# CHEMICAL ANALYSIS

Vol. 1. **The Analytical Chemistry of Industrial Poisons, Hazards, and Solvents.** *Second Edition.* By the late Morris B. Jacobs

Vol. 2. **Chromatographic Adsorption Analysis.** By Harold H. Strain (*out of print*)

Vol. 3. **Colorimetric Determination of Traces of Metals.** *Third Edition.* By E. B. Sandell
**Photometric Determination of Traces of Metals: General Aspects.** *Fourth Edition of Part 1 of Colorimetric Determination of Traces of Metals.* By E. B. Sandell and Hiroshi Onishi

Vol. 4. **Organic Reagents Used in Gravimetric and Volumetric Analysis.** By John F. Flagg (*out of print*)

Vol. 5. **Aquametry: A Treatise on Methods for the Determination of Water.** *Second Edition* (*in three parts*). By John Mitchell, Jr. and Donald Milton Smith

Vol. 6. **Analysis of Insecticides and Acaricides.** By Francis A. Gunther and Roger C. Blinn (*out of print*)

Vol. 7. **Chemical Analysis of Industrial Solvents.** By the late Morris B. Jacobs and Leopold Scheflan

Vol. 8. **Colorimetric Determination of Nonmetals.** *Second Edition.* Edited by the late David F. Boltz and James A. Howell

Vol. 9. **Analytical Chemistry of Titanium Metals and Compounds.** By Maurice Codell

Vol. 10. **The Chemical Analysis of Air Pollutants.** By the late Morris B. Jacobs

Vol. 11. **X-Ray Spectrochemical Analysis.** *Second Edition.* By L. S. Birks

Vol. 12. **Systematic Analysis of Surface-Active Agents.** *Second Edition.* By Milton J. Rosen and Henry A. Goldsmith

Vol. 13. **Alternating Current Polarography and Tensammetry.** By B. Breyer and H. H. Bauer

Vol. 14. **Flame Photometry.** By R. Herrmann and J. Alkemade

Vol. 15. **The Titration of Organic Compounds** (*in two parts*). By M. R. F. Ashworth

Vol. 16. **Complexation in Analytical Chemistry: A Guide for the Critical Selection of Analytical Methods Based on Complexation Reactions.** By the late Anders Ringbom

Vol. 17. **Electron Probe Microanalysis.** *Second Edition.* By L. S. Birks

Vol. 18. **Organic Complexing Reagents: Structure, Behavior, and Application to Inorganic Analysis.** By D. D. Perrin

Vol. 19. **Thermal Methods of Analysis.** *Second Edition.* By Wesley Wm. Wendlandt

Vol. 20. **Amperometric Titrations.** By John T. Stock

Vol. 21. **Reflectance Spectroscopy.** By Wesley Wm. Wendlandt and Harry G. Hecht

Vol. 22. **The Analytical Toxicology of Industrial Inorganic Poisons.** By the late Morris B. Jacobs

Vol. 23. **The Formation and Properties of Precipitates.** By Alan G. Walton

Vol. 24. **Kinetics in Analytical Chemistry.** By Harry B. Mark, Jr. and Garry A. Rechnitz

Vol. 25. **Atomic Absorption Spectroscopy.** *Second Edition.* By Morris Slavin

Vol. 26. **Characterization of Organometallic Compounds** (*in two parts*). Edited by Minoru Tsutsui

Vol. 27. **Rock and Mineral Analysis.** *Second Edition.* By Wesley M. Johnson and John A. Maxwell

| Vol. 28. | The Analytical Chemistry of Nitrogen and Its Compounds (*in two parts*). Edited by C. A. Streuli and Philip R. Averell |
|---|---|
| Vol. 29. | The Analytical Chemistry of Sulfur and Its Compounds (*in three parts*). By J. H. Karchmer |
| Vol. 30. | Ultramicro Elemental Analysis. By Günther Tölg |
| Vol. 31. | Photometric Organic Analysis (*in two parts*). By Eugene Sawicki |
| Vol. 32. | Determination of Organic Compounds: Methods and Procedures. By Frederick T. Weiss |
| Vol. 33. | Masking and Demasking of Chemical Reactions. By D. D. Perrin |
| Vol. 34. | Neutron Activation Analysis. By D. De Soete, R. Gijbels, and J. Hoste |
| Vol. 35. | Laser Raman Spectroscopy. By Marvin C. Tobin |
| Vol. 36. | Emission Spectrochemical Analysis. By Morris Slavin |
| Vol. 37. | Analytical Chemistry of Phosphorus Compounds. Edited by M. Halmann |
| Vol. 38. | Luminescence Spectrometry in Analytical Chemistry. By J. D. Winefordner, S. G. Schulman, and T. C. O'Haver |
| Vol. 39. | Activation Analysis with Neutron Generators. By Sam S. Nargolwalla and Edwin P. Przybylowicz |
| Vol. 40. | Determination of Gaseous Elements in Metals. Edited by Lynn L. Lewis, Laben M. Melnick, and Ben D. Holt |
| Vol. 41. | Analysis of Silicones. Edited by A. Lee Smith |
| Vol. 42. | Foundations of Ultracentrifugal Analysis. By H. Fujita |
| Vol. 43. | Chemical Infrared Fourier Transform Spectroscopy. By Peter R. Griffiths |
| Vol. 44. | Microscale Manipulations in Chemistry. By T. S. Ma and V. Horak |
| Vol. 45. | Thermometric Titrations. By J. Barthel |
| Vol. 46. | Trace Analysis: Spectroscopic Methods for Elements. Edited by J. D. Winefordner |
| Vol. 47. | Contamination Control in Trace Element Analysis. By Morris Zief and James W. Mitchell |
| Vol. 48. | Analytical Applications of NMR. By D. E. Leyden and R. H. Cox |
| Vol. 49. | Measurement of Dissolved Oxygen. By Michael L. Hitchman |
| Vol. 50. | Analytical Laser Spectroscopy. Edited by Nicolo Omenetto |
| Vol. 51. | Trace Element Analysis of Geological Materials. By Roger D. Reeves and Robert R. Brooks |
| Vol. 52. | Chemical Analysis by Microwave Rotational Spectroscopy. By Ravi Varma and Lawrence W. Hrubesh |
| Vol. 53. | Information Theory As Applied to Chemical Analysis. By Karel Eckschlager and Vladimir Štěpánek |
| Vol. 54. | Applied Infrared Spectroscopy: Fundamentals, Techniques, and Analytical Problem-solving. By A. Lee Smith |
| Vol. 55. | Archaeological Chemistry. By Zvi Goffer |
| Vol. 56. | Immobilized Enzymes in Analytical and Clinical Chemistry. By P. W. Carr and L. D. Bowers |
| Vol. 57. | Photoacoustics and Photoacoustic Spectroscopy. By Allan Rosencwaig |
| Vol. 58. | Analysis of Pesticide Residues. Edited by H. Anson Moye |
| Vol. 59. | Affinity Chromatography. By William H. Scouten |

# Potentiometry and Potentiometric Titrations

# CHEMICAL ANALYSIS

## A SERIES OF MONOGRAPHS ON ANALYTICAL CHEMISTRY AND ITS APPLICATIONS

*Editors*
**P. J. ELVING, J. D. WINEFORDNER**
*Editor Emeritus:* **I. M. KOLTHOFF**

*Advisory Board*

Fred W. Billmeyer, Jr.   Victor G. Mossotti
Eli Grushka              A. Lee Smith
Barry L. Karger          Bernard Tremillon
Viliam Krivan            T. S. West

**VOLUME 69**

A WILEY-INTERSCIENCE PUBLICATION

**JOHN WILEY & SONS**

New York / Chichester / Brisbane / Toronto / Singapore

# Potentiometry and Potentiometric Titrations

**E. P. SERJEANT**

*University of New South Wales*
*Duntroon, Australia*

A WILEY-INTERSCIENCE PUBLICATION

**JOHN WILEY & SONS**

New York  /  Chichester  /  Brisbane  /  Toronto  /  Singapore

Copyright © 1984 by John Wiley & Sons, Inc.

All rights reserved. Published simultaneously in Canada.

Reproduction or translation of any part of this work beyond that permitted by Section 107 or 108 of the 1976 United States Copyright Act without the permission of the copyright owner is unlawful. Requests for permission or further information should be addressed to the Permissions Department, John Wiley & Sons, Inc.

*Library of Congress Cataloging in Publication Data:*
Serjeant, E. P.
  Potentiometry and potentiometric titrations.

  (Chemical analysis, ISSN 0069-2883; v. 69)
  "A Wiley-Interscience Publication."
  Includes index.
  1. Electrochemical analysis.  I. Title.  II. Series.

  QD115.S39  1984    543'.08712    83-21903
  ISBN 0-471-07745-3

Printed in the United States of America

10 9 8 7 6 5 4 3 2 1

*Dedicated to my Lord,*
*Creator of energy and matter*

# PREFACE

This work was undertaken in response to a request for a book that would replace the classic Kolthoff–Furman *Potentiometric Titrations* published in 1931. In recent years there has been a resurgence of interest in the applications of potentiometry to analysis as a result of the development of ion selective and associated membrane electrodes. The literature covering the theory and application of these electrodes is now large, and there are a number of good texts devoted exclusively to this aspect of potentiometry. At the same time there has also been a steady, though less spectacular, development in the applications of electrodes and techniques that were already established in 1931. These include the applications of potentiometry both to titrimetric analysis and to the investigation of the solution equilibria upon which these analyses are based. It was felt necessary, therefore, that this book should survey the endeavors in all three areas and, accordingly, it is divided into three parts. The first part deals with cells, electrodes, and the basic procedures of direct potentiometry together with a description of electrodes and techniques used for equivalence point detection in titrimetric analysis. The second part deals with the applications of potentiometry to the determination of solution equilibrium data, concentrating mainly upon the determination of ionization constants of acids and bases in aqueous and nonaqueous solution, and also upon the determination of stability constants. The third part is devoted to titrimetric analysis and covers the four main branches; acid-base, complexometric, oxidation–reduction, and precipitation titrations. The main thrust of the literature search into these topics was concentrated upon the two decades from 1958 to 1978, although some reference is made to important work prior to and subsequent to this period.

Within the three subdivisions mentioned, the measurement of a cell potential can be performed to fulfill different requirements ranging from measurements having thermodynamic significance performed at "zero current" in carefully calibrated cells without liquid junction to the monitoring of a potential between two electrodes immersed in a titrant solution through which a current is deliberately impressed during the course of a titration. In the first case the object of the measurement of potential is to interpret it in terms of the activity of an electroactive electrolyte present in the cell solution by means of the thermodynamically derived Nernst equation. In the second case the monitoring of the potential is performed only to establish that point in the titration at which it is hoped a large change in potential can be interpreted as signifying the point of completion of the titration reaction. The volume (or weight) of titrant required to reach this point is of prime interest to the analyst, and the actual values of the potentials have no thermodynamic or even quasi-thermodynamic significance. It is difficult, therefore, to preserve the integrity of the word "potentiometry" if both these applications are described equally as potentiometric. To distinguish unequivocally between these two types of applications, the use of the term "potentiometry" in this book is taken to mean the measurement of a cell potential $E_{cell}$ which can be interpolated in terms of activity or concentration by means of a form of the Nernst equation written as

$$E_{cell} = \text{constant} \pm \text{slope} \log a_x$$

where $a_x$ is the activity of an electroactive electrolyte, or a single electroactive species such as an ion. In contradistinction to this, the term "potentiotitrimetry" is used to describe a traditional method of locating the equivalence volume of a titration by the monitoring of cell potentials during the course of the reaction. Following upon these definitions, a "*potentiometric titration*" is one in which the measured potentials are used to obtain the activity of an electrolyte or a species at particular points during a titration. To this end the cell must be carefully calibrated before the titration. The determination of an ionization constant by $p$H titration is an example of a potentiometric titration, and so is the titrimetric application when computer-generated curves are used to fit experimental data to theoretical curves in order to determine the equivalence volume.

There are now many fields of scientific endeavor in which either direct potentiometry or potentiotitrimetry make a useful contribution as a powerful analytical tool. It is not always evident, however, that workers without formal training in the analytical chemistry of these techniques fully appreciate the theoretical basis for their application, and their limitations. It has been the aim of the author, therefore, to present the necessary basic theory in a form which is assimilable to those with only an elementary knowledge of chemistry, and this theory is then developed further at appropriate points in the book to provide a basis for the understanding of the practical applications of the techniques. The emphasis has been towards these applications, and to this end much practical information is contained in over 60 tables presented in the book. Some detailed practical instructions are included in the text but, unfortunately, these inclusions have had to be curtailed in order to present a comprehensive treatment of the subject as a whole in the space available. It is hoped this deficiency is remedied to a degree by the inclusion of many references to the original work. Detailed descriptions of general purpose apparatus is not a feature of the book since these are available from the manufacturers, and only passing references are made to the important field of automation. It is worth noting that some books devoted to ion selective electrodes give details of automation techniques and, of course, Svehla's *Automatic Potentiometric Titrations* fulfills a need in this direction.

It is hoped that this book will be useful also in formulating courses in potentiometry and in titrimetry at the undergraduate and the beginning graduate level. Indeed, parts of it have been derived from or have formed a basis for undergraduate courses given to chemistry majors at the second and third-year levels. Thus parts of Chapter 1 were derived from a 12-lecture course given to second-year students as part of a topic dealing with electrochemistry and thermodynamics. This was backed by a practical course that included some of the $p(a_H \gamma_{Cl})$ measurements described in Chapter 6. Parts of Chapters 3 and 4 have formed the basis for a course in potentiometry given to students in second and third-year analytical chemistry courses. However, the primary purpose of the book is to provide a source of reference for practicing analytical chemists and those engaged in allied fields.

The guidance and patience of Dr. P. J. Elving, University of Michigan, who suggested the project, and the comments of his editorial colleague, Dr. J. D. Winefordner, University of Florida, have been particularly appreciated over the years taken in compiling this book. The author gratefully acknowledges the help given by his wife, Joan, in processing many of the thousands of abstracts collected during the literature search, and also thanks Dr. R. G. Bates, Emeritus Professor, University of Florida, for reading Chapter 4 while it was still in the handwritten state. Dr. D. D. Perrin, Professorial Fellow, Australian National University, is thanked for his useful comments on Chapter 8, and Dr. A. Albert, Emeritus Professor, Australian National University, and Research Professor, State University of New York, Stonybrook, is thanked for his encouragement and for reading Chapters 6 and 7. The use of library facilities and the help given by the staffs of the Chemistry Department and Health Center libraries in the University of Florida and the John Curtin School of Medical Research and Research School of Chemistry libraries in the Australian National University are also acknowledged with gratitude. The staffs of the Bridges Library,

Duntroon, and the Defence Central Library, Canberra, are thanked for initiating the computer searches that produced a large number of references to the more recent literature. The author also wishes to thank Mrs. Helen Mann and Mrs. Margaret Bacon who typed the bulk of the manuscript.

E. P. SERJEANT

*Duntroon, Australian Capital Territory*
*March 1984*

# ACKNOWLEDGMENTS

Permission from the following publishers to reproduce copyright material is gratefully acknowledged:

The American Chemical Society for Figures 3.4 and 3.5c, which appeared in *Analytical Chemistry* [Figure 3.4 from L. W. Marple and J. S. Fritz, *Anal. Chem.* **34**, 796 (1962), and Figure 3.5c from T. Fujinaga, S. Okazaki, and H. Freiser, *Anal. Chem.* **46**, 1842 (1974)].

Elsevier Scientific Publishing Company for Figures 3.5d, 3.10b and 9.3, which appeared in *Analytica Chimica Acta* [Figure 3.5d from J. Ruzicka, C. G. Lamm, and J. C. Tjell, *Anal. Chim. Acta* **62**, 15 (1972), Figure 3.10b from E. H. Hansen and J. Ruzicka, *Anal. Chim. Acta* **72**, 353 (1974), and Figure 9.3 from J. M. H. Fortuin, *Anal. Chim. Acta* **24**, 175 (1961)].

Instrument Society of America for Figure 3.2b, which appeared in *Advances in Instrumentation, Part 3* as Figure 1 of paper 72-719 by H. L. Koppel and A. G. Gealt, *Advan. Instrum.* **27** (3), 719 (1972).

*Journal of Chemical Education* for Figure 3.8 taken from a paper by A. Craggs, G. J. Moody, and J. D. R. Thomas, *J. Chem. Educ.* **51**, 541 (1974).

McGraw-Hill Book Company Inc. for Figure 3.3(VI), which appeared in *The Determination of Stability Constants* by F. J. C. Rossotti and H. Rossotti, McGraw-Hill, New York, 1961, as Fig. 7-1.

Pergamon Press Ltd. for Figure 7.1, which appeared in *Talanta* as Fig. 2 of a paper by A. P. Kreshkov, *Talanta* **17**, 1029 (1970), and for Figures 3.7a and 9.2, which appeared in *Automatic Potentiometric Titrations* by G. Svehla, Pergamon, Oxford, 1978 as Fig. 4.4 and Fig. 8.11, respectively.

Plenum Publishing Company for Figure 12.1, which appeared in the *Journal of Analytical Chemistry of the USSR* as Fig. 1 of a paper by A. I. Busev and Li Gyn, *J. Anal. Chem. USSR* **14**, 741 (1959); *Zh. Anal. Khim.* **14**, 688 (1959).

John Wiley and Sons, Inc. for Table 3.1 and Figure 3.1, which appeared in *Determination of pH* by R. G. Bates, 2nd ed., Wiley, New York, 1973 as Table 10-1 and Fig. 10-4, respectively; for Figures 3.2a, 3.5a, 3.5b, and 3.7b, which appeared in *Analysis with Ion Selective Electrodes* by P. L. Bailey, Heydon, London, 1976 as Figures 2.1(a), 5.1(a), 5.1(b), and 6.1(c), respectively; for Figure 3.10*a*, which appeared in *Potentiometric Water Analysis* by D. Midgley and K. Torrance, Wiley, Chichester, 1978 as Figure 3.1; for Figure 3.6*a*, which appeared in *Experimental Electrochemistry for Chemists* by D. T. Sawyer and J. L. Roberts, Jr., Wiley, New York, 1974 as Figure 6-1(a).

E. P. S.

# CONTENTS

## PART 1: CELLS, ELECTRODES, AND BASIC PROCEDURES

**Chapter 1. Rudiments of Potentiometry**    3

     1.1. Direct Potentiometry 3
     1.2. Activity–Concentration Relations 14
     1.3. Reduction Potentials 29
     1.4. The Cell Notation 47

**Chapter 2. Classifications of Potentiometry and Titrimetry**    53

     2.1. Types of Titrations 53
     2.2. Types of Cells 56
     2.3. Titrimetry Using Potential Monitoring 61
     2.4. Electrolytic Potentiotitrimetry and Differential Electrolytic Potentiotitrimetry 69

**Chapter 3. Classifications and Descriptions of Electrodes**    79

     3.1. Categories of Oxidation–Reduction Electrodes 80
     3.2. Reference Electrodes in Aqueous Solution 81
     3.3. Design of Reference Half-Cells 104
     3.4. Reference Half-Cells in Nonaqueous Solvents 113
     3.5. Categories and Descriptions of Membrane Electrodes 118

**Chapter 4. Procedures of Analytical Potentiometry**    191

     4.1. Methods of Calibration 192
     4.2. Standard Addition/Subtraction Methods 222
     4.3. Some Relevant Properties of Ion Selective Electrodes 232

## Chapter 5. Electrodes for Potentiotitrimetry 253

5.1. Electrode Requirements and Advantages of Potentiotitrimetry 253
5.2. Electrodes for Acid-Base Titrimetry 255
5.3. Electrodes for Complexometric Titrations 264
5.4. Electrodes for Precipitation Titrations 278
5.5. Electrodes for Oxidation-Reduction Titrations 292

## PART 2: APPLICATIONS TO THE DETERMINATION OF SOLUTION EQUILIBRIUM DATA

## Chapter 6. The Determination of Ionization Constants in Aqueous Solution 305

6.1. Introduction 305
6.2. Methods Based upon Measurements of $p(a_H\gamma_{Cl})$ 310
6.3. The $p$H Titration Method 326
6.4. The Determination of the Ionic Product of Water 332
6.5. Polyfunctional Acids and Bases 337

## Chapter 7. Acid-Base Equilibria in Nonaqueous Solvents 363

7.1. Introduction 363
7.2. General Properties of Amphiprotic Solvents 368
7.3. Potentiometric Measurements with Cells Containing a Glass Electrode in Selected Nonaqueous Solvents 376
7.4. Measurements in Mixed Aqueous Solvents 409
7.5. Half-Neutralization Potentials 418

## Chapter 8. The Determination of Stability Constants by Potentiometric Titration 431

8.1. Introduction 431
8.2. The $p$H Titration Method 434

## PART 3: APPLICATIONS TO TITRIMETRIC ANALYSIS

**Chapter 9. The Basics of Titrimetry** — 445

    9.1. General Principles 445
    9.2. Methods of Assessing the Equivalence Volume, $V_E$ 457

**Chapter 10. Acid-Base Titrimetry** — 475

    10.1. Introduction 475
    10.2. Titrations in Aqueous Solutions 476
    10.3. Titrations in Nonaqueous Solvents 481

**Chapter 11. Complexometric Titrations** — 501

    11.1. A Survey of Methods 501
    11.2. The Preparation of Titrants and Other Reagents 525

**Chapter 12. Oxidation-Reduction Titrations** — 531

    12.1. General Considerations and Applications 534
    12.2. Some Oxidants Used as Titrants 570
    12.3. Some Reductants Used as Titrants 584
    12.4. Oxidation-Reduction Titrations of Some Organic Compounds in Various Media 594

**Chapter 13. Precipitation Titrations** — 609

    13.1. Some Basic Considerations 609
    13.2. Titration of Inorganic Anions 631
    13.3. Titration of Inorganic Cations 642
    13.4. Applications to Organic Analysis 654

## APPENDIX

Table A.1   Values of the Nernst Factor $k$ in Volts from 0 to 100°C   675

Table A.2   Values of A and B in the Debye-Huckel Equation from 0 to 100°C for Water   676

Table A.3  Values of the Ion Size Parameter $a_i$ and Ion Activity
Coefficients in Water at 25°   677

Table A.4  Thermodynamic Values of the Ionic Product of Water $K_w$
from 0 to 60°C   678

Index   679

# PART 1
# CELLS, ELECTRODES, AND BASIC PROCEDURES

CHAPTER
1
# RUDIMENTS OF POTENTIOMETRY

In analytical chemistry there is a dichotomy of interest in the applications of measuring or monitoring cell potentials that develop between two suitable electrodes when they dip into a solution containing a species whose concentration is to be measured. One application aims at deducing the concentration of the species (called *the determinand*) directly from the value of the measured potential; this has become known as *analysis by direct potentiometry*. The other sphere of interest is centered upon the determination of the equivalence volume for a titrimetric reaction by the monitoring of potentials that develop between the electrodes during the course of the titration. An estimate of the equivalence volume, and thence the concentration of determinand, is made from the comparatively large change in potential which is observed when all the determinand has reacted with the titrant according to a known stoichiometry (see Figure 1.4). This is only one method among the many that are available to the analyst for the detection of equivalence points. Therefore, to better reflect that the actual potentials obtained often have no theoretical interest and are thus subsidiary to the main objective of determining an equivalence volume, this application is known in this book as *potentiotitrimetry*. This chapter serves as an introduction to both of these applications, but the main emphasis here is upon enumerating the general principles upon which analysis by direct potentiometry is based.

## 1.1 DIRECT POTENTIOMETRY

A simple oxidation-reduction reaction is taken to illustrate the rudiments of potentiometry with the purpose of deriving equations that are pertinent to the analyses of ion concentrations regardless of whether the method relies upon the potentials produced by an oxidation-reduction electrode or upon those produced by a more recently developed ion selective electrode versus a suitable reference half-cell. These equations are similar in form to equation (1.8) [see also equation (3.2)].

# RUDIMENTS OF POTENTIOMETRY

## What Is Meant by Potentiometry

In this book potentiometry is defined as the measurement of the electromotive force of a reversible cell. This electromotive force, commonly referred to as the cell e.m.f., the cell potential or $E_{cell}$, arises from a spontaneous process that has the potential or capacity to occur in the cell. The object of performing potentiometric measurements in chemistry is to operate the cell under such conditions that the spontaneous process does not occur and, under this condition to interpret the measured $E_{cell}$ in terms of a unit of concentration for one of the components present in the cell solution.

## Reversible Cells

The reversible cell and its e.m.f. are at the core of potentiometry, and to illustrate the terms, the spontaneous reaction produced by the dissolution of zinc in acid is used as an example. Thus

$$Zn_{(s)} + 2H^+ \rightarrow Zn^{2+} + H_{2(g)}$$

The overall free energy change ($\Delta G$) is negative for this oxidation-reduction reaction in which solid pure zinc is oxidized by hydrogen ion to form zinc ion, the hydrogen ion being simultaneously reduced to hydrogen gas. For any chemical process it can be deduced from basic thermodynamic definitions that $\Delta G$ (in joules, J) at constant temperature and pressure is related to the activity quotient $Q$ by

$$\Delta G = \Delta G^\circ + RT \ln Q \qquad (1.1)$$

The terms of this equation are: $T$ the absolute temperature (°K), $R$ the ideal gas constant (8.3143 J °K$^{-1}$ mol$^{-1}$), and $\Delta G^\circ$ the standard free energy change for the reaction. The activity quotient for a reaction of the type

$$lL + mM \cdots \rightarrow xX + yY \cdots$$

is defined as

$$Q = \frac{a_X^x \cdot a_Y^y}{a_L^l \cdot a_M^m}$$

The activity terms $a_X$, $a_Y$, $a_L$, and $a_M$ are usually expressed either as molality (moles kg$^{-1}$) or molarity (moles L$^{-1}$) for solutions; the corresponding term for gaseous reactants or products is the active pressure or fugacity ($f$ atm). When all the substances are in their standard states and each of their activities is unity, the free energy change accompanying the reaction becomes equal to the standard free change $\Delta G°$ (i.e., $Q = 1$). The application of equation (1.1) to the example reaction

$$Zn_{(s)} + 2H^+ \rightarrow Zn^{2+} + H_{2(g)}$$

yields

$$\Delta G = \Delta G° + RT \ln \frac{a_{Zn^{2+}} \cdot P_{H_2}}{a_{H^+}^2} \qquad (1.2)$$

There is no term for $a_{Zn(s)}$ in this equation because the activity of a pure solid substance is conventionally taken as unity. It is also assumed that the stoichiometric pressure of hydrogen gas, ($P_{H_2}$) is equal to the required thermodynamic quantity, fugacity. If the reaction is carried out in a test tube open to the atmosphere, the pressure of hydrogen required to escape from the solution phase is only about 1 atm and at this low pressure fugacity tends to equal the stoichiometric pressure.

The basic philosophy of potentiometry is that we measure the free energy change, classically associated with oxidation-reduction reactions, by measuring the amount of electrical energy (volt) to stop the reaction. The necessary interconversion of units for this purpose is given by

$$\Delta G = -nFE_{cell} \qquad (1.3a)$$

and

$$\Delta G° = -nFE°_{cell} \qquad (1.3b)$$

where $n$ is the number of moles of electrons transferred in the overall reaction, $F$ is the Faraday constant (96,487 coulombs), and $E_{cell}$ is the e.m.f. which must be applied to stop the reaction. Thus in the example the aim is to convert the fairly vigorous reaction between zinc and acid into the delicately balanced equilibrium

$$Zn_{(s)} + 2H^+ \rightleftharpoons Zn^{2+} + H_{2(g)}$$

This requires that the oxidation process

$$Zn_{(s)} \rightleftharpoons Zn^{2+} + 2e$$

and the reduction process $2H^+ + 2e \rightleftharpoons H_{2(g)}$ should each have the potential or the capacity to occur simultaneously but separately. Thus two separate compartments or *half-cells* are necessary; one in which solid pure zinc and zinc ion are coexistent at an interface, $Zn_{(s)}|Zn^{2+}$, and the other in which hydrogen ion and hydrogen gas coexist, also at an interface $H^+|H_{2(g)}$. The latter interface consists of finely divided platinum, deposited electrolytically on a platinum plate, which allows an exchange between hydrogen gas passing over the plate and hydrogen ion in the solution through which the gas ultimately flows.

Having separated the two half-reaction centers, the only way that the overall reaction could occur is to conduct the flow of electrons from the electron source at the $Zn_{(s)}|Zn^{2+}$ interface to the region relatively lower in electron density at the $H^+|H_{2(g)}$ interface by means of an electronic (or ohmic) conductor such as a copper wire. It must be stipulated also that there should be a means of conducting the electron flow within and between the solution phases of the two half-cells. This is achieved by means of an ionic conductor, often a salt bridge (Figure 1.1). The cell depicted in Figure 1.1 can also be represented by a schematic representation called *cell notation* in which only the interfaces are shown. Using this notation, the cell under discussion is represented as

$$Zn|ZnSO_4\|HCl|H_{2(g)} \text{ (1 atm); Pt}$$

The double verticle lines ($\|$) represent the two liquid–liquid junctions at either end of the salt bridge.

If the cell is short-circuited by connecting the $Zn|ZnSO_4$ interface to the $HCl|H_2$ interface externally to the cell, electrons can flow from the zinc half-cell to the hydrogen half-cell and the spontaneous overall oxidation-reduction reaction will occur. The net result is that an electric current is produced. This negates the purpose of potentiometry, however, which has as its aim the measurement of $E_{cell}$ under equilibrium conditions. Two methods are available for such measurements and both methods have as their objective the minimization of the electron flow, and hence the current, in the conductor external to the cell.

### Measurement of Cell E.M.F.

The classical method is to measure the amount of electrical energy required to exactly balance the electron flow emanating from the cell reaction by a flow equal in magnitude but opposite in the direction of flow from an external source of e.m.f. This concept forms the basis for the simplified potentiometer circuit (Figure 1.1) which is connected across the terminals

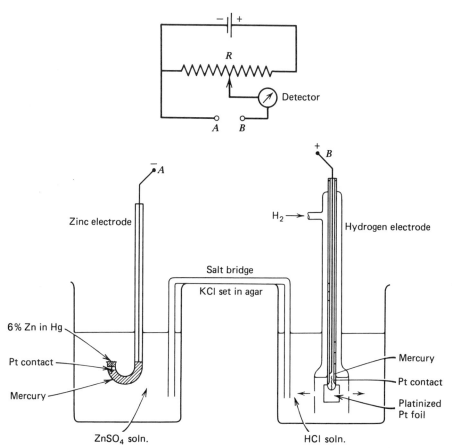

**Figure 1.1.** Diagrammatic representation of a simplified potentiometer circuit and a cell in which the reaction $Zn_{(s)} + 2H^+ \rightarrow Zn^{2+} + H_2$ would proceed if current were drawn. The cell is connected across the terminals $AB$ as shown.

$AB$. The potential applied to the cell from the external source of e.m.f. is determined by the value of the adjustable resistance $R$ which has been previously calibrated in terms of volts. The detector $D$, usually a sensitive galvanometer or (better still) an electrometer, records a zero deflection when the applied e.m.f. is equal to the cell e.m.f. If it now be imagined that the adjustable resistance is controlled by the rotation of a dial calibrated directly in volts, the value of $E_{cell}$ can be read directly from the dial when the detector is showing a zero deflection.

The second, and the more modern, method is to use a voltmeter with a high input impedance ($> 10^{12}$ Ω), such as a $p$H meter, which when

# 8 RUDIMENTS OF POTENTIOMETRY

connected across $AB$, imposes such a high resistance in the cell circuit that the flow of electrons, and thus the current flowing, becomes negligible (usually $< 10^{-12}$ A). The value of $E_{cell}$ is recorded directly on either an analogue or digital scale usually in millivolts (mV).

### The Nernst Equation

The purpose of measuring $E_{cell}$ is to interpret this value in terms of a unit of concentration for one of the electroactive species contained in the cell. The basis for this interpretation is the Nernst equation which is obtained by combining equation (1.1) with equations (1.3a) and (1.3b). Thus

$$E_{cell} = E^{\circ}_{cell} - \frac{RT}{nF} \ln Q \quad (1.4)$$

Measurements of $E_{cell}$ must be made at constant temperature and it is customary to combine the constants $R$, $T$, and $F$ into a single constant $k$, $(RT\ln 10)/F$. Values of $k$ at various temperatures are to be found in the appendix (Table A.1). The Nernst equation, when applied to the equilibrium

$$Zn_{(s)} + 2H^+ \rightleftharpoons Zn^{2+} + H_2 \; (P = 1 \text{ atm})$$

becomes

$$E_{cell} = E^{\circ}_{cell} - \frac{k}{2} \log \frac{a_{Zn^{2+}}}{a^2_{H^+}} \quad (1.5)$$

The value of $n$ is 2 because two moles of electrons would be transferred from $Zn \rightleftharpoons Zn^{2+} + 2e$ to $2H^+ + 2e \rightleftharpoons H_2$ if the reaction were allowed to proceed.

### The Practical Condition of Reversibility

The Nernst equation is applicable for the interpretation of $E_{cell}$ in terms of an activity only if the cell is operating under reversible conditions. To appreciate this, imagine what would happen within the cell if a detector requiring a comparatively large current to activate it were used in the potentiometer (Figure 1.1). If the applied e.m.f. were less than the cell e.m.f., the detector would sluggishly indicate a deflection and in doing so would draw current which would emanate from the spontaneous cell reaction. The activities or concentrations of the species in the half-cells would be altered by zinc dissolving and increasing the concentration of zinc ion in the zinc half-cell accompanied by an equivalent diminution in hydrogen ion concentration in the other half-cell. Attempts to increase the applied e.m.f. to find a balance might or might not be successful depending upon the kinetics of the electrode reactions. For a cell to be truly thermodynamically reversible, no current should ever be drawn from it nor applied to it during the course of the measurement. This is clearly impossible because the resistance of the detector would have to be infinitely large. For this reason statements such as "potentiometry at zero current" have no place in the real world. However, modern instruments draw such small currents that reversibility in the actual measurement of $E_{cell}$ is no longer a problem; the problem is in the reversibility of the electrode. For example, had a solid rod of pure zinc been used in the cell of Figure 1.1, it is very unlikely that it would behave reversibly no matter how refined the equipment used for the measurement of $E_{cell}$. In common with most other metals zinc does not give the reversible exchange

$$Zn_{(s)} \rightleftharpoons Zn^{2+} + 2e$$

when the solid metal is immersed in an aqueous solution containing the metal ion. Strains in the solid or surface impurities often cause unsteady or nonreproducible values of the e.m.f. and even if the equilibrium value is attained eventually, it is reached only very slowly. In some cases, however, this problem can be minimized by dissolving the carefully cleaned pure metal in freshly redistilled mercury. Fortunately, zinc is an example of such a metal and can be used as an amalgam electrode. A concentration in range 2–6% W/V zinc in mercury is sufficient to give a reasonable rate of response; that is, the exchange equilibrium

$$Zn_{(s)} \rightleftharpoons Zn^{2+} + 2e$$

at the interface is established within 20 min.

**The Liquid Junction Potential**

There is yet one more factor that prevents the example cell

$$Zn(Hg)|ZnSO_4\|HCl|H_{2(g)} \text{ (1 atm)}; Pt$$

from satisfying the criteria for thermodynamic reversibility. This factor is the electrical potential set up as a result of the liquid–liquid junctions established by the salt bridge. *The liquid junction potential*, as it is called, is present in most of the cells used in the analytical applications of direct potentiometry. An acknowledgement that it exists is made by including a term $E_j$ in the Nernst equation which, for the example cell, now becomes

$$E_{cell} = E_{cell}^{\circ} - \frac{k}{2} \log \frac{a_{Zn^{2+}}}{a_{H^+}^2} + E_j \qquad (1.6)$$

This simple expedient does not now make the equation (1.6) strictly valid, for $E_j$ is not thermodynamically defined. However, by a judicious choice of the type of electrolyte solution used in the salt bridge, it is possible to minimize the liquid junction potential $E_j$ to a value which is usually assumed to be constant. A cell with liquid junction would never be used if the results obtained from measurement of $E_{cell}$ are required to be interpreted in terms of thermodynamic quantities. *A cell without liquid junction* (no salt bridge) is preferred for these applications.

### Ion Activity-Concentration Measurements

The use of the example cell for the determination of the concentration of zinc ion in a pure solution of zinc sulfate is now outlined as an illustration of the factors that must be considered in the interpretation of the measured value of $E_{cell}$ in terms of a unit of concentration for one of the components in the cell. The assumption is made that $E_j$ is constant and it is specified that the temperature will be 298 °K. Equation (1.6) can thus be written as

$$E_{cell} = \text{constant} - \frac{0.05916}{2} \log \frac{a_{Zn^{2+}}}{a_{H^+}^2} \qquad (1.7)$$

According to equation (1.7), the prospect of the determination is not encouraging for there are too many unknowns. The solution to the problem, however, is to be found in the technique of *cell calibration*. This is a recurrent theme in the field of direct potentiometry and standards are required for this purpose. With the notable exception of the arbitrarily defined potential for the standard hydrogen electrode, these standards are never based upon the potentials of cells or electrodes for, no matter how carefully these are prepared, it appears difficult to obtain electrodes that are exactly reproducible throughout the scientific world (1). Thus solutions are specified as calibration standards for these can be prepared more reproducibly than electrodes can. This topic is not outlined here but it is described in Chapter 4. For the purpose of this introductory section the erroneous assumption is made (see Section 1.2) that the stoichiometric concentration of a substance in a solution is equal to the activity of the substance in that solution. Thus equation (1.7) is written as

$$E_{cell} = \text{constant} - \frac{0.059}{2}\log\frac{c_{Zn^{2+}}}{c_{H^+}^2} \quad (1.8)$$

### Calibration and Analysis

The calibration procedure is now straightforward. The value $c_{H^+}$ could be kept constant at 0.01 $M$ in the hydrogen ion–hydrogen gas half-cell and a series of standard zinc sulfate solutions could be prepared covering the concentration range of interest (preferably in the range 0.01–0.002 $M$). Thus for each one of these known concentrations the cell can now be represented as

$$\text{Zn(Hg)} | \text{ZnSO}_4 \text{ (known molarity)} \| \text{HCl}(0.01\ M) | \text{H}_2\ (1\ \text{atm}); \text{Pt}$$

and equation (1.8) becomes

$$E_{cell} = \text{constant} - \frac{0.059}{2}\log c_{Zn^{2+}} - \frac{0.059}{2}\log\frac{1}{(0.01)^2} \quad (1.9a)$$

or

$$E_{cell} = (\text{constant} - 0.118) - \frac{0.059}{2}\log c_{Zn^{2+}} \quad (1.9b)$$

If the measured values of $E_{cell}$ are plotted against the corresponding known values of $-\log c_{Zn^{2+}}$, a slightly curved calibration graph would be obtained (Figure 1.2). The reason for the departure from the linear relation predicted by equation (1.9b) is primarily associated with the erroneous assumption that $a_{Zn^{2+}} = c_{Zn^{2+}}$, although in some instances $E_j$ can also vary. Provided that the unknown concentration of zinc ion in a pure zinc sulfate solution is such that its measured value of $E_{cell}$ is within the range covered by the calibration curve, its concentration can be interpolated from the graph.

### Comments on Procedure

It is worthwhile at this stage to consider the procedure adopted to achieve the objective which was the determination of zinc ion in a pure zinc sulfate solution. The major point of interest in the technique outlined is that the concentration of hydrogen ion in the $H^+ | H_2\ (1\ \text{atm})$; Pt half-cell has been

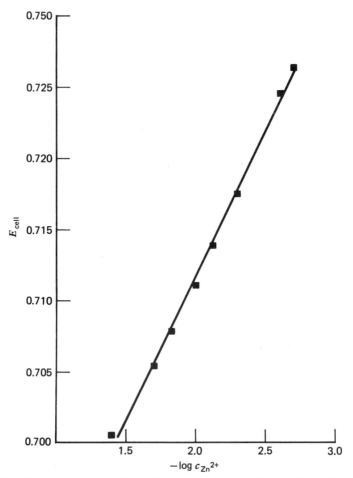

**Figure 1.2.** Calibration curve which could be used for the determination of zinc(II) concentration in an aqueous solution of pure zinc(II) sulfate.

kept constant and so has the pressure of hydrogen gas throughout the calibration and the determination. Regardless of whether $a_{H^+}$ or $c_{H^+}$ is used, this expedient permitted the expansion of equation (1.8) to yield equation (1.9a) in which the term for the hydrogen half-cell has a constant value. The exact value of this term need not be known (0.118 V is an approximation) for it has been incorporated into a constant term enclosed in brackets in equation (1.9b). Recall that this constant also includes a term for $E_{cell}^{\circ}$ which is a thermodynamically defined constant whose value is unknown (for the actual electrodes used) and the liquid junction potential $E_j$ which is also unknown but assumed to be constant. Provided, therefore, that provision is made in the design of the hydrogen half-cell to ensure that $c_{H^+}$ remains constant throughout the calibration and the determination, the constant terms should remain constant during the time taken for the analysis. If this condition is met, then the hydrogen half-cell $H^+(0.01\ M)|H_2$ (1 atm); Pt has been used as the *reference* half-cell. Such half-cells are commonly termed *reference electrodes*. The appropriate term for the $Zn(Hg)|Zn^{2+}$ half-cell is the *indicator electrode* because, after suitable calibration, it has been possible to deduce the concentration of the determinand (the zinc ion) from the indicated potential of the zinc half-cell with respect to the reference electrode, that is, from the measured value of $E_{cell}$.

**Practical Impediments and How They Are Overcome**

Having described the basic definitions and concepts associated with direct potentiometry it is now pertinent to enquire whether the example cell could be used for the routine determination of zinc ion in samples likely to be submitted for analysis. Unfortunately, the method previously described is not suitable for two reasons:

1. The indicator electrode is too reactive.
2. The reference electrode is too cumbersome.

These objections are discussed below.

1. The fact that zinc (a reductant) reacts with hydrogen ion (an oxidant) is the portent that zinc will also react with oxidants having a less negative standard reduction potential than that listed for $Zn^{2+} + 2e \rightleftharpoons Zn$ in Table 1.3. Although it is theoretically possible to devise an electrode for every process for which a stoichiometrically balanced half-reaction of the type

$$\text{oxidant} + ne \rightleftharpoons \text{reductant}$$

can be written, the use of such electrodes in analysis by direct potentiometry is almost totally inhibited by their reactivity. Fortunately, *ion selective electrodes*, the potentials of which develop mainly as a result of mass exchange across an interface in contrast to an electron exchange across an interface, are much more specific for the electroactive species to which they respond. This type of electrode also develops a potential which is linearly dependent on the logarithm of the concentration of a given electroactive species in solution so that the basic principles governing calibration and use are similar to those already outlined in this chapter. The period since their development in the 1960s has been notable for a resurgence of interest in the application of direct potentiometry to analytical chemistry. Ion selective electrodes are discussed in Chapters 3 and 4 and feature prominently throughout the subsequent portions of this book.

2. In addition to making a constant contribution to the value of $E_{cell}$ (see equation 1.9b), a reference electrode should be compact and, for preference, easily transportable. The hydrogen reference electrode with its accompanying gas cylinder, gas purification, and presaturation train, although very suitable as an electrode for some types of work requiring a high degree of precision, is too cumbersome for routine applications. One of the attractive features of analysis by direct potentiometry is that the equipment can be taken to perform the determination at the place where the sample exists in its unique environment. These determinations are performed using ion selective electrodes that are easily transportable and the reference electrode against which their potential is measured must be equally robust and transportable. Electrodes suitable as reference electrodes are described in Chapter 3.

## 1.2 ACTIVITY–CONCENTRATION RELATIONS

In the preceding section it was necessary to circumvent a seeming impasse by making a simplifying, but erroneous, assumption that the activity of an ion (its effective concentration) was equal to the stoichiometric concentration of that ion for identical units of concentration. This assumption was, in effect, imposing a condition of ideality on the solution (zinc sulfate) by supposing that each type of ion ($Zn^{2+}$ and $SO_4^{2-}$) exists in the solution as a discrete entity and that no forces of attraction exist between these ions of opposite charge nor between the ions and the solvent. Such a condition would be approached only if the solution were diluted so that the ionic concentration tended towards an infinitely small value. The divergence between ideal and real behavior for solutions of finite concentrations is allowed for by a factor termed the *activity coefficient* which relates the activity of a species $i$, in the example an ion, to its stoichiometric concentration (usually molal or molar) by

$$a_i = m_i \gamma_i \, (\text{mole kg}^{-1}); \quad a_i = c_i y_i \, (\text{mole L}^{-1}) \quad (1.10)$$

The terms $\gamma_i$ and $y_i$ are the activity coefficients corresponding to the respective concentration scales molal ($m$) and molar ($c$). Interconversion of activity coefficients between the two scales of concentration is achieved by

$$y = \gamma d_0 \frac{m}{c} \quad (1.11)$$

where $d_0$ is the density of the solvent. The activity coefficient is a dimensionless quantity the value of which becomes equal to unity at infinite dilution, and often at another concentration.

### Definition of Activities

When a solution of strong electrolyte, represented here as $A_rB_s$, is prepared in the laboratory so that its concentration is known to be $m$ (mole kg$^{-1}$), there are two methods by which its activity can be expressed in terms of a measurable quantity known as the mean *activity coefficient*. On dissolving the solid the process is

$$A_rB_s \rightarrow rA^{z+} + sB^{z-}$$

and in the solution, therefore, the concentration of positive $(m_+)$ and negative $(m_-)$ ions are related to the total concentration, $m$ by $m_+ = rm$ and $m_- = sm$. Using the appropriate definition of an activity coefficient given by equation (1.10), the respective activities of the ions in the solution are

$$a_+ = rm\gamma_+ \quad \text{and} \quad a_- = sm\gamma_-$$

The *activity of the electrolyte* as a whole, $a_2$, is defined as

$$a_2 = a_+^r \cdot a_-^s$$
$$= (rm\gamma_+)^r \cdot (sm\gamma_-)^s$$

or

$$a_2 = (r^r \cdot s^s) m^{r+s} \cdot \gamma_+^r \gamma_-^s \qquad (1.12)$$

It is often more convenient, however, to express the activity of the electrolyte in terms of the geometric mean activity of the positively and negatively charged ions it contains with the object of converting the activity coefficient terms in equation (1.12) into the mean activity coefficient. To this end the total number of moles of ions $(r + s)$ derived from 1 mole of $A_rB_s$ is defined by $\nu$ and the *mean activity* of the solution $a_\pm$ is given by

$$a_\pm = (a_+^r \cdot a_-^s)^{1/\nu} = (a_2)^{1/\nu}$$

Combining this definition with equation (1.12) yields

$$a_\pm = (r^r \cdot s^s)^{1/\nu} m (\gamma_+^r \cdot \gamma_-^s)^{1/\nu}$$

# 16  RUDIMENTS OF POTENTIOMETRY

The term $(\gamma_+^r \gamma_-^s)^{1/\nu}$ is the *mean activity coefficient* and is given the symbol $\gamma_\pm$. Hence

$$a_\pm = (r^r \cdot s^s)^{1/\nu} m \cdot \gamma_\pm \quad (1.13)$$

where the quantity $(r^r \cdot s^s)^{1/\nu} m$ is defined as the *mean ionic molality*, $m_\pm$ for the electrolyte solution. Thus

$$a_\pm = m_\pm \cdot \gamma_\pm$$

Sometimes an expression containing the activity of an electrolyte is obtained and it is useful also to express equation (1.12) in terms of the mean activity coefficient. Thus from the preceding definitions the appropriate equation is

$$a_2 = (r^r \cdot s^s) m^\nu \gamma_\pm^\nu \quad (1.14)$$

The results of these definitions, when applied to solutions of some electrolytes, are summarized in Table 1.1. The potentiometric method used for the determination of a mean activity coefficient $\gamma_\pm$ for an electrolyte is discussed later.

Table 1.1. Relation of Activity ($a_2$), Mean Activity ($a \pm$), Mean Activity Coefficient ($\gamma \pm$), and Ionic Strength ($I$) to Concentration[a] ($m$) for Some Electrolytes

| Electrolyte $A_r B_s$ | $r$ | $s$ | $\nu$ | $a \pm$ (Eq. 1.13) | $a_2$ (Eq. 1.14) | $I$ (Eq. 1.15) |
|---|---|---|---|---|---|---|
| KCl | 1 | 1 | 2 | $m\gamma_\pm$ | $m^2 \gamma_\pm^2$ | $m$ |
| ZnSO$_4$ | 1 | 1 | 2 | $m\gamma_\pm$ | $m^2 \gamma_\pm^2$ | $4m$ |
| AlPO$_4$ | 1 | 1 | 2 | $m\gamma_\pm$ | $m^2 \gamma_\pm^2$ | $9m$ |
| Na$_2$SO$_4$ | 2 | 1 | 3 | $1.5874\, m\gamma_\pm$ | $4m^3 \gamma_\pm^3$ | $3m$ |
| CaCl$_2$ | 1 | 2 | 3 | $1.5874\, m\gamma_\pm$ | $4m^3 \gamma_\pm^3$ | $3m$ |
| Na$_3$PO$_4$ | 3 | 1 | 4 | $2.2795\, m\gamma_\pm$ | $27m^4 \gamma_\pm^4$ | $6m$ |
| La(NO$_3$)$_3$ | 1 | 3 | 4 | $2.2795\, m\gamma_\pm$ | $27m^4 \gamma_\pm^4$ | $6m$ |
| Ca$_3$(PO$_4$)$_2$ | 3 | 2 | 5 | $2.5509\, m\gamma_\pm$ | $108m^5 \gamma_\pm^5$ | $15m$ |
| La$_2$(SO$_4$)$_3$ | 2 | 3 | 5 | $2.5509\, m\gamma_\pm$ | $108m^5 \gamma_\pm^5$ | $15m$ |
| Na$_6$A$^{6-b}$ | 6 | 1 | 7 | $4.6450\, m\gamma_\pm$ | $46656m^7 \gamma_\pm^7$ | $21m$ |

[a] The same relations are valid for concentrations in mole L$^{-1}$ if the mean activity coefficient is represented by $y \pm$ (equation 1.10).
[b] Represents the hexasodium salt of a hexaprotic acid such as benzenehexacarboxylic acid.

## Calculation of Activity Coefficients

The general formulation of the relationship between the concentration of an ion, $m_i$ (molal) or $c_i$ (molar) and its corresponding activity coefficient ($\gamma_i$ or $y_i$) is based upon the Debye–Huckel equation (2). An important term in this formulation is the ionic strength $I$ which is a measure of the intensity of the electrical field due to the ions in a solution and is defined as

$$I = 0.5 \Sigma m_i z_i^2 \; (\text{mole kg}^{-1}); \quad I = 0.5 \Sigma c_i z_i^2 \; (\text{mole L}^{-1}) \quad (1.15)$$

where $z_i$ is the valency of the ion. The ionic strength is related to the activity coefficient of an ion by the Debye–Huckel equation

$$-\log \gamma_i = \frac{A z_i^2 (I)^{1/2}}{1 + B a_i (I)^{1/2}} \quad (1.16)$$

The terms $A$ and $B$ are constants which vary with the dielectric constant and temperature of the pure solvent. For water the values of $A$ are approximately 0.511 and 0.507 at 25 and 20°C, respectively and $B \cong 0.329$ and $0.328 \times 10^8$ cm$^{-1}$ mole$^{1/2}$ (kg or L)$^{1/2}$ at the corresponding temperatures for both the molal and molar scales of concentration. The term $a_i$ is the "ion size parameter" or "the mean distance of closest approach" of the ions and is usually expressed in angstrom units. More accurate values of $A$ and $B$ in water medium for both scales of concentrations at various temperatures are listed in the appendix (Table A.2). Estimates of the values for $a_i$ made by Kielland (7) for various inorganic and organic ions obtained from data for the ionic mobilities, radii in the crystalline solid, deformability, and hydration numbers are also given in the appendix (Table A.3). For most practical purposes, however, solutions contain at least two ionic substances and the assignation of a value to $a_i$ becomes a problem. In some instances $a_i$ can be regarded as an adjustable parameter which allows the Debye–Huckel equation to be used for the linear extrapolation of an experimentally derived quantity to zero ionic strength. In other determinations the lack of definition for $a_i$ imposes a limiting condition for the application of a method.

## Forms of the Debye–Huckel Equation

The simplest form of the Debye–Huckel equation is obtained by assuming that the ion exists as a point charge for which $a_i = 0$. In this form equation (1.16) is known as the Debye–Huckel limiting law and at 25°, when the constant $A$ is 0.511, is approximated to

$$-\log \gamma_i = 0.5 z_i^2 (I)^{1/2}$$

For an electrolyte,

$$A_r B_s \to rA^{z+} + sB^{z-}$$

both this equation and equation (1.16) can be adapted to calculate the mean activity coefficient $\gamma_\pm$. Using the definition that

$$\gamma_\pm = (\gamma_+^r \gamma_-^s)^{1/\nu}$$

it follows that

$$\nu \log \gamma_\pm = r \log \gamma_+ + s \log \gamma_- \tag{1.17}$$

Substituting the Debye–Huckel limiting law for the ion activity coefficients on the right-hand side of this equation yields

$$-\nu \log \gamma_\pm = [rz_+^2 + sz_-^2] 0.5 (I)^{1/2}$$

The condition of electroneutrality for the electrolyte as a whole demands that $rz_+ = sz_-$ and, since $\nu = r + s$, $\nu$ will also equal $(rz_+^2 + sz_-^2)/z_+ z_-$. Thus

$$-\log \gamma_\pm = 0.5 z_+ z_- (I)^{1/2}$$

for which the values of $z_+$ and $z_-$ are absolute integers; that is, they are taken without regard to sign. Thus there are two principal forms of the Debye–Huckel equation (25°): one for the mean activity coefficient of an electrolyte and the other for the activity coefficient of an ion. These are

$$-\log \gamma_\pm = \frac{0.511 z_+ z_- (I)^{1/2}}{1 + 0.328 a_i (I)^{1/2}} \quad \text{and} \quad -\log \gamma_i = \frac{0.511 z_i^2 (I)^{1/2}}{1 + 0.328 a_i (I)^{1/2}}$$

$$\tag{1.18}$$

for values of $a_i$ in angstrom units. There are, however, empirical variants of these equations and some of the more commonly used forms are summarized in Table 1.2. It should be noted that the upper limit of validity for

## ACTIVITY–CONCENTRATION RELATIONS

**Table 1.2. Commonly used Equations for the Calculation of Single Ion Activity Coefficients (Molal Scale)**

| Equation | Range of Validity | Author | Ref. |
|---|---|---|---|
| $-\log \gamma_i = Az_i^2 I^{1/2}$ | $I < 10^{-3}$ | Debye and Huckel | 2 |
| $-\log \gamma_i = Az_i^2 I^{1/2}/(1 + Ba_i I^{1/2})$ | $I < 10^{-2}$ | Debye and Huckel | 2[a] |
| $-\log \gamma_i = Az_i^2 I^{1/2}/(1 + I^{1/2})$ | $I < 0.1$ | Guntelberg | 3 |
| $-\log \gamma_i = Az_i^2 I^{1/2}/(1 + 1.5 I^{1/2})$ | $I < 0.1$ | Scatchard | 4 |
| $-\log \gamma_{Cl^-} = A I^{1/2}(1 + 1.5 I^{1/2})$ | $I < 0.1$ | Bates and Guggenheim | 5[b] |
| $-\log \gamma_i = Az_i^2 \{[I^{1/2}/(1 + I^{1/2})] - 0.2I\}$ | $I < 0.1$ | Davies | 6 |

[a] Values for $A$ and $B$ for both molal and molar scales of concentration in aqueous solutions are given in the appendix, Table A.2; Kielland's estimate of $a_i$ for various ions are given in Table A.3.
[b] Known as the Bates–Guggenheim Convention.

an equation depends upon the ionic strength of a solution rather than upon its concentration. For example, the ionic strength for a 0.05 $m$ solution of potassium chloride is given by the application of equation (1.15) as

$$I = 0.5(0.05 \times 1^2 + 0.05 \times 1^2) = 0.05\ m,$$

whereas for a 0.05 $m$ solution of lanthanum(III) nitrate it is

$$I = 0.5(1 \times 0.05 \times 3^2 + 3 \times 0.05 \times 1^2) = 0.3\ m$$

If a given form of the Debye–Huckel equation is said to be valid up to $I = 0.1$, then that equation could be applied to 0.05 $m$ potassium chloride ($I = 0.05$) but not to 0.05 $m$ lanthanum nitrate ($I = 0.3$). Consider now the electrolyte $Na_6A$. Reference to column 7 of Table 1.1 reveals that the factor relating ionic strength to concentration is $I = 21\ m$. The maximum value of the concentration of $Na_6A$ for which such a form of the Debye–Huckel equation ($I \leq 0.1$) is valid will be only $0.1/21 = 0.0048\ m$.

### Determination of Mean Activity Coefficients

The method described here requires the design and construction of a reversible cell so that the measured e.m.f. of the cell will be related to the activity of the electrolyte as a whole ($a_2$) by

$$E_{cell} = E_{cell}^o - \frac{k}{n} \log a_2 \qquad (1.19)$$

This requirement is met for an aqueous solution of pure electrolyte $A_rB_s$ of known concentration ($m$) by a cell that can be represented as

$$\text{electrode } A | A^{z+} B^{z-} | \text{electrode } B$$

It is implicit in the cell notation that electrode $A$ is reversible to species $A^{z+}$ and the electrode $B$ to $B^{z-}$ and, in contrast to the example cell of Section 1.1, the design of this cell requires the elimination of the salt bridge. Recall that in the cell

$$\text{Zn} | \text{Zn}^{2+} \| \text{H}^+ | \text{H}_2 ; \text{Pt}$$

the salt bridge prevented reaction between zinc and hydrogen ion under the operational condition that negligible current was drawn from the cell. It is also equally important in the design of a cell without liquid junction to prevent any possible reaction between the materials used in the construction of electrode $A$ with species $B^{z-}$.

Given that the conditions of thermodynamic reversibility are approached not only by the electrodes used in the construction of the cell, but also by the method used for the measurements of $E_{\text{cell}}$, the determination of the mean activity coefficient ($\gamma_\pm$) for the electrolyte solution containing $rA^{z+}$ and $sB^{z-}$ requires the usual stages:

1. Calibration of the cell.
2. Use of the calibrated cell for the determination.

1. The quantity required for the calibration is $E°_{\text{cell}}$ at the carefully controlled temperature $T$. To this end equation (1.19) is expanded for the term $a_2$ ($a_2 = r^r \cdot s^s \cdot m^\nu \gamma_\pm^\nu$) and then rearranged to yield

$$E_{\text{cell}} + \frac{k}{n} \log r^r \cdot s^s \cdot m^\nu = E°_{\text{cell}} - \frac{k}{n} \log \gamma_\pm^\nu \quad (1.20)$$

The terms on the left-hand side of this equation are all known because $E_{\text{cell}}$ is the measured quantity for a known concentration ($m$) of electrolyte, $n$ is the number of moles of electrons ($r_{z+}$ or $s_{z-}$), and the remaining terms arise from the definition of $a_2$. If it is now stipulated that the concentrations of the solutions of electrolyte to be used for the calibration are such that none will give rise to an ionic strength greater than 0.1 $m$, the term $\gamma_\pm$ can be approximated by

$$-\log \gamma_\pm = \frac{A z_+ \cdot z_- I^{1/2}}{1 + B a_i I^{1/2}}$$

whereupon equation (1.20) becomes

$$E_{cell} + \frac{k}{n}\log r^r \cdot s^s \cdot m^\nu = E_{cell}^\circ + \frac{k}{n}\nu Az_+ \cdot z_- \frac{I^{1/2}}{1 + Ba_i I^{1/2}}$$

$$= E_{cell}^\circ + \text{constant} \frac{I^{1/2}}{1 + Ba_i I^{1/2}} \quad (1.21)$$

At the given temperature for the given electrolyte the numerical value of "constant" is known and so is $B$ in the Debye–Huckel equation. The aim is now to use $a_i$ as an adjustable parameter such that when the values of the left-hand side of the equation are plotted against the corresponding values of $I^{1/2}/(1 + Ba_i I^{1/2})$, a linear relation is obtained the slope of which is numerically equal to the value of constant in equation (1.21). Such a series of calculations can be programmed to be performed by a digital computer, or in some instances by desk calculator using regression analysis. The value of the intercept ($I = 0$) when the calculated slope becomes equal to constant is the required quantity of $E_{cell}^\circ$. It may not always be necessary to adopt this type of procedure, however, and a plot of the experimental values on the left-hand side of equation (1.21) versus $I^{1/2}/(1 + I^{1/2})$ could yield a value for $E_{cell}^\circ$ that does not differ significantly from the value that would be obtained by the more elaborate method of calculation outlined previously.

2. The evaluation of the mean activity coefficient for each of the solutions used in the calibration can be performed using equation (1.19) since the only unknown in that equation is now $a_2$ and this has already been defined in terms of the known concentration $m$ and the required mean activity coefficient. Unlike the calibration procedure outlined in Section 1.1, however, there is no need to be restricted in this determination to the upper limit of concentration for which the calibration has been performed because no extra-thermodynamic assumptions have been made.

### Design Features of a Cell Without Liquid Junction

In the preceding description of the general method for the determination of mean activity coefficients the cell without liquid junction was represented as

$$\text{electrode } A | A^{z+} B^{z-} | \text{electrode } B$$

It is now necessary to be more specific as to what is meant by the term a *cell without liquid junction* and to consider the essential design features of this type of cell. Consider, for example, a cell that could be used for the determination of the mean activity coefficients of zinc sulfate solutions:

$$\text{Zn electrode}|\text{ZnSO}_4|\text{sulfate electrode}$$

A zinc amalgam electrode of the type described in Section 1.1 would be a suitable indicator electrode for zinc ion $[\text{Zn(Hg)}|\text{Zn}^{2+}]$ and its half-reaction can be represented by the equilibrium

$$\text{Zn}^{2+} + 2e \rightleftharpoons \text{Zn(Hg)}_{(l)}$$

The only appreciable soluble species that must appear in the other half-cell reaction is the sulfate ion and this restricts the choice of suitable half-reactions to those that include a sparingly soluble sulfate such that

$$\text{M}_x(\text{SO}_4)_{y(s)} + (2y)e \rightleftharpoons x\text{M}_{(s)} + y\text{SO}_4^{2-}$$

Furthermore, it must be specified that the pure metal M must not react with the solvent, in this case water, and the solubility of the metal sulfate should be as low as possible. A possible half-reaction that partially fulfils these specifications is

$$\text{PbSO}_{4(s)} + 2e \rightleftharpoons \text{Pb}_{(s)} + \text{SO}_4^{2-}$$

Imposing the condition that all species represented by the half-reaction must coexist at the interface of the corresponding half-cell leads to $\text{SO}_4^{2-}|\text{PbSO}_{4(s)}; \text{Pb}_{(s)}$ as a half-cell corresponding to $B^{z-}|$electrode $B$ in the general notation. If the half-cell is to be a practical proposition, lead must be present as an amalgam for the same reasons as those already outlined for metallic zinc. Thus the cell reversible to the activity of zinc sulfate can be represented as

$$\text{Zn(Hg)}|\text{Zn}^{2+}\text{SO}_4^{2-}|\text{PbSO}_4; \text{Pb(Hg)}$$

and the equilibrium cell reaction as

$$\text{Zn(Hg)}_l + \text{PbSO}_{4(s)} \rightleftharpoons \text{Pb(Hg)}_l + \text{Zn}^{2+} + \text{SO}_4^{2-}$$

The Nernst equation, corresponding to equation (1.19), for this reaction is, therefore,

$$E_{\text{cell}} = E_{\text{cell}}^\circ - \frac{k}{2}\log a_{\text{Zn}^{2+}} \cdot a_{\text{SO}_4^{2-}} = E_{\text{cell}}^\circ - \frac{k}{2}\log a_2 \quad (1.22a)$$

which leads to

$$E_{cell} + \frac{k}{2}\log m^2 = E_{cell}^\circ - \frac{k}{2}\log \gamma_\pm^2 \qquad (1.22b)$$

as the expression corresponding to equation (1.20). The validity of equation (1.22), however, depends upon the validity of the reaction which is its basis. The cell equation assumes lead sulfate to be present wholly as a solid and this is also assumed to be the case in the representation of the cell by the notation. Both ignore the fact that lead sulfate is slightly soluble, its solubility product $K_{sp}$ being about $2 \times 10^{-8}$ at 25°C. Thus the activity of lead ion in the cell when it contains a 0.01 $m$ solution of zinc sulfate will be about $2 \times 10^{-8}/10^{-2} = 2 \times 10^{-6}$ mole kg$^{-1}$. A tenfold dilution of the zinc sulfate in the cell solution will increase the concentration of lead ion by approximately tenfold. For these two conditions the cell can be represented by the notations

$$\text{Zn(Hg)}|\text{Zn}^{2+}(0.01\ m), \text{SO}_4^{2-}(ca.\ 0.010002\ m),$$

$$\text{Pb}^{2+}(ca.\ 2 \times 10^{-6})|\text{PbSO}_4; \text{Pb(Hg)}$$

and

$$\text{Zn(Hg)}|\text{Zn}^{2+}(0.001\ m), \text{SO}_4^{2-}(ca.\ 0.00102\ m),$$

$$\text{Pb}^{2+}(ca.\ 2 \times 10^{-5})|\text{PbSO}_4; \text{Pb(Hg)}$$

respectively. Although the cell notations contain no indication of the presence of a salt bridge, there is nonetheless a liquid junction or diffusion potential whose magnitude depends on the differences in concentration existing within the two half-cells. The cell can, therefore, be represented as

$$\text{Zn(Hg)}|\text{ZnSO}_4(m) \vdots\vdots \text{Zn}^{2+}(m), \text{Pb}^{2+}\left(ca.\ \frac{K_{sp}}{m}\right),$$

$$\text{SO}_4^{2-}\left[ca.\ \left(m + \frac{K_{sp}}{m}\right)\right]|\text{PbSO}_4; \text{Pb(Hg)}$$

The more dilute the concentration of zinc sulfate in the cell, the greater the solubility of an interfering ion (Pb$^{2+}$)*, and the greater becomes the depar-

―――――――
*Interfering reaction is Pb$^{2+}$ + Zn$_{(s)}$ → Pb$_{(s)}$ + Zn$^{2+}$.

ture of the actual sulfate ion concentration in the cell from the value of its stoichiometric concentration in the prepared zinc sulfate solution. Thus the diffusion potential can be expected to increase across the interface $PbSO_4|$ solution, represented by $\vdots$, as the solution of zinc sulfate is made more dilute. Whether the two half-cells in a cell "without" liquid junction need to be physically separated (as in Figure 1.3) depends upon the solubility of the materials used in the construction of the electrodes and upon the concentrations of the two electroactive species contained in the cell solution. This applies particularly to the electrodes that are reversible to anions in the commonly used types of cells "without" liquid junction. These are of the type

anion|metal–anion salt; metal

and are referred to as *electrodes of the second kind* (two solids in the half-reaction). It is the solubility of the metal–anion salt ($PbSO_4$ in the preceding example), as reflected in the magnitude of its solubility product, that dictates the lower limit of concentration for which these electrodes can be used, and ideally $K_{sp} \rightarrow 0$. It is now necessary to state that for most anion-reversible electrodes of this type in regular use, the solubility of the

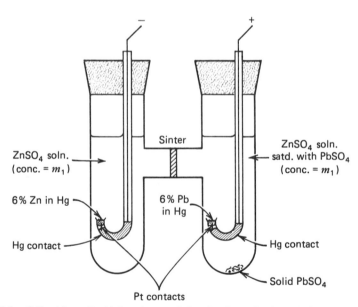

Figure 1.3. Cell without liquid junction used for the determination of the mean activity coefficients of zinc(II) sulfate solutions.

metal–anion salt is much less than for the example $SO_4^{2-}\,|\,PbSO_4;\,Pb(Hg)$. The most commonly used electrode of the second kind employed in cells without liquid junction is the silver–silver chloride electrode ($Cl^-\,|\,AgCl;\,Ag$). The value of $K_{sp}$ for silver chloride ($1.5 \times 10^{-10}$ at 25°) is about 100 times less than that for lead sulfate. The solubility product for a salt is always finite, however, and when using cells without liquid junction of the type under discussion it is necessary to design the conditions of the experiment so that it is possible to assume that the diffusion potential will be very small and will make a negligible contribution to the calibration factor, $E_{cell}^{\circ}$. In the calibration of a cell the best experimental conditions for the procedure are usually a compromise between conflicting factors:

1. The desire to work at low ionic strengths to prevent a long, and possibly uncertain, extrapolation of experimental quantities (see equation 1.21) to a condition where $I = 0$.
2. The realization that at low concentrations the solubility of the electrode material can become a problem for reasons stated previously.

Returning now to the determination of the mean activity coefficients for zinc sulfate solutions, equation (1.22) can be expanded to

$$E_{cell} + k \log m = E_{cell}^{\circ} + k \cdot \frac{4A \cdot I^{1/2}}{1 + Ba_i I^{1/2}}$$

Specifying that the experiment will be performed at a constant temperature of 20°, the numerical values of the constants appropriate to this temperature are $k = 0.05817$, $A = 0.507$, and $B = 0.329$ whereupon the equation becomes

$$E_{cell} + 0.05817 \log m = E_{cell}^{\circ} + 0.118\left(\frac{I^{1/2}}{1 + 0.329 a_i I^{1/2}}\right) \quad (1.23)$$

The upper limit of concentration for zinc sulfate ($m$) that can be used for the calibration is imposed by the upper limit of the ionic strength for which equation (1.23) is likely to yield a linear relation. This is assumed to be $I = 0.1$ which imposes the condition that the concentration of zinc sulfate cannot exceed $0.1/4 = 0.025$ mole kg$^{-1}$.

The lower limit of concentration is determined ultimately by the precision of the instrument used to measure $E_{cell}$. If it is desired to measure $E_{cell}$ to $\pm 0.01$ mV, very approximate calculations using $K_{sp}^{*}$ for lead sulfate

*Relative error (mV) $\cong \frac{58}{2} \log[(m + K_{sp}/m)/m]$

suggest that the lower limit of concentration will be about 0.005 $m$, with a concentration of 0.002 $m$ as an approximate limit if the instrument is calibrated to read 0.1 mV. In either case, therefore, the calibration can be performed only over a limited range of concentration, the limits being 0.005–0.025 $m$ ($I = 0.02 - 0.1$ $m$) for the more precise instrument ($\pm 0.01$ mV).

The calibration procedure requires the preparation of at least six solutions of pure zinc sulfate prepared in pure (deoxygenated) water, the known concentrations of which are within the range 0.005–0.025 $m$. $E_{cell}$ is measured and the value of the left-hand side of equation (1.23) calculated for each of these concentrations. Thereafter, the calculation by the method of successive approximations which is applied to the adjustable parameter $a_i$ can be performed as outlined elsewhere in this book. An alternative method in this particular case is to replace the bracketed term on the right-hand side of equation (1.23) with $I^{1/2}/(1 + I^{1/2}) - 0.21$ (Table 1.2) because the range of ionic strength is necessarily small and the values relatively high. The individual values of ($E_{cell} + 0.05817 \log m$) when plotted against the corresponding value of $\{(4m)^{1/2}/[1 + (4m)^{1/2}] - 0.2(4m)\}$ should yield a straight line of intercept $E_{cell}^\circ$.

The value of $E_{cell}^\circ$ can then be used to calculate the mean activity coefficient for a solution of zinc sulfate more concentrated than 0.005 $m$ by measuring $E_{cell}$ at 20° for the known concentration $m$ and using the expression (see equation 1.22)

$$E_{cell} + 0.05817 \log m = E_{cell}^\circ + 0.05817 \log \gamma^\pm$$

For a solution more dilute than 0.005 $m$, however, use must be made of the extrapolated portion of the calibration graph and the mean coefficient can be interpolated from this.

### Quasi-Thermodynamic Cells

A systematic study of the values of the mean activity coefficients for different electrolytes forms an important basis for the establishment of the ion activity standards discussed in Chapter 4. However, the values of the mean activity coefficients for only one specific electrolyte (e.g., $ZnSO_4$) are of little use in the calibration of a cell with liquid junction. Imagine now that the cell

$$Zn | ZnSO_4 \| HCl\ (0.01\ M) | H_2\ (1\ atm); Pt$$

used as the example cell in Section 1.1, is transformed so that it consists of a perfect zinc electrode and a perfect reference electrode with such an inbuilt salt bridge that no liquid junction potential is induced. That is,

zinc electrode | $ZnSO_4$ || reference electrode

Such a cell, of course, does not exist, but if it did, the measured $E_{cell}$ would still be related to the activity of zinc ion by an equation analogous to (1.9b). The correct form of this equation would be

$$E_{cell} = \text{constant} - \frac{k}{2} \log a_{Zn^{2+}} = \text{constant} - \frac{k}{2} \log m_{Zn^{2+}} \cdot \gamma_{Zn^{2+}}$$

(1.24)

in which the value of constant includes the value for the potential of the "perfect" reference electrode. The reason why it is not thermodynamically valid to use known values of the mean activity coefficients for zinc sulfate in the calibration of the cell is deducible from the definition of mean activity coefficients (equation 1.17), which for zinc sulfate is

$$2 \log \gamma_\pm = \log \gamma_{Zn^{2+}} + \log \gamma_{SO_4^{2-}}$$

(1.25)

Calculation of $\gamma_{Zn^{2+}}$ is not possible from this equation unless $\gamma_{SO_4^{2-}}$ is known and the latter quantity can neither be determined experimentally nor can it be calculated from the Debye–Huckel equation without recourse to extra-thermodynamic assumptions. Thus the results obtained with this hypothetically perfect cell would only have thermodynamic validity when the concentration of zinc sulfate is very close to zero. By analogy it is possible to extend this argument to the most commonly used type of cell in analysis by direct potentiometry. These are of the type

indicator electrode $A$ | species A || reference electrode

in which the liquid junction potential is assumed to be constant. For solutions of finite concentrations it is neither possible to calibrate these cells nor to interpret subsequently measured $E_{cell}$ values in terms of activities with strict thermodynamic validity. It is, however, possible to use such a cell for the measurement of single ion activities provided that experimental conditions are selected such that there are good reasons for supposing that the results will bear a very close resemblance to those which would be obtained under rigorous thermodynamic conditions. It is to emphasize this supposition that cells used for the measurement of single ion activities are described as *quasi-thermodynamic*. Thus the most commonly measured single ion activity, the quantity $p\mathrm{H}(-\log a_{\mathrm{H}^+})$, is described as a quasi-thermodynamic quantity. By contrast, the closely related quantity $p(a_\mathrm{H}\gamma_\mathrm{Cl})$ described elsewhere is a thermodynamic quantity because it is derived from the measurement of the e.m.f. of a cell in which both electrodes act as indicator electrodes. This type of cell, represented as

$$\text{electrode } A \,|\, \mathrm{A}^{z+}\,\mathrm{B}^{z-} \,|\, \text{electrode } B$$

can be described as a thermodynamic cell since, under the condition of good experimental design, no extra thermodynamic assumptions are necessary in the interpretation of $E_\mathrm{cell}$ in terms of the activity of the electrolyte as a whole.

## The Concept of Constant Ionic Strength

In spite of the difficulties outlined, it is advisable now to recall that the example cell of Section 1.1 could be calibrated and used under very restricted conditions to determine the concentration of zinc ion in a solution of pure zinc sulfate. There is, in fact, a dichotomy of purpose in performing potentiometric measurements using cells with liquid junction. First, there are those types of measurements for which the activity of an ion is a more fundamental quantity as in the determination of equilibrium constants (e.g., see Chapter 6). The second type of measurement, more important in many routine analytical applications, requires the determination of concentrations. For the latter methods the ionic strength of a sample solution is unknown and this creates a problem in the preparation of suitable standards with which to calibrate the cell. One of the methods used to solve this problem is to create a medium of high and "constant" ionic strength by the addition of an inactive electrolyte such that the standards and the suitably diluted samples are all at the same ionic strength. When the electroactive species is present as a small fraction of the total electrolyte concentration (i.e., inactive electrolyte + electroactive species), it is assumed that the activity coefficients of the electroactive species are constant over a limited range of concentrations for the electroactive species. As an illustration of what is meant by this assumption, consider the standardization of the previous imaginary cell, zinc electrode|$\mathrm{ZnSO_4}$||reference electrode, in the presence of a 3 $M$ solution of sodium perchlorate as the inactive electrolyte, and the subsequent determination of zinc ion in a sample solution. For the calibration the necessary cells can be represented as

$$\text{zinc electrode}|\mathrm{ZnSO_4}(c_1\ldots c_n), \mathrm{NaClO_4}(3\;M)\,\|\,\text{reference electrode}$$

where $c_1 \ldots c_n$ represent known molar concentrations of each standard solution. For a solution of concentration $c_1$ ($c_1 > c_n$) from equation (1.24),

$$E_{\text{cell}_1} = \text{constant} - \frac{k}{2} \log c_1 - \frac{k}{2} \log y_{\text{Zn}^{2+}}$$

and for solution of concentration $c_n$ the corresponding equation is

$$E_{\text{cell}_n} = \text{constant} - \frac{k}{2} \log c_n - \frac{k}{2} \log y_{\text{Zn}^{2+}}$$

The basis of the assumption is that the activity coefficients $y_{\text{Zn}^{2+}}$ (molar scale) will remain constant if the contributions of the concentrations of zinc sulfate to the total ionic strengths are very small in comparison to the total ionic strengths; that is, $I = 3.00 + 4c_{\text{ZnSO}_4} \cong 3.0$. Under this circumstance a plot of the measured values $E_{\text{cell}_1} \ldots E_{\text{cell}_n}$ versus the logarithm of the corresponding concentrations $c_1 \ldots c_n$ should be linear if $c_1$ does not exceed 0.01 $M$.

The validity of the subsequent interpretation of a measured $E_{\text{cell}}$ in terms of $c_{\text{Zn}^{2+}}$ from the calibration graph depends upon some prior knowledge of the likely ionic strength in the sample solution. If it is possible to add sodium perchlorate to the sample solution so that $E_{\text{cell}}$ remains within the range of those values covered by the calibration and, at the same time, the total ionic strength remains at 3.0 $M \pm 2\%$, then it is valid to interpolate $c_{\text{Zn}^{2+}}$ from the calibration graph. This type of procedure might be acceptable in samples of potable waters, for example, since these are generally of low ionic strength. The procedure, however, would be unacceptable in samples of seawater for which the ionic strength is inherently high ($I \cong 0.7\ M$).

The concept of constant ionic strength is used in the formulation of total ionic strength adjustment buffers and in the determination of stability constants described in Chapter 8.

## 1.3 REDUCTION POTENTIALS

### The Significance of $E^{\circ}_{\text{cell}}$

Hitherto in this chapter it has been convenient to regard the term $E^{\circ}_{\text{cell}}$ as a calibration factor. This has been done in order to emphasize that in practice it is always necessary to calibrate the electrodes before any subsequent measurement of the cell e.m.f. can be interpreted in terms of an unknown ion activity. However, as is evident from the equation $\Delta G° = -nFE°_{cell}$ (equation 1.3b), the quantity $E°_{cell}$ is related to the standard free energy change for the cell reaction at a given temperature. Thus it applies specifically to the cell e.m.f. that would be obtained under conditions of unit activity for electroactive species in solution or to unit pressure for any gaseous species appearing in the cell reaction. For example, $E°_{cell}$ for the reaction discussed earlier applies only to the condition

$$PbSO_{4(s)} + Zn_{(s)} \rightleftharpoons Pb_{(s)} + Zn^{2+}\,(a=1) + SO_4^{2-}\,(a=1)$$

in which the activities of both zinc ion and sulfate ion are unity at a given temperature. Reference to Figure 1.3 acts as a reminder that the whole cell consists of two half-cells. The overall value of $E°_{cell}$ is, therefore, a composite of the two unique half-cell standard potentials or $E°$ values at the given temperature. It is customary to associate these two $E°$ values with the constituent half-reactions each expressed as a reduction reaction. Thus the two half-reactions under standard conditions at a given temperature can be represented as

$$Zn^{2+}\,(a=1) + 2e \rightleftharpoons Zn_{(s)}$$

giving rise to the standard reduction potential $E°_{Zn^{2+};Zn}$ and

$$PbSO_{4(s)} + 2e \rightleftharpoons Pb_{(s)} + SO_4^{2-}\,(a=1)$$

for which the standard reduction potential is $E°_{PbSO_4;Pb}$.

In the absence of any other information it can be deduced that $E°_{Zn^{2+};Zn}$ will be more negative than $E°_{PbSO_4;Pb}$ because in order to obtain a positive and meaningful value of the cell potential it is found necessary to connect the zinc half-cell to the negative terminal of the potentiometer. It follows that if values for two standard reduction potentials are available, $E°_{cell}$ is the difference between the more positive (or less $^-ve$) $E°$ value and the more negative (or less $^+ve$) $E°$ value. That is,

$$E°_{cell} = \overset{+}{E}° - \bar{E}°, \qquad (1.26)$$

which becomes

$$E°_{cell} = E°_{PbSO_4;Pb} - E°_{Zn^{2+};Zn}$$

for the example reaction. For the general case when the activities of the electroactive species are not unity, equation (1.26) can be written as

$$E_{\text{cell}} = \overset{+}{E} - \overset{-}{E} \tag{1.27}$$

In this equation both $\overset{+}{E}$ and $\overset{-}{E}$ can be defined in terms of $\overset{+}{E}{}^\circ$ and $\overset{-}{E}{}^\circ$, and the activities of the participating electroactive species in each half-reaction by means of the Nernst equation. Applying this approach to a measured potential of the example cell $\text{Zn(Hg)} | \text{Zn}^{2+}\text{SO}_4^{2-} \overset{+}{|} \text{PbSO}_4; \text{Pb(Hg)}$ for which the constituent half-reactions, each expressed as reductions, are

$$\text{Zn}^{2+} + 2e \rightleftharpoons \text{Zn}_{(s)} \tag{1.28}$$

and

$$\text{PbSO}_{4(s)} + 2e \rightleftharpoons \text{Pb}_{(s)} + \text{SO}_4^{2-} \tag{1.29}$$

yields

$$E_{\text{cell}} = \left( E^\circ_{\text{PbSO}_4; \text{Pb}} - \frac{k}{2}\log a_{\text{SO}_4^{2-}} \right) - \left( E^\circ_{\text{Zn}^{2+}; \text{Zn}} - \frac{k}{2}\log \frac{1}{a_{\text{Zn}^{2+}}} \right) \tag{1.30}$$

The definition of the activity quotient $Q$ remains unaltered by this treatment being $Q = a_{\text{Pb}_{(s)}} \cdot a_{\text{SO}_4^{2-}} / a_{\text{PbSO}_{4(s)}} = a_{\text{SO}_4^{2-}}$ for half-reaction (1.29) and $Q = a_{\text{Zn}_{(s)}} / a_{\text{Zn}^{2+}} = 1/a_{\text{Zn}^{2+}}$ for half-reaction (1.28). Equation (1.30) can also be written as

$$E_{\text{cell}} = \left( E^\circ_{\text{PbSO}_4; \text{Pb}} - E^\circ_{\text{Zn}^{2+}; \text{Zn}} \right) - \frac{k}{2}\log a_{\text{SO}_4^{2-}} \cdot a_{\text{Zn}^{2+}}$$

which, from the definition of $E^\circ_{\text{cell}}$ given by equation (1.26), becomes identical to equation (1.22a); that is,

$$E_{\text{cell}} = E^\circ_{\text{cell}} - \frac{k}{2}\log a_{\text{Zn}^{2+}} \cdot a_{\text{SO}_4^{2-}}$$

**Standard Reduction Potentials of Single Half-Cells**

In the absence of an arbitrarily defined standard reduction potential neither $E^\circ_{\text{PbSO}_3; \text{Pb}}$ nor $E^\circ_{\text{Zn}^{2+}; \text{Zn}}$ could be determined experimentally. This difficulty is surmounted by designating that the half-reaction

$$2\text{H}^+ \,(a = 1) + 2e \rightleftharpoons \text{H}_2 \,(1 \text{ atm})$$

shall be the arbitrary standard and specifying that the standard reduction potential for this half-reaction shall be zero at all temperatures; that is,

$$2H^+ + 2e \rightleftharpoons H_2; \qquad E^\circ = 0.0000 \text{ V at all temperatures}$$

This internationally accepted definition allows quantities such as $E^\circ_{Zn^{2+};Zn}$ and $E^\circ_{PbSO_4;Pb}$ to be expressed in terms of a hypothetical perfect cell that contains a standard hydrogen electrode as a notional reference electrode.* For example, the hypothetical cell

$$Zn|Zn^{2+} \; (a=1) \vdots H^+ \; (a=1)|H_2 \; (1 \text{ atm}); \text{Pt}$$

in which $\vdots$ represents a perfect salt bridge, allows $E^\circ_{Zn^{2+};Zn}$ to be defined at a given temperature. This cell is similar to the example cell of Section 1.1, the zinc half-cell being negative with respect to the hypothetical standard hydrogen half-cell. If we are now told that $E^\circ_{cell}$ is 0.763 V at 25°C, it follows from equation (1.26) that the standard reduction potential $E^\circ_{Zn^{2+};Zn}$ is −0.763 V. A few standard reduction potentials are presented in Table 1.3 and a wide selection is given in Reference 7a. These values comply with the conventionally accepted form (8) oxidant + $ne \rightleftharpoons$ reductant. Reference to the tabulated entry $PbSO_{4(s)} + 2e \rightleftharpoons Pb_{(s)} + SO_4^{2-}$, which for $E^\circ$ is given as −0.356 V, provides the additional data that allows $E^\circ_{cell}$ for the example cell $Zn|Zn^{2+}(a=1), SO_4^{2-}(a=1)|PbSO_4; Pb$ to be calculated at 25°. Application of equation (1.26) yields

$$E^\circ_{cell} = (-0.356) - (-0.763) = 0.407 \text{ V}$$

It should be noted that this value is specific to 25° and it is impossible to predict, for example, what the value of $E^\circ_{cell}$ at 20° would be without reference to other thermodynamic data. A warning is given at this stage that standard reduction potentials used throughout this book should not be confused with the oxidation potentials found in some of the older literature (9). For example, the standard oxidation potential (25°) for zinc is +0.763 V. This is applicable to the oxidation half-reaction $Zn_{(s)} \rightleftharpoons Zn^{2+} + 2e$ and the potential is positive with respect to $H_2 \rightleftharpoons 2H^+ + 2e$. Interconversion of such values to the accepted standard reduction potentials is achieved by writing the oxidation half-reaction in the reverse direction and changing the sign of the potential. Thus $Zn_{(s)} \rightleftharpoons Zn^{2+} + 2e$ (+0.763 V) becomes $Zn^{2+} + 2e \rightleftharpoons Zn_{(s)}$; $E^\circ = -0.763$ V.

**Predictive Uses of Standard Reduction Potentials**

The basic rule for the prediction of the direction of spontaneity for any chemical reaction is that the free energy change associated with the reaction

―――――――――
*Such an electrode has no basis in reality (see cell (1.37)).

Table 1.3. Some Standard Reduction Potentials

| Oxidant + $ne \rightleftharpoons$ Reductant | $E°$, V |
|---|---|
| $Na^+ + e \rightleftharpoons Na$ | −2.713 |
| $Al^{3+} + 3e \rightleftharpoons Al$ | −1.66 |
| $Zn^{2+} + 2e \rightleftharpoons Zn$ | −0.763 |
| $Zn^{2+} + 2e \rightleftharpoons Zn(Hg)$ | −0.763 |
| $Fe^{2+} + 2e \rightleftharpoons Fe$ | −0.44 |
| $Cr^{3+} + e \rightleftharpoons Cr^{2+}$ | −0.38 |
| $Cd^{2+} + 2e \rightleftharpoons Cd$ | −0.403 |
| $PbSO_4 + 2e \rightleftharpoons Pb + SO_4^{2-}$ | −0.356 |
| $PbSO_4 + 2e \rightleftharpoons Pb(Hg) + SO_4^{2-}$ | −0.351 |
| $V^{3+} + e \rightleftharpoons V^{2+}$ | −0.255 |
| $Sn^{2+} + 2e \rightleftharpoons Sn$ | −0.14 |
| $Pb^{2+} + 2e \rightleftharpoons Pb$ | −0.126 |
| $2H^+ + 2e \rightleftharpoons H_2$ | 0.0000 |
| $S_4O_6^{2-} + 2e \rightleftharpoons 2S_2O_3^{2-}$ | 0.09 |
| $AgCl + e \rightleftharpoons Ag + Cl^-$ | 0.222 |
| $Hg_2Cl_2 + 2e \rightleftharpoons 2Hg + 2Cl^-$ | 0.268 |
| $Cu^{2+} + 2e \rightleftharpoons Cu$ | 0.34 |
| $Fe(CN)_6^{3-} + e \rightleftharpoons Fe(CN)_6^{4-}$ | 0.36 |
| $O_2 + 2H_2O + 4e \rightleftharpoons 4OH^-$ | 0.401 |
| $Cu^+ + e \rightleftharpoons Cu$ | 0.52 |
| $MnO_4^{2-} + 2H_2O + 2e \rightleftharpoons MnO_2 + 4OH^-$ | 0.50 |
| $I_3^- + 2e \rightleftharpoons 3I^-$ | 0.545 |
| $MnO_4^- + e \rightleftharpoons MnO_4^{2-}$ | 0.57 |
| $I_2 + 2e \rightleftharpoons 2I^-$ | 0.621 |
| $O_2 + 2H^+ + 2e \rightleftharpoons H_2O_2$ | 0.69 |
| $OBr^- + H_2O + 2e \rightleftharpoons Br^- + 2OH^-$ | 0.76 |
| $Fe^{3+} + e \rightleftharpoons Fe^{2+}$ | 0.771 |
| $Ag^+ + e \rightleftharpoons Ag$ | 0.799 |
| $OCl^- + H_2O + 2e \rightleftharpoons Cl^- + 2OH^-$ | 0.89 |
| $Br_2 + 2e \rightleftharpoons 2Br^-$ | 1.08 |
| $2IO_3^- + 12H^+ + 10e \rightleftharpoons I_2 + 6H_2O$ | 1.19 |
| $MnO_2 + 4H^+ + 2e \rightleftharpoons Mn^{2+} + 2H_2O$ | 1.23 |
| $O_2 + 4H^+ + 4e \rightleftharpoons 2H_2O$ | 1.229 |
| $Cr_2O_7^{2-} + 14H^+ + 6e \rightleftharpoons 2Cr^{3+} + 7H_2O$ | 1.33 |
| $Cl_2 + 2e \rightleftharpoons 2Cl^-$ | 1.358 |
| $Mn^{3+} + e \rightleftharpoons Mn^{2+}$ | 1.51 |
| $MnO_4^- + 8H^+ + 5e \rightleftharpoons Mn^{2+} + 4H_2O$ | 1.51 |
| $2BrO_3^- + 12H^+ + 10e \rightleftharpoons Br_2 + 6H_2O$ | 1.50 |
| $Bi_2O_4 + 4H^+ + 2e \rightleftharpoons 2BiO^+ + 2H_2O$ | 1.59 |
| $Ce^{4+} + e \rightleftharpoons Ce^{3+}$ | 1.61 |
| $MnO_4^- + 4H^+ + 3e \rightleftharpoons MnO_2 + 2H_2O$ | 1.68 |
| $F_2 + 2e \rightleftharpoons 2F^-$ | 2.87 |

should be negative. The accuracy of the prediction and its relevance to the development of a possible laboratory reaction depends upon the selection of thermodynamic data that is completely appropriate to the course of the reaction. When used within this qualification, standard electrode potentials are valuable data that make possible the prediction of the direction of spontaneity for oxidation–reduction reactions. For these reactions to be potentially spontaneous when the reactants are at unit activity, $E^\circ_{cell}$ must be positive since this condition will correspond to a negative value of the standard free energy change, $\Delta G^\circ$ for the reaction (see equation 1.3b). Therefore, if any pair of half-reactions is selected and arranged as represented by the following order:

$$\text{oxidant}_1 + n_1 e \rightleftharpoons \text{reductant}_1 \quad \overline{E}^\circ = \text{more } ^-ve \text{ or less } ^+ve$$

$$\text{oxidant}_2 + n_2 e \rightleftharpoons \text{reductant}_2 \quad \overset{+}{E}^\circ = \text{less } ^-ve \text{ or more } ^+ve$$

a spontaneous reaction is feasible between reductant$_1$ and oxidant$_2$. The reaction is represented by

$$n_2 \text{reductant}_1 + n_1 \text{oxidant}_2 \rightarrow n_1 \text{reductant}_2 + n_2 \text{oxidant}_1$$

as in the following examples:

1. $$Zn^{2+} + 2e \rightleftharpoons Zn_{(s)} \quad E^\circ = -0.76 \text{ V}$$

$$Cu^{2+} + 2e \rightleftharpoons Cu_{(s)} \quad E^\circ = +0.34 \text{ V}$$

Reaction $Zn_{(s)} + Cu^{2+} \rightarrow Cu_{(s)} + Zn^{2+}$; $E^\circ_{cell} = (0.34) - (-0.76)$
$= 1.10 \text{ V}$

2. $$Fe^{3+} + e \rightleftharpoons Fe^{2+} \quad E^\circ = 0.77 \text{ V}$$

$$Cr_2O_7^{2-} + 14H^+ + 6e \rightleftharpoons 2Cr^{3+} + 7H_2O \quad E^\circ = 1.33 \text{ V}$$

Reaction $6Fe^{2+} + Cr_2O_7^{2-} + 14H^+ \rightarrow 2Cr^{3+} + 7H_2O + 6Fe^{3+}$;
$E^\circ_{cell} = 1.33 - 0.77 = 0.56 \text{ V}$

3. $$Cl_2 + 2e \rightleftharpoons 2Cl^- \quad E^\circ = 1.36 \text{ V}$$

$$MnO_4^- + 8H^+ + 5e \rightleftharpoons Mn^{2+} + 4H_2O \quad E^\circ = 1.51 \text{ V}$$

Reaction $10Cl^- + 2MnO_4^- + 16H^+ \rightleftharpoons 2Mn^{2+} + 8H_2O + 5Cl_2$;
$E^\circ_{cell} = 1.51 - 1.36 = 0.15 \text{ V}$

## REDUCTION POTENTIALS

Notice that in all these examples the number of moles of electrons transferred in a given reaction cancel and thereby do not appear in the overall stoichiometric equation. Although it has been necessary in examples 2 and 3 to multiply a given half-reaction in order to achieve this cancellation, the relevant value of $E°$ is not affected. For example, the $E°$ value for $5Cl_2 + 10e \rightleftharpoons 10Cl^-$ is the same as the given value for $Cl_2 + 2e \rightleftharpoons 2Cl^-$. These balanced chemical equations not only represent the spontaneously occurring "test tube" reactions, but they also represent the overall cell reaction. Any one of these reactions can form the basis for the design of a cell the potential of which could, in theory, be dependent upon the activities of the participating electroactive species. The data of example 2 are used as an illustration first of the basic concept of cell design and, second, of how the data are used in the formulation of the appropriate Nernst equation. The half-reaction $Fe^{3+} + e \rightleftharpoons Fe^{2+}$ has a less positive $E°$ value than the $E°$ value for $Cr_2O_7^{2-} + 14H^+ + 6e \rightleftharpoons 2Cr^{3+} + 7H_2O$. In the standard state of unit activities at 25° the $Fe^{3+}$; $Fe^{2+}$ couple will, therefore, be negative with respect to the $Cr_2O_7^{2-}$; $H^+$; $Cr^{3+}$ triad. The corresponding half-cells will be formed if the soluble electroactive species specified by each half-reaction coexists at the interface between the solution and a chemically inert conductor. A platinum electrode is often assumed to be sufficiently inert to provide this nonreactive interface, and thus the cell can be represented by the notation

$$Pt|Fe^{3+}(a=1), Fe^{2+}(a=1) \| Cr_2O_7^{2-}(a=1),$$
$$Cr^{3+}(a=1), H^+(a=1)|Pt \qquad (1.31a)$$

An electrode constructed of a material which is not a species specified in a half-reaction and which provides this "inert" interface is called a *zeroth electrode*. The term arises because such electrodes are used in half-cells for which the specified half-reactions contain *zero* solid phase. In other words, all the participating electroactive species are either present in solution as in the example half-reactions, or a gas in equilibrium with a conjugate solution species as in the couple $H^+$; $H_{2(g)}$; the latter couple is, of course, the basis for the hydrogen electrode.

Recourse to the use of the data contained in tables of $E°$ values can be of assistance in the correct formulation of the Nernst equation either in terms of the overall cell reaction or in terms of each half-reaction. The Nernst equation pertaining to cell (1.31), for example, can be derived by either of these methods as follows:

1. The application of the Nernst equation to the overall cell reaction

$$6Fe^{2+} + Cr_2O_7^{2-} + 14H^+ \rightleftharpoons 2Cr^{3+} + 7H_2O + 6Fe^{3+} \qquad (1.31b)$$

which yields at 25° (Nernst factor $k = 0.0592$ and $n = 6$)

$$E_{cell} = E°_{cell} - \frac{0 \cdot 0592}{6} \log\left(\frac{a^2_{Cr^{3+}} \cdot a^6_{Fe^{3+}}}{a_{Cr_2O_7^{2-}} \cdot a^{14}_{H^+} \cdot a^6_{Fe^{2+}}}\right) \quad (1.32a)$$

2. The application of the Nernst equation separately to the half-reactions

$$Fe^{3+} + e \rightleftharpoons Fe^{2+}; \quad E° = 0.77 \text{ V}$$

$$Cr_2O_7^{2-} + 14H^+ + 6e \rightleftharpoons 2Cr^{3+} + 7H_2O; \quad E° = 1.33 \text{ V}$$

$E_{cell}$ is defined as

$$E_{cell} = \overset{+}{E}_{Cr_2O_7^{2-}; Cr^{3+}} - \overset{+}{E}_{Fe^{3+}; Fe^{2+}}$$

which yields (25°)

$$E_{cell} = \left(\overset{+}{E}° - \frac{0.0592}{6} \log \frac{a^2_{Cr^{3+}}}{a_{Cr_2O_7^{2-}} \cdot a^{14}_{H^+}}\right) - \left(\overset{-}{E}° - \frac{0.0592}{1} \log \frac{a_{Fe^{2+}}}{a_{Fe^{3+}}}\right)$$

or

$$E_{cell} = E°_{cell} - \frac{0.0592}{6} \log\left(\frac{a^2_{Cr^{3+}} \cdot a^6_{Fe^{2+}}}{a_{Cr_2O_7^{2-}} \cdot a^{14}_{H^+} \cdot a^6_{Fe^{2+}}}\right) \quad (1.32b)$$

Both methods lead to the identical equations (1.32) as is to be expected. Superficially, therefore, it appears quite simple to devise a cell apparently endowed with a Nernstian response based upon any pair of half-reactions appearing in the table of $E°$ values. Such an impression, however, is entirely erroneous because the entry for an $E°$ value does not necessarily imply that the value was determined as a result of e.m.f. measurements or indeed that the half-cell reaction is thermodynamically reversible. In fact, most of the values quoted are calculated from equilibrium data derived from thermochemical or equilibrium studies of oxidation-reduction systems (9) rather than from the direct determination of the potential of an electrode system. Thus although reaction (1.31a) represents the correct stoichiometry, equations (1.32) will not account for the observed variation of $E_{cell}$ with changes in the activities of the participating electroactive species. This is because the half-reaction

$$Cr_2O_7^{2-} + 14H^+ + 6e \rightleftharpoons Cr^{3+} + 7H_2O$$

is irreversible and the standard potential is not derived from direct measurement. The reaction mechanism is thought to involve the transistory existence of active intermediates (see Chapter 12) and actually the expression for $E^+$ contained in equations (1.32) is found to be practically independent of the concentration of Cr(III). Furthermore, the potential is also strongly dependent upon the hydrogen ion concentration as is also the case with other reactions that include a large number of moles of hydrogen ion in their stoichiometry. This dependence cannot be predicted from the coefficients in the balanced equation. It is these types of factors that restrict the use of half-cells based on oxidation-reduction systems to a very small number in comparison with the large number that superficially appear possible. For those few that are available, the use of standard reduction potentials is complicated by the fact that concentrated solutions containing more than one electrolyte are encountered frequently in practice. A well-studied example is the half-cell $Pt|Fe^{3+}, Fe^{2+}$. In view of the problem associated with the activity coefficient of a single ionic species, it can be appreciated that the interpretation of the potential of this half-cell in terms of the activity of each of the species $Fe^{3+}$ and $Fe^{2+}$ would have no practical thermodynamic significance. Nevertheless, the relationship can be formulated as

$$E_{Fe^{3+}, Fe^{2+}} = E^\circ_{Fe^{3+}, Fe^{2+}} - \frac{k}{1} \log \frac{c_{Fe^{2+}} \cdot y_{Fe^{2+}}}{c_{Fe^{3+}} \cdot y_{Fe^{3+}}} \quad (1.33)$$

where $c$ represents molar concentrations and $y$ represents the molar activity coefficients. In practice this half-cell is used under conditions in which it is possible to assume that the total ionic strength is constant whereupon the ratio of the molar activity coefficients, $y_{Fe^{2+}}/y_{Fe^{3+}}$ can also be assumed to be a constant. If this constant, specific to a given set of conditions, is incorporated with the standard potential, then equation (1.33) can be written as

$$E_{Fe^{3+}, Fe^{2+}} = (E^\circ + \text{constant}) - k \log \frac{c_{Fe^{2+}}}{c_{Fe^{3+}}} \quad (1.34)$$

or

$$E_{Fe^{3+}, Fe^{2+}} = E^{\circ\prime}_{Fe^{3+}, Fe^{2+}} - k \log \frac{c_{Fe^{2+}}}{c_{Fe^{3+}}} \quad (1.35)$$

Since "constant" of equation (1.34) is specific to a given set of conditions, $E^{o\,\prime}_{Fe^{3+},Fe^{2+}}$ will also be a conditional constant and is known as *the formal reduction potential* which is usually abbreviated to *the formal potential.*

**Formal Potentials**

Superficially, it would appear that equation (1.35), being derived assuming the constancy of a ratio of activity coefficients, is totally descriptive of all formal potentials. This is not the case, for the formal potential not only incorporates in one figure these activity ratios but in many instances can also include effects which may result from acid-base and complexation equilibria, liquid junction potentials, and others such as solvolysis and association effects which may be specific to a given set of experimental conditions. The term *formal* encompasses these conditions because it is descriptive of the concentration term *formality* F, the number of formula weights of the substance per liter of solution. This concentration scale has the advantage over molality or molarity in that it defines unambiguously the composition of a solution without commitment as to the actual ionic or molecular species present. For example, an equiformal mixture of iron(III) and iron(II) in 1F perchloric acid exhibits a formal potential of 0.73 V at 25° with respect to the standard hydrogen half-cell [i.e., $2H^+$ ($a = 1$) + $2e$ ⇌ $H_2$ (1 atm); $E° = 0.000$ V]. The formal potential of the same couple is 0.70 V in 1F hydrochloric acid and a value of 0.6 V is observed in 1F phosphoric acid. These differences arise because the degree to which complex formation occurs between iron(III) and the chloride and phosphate anions is greater than the degree to which the corresponding complexes are formed between iron(II) and these anions. Thus although the hydrochloric acid and phosphoric acid solutions contain the same formal concentrations of both iron(III) and iron(II), the concentration ratio of these electroactive species actually present in each solution is not the same and the net effect is a shift in potential.

Oxidation-reduction couples such as Fe(III); Fe(II), Tl(III); Tl(I), and Ce(IV); Ce(III), represented by half-reactions of the type

$$M^{z+} + ne \rightleftharpoons M^{(z-n)+} \quad (z > n) \quad (1.36)$$

in which hydrogen ion is not an electroactive species are assumed to have formal potentials derived from solutions containing equal formal concentrations of each species rather than that each species should be present at a concentration of 1F. The solution is understood to contain a specified electrolyte in such a comparatively large concentration that the liquid junction potential and the activity coefficients are made independent of variations in the magnitude of the equiformal concentrations of oxidized and reduced species. Although hydrogen ion is not an electroactive participant in half-reactions of this type, it must be considered as a potential-determining ion because the actual concentrations of the specified electroactive species present in the solution depend in most cases upon the $p$H of the solution. This type of dependence is illustrated by the half-reaction

$$[Fe(CN)_6]^{3-} + e \rightleftharpoons [Fe(CN)_6]^{4-}; \quad E^\circ = 0.356 \text{ V}$$

for which the standard potential might appear to be independent of the hydrogen ion concentration. However, the hexacyanoferrate(III) and the hexacyanoferrate(II) anions are conjugate bases to the corresponding hexacyanoferric(III) and hexacyanoferric(II) acids. If these acids were of equal strength, then the reduction potential would be independent of hydrogen ion concentration. In fact, the mono and dihydrogen hexacyanoferrate(II) anions are weaker acids than the corresponding iron(III) complex acids with the result that the concentration of the hexacyanoferrate(II) is decreased relative to the hexacyanoferrate(III) as the acid concentration increases. The values of the formal potentials 0.48, 0.56, and 0.71 V corresponding to the respective concentrations of hydrochloric acid 0.01F, 0.1F, and 1F are indicative of the magnitude of this effect.

The acid formality is also specified even for those half-reactions such as $Cr_2O_7^{2-} + 14H^+ + 6e \rightleftharpoons 2Cr^{3+} + 7H_2O$ in which the hydrogen ion appears as an electroactive species. The actual Nernst expression used for these types of half-reactions does not contain a term in hydrogen ion because the theoretical Nernstian behavior predicted by the inclusion of a hydrogen ion concentration or activity term is not observed in practice. The equations relating an experimentally derived potential $E$ to the formal potential $E^{\circ\prime}$ and the ratio of reductant to oxidant concentrations for both types of half-reactions (i.e., those with and without $H^+$) are therefore identical, and each is written as

$$E = E^{\circ\prime} - \frac{k}{n} \log \frac{c_{\text{reductant}}}{c_{\text{oxidant}}}$$

In this equation $c_{\text{oxidant}}$ is the formality of the soluble ion in the higher oxidation state [e.g., Fe(III), Ce(IV), Cr(VI), Tl(III), etc.] and $c_{\text{reductant}}$ is the corresponding formality for the lower oxidation state [e.g., Fe(II), Ce(III), Cr(III), Tl(I), etc.]. When the ratio of the formalities is equal to unity, $E$ becomes equal to the formal potential for the carefully specified experimental conditions. Some experimentally derived values (10) typical of these definitions are given in Table 1.4 for the Fe(III); Fe(II) and the Cr(VI); Cr(III) couples (see also Table 12.2).

A slightly different definition pertains to a half-reaction of the type

$$M^{z+} + ne \rightleftharpoons M_{(s)} \quad (\text{i.e., } z = n)$$

The formal potentials for these are defined as the potentials that are obtained from cells in which the formalities of the metal ions, $M^{z+}$ are unity. Once again the experimental conditions must be specified as in the following:

$$Ag^+ + e \rightleftharpoons Ag_{(s)}; \quad E^{\circ\prime} = 0.792(1F\ HClO_4);$$

$$E^{\circ\prime} = 0.770(1F\ H_2SO_4)$$

All formal reduction potentials are expressed on the same scale as the thermodynamically derived standard reduction potentials but differ from the latter in that each value of a formal potential includes a contribution from some sort of liquid junction potential. The magnitude of this contribution depends largely upon the type of reference electrode used in the determination and, strictly speaking, this half-cell should be specified. Thus although the value of $E^{\circ\prime} = 0 \cdot 792$ for the $Ag^+$; Ag couple in 1F perchloric

Table 1.4. Formal Potentials of $Fe^{3+}$; $Fe^{2+}$ and $Cr_2O_7^{2-}$; $Cr^{3+}$ Half Reactions[a]

| Acid Present | $E^{\circ\prime}$ Iron (V) | $E^{\circ\prime}$ Dichromate (V) |
|---|---|---|
| 0.1 $M$ HCl | 0.73 | 0.93 |
| 0.5 $M$ HCl | 0.72 | 0.97 |
| 1 $M$ HCl | 0.70 | 1.00 |
| 2 $M$ HCl | 0.69 | 1.05 |
| 3 $M$ HCl | 0.68 | 1.08 |
| 0.1 $M$ $H_2SO_4$ | 0.68 | 0.92 |
| 0.5 $M$ $H_2SO_4$ | 0.68 | 1.08 |
| 4 $M$ $H_2SO_4$ | 0.68 | 1.15 |
| 0.1 $M$ $HClO_4$ | 0.735 | 0.84 |
| 1 $M$ $HClO_4$ | 0.735 | 1.025 |

[a] Values include a saturated calomel electrode liquid junction potential.

REDUCTION POTENTIALS 41

acid implies that this is the potential of the hypothetical cell

$$\text{Pt}; \text{H}_2\,(1\,\text{atm})|\text{H}^+\,(a = 1)\|\text{Ag}^+(1\text{F}), \text{HClO}_4(1\text{F})|\text{Ag} \quad (1.37)$$

it is impossible to use the standard hydrogen reduction potential as being representative of a practical reference electrode because single ion activities (e.g., $a_\text{H} = 1$) cannot be determined nor can they be adjusted independently. The calomel electrode is the most popular type of reference electrode and unless otherwise stated it is generally assumed that this is the reference electrode used for the determination of formal potentials. The liquid junction potential introduced by this type of reference electrode is generally taken as being reproducibly small when the other stated experimental conditions are adhered to. A condition implicit in the use of a reference electrode is that its potential, represented as $E_\text{ref}$, is known. Thus the practical equivalent to cell (1.37) is

$$\text{calomel reference electrode} \| \text{Ag}^+(1\text{F}), \text{HClO}_4(1\text{F}) | \text{Ag} \quad (1.38)$$

and the measured value of the cell potential $E_\text{cell}$ is related to the formal potential $E^{\circ\prime}$ by rearrangement of equation (1.27), $E_\text{cell} = \overset{+}{E} - \overset{-}{E}$, to yield

$$E^{\circ\prime}_{\text{Ag}^+;\,\text{Ag}} = E_\text{cell} + E_\text{ref} \quad (1.39)$$

In practice, a cell used for the determination of formal potentials must be calibrated before use against a standard solution having a known potential as described elsewhere in this book.

### The Use of Formal Potentials

These potentials have greater application to experimental work than the corresponding standard reduction potentials because formal potentials are subject to direct measurement. In contrast, recall that the determination of the thermodynamic quantity $E^\circ_\text{cell}$, described in the calibration procedure, was obtained only by the extrapolation of experimentally derived data to zero ionic strength, a condition of no direct practical application. It follows that, provided experimental conditions are chosen which approximate those for which the formal potentials were measured, predictions as to the direction of spontaneity for a reaction or to the potentials likely to be encountered during the course of a titration are more reliably derived from formal potentials than from the corresponding standard reduction potentials. For example, the formal potentials for the half-reactions $I_2 + 2e \rightleftharpoons 2I^-$ and $\text{Fe(CN)}_6^{3-} + e \rightleftharpoons \text{Fe(CN)}_6^{4-}$ are 0.54 V and 0.71 V, respectively in 1F sulfuric acid. Hence the reaction observed in dilute sulfuric acid is

$$2I^- + 2Fe(CN)_6^{3-} \rightarrow 2Fe(CN)_6^{4-} + I_2$$

as would be predicted from these data. This reaction is the reverse of that which would be predicted from the standard potentials

$$Fe(CN)_6^{3-} + e \rightleftharpoons Fe(CN)_6^{4-}; \quad E^\circ = 0.36 \text{ V}$$

and

$$I_{2(s)} + 2e \rightleftharpoons 2I^-; \quad E^\circ = 0.54 \text{ V}$$

According to the stoichiometry of the half-reaction, the $E$ value for the hexacyanoferrate(III); hexacyanoferrate(II) couple is apparently independent of the relative strengths of the corresponding iron(III) and iron(II) acids. However, as was pointed out earlier, the value of a formal potential for this couple includes the effect of the difference in the acidities of the electroactive species and hence furnishes more reliable data for predictions under a given set of conditions. Formal potentials are also applied to the prediction of the shape of a potentiometric titration curve for an oxidation-reduction reaction. Such curves can be used as a basis for the selection of devices specific to the detection of the equivalence points for these reaction. The most popularly used of these devices are the visual indicators, but electrode systems, classified for convenience as potentiometric, also find application (see Section 2.3).

### The Shape of a Potentiometric Titration Curve

Thus far in this introductory chapter the emphasis has been towards describing the use of potentiometry for the direct determination of ion activities. From this description basic concepts have been developed. For example, the purpose of the present discussion on reduction potentials has been to define and describe a term that has arisen from these general considerations. In Section 1.1 the basic procedure was described whereby the cell

$$Zn(Hg)|ZnSO_4\|HCl\ (0.01\ M)|H_2\ (1\ atm);\ Pt$$

could be used under restricted conditions to determine the concentration of zinc ion (the determinand) in a solution. It was also implied that the cell could be represented by a generalized notation:

indicator electrode $(Zn)|$determinand $(Zn^{2+})\|$reference electrode.

REDUCTION POTENTIALS 43

We can now compare this cell with cell (1.38) and it can be concluded that if the concentration of zinc sulfate is 1F, the measurement of $E_{cell}$ could yield the formal potential, the equation analogous to (1.39) being

$$E^{\circ\prime}_{Zn^{2+};Zn} = E_{ref} - E_{cell}$$

The use of these types of cells in routine chemical analysis by direct potentiometry, however, is inhibited by the reactivity of oxidation-reduction electrodes. Prior to the development of the less reactive ion selective electrodes in the 1960s, by far the most popular application of potentiometry to analysis was the detection of equivalence points in titrimetry. The monitoring of the potentials between two suitably chosen electrodes provides data from which it is possible to deduce these stoichiometric equivalence points for many acid-base, complexometric, oxidation-reduction, and precipitation titrations. Often the sole purpose of monitoring potentials for these applications is directed toward observing a sharp change of potential when the reaction between the titrant and the solution being titrated (the titrand) is complete. In many cases it is neither desirable, nor indeed possible, to interpret the potentials in terms of an active concentration by means of the Nernst equation. If, however, a cell of the type

reference electrode‖titrand solution|indicator electrode

is used, the potentials observed during the course of a titration can be predicted from a knowledge of the stoichiometry of the reaction, which may require calculations based upon the relevant equilibrium constants, together with the formal potential of the indicator electrode and the potential of the reference electrode. This cell notation implies that the indicator electrode is reversible to the determinand present in the half-cell solution being titrated. A zeroth indicator electrode immersed initially in a sample solution of a reductant which is then made to undergo a reaction with an oxidant as titrant is an example of such a half-cell. If the titrand solution initially contains iron(II), a reductant, and this solution is titrated with cerium(IV), an oxidant, the reaction deducible from the following formal potentials in 1F sulfuric acid,

$$Fe^{3+} + e \rightleftharpoons Fe^{2+}; \quad E^{\circ\prime} = 0.68 \quad \text{and}$$

$$Ce^{4+} + e \rightleftharpoons Ce^{3+}; \quad E^{\circ\prime} = 1.44$$

is

$$Fe^{2+} + Ce^{4+} \rightarrow Fe^{3+} + Ce^{3+}$$

Assuming that the reference electrode is a calomel electrode, the potential of which is 0.246 V at 25°, the cell at any point before the equivalence volume can be represented as

$$\text{calomel reference electrode} \| Fe^{2+}, Fe^{3+}(1F\ H_2SO_4) | Pt$$

in which the platinum interface is the zeroth electrode. The cell notation implies that this indicator electrode is positive with respect to the calomel reference electrode and thus, from equation (1.27),

$$E_{cell} = \overset{+}{E}_{ind} - \overset{-}{E}_{ref}$$

The expression relating $E_{cell}$ to the ratio of the concentrations of iron(II) to iron(III) at 25° in 1F sulfuric acid is, therefore,

$$E_{cell} = \left( E^{\circ\prime}_{Fe(III); Fe(II)} - \frac{0.059}{1} \log \frac{c_{Fe(II)}}{c_{Fe(III)}} \right) - 0.246 \quad (1.40)$$

in which $E^{\circ\prime} = 0.68$ V. This equation allows calculation of $E_{cell}$ for any concentration ratio of iron(II)/iron(III). Now in the titration, represented by the overall reaction

$$Fe^{2+} + Ce^{4+} \rightarrow Fe^{3+} + Ce^{3+} \quad (1.41)$$

the ratio iron(II)/iron(III) will depend upon the amount of cerium(IV) titrant added. Although detectable amounts of iron(II) remain in the solution, the cerium(IV) titrant is reduced to cerium(III). Thus if $x$ is the fraction of the stoichiometric amount of oxidant added, for $0 < x < 1$ equation (1.40) can be written as

$$E_{cell} = \left( 0.68 - \frac{0.059}{1} \log \frac{1-x}{x} \right) - 0.246 \quad (1.42)$$

which now becomes the equation for the titration curve. After the equivalence point has been reached, the cell potential is determined by the ratio of the excess concentration of cerium(IV) titrant to cerium(III) concentration. The cerium(III) concentration is equal to the concentration of iron(II) originally present and thus the cell potential after the equivalence point is given by

$$E_{cell} = \left( E^{\circ\prime}_{Ce(IV); Ce(III)} - \frac{0.059}{1} \log \frac{c_{Ce(III)}}{c_{Ce(IV)}} \right) - 0.246 \quad (1.43)$$

The equation analogous to equation (1.42) for this condition is

$$E_{cell} = \left(1.44 - \frac{0.059}{1}\log\frac{1}{x-1}\right) - 0.246 \qquad (1.44)$$

in which $x > 1$.

At the equivalence point

$$\frac{c_{Fe(II)}}{c_{Fe(III)}} = \frac{c_{Ce(IV)}}{c_{Ce(III)}} \qquad (1.45)$$

Adding equations (1.40) and a form of equation (1.43) made specific to the inverted cerium concentrations ratio given by (1.45), the logarithmic terms in each of these equations cancel and thus the cell potential at the endpoint is given by

$$E_{cell} = \tfrac{1}{2}\left(E^{\circ\prime}_{Fe(III);\,Fe(II)} + E^{\circ\prime}_{Ce(IV);\,Ce(III)} - 2(0.246)\right) = 0.81 \text{ V} \qquad (1.46)$$

The values of $E_{cell}$ to be expected during the course of the titration can, therefore, be calculated by the use of equation (1.42) up to the equivalence point, equation (1.46) at the equivalence point, and equation (1.43) after the equivalence point. Specimen results of these calculations are presented in Table 1.5 and expressed graphically as a typical titration curve in Figure 1.4.

Table 1.5. Calculated Cell Potentials for an Oxidation-Reduction Titration

| | Values $E_{cell}$ for: calomel electrode $\|$ Fe(II) + Ce(IV) titration $\|$ Pt | | |
|---|---|---|---|
| Fraction $x^a$ | Before e.p.[b] (1.42)[c] (V) | At e.p. (1.46) (V) | After e.p. (1.44) (V) |
| 0.1 | 0.38 | — | — |
| 0.3 | 0.41 | — | — |
| 0.5 | 0.43 | — | — |
| 0.7 | 0.46 | — | — |
| 0.9 | 0.49 | — | — |
| 1.0 | — | 0.81 | — |
| 1.1 | — | — | 0.14 |
| 1.3 | — | — | 1.16 |
| 1.5 | — | — | 1.18 |
| 2.0 | — | — | 1.19 |

[a] $x$ is the fraction of the stoichiometric amount of oxidant [Ce(IV)] added.
[b] e.p. is the equivalence point.
[c] Equation number, see text.

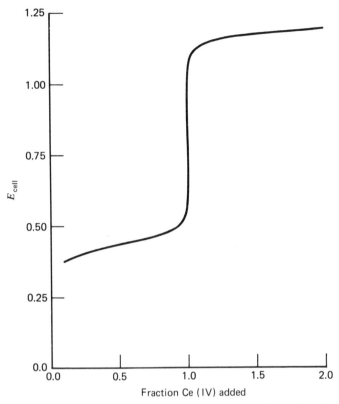

**Figure 1.4.** The shape of a typical titration curve. This curve was obtained by plotting values calculated by means of equations (1.42), (1.46), and (1.43) for the titration of iron(II) with cerium(IV).

Although the example titration curve has been calculated, it has the typical sigmoid shape characteristic of potentiometric titration curves observed not only for oxidation-reduction reactions, but also for many acid-base, complexometric, and precipitation reactions. The most prominent feature of these curves is the sharp change of potential observed in the region of the stoichiometric equivalence point. A study of how the potentials change with incremental additions of titrant in this region allows the volume of titrant required to reach the equivalence point to be deduced. This topic is described in Chapter 9.

Two other features pertinent to oxidation-reduction titrations and their application to routine analysis are briefly outlined continuing with the reaction $Fe^{2+} + Ce^{4+} \rightarrow Fe^{3+} + Ce^{3+}$ as the example.

1. *Visual indication* of the equivalence point is possible if an oxidation-reduction indicator having a formal potential approximately equal to the potential of the indicator electrode at the equivalence point is selected. For example, equation (1.46) yields $E_{cell} = 0.81$ V for the equivalence point of the titration. The potential of the indicator electrode is calculated from $E_{ind} = E_{cell} + 0.246 = 1.06$ V on the standard hydrogen electrode scale. An indicator suitable for use would have a formal potential of approximately 1.06 V in the titration medium.

2. *Potentiometric indication* of the equivalence point is also possible by using two indicator electrodes each responding to changes in activities of the electroactive species but each responding at a different rate. If a platinum electrode and a tungsten electrode are both immersed in the iron(II) titrand solution, the platinum electrode responds in a rapid and Nernstian fashion to changes in the ratio Fe(II)/Fe(III) as the titration proceeds. The tungsten electrode, however, tends to be comparatively irreversible and unresponsive to changes in this ratio. The sharp change of potential which has been predicted in the region of stoichiometric equivalence point is generally commensurate with a large change in potential developing between the two indicator electrodes. In this application the potential is used merely to monitor the titration, and it would not be possible to interpret the cell potential at any point during the titration in terms of a unit of concentration by means of the Nernst equation. Thus the entire emphasis of this application not only contravenes the definition of potentiometry given earlier, but the measurement of potential also assumes a subsidiary role to the primary purpose which is the measurement of a titration volume. The term *potentiotitrimetry* is used in this book to reflect this change of emphasis.

Potentiotitrimetry is applicable to all types of reactions and, in contrast to titrimetry using visual indicators, the technique can be used with colored or opaque solutions. Principal among the disadvantages is the likelihood that a potentiotitrimetric method will be more time-consuming than the comparable indicator procedure although this disadvantage is frequently offset by improved accuracy.

## 1.4 THE CELL NOTATION

In previous discussions the cell notation has been used as an expedient by which cells are represented diagrammatically rather than by a full illustration. Essentially these notations have conformed to the International Union of Pure and Applied Chemistry convention (11) pertaining to the diagrams for galvanic cells. However, important details and implications have been omitted from the discussion so far. Thus in the notation used for the cell

$$\text{Zn(Hg)} | \text{ZnSO}_4 \| \text{HCl } (0.01 \, M) | \text{H}_2 \, (1 \text{ atm}); \text{Pt} \qquad (1.47)$$

no information is given as to the materials used interconnecting the zinc amalgam (liquid) interface to the negative terminal of the potentiometer and for the connection of the platinum interface to the postive terminal. It has been stated, quite correctly, that the single vertical lines represent phase boundaries and thus the interface between solid platinum and hydrogen gas should also be represented in this manner rather than by a semicolon. Incorporating these two correcting modifications to the cell notation (1.47) yields

$$\text{Cu}_{(s)} | \text{Pt}_{(s)} | \text{Hg}(6\% \text{ Zn})_{(l)} | \text{ZnSO}_4(\text{aq}) \| \text{HCl } (0.01 \, M) | \text{H}_2 \, (1 \text{ atm}) | \text{Pt}_{(s)} | \text{Cu}_{(s)}$$

in which the double vertical lines represent the salt bridge. Notice that the phases of the species are all specified by the subscript (s), (l), or (g) corresponding to solid, liquid, or gas, respectively, and the solutions as aqueous either by concentration or by the abbreviation (aq). The connection of the 6% zinc amalgam interface is here shown as requiring a connection by means of a platinum wire to the copper wire in order to connect this liquid interface to the negative terminal of the potentiometer. Although most cell notations given in textbooks and in the research literature use the abbreviated form of the notation as given by cell (1.47), it is very important that the method used to connect the electroactive interfaces to the measuring device should be borne in mind. Bimetallic junctions exposed to variations in atmospheric temperature give rise to thermal effects that can cause errors and variations in the measurement of cell e.m.f. no matter how carefully the temperature of the solution phase in the cell itself is controlled. For example, a commercially available electrode having the following hexametallic junctions,

$$\text{Cu}_{(s)} | \text{stainless steel}_{(s)} | \text{Ag}_{(s)} | \text{spring}_{(s)} | \text{Pt}_{(s)} | \text{Ag}_{(s)} | \text{AgCl}_{(s)} | \text{aq phase} \dots$$

five of which protruded above the water level in a thermostatically controlled bath, proved to be an admirable indicator of changes in laboratory temperature when the electrode was incorporated in a cell the e.m.f. of which was monitored continuously.

Cells without liquid junction can be represented by two methods, one of which ignores the possibility of a junction potential being present in the cell, and the more descriptive notation which acknowledges this possibility. Thus the cell can be represented as

$$Cu_{(s)}|Pt_{(s)}|Hg(6\%Zn)_{(l)}|ZnSO_4(aq)|PbSO_{4(s)}|Hg(6\%Pb)_{(l)}|Pt_{(s)}|Cu_{(s)}$$

or, more fully as

$$Cu_{(s)}|Pt_{(s)}|Hg(6\%Zn)_{(l)}|ZnSO_4(aq) \vdots ZnSO_4(aq),$$

$$PbSO_4(aq)|PbSO_{4(s)}|Hg(6\%Pb)|Pt_{(s)}|Cu_{(s)} \qquad (1.48)$$

in which the diffusion potential established across a liquid junction $\vdots$ is acknowledged but is usually assumed to be eliminated. The reason for this assumption was discussed earlier.

Regardless of what detail is given by the notation, these cell diagrams all imply the same type of information: under the conditions of a potentiometric measurement (current $\to$ 0) the e.m.f. of a cell is equal in sign and magnitude to the potential of the right-hand terminal when that of the left is taken as zero. The only condition of practical application, however, is when the value of $E_{cell}$ is positive corresponding to a spontaneous cell reaction ($\Delta G$ negative) which would occur only if current were drawn from the cell. Thus according to the convention, the right-hand terminal will always be positive with respect to the left-hand terminal, and the cell notations given not only in this chapter but also throughout this book comply with this convention. The spontaneous oxidation-reduction cell reaction can, therefore, be deduced from the cell notation by the following argument: since electrons are negatively charged there will be a potentially spontaneous flow of electrons from the left-hand terminal to the right-hand half-cell that will tend to increase the number of reduced entities in this positive half-cell. There will, of course, be a corresponding increase in the number of oxidized species in the left-hand half-cell.

Because a multitudinous array of vertical bars in the full notation can complicate the interpretation of the essential information conveyed by a cell diagram, the abbreviated form used earlier in this chapter is retained. This form is similar to that given by Bates (12) and can be summarized as follows:

1. Interfaces between materials used in the construction of oxidation-reduction electrodes are represented by a semicolon (;).
2. Interfaces between an oxidation-reduction electrode and a solution to form a half-cell are represented by a single vertical line (|).
3. Closely spaced and double vertical lines represent a salt bridge (||).

4. Two or more solutes in the same solution are separated by commas (,).
5. The conductors (e.g., copper wires) attached to the electrodes during the measurement of the e.m.f. of the cell are omitted from the notation. It will usually be assumed that thermal effects are minimized in the design of the electrode.

The glass-saturated calomel electrode system commonly used for the measurement of $p$H, when immersed in a standardizing $p$H buffer solution, would be represented by this notation as

$$Ag; AgCl|H^+, Cl^-|glass|KH_2PO_4(0.025 \ m),$$

$$Na_2HPO_4(0.025 \ m)||sat.KCl|Hg_2Cl_2; Hg$$

Here the silver–silver chloride (oxidation-reduction) electrode dipping into an aqueous solution of hydrogen ion and chloride ion forms a reference half-cell contained inside a hydrogen-ion responsive glass membrane. The external surface of this membrane dips into a 0.025 *equimolal* solution of potassium dihydrogen phosphate and disodium hydrogen phosphate forming the complete half-cell which is reversible to hydrogen ion. The other half-cell, a saturated calomel reference electrode, is separated from the phosphate buffer solution by a salt bridge which gives rise to a liquid junction potential. The notation shows the calomel reference electrode is positive with respect to the glass electrode.

### REFERENCES

1. R. G. Bates and J. B. Macaskill, *Pure Appl. Chem.* **50**, 1701 (1978).
2. P. Debye and E. Huckel, *Physikal. Z.* **24**, 185, 305, 334 (1923); **25**, 145 (1924).
3. E. Guntelberg, *Z. Phys. Chem.*, **123**, 199 (1926).
4. G. Scatchard, *Chem. Rev.* **19**, 309 (1936).
5. R. G. Bates and E. A. Guggenheim, *Pure Appl. Chem.* **1**, 163 (1960).
6. C. W. Davies, *J. Chem. Soc.* **1938**, 2093; *Ion Association*, Butterworths, London, 1962, p. 41.
7. J. Kielland, *J. Amer. Chem. Soc.*, **59**, 1675 (1937).
7a. G. Milazzo and S. Caroli, *Tables of Standard Electrode Potentials*, Wiley, New York, 1978.
8. IUPAC, 17 Conference (Stockholm), 1953; T. S. Licht and A. J. Bethune, *J. Chem. Educ.*, **34**, 433 (1957); A. J. Bethune, *J. Electrochem. Soc.* **102**, 288C (1955).

# REFERENCES

9. W. M. Latimer, *The Oxidation State of the Elements and Their Potentials in Aqueous Solution*, 2nd ed., App. 1, Prentice-Hall, Englewood Cliffs, N.J., 1952; G. Charlot, *Oxidation-Reduction Potentials*, Pergamon Press, London, 1958; G. Charlot and others, *Selected Constants*, Supplement to Pure Appl. Chem. IUPAC, Butterworth, London, 1971.
10. G. F. Smith, *Anal. Chem.* **23**, 925 (1951).
11. IUPAC, Manual of Symbols and Terminology, App. III Electrochemical Nomenclature, *Pure Appl. Chem.* **37**, 501 (1974).
12. R. G. Bates, *Determination of pH Theory and Practice*, 2nd ed., Wiley, New York, 1973, p. 12.

CHAPTER
2
# CLASSIFICATIONS OF POTENTIOMETRY AND TITRIMETRY

## 2.1 TYPES OF TITRATIONS

### General Classifications

In Chapter 1 it was mentioned that the development of potentiometry as a method used for the location of the equivalence point of a titration predates the development of the widespread application of direct potentiometry as an analytical technique. The monitoring of a potential during the course of a titration is only one method among many that can be used to detect the stoichiometric completion of a reaction suitable for titrimetry. Replacing the older term *volumetric analysis, titrimetry* is the word used to describe the quantitative measure (volume or weight) of a solution of accurately known concentration which is required to react stoichiometrically with all of the determinand contained in a solution. It is the type of reaction that establishes the traditional classifications within titrimetry rather than the method used to determine the equivalence volume (or weight). These classifications are retained in Part 3 of this book and are described as:

Acid-base titrimetry
Complexometric titrimetry
Oxidation-reduction titrimetry
Precipitation titrimetry

Confusion as to what actually constitutes a titration can be caused by two of the techniques now generally employed in analysis by direct potentiometry using cells with liquid junctions. These techniques are described as the following:

1. *The Standard Addition Method.* The known addition method in which the concentration of a particular electroactive species in a

sample is determined by adding known amounts of that species and recording the change in $E_{cell}$. An example of this method is given in Section 2.2 of this chapter under *differential potentiometry* and the technique is also described as a *direct potentiometric* technique in Chapter 4.

2. *The Standard Subtraction Method.* The known subtraction method in which changes in e.m.f. resulting from the addition of a known amount of a species that reacts with the species of interest (usually by complexation or by precipitation) are employed to determine the original activity or concentration of the species (see Chapter 4).

Both methods are sometimes classified as titrimetric, but such a classification is erroneous. In the standard addition method no reaction occurs and this lack of reaction contravenes the basic definition for titrimetry. Superficially, the standard subtraction method might appear to conform with our definition of titrimetry in that a reaction occurs. However, only a quantity of reagent sufficient to reduce the concentration of the species of interest to about one-half of its original concentration is usually added. By contrast in the classical applications of titrimetry the concentration of the determinand is assumed to be negligible at the equivalence point and on this ground the standard subtraction method cannot be regarded as titrimetric.

A second source of confusion in the nomenclature of that branch of titrimetry which relies upon the monitoring of potentials as the method of indication is caused by the use of the words *differential* and *differentiating*. In this book *differential* is used to describe a titration that involves the monitoring of the difference between the potentials of two electrodes forming part of a concentration cell (with or without diffusion; see Section 2.2). Measurements with these cells form a category of potentiometry called *differential potentiometry*. When used within the context of titrimetry the closely related word *differentiating* is used to describe a titration performed under conditions that enable a distinction to be made between the amounts of two or more closely related compounds or ions present in the same solution. Sometimes these *differentiating titrations* are mistakenly called *differential titrations* which adds to the confusion. In this book the word *discriminative* is used to resolve this ambiguity and a *discriminative titration* is one that allows a distinction to be made between two or more closely related species.

A further source of confusion also arises from the word *differential* when what is meant is *derivative*. Thus the term $\Delta E_{cell}/\Delta V$, meaning the change in a potential per unit change in the amount of titrant, is the first derivative of the potential of an indicator electrode with respect to the amount (weight or volume) of added reagent. A *derivative titration* (1) is one that involves measuring, recording, or computing this first derivative.

## Potentiometric Titrations and Potentiotitrimetry

Two general applications of potentiometric methods have been outlined in Chapter 1. In the first method a suitably designed electrode system is calibrated in terms of a given electroactive species by measuring the values of $E_{cell}$ for a series of standard solutions; the calibration data then permits the interpolation of $E_{cell}$ directly in terms of concentration or activity for a solution of a sample. This application is known as *direct potentiometry*. The basis for direct potentiometry is that the Nernst equation should be applicable which, in turn, demands that $E_{cell}$ is measured under the truly potentiometric condition of negligible current flow.

In the second method *the monitoring* of a potential, as opposed to the implications of potential *measurement* as given previously, serves to locate the equivalence point of a titration. It was shown earlier that changes in the potential set up between two indicator electrodes constructed of *different* materials, but each responding at a different rate to the same electroactive species, can provide a sufficiently good monitor for equivalence point indication. Furthermore, it is shown later that the response of a pair of electrodes constructed of the *same* materials can be similarly modified by the deliberate application of a small current between them during the course of a titration. This modification of response can also be achieved by a different chemical or physical pretreatment for each electrode in some instances. Bearing in mind that the definition of potentiometry is thermodynamically based, it would be a travesty of this definition to describe a monitoring role, in which the measured potential cannot be interpreted by the Nernst equation, as potentiometric. It is important to realize that direct potentiometry provides equilibrium information regarding the system, whereas the titrimetric monitoring role yields primarily stoichiometric information. These two functions are distinguished by retaining the rigid definition of potentiometry and reserving the word *potentiotitrimetry* for the stoichiometric application. For example, the ionization constant for a weak acid can be determined by a titration in which the measured value of $E_{cell}$ is interpreted in terms of $p$H at various stages during the titration of the weak acid with strong base (see Chapter 6). This technique is described as the determination of an ionization constant by *potentiometric titration* because potentiometry is assumed to comply with the fundamental thermodynamic definition. A potentiometric titration therefore is an application of *direct potentiometry* and, as such, entails careful attention to those details specific to potentiometry mentioned in Chapter 1. Paramount among these are the requirements implicit to cell calibration under conditions similar to those encountered in the titration. Usually in a potentiometric titration, as defined here, the original stoichiometric concentrations of both titrant and titrand are known and the purpose of performing the titration is to determine some physical constant that pertains to a solution equilibrium process. Thousands of such constants for the processes described in Part 2 of this book are determined each year.

The emphasis placed upon cell calibration is often the distinguishing feature between potentiometric titrations and potentiotitrimetry for there are occasions when the same cell can be used for either technique. In potentiotitrimetry, however, the change in potential per unit volume of titrant is of greater importance than the actual value of the potential and thus the need for careful cell calibration is diminished.

The nomenclature recommended by the International Union of Pure and Applied Chemistry (1) does not distinguish between the two roles fulfilled by the measurement of $E_{cell}$ as opposed to the monitoring of $E_{cell}$ during the course of a titration. Both applications are called *potentiometric titrations* and traditionally there has been no necessity to make such a distinction as is made in this book because a given work has usually dealt exclusively with either the titrimetric application of potentiometry or the application of potentiometry to some aspect of ion activity measurements. Consistency for the definition of potentiometry presents no problem if only one of these applications is described.

## 2.2 TYPES OF CELLS

### Chemical Cells and Concentration Cells

In addition to the dichotomy of functions (i.e., ion activity measurements and titrimetric indicator role) outlined, there are two general categories of cells available to fulfill these functions. The category to which a given cell belongs is decided by the answer to the question. Does the cell contain a salt bridge? As was pointed out in Chapter 1, the measured e.m.f. of a cell that contains a salt bridge will include a contribution from a diffusion or liquid junction potential $E_j$. Such a cell is therefore described as *a cell with diffusion*. The other category is a *cell without diffusion* as exemplified by a cell such as

$$Zn(Hg)|ZnSO_4(aq)|PbSO_4; Pb(Hg)$$

in which the diffusion potential can be assumed to be negligible under the conditions of good experimental design. All the cells so far described that fall into these two categories have had a different oxidation-reduction system in each of the constituent half-cells. Their e.m.f. is derived from the free-energy change of a balanced oxidation-reduction reaction in which an electron transfer occurs from the reduced form of one couple to the oxidized form of the other. Because they are based upon a potentially spontaneous chemical reaction these cells are known generically as *chemical cells*. When a chemical cell is to be used for ion-activity/concentration measurements, the term *direct potentiometry* is appropriate. When the prime motive in using a chemical cell is as an aid in detecting the equivalence point of a titration, the method is described here as *potentiotitrimetric*.

Another generic classification arises to describe those cells in which the same oxidation-reduction couple is present in each half-cell but the activity of one or more of the participating species differs from one half-cell to the other. These are known as *concentration cells* which, in common with chemical cells, can be subdivided into two categories according to whether a cell contains a salt bridge. Concentration cells with and without diffusion are described in more detail later and their use gives rise to the techniques described as *differential potentiometry* and *differential potentiotitrimetry* in which the term *differential* can be taken as implying that a concentration cell is used.

Despite the fact that in recent applications of potentiometry oxidation-reduction electrode processes are not necessarily involved, the foregoing classifications of chemical cells and concentration cells are conveniently used to subdivide chemical potentiometry into the two major branches of *direct potentiometry* and *differential potentiometry*, respectively. An outline of direct potentiometry and its application to ion activity/concentration determinations was given in Chapter 1.

## Concentration Cells with and without Diffusion

There are two possible methods by which a differential potentiometric cell can be devised so that it will be consistent with the general definition of a concentration cell. The essentials of these methods are based upon cells that contain the following components.

1. *Two different electrodes* each reversible to the same electroactive species dipping into *one solution*. The type or the quantity of electroactive material of each electrode is different and the potential between the two develops as a result of this difference. Sometimes the potential can be described by the Nernst equation as in the *concentration cell without diffu-*

*sion* represented by

$$\text{Hg}(\text{Zn}, a = a_1)|\text{Zn}^{2+}|(\text{Zn}, a = a_2)\text{Hg}$$

In this cell two zinc amalgam electrodes having different active concentrations of zinc contained in the mercury ($a_1$ and $a_2$) dip into a single solution which contains zinc ion. The cell reaction involves only the two different phase activities of the constituent electrodes,

$$\text{Zn}(a_1) \rightleftharpoons \text{Zn}(a_2)$$

The Nernst equation can be formulated as

$$E_{\text{cell}} = E_{\text{cell}}^{\circ} - \frac{k}{2}\log\frac{a_2}{a_1} = \frac{k}{2}\log\frac{a_1}{a_2} \qquad (2.1)$$

The quantity $E_{\text{cell}}^{\circ} = 0$ because the standard reduction potentials for each half cell ($\text{Zn}^{2+} + 2e \rightleftharpoons \text{Zn}$) are identical.

2. *Two identical electrodes* each reversible to the same electroactive species contained in *two solutions* having different active concentrations of electroactive species. The only method by which the resulting half-cells can be interconnected is by means of a salt bridge. For example, the cell described by the notation

$$\text{Pt}; \text{H}_2\,(1\text{ atm})|\text{H}^+(a_1)||\text{H}^+(a_2)|\text{H}_2\,(1\text{ atm}); \text{Pt}$$

in which the activities of hydrogen ion $a_1$ and $a_2$ are not equal, represents a *concentration cell with diffusion*. The cell reaction for this cell is

$$\text{H}^+(a_2) \rightleftharpoons \text{H}^+(a_1)$$

and the Nernst expression is

$$E_{\text{cell}} = E_{\text{cell}}^{\circ} + E_j - k\log\frac{a_1}{a_2} = E_j + k\log\frac{a_2}{a_1} \qquad (2.2)$$

in which $E_j$ represents the junction or diffusion potential. Under the conditions of constant ionic strength, brought about by identically high salt concentrations in each of the half-cell solutions, this equation can be written as

$$E_{\text{cell}} = k\log\frac{c_2}{c_1} \qquad (2.3)$$

The value of $E_{cell}$, therefore, depends upon the ratio of the two concentrations of the hydrogen ion, the assumption being that the condition of constant ionic strength not only causes the activity coefficients to become equal for the two solutions, but is also effective in eliminating the diffusion potential $E_j$ between the two half-cells.

Of these two types of cells only concentration cells with diffusion find any application in analytical chemistry that is consistent with the previously stated definition of potentiometry. Differential potentiometry is therefore synonymous with the use of these cells. On the other hand, the application of concentration cells without diffusion is confined usually to the detection of endpoints in titrimetric analysis and is thus described as a differential potentiotitrimetric technique.

**Differential Potentiometry**

The general form of equation (2.3)

$$E_{cell} = \frac{k}{n} \log \frac{c_2}{c_1} \qquad (2.4)$$

forms the basis for differential potentiometry. In theory, if the concentration of the determinand ($c_1$) in the solution contained in one half-cell is known, then the unknown concentration ($c_2$) in the other may be calculated from the measured value of $E_{cell}$ provided both half-cells are at the same temperature. The usual procedure, however, consists of adjusting the composition of one half-cell (the reference half-cell) to match that of the other (the sample half-cell) as evidenced by a zero cell potential between identical electrodes. The minimization of the liquid junction potential and of activity effects is achieved by a careful selection of an inert electrolyte solution to ensure that an identical condition of constant ionic strength (constant $p$H, etc.) exists between the two half-cells. If the known concentration in the reference half-cell is adjusted by either the *known addition* (see Section 2.1) of a standard solution of the determinand prepared in the inert electrolyte medium, or by the known addition of this electrolyte medium alone until $E_{cell} = 0$, the concentration of the determinand in the sample half-cell may be calculated after due allowance has been made for dilution effects.

The zeroth electrode system Pt|Ce(IV), Ce(III), mentioned in Chapter 1, when used in a concentration cell with diffusion provides an example of the differential potentiometric determination of fluoride (2). The potential of a Ce(III)–(IV) system is affected by fluoride due to complexation of cerium(IV) thus altering, in a manner predictable from the respective stability constants, the potential determining ratio $c_{Ce(III)}/c_{Ce(IV)}$. The cell can be represented initially as

Pt|Ce(IV), Ce(III), $H_2SO_4$(0.25 $M$)||Ce(IV), Ce(III), $H_2SO_4$(0.25 $M$)|Pt
half-cell A (sample)    half-cell B (reference)

(2.5)

in which half-cell A and half-cell B contain equal volumes of the same solution of cerium(IV), cerium(III) and 0.25 $M$ sulfuric acid. The initial potential of this cell should then be zero, but sometimes a bias potential may be observed due to slight differences in the manufacture of the electrodes. The sample solution of fluoride is added to half-cell A and an equal volume of water to half-cell B. Incremental additions of a standard fluoride solution are made to half-cell B and, in order to maintain identical conditions of acidity and ionic strength in both half-cells, the same incremental additions of water are made to half-cell A. If $E_{cell}$ is measured after each of the respective known volumes have been added to the two half-cells and this e.m.f. plotted against the corresponding volume, the concentration of fluoride can be interpolated from the volume of the standard fluoride added to cell B which is required to yield $E_{cell} = 0$. If an initial bias potential is observed, then the volume added to reach this potential (as opposed to $E_{cell} = 0$) is the correct volume to use in calculating the fluoride concentration.

This null point technique, first performed by de Brouckere in 1928 (3) and refined about 1960 by Malmstadt and co-workers (4)–(12) using oxidation-reduction electrodes, has now been extensively used with ion selective electrodes. The basic principle of the method, however, remains identical regardless of the type of electrodes used, provided they form part of a concentration cell with diffusion. *Null point potentiometry* is a name sometimes used to describe the method but the International Union of Applied Chemistry has recommended (1) the discontinuation of this name and its replacement by *differential potentiometry*.

### Differential Potentiotitrimetry

At first sight an obvious method for the detection of an equivalence point would be to use a *concentration cell with diffusion* similar to cell (2.5) in which the reference half-cell contains a solution that has already been titrated to the exact equivalence potential using a chemical cell. If the solution contained in the sample side of the concentration cell is then titrated, the equivalence point will be commensurate with a zero value for

$E_{cell}$. Thus in the example oxidation-reduction titration of Chapter 1,

$$Fe^{2+} + Ce^{4+} \to Fe^{3+} + Ce^{3+}$$

the reference half-cell would consist of a solution composed of the species present at the equivalence point. The potential of this solution should be about 1.06 V with respect to the standard hydrogen electrode as calculated by equation (1.46) and would contain only the electroactive species iron(III) and cerium(III) in 1F sulfuric acid. The sample half-cell, formed by immersing a platinum electrode in a solution of iron(II) prepared in 1F sulfuric acid, would need to be connected to the reference half-cell by means of a salt bridge. Initially, this cell can be represented as

$$\underset{\text{sample half-cell}}{Pt|Fe^{2+}, 1F\ H_2SO_4} \| \underset{\text{reference half-cell}}{Fe^{3+}, Ce^{3+}, 1F\ H_2SO_4|Pt} \qquad (2.6)$$

If the cerium(IV) titrant solution is added to the sample half-cell, the volume of titrant required to yield $E_{cell} = 0$ is the equivalence volume. This method is rarely used in practice because the solution contained in the reference half-cell is unbuffered with respect to small quantities of interfering species which will cause the potential of this half-cell to change. These interferences are sometimes of atmospheric origin. In oxidation-reduction reactions absorption of atmospheric oxygen can cause a drift in the reference potential; in acid-base titrations a similar error in the value of the reference potential may be caused by the absorption of carbon dioxide. For these types of reasons differential potentiotitrimetry is confined to the use of *concentration cells without diffusion*.

## 2.3 TITRIMETRY USING POTENTIAL MONITORING

In the preceding paragraph it was shown by a process of elimination that of the two types of *concentration cells* associated with differential potentiometry only those *without diffusion* find application in titrimetric analysis. Conversely, of the two types of chemical cells associated with direct potentiometry only *chemical cells with diffusion* find common use as potential monitors for equivalence point location in titrimetry. These applications to potentiotitrimetry using chemical cells with diffusion and to differential potentiotitrimetry using concentration cells without diffusion are outlined in this section.

## Applications of Chemical Cells with Diffusion to Titrimetry

The cells commonly used for potentiotitrimetry can be represented by the notation

indicator electrode | determinand ‖ reference electrode

in which the reference electrode is often a calomel reference electrode. For all types of titrations a plot of $E_{cell}$ versus the volume of titrant usually has a sigmoid shape similar to that of Figure 1.4. The methods used to deduce the equivalence volume of titrant from curves of this shape are discussed in Chapter 9. For the purposes of the generalizations made in this section it is sufficient to state that the stoichiometric equivalence volume is often located from the volume of titrant required to give a maximum in the values of $\Delta E_{cell}/\Delta V$, the change of $E_{cell}$ per incremental unit volume of titrant. The indicator electrode selected must respond to changes in concentration of either a species generated by or consumed in the titration reaction. Under circumstances in which a titration has to be performed routinely, it is often advantageous to devise a reference electrode the potential of which is identical to the potential that the indicator electrode will assume at the equivalence point. This is a variant of the concept discussed using the example concentration cell (2.6), except that the objection of possible variations in the value of the potential of the reference half-cell can be overcome in some instances by using an electrode reversible to hydrogen ion immersed in a solution buffered with respect to $p$H. When such a reference electrode is incorporated to form the titration cell which is connected to a detector such as a galvanometer or even a small battery operated digital multimeter, the equivalence volume is that volume of titrant required to give a zero reading on the detector.

A quinhydrone electrode (see Reference 13 for details) when incorporated into the reference half-cell of Figure 2.1 has a sufficiently low resistance to be used with the types of detectors mentioned and offers quite a wide range of reference potentials which can be selected by judicious choice of a $p$H buffer solution. Quinhydrone is an equimolecular mixture of 1,4-cyclohexanedione (benzoquinone, represented here as Q) and 1,4-benzenediol (hydroquinone, represented as $H_2Q$) both of which are slightly soluble in water forming an oxidation-reduction system represented by $Q_{(s)} + 2H^+ + 2e \rightleftharpoons H_2Q_{(s)}$ for which $E° = 0.699$ V at 25°. The potential of a zeroth electrode such as platinum or gold dipping into an aqueous solution containing quinhydrone depends upon the activity of hydrogen ion as is shown by the equation (25°)

$$E_{Q;H_2Q} = 0.699 - 0.0592 \log \frac{1}{a_{H^+}}$$

or

$$E = 0.699 - 0.0592 \, p\text{H} \qquad (2.7)$$

# TITRIMETRY USING POTENTIAL MONITORING

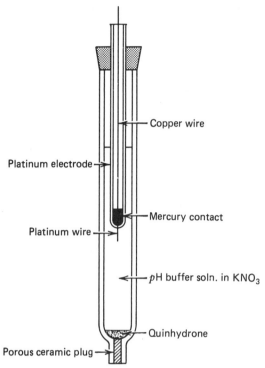

**Figure 2.1.** Quinhydrone reference half-cell. The $p$H of the buffer solution is adjusted so that the potential of the quinhydrone electrode will be equal to the potential of the indicator electrode at the equivalence point [see cell notation (2.9)].

The useful range of the electrode extends from $p$H 0 to 8 allowing a range of reference potentials of from 0.225 to 0.599 V to be obtained.

In practice the potential of the titration cell, represented at the equivalence point as

calomel electrode ‖ solution at the equivalence point | indicator electrode

(2.8)

is determined. From this value of $E_{cell}$ and from a knowledge of the approximate value of the potential of the calomel electrode (about 0.24 V), the potential of the indicator electrode $E_{ind}$ at the equivalence point can be calculated very approximately on the standard hydrogen electrode scale using an equation of the type $E_{ind} = E_{cell} + 0.24$. This value of $E_{ind}$ is substituted for $E$ in equation (2.7) which allows the calculation of a very approximate value for the $p$H of the solution that must be prepared for use in the quinhydrone half-cell. This value of $p$H is used merely for guidance in the selection of a suitable substance from which to prepare an appropriate buffer solution. Solutions of citric acid ($pK_1 = 3.1$, $pK_2 = 4.8$, $pK_3 = 6.4$) from which it is possible to prepare $p$H buffer solutions in the range 2.5–7.0 are suitable for this purpose. The acid is dissolved in a solution of 1 $M$ potassium nitrate. To a convenient volume of the solution a few milligrams of solid quinhydrone is added and a half-cell formed by dipping a platinum electrode in the solution. The calomel electrode that was used in cell (2.8) is combined with this quinhydrone half-cell to form the complete cell

calomel electrode∥citric acid solution in 1 $M$ KNO$_3$|Q; H$_2$Q; Pt  (2.9)

The potential of this cell is adjusted to the exact potential of cell (2.8) at the equivalence point by the dropwise addition of a concentrated solution of sodium hydroxide. The citrate buffer solution prepared by this procedure is then poured into the body of the electrode shown in Figure 2.1. Insertion of the platinum electrode used in cell (2.9) completes the preparation of a reference electrode the potential of which will be commensurate with the potential of the indicator electrode [cell (2.8)] at the equivalence point. Before proceeding to the determination of sample solutions it is advisable to restandardize the titrant using the cell into which the reference electrode is incorporated.

A reference electrode prepared in this manner can be used for periods of up to six months without renewal of the buffer solution provided it is stored in the dark when not in use. Benzene hexacarboxylic acid (mellitic acid $pK_1 \sim 0.8$, $pK_2$ 2.3, $pK_3$ 3.5, $pK_4$ 5.2, $pK_5$ 6.5, $pK_6$ 7.7) is sufficiently soluble to be used in the preparation of suitable $p$H buffers by which it is possible to cover the entire $p$H range for which the quinhydrone electrode can be used. This method is sometimes known as the *Pinkhof–Treadwell method* (14).

Another type of reference half-cell, the potential of which will remain constant throughout the course of a titration, can be formed by sealing an electrode responding to a species contained in the titrant solution into an extended portion of the delivery tube below the stopcock of a burette. The indicator electrode is immersed in the titrand solution contained in a beaker and electrical contact between this indicator half-cell and the reference half-cell is made by means of the titrant solution by dipping the extended burette tip into the titrand solution. The equivalence point is located from the maximum change in potential that occurs in the vicinity of the equivalence volume. This method, which can be described as the *titrant stream monitor method* (15), is similar in concept to the *retarded electrode method* described in the next subsection. However, in the titrant stream monitor method the reference half-cell potential remains constant throughout the titration and the reference electrode need not respond to the same electroactive species as the indicator electrode. For this reason the cell has been included as a chemical cell with diffusion rather than as a concentration cell.

## Applications of Concentration Cells without Diffusion to Titrimetry

Equivalence point detection by differential potentiotitrimetry depends upon the difference in the potential established between two electrodes that respond to the same electroactive species at different rates. This difference in cell potential becomes suddenly accentuated in the vicinity of the equivalence point. The concentration cell without diffusion can be represented as

electrode A | titrand solution | electrode B

in which the differential rate of response depends upon the methods used in the preparation of the two electrodes A and B. These can be constructed of the following.

1. *Different materials* each of which will respond to the same electroactive species present in either the titrand or the titrant solutions. The most widely used of these *bimetallic systems*, as they are called, is the platinum–tungsten pair developed by Willard and Fenwick in 1922 (16). More recent applications (17) extend their use in routine analysis requiring acid-base, complexometric, or precipitation titrimetry. The principle is not confined to oxidation-reduction electrode systems and the use of ion selective electrodes each having a different selectivity for the same species is a variant of the method (18).

2. *Identical materials* pretreated so as to produce a different type of surface–solution interface. The pretreatment might cause a difference in oxide layer on the metals, a difference in the surface areas of the materials, or some other modification in the nature of the surfaces. The technique was first applied in 1928 (19) when it was found possible to use the difference in potential which develops between a platinum electrode and a platinized platinum electrode (i.e., a platinum electrode coated electrolytically with finely divided platinum) to detect the equivalence point of a titration of iron(II) with potassium dichromate. The potential between both electrodes is approximately constant in iron(II) sulfate, but in the vicinity of the equivalence point a potential develops as a result of a difference in response to the dichromate ion. Differences in the stress corrosion of two silver-plated platinum microelectrodes of different sizes have been claimed (20) to produce a differential response which allows the equivalence point in a titration of silver nitrate with potassium iodide to be located by potential

monitoring. More recently, adsorption polarized electrodes have been used to monitor this titration (21).

3. *Identical materials* separated in such a manner that the response of one is physically retarded with respect to the other. A simple version of this *retarded electrode method*, originally developed in 1928 (22), is shown in Figure 2.2. Electrode A, in the example a platinum electrode, dips into the titrand solution. The retarded electrode, electrode B, is an identical electrode which is sealed inside a small medicine dropper fitted with a rubber bulb that allows the flushing and refilling of the retarded electrode compartment. At the outset of the titration a small volume (negligible in comparison with the bulk of the solution) of titrand solution is contained in the dropper. A potential between A and B will develop when an incremental addition of the titrant is made to the bulk of the solution. By flushing and refilling the retarded electrode compartment before and after each incremental addition of titrant, the potential assumed by electrode B can be regarded as a memory of the potential that electrode A possessed at the previous addition. The equivalence point is, therefore, commensurate with the largest value of the potential recorded during the course of the titration.

It is not possible in any of these three methods to interpret the value of the measured cell potential by means of the Nernst equation. Such electrode systems are often described as being *polarized* which is the term used to imply non-Nernstian behaviour. In some instances selection of suitable electrode systems for differential potentiotitrimetry can be aided by electrokinetic data as outlined below.

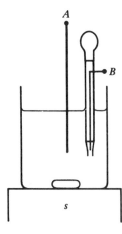

**Figure 2.2.** Example of a cell used for differential potentiotitrimetry. Electrodes *A* and *B* are made of identical materials, but *B* dips into a portion of the solution to which the indicator electrode *A* had already responded at the previous addition of titrant. After each addition and reading, the solution in the dropper is exchanged with the bulk solution.

## The Equilibrium Exchange Current Density

It was stated earlier that a negligible potential exists between bright platinum and platinized platinum electrodes in iron(II) sulfate solution, but a finite potential is established when the two electrodes are immersed in a dichromate solution. The latter potential reflects the difference in the rates at which the equilibrium process

$$Cr_2O_7^{2-} + 14H^+ + 6e \rightleftharpoons 2Cr^{3+} + 7H_2O$$

is established at the two electrodes. An equilibrium potential for a given *single electrode* is established only when the rates for the reduction reaction (oxidant + $ne \rightarrow$ reductant) and the oxidation reaction (reductant $\rightarrow$ oxidant + $ne$) become equal. Even when this equality of rates is established there will be a continual exchange of material and charge across the electrode interface. If material is transferred, then according to the laws of electrolysis a current must be associated with each of oxidation and reduction processes and, at equilibrium, the magnitude of these currents must be equal. Such a magnitude is designated the *equilibrium exchange current density* $j_0$, which reflects the rate at which a given electrode will achieve an equilibrium potential with respect to an electroactive species to which it is reversible. It is only when the equilibrium potential is established that the value of the potential has thermodynamic significance in terms of the Nernst equation and, as was stated in Chapter 1, this only occurs if the net overall current flowing in the system is negligibly small. Exchange current densities are, therefore, pertinent to the kinetic properties (23) of a single electrode for a given set of conditions and values of $j_0$ vary with concentration. Furthermore, any modification in the surface properties of this electrode whether caused, for example, by deliberate surface abrasion or by chemical means will alter the rate at which the electrode attains an equilibrium potential in a solution of a particular concentration. This alteration in rate is reflected by a change in the equilibrium exchange current density. A large value of the exchange current density is consistent with a comparatively rapid attainment of equilibrium potential and thus it would be expected that if platinized platinum responds more rapidly to the dichromate ion than bright platinum does, the value of $j_0$ for platinized platinum in dichromate will be greater than the value of $j_0$ for bright platinum in the same solution. Similarly in the concentration cell without diffusion

$$Pt|H^+|W$$

Table 2.1. Estimates of Exchange Current Densities ($j_0$)

| Electrode | Solution | $j_0$ (A cm$^{-2}$) |
|---|---|---|
| Pd[a] | 1 M $H_2SO_4$ | $1 \times 10^{-3}$ |
| Pt[a] | 1 M $H_2SO_4$ | $7.9 \times 10^{-4}$ |
| Ni[a] | 1 M $H_2SO_4$ | $6.3 \times 10^{-6}$ |
| Au[a] | 1 M $H_2SO_4$ | $4.0 \times 10^{-6}$ |
| W[a] | 1 M $H_2SO_4$ | $1.3 \times 10^{-3}$ |
| Cd[a] | 1 M $H_2SO_4$ | $1.6 \times 10^{-11}$ |
| Tl[a] | 1 M $H_2SO_4$ | $1 \times 10^{-11}$ |
| Hg[a] | 1 M $H_2SO_4$ | $5 \times 10^{-13}$ |
| Ag[b] | 1 M $AgClO_4$ | 1.0 |
| Tl[b] | 1 M $TlClO_4$ | $1 \times 10^{-3}$ |
| Pb[b] | 0.5 M $Pb(ClO_4)_2$ | $8 \times 10^{-4}$ |
| Zn[b] | 0.5 M $Zn(ClO_4)_2$ | $3 \times 10^{-8}$ |
| Zn[b] | 0.5 M $ZnSO_4$ | $3 \times 10^{-5}$ |
| Zn[b] | 0.5 M $ZnSO_4$ | $3 \times 10^{-4}$ |
| Ag metal[c] | 1 M $AgNO_3$ | 0.1 |
| $Ag_2S$[c] | 1 M $AgNO_3$ | 1 |
| Fluoride ($LaF_3$)[c] | 1 M KF; 0.7 M $K_2SO_4$ | $5 \times 10^{-5}$ |
|  | 1 M KOH; 0.7 M $K_2SO_4$ | $5 \times 10^{-6}$ |
|  | 0.01 M KF; 0.7 M $K_2SO_4$ | $10^{-5}$ |
|  | 0.025 M $La(NO_3)_3$ | $7 \times 10^{-7}$ |
| $6 \times 10^{-7}$ M valinomycin/ n-decanol[c] | 1 M KCl | $3 \times 10^{-7}$ |
| $6 \times 10^{-4}$ M valinomycin/ n-decanol[c] | 1 M KCl | $10^{-7}$ |
| $6 \times 10^{-4}$ M valinomycin/ n-decanol[c] | 1 M KCl | $10^{-5}$ |
| $6 \times 10^{-3}$ M valinomycin/ n-decanol[c] | 1 M KCl | $2 \times 10^{-4}$ |
| $2.7 \times 10^{-3}$ M valinomycin/ dipenylether[c] | 1 M KCl | $6 \times 10^{-4}$ |
|  | 1 M K picrate | $9 \times 10^{-3}$ |
|  | 1 M RbCl | $2.7 \times 10^{-3}$ |
|  | 1 M CsCl | $2 \times 10^{-4}$ |
|  | 1 M $NH_4Cl$ | $3.4 \times 10^{-6}$ |
|  | 1 M NaCl | $10^{-6}$ |
|  | 1 M LiCl | $8 \times 10^{-7}$ |
|  | 0.1 M $MgCl_2$ | $2 \times 10^{-7}$ |

[a] Data from J. O/M Bockriss and A. K. N. Reddy, *Modern Electrochemistry*, Vol. 2, Plenum, New York, 1970, p. 1238. For hydrogen evolution reaction.
[b] Data from J. M. West, *Electrodeposition and Corrosion Processes*, 2 ed., Van Nostrand Reinhold, London 1970, p. 17. Values at 25°.
[c] Data from J. Koryta, *Anal. Chim. Acta* **111**, 1 (1979). Values are for ion selective electrodes (inclusive of constituent interfaces) with the exception of the value for Ag|AgNO$_3$(1 M).

which can be used for acid-base analysis using differential potentiotitrimetry, the observation that the rate of response of platinum to changes in hydrogen ion concentration is faster than that of tungsten immersed in the same solution is qualitatively predictable from the values of the following equilibrium exchange current densities at 25°C:

$$\text{Platinum in 1 } M \text{ H}_2\text{SO}_4, \quad j_0 = 794 \times 10^{-6} \text{ A cm}^{-2}$$

$$\text{Tungsten in 1 } M \text{ H}_2\text{SO}_4, \quad j_0 = 1.3 \times 10^{-6} \text{ A cm}^{-2}$$

Values of exchange current densities are, therefore, data that can provide guidance in the selection of electrode systems that may be suitable for differential potentiotitrimetry and some values are given in Table 2.1. A good summary of the basic theory governing the concept of the exchange current density is to be found in Reference (24).

## 2.4 ELECTROLYTIC POTENTIOTITRIMETRY AND DIFFERENTIAL ELECTROLYTIC POTENTIOTITRIMETRY

In this book *electrolytic potentiotitrimetry* is defined as the monitoring of the potential of a chemical cell with liquid junction during the course of a titration when a current of a few microamps is made to flow through the cell. The method was first proposed in 1911 (25). Sometimes in order to avoid the accumulative and possibly deleterious effects which might be caused by repeated passages of current through the reference electrode, a third or *auxiliary electrode* is used merely to conduct the current. In these instances the value of the current flowing through the indicator electrode and the auxiliary electrode is adjusted to the desired value but the potential is monitored between the indicator and reference electrode as is usual for a chemical cell with liquid junction.

The corresponding technique of applying a small current through a concentration cell without liquid junction during the course of a titration is termed *differential electrolytic potentiotitrimetry*, and was introduced in 1922 (26). Often the two electrodes are made of the same material and the current flowing through each causes a modification in the response of one electrode to a given electroactive species with respect to the response of the other. As a result a large change in potential is observed in the vicinity of the equivalence point.

The recommended IUPAC nomenclature corresponding to the terms electrolytic potentiotitrimetry and differential electrolytic potentiotitrimetry are *controlled current potentiometric titrations* and *controlled current potentiometric titrations with two indicator electrodes*, respectively. The variety of other names found in the literature include *potentiometric titrations with one polarized electrode*, *potentiometric titrations with two polarized electrodes*, and variants of these in which the term *constant current* is included. Another widespread term used to describe differential electrolytic potentiotitrimetry is *bipotentiometric titration*. The terms used in this book are derived from *potentiometry* and *differential potentiometry* and the applications of these techniques to *titrimetry*. The word *electrolytic* is used in the same sense as that suggested by Bishop (28) as implying that electrolysis is proceeding under the influence of a minute heavily stabilized current. This electrolysis current is less than the diffusion current of the electroactive species and the solutions are normally stirred.

The general theory relating the magnitude of a current that flows through any cell when a potential is applied to it and the relevance of this theory to the two types of electrolytic potentiotitrimetry is discussed in terms of the resulting voltammetric curves in Reference 27. For the purpose of this book it is sufficient to set forth the following points:

1. Electrode reactions are special types of heterogeneous chemical reactions in which there is a net charge transfer and hence a current flow across an interface which makes the rate of reaction dependent upon the applied potential.

2. For a current to flow through a cell, the potential applied across the cell must depart from either the potentiometrically determined or the predicted value of $E_{cell}$. A condition of thermodynamic reversibility pertains only if a microscopic decrease in applied potential will cause the spontaneous cell reaction to occur and, conversely, a microscopic increase in potential above that of $E_{cell}$ causes the reverse reaction to proceed. In either of these two cases a microscopic current will flow.

3. In practice very few electrode reactions are thermodynamically reversible and hence the difference between $E_{cell}$ and the applied potential has to be much greater than is implied by the criterion for thermodynamic reversibility in order that a current should flow.

4. The greater the difference between $E_{cell}$ and the potential that must be applied before a current of even a few microamps is detected in the cell, the higher the activation energy for the reaction occurring in the cell.

The electrolytic methods, therefore, have particular application to those electrode reactions in which attainment of an equilibrium potential is slow under the usual potentiometric condition of "zero" current. These electrode reactions are usually characterized by a low value of the exchange current density.

The methods of electrolytic potentiotitrimetry and differential electrolytic potentiotitrimetry have been used to locate the equivalence points in acid-base, complexometric, precipitation, and oxidation-reduction titrations. In the latter application the methods can be applied to those oxidation-reduction titrations in which one of the reagents (the titrant or the titrand) exhibits slow kinetics at the interface of a zeroth electrode such as platinum. Commonly employed reagents in this category are the oxidant potassium dichromate, and the reductants sodium thiosulfate and arsenic(III) oxide. The half-reactions appropriate to these reagents are described as *irreversible* and can be represented as follows:

the irreversible dichromate/chromium(III) couple

$$Cr_2O_7^{2-} + 14H^+ + 6e \rightleftharpoons 2Cr^{3+} + 7H_2O$$

the irreversible tetrathionate/thiosulfate couple

$$S_4O_6^{2-} + 2e \rightleftharpoons 2S_2O_3^{2-}$$

and the irreversible arsenic(V)/arsenic(III) couple emanating from

$$H_3AsO_4 + 2H^+ + 2e \rightleftharpoons H_3AsO_3 + H_2O$$

in which the reductant $H_3AsO_3$ is formed from the reagent $As_2O_3$ by the acid-base reactions $As_2O_3 + H_2O + 2OH^- \rightarrow 2H_2AsO_3^-$:

$$2H_2AsO_3^- + H_{(HCl)}^+ \rightarrow H_3AsO_3$$

**Electrolytic Potentiotitrimetry**

This technique is advantageously applied when a reagent forming an irreversible couple is reacted with a titrant forming a reversible couple such as iodine, iron(II), or cerium(IV). In these instances the passage of a small current ($< 10 \mu A$) through a chemical cell consisting of a platinum indicator electrode and a reference electrode produces a considerable enhancement in the change of potential circumjacent to the equivalence point in comparison with the change that would be observed using the same cell under the conditions of "zero" current. Even under the influence of a large value of the applied potential, a highly irreversible couple will not form a potential-determining state of coexistence at the surface of a platinum electrode. Any current that flows as a result of the potential being applied to the cell stems from either the oxidation or reduction of some other species in the solution or of the solvent itself. If, for example, thiosulfate solution is titrated with iodine in aqueous solution, $2S_2O_3^{2-} + I_2 \rightarrow S_4O_6^{2-} + 2I^-$, either of these possibilities can occur depending upon the polarity of the platinum indicator electrode.

1. *Platinum electrode polarized cathodically.* Water or the dissolved oxygen contained in it will be reduced prior to the equivalence point. Thereafter, reduction of iodine occurs which results in a substantial potential shift.

2. *Platinum electrode polarized anodically.* The oxidation of the iodide ion will occur before the equivalence point and, of course, this ion will still be present after the equivalence point. Little change in potential occurs, therefore, when excess iodine is added.

A comparison between these two conditions (curves $a$ and $b$) and the titration curve (curve $c$) obtained under the potentiometric condition ($i \rightarrow 0$) is given in Figure 2.3. The potentials observed for curve $a$ before the

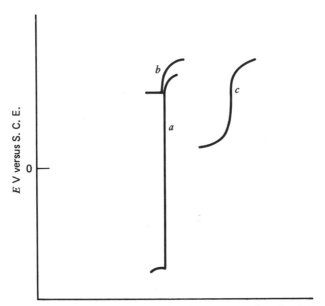

**Figure 2.3.** Titration curves of thiosulfate with iodine obtained when the platinum electrode was polarized cathodically (curve $a$), and when it was polarized anodically (curve $b$) with respect to the saturated calomel reference half-cell. Curve $c$ is the curve obtained when $i \rightarrow 0$.

equivalence point bear no relation to the activity ratio thiosulfate/tetrathionate, but after the equivalence point the potential becomes dependent upon the activity ratio iodide/iodine. By contrast, for titrations in which both the titrand and the titrant solutions contain electroactive species that can be regarded as reversible, enhancement of the inflection at the equivalence point is not markedly improved by passage of current through the indicator electrode. For example, only a marginal improvement is observed when electrolytic potentiotitrimetry is applied to the titration of iron(II) with cerium(IV).

## Differential Electrolytic Potentiotitrimetry

The term implies that the current flows through two electrodes each of which, in the absence of any kinetic data to the contrary, might be thought of as being reversible under conditions of $i \rightarrow 0$ to the same electroactive species in the solution. In practice this technique, in which current flows usually through two identical platinum electrodes, can be applied to either of the following cases:

1. A reversible couple titrated with a reversible couple, as in the example:

    iron(II) with cerium(IV)

2. An irreversible couple titrated with a reversible couple or vice versa, as in the following examples:

    iron(II) (reversible) with dichromate (irreversible)

    arsenite (As(III), irreversible) with iodine (reversible)

Each of the three example titrations yields dissimilar types of differential curves (Figure 2.4) from which, in each case, it is possible to deduce the equivalence volume. The low potentials obtained in the early stages of the titration of iron(II) with cerium(IV) are explained by the cathodic reduction of iron(III) at the surface of one platinum electrode and the simultaneous oxidation of iron(II) at the anodic platinum electrode. At the equivalence point, however, only iron(III) and cerium(III) are present and thus the potential of the anode changes so that cerium(III) can be oxidized at its surface. After the equivalence point, excess cerium(IV) is present and the cathode potential will, once again, be similar in magnitude to that of the anode. Provided that the current flowing is small, the equivalence volume is

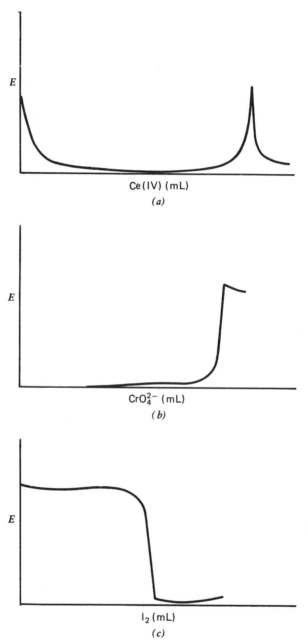

**Figure 2.4.** Differential electrolytic potentiotitrimetric curves obtained with two platinum electrodes when: (a) Fe(II) was titrated with Ce(IV); (b) Fe(II) (reversible) was titrated with $Cr_2O_7^{2-}$ (irreversible); (c) As(III) (irreversible) was titrated with $I_2$ (reversible).

deducible from the maximum difference in the potential that develops between the two electrodes.

The change in potential commensurate with the equivalence volume in the case of titration reactions between reversible and irreversible couples is due to a cathodic decomposition of the solvent at some stage during the titration. In the example of a reversible couple being titrated with an irreversible couple, a large potential develops as a result of solvent reduction at and after the equivalence volume. For the converse case a large potential emanating from this cause develops before the equivalence point.

## Apparatus and General Conditions

The apparatus required for the routine application of the technique is quite easy to assemble. The electrodes through which the current is to flow are connected through a high resistance to a source of constant d.c. voltage. Typically, this ballast resistance should have a value of no less than 100 M$\Omega$ and the voltage source of such a value that a current of less than 1 $\mu$A could be obtained. A resistance of 200 M$\Omega$ and a battery of 2–120 V has been found suitable (28). The potential that exists across the electrodes during a titration is conveniently monitored by a $p$H meter. It is also worth noting that almost all $p$H meters of modern design have provision for imposing a polarizing current of 5 or 10 $\mu$A through the electrodes. Suitable adaptation of this facility can obviate the requirement for the separate type of constant current source mentioned. The overall resistance of the solution in the titration cell is usually kept small by maintaining a condition of high ionic strength. In many oxidation-reduction titrations this condition is inherent in the high acid concentrations necessary for the correct stoichiometry of the reaction. By comparison, the concentration of the electroactive determinand makes only a small contribution to the total ionic strength. In these circumstances the fraction of the current carried through the cell by an electroactive species upon which the quantitative effectiveness of the method depends becomes a negligible factor. In some titrations, however, a condition of high ionic strength must be created by the addition of a supporting electrolyte of nonreactive ions. In this case the ions of the added salt carry practically all the current and the potential gradient is established so very close to the electrode surface that the gradient cannot cause the electroactive titrand ions to migrate from the bulk of the solution. Thus the current flowing can be regarded as being due primarily to diffusion although an effect due to the mechanical stirring of the solution is always evident. In the routine applications of the method the concentration of electroactive species is usually greater than 0.001 $M$ and at such concentrations the current flowing in the cell is much less than the limiting current which could be made to flow by increasing the applied potential. For accurate results the ratio of the current flowing to the limiting current is often about $10^{-3}$. At such low ratios, losses of electroactive material due to chemical changes resulting from electrolysis are negligible. Electrolysis, of course, need not be considered as a possible source of error if the titration reaction between two reversible couples is performed in a cell containing two polarized electrodes; the product reduced on the cathode is oxidized on the anode. Small systematic errors are, nonetheless, introduced by the current flowing through two electrodes of different areas producing differences in the current density and these errors are exacerbated if there is also a difference in stirring rate at the electrodes. Typically, at concentrations of about 0.1 $M$ there is a discrepancy between the endpoints and equivalence points of about 0.01–0.02 mL for the titrations of iron(II) with cerium(IV) or dichromate. Errors are also produced by the fouling and deactivation of the electrodes in some cases. Despite these, the accuracy compares favorably with the classical ($i \rightarrow 0$) potentiotitrimetric method showing also a considerable improvement (two to tenfold) in both the sharpness of the endpoint and in the attainment of equilibrium values of the electrode potentials (28).

Examples of these methods are exemplified by the microanalytical determination of nanogram quantities of halides (29), and a more recent adaptation of the method in which the two electrodes are polarized by an alternating current (30–32). These current densities are somewhat larger than those used in the d.c. method which still further accelerates the electrode response, whereas the alternating anodization and cathodization of the electrode minimizes the fouling and deactivation effects mentioned previously.

**The Karl Fischer Titration**

This method, used for the determination of water in organic materials, is a notable example of the application of differential electrolytic potentiotitrimetry to equivalence point detection in nonaqueous systems. An entire book (33) has been devoted largely to the analytical applications of the method in which the titrant is an anhydrous solution containing iodine, sulfur dioxide, and excess pyridine usually prepared in methanol. The titration reaction between the titrant and water can be represented by

$$I_2 Pyr. + SO_2 Pyr. + H_2O \xrightarrow{Pyr.} SO_3 Pyr. + 2 Pyr.H^+ I^-$$

where Pyr. represents pyridine. The product $SO_3 Pyr.$ reacts further with methanol to form pyridinium methyl sulfate

$$SO_3 Pyr. + CH_3OH \rightarrow Pyr.H^+ CH_3 SO_4^-$$

From these reactions it is evident that 1 mole of water requires 1 mole of iodine. When differential electrolytic potentiotitrimetry is used to detect the equivalence point, the titration curve is similar to that shown in Figure 2.4b.

An impediment to the method is caused by the instability of the iodine–sulfur dioxide–pyridine–methanol reagent which, even in the absence of water, undergoes a rapid initial loss in strength followed by slower change. Remedies have included the introduction of the sample into pyridine–methanol–sulfur dioxide followed by titration with the more stable solution of iodine (alone) in methanol (34, 35). The use of a stabilized reagent with 2-methoxyethanol in place of methanol has been proposed (36) and titrations have also been performed in dimethylformamide (37). More recently, coulometrically generated iodine reagent has been used in conjunction with a potentiometrically ($i \to 0$) indicated equivalence point (38). A comparison between this method of equivalence point detection and amperometric detection was made subsequently (39). A kinetic study of the reaction which utilized results obtained by potentiometric measurements has also been reported (40).

## REFERENCES

1. IUPAC, "The Classification and Nomenclature of Electroanalytical Techniques," *Pure Appl. Chem.* **45**, 81 (1976).
2. T. A. O'Donnell and D. F. Stewart, *Anal. Chem.* **33**, 337 (1961).
3. L. de Brouckere, *Bull. Soc. Chim. Belg.* **37**, 103 (1928).
4. H. V. Malmstadt and J. D. Winefordner, *Anal. Chim. Acta*, **30**, 283 (1959).
5. H. V. Malmstadt and J. D. Winefordner, *J. Am. Water Works Assoc.* **51**, 733 (1959).
6. H. V. Malmstadt and J. D. Winefordner, *Anal. Chem.* **32**, 281 (1960).
7. H. V. Malmstadt and H. L. Pardue, *Anal. Chem.* **32**, 1034 (1960).
8. H. V. Malmstadt, T. P. Hadjiioannou, and H. L. Pardue, *Anal. Chem.* **32**, 1039 (1960).
9. H. V. Malmstadt and H. L. Pardue, *Anal. Chem.* **33**, 1040 (1961).
10. H. V. Malmstadt and J. D. Winefordner, *Anal. Chim. Acta* **24**, 91 (1961).
11. J. D. Winefordner and M. Tin, *Anal. Chem.* **35**, 382 (1963).
12. J. D. Winefordner and G. A. Davison, *Anal. Chim. Acta* **28**, 480 (1963).
13. R. G. Bates, *Determination of pH Theory and Practice*, 2nd ed., Wiley, New York, 1973, p. 294.
14. J. Pinkhof, *Dissertation*, Amsterdam, 1919; W. D. Treadwell and L. Weiss, *Helv. Chim. Acta* **2**, 680 (1919).
15. H. H. Willard and A. W. Boldyreft, *J. Am. Chem. Soc.* **51**, 471 (1929).
16. H. H. Willard and F. Fenwick, *J. Am. Chem.* **44**, 2504 (1922).
17. S. A. Darwish and R. Salim, *Microchem. J.* **18**, 670 (1973).

18. I. C. Popescu, C. Liteanu, and A. Mocanu, *Rev. Roum. Chim.* **20**, 397 (1975).
19. E. Muller and H. Kogert, *Z. Physik. Chem.* **136**, 437 (1928).
20. R. B. Hanselman, G. H. Schade, Jr., and L. B. Rogers, *Anal. Chem.* **32**, 761 (1960).
21. E. Schumacker and F. Umland, *Fresenius' Z. Anal. Chem.* **272**, 257 (1974).
22. N. F. Hall, M. A. Jensen, and S. E. Backström, *J. Am. Chem. Soc.* **50**, 2217 (1928).
23. J. O'M. Bockris and A. K. N. Reddy, *Modern Electrochemistry*, Vol. 2, Plenum, New York, 1970, p. 876.
24. C. W. Davies and A. M. James, *A Dictionary of Electrochemistry*, Macmillan Press, London, 1976, p. 121.
25. P. Dutoit and G. Weise, *J. Chem. Phys.* **578**, 608 (1911).
26. H. H. Willard and F. Fenwick, *J. Am. Chem. Soc.* **44**, 2504, 2515 (1922).
27. C. N. Reilley and R. W. Murray in *Treatise on Analytical Chemistry*, I. M. Kolthoff and P. J. Elving, Eds., Part 1, Vol. 4, Chapter 43, Interscience, New York, 1963.
28. E. Bishop, *Analyst (London)* **83**, 212 (1958); **85**, 422 (1960).
29. E. Bishop and R. G. Dhaneshwar, *Anal. Chem.* **36**, 726 (1964).
30. E. Bishop and T. J. N. Webber, *Analyst (London)* **98**, 697 (1973).
31. E. Bishop and T. J. N. Webber, *Analyst (London)* **98**, 769 (1973).
32. E. Bishop and P. Cofre, *Analyst (London)* **103**, 162 (1978).
33. J. Mitchell, Jr. and D. M. Smith, *Aquametry*, Interscience, New York, 1948.
34. A. Johansson, *Acta Chem. Scand.* **3**, 1058 (1949).
35. W. Seaman, W. H. McComas, Jr., and G. A. Allen, *Anal. Chem.* **21**, 510 (1949).
36. E. D. Peters and J. L. Jungnickel, *Anal. Chem.* **27**, 450 (1955).
37. R. F. Swensen and D. A. Keyworth, *Anal. Chem.* **35**, 863 (1963).
38. A. Cedergren, *Talanta* **21**, 367 (1974).
39. A. Cedergren, *Talanta* **21**, 553 (1974).
40. J. C. Verholf and E. Barendrecht, *J. Electroanal. Chem. Interfacial Electrochem.* **71**, 305 (1976).

CHAPTER

3

# CLASSIFICATIONS AND DESCRIPTIONS OF ELECTRODES

The most useful cell in the applications of direct potentiometry to chemical analysis is the quasi-thermodynamic type represented as

indicator electrode A | species A (constant $I$) || reference electrode  (3.1)

It was mentioned in Chapter 1 that careful calibration of the cell under conditions of constant temperature and constant ionic strength ($I$) leads to a linear relationship which can be expressed as

$$E_{cell} = constant - slope \cdot \log c_A \quad (3.2)$$

The two major contributing terms in the magnitude of *constant* are the standard potential of the indicator electrode and the potential of the reference electrode which are unequivocal constants at a given temperature. The other two contributors to this term are the junction potential and the activity coefficient of species A both of which are assumed to be constant if all solutions have the same ionic strength. For an indicator electrode of exact Nernstian response, the value of the *slope* is deducible from the value of the Nernst factor $k$ (see appendix) at the required temperature divided by the charge on the determinand ion. Nevertheless, it is not a prerequisite to the success of the method that the value of the *slope* should be exactly equal to the appropriate fraction of the Nernst factor $k$. It is sufficient that a linear relation between $E_{cell}$ and $\log c_A$ should exist and that this relation should be reproducible over the period of time required for the calibration and the analysis. A factor that must be considered, however, is the type of extraneous species to which the indicator electrode may respond or which may react chemically with the materials used in its construction. It is on the basis of such considerations that oxidation-reduction electrodes are almost totally excluded as indicator electrodes in cells that are used in analysis by direct potentiometry. By comparison, *membrane electrodes*, as the other class of electrodes are called, are less reactive than oxidation-reduction electrodes and since the 1960s have found increasing application in potentiometric analysis. Unlike oxidation-reduction half-cells, the potentials of which are derived from an electron exchange process, oxidant + $ne \rightleftharpoons$ reductant, membrane potentials develop as a result of the equilibrium established between the constant activity of a species held in a membrane and a lower activity of the same species present in the solution under test. The potential that develops is proportional to the logarithm of the activity of the species in solution. Thus when a membrane electrode is used as the indicator electrode of cell (3.1), the relation between the potential of that cell and the concentration of species A is described by equation (3.2). It is usual for the reference electrode of quasi-thermodynamic cells, conforming to the notation of cell (3.1), to be derived from half-cells described as *oxidation-reduction electrodes*. On the other hand, the indicator electrodes employed in the direct potentiometric applications of these cells generally belong to the class of electrodes described as *membrane electrodes*. *Ion selective electrodes* constitute a category within the classification of membrane electrodes and these are important examples of the type of indicator electrodes used for the applications of direct potentiometry to analysis. In potentiotitrimetry, however, both oxidation-reduction and membrane electrodes appear to find an equal application as indicator electrodes.

## 3.1 CATEGORIES OF OXIDATION-REDUCTION ELECTRODES

There are subdivisions within the broad classification of oxidation-reduction electrodes that are made according to the number of ion-crossing interfaces present in the solid materials from which the electroactive part of the electrode is formed. These subdivisions are described by the equivalent terms *type*, *class*, or *kind*, and of these terms, *kind* is the most commonly used. The electrode kind can often be assigned according to the number (if any) of solid phases that appear in the balanced half-reaction. If this number is zero, as in $Fe^{3+} + e \rightleftharpoons Fe^{2+}$ or $2H^+ + 2e \rightleftharpoons H_{2(g)}$, the electrode that responds in a predictable manner to changes in the activity ratio reductant/oxidant is called a *zeroth kind electrode*. Implicit in this definition is the unreactive nature of the conducting solid material when it is immersed in the solution containing the oxidation-reduction couple. Platinum metal is the material most frequently used to provide this type of interface and the electrodes so formed are sometimes referred to as *inert electrodes* or as *redox electrodes*. These terms should not be used. *Inert* implies that the electrode material is unreactive under all conditions of normal use which negates the possibility of such observed behavior as is inferred by $PtO_2 + 2H^+ + 2e \rightleftharpoons Pt(OH)_2(1,2)$; the word *redox* could be taken as describing the whole classification of oxidation-reduction electrodes.

Electrodes of the first, second, and third kinds have one, two, or three ion-crossing interfaces, respectively, as shown by the following examples:

1. Often metal or metal-amalgam electrodes.

$$Ag_{(s)}|Ag^+ \quad \text{and} \quad Zn(Hg)_{(l)}|Zn^{2+}$$

   typical half-reaction $Ag^+ + e \rightleftharpoons Ag_{(s)}$, that is, one solid.

2. Metal-sparingly soluble metal salt electrodes.

$$Ag_{(s)}; AgCl_{(s)}|Cl^- \quad \text{and} \quad Pb(Hg)_{(l)}; PbSO_{4(s)}|SO_4^{2-}$$

   typical half-reaction $AgCl_{(s)} + e \rightleftharpoons Ag_{(s)} + Cl^-$, that is, two solids.

3. Metal-two sparingly soluble metal salts having a common anion

$$Ag_{(s)}; Ag_2C_2O_{4(s)}; CaC_2O_{4(s)}|Ca^{2+} \quad \text{and} \quad Pb(Hg)_{(l)}; PbC_2O_{4(s)}|Ca^{2+}$$

   typical half-reaction $PbC_2O_{4(s)} + Ca^{2+} + 2e \rightleftharpoons Pb_{(s)} + CaC_2O_{4(s)}$, that is, three solids.

It should be noted that for all three kinds of electrodes an example has been given in which a metal has been specified by the notation as being present as a metal amalgam. In these instances the mercury can be regarded as a zeroth interface because, ideally, it is not essential to the operation of the half-cell as specified by the half-reaction. Despite the presence of mercury as a zeroth interface in the appropriate examples of the second and third kinds of electrodes, such electrodes are not classified as amalgam electrodes. Common usage of the term *amalgam electrode* confines this description to electrodes of the first kind in which a metal susceptible to oxidation is used in the form of a dilute amalgam.

## 3.2 REFERENCE ELECTRODES IN AQUEOUS SOLUTIONS

### General Properties

In its most common application the potential of a reference electrode must remain constant during the calibration and the subsequent use of a cell. This function is implicit in the general equation (3.2) and was fulfilled by the use of a hydrogen reference electrode in the cell $Zn(Hg)|Zn^{2+}\|H^+(0.01\ M)|H_2$ (1 atm); Pt described earlier. The provision of an invariant potential is not necessarily the sole requirement for an electrode to be described as a reference electrode, however, as was exemplified by the concentration cell with diffusion (cell 2.5). A general form of this type of cell is

X responsive electrode | unknown conc X ‖ known conc X | X responsive electrode
half-cell A (sample)                                   half-cell B (reference)

In the example used, the potentials of both the sample half-cell and the reference half-cell would vary during the adjustment of the concentrations of the two half-cell solutions so that a zero value of $E_{cell}$ was ultimately obtained. Another application in which the potential of a reference electrode varies is in the checking of the response of a newly developed electrode against that of an older type having a proven performance. For example, the response of a hydrogen ion selective electrode might require checking against that of a hydrogen electrode in the chemical cell without diffusion

$$H^+ \text{ selective electrode} | H^+ | H_2 \text{ (1 atm)}; Pt \qquad (3.3)$$

In this instance it would be hoped that the value of $E_{cell}$ would remain constant over a range of hydrogen ion activities because this would indicate that the electrode under test has the same response as the hydrogen reference electrode.

Chemical cells without liquid junction represented by the type

$$Pt; H_2 \text{ (1 atm)} | H^+ \text{ buffer solution}, Cl^- \text{ (known molality)} | AgCl; Ag$$

have also been used to establish a practical scale of $p$H (see Chapter 4). The silver–silver chloride electrode is the reference electrode in these cells. Nevertheless, it is the variation of the known concentrations of chloride ion over a substantial range that allows the ultimate assignation of a $p$H value to the buffer solutions. The stoichiometric concentrations of the buffer solution components are kept constant throughout the series of measurements of $E_{cell}$, and thus the variation in the potentials of the hydrogen ion indicator electrode is small in comparison with the variation in the potentials of the reference electrode.

The foregoing examples serve to illustrate that the role of the reference electrode is unique to a particular classification of cell. These roles can be summarized as follows:

1. *In chemical cells with diffusion*: to provide a reference half-cell of fixed potential.
2. *In chemical cells without diffusion*: to provide half-cell potentials that respond reproducibly to variations in the known concentrations of a species to which it is reversible.

3. *In concentration cells with diffusion*: to provide a means of monitoring the potential of a half-cell having a known solution composition with respect to an identical electrode system immersed in a solution of unknown composition contained in the other half-cell.

A fourth and more recent application of reference electrodes is the provision of a constant internal reference potential in some types of membrane electrodes.

The essential characteristics of an ideal reference electrode are reversibility, stability, and reproducibility (3, 4). It was mentioned in Chapter 2 that good reversibility is associated in the electrokinetic sense with a high value of the exchange current density which, in the thermodynamic sense, implies that the potential-activity relation is described by the Nernst equation. A high value of the exchange current density for a reference electrode is also consistent with a rapid attainment of an equilibrium potential together with a negligible shift in potential should a small current flow through the electrode. If $E_{cell}$ is measured using an instrument having a high input impedance, as is usual in the modern practice of direct potentiometry, the magnitude of this current will be negligible.

A rapid equilibration to a stable and consistent value of potential after a sudden temperature change is one aspect of the reproducibility of an electrode. Another aspect of reproducibility is that in the preparation of a batch of electrodes by an identical method, the potential difference between any pair of them should be negligible when immersed in the same solution of electroactive species. That is, the *bias potential* should ideally be zero. Good reproducibility within a given batch, however, does not appear to imply that the potentials for all reference electrodes prepared by the same method at different times or at different places will have the same value. Indeed, evidence is to the contrary. For example, the standard potential for the silver–silver chloride electrode, one of the most widely used reference electrodes, was found to vary by as much as 0.2 mV in a collaborative study that compared data obtained for this electrode in different laboratories (5). Paradoxically, values of other constants or values of other properties derived from the use of cells in which a particular type of reference electrode is used may agree consistently from laboratory to laboratory. Periodic cell calibration, by an adaptation of the method described for the determination of the mean activity coefficients of zinc sulfate solutions, is responsible for the resolution of this apparent paradox. The frequency with which cell calibration should be performed is dictated by the relative reproducibilities of the potentials that are established by the two electrodes. It is the period over which the less stable electrode can be expected to reproduce a constant potential that establishes this frequency. Usually the indicator electrode in a chemical cell with diffusion is less stable in this respect than the reference electrode. The change of the electrode potential over a period of time is only one aspect of stability. Another factor pertaining particularly to chemical cells without diffusion is the stability of the solid phase of the electrode in the solution phase in which it is immersed. This aspect of stability can either be associated with the physical or the chemical properties of the material. The solubility of the solid phase of an electrode of the second kind is relevant to the physical properties of the material as was discussed for the Pb(Hg); $PbSO_4$ electrode. The ability of the electroactive materials to withstand attack, particularly by complexants or oxidation–reduction reagents, are aspects of the chemical properties that must be considered. For example, a silver–silver chloride electrode would not be suitable as a reference electrode in a solution that contains a high concentration of an amine.

## The Hydrogen Electrode

The cell depicted in Figure 1.1 contained a hydrogen electrode acting as a reference electrode; a brief description of the electrode was given and the impediments to its use as a common reference electrode in chemical cells with diffusion were discussed. In practice, however, the hydrogen electrode can be regarded as the reference electrode for all measurements of the $p$H values of aqueous solutions. In this respect it is a notional reference electrode because, although it is not popularly used for $p$H measurements, the $p$H values of the standard buffer solutions used to calibrate more convenient electrode combinations were all derived by extrapolation of data derived from cells in which the hydrogen electrode was the indicator electrode. For example, the response of the glass electrode is adjusted so that it becomes coincident with the response to change in $p$H that would be observed if the hydrogen electrode were used as the indicator electrode in the same cell.

The standard potential of the hydrogen electrode is defined arbitrarily to be zero at all temperatures and this definition forms the basis for the standard oxidation–reduction potentials and the formal potentials discussed in Chapter 1. There is a difference between this definition and the statement that the standard hydrogen electrode has a potential of zero at all temperatures. The concept that it is possible to construct a standard hydrogen electrode (Pt; $H_2$ (1 atm)| $H^+(a = 1)$) is a misrepresentation (6) because it is impossible to determine the activity coefficient of a single ion, neither is it possible to calculate this quantity even approximately for such a concentrated solution. Care must therefore be taken in the terminology. For example, the information that the standard potential of the silver–silver chloride electrode was determined to be 0.23142 V at 10° using the cell

$$\text{Pt}; \text{H}_2 \, (1 \text{ atm}) | \text{HCl}(0.01 \, m) | \text{AgCl}; \text{Ag} \qquad (3.4)$$

really means that standard potential of the cell ($E^{\circ}_{\text{cell}}$) was found to be 0.23142 V. It is because the standard potential of the hydrogen electrode is defined as being zero at 10° that allows the assignation of this value for $E^{\circ}_{\text{cell}}$ as the standard potential for $\text{AgCl}_{(s)} + e \rightleftharpoons \text{Ag}_{(s)} + \text{Cl}^-$ at 10°. Similarly, if $E^{\circ}_{\text{cell}}$ is determined as 0.22234 V at 25°, the difference of $-9.08$ mV $(0.22234 - 0.23142)$ is completely attributed to the effect of change in temperature upon the standard potential of the silver–silver chloride couple despite the inevitable variation of the absolute standard potential of the hydrogen electrode with temperature. Neither the hydrogen electrode nor the silver–silver chloride electrode can be regarded as standard electrodes in these instances and the concept of any *standard electrode* as defined in some textbooks has no basis in reality.

When prepared following the instructions summarized by Bates (7), the hydrogen electrode is one of the most reproducible electrodes available as is evident from the low bias potentials between electrodes which are less than 10 μV. The base of the hydrogen electrode is a piece of platinum foil about 0.125 mm thick and about 1 cm². A length of platinum wire (26 gauge) is spot welded near the center of one edge of the square and this length is used to attach the foil to a piece of flint glass tubing about 5 mm external diameter and about 8 cm long. The seal of the wire into the tube is accomplished by melting a bead of glass over the entire wire-foil junction before finally sealing the remaining length of wire into the glass tube (Figure 3.1). In order to avoid any possibility of the manifestation of the thermal effects, it is advisable to ensure that the junction between the copper wire (used as a connection for the electrode) and the platinum wire inside the tube is made so that when the electrode is in use this junction will be below the surface of the solution. The connection can be made by a spot weld prior to sealing the platinum wire into the glass tube or by the use of a small volume of molten Wood's metal which is poured into the body of the electrode after the glass–platinum wire seal has been made. In the latter case the copper connecting wire is dipped into this low melting alloy while it is still liquid. The connection can also be made through mercury but this introduces a risk that mercury will work through the seal. Whatever method is used, the connection should be protected against undue strain by providing a suitable collar at the top of the electrode body.

The platinum foil base of the electrode is usually coated with finely divided platinum (*platinum black*), which is able to catalyze the exchange $2\text{H}^+ + 2e \rightleftharpoons \text{H}_2$, although the electrode will also function correctly if this

Figure 3.1. Base for the hydrogen electrode (7).

platinum black is dispersed in the solution (8). The platinum black is generally deposited electrolytically onto the surface of the foil once this surface has been prepared by brief immersion in 50% aqua regia followed by washing first with concentrated nitric acid and finally with water. Immediately before the deposition process, surface oxides are removed by electrolysis as a cathode in an electrolyte of 0.01 $M$ sulfuric acid. After washing with water, the electrode is made to form a cathode in a compartment separated from a platinum anode compartment by means of a salt bridge or by the use of a H-shaped cell in which the two limbs are separated with a glass-frit cross tube. The recommended procedure (9) uses an aqueous electrolyte containing 3.5% chloroplatinic acid and 0–0.005% lead acetate and an electrolysis time of up to 10 min at a current density of 30 mA cm$^{-2}$ in a well-stirred solution. Usually a deposition time of 5 min is adequate since smaller quantities of deposit are concomitant with a faster equilibration time and a lower adsorptivity of material from a solution. After their preparation the electrodes are washed with water and when not in use are stored in water.

If a hydrogen electrode is to be incorporated into a cell from which accurate values of e.m.f. are to be obtained, the cell configuration typified by Figure 1.1 would not be suitable. However, chemical cells with liquid junction in which that type of *Hildebrand* hydrogen electrode is used as an indicator electrode have been applied for many years (9) to obtain results of medium precision. Nowadays, the hydrogen electrode is almost exclusively used in a cell without diffusion. The cell compartment is either of the H-type (see Figure 1.3) or of the U-type in order that the constituent half-cells may be established in separate limbs. It is sometimes necessary to interpose a stopcock or a glass frit between the two half-cells, and it is usually advisable to incorporate a facility in the cell compartment that permits the two half-cells to be filled or emptied without the removal of the cell from the thermostatically controlled fluid bath. The device that holds the cell in this bath should be such that the depth of submersion of the cell in the fluid excludes the possibility of loss of solvent from the cell solution through condensation. It is necessary to make provision for the attachment of a presaturator to the limb in which the hydrogen half-cell is to be established and this vessel is maintained at the same temperature as the cell. Its purpose is to prevent change in composition of the half-cell solution due to the flow of hydrogen gas. The presaturator has a volume about one-third that of the hydrogen half-cell and is usually manufactured as an integral part of this half-cell. It contains a portion of the cell solution through which hydrogen gas is bubbled before the gas enters the actual hydrogen half-cell solution at the rate of one to two bubbles per second through a jet of about 1 mm diameter. Provision can also be made for the flow of gas to be diverted so that it can be used as part of the fill–empty control system if this is to be incorporated into the overall cell design. It is axiomatic that atmospheric oxygen should not be allowed in contact with the platinum black surface of the electrode for any prolonged period. Electrolytic grade hydrogen, commercially available in cylinders, contains 0.15–0.20% oxygen together with small amounts of other impurities and these must be removed by the means of a purification train installed in the gas line (10). It should be noted that palladium membrane electrolytic generators are available for the *in situ* production of hydrogen that does not require further purification.

The notation (3.4) specifies that the partial pressure of hydrogen gas at the electrode surface is 1 atm. The cell reaction, deducible from the notation, is

$$H_2 \, (1 \, atm) + 2AgCl_{(s)} \rightleftharpoons 2H^+ + 2Cl^- + Ag_{(s)} \quad (3.5)$$

and the appropriate expression for the Nernst equation is

$$E_{cell} = E^\circ_{cell} - \frac{k}{2} \log \frac{a^2_{H^+} \cdot a^2_{Cl^-}}{1} \quad (3.6)$$

This equation is valid only if the partial pressure of hydrogen gas ($p_{H_2}$) at the electrode is 1 atm. The more general equation, however, is

$$E_{cell} = E_{cell}^{\circ} - \frac{k}{2}\log\frac{a_{H^+}^2 \cdot a_{Cl^-}^2}{p_{H_2}} \qquad (3.7)$$

and this is the equation which must be applied when, as in a typical experiment, the partial pressure of hydrogen is not 1 atm. Provided the aqueous solution contains no volatile solutes such as ammonia or carbon dioxide, the value of $p_{H_2}$ (atm) to use in equation (3.7) is given by

$$p_{H_2} = \frac{P - p_{H_2O}}{760} \qquad (3.8a)$$

where $P$ is the barometric pressure and $p_{H_2O}$ is the vapor pressure of water in the solution* at the given temperature both expressed in mm Hg. A refinement of this equation allows for a hydrostatic contribution to the pressure, the magnitude of which depends upon the depth (in mm) of immersion of the hydrogen gas delivery jet in the solution (11). Equation (3.8a) is then written as

$$p_{H_2} = \frac{P - p_{H_2O} + 0.03h}{760} \qquad (3.8b)$$

where $h$ is the depth of immersion (mm). This correction is superfluous if, as is usual, the depth of immersion of the jet is kept constant throughout the cell calibration and the subsequent measurements. The tedium of calculations can be reduced by allowing for the effects implicit in equations (3.7) and (3.8) in terms of an overall correction to be added directly to the experimental values of $E_{cell}$. This is done by considering the theoretical concentration cell without diffusion

$$\text{Pt; H}_2 \text{ (1 atm)} | \text{H}^+ | \text{Pt; H}_2(p_{H_2})$$

which summarizes the state of a cell operating at an experimental pressure of $p_{H_2}$ with respect to the same cell ideally operating at a standard pressure of 1 atm. The cell reaction for this cell is

$$\text{H}_2(760 \text{ mm}) \rightleftharpoons \text{H}_2(p_{H_2} \text{ in mm})$$

---

*Strictly this should also be corrected for Raoult's law and for the osmotic coefficient.

and the difference in potential between the electrodes ($\Delta E$) is given by

$$\Delta E = -\frac{k}{2}\log\frac{p_{H_2}}{760}$$

In terms of the barometric pressure $P$ and the vapor pressure of water $p_{H_2O}$ this equation becomes

$$\Delta E = \frac{k}{2}\log\frac{760}{P - p_{H_2O}} \tag{3.9}$$

which gives the correction in volts which must be added to $E_{cell}$ to allow for the effect of the actual pressure of hydrogen in the operational half-cell. Values of these corrections (in mV) for temperatures 0–95°C and barometric pressures 720–775 mm are given in Table 3.1.

The values of the potentials developed by the hydrogen electrode are greatly affected by such catalytic poisons as sulfide and cyanide, and gross deviations from the range of expected values are symptomatic of this effect. Drifting potentials, on the other hand, can be attributable to the chemical reduction of oxidant species such as the benzoic acids and nitrophenols on the catalytic platinum black surface. The latter effect can sometimes be overcome by using a palladium black deposit which has a lower catalytic activity. A diagnostic test (12) to detect the reduction of a particular species can be made by comparing the potential between two electrodes of differing catalytic activity. In the absence of any oxidant species the bias potential (if any) between the electrodes should be constant. If the magnitude of the bias potential increases or varies with time in the presence of a suspected oxidant species, it is very probably that reduction of that species is occurring on the more active surface. Such a test has been used to confirm that a hydrogen electrode can be used in 5 $M$ perchloric acid. Electrodes plated with palladium black are to be preferred, however, only when the presence of oxidants in the solution excludes the use of the more active platinum catalyst. The response and reproducibility of the palladium plated electrodes are generally inferior to the platinized type.

Palladium can be deposited on a platinum cathode from a 1–2% solution of palladous chloride in 1 $M$ hydrochloric acid in a cell containing a palladium anode. Bates (13) has prepared the plating solution containing the equivalent of about 0.5 g metal by the electrolytic oxidation of palladium foil in 12 M hydrochloric acid (50 mL) at a current of 500 mA which is made to flow with frequent reversals of polarity between two strips of the metal. Subsequent evaporation of this solution to a volume within the range 2–5 mL followed by the addition of water (50 mL) and lead acetate (40 mg) yields a suitable plating solution.

Table 3.1. Pressure Corrections for the Hydrogen Electrode in Millivolts[a]

| Barometric Pressure (mm Hg) | Temperature (°C) | | | | | | | | | | | | | | | | | | | |
|---|---|---|---|---|---|---|---|---|---|---|---|---|---|---|---|---|---|---|---|---|
| | 0 | 5 | 10 | 15 | 20 | 25 | 30 | 35 | 40 | 45 | 50 | 55 | 60 | 65 | 70 | 75 | 80 | 85 | 90 | 95 |
| 720 | 0.71 | 0.76 | 0.82 | 0.89 | 0.99 | 1.13 | 1.30 | 1.52 | 1.81 | 2.18 | 2.67 | 3.30 | 4.12 | 5.18 | 6.60 | 8.51 | 11.17 | 15.06 | 21.35 | 34.56 |
| 725 | 0.63 | 0.67 | 0.73 | 0.81 | 0.91 | 1.03 | 1.20 | 1.42 | 1.71 | 2.08 | 2.56 | 3.18 | 3.99 | 5.05 | 6.45 | 8.34 | 10.96 | 14.79 | 20.95 | 33.66 |
| 730 | 0.55 | 0.59 | 0.65 | 0.72 | 0.82 | 0.94 | 1.11 | 1.32 | 1.61 | 1.97 | 2.45 | 3.06 | 3.87 | 4.91 | 6.30 | 8.17 | 10.76 | 14.53 | 20.57 | 32.81 |
| 735 | 0.47 | 0.51 | 0.56 | 0.63 | 0.73 | 0.85 | 1.02 | 1.23 | 1.51 | 1.87 | 2.34 | 2.95 | 3.74 | 4.78 | 6.15 | 8.00 | 10.56 | 14.27 | 20.19 | 32.01 |
| 740 | 0.39 | 0.43 | 0.48 | 0.55 | 0.64 | 0.76 | 0.92 | 1.13 | 1.41 | 1.77 | 2.23 | 2.83 | 3.62 | 4.65 | 6.01 | 7.83 | 10.36 | 14.01 | 19.82 | 31.24 |
| 745 | 0.31 | 0.34 | 0.39 | 0.46 | 0.55 | 0.67 | 0.83 | 1.04 | 1.31 | 1.66 | 2.12 | 2.72 | 3.50 | 4.52 | 5.86 | 7.67 | 10.16 | 13.77 | 19.46 | 30.51 |
| 746 | 0.29 | 0.33 | 0.38 | 0.45 | 0.53 | 0.65 | 0.81 | 1.02 | 1.29 | 1.64 | 2.10 | 2.70 | 3.48 | 4.49 | 5.83 | 7.63 | 10.12 | 13.72 | 19.39 | 30.37 |
| 747 | 0.28 | 0.31 | 0.36 | 0.43 | 0.52 | 0.63 | 0.79 | 1.00 | 1.27 | 1.62 | 2.08 | 2.68 | 3.45 | 4.46 | 5.80 | 7.60 | 10.08 | 13.67 | 19.31 | 30.23 |
| 748 | 0.26 | 0.30 | 0.34 | 0.41 | 0.50 | 0.62 | 0.78 | 0.98 | 1.25 | 1.60 | 2.06 | 2.65 | 3.43 | 4.44 | 5.77 | 7.57 | 10.04 | 13.62 | 19.24 | 30.09 |
| 749 | 0.24 | 0.28 | 0.33 | 0.40 | 0.48 | 0.60 | 0.76 | 0.96 | 1.23 | 1.58 | 2.04 | 2.63 | 3.40 | 4.41 | 5.74 | 7.54 | 10.00 | 13.57 | 19.17 | 29.95 |
| 750 | 0.23 | 0.26 | 0.31 | 0.38 | 0.47 | 0.58 | 0.74 | 0.94 | 1.21 | 1.56 | 2.02 | 2.61 | 3.38 | 4.39 | 5.72 | 7.50 | 9.96 | 13.52 | 19.10 | 29.81 |
| 751 | 0.21 | 0.25 | 0.30 | 0.36 | 0.45 | 0.57 | 0.72 | 0.92 | 1.19 | 1.54 | 2.00 | 2.58 | 3.35 | 4.36 | 5.69 | 7.47 | 9.93 | 13.47 | 19.03 | 26.68 |
| 752 | 0.20 | 0.23 | 0.28 | 0.34 | 0.43 | 0.55 | 0.70 | 0.90 | 1.17 | 1.52 | 1.97 | 2.55 | 3.33 | 4.33 | 5.66 | 7.44 | 9.89 | 13.42 | 18.96 | 29.54 |
| 753 | 0.18 | 0.22 | 0.26 | 0.33 | 0.42 | 0.53 | 0.68 | 0.89 | 1.15 | 1.50 | 1.95 | 2.53 | 3.30 | 4.31 | 5.63 | 7.40 | 9.85 | 13.37 | 18.90 | 29.41 |
| 754 | 0.16 | 0.20 | 0.24 | 0.31 | 0.40 | 0.51 | 0.67 | 0.87 | 1.13 | 1.48 | 1.93 | 2.51 | 3.28 | 4.28 | 5.60 | 7.37 | 9.81 | 13.33 | 18.83 | 29.28 |
| 755 | 0.15 | 0.18 | 0.23 | 0.29 | 0.38 | 0.49 | 0.64 | 0.85 | 1.12 | 1.46 | 1.91 | 2.49 | 3.25 | 4.26 | 5.57 | 7.34 | 9.77 | 13.28 | 18.76 | 29.14 |
| 756 | 0.13 | 0.17 | 0.21 | 0.28 | 0.36 | 0.48 | 0.63 | 0.83 | 1.10 | 1.44 | 1.89 | 2.47 | 3.23 | 4.23 | 5.55 | 7.31 | 9.74 | 13.23 | 18.69 | 29.01 |
| 757 | 0.12 | 0.15 | 0.20 | 0.26 | 0.35 | 0.46 | 0.61 | 0.81 | 1.08 | 1.42 | 1.87 | 2.45 | 3.20 | 4.20 | 5.52 | 7.28 | 9.70 | 13.18 | 18.62 | 28.83 |
| 758 | 0.11 | 0.14 | 0.18 | 0.24 | 0.33 | 0.44 | 0.59 | 0.79 | 1.06 | 1.40 | 1.85 | 2.42 | 3.18 | 4.18 | 5.49 | 7.24 | 9.66 | 13.13 | 18.56 | 28.76 |

| | | | | | | | | | | | | | | | | |
|---|---|---|---|---|---|---|---|---|---|---|---|---|---|---|---|---|
| 759 | 0.09 | 0.12 | 0.16 | 0.23 | 0.31 | 0.42 | 0.58 | 0.78 | 1.04 | 1.38 | 1.83 | 2.40 | 3.17 | 4.16 | 5.46 | 7.21 | 9.62 | 13.09 | 18.49 | 28.63 |
| 760 | 0.07 | 0.10 | 0.15 | 0.21 | 0.30 | 0.41 | 0.56 | 0.76 | 1.02 | 1.36 | 1.79 | 2.39 | 3.14 | 4.13 | 5.43 | 7.18 | 9.58 | 13.04 | 18.41 | 28.50 |
| 761 | 0.06 | 0.09 | 0.13 | 0.20 | 0.28 | 0.39 | 0.54 | 0.74 | 1.00 | 1.34 | 1.77 | 2.36 | 3.12 | 4.10 | 5.40 | 7.15 | 9.55 | 12.99 | 18.35 | 28.38 |
| 762 | 0.04 | 0.07 | 0.12 | 0.18 | 0.26 | 0.37 | 0.52 | 0.72 | 0.98 | 1.32 | 1.76 | 2.34 | 3.09 | 4.08 | 5.38 | 7.12 | 9.51 | 12.95 | 18.29 | 28.25 |
| 763 | 0.03 | 0.06 | 0.10 | 0.16 | 0.24 | 0.36 | 0.50 | 0.70 | 0.96 | 1.30 | 1.74 | 2.32 | 3.07 | 4.05 | 5.35 | 7.08 | 9.47 | 12.90 | 18.22 | 28.13 |
| 764 | 0.01 | 0.04 | 0.08 | 0.15 | 0.23 | 0.34 | 0.49 | 0.69 | 0.94 | 1.28 | 1.72 | 2.30 | 3.05 | 4.03 | 5.32 | 7.05 | 9.44 | 12.85 | 18.16 | 28.00 |
| 765 | −0.01 | 0.02 | 0.07 | 0.13 | 0.21 | 0.32 | 0.47 | 0.67 | 0.92 | 1.26 | 1.70 | 2.28 | 3.02 | 4.00 | 5.29 | 7.02 | 9.40 | 12.80 | 18.09 | 27.89 |
| 766 | −0.02 | 0.01 | 0.05 | 0.11 | 0.19 | 0.31 | 0.45 | 0.65 | 0.91 | 1.24 | 1.68 | 2.26 | 3.00 | 3.98 | 5.26 | 6.99 | 9.36 | 12.76 | 18.03 | 27.76 |
| 767 | −0.04 | −0.01 | 0.04 | 0.10 | 0.18 | 0.29 | 0.43 | 0.63 | 0.89 | 1.22 | 1.66 | 2.23 | 2.98 | 3.95 | 5.24 | 6.96 | 9.32 | 12.71 | 17.96 | 27.64 |
| 768 | −0.05 | −0.02 | 0.02 | 0.08 | 0.16 | 0.27 | 0.42 | 0.61 | 0.87 | 1.20 | 1.64 | 2.21 | 2.95 | 3.93 | 5.21 | 6.93 | 9.29 | 12.66 | 17.90 | 27.53 |
| 769 | −0.07 | −0.04 | 0.00 | 0.06 | 0.14 | 0.26 | 0.40 | 0.59 | 0.85 | 1.18 | 1.62 | 2.19 | 2.93 | 3.90 | 5.18 | 6.90 | 9.25 | 12.62 | 17.83 | 27.41 |
| 770 | −0.08 | −0.05 | −0.01 | 0.05 | 0.13 | 0.24 | 0.38 | 0.57 | 0.83 | 1.16 | 1.60 | 2.17 | 2.91 | 3.88 | 5.16 | 6.87 | 9.21 | 12.57 | 17.77 | 27.29 |
| 771 | −0.10 | −0.07 | −0.03 | 0.03 | 0.11 | 0.22 | 0.36 | 0.56 | 0.81 | 1.14 | 1.58 | 2.14 | 2.88 | 3.85 | 5.13 | 6.84 | 9.18 | 12.53 | 17.70 | 27.17 |
| 772 | −0.11 | −0.08 | −0.04 | 0.01 | 0.09 | 0.20 | 0.34 | 0.54 | 0.79 | 1.12 | 1.56 | 2.12 | 2.86 | 3.83 | 5.10 | 6.80 | 9.14 | 12.48 | 17.64 | 27.06 |
| 773 | −0.13 | −0.10 | −0.06 | 0.00 | 0.08 | 0.19 | 0.32 | 0.52 | 0.77 | 1.10 | 1.54 | 2.10 | 2.84 | 3.80 | 5.07 | 6.77 | 9.10 | 12.44 | 17.58 | 26.95 |
| 774 | −0.14 | −0.11 | −0.08 | −0.02 | 0.06 | 0.17 | 0.31 | 0.50 | 0.75 | 1.09 | 1.52 | 2.08 | 2.81 | 3.78 | 5.04 | 6.74 | 9.07 | 12.39 | 17.51 | 26.83 |
| 775 | −0.16 | −0.13 | −0.09 | −0.04 | 0.04 | 0.15 | 0.29 | 0.48 | 0.74 | 1.07 | 1.50 | 2.06 | 2.78 | 3.75 | 5.02 | 6.71 | 9.03 | 12.35 | 17.45 | 26.72 |

*Source*: Reference 11.
[a]See equation (3.7).

## Silver–Silver Chloride Electrodes

This electrode is an example of an electrode of the second kind which is reversible to an anion. Experimental applications of this kind of electrode are restricted not only by the reactivity of the electrode material, as was discussed for the hydrogen electrode, but also by the solubility of the solid metal–anion salt in the solvent of interest. Thus, as was mentioned for the lead–lead sulfate electrode, the ultimate dilution at which an electrode of this type is effective depends on both the value of the solubility product for the metal salt and the precision with which it is intended to measure $E_{cell}$. In the absence of any interfering species that may react with the electrode material, the relative error attributable to the solubility effect is given approximately by the equation

$$\text{Relative error (mV)} \simeq \frac{1000k}{n} \log\left[\frac{(m + K_{sp}/m)}{m}\right] \quad (3.10)$$

where $k$ and $n$ are the familiar terms in the Nernst equation, $K_{sp}$ is the molal solubility product at a particular temperature and $m$ is the molal concentration of the soluble electroactive species.

The half-reaction for the silver–silver chloride electrode is

$$AgCl_{(s)} + e \rightleftharpoons Ag_{(s)} + Cl^-$$

from which it is deduced that the potential assumed by the electrode at a given temperature depends solely upon the activity of the chloride ion in the solution. The standard potential for the electrode is about 0.2225 V ± 0.2 mV (5) at 25° and the value of the solubility product is $1.5 \times 10^{-10}$ at this temperature. The appropriate form of equation (3.10) is, therefore, relative error (mV) $\simeq 59.2 \log[(m + 1.5 \times 10^{-10}/m)/m]$. It follows that if $E_{cell}$ is to be measured to $\pm 0.1$ mV, the minimum permitted chloride ion concentration is about $2 \times 10^{-4}$ mole kg$^{-1}$ with a minimum concentration of about $6 \times 10^{-4}$ mole kg$^{-1}$ if the precision of measurement for $E_{cell}$ is to be $\pm 0.01$ mV. In practice the concentrations of chloride solutions prepared for use in chemical cells without diffusion for which the silver–silver chloride electrode is used as a reference electrode normally exceed 0.001 mole kg$^{-1}$.

Although the lower limit of chloride ion concentration depends upon the solubility of silver chloride in water, the upper limit of concentration is determined by the dissolution of silver chloride as the anionic complex $AgCl_2^-$. For example, the solubility of silver chloride in a saturated solution of potassium chloride is about $6 \times 10^{-3} M$ at 25° as compared with its solubility in water of about $1.2 \times 10^{-5}$. If the electrode is to be immersed in a concentrated potassium chloride solution to form a reference half-cell suitable for use in a chemical cell with diffusion, the solution should be presaturated with solid silver chloride, otherwise dissolution of the silver chloride coating will occur. This effect is not observed when the silver–silver chloride electrode is used as a reference electrode in a cell without diffusion that contains synthetic seawater. It has been described as an ideal reference electrode for this application (14).

The precision of a silver–silver chloride electrode is severely impaired by the presence of bromide impurity in the chloride solutions and slightly impaired by the presence of dissolved oxygen when chloride is present in acid solutions. The bias potential of an electrode immersed in a solution containing 0.064 mole percent potassium bromide was found to be 1 mV relative to an electrode immersed in a bromide-free solution (15). By comparison, the oxygen effect is very small and can be neglected altogether if a precision of less than 0.3 mV is sufficient in acid solution, or if the solution is neutral. Dissolved oxygen can be removed from the solution by bubbling nitrogen saturated with water vapor through it. A method is also available for the preparation of chlorides free from bromide impurities (15) and this reference also includes a sensitive colorimetric test for the detection of traces of bromide.

The calibration of chemical cells without diffusion has been mentioned in Chapter 1 with respect to the cell

$$Zn(Hg)|ZnSO_4|PbSO_4; Pb(Hg)$$

and it was pointed out that $E^\circ_{cell}$ is the calibration factor rather than the difference between the published $E^\circ$ values of the constituent electrodes. An exactly analogous method of calibration is used for cells in which the silver–silver chloride electrode is the reference electrode. For example, calibration of the type of cell represented as

$$Pt; H_2 (1\ atm)|H^+(pH\ buffer\ solution); Cl^-(known\ molality)|AgCl; Ag$$

and used in the establishment of the NBS scale of $pH$ is performed by the determination of the quantity $E^\circ_{cell}$ for

$$Pt; H_2 (1\ atm)|HCl\ solutions(known\ molality)|AgCl; Ag$$

The Nernst equation for this cell is given by equation (3.6) which can be written as

$$E_{cell} = E^\circ_{cell} - k \log a_2 \qquad (3.11)$$

where $a_2$ is the activity of the electrolyte as a whole which, as is evident from equation (1.14; see also Table 1.1), allows equation (3.11) to be written in conformity with equation (1.20) as

$$E_{cell} + k \log m_{HCl}^2 = E_{cell}^\circ - k \log \gamma_\pm^2 \tag{3.12}$$

The solution of this equation by a least squares routine to yield the calibration factor $E_{cell}^\circ$ is described elsewhere. Although a recent study (5) has shown that there are variations of $\pm 0.2$ mV between seemingly identical determinations of $E_{cell}^\circ$, as performed by different workers, there is good agreement for the mean activity coefficients of hydrochloric acid solutions calculated from these data. The known activity coefficient of 0.01 $m$ hydrochloric acid at a given temperature is therefore used which allows rearrangement of equation (3.11) to

$$E_{cell}^\circ = E_{cell} + 2k \log 0.01 \gamma_\pm \tag{3.13}$$

in which all the terms on the right-hand side are known. This expediency allows a rapid and accurate method of cell calibration. Values of the mean activity coefficients* for 0.01 $m$ hydrochloric acid solutions at various temperatures are as follows:

0°, 0.9078; 10°, 0.9068; 20°, 0.9057; 25°, 0.9047; 30°, 0.9036;

40°, 0.9026; 50°, 0.9005; 60°, 0.8995

In addition to providing a reference electrode for the type of chemical cell without diffusion referred to previously, silver–silver chloride electrodes are used as reference half-cells in chemical cells with diffusion represented, for example, as

indicator electrode A | Species A ‖ 3.5 $M$ KCl | AgCl; Ag

and also as internal reference electrodes in ion selective electrodes of which the glass electrode is an example. The method of preparation and the design of the electrode is often dictated by the intended application of the electrode. Generally, the three types of applications can be associated with

---

*Calculated from $p(a_H \gamma_{Cl})$ values given by Bates (16).

the following commonly used methods of preparation:

1. Chemical cells without diffusion: reference electrode prepared by the *thermal-electrolytic method*.
2. Chemical cells with diffusion: reference half-cell prepared by the *bielectrolytic method*.
3. Ion selective electrode application: internal reference electrode prepared by the *electrolytic method*.

**The Thermal-Electrolytic Method.** Bates (17) gives the experimental details required to prepare this type of silver–silver chloride electrodes that are characterized by a very small bias potential ($\pm 20$ $\mu$V) between any two within a given batch. Silver oxide, prepared from the addition of sodium hydroxide to silver nitrate solution, is washed many times with water until the conductivity of the wash liquor is constant. A portion of the oxide is applied as a paste in water to a helix formed from platinum wire which has been sealed through a glass tube by a method similar to that described for the hydrogen electrode. This stage is usually repeated a number of times until spherical electrodes are obtained. The electrodes are suspended in a crucible furnace heated to about 500°C and allowed to remain there for 10 min or until they are completely white. The silver electrode so formed, containing about 150–200 mg silver, is made the anode in an H-shaped electrolysis cell that contains a platinum cathode and 1 $M$ hydrochloric acid (prepared from the doubly distilled acid) as the electrolyte. An electrolysis current of 10 mA passing for 45 min will convert about 15–20% of the silver to silver chloride. Each electrode is soaked overnight by immersion in a 0.05 $M$ solution of hydrochloric acid and then it is stored in water for a few days before use. This process, known as the *conditioning process*, plays an important part in the preparation of any newly made electrode, including the commercially available ones, for regular use. The soaking of the electroactive part of the electrode in a suitable solution promotes an ageing process which is necessary for the establishment of a reproducible potential. The type of conditioning varies with the type of electrode.

Silver–silver chloride electrodes made by the thermal-electrolytic process are light grey to white in color. After the initial overnight soaking in 0.05 $M$ hydrochloric acid solution a batch of new electrodes should be tested in a systematic way for the existence of a bias potential between any pair of them. If the bias potential for any individual electrode deviates by more than 0.1 mV from the values of the remainder it should be rejected. Electrodes prepared by the thermal-electrolytic method, by reason of the spongy and porous nature of the silver–silver chloride mass, are somewhat slower to establish an equilibrium potential than electrodes prepared by other methods. Nonetheless, in the author's experience the equilibration time for a conditioned electrode is usually less than 15 min at 25°.

**The Bielectrolytic Method.** The electrode is formed by an initial cathodic deposition of silver on to a piece of platinum (wire, foil, or gauze) followed by the anodic oxidation of portion of this deposit to form silver chloride by electrolysis in a hydrochloric acid solution.

A plating solution containing 10 g per liter potassium silver cyanide is prepared from the solid (C.P. grade) which has been doubly recrystallized from water and dried over calcium chloride. Immediately before the first stage of the electrolysis process, traces of excess cyanide should be removed from the plating solution by the careful addition of a dilute solution of silver nitrate until the solution is slightly turbid. The clear supernatant liquor which appears once the solid silver cyanide has settled is removed by decantation and a portion of this solution transferred to an H-type electrolytic cell that is fitted with a platinum anode. The carefully cleaned platinum base for the electrode is made the cathode of this cell and is then plated with silver from the well-stirred solution at a current density of 0.4 mA cm$^{-2}$ for about 6 h. The silver deposit should be white and velvetlike in appearance and should wet uniformly. Nonuniformity in the wettability of the surface is indicative that the behavior of the silver–silver chloride electrode produced subsequently from this surface will be erratic. Before chloridizing the silver base of the electrodes, they are soaked in concentrated ammonia solution for 1–6 h and then washed frequently with water over a period of 1–2 days. The conversion of a portion of the silver to form silver chloride is accomplished by anodic oxidation, carried out in the H-shaped cell containing 0.1 M hydrochloric acid and fitted with a platinum cathode, at a current density of 0.4 mA cm$^{-2}$ for about 30 min. The electrodes are conditioned in 0.05 M hydrochloric acid followed by immersion in water as described for the thermal-electrolytic type. The bias potentials between electrodes prepared by this method are less than 0.1 mV and variants of the method described are summarized in a table contained in Reference 17.

The response times of the bielectrolytic type of silver–silver chloride electrodes are shorter than those of the thermal-electrolytic type which makes them more suited to titrimetric applications.

**The Electrolytic Method.** The base for electrodes prepared by this method is a piece of silver wire which, after cleaning in 3 *M* nitric acid followed by copious washings with water, is anodically chloridized by the method described for the bielectrolytic type. Although the potential for a given electrode will remain stable with time, the bias potential between electrodes tends to be high ($\pm 5$ mV) which is attributable to stresses and strains introduced by the drawing process used in the production of the wire and also by impurities contained in the metal. The electrodes, however, are suitable as internal reference electrodes for ion selective electrodes and their ease of preparation is convenient whenever a silver–silver chloride electrode is required as an indicator electrode in the potentiotitrimetric determination of chloride.

Silver–silver chloride electrodes are available from all of the principal manufacturers and suppliers of ion selective electrodes. They can be supplied either in a form suitable for use as an indicator electrode or supplied as part of a complete reference half-cell designed for use in a cell with liquid junction. For some industrial applications these reference half-cells are preferred to the mercurous–mercurous chloride type of reference half-cell described in the next subsection. Although the latter type of reference half-cell is generally acceptable for laboratory service, the risk of using potentially dangerous mercury compounds (e.g., calomel) is inhibiting the use of calomel half-cells for process-control applications and has resulted in an industrial preference for the silver–silver chloride half-cell (18). The silver–silver chloride reference half-cell is also less subject to hysteresis as a function of temperature.

The laboratory and industrial versions of the silver–silver chloride reference half-cells (Figure 3.2) are used in cells of the type

indicator electrode A | species A ∥ KCl($3.5\ M$) saturated with AgCl | AgCl; Ag

in which they are required to provide a constant potential so that changes in $E_{\text{cell}}$ are related directly to changes in the concentration of species A. This constant reference potential is, of course, dependent upon the constancy of the chloride ion concentration in the reference half-cell and, at one time, a saturated solution of potassium chloride was considered the most satisfactory source of chloride ion. This solution is no longer considered suitable, however, because the effect of lowering the temperature can cause the liquid junction to become clogged with potassium chloride crystals which sometimes result in the high resistance of this junction. The use of 3.5 $M$ potassium chloride solution, presaturated with silver chloride for the desired temperature range, avoids this problem yet does not appear to appreciably affect the junction potential (19). The approximate values (in mV) of the half-cell potentials for the silver–silver chloride reference half-cell with respect to $2\text{H}^+ + 2e \rightleftharpoons \text{H}_2$ (1 atm), $E° = 0.0$ at the given temperatures are:

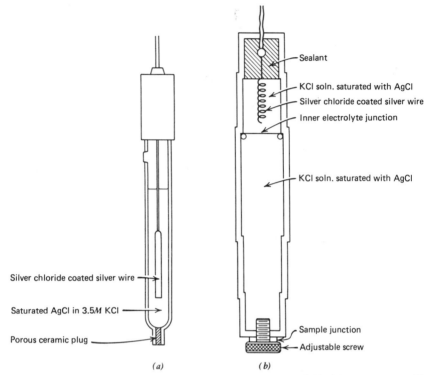

**Figure 3.2.** Silver–silver chloride reference half-cells: (*a*) refillable laboratory type; (*b*) industrial type (18).

10°, 215; 15°, 212; 20°, 208; 25°, 205; 30°, 201; 35°, 197; 40°, 193. These values include the contribution made by the liquid junction potential. The methods available for the establishment of a liquid junction for a reference half-cell are discussed elsewhere in this book.

Other types of silver–silver chloride electrodes include a flow-through porous electrode prepared by compressing a mixture 50% silver, 20% silver chloride, and 30% sodium bicarbonate into cylinders (1.7 × 0.9 cm) followed by sintering at 450° (20), a stable electrode for potential measurements in sea water (21), and an improved reference electrode (22).

## The Mercury–Mercurous Chloride Reference Half-Cell (Calomel Electrode)

The mercury–mercurous chloride electrode is another example of an electrode of the second kind that is reversible to chloride ion according to the electrode reaction

$$Hg_2Cl_{2(s)} + 2e \rightleftharpoons 2Hg_{(l)} + 2Cl^-  \qquad (3.14)$$

However, as a reference electrode in a cell without diffusion it is inferior to the silver–silver chloride electrode in its response to chloride ion. It is more susceptible to the effect of dissolved oxygen and can exhibit a marked temperature hysteresis effect. The hysteresis effect has been traced to the slow disproportionation reaction

$$Hg_2Cl_2 \rightarrow Hg + HgCl_2$$

with the resulting formation of mercuric complexes in solution which cause a slow drift in potential. When the temperature is lowered the solution will still contain the dissolved mercuric species formed at the higher temperature and thus the retardation in the establishment of equilibrium conditions is accentuated. The amplitude of the effect depends upon the chloride ion concentration, being very small in 0.1 $M$ potassium chloride solution but most pronounced in saturated potassium chloride.

The use of the calomel electrode is almost totally restricted to the provision of a constant potential established when it is incorporated into a reference half-cell that contains a solution of potassium chloride. The concentrations commonly used in the preparation of this solution are 3.8, 3.5, or 3.0 $M$ which seem to be replacing the saturated solution used in the hitherto ubiquitous saturated calomel electrode. The reasons for this change, given in the previous description of the silver–silver chloride reference half-cell, apply also to the calomel half-cell. In design, the calomel reference half-cell is similar to the silver–silver chloride reference half-cell shown in Figure 3.2 except that the silver–silver chloride element is replaced by a calomel element. Calomel paste is formed by grinding together a mixture of mercurous chloride and mercury moistened with a few drops of the appropriate potassium chloride solution (i.e., 3.8, 3.5, or 3.0 $M$). This paste is packed to a depth of about 1 cm in a glass tube of about 5–6 mm dia. and thus forms the calomel element. This element is positioned (e.g., by means of a rubber bung) in a wider diameter tube containing the potassium chloride solution which has been fitted with a means of establishing a liquid junction. Connection to the calomel element is made by placing a few drops of mercury above the calomel paste followed by insertion of the connecting wire into the mercury.

The calomel electrode has only a limited life at high temperatures and it is inadvisable to use a calomel reference half-cell at temperatures greater than 70° for long periods. The slow establishment of an equilibrium

# CLASSIFICATIONS AND DESCRIPTIONS OF ELECTRODES

**Table 3.2. Approximate Potentials of the Calomel Reference Half-Cell**

| Temperature (°C) | Half-Cell Potential (mV) at KCl Concentrations[a] | | | | |
|---|---|---|---|---|---|
| | 0.1 | 3.0 | 3.5 | 4.0 | Saturated |
| 10 | 336 | 260 | 256 | — | 254 |
| 15 | 336 | — | — | — | 251 |
| 20 | 336 | 257 | 252 | — | 248 |
| 25 | 336 | 255 | 250 | 246 | 244 |
| 30 | 335 | 253 | 248 | 244 | 241 |
| 35 | 334 | — | — | — | 238 |
| 40 | 334 | 249 | 244 | 239 | 234 |

[a] Concentrations expressed in mol $L^{-1}$ at 25°C.

potential after a temperature change due to the hysteresis effect is less marked the greater the dilution of the solution in the half-cell. It is for this reason that the approximate potentials for the reference half-cells given in Table 3.2 include values for 0.1 $M$ solution of potassium chloride obtained at various temperatures.

### The Thallium Amalgam–Thallous Chloride Reference Half-Cell ("Thalamid" Electrode)

The electrode is another example of a metal–metal chloride electrode but in this case the metal is present as a 40 weight percent amalgam. The potential of the electrode depends upon the activity of chloride in the solution as is evident from the half-reaction

$$TlCl_{(s)} + e \rightleftharpoons Tl(40\% \text{ in } Hg_{(l)}) + Cl^-$$

Thallium(I) chloride is too soluble, however, for the electrode to be used as the reference electrode reversible to chloride ion in a chemical cell without diffusion. It is, therefore, used either as the electrode in a reference half-cell containing a potassium chloride solution or as the internal reference electrode in some types of hydrogen ion selective glass electrodes. Although rarely used in comparison with the silver–silver chloride and the calomel types of reference half-cells, the electrode is remarkably free of thermal hysteresis and may be used over the temperature range 0–135°. The potentials of the reference half-cells represented as

$$KCl(\text{known concentration}), TlCl(\text{sat}) | TlCl; Tl(Hg)$$

are above 800 mV more negative than the corresponding silver–silver chloride or mercury–mercurous chloride reference half-cells that contain identical concentrations of potassium chloride. This may cause difficulties in obtaining on-scale readings with some types of ion-activity meters that are designed for use with the silver or mercury types of reference half-cells. This difficulty can sometimes be overcome by using indicating electrodes that have Thalamid® (23) internal reference electrodes. Such electrodes are available from Jenaer Glaswerk Schott und Gen. (Mainz) for both laboratory and industrial use. The high toxicity of the electroactive components, however, can present problems in the event of accidental breakage and also in the disposal of worn electrodes.

### The Mercury–Mercurous Sulfate Reference Half-Cell

This electrode, like the calomel electrode, is an electrode of the second kind based upon the mercury(I)–mercury oxidation-reduction system, but in this case the sparingly soluble salt is mercurous sulfate. Thus the reaction corresponding to (3.14) is

$$Hg_2SO_{4(s)} + 2e \rightleftharpoons 2Hg_{(l)} + SO_4^{2-}$$

for which the standard reduction potential (24) is 0.612 V at 25°. The comparatively high aqueous solubility of mercurous sulfate ($ca.$ 0.0012 mole $kg^{-1}$ at 25°) restricts the use of the electrode as a reference electrode in chemical cells without diffusion to sulfate ion concentrations greater than 0.01 molal. In spite of this solubility, and also a tendency of the salt to hydrolyze, the electrode has the desirable features of outstanding reproducibility and an insensitivity to polarization (high exchange current density), properties that have led to its inclusion as an electrode in the standard Weston cell.

The electrode is available commercially supplied as part of a reference half-cell that looks exactly like a calomel reference half-cell. The mercury–mercurous sulfate element in this type of half-cell is prepared by a method similar to the one used for the calomel element. These reference half-cells are of two types: one in which the mercury–mercurous sulfate element dips into a saturated solution of potassium sulfate, and the other in which this element dips into a 1 $M$ solution of sodium sulfate. The potentials of these two types are 0.658 V for the saturated potassium sulfate type (25), and 0.648 V for the 1 $M$ sodium sulfate type (26), both with respect to the standard $2H^+ + 2e \rightleftharpoons H_2$ at 22°. These values include the liquid junction potentials which are greater than those for the calomel types of reference half-cells. When calibrating chemical cells with diffusion into which mercury–mercurous sulfate electrodes are incorporated as reference half-cells, added emphasis must be placed, therefore, upon ensuring that the conditions in the calibrating solutions parallel very closely those likely to be encountered in the test solutions. Generally, these types of reference half-cells are used only for those determinations in which diffusion of chloride ion from either silver–silver chloride or calomel types of reference half-cells could prove troublesome. However, in such cases it is also possible to use *double-junction reference half-cells*. These are conventional reference half-cells, usually of the chloride type, fitted with an extra compartment filled with an electrolyte that is compatible not only with the electroactive species of the reference half-cell, but also with all the electroactive species (solution and solid) that are present in the indicator half-cell.

### Other Types of Reference Half-Cells

Reference half-cells containing *electrodes reversible to hydrogen ion or hydroxyl ion* do not appear to be widely used in practice. Mention has already been made of the quinhydrone reference electrode which, although susceptible to oxidation and acidic dissociation effects above $p$H 7–8 when used as an indicator electrode for $p$H measurements, provides a very easily made and reproducible reference half-cell. Quinhydrone is commercially available and should be recrystallized from water at 70° and dried at room temperature before use. A zeroth electrode such as platinum or gold (previously cleaned in hot chromic acid followed by washing with water and absolute alcohol) when dipped into a $p$H buffer solution saturated with respect to quinhydrone will assume a potential that depends upon the $p$H of the solution according to the approximate equation

$$E = 0.7 - kp\text{H}$$

in which $k$ is the Nernst factor. A reference half-cell can be formed by pouring the buffer solution into a small vessel fitted with a ceramic plug similar to those used in the construction of calomel or silver–silver chloride reference half-cells, adding a further few milligrams of solid quinhydrone to the solution, and positioning the zeroth electrode in the vessel. The use of reference half-cells prepared by this method is worth considering if the indicator electrode has also to be used in a solution that is buffered with respect to $p$H. In such cases it is possible that the same buffer solution could be used in both half-cells. A properly conditioned glass electrode will also maintain a constant potential for a number of months when immersed in a $p$H buffer solution and can be used in place of a quinhydrone electrode. For example, a hydrogen ion selective glass electrode has been used as a reference electrode in the measurement of sodium ion activities with a sodium ion-sensitive glass electrode as the indicator electrode (27). The liquid junction potential was eliminated in these measurements. The input impedance of most modern ion activity meters is sufficiently large to allow the connection of two electrodes of these types each having a high resistance (*ca.* $10^8$–$10^{10}$ Ω). However, should this be desired, it is better that the two electrodes are connected to the meter by means of a suitably designed screened terminal box rather than to extensively modify the terminals of the meter.

A number of metal–metal oxide electrodes have been found to be reversible to hydroxyl ion (28, 29). Of these, the *mercury–mercuric oxide electrode* has the advantage of a highly reproducible standard state for the metallic phase and complete freedom from any disturbing effect due to the variable valency of the oxide (30). The half-reaction is

$$HgO_{(s)} + H_2O + 2e \rightleftharpoons Hg_{(l)} + 2OH^-$$

and when the electrode is incorporated in the cell Pt; $H_2$ (1 atm)|NaOH (known molality)|HgO; Hg, the cell potential responds in a predictable manner for sodium hydroxide molalities of between 0.001 and 0.3 mole kg$^{-1}$. The standard potential for this cell is 0.926 V at 25°. The electrode is available commercially as part of the reference half-cells:

Hg; HgO|Ba(OH)$_2$ (saturated solution), and

Hg; HgO|Ca(OH)$_2$ (saturated solution),

both of which are well suited for measurements in alkaline solutions. The potentials (25°) of these half-cells with respect to the standard $2H^+ + 2e \rightleftharpoons H_2$ (1 atm) are about 0.146 V for the barium hydroxide type and about 0.192 V for the calcium hydroxide type (31). These are experimental values determined against a saturated calomel half-cell and thus they include a contribution from the liquid junction potentials.

The mercury–mercurous carboxylate types of electrodes appear to be feasible for use as reference half-cells using *p*H buffers prepared from the appropriate carboxylic acid–carboxylate solutions. Covington (32) lists the mercury–mercurous acetate and the corresponding formate, propanoate, oxalate, and benzoate electrodes which are prepared by a method analogous to that used for the calomel electrode except that the appropriate mercurous carboxylate solution is used. A mercury–mercurous picrate electrode has also been described (33).

## 3.3 THE DESIGN OF REFERENCE HALF-CELLS

A distinction has been made in the preceding section between the terms *reference electrode* and *reference half-cell*. The latter term has been applied specifically to chemical cells with liquid junction in which the role of the reference half-cell is to provide an invariant potential over the period required for the calibration and the subsequent activity determinations made with the cell. The frequency of the calibration, however, is usually determined by variations in the potential of the indicator electrode rather than by any instability in the potential of the commonly used reference half-cells. This is particularly true if, as is normal, the indicator electrode in the cell is of the membrane type. Despite the relatively stable potentials of reference half-cells, care must be exercised in choosing a type of half-cell that will be compatible with the indicator electrode. This compatibility depends upon the nature of the solution in the reference half-cell, for this solution leaks through the liquid junction into the test solution, which can cause the response of the indicator electrode to be adversely affected. In this discussion it is assumed, of course, that the range of possible $E_{cell}$ values are within the range of potentials for which the measuring device is calibrated.

The electrolyte solution in a reference half-cell usually has two functions. First, it provides a medium having a fixed activity of the ion to which the electrode is reversible thereby ensuring that the complete half-cell maintains a fixed potential. Second, the electrolyte is the ionic conductor or *bridging solution* through which a conducting pathway is established to the test solution and thus to the indicator electrode. In most applications the cell and the bridge solutions are the same, but in those cases where the cell solution is incompatible with either the test solution or the indicator electrode, a separate bridging solution is required. The design of the reference half-cell must be modified to accommodate this separate bridging solution and, of necessity, this design must introduce an extra liquid junction. Such half-cells are called *double-junction reference half-cells*, often loosely referred to as *double-junction reference electrodes*. Whenever possible, however, it is better to avoid the use of these modified half-cells by choosing a different type of reference half-cell so that the functions of the half-cell and bridge solutions are provided by the one solution. It is axiomatic that this solution should not contain the determinand species, neither should it react with any species in the determinand solution nor release into it a species that will react with or otherwise cause a modification to the response of the indicator electrode. Ideally, the reference half-cell solution should be of such a nature that its flow across the liquid junction between this half-cell and the test solution does not result in the establishment of a *liquid junction* potential. However, this ideality is not attainable in practice. Thus the aim is to reduce the magnitude of the liquid junction potential and to make its value reproducible from one measurement to the next. It is only when the latter condition is met that equation (3.2) is valid. This is because the value of the liquid junction potential has been assumed to make an invariant contribution to the value of *constant* in the equation.

## The Liquid Junction Potential

When two dissimilar electrolytes are separated by a permeable interface, a junction potential is set up as a result of the unequal rates of diffusion of cations and anions across the interface. The boundary on one side of the interface tends to assume the charge of the slower moving ion and thus an electric field gradient is established that counteracts the further diffusion of the faster moving ion from the other side of the interface. The size of the equilibrium junction potential depends upon the mobilities, concentrations, charges, and the individual ion activity coefficients of all the ions on either side of the interface and upon the physical form of the junction. It is not possible to determine the magnitude of the liquid junction potential established between two dissimilar electrolytes, neither is it possible to compute its unequivocal value. Reasonable extrathermodynamic estimates, however, can be made (34) provided that the overall compositions of the solutions are known, but these estimates cannot be applied directly to any routine analytical measurement made using a chemical cell with diffusion. The aim for such measurements, therefore, is to reduce the magnitude of the liquid junction potential to a value that will be reproducible from one measurement to the next. The design of the reference half-cell is usually such that the flow of the bridge solution is unidirectional provided it is positioned correctly so that the level of this solution in the half-cell is above the level of the test solution. A bridge solution that contains cations and anions of similar mobilities will give rise to a smaller liquid junction potential than one in which the difference in mobilities is greater. For example, a bridge solution formed from $H^+Cl^-$, in which the mobility of the hydrogen ion is about five times that of the chloride ion, will create a liquid junction potential in the decades of millivolts, whereas a solution of $K^+Cl^-$, for which the difference is only about 1.4%, will cause a liquid junction potential which is usually less than 5 mV. A change in the liquid junction potential when the electrodes are transferred from a calibrating solution to a test solution will cause an error in the analytical measurement. The magnitude of the relative error caused by a change in the liquid junction potential between a calibrating solution and a test solution for which identical values of $E_{cell}$ are obtained at constant ionic strength is given by differentiation of the equation (see equation 3.2)

$$-\log c_A = \frac{E_{cell} - constant}{slope}$$

At the same potential for the two cells the change in liquid junction potential will cause a change $\Delta E$ in the value of *constant*. Hence:

$$0.434 \frac{\Delta c_A}{c_A} = \frac{\Delta E}{slope}$$

For a slope of 59 mV decade$^{-1}$, the relative error ($\Delta c_A/c_A$) in the concentration of species A will be given by

$$\text{percent relative error} \sim 4\Delta E$$

Thus a difference of 1 mV in the liquid junction potential between the two solutions will cause a relative error of about 4% which will be inherent even if the $E_{cell}$ values are measured with a precision of $\pm 0.01$ mV.

Usually, the bridge solution is chosen so that the liquid junction potential will be minimized. To this end the nature of the cations and anions forming the bridge solution should be such that the summation of the products of the charge $z$ (positive for cations, negative for anions) activity $a$, and mobility $u$ for each type of ion will yield a value close to zero. Thus ideally

$$\Sigma_i z_i a_i u_i = 0 \tag{3.15}$$

and when this condition is met the bridge solution is said to be *equitransferent*. Potassium chloride solutions are almost equitransferent and it is, of course, for this reason that the salt is used, not only to provide a solution of known concentration of chloride ion (3.0 $M$, 3.5 $M$, 3.8 $M$, or saturated KCl) for the calomel, silver–silver chloride, and Thalamid reference half-cells, but also to provide a bridge solution with an inherently low junction potential. It is possible to formulate bridge solutions that are even more closely equitransferent than potassium chloride (35), but under operational conditions these solutions do not show any real improvement over the ubiquitous potassium chloride.

The size of the liquid junction formed at the interface between a concentrated solution of potassium chloride and an aqueous test solution is also affected by the nature of the test solution. In general it appears, from the cases studied using saturated potassium chloride solutions (36), that the liquid junction potential increases as the test solution is made more dilute and its magnitude depends also upon the relative mobilities of cations and anions in the test solution. Thus for an identical condition of constant ionic strength, the sodium chloride solution yields about twice the junction potential that is obtained for the potassium chloride solution (37). Nonetheless, the presence of a relatively high concentration of an electrolyte, as is usual in the provision of a condition of constant ionic strength, will tend to improve the constancy of the liquid junction potential. However, in these instances it is advisable that the ionic strength of the bridge solution is at least 10 times that of the test solution.

Sample solutions containing large concentrations of the very mobile hydrogen ion or the slightly less mobile hydroxyl ion give rise to large values of the liquid junction potential at the interface with a concentrated solution of potassium chloride. Outside the range $p$H 2–12 the value of the junction potential will vary markedly as the concentrations of these ions are changed. Bates (38) discusses the effects in relation to $p$H measurements. Errors due to this cause will not be confined to $p$H measurements, however, but will be included in all measurements made using chemical cells with diffusion. If conditions of comparatively high concentrations of acid or alkali are an unavoidable prerequisite for the application of a method, consideration should be given to changing the type of reference half-cell. Presumably, at such extremes of $p$H the test solution would need to be buffered, and thus a reference half-cell formed from an identical buffer solution with a $p$H responsive electrode might ensure a constant and low value of the liquid junction potential. At values of $p$H $> 12$ the mercury–mercuric oxide electrode would be satisfactory as the "$p$H responsive electrode," and at values of $p$H $< 2$ consideration could be given to either a quinhydrone or a glass electrode. A completely different alternative is to use double junction reference half-cell of the chloride type (e.g., the silver–silver chloride or mercury–mercurous chloride types). At the extremes of $p$H the extra bridge solution would need to closely resemble the sample solution and the whole reference half-cell assembly would need to be maintained at a constant temperature. The latter condition is sometimes difficult to achieve, and, as a general rule, it is better to avoid the use of double junctions whenever possible.

Sample solutions containing colloidal and particulate matter can cause a destablization of the liquid junction potential which, for $p$H measurements, gives rise to a suspension or Pallman effect (39). Bates (40) discusses this effect in relation to the measurement of $p$H in suspensions of soils, and a similar effect causing wide variations in the liquid junction potential in blood samples has also been observed (41). Suspension effects have also been reported in the potentiometric determination of chloride ion in soil

(42) causing large liquid junction potentials which were attributed to the use of the reference half-cell $K_2SO_4|Hg_2SO_4$; Hg. A calomel reference half-cell, in which the diffusion of potassium chloride into the soil sample was minimized by the use of a junction of low flow-rate, was found to be satisfactory for this determination.

The major cause of a liquid junction potential in the types of chemical cells with diffusion used in routine analysis is associated, therefore, with variability in type, quantity, and characteristics of each individual species present in both the bridge and the test solutions. A further, but comparatively minor, contributing cause is the method by which the junction is established and this aspect is discussed in the next subsection. Here, however, it is advisable to outline the conditions whereby the liquid junction potential formed by a given reference half-cell, selected for a particular analytical determination, can be assumed to maintain a constant value for both the calibrating and the analyte solutions. Clearly, the components of the calibrating solutions need to closely resemble those of the analyte solutions both in respect to quantity and type. One approach is to ensure a condition of identical ionic strength and often $p$H between the two series of solutions. This is achieved by the addition of swamping concentrations of electrolyte which may also contain $p$H buffer components. In addition to establishing conditions by which the liquid junction potential is likely to remain constant, this expediency also creates conditions whereby it is often assumed that the activity coefficient of the determinand remains constant for both series of calibration and test solutions (see equation 3.2). This method, however, is impracticable for samples that contain high but variable concentrations of electrolytes, and in these instances it is advisable to employ methods based upon standard addition or subtraction techniques. These techniques, discussed in Chapter 4, are useful in creating conditions under which it is reasonable to assume that both the activity coefficient of the determinand and the liquid junction potential will remain constant throughout the determination. Either direct potentiometry using chemical cells with diffusion, or differential potentiometry using concentration cells with diffusion can be used with these techniques.

### Common Methods of Forming the Liquid Junction

The geometry of the interface between two electrolyte solutions affects the stability of the junction potential that is established across this interface. Ideally, the shape of the junction should be such that ions from the denser solution diffuse upwards so that mixing by gravitational flow is prevented. Invariably, this condition is not met, but methods (43) are available whereby liquid junction potentials of high stability and reproducibility can be obtained if the interface is formed within a cylindrical tube. Such junctions are categorized as the constrained diffusion type with cylindrical symmetry. For example, junctions formed within a vertical length of 1 mm precision bore capillary tubing seem to be particularly stable (44, 45). However, these junctions are somewhat inconvenient to form and thus the more convenient unidirectional flow types of junctions, the designs of which contravene the basic principles for reproducibility and stability of the junction potential, are commonly used in the commercially available reference half-cells. Despite these deficiencies, adequate stability and reproducibility can be obtained with such half-cells provided that the type of junction is compatible with the type of application. Effectively, the junction is an impeding device that allows a slow but steady flow of the bridge solution into the sample solution. Choice of a suitable junction device allows flow-rates ranging from approximately 0.01 to 1–2 ml per 5 cm head of bridge solution per 24 h period to be selected. Basically, there are four types of junctions used:

1. *The ceramic plug type.* The use of this type of porous plug allows a very slow leakage of the bridge solution into the sample solution; typically about 0.01–0.1 mL per 5 cm head per day. For reference half-cells fitted with a single junction, the ceramic plug is sealed into the lower end of the tube containing the reference solution. For "double junction electrodes," however, the ceramic plug connecting the reference solution to the bridge solution can be set into the wall of the reference half-cell. Connection of the bridge solution to the sample solution need not necessarily be made through a ceramic plug sealed to the bottom of the overall housing (see Figure 3.2). In general, this type of junction is robust and trouble-free when used in clear solutions and has a reproducibility of $\pm 0.1$ mV in aqueous $p$H buffer solutions. However, it should not be used in solutions that contain particulate material (including precipitates) or colloids. Once the junction becomes fouled it is very difficult to clean.

2. *The ground-sleeve type.* Seepage of the bridge solution between the two surfaces forming a ground-glass joint is the general concept upon which the design of this type of junction is based. An important requirement is that the fit between the ground-glass portions of the joint should be good otherwise the flow of electrolyte will be irregular and the potential will be unstable. When fitted with this type of junction, it is the quality of the ground-glass joint that determines the quality of the whole reference half-cell. The best quality ones are very satisfactory but there is a wide variation in the quality of the half-cells that are supplied commercially. It has been found (46) that these types of reference half-cells can give signals with four to five times the fluctuations given by the ceramic plug type. Two forms of the ground-sleeve type of junction are given in Figure 3.3. In one, the

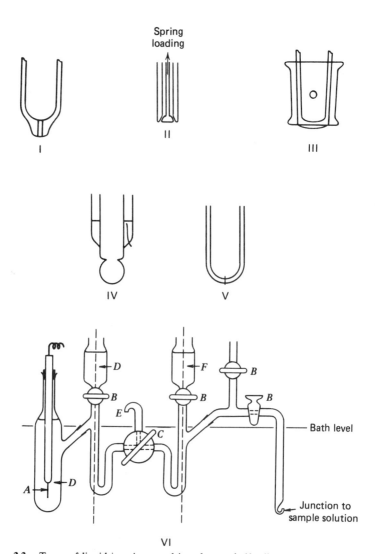

**Figure 3.3.** Types of liquid junctions used in reference half-cells: I, ceramic frit; II and III, ground sleeve types; IV, wick or fibre type; V, imperfect glass to platinum or palladium seal; VI, "Wilhelm" type reference half-cell fitted with J-shaped liquid junction to sample solution (not shown). The parts outside the dashed lines may be bent perpendicular to the plane of the paper. $A$, silver electrode; $B$, two-way stopcocks; $C$, three-way stopcocks; $D$, 0.01 M $AgClO_4$ + 2.99 M $NaClO_4$; $E$, to waste; $F$, 3 M $NaClO_4$ (51).

ground-glass joint is modified by drilling a hole through the inner or cone portion of the joint. In the other, a comparatively small contact area of joint is used, the cone portion being secured in the socket by means of a spring which is attached to the top of a piece of glass rod sealed to the cone. The flow-rate of the bridge solution through this type of junction is often about a hundred times greater than that through the ceramic plug type. Typically, flow-rates of 1–2 ml per 5 cm head per day are recorded (47) and this necessitates frequent replenishment of the bridge solution. One great advantage of the general form of the junction is the ease with which it can be dismantled for cleaning. A concomitant advantage is that a reference half-cell fitted with this type of junction can be used in soap and other surfactant solutions, emulsions, and also in solutions containing particulate material.

3. *The wick or fiber type.* This junction is formed through an asbestos or quartz fiber sealed into the bottom of the tube that contains the reference solution for the half-cell. It has also been used to provide the junction for combination electrodes. A typical configuration for one of these is shown in Figure 3.3. A combination electrode is one that contains the indicator electrode and the reference half-cell manufactured as a single assembly. In the diagram, connection of the reference half-cell to the test solution is made by means of a fiber sealed through a silicone grommet. In general use the reproducibility of this type of junction is inferior to the ceramic type having a day to day stability of about $\pm 2$ mV under favorable conditions. The flow-rates through different asbestos wick junctions may vary by as much as an hundredfold (46) and they are particularly prone to clogging when used in samples with a high suspended solids content or in solutions containing colloidal material. Wicks formed of cellulose or of soft woods such as bamboo have also been used.

4. *The imperfect seal type.* A crack deliberately made in the bottom of a cylindrical vessel that will hold the bridge solution provides a satisfactory means of forming a junction. Another pattern is an imperfect glass to metal seal that can be made when palladium or platinum is sealed through glass. In the latter type, the annulus formed around the metal is sufficient to allow an adequate flow of bridge solution through the seal. The metal, however, will respond as a zeroth electrode (48) if the reference half-cell is used in solutions containing strong oxidants (0.2 $M$ $KMnO_4$ in 0.05 $M$ $H_2SO_4$) or reductants (0.5 $M$ $SnCl_2$ in 1 $M$ HCl). Reference half-cells are available commercially in which both these patterns of imperfect seals are used as a means of forming the junction.

There are, of course, other methods of establishing conducting pathways between the reference and sample solutions (49). Noteworthy among these is the inverted U-tube of small diameter that can be used to connect the reference half-cell to the sample solution through an intermediate vessel containing the bridge solution. When filled with the bridge solution set in agar, capillary U-tubes are satisfactory for some types of semimicro measurements (50). If the intermediate bridge solution differs from the reference solution, a liquid junction between the vessels containing these solutions is sometimes made through a closed stopcock. The "Wilhelm" type of reference half-cell (51) shown in Figure 3.3 is an example in which this type of connection is made. In essence the use of closed stopcocks as junctions is analogous to the ground-sleeve type mentioned in paragraph 2. The J-shaped junction fitted to the "Wilhelm" assembly provides the interface between the bridge solution and the sample solution (not shown). The design of this interface is dictated by the requirement that the shape of the junction should be such as to prevent mixing by gravitational flow and to allow transfer of ions by diffusion only. As shown, the whole assembly would be used as the reference half-cell and salt bridge for the potentiometric determination of stability constants using 3 $M$ sodium perchlorate as the medium of constant ionic strength (see Chapter 8). It should be noted that the reference electrode in this assembly is silver and the reference half-cell is formed with this electrode immersed in a dilute (0.01 $M$) solution of silver perchlorate prepared in a concentrated (2.99 $M$) solution of sodium perchlorate which yields an overall ionic strength of 3.00 mol $L^{-1}$. The bridge solution is prepared from 3.00 $M$ sodium perchlorate and the J-shaped junction dips into a dilute solution (perhaps 0.01 $M$) of metal complex also prepared in 3.00 $M$ sodium perchlorate. Under these circumstances the liquid junction potential can be expected to be negligible above $p$H 3. There is no reason why this type of enterprising approach cannot be applied to modify commercially available "single vessel" reference half-cells so that they become more specific to a particular analytical determination.

The J-shaped junction is rarely used in routine analysis principally because the design of the reference half-cell that employs this type of junction would be cumbersome in comparison with the compact designs that become possible when the porous ceramic plug or ground-sleeve type of junctions are incorporated. An hydrostatic head of bridge solution must be maintained when using these popular types of restricted flow junctions or there is a possibility of back-diffusion of a sample solution into the reference half-cell solution. In normal laboratory use an hydrostatic head of about 1 cm above the level of the sample solution is sufficient to ensure the required unidirectional flow of the half-cell solution. When used in-line, however, as in some types of industrial process control applications, an hydrostatic head of 10–50 cm of bridge solution is common, whereas even more elaborate mechanical pressurization may sometimes be needed. The advantages and disadvantages accompanying the use of pressurized reference half-cells in process $p$H measurements have been summarized (18) and an ingenious screw-type of liquid junction is described which provides an alternative to mechanical pressurization. The industrial version of the silver–silver chloride reference half-cell shown in Figure 3.2 is fitted with this kind of screw junction. The design of the half-cell shows an upper and detachable compartment separated by a porous membrane from a lower compartment that terminates in the screw junction at the bottom. The upper compartment contains the silver–silver chloride electrode immersed in a solution of potassium chloride (1.0, 3.0, 3.5, or 4.0 $M$) saturated with respect to silver chloride. The purpose of the identical solution in the lower compartment is to minimize the possibility of process contaminants fouling the reference half-cell compartment. The screw junction is formed from a porous fluorocarbon washer that allows a flow of the bridge solution from a center hole outward through a web toward its circumference. The pressure exerted by the screw on this washer determines not only the rate of flow of solution through the junction, but also the electrical resistance across the whole assembly. The resistance when the screw is only finger-tight is about 1000 $\Omega$ and the device will operate satisfactorily under this condition for long periods at low and relatively constant temperatures. However, by tightening the screw so that a resistance of about 50,000 $\Omega$ is obtained, the assembly can be used at high temperatures such as those which are produced by boiling solutions or by steam sterilization. The whole device is conveniently designed to ensure ease of servicing.

A resistance of 50,000 $\Omega$ across a reference half-cell is far in excess of the resistance across the usual type of junctions for these are typically less than 5,000 $\Omega$ even when the filling solution is 1 $M$ potassium chloride (52). With modern instrumentation, however, such high resistances are not the encumbrance they once were.

### 3.4 REFERENCE HALF-CELLS IN NONAQUEOUS SOLVENTS

Analysis by direct potentiometry is confined almost entirely to the use of aqueous or predominantly aqueous solutions. Although the useful ranges of the sodium ion and the iodide ion membrane electrodes in the lower straight-chain alcohols have been investigated (53), the general use of membrane indicator electrodes is inhibited by the poor performance of most of these types of electrodes in nonaqueous media (54). This restriction is imposed by such effects as dissolution of the membrane material, or attack by the solvent upon the membrane matrix or on the seal between the membrane and the electrode body. Notably exempt from these restrictions, however, is the glass electrode which retains a reproducible and rapid response in a wide variety of solvents. The topic of relevance to analytical chemistry is primarily centered, therefore, upon the potentiometric investigation of acid-base equilibria in nonaqueous solvents (see Chapter 7), and upon the important associated technique of acid-base potentiotitrimetry in pure and mixed organic solvents (see Chapter 10). The same type of reference half-cell can be used with either of these applications.

The most widely used reference half-cells are those, such as the calomel and silver–silver chloride types, that are supplied commercially for use in aqueous solution. The type of liquid junction preferred is the ceramic plug type (55). These half-cells can be modified for use in nonaqueous solvents either (1) by retaining the aqueous phase in the half-cell and fitting an extra salt bridge, or (2) by replacing the aqueous half-cell solution with one prepared in an appropriate solvent.

1. Modification of an aqueous type of reference half-cell by fitting a salt bridge that contains an electrolyte prepared in the solvent of interest can cause large junction potentials that vary markedly with the nature of the bridge solution. For example, the potentials between a saturated aqueous solution of potassium chloride and ethanol–water mixtures show the following large variations (56): 50% ethanol, 25 mV; 80% ethanol, 75 mV; 100% ethanol, 140 mV. Large variations in the cell potentials between a saturated calomel reference half-cell in water and a $Ag|Ag^+$ couple in acetonitrile can also be induced merely by changing the composition of the bridge solution between the two solvent systems (57). Precipitation of potassium chloride at the aqueous–acetonitrile interface can also contribute to a fairly rapid drift in potential that has been observed when using an aqueous calomel reference half-cell in this solvent (58).

The solution used for the salt bridge should, for preference, be prepared in a nonvolatile highly polar solvent that will not cause precipitation either at the aqueous reference half-cell solution interface or at the interface with the titrand solution. Similarly, the bridge electrolyte itself should not precipitate at either of these interfaces. The nature of the bridge solution should be such that neither the solvent nor the solute will cause a modification to the general properties of the titrand solvent system. For example, diffusion of traces of water can sometimes alter the acid-base characteristics of the solvent used in the preparation of the titrand solution and so can diffusion of potassium or sodium ion into this solution. Absence of chloride ion in the bridge solution is also advantageous in some circumstances. Several types of conducting pathways have been used to connect an aqueous calomel reference half-cell to titrand solutions prepared in acetone, 2-methyl-2-propanol, 2-propanol, or pyridine (59). Sometimes these need to be of the triple junction type shown in Figure 3.4 which is suitable for use in 2-methyl-2-propanol. With modified saturated calomel half-cells of the dual

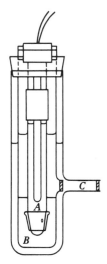

Figure 3.4. Triple junction calomel reference half-cell suitable for use in solutions prepared in 2-methyl-2-propanol. The junctions are made through: $A$, saturated KCl solution in water; $B$, a two-phase system of 2-methyl-2-propanol/water saturated with KCl; $C$, 2-methyl-2-propanol saturated with tetrabutylammonium chloride. The first junction is of the sleeve type, but the remaining two are through sintered glass (shaded areas). Reprinted by permission from Reference (59).

or triple junction types it is possible to maintain reproducibility of the half-cell potential to within 2 mV for long periods. However, if it is intended to report values of the potentials of cells that use these modified reference half-cells, it is important that the exact composition of the bridge solution(s) should be specified because the reproducibility of the measurement will be dependent upon the nature of these solutions.

2. A satisfactory modification to a commercially supplied calomel or silver–silver chloride reference half-cell can sometimes be made by immersing the reference element (Hg; $Hg_2Cl_2$ or Ag; AgCl) into a solution of chloride ion prepared in a suitable solvent. This solvent need not necessarily be identical to that of the titrand medium, although it is, of course, preferable that the same medium be used in the preparation of the reference electrolyte and the titrand solution. The solid phase of the reference element must not be appreciably soluble in or react with the reference solution: reasons that exclude both calomel and silver–silver chloride elements from being suitable for direct use in aprotic solvents. For example, calomel disproportionates in a number of these solvents such as acetonitrile, 4-methyl-1,3-dioxolan-2-one (propylene carbonate), dimethylformamide, and dimethylsulfoxide. Silver chloride, on the other hand, becomes appreciably soluble in these solvents as a result of the formation of anionic complexes with chloride ion. Nonetheless, both types of reference half-cells have been used in a wide variety of other solvents. For example, a form of the calomel reference half-cell commonly used in nonaqueous titrations contains a reference solution of potassium chloride in methanol, ethanol, or

Table 3.3. Reference Half-Cells for Nonaqueous Solvents

| Solvent | Reference Half-Cell | Ref. | Some Soluble Electrolytes[a] |
|---|---|---|---|
| Acetic acid, HOAc | Hg; Hg$_2$Cl$_2$∣NaCl + NaClO$_4$ satd in HOAc | (61) | LiCl, KCl, ZnCl$_2$, NaClO$_4$, Mg(ClO$_4$)$_2$, NH$_4$OAc, NaOAc |
| | Hg; Hg$_2$Cl$_2$∣satd LiCl in HOAc | (64) | |
| | Hg∣satd Hg$_2$(OAc)$_2$, satd NaClO$_4$ in HOAc | (65) | |
| | Ag; AgCl∣satd KCl in HOAc | (66, 67) | |
| Acetic anhydride, Ac$_2$O | Ag; AgCl∣LiCl in Ac$_2$O | (68) | NaClO$_4$, LiClO$_4$ |
| | Ag∣0.01 M AgClO$_4$, 0.1 M LiClO$_4$ in Ac$_2$O | (69) | |
| | Hg∣satd Hg$_2$(OAc)$_2$, satd NaClO$_4$ in HOAc | (65) | |
| Acetone, Me$_2$CO | Hg; Hg$_2$Cl$_2$∣satd LiCl in Me$_2$CO | (70) | LiNO$_3$, LiClO$_4$, NaClO$_4$, TEAP[b] |
| 1:1 Me$_2$CO + 1,2-ethanediol | Hg; Hg$_2$Cl$_2$∣satd KCl in the mixed solvent | (71) | |
| Acetonitrile, CH$_3$CN | Ag∣0.01 M AgNO$_3$ in CH$_3$CN | (72) | LiClO$_4$, NaClO$_4$, tetraalkyammonium salts up to 0.1 M e.g., TMAC, TEAB, TBAI[b] |
| | Hg; Hg$_2$Cl$_2$ ground∣0.1 M TEAP[b] in CH$_3$CN with KCl and KClO$_4$ | (73) | |
| Dimethylformamide, DMF | Tl(Hg); TlCl∣0.1 m LiCl in DMF | (74) | LiCl, NaClO$_4$, tetraalkyammonium halides, perchlorates and fluoroborates; TBAI, TEAP[b] up to 0.1 M are convenient |
| | Pb(Hg); PbCl$_2$∣0.1 m LiCl in DMF | (74) | |
| | Cd(Hg); CdCl$_2$∣0.1 m LiCl in DMF | (74, 75) | |
| Dimethylsulfoxide, DMSO | Ag; AgCl∣satd KCl in DMSO | (76) | LiCl, NaNO$_3$, NaClO$_4$, NaOAc, TEAB, TBAP, TEAN[b] |
| | Ag∣AgCl$_2^-$ in DMSO | (77) | |
| | Tl(Hg); TlCl∣0.1 m LiCl in DMSO | (74) | |
| Ethanol, EtOH (also 2-propanol) | Ag; AgCl∣in cells without diffusion | (78) | LiCl, LiClO$_4$, LiBr, KOEt |
| | Hg; Hg$_2$Cl$_2$∣satd KCl in EtOH | (60) | |
| 1,2-Ethane diamine, C$_2$H$_8$N$_2$ | Hg; Hg$_2$Cl$_2$∣satd LiCl in C$_2$H$_8$N$_2$ | (79) | LiCl, NaNO$_3$, TMAB, TEAN, TEAC, TEAP, TPAI[b] and NNN-triethyl 3-methylbutylammonium iodide |
| | Zn(Hg); ZnCl$_2$∣0.25 M LiCl in C$_2$H$_8$N$_2$ | (80) | |

| Solvent | Electrode | Ref. | Electrolyte |
|---|---|---|---|
| Methanol, MeOH | Ag; AgCl\| in cells without diffusion | (78) | LiCl; $NH_4Cl$, KOMe, NaOMe, TEAB, TEAI[b] |
| | Hg; $Hg_2Cl_2$\| satd KCl in MeOH | (60) | |
| N-Methylacetamide, MeCONHMe | Ag; AgCl\| satd NaCl in MeCONHMe | (81) | KCl, KI, $NH_4Cl$, TEAB, TEAP[b] |
| Nitromethane, $MeNO_2$ | Ag; AgCl\| satd TMAC,[b] 0.09 M TMAP[b] in $MeNO_2$ | (82) | $LiClO_4$, $Mg(ClO_4)_2$, TMAC, TEAP,[b] tributylethylammonium nitrate and iodide |
| Propylene carbonate, $C_4H_6O_3$ (4-methyl-1,3-dioxolan-2-one) | Pt black\|$I_3^-$, $I^-$ in $C_4H_6O_3$ | (83) | $LiClO_4$ (0.83 M), TEAP[b] (0.25 M) |
| | Li ribbon\|1 M $LiClO_4$ in $C_4H_6O_3$ | (84) | |
| | Ag\|0.02 M $AgClO_4$, 0.2–1.0 M $LiClO_4$ in $C_4H_6O_3$ | (85) | |
| | Hg; $Hg_2Cl_2$\| satd TEAC[b] in $C_4H_6O_3$ | (86) | |
| | Hg; $Hg_2Cl_2$\| satd KCl, 0.1 M TEAP[b] in $C_4H_6O_3$ | (87) | |
| | Tl(Hg); TlCl\| alkali metal halide in $C_4H_6O_3$ | (88) | |
| Pyridine | Ag\|1 M $AgNO_3$ in pyridine | (89) | LiCl, $LiClO_4$, $LiNO_3$, TBAI, TEAC, TEAB, TPAB[b] |
| | Pt\|0.0025 M CuCl, 0.025 M $CuCl_2$, 0.1 M TEAC[b] in pyridine | (90) | |
| Sulfolane, $C_4H_8SO_2$ (tetrahydrothiophene-S, S-dioxide) | Ag; AgCl\| satd TEAC, 0.1 M TEAP[b] in $C_4H_8SO_2$ | (63) | $LiClO_4$, $NaClO_4$, TEAP,[b] all at 0.1 M |
| Tetrahydrofuran, THF | Ag\| satd $AgClO_4$, 0.3 M $LiClO_4$ in THF | (91) | $LiClO_4$, $NaClO_4$, TBAI[b] |

[a] From Reference 63.
[b] Abbreviations: TBAI, tetrabutylammonium iodide; TBAP, tetrabutylammonium perchlorate; TEAB, tetraethylammonium bromide; TEAC, tetraethylammonium chloride; TEAI tetraethylammonium iodide; TEAN, tetraethylammonium nitrate; TEAP, tetraethylammonium perchlorate; TMAB tetramethylammonium bromide; TMAC, tetramethylammonium chloride; TMAP, tetramethylammonium perchlorate; TPAB tetrapropylammonium bromide; TPAI, tetrapropylammonium iodide.

2-propanol (60). It should be noted, however, that the magnitude of the liquid junction potential has been found to vary by as much as 50 mV when using a reference half-cell of this type (59). Another form of the calomel half-cell uses a saturated solution of sodium chloride and sodium perchlorate in glacial acetic acid (61). A saturated solution of sodium chloride, this time prepared in dimethyl sulfoxide, can be used also as the reference solution in a silver–silver chloride half-cell (62).

The use of the calomel and silver–silver chloride reference half-cells together with the applications of other types of reference half-cells to measurements in nonaqueous solvents are summarized in Table 3.3. The remarks concerning the liquid junction potentials and precipitation effects made in the foregoing descriptions are also pertinent to all the half-cells listed.

## 3.5 CATEGORIES AND DESCRIPTIONS OF MEMBRANE ELECTRODES

### Membranes and Their Incorporation into Electrodes

The first membrane electrode to be developed and to be used extensively was the hydrogen ion selective glass membrane electrode discovered by Cremer in 1906 (92). For some time it was commonly believed that the response of the glass electrode to changes in $p$H resulted from the selective permeation of a thin glass membrane by the hydrogen ion. However, the insight provided comparatively recently by the work of Karreman and Eisenman (93) and Nicolsky et al. (94) by which it was possible to attribute this $p$H response to an ion-exchange process across a hydrated phase boundary helped in understanding the principles upon which the development of other types of membrane electrodes could be based. Although some success in the technology of membrane electrodes had been reported in the three decades subsequent to the development of the compressed silver chloride, silver bromide, and silver iodide disc electrodes by Kolthoff and Sanders in 1937 (95), the current resurgence of interest in the techniques of direct and differential potentiometry as feasible routine analytical methods really dates from 1966 when Frant and Ross invented the lanthanum fluoride membrane electrode which is reversible to fluoride ion (96). Within one decade membranes had been developed from which it was possible to construct and market electrodes that would respond in a reproducible manner to the ions $Na^+$, $K^+$, $Mg^{2+}$, $Ca^{2+}$, $Cd^{2+}$, $Cu^{2+}$, $Ag^+$, $NH_4^+$, $S^{2-}$, $I^-$, $Br^-$, $Cl^-$, $CN^-$, $SCN^-$, $NO_3^-$, $ClO_4^-$, $BF_4^-$. The electroactive materials used in these membranes can be broadly classified (97) into glass, insoluble inorganic salts such as lanthanum fluoride, and large water-insoluble organic molecules some of which are long-chain ion-exchange materials, whereas others are charged or uncharged complexing agents. Although there is no common mechanism available to describe the way this variety of materials will respond to a pertinent species in an electrolyte solution, they all set up an exchange equilibrium or an ion-exchange process across a phase boundary, and they all can conduct electricity by either an ionic, electronic, or defect mechanism or by some combination of these. The membrane contains a constant activity of immobilized species held in a phase ideally immiscible with the electrolyte solution so that this species may enter into rapid equilibrium with the same species contained in the solution. The potential that develops depends principally upon the free energy change associated with mass transfer across the phase boundary. When the membrane is used as the sensing component for an indicator electrode which is then incorporated into a cell represented, for example, as

$$\text{Hg; Hg}_2\text{Cl}_2 | 3.5 \ M \ \text{KCl} \| \text{species A } (a_A = a_1) | \text{membrane } (a_A = \text{constant}) \text{electrode} \quad (3.16)$$

$E_{\text{cell}}$ is related to the activity of the species in the indicator half-cell solution $a_1$ by an equation similar to (3.2); that is,

$$E_{\text{cell}} = constant \pm slope \log a_1 \quad (3.17)$$

The sign in this equation depends upon the relative polarities of the two half-cells, and it is assumed that the activity of species A($a_A$) in the membrane remains constant.

The cell notation (3.16) does not specify how a conducting pathway is made between the sensor membrane and the instrument used to measure the cell potential. In many commercially available membrane electrodes the necessary conducting medium is provided by an electrolyte solution containing a fixed concentration of a species (e.g., Cl$^-$) in which an appropriate reference electrode (e.g., Ag-AgCl) is immersed. If this method of connection is used, then cell (3.16) could be more fully represented by the notation

$$\text{Hg; Hg}_2\text{Cl}_2 | 3.5 \ M \ \text{KCl} \| \begin{array}{|c|} \text{analyte soln} \\ \text{of A } (a_A = a_1) \end{array} \begin{array}{|c|} \text{Sensor membrane} \\ \text{responding to A} \end{array} \begin{array}{|c|} \text{fixed conc A,} \\ \text{fixed conc Cl}^- \end{array} | \text{AgCl; Ag}$$

← external reference →          ← internal reference →
half cell                                             half cell

(3.18)

The role of the internal reference half-cell is, therefore, to provide a connecting medium of invariant potential between the sensor membrane and the terminal of the instrument (Figure 3.5a). An alternative to this indirect method, however, is to attach the membrane material directly to a solid conductor and thence to the measuring instrument. Since the solution phase of the internal reference half-cell becomes superfluous in such circumstances, these types of membrane electrodes are sometimes referred to as *solid state electrodes*. The direct attachment of a wire contact to the membrane, as shown in Figure 3.5b is sometimes used when the physical properties of the membrane material are suitable. Of wider application, however, are electrodes prepared by coating a zeroth electrode surface with the membrane material. Commonly used zeroth interfaces are provided by platinum wire (Figure 3.5c) and hydrophobized graphite (Figure 3.5d).

The first coated platinum wire electrodes were reported in 1971 by Cattrall and Freiser (100) who prepared a calcium ion responsive electrode by coating a platinum wire, attached to the central conductor of a coaxial cable, with a calcium ion exchanger in PVC dissolved in cyclohexanone. A responsive coating was formed when the cyclohexanone evaporated. By extending the principle of this simple method of manufacture to other types of organic membrane materials, it has been possible to prepare many types of anion-responsive coated wire electrodes and some cation-responsive ones (101). When used as indicator electrodes in cells with liquid junction, the observed potential-concentration relation is sufficiently reproducible to be described by equation (3.17), and they have also been used as indicator electrodes in potentiotitrimetry.

When graphite is made water-repellent (i.e., hydrophobic) with Teflon, the graphite retains its conductivity and is easily compressed at room temperature into virtually any shape. Rods fashioned from this material form the zero interface for the Selectrode® type of membrane electrodes developed by Ruzicka (99) and shown in Figure 3.5d. The rod is housed in a Teflon tube and electrical contact to it is made by means of an electrically screened stainless steel wire screwed into the upper surface of the rod. The membrane material is applied to the lower surface of the rod by rubbing it with about 2–4 mg of the finely divided or colloidal material. After removing the excess loosely attached material, the active surface is hand polished at room temperature with the rounded end of a glass rod. A scant coating of material forms a more satisfactory surface than an excess since the latter often adds to the porosity of the surface causing a retention of the sample solution which, in turn, slows the response of the electrode. When the response of the electrode deteriorates after prolonged use, the membrane is removed with the blade of a scalpel. This exposes a new surface of rod which can then be recoated with a fresh portion of a membrane material.

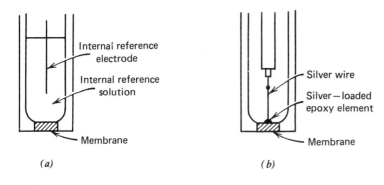

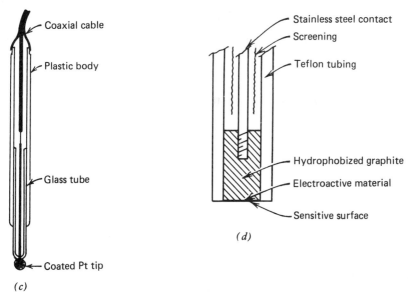

**Figure 3.5.** Methods of connecting the sensor membrane: (a) by an internal reference half cell; (b) by a solid internal contact; (c) by direct immersion of a platinum wire in a solution of electroactant which is then allowed to evaporate forming a coated wire electrode (reprinted by permission from Reference 98); (d) by rubbing the finely divided electroactant with graphite to form a Selectrode ®(99).

The latter need not necessarily be the same type of material as that which was removed.

The mechanism by which individually reproducible potentials are obtained with these solid state electrodes depends upon the establishment of some form of reference potential at the interface between the membrane material and the graphite or wire conductor. For example, in the case of the coated wire electrodes, Buck (102, 103) has suggested that the plastic membrane–metal interface is a capacitor and the inner potential of the metallic conductor is determined by capacitative coupling to the inner potential of the membrane phase. It has also been suggested (104) that since a coating agent such as PVC is permeable to both oxygen and water, an oxygen electrode is set up at the Pt–PVC interface and this may function as an internal reference electrode.

### Classifications of Membrane Materials

Regardless of whether a complete internal reference half-cell [see cell (3.18)] or, in the case of the solid-state membrane electrodes, a quasi-reference electrode provides the internal reference potential, the whole electrode is classified according to the nature of the membrane material. In the nomenclature recommended by IUPAC (105) the classification is based upon the form of membrane fabrication. Within this classification there are distinct categories:

1. Solid membranes.
2. Liquid membranes.
3. Special electrodes.

There are further subdivisions within each of these as follows:

**Solid Membranes (Homogeneous).** Included in this category are *glass membrane electrodes, nonporous membrane electrodes* formed from an homogeneous mixture of electroactive material and inert polymeric support which can be "solidified," and *crystal membrane electrodes* prepared from either a single compound or an homogeneous mixture of compounds (e.g., $Ag_2S$, $AgI/Ag_2S$).

**Solid Membranes (Heterogeneous).** An existing inert matrix such as is provided by materials such as PVC, silicone rubber, parchment, or hydrophobized graphite is impregnated with electroactive species which becomes distributed nonuniformly within the matrix.

**Liquid Membranes.** These are formed from a porous material such as PVC or cellulose acetate filter which acts as support for an electroactive substance of very low aqueous solubility which is applied to the support as a solution. These substances can be bulky *cations* (e.g., quaternary ammonium salts) which provide anionic-responsive membranes. Conversely, bulky *anions* [e.g., tetrakis(*p*-chlorophenyl)borate anions] can be used to form membranes that are responsive to cations. Cationic-responsive liquid membranes can also be formed from *neutral* molecular carriers such as macrocyclic compounds and antibiotics.

**Special Electrodes.** These include various types of *gas-sensing electrodes*, *enzyme substrate electrodes*, and, presumably, solid-state devices such as the *ion-sensitive field effect transistors* (ISFETS) would be included in this category.

A schematic representation derived from this type of classification has been suggested as a teaching aid (106). These classifications that are based upon the compositions of membranes rather than upon the nature of the electroactive material are no longer entirely appropriate, however, because in some cases the same electroactive material can be used in two or more categories within the classification. As the technology of membrane fabrication continues to improve it may become increasingly difficult, therefore, to classify the whole membrane. By comparison, the nature of the electroactive material, regardless of the way it is supported in the membrane, is easy to define and it is for these reasons that the following type of classification, based upon that proposed originally by Covington (97), is used here.

There are several main categories into which the electroactive materials for membrane electrodes fall. These are:

1. Glass.
2. Insoluble inorganic salts.
3. Large organic molecules such as organic ion exchangers and neutral carriers.

The majority of cation and anion selective membrane electrodes fall unambiguously into these categories. The exceptions are the solid-state devices (e.g., ISFETS) that are ion selective. Also excluded from this classification are sensors, the response of which depends indirectly upon changes in the concentration of a nonionic determinand. The enzyme electrodes and gas sensing "electrodes" are examples of these types of sensors. These are, in fact, based upon membrane electrodes prepared from one of the three given types of electroactive materials. In these cases the membrane material may be *modified* so that it responds indirectly to a different species by interposing a selective chemical reaction that yields the species usually sensed by the membrane. Thus the basic mode of action of the enzyme substrate electrode depends upon sensitizing a membrane with an enzyme that will react with the determinand (a substrate) yielding a product to which the membrane electrode is selective. The mode of action of gas sensing "electrodes" depends upon a similar principle. For example, a hydrogen ion-responsive glass membrane electrode can be made into a carbon dioxide sensor by surrounding the glass membrane with a thin film of a dilute solution of sodium bicarbonate held in position against the membrane with microporous PTFE. Diffusion of carbon dioxide through the PTFE causes a diminution of $p$H which is related to the partial pressure of the gas in the sample. The description of these gas sensors, however, as "electrodes" is erroneous because as commonly used they are complete chemical cells.

The incorporation of these types of enzyme substrate electrodes and gas sensors into the overall classification is conveniently achieved by including a fourth category; the fifth category identifies the remainder as follows:

4. Sensors with modified membranes.
5. Miscellaneous electrodes (e.g., ISFETS).

### Glass Electrodes

This category of ion selective membrane electrodes was in use and had been extensively studied and reviewed (107, 108) before the development of other classes of membrane electrodes. All glass electrodes respond to changes in hydrogen ion activity and the response of all are modified to some degree by the presence of such monovalent cations as sodium and potassium ions and to a lesser extent by silver and lithium ions. The enhancement of the response to monovalent cations other than hydrogen ion in the higher $p$H region is made possible by changing the composition of the glass membrane, the system $Al_2O_3, Na_2O, SiO_2$ being particularly versatile in this respect. Glasses with a low content of alumina, together with a ratio of $SiO_2$ to $Na_2O$ that is sufficiently high so that hydrolytic stability is ensured, can be used to prepare essentially $p$H-responsive membranes. A composition with a higher content of alumina, for example, 5% $Al_2O_3$, 27% $Na_2O$, and 68% $SiO_2$,* yields a membrane with a higher selectivity for potassium (and silver) ion than sodium ion. Increasing the alumina content still higher to 18% $Al_2O_3$, 18% $Na_2O$, and 71% $SiO_2$* yields a membrane in which the order of selectivity is $Na^+ > Li^+ > K^+$ at $p$H 7. In all cases, however, the hydrogen ion response will predominate in the lower $p$H range.

*All percentages are mole percent.

A glass electrode is formed by blowing a bulb of the molten membrane material on to the end of a piece of glass tubing. An internal reference half-cell, which is usually a silver–silver chloride electrode dipping into a solution of invariant composition containing chloride ion (Solution A in Figure 3.6), is used as a means of connecting the internal surface of the membrane by way of a suitably screened cable to the ion activity meter. The potential difference between the internal and external surfaces of the membrane when the latter is immersed in the solution under test is measured against an external reference half-cell. This is often the 3.5 $M$ calomel reference half-cell given in the cell notation of Figure 3.6. Since both the potentials of the internal and external reference half-cells remain constant, any variation in the activity of the cation contained in the solution under test will be reflected in a change in the potential that develops across the membrane. This will cause a corresponding change in the value of $E_{cell}$ as is predicted by equation (3.18). A glass $p$H-responsive membrane electrode in

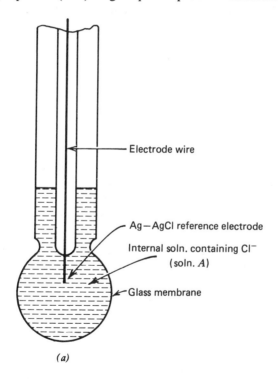

**Figure 3.6.** (*a*) The glass electrode. (*b*) The notation of a cell containing this electrode used in conjunction with a calomel reference half-cell:

$$\text{Hg; Hg}_2\text{Cl}_2 \big| 3.5\text{M KCl} \big\| \text{Test soln.} \big| \genfrac{}{}{0pt}{}{\text{Glass}}{\text{membrane}} \big| \text{Soln. A} \big| \text{AgCl; Ag}$$

which the internal reference solution was replaced by mercury can be regarded as a solid-state version of the electrode and has been used for the measurement of $p$H at temperatures down to $-30°$ (109).

The theory of how a potential develops across a thin glass membrane is not discussed in detail here, but can be found in References 107 and 108. Reference to more recent work can be traced from the reviews of R. P. Buck (102, 103, 110). A glass membrane after it has been soaked in an aqueous solution can be regarded as consisting of at least three distinct regions:

| test solution | hydrated glass layer | dry glass layer | hydrated glass layer | internal solution |

The thickness of the dry glass layer is about 50 $\mu$m, whereas the thickness of the hydrated layers is typically within the range $10^{-2}$–10 $\mu$m and depends primarily upon the type of glass from which the membrane is fabricated. Generally, electrodes made from glasses that form thin hydrated layers have superior electrode functions to those produced from glasses that form thick layers (111), and this has been attributed to the more cohesive structure present in the thinner hydrated layers (112). The hydrated layers formed on either side of the membrane act as immobilized solutions and for this reason they are sometimes referred to as *gel layers* (111). They are essential to the operation of a glass electrode and effects such as chemical degradation and hydration which affect the structure, and often the thickness of a layer, will cause a change in the response of the electrode. The gel layer provides an interface with the solution at which monovalent cations of the glass are exchanged and the anionic sites are involved in an exchange equilibrium with the determinand ions in the test solution. Soluble anions, however, play no role in the processes occurring within the gel layer. The effect of increasing the alumina content of the glass membrane is to reduce the degree to which the $H^+-M^+$ exchange process occurs. Presumably, this is caused by the replacement of silicon by aluminum in $\equiv$ SiONa and this favors the alkali–metal selectivity of the glass. The composition of the gel layers is not homogeneous and neither is there the sharp definition between these layers and the dry glass that is implied by the preceding representation. There is, for example, a gradation in the concentration of alkali metal ion contained in the gel layer that increases with the distance from the solution interface in toward the dry glass. In the more $p$H-sensitive glasses this concentration gradient is stepwise, whereas in glasses that exhibit a greater response to monovalent metal cations a smooth concentration gradient is established. At the periphery, alkali–metal cation and hydrogen ion interdiffuse readily in a silicate network which is deficient in the metal ion and partially hydrolyzed. In contrast with the concentration gradient of the alkali–metal ion, however, there is a gradual diminution in the concentration of water toward a transition or barrier region between the gel layer and the dry glass. Here the interdiffusion between proton and metal ion reaches a minimum and this barrier region inhibits the exchange of species between the dry glass and the gel layer. The conductivity of this region is low and, therefore, the predominant species will be $\equiv$Si—OH. Nonetheless, there will be a flux of cations from the dry glass to replace those dissolved away at the outer surface of the gel layer. A defect or jump mechanism involving the interstitial cations possibly accounts for this diffusion in the dry glass (113).

The magnitude of the potential established by a glass electrode depends predominantly upon the relative ion exchange processes that occur at the test solution–external gel layer interface, and the internal reference solution–internal gel layer interface. Once the internal gel layer has been fully developed, the potential at the internal boundary can be expected to remain invariant provided that the composition of the internal reference half-cell solution remains constant. The manner in which the overall electrode potential changes depends primarily, therefore, upon variations that occur in the activity of the determinand ion at the surface between the gel layer and the test solution. There are, however, a number of other contributions to the overall potential that arise as a result of the concentration gradients established within the gel layer itself and the diffusion of ions within this layer. For example, diffusion causes a charge separation the effect of which is to produce a counteracting potential similar in nature to the liquid junction potential. There are also contributions from the diffusion and phase boundary potentials of the inner surface of the glass that can be regarded as constant, and a time-dependent potential due to the movement of monovalent cations from the dry glass. If the sum of all these contributions were identical for the internal and external profiles of the membrane surfaces, then they would cancel when the membrane separated two solutions of identical composition. For a glass electrode this ideality would be identified by a zero electrode potential when the test and internal reference solutions had the same compositions. Inevitably, a small residual potential is observed and this is known as the *asymmetry potential*.

The major cause of the asymmetry potential is attributed to chemical and physical differences between the external and internal glass surfaces produced by the manufacturing process. For example, volatilization of some sodium from the external surface of the bulb will cause a slight difference in chemical composition; slight differences in the rates of cooling between the two surfaces will cause an unequal distribution of stress. The glasses now used in the manufacture of glass electrodes produce membranes characterized by small asymmetry potentials. Once they are conditioned by prolonged soaking in an aqueous solution, the potentials of these electrodes are very stable. A minimum conditioning period of about 14 days is recommended for accurate work (114), the maximum stability being achieved after several months (115).

Chalcogenide glasses have been reported (116) from which it has been possible to prepare membranes responsive to iron(III) and copper(II). The iron(III) responsive membrane (117) was prepared from the glass Fe-1173 ($Ge_{28}Sb_{12}Se_{60}$) and has been found to respond to uncomplexed iron(III) in sulfate solution. It was used to evaluate formation constants of iron(III) sulfate and as an indicator for titrations of sulfate with barium (118). The copper(II) responsive membrane (119) was made from a glass in which the active compound is reported as $Cu_6As_4S_9$. Both electrodes have disadvantages that exclude them from being useful in routine analysis. Both are susceptible to oxidation, both require frequent reactivation, and both have long equilibration times when compared with the standard established by other types of membrane electrodes.

## Electrodes Based on Membranes Prepared from Insoluble Inorganic Salts

The solubility of the electroactive material in the solvent, usually water, should be less than $10^{-6}$ mole $L^{-1}$ and when formed into a membrane, the material should act as an ionic conductor having a resistance small in comparison with the input impedance of the ion activity meter. In order to develop a reversible potential, the exchange current density of the material should be high in comparison with the current flowing in the meter circuit which is typically in the range $10^{-11}$ to $10^{-15}$ A (see Table 2.1). In order to ensure the necessary electrical conductance there must be cohesive contact between particles of the electroactive material. One method of fulfilling this requirement is to machine or compress the pure electroactive material either alone or as a homogeneous mixture, often with silver sulfide, to form a *homogeneous membrane* in the form of a disc. The alternative method, frequently applied to the same electroactive material, is to disperse the material into an inert matrix such as PVC, polythene, or silicon rubber to form a *heterogeneous membrane*. In terms of potentiometric response, there appears little difference between the same material whether it is used in the form of a homogeneous or as a heterogeneous membrane electrode. The latter membranes are more easily manufactured and are more flexible.

The membrane is housed in the body of the electrode assembly by a liquid-tight seal, and electrical contact between the internal surface of the membrane and the ion activity meter can be made through an internal reference half-cell and then by screened cable to the terminal of the meter.

Frequently, as was also mentioned for glass electrodes, a silver–silver chloride electrode is used. However, for these types of membrane electrodes the internal reference solution must contain, in addition to the electroactive chloride ion, an ion to which the membrane responds. A similar condition pertains, of course, also to the glass electrode since the aqueous internal reference solution contains hydrogen ion. The alternative method of making a direct electrical connection to the internal surface of an inorganic salt disc by solid-state contact is being used more frequently in recent years. Many of these types of membranes (see Table 3.4) contain silver salts and hence connection to them can be made by either embedding a silver wire into the membrane, or by sealing the internal contact to the membrane by means of an epoxy cement that is loaded with finely divided silver. Satisfactory solid-state connections have also been made using other materials such as platinum, mercury, and carbon.

Homogeneous membranes can sometimes be produced from a single crystal. They can also be formed by a sintering process or by compressing the electroactive material into a disc. For example, the fluoride ion selective membrane electrode can be made from a single crystal of lanthanum fluoride which has been doped with europium(II). The latter is used to increase the electrical conductivity (120) and to aid the ionic transport essential for a reasonably rapid potentiometric response. Alternatively, a fluoride responsive ceramic membrane may be prepared by sintering a mixture of lanthanum fluoride, europium fluoride, and calcium fluoride at temperatures greater than 1200° for 3–15 h in a controlled atmosphere (121). The response characteristics of electrodes based on the ceramic type of membrane appear similar to those based on the doped single crystal type which is the more commonly used version. Homogeneous membranes prepared by compressing the electroactive materials into discs are often based on silver salts. Although some single salts such as silver chloride, silver bromide, and silver sulfide form cohesive discs by compression, it is more usual for the majority of salts, such as silver iodide, for example, to crumble once the pressure on the disc is released. The addition of a binder to the electroactive material counteracts this effect. In these instances, the solubility of the binder must be less than the solubility of the electroactive salt and this requirement limits the selection to highly insoluble salts such as silver sulfide, silver selenide, and mercuric sulfide. After compression these salts form cohesive membranes in which the necessary low solubility is combined with a suitably low electrical resistance.

An electrode containing a membrane prepared by the compression of a pure inorganic salt exhibits two main alternative responses; one to the cation of the salt and the other to the anion. For example, a silver sulfide membrane electrode will respond to changes in the activity of silver ion or sulfide ion. The response to silver ion is termed the *primary response* and the response to sulfide ion is the indirect or *secondary response* because it is essentially a response to free silver ions in solution fixed by the solubility product of silver sulfide (*ca.* $10^{-50}$ at 25°) and the activity of sulfide ion. It follows form the definition of the solubility product that if the electrode is immersed in a solution that contains sulfide ion of unit activity, the activity of silver ion at equilibrium will be about $10^{-25}$ mole L$^{-1}$ at 25°; if present at $10^{-4}$ mole L$^{-1}$, the silver ion activity will increase to about $10^{-23}$ mole L$^{-1}$. These ion exchange equilibria are established rapidly, and result in a Nernstian response for solutions containing silver ion activities ranging from $10^{-25}$ to $10^{-1}$ mole L$^{-1}$. If, however, electrodes are fabricated from compressed membranes containing equimolar amounts of Ag$_2$S and AgX (X = Cl, Br, or I), a modification to the original response of the silver sulfide alone is observed. Once again the primary response depends upon the activity of silver ion in the solution, but it is now the solubility of the silver halide that controls the secondary response because this salt is more soluble than silver sulfide. Thus the silver ion activity in the solution is fixed by the solubility equilibrium:

$$AgX_{(s)} \rightleftharpoons Ag^+_{(aq)} + X^-_{(aq)} \tag{3.19}$$

Although cohesive discs can be formed from pure silver chloride or pure silver bromide, the electrodes derived from these do not function as well as those derived from the mixtures formed with silver sulfide. The pure halide discs are very sensitive to light and have a high electrical resistance. These undesirable characteristics are much diminished by the presence of silver sulfide which also improves the compressibility of the electroactive material and the density and strength of the resulting disc. Similarly, some compressed metal sulfides in isolation can be used to produce responsive electrodes, but those formed from mixed membranes with silver sulfide yield a superior electrode response. The primary response of these mixed membrane metal sulfide–silver sulfide electrodes is also toward the silver ion, but the secondary response is due in these cases to the aqueous metal ion. This increases the silver ion activity in the test solution by a similar type of equilibrium to (3.19). Thus the mixed membrane CdS–Ag$_2$S can be used to produce an electrode reversible to cadmium ion by virtue of the rapidly established equilibrium

$$Cd^{2+}_{(aq)} + Ag_2S_{(s)} \rightleftharpoons CdS_{(s)} + 2Ag^+_{(aq)}$$

The electrode response to silver ion is therefore a measure of the cadmium ion activity. Electrodes that respond to copper(II) and lead(II) by an analogous process can be made from the mixed membranes formed by the compression of the appropriate metal sulfide with silver sulfide. The copper electrode produced by this method is particularly prone to interference from chloride and bromide ion. It has been suggested that interference by chloride ion (122) is due to the reaction

$$Ag_2S_{(s)} + Cu^{2+}_{(aq)} + 2Cl^-_{(aq)} \rightleftharpoons 2AgCl_{(s)} + CuS_{(s)}$$

which appears to be confirmed (123), since after exposure to a high concentration of chloride the electrode becomes responsive to chloride ion. An extraneous but reproducible response caused by an interferent can sometimes increase the versatility of an electrode. An electrode based upon the mixed membrane silver iodide–silver sulfide, for example, in addition to its primary and secondary responses can be made responsive to mercury(II) or to cyanide ion by the selection of experimental conditions that will favor the formation of stable complexes with the electroactive material AgI. Thus in the $p$H range 11–13 the electrode becomes responsive to cyanide ion due to the consumption of silver iodide from the membrane by the almost stoichiometric reaction

$$AgI_{(s)} + 2CN^- \rightarrow Ag(CN)_2^- + I^-$$

The activity of the primary ion $Ag^+$ in the solution is therefore determined by cyanide activity represented by

$$Ag(CN)_2^- \rightleftharpoons Ag^+ + 2CN^-$$

In the lower $p$H range 4–5 the electrode is responsive to mercury(II) as a result of the reaction between this ion and silver iodide which also involves complex formation (124)

$$Hg^{2+} + nAgI_{(s)} \rightleftharpoons HgI_n^{2-n} + nAg^+$$

If the concentration of mercury(II) is less than $10^{-4}$ mole $L^{-1}$, $n = 1$, and a response slope of about 59 mV per decade is observed. At higher concentrations $n$ tends toward 2 and the slope is about 29 mV per decade. A crystal membrane formed directly from solid $Ag_2HgI_4$ has been used to produce an electrode which has been reported (125) as exhibiting a linear response for silver ion, iodide ion, and mercury(II) in the range $10^{-6.5}$–$10^{-1}$ mole $L^{-1}$.

The response of these types of precipitate-based membrane electrodes depends greatly upon conditions used in the precipitation of the membrane materials. Silver sulfide and some other metal sulfides, for example, may be precipitated from a solution containing the appropriate pure metal salt by any one of the following methods:

1. By bubbling hydrogen sulfide gas through an acidified solution.
2. By an homogeneous precipitation using an excess of thioacetamide.
3. By adding an excess of sodium sulfide to an alkaline solution.
4. By adding sodium thiosulfate to a strong acid solution.

Method 1 appears to be the best method for the preparation of the sulfides of silver, cadmium, copper, and lead provided that precipitation is carried out under stoichiometric conditions such that only a minimal excess of the soluble salt remains and yet excess hydrogen sulfide is entirely absent. Homogeneous precipitation also produced satisfactory materials from which to form silver sulfide–copper sulfide and silver sulfide–cadmium sulfide membrane electrodes with a Nernstian response that contrasted with the unsatisfactory response obtained when Method 3 was used (126, 127). Presumably, coprecipitation of the hydroxides is a cause contributing to the failure of the latter membranes. However, a variant of Method 3 has been used to prepare silver sulfide and lead sulfide either individually or as a coprecipitated mixture (128). The metal nitrate solutions ($1\ M$) were added slowly to a 20% excess of sodium sulfide ($0.3\ M$). After a series of elaborate washing procedures, which are necessary to remove excess reagent and to optimize the physical properties of the precipitate particles, the precipitates were dried at 110° for 24 h. The compression of the dried homogenized materials to produce cohesive membranes can be carried out using a laboratory type press in an evacuable die designed originally for the preparation of potassium bromide discs used in infrared spectroscopy. If elevated temperatures are required, the pressing die is preheated to the desired temperature using electrical heating tape. Once formed, electrical contact to the membrane is conveniently made through a silver wire sealed directly with a silver based epoxy onto one face of the disc. By means of a suitable plastic cement the perimeter of this face of the disc can then be sealed onto the end surface of a rigid piece of thick-walled plastic tubing. A thin-walled metal tube that fits securely against the inner surface of the plastic and which extends along about 75% of its length provides suitable screening when attached to the external shield of the connecting cable.

The type of solid materials possessing the necessary properties from which to prepare suitable membranes appears to be restricted (129) to the following types of inorganic salts: Group II and rare earth fluorides; halides of silver, lead, mercury, and thallium(I); sulfides of silver, lead, mercury, zinc, copper(I) and (II), and cadmium; selenides and tellurides of these metals; silver thiocyanate, azide, chromate, cyanide, and phosphate; bismuth phosphate, lead phosphates, and lead sulfates.

Heterogeneous membranes can utilize the same ion-exchanging electroactive solids as those used to form homogeneous membrane electrodes. Electrodes based on heterogeneous membranes were first described by Pungor and Hallos-Rokosinyi in 1961 (130). They used ion-exchanging silver halides supported on a matrix provided by silicone rubber which acts as an inert binder. These membranes were formed (131) by homogenizing a mixture of the precipitate (50% by total weight) and polysiloxane, which is then treated with a silane cross-linking agent and a catalyst. The degree of the cross-linkage determines the manner in which the electroactive particles are incorporated into the membrane surface which, in turn, affects the quality of the membrane function. Once cured, the membrane material is cut into discs 0.3–0.5 mm thick, which can then be affixed to the electrode body by means of silicone rubber glue. Heterogeneous membranes using polyethylene as the inert matrix have been prepared by mixing the dried and sieved precipitate with polythene powder and then subjecting this mixture to high pressures (100–300 atm = $10^7$ to $3 \times 10^7$ Pa) at temperatures in the range 100–300°C (132). The resulting discs (1–2 mm thick) are subsequently thermosealed in a suitable mold to a rigid polythene tube producing electrodes with no leakage and great mechanical strength. Moody and Thomas (133) have summarized the properties of the matrix support and the solids to be incorporated as follows:

A suitable matrix must:

1. be chemically inert and provide good adhesion for the sensor particles;
2. be hydrophobic;
3. be tough, flexible, yet nonporous and crack resistant to prevent leakage of internal solutions;
4. not swell in sample solutions. This is one reason for the wide use of silicone rubber, which takes up < 0.1 weight percent of water. Any extensive swelling disrupts the active chain of sensor particles.

In turn, the active sensor material (obtained, for example, by precipitation methods) should:

1. be physically compatible with the matrix;
2. have a low solubility product;
3. be of the right grain size (1–15 nm) which is a function of precipitation technique;

4. be mixed with the matrix support in the right proportion, usually 50 weight percent. This is essential in order to maintain physical contact between sensor particles and provide electrical conduction through the membrane, otherwise the resistance may be too great;
5. undergo rapid ion-exchange at the membrane-sample interface.

Some membranes based on insoluble inorganic salts are listed in Table 3.4, and these are arranged alphabetically on the first letter of the formula (inorganic salts), or common name (polymers) of the material used as a matrix for the membrane. Once assigned alphabetically, all single component membranes are given before those containing two components; all the two components are given before three components, and so on.

The relative merits of silver and mercury chalcogenides (i.e., sulfide, selenide, telluride) as matrices for the electroactive chalcogenides of bismuth, cadmium, cobalt, copper, lead, nickel, and zinc for cation-responsive electrodes, and silver or mercury(I) halides for anion-responsive electrodes have been assessed (180). For this evaluation some 235 electrodes were prepared by compression (8000 Kg cm$^{-2}$, ~ 8 × 10$^8$$^8$Pa) in the temperature range 25–200°C (162), and electrodes with satisfactory responses for copper(II), mercury(II), and lead(II) were obtained. However, all electrodes based upon membranes in which bismuth(III), cobalt(II), and zinc(II) chalcogenides were the active materials performed poorly. For the cation-responsive electrodes the silver sulfide matrix had the advantage of a lower oxidation-reduction sensitivity, whereas electrodes reversible to halides that were based upon mercury(II) sulfide–mercury(I) halide were more sensitive than those based upon the corresponding silver salts (see also References 159, 160, 181). Halide-responsive electrodes of relatively low impedance have also been prepared (182) by the precipitation of silver chloride, silver bromide, or silver iodide on to finely divided gold particles ( < 3 $\mu$m), the optimum AgCl: Au mole ratio, for example, being 9:1. These electrodes have a low oxidation-reduction sensitivity and a short response time (0.1–10 sec). The silver chloride–gold membrane has been used in a flow-through cell (183) for the continuous monitoring of water (184).

The low oxidation-reduction sensitivity of the halide membrane electrodes in comparison with that shown by the corresponding second kind oxidation-reduction electrodes was a property noted in 1937 (95). In oxidation-reduction electrodes of the type AgX; Ag| the deposit of the salt is porous and thus the test solution comes into direct contact with the silver metal. The metal can therefore act as a zeroth interface which will respond to any oxidation-reduction system present in the test solution, and, of course, such an interface does not exist in a membrane that contains only

Table 3.4. Some Electrodes Based on Membranes Prepared from Inorganic Salts

| Membrane | Type | Sensitive to | Comments | Refs. |
|---|---|---|---|---|
| AgCl (cast) | Homogeneous | $Cl^-$ | Not affected by $KMnO_4$, AgBr and AgI disks described | (95) |
| $Ag_2HgI_4$ | Homogeneous | $Ag^+, Hg^{2+}, I^-$ | Crystal membrane | (125) |
| $Ag_{19}I_{15}P_2O_7$–$Ag_3PO_4$ | Heterogeneous | $PO_4^{3-}$ | Thin layer of electroactive $Ag_3PO_4$ formed on $Ag_{19}I_{15}P_2O_7$ support | (134) |
| $Ag_2S$ | Homogeneous | $Ag^+, S^{2-}$ | Ceramic membrane | (135, 136) |
|  |  |  |  | (137) |
|  |  |  | Micro electrode | (138) |
| $Ag_2S$–AgI | Homogeneous | $I^-, CN^-$ |  | (139, 140) |
| $Ag_2S$–Ag X | Homogeneous | $X = Cl^-, Br^-, I^-$ |  | (141) |
|  |  |  | Micro electrode | (138) |
| (also $Ag_3SBr$, $Ag_3SI$, and $Ag_{19}I_{15}P_2O_7$ supports) | Heterogeneous |  | Thin layer electroactive AgX formed by reaction of $X_2$ gas | (134) |
| $Ag_2S$–CdS | Homogeneous | $Cd^{2+}$ | Micro electrode | (142, 143) |
|  |  |  |  | (138) |
|  |  |  | Made by crystn at 600–1100° in sealed silica ampoules | (144) |
|  | Heterogeneous |  | Thin layer CdS deposited on $Ag_2S$ support | (145) |
| $Ag_2S$–CuS | Homogeneous | $Cu^{2+}$ | Several methods of preparation are discussed | (146) |
|  |  |  |  | (147) |
|  |  |  | Micro electrode | (138) |
|  | Heterogeneous |  | Thin layer CuS deposited on $Ag_2S$ support | (145) |

(*continued*)

Table 3.4. (Continued)

| Membrane | Type | Sensitive to | Comments | Refs. |
|---|---|---|---|---|
| $Ag_2S$–CuSe | Homogeneous | $Cu^{2+}$ | Comparison made with $Ag_2S$–CuS electrode | (148) |
| $Ag_2S$–PbS | Homogeneous | $Pb^{2+}$ | Micro electrode | (143, 149) (138) |
|  | Heterogeneous |  | Thin layer PbS deposited on $Ag_2S$ support | (145) |
| $Ag_2S$–PbSe PbTe | Homogeneous | $Pb^{2+}$ | $Ag^+$, $Cu^{2+}$, $Fe^{3+}$, $Hg^{2+}$, $Cl^-$, and $S^{2-}$ interfere Prepared by sintering | (150) |
| $Ag_2S$ + MS | Homogeneous | M = $Cd^{2+}$, $Co^{2+}$ $Cu^{2+}$, $Hg^{2+}$ $Ni^{2+}$, $Pb^{2+}$, $Zn^{2+}$ | On the surface of a silver wire | (151) |
| $Ag_2Se$–CdSe $Ag_2Te$–CdTe | Homogeneous | $Cd^{2+}$ | Made by crystn at 600–1100° in sealed silica ampoules | (144) |
| $Ag_2S$–CdS–$Cu_2S$ | Homogeneous | $Cd^{2+}$ | Ceramic membrane $S^{2-}$ and $I^-$ interfered seriously | (152) |
| $Ag_2S$–$Cu_2S$–PbS | Homogeneous | $Pb^{2+}$ | Ceramic membrane Response time less than $Ag_2S$ – PbS type | (153) |
| $Ag_2S$–$Cu_2S$–PbS–$PbSO_4$ | Homogeneous | $SO_4^{2-}$ | Hot pressed pellet | (128, 154) |
| $Cu_2S$ | Homogeneous | $Cu^{2+}$ | Ceramic membrane | (155) |
| $Cu_{1.8}Se$ | Homogeneous | $Cu^{2+}$ | Single crystal; $Ag^+$ $Hg^{2+}$ interfere | (156) |

| Material | Type | Ion | Notes | Ref |
|---|---|---|---|---|
| Graphite, hydrophobized-$Ag_2S$ | Homogeneous | $Ag^+$, $S^{2-}$ | Selectrodes® | (157) |
| AgX | | $X = Cl^-, Br^-, I^-$ | | |
| CdS | | $Cd^{2+}$ | | |
| CuS | | $Cu^{2+}$ | | |
| HgS | | $Hg^{2+}$ | | |
| PbS | | $Pb^{2+}$ | | |
| Graphite, hydrophobized-$Cu_2S$-cupric tartrate | Homogeneous | tartrate | Based on a Selectrode® | (158) |
| $HgS$–$Hg_2Br_2$ | Homogeneous | $Br^-$ | | (159) |
| $HgS$–$Hg_2Cl_2$ | Homogeneous | $Cl^-$ | Sensitivity greater than Ag–AgCl | (160) |
| | | | Compressed at 8000 kg cm$^{-2}$ at temperatures 20–200° | (161, 162) |
| $LaF_3$ | Homogeneous | $SO_3^{2-}$ | | (163) |
| | | $F^-$ | Doped with Eu, Pr, Sm, Ho, and Nd | (120) |
| Rare earth fluorides | | | Single crystal | (96) |
| $LaF_3$ | | | Ceramic membrane formed with europium and calcium fluorides | (121) |
| | | | Solid state electrode Silver attached to membrane | (164) |
| Polythene–AgX | Heterogeneous | $X = Cl^-, Br^-, I^-$ | | (165) |
| Polythene–$Ag_2S$ | Heterogeneous | $S^{2-}$ | | (132) |
| Polythene–$Ag_2S$–CdS | Heterogeneous | $Cd^{2+}$ | $Ag^+$, $Hg^{2+}$, $Cu^{2+}$ interfere | (127) |
| Polythene–$Ag_2S$–CuS | Heterogeneous | $Cu^{2+}$ | | (166) |
| Polythene–$Ag_2S$–PbS | Heterogeneous | $Pb^{2+}$ | $Ag^+$, $Cu^{2+}$, $Hg_2^{2+}$, $Hg^{2+}$ interfere | (167) |

(*continued*)

Table 3.4. (Continued)

| Membrane | Type | Sensitive to | Comments | Refs. |
|---|---|---|---|---|
| PVC–BaSO$_4$ | Heterogeneous | SO$_4^{2-}$ | PO$_4^{3-}$, CO$_3^{2-}$, HPO$_4^{2-}$, CH$_3$COO$^-$ interfere | (168) |
| PVC–K$_2$Zn$_3$[Fe(CN)$_6$]$_2$ | Heterogeneous | Cs$^+$ | | (169) |
| Silicone rubber–AgI | Heterogeneous | I$^-$ | 0.1 M KCl does not interfere | (170) |
| Silicone rubber–AgX | Heterogeneous | Ag$^+$, X$^-$ = Cl$^-$, Br$^-$, I$^-$ | | (171) |
| Silicone rubber–Ag$_2$S–CdS | Heterogeneous | Cd$^{2+}$ | Ag$^+$, Cu$^{2+}$, Fe$^{3+}$, Hg$^{2+}$, S$^{2-}$, and I$^-$ interfere | (172) |
| Silicone rubber–BaSO$_4$ | Heterogeneous | SO$_4^{2-}$ | Phosphate interferes | (171) |
| Silicone rubber–CaF$_2$ | Heterogeneous | F$^-$ | | (173) |
| Silicone rubber–Cs12–molybdophosphate | Heterogeneous | Cs$^+$ | | (174) |
| Silicone rubber–Cu$_2$S | Heterogeneous | Cu$^{2+}$ | | (175) |
| Silicone rubber–Cu$_{2-x}$S | Heterogeneous | Cu$^{2+}$ | High selectivity for Cu$^{2+}$ | (176) |
| Silicone rubber–K$_2$Zn$_3$[Fe(CN)$_6$]$_2$ | Heterogeneous | K$^+$ | Selectivity comparable to cationic glass electrodes | (177) |
| Silicone rubber–LaF$_3$ | Heterogeneous | F$^-$ | | (173) |
| Silicone rubber–PbS | Heterogeneous | Pb$^{2+}$ | Cu$^{2+}$, Ag$^+$, Hg$^{2+}$, SO$_4^{2-}$ interfere | (178) |
| Silicone rubber–phosphate salts | Heterogeneous | PO$_4^{3-}$ | | (179) |

insoluble salts. A great difference in the reactivities of the two types has been noted (97). For example, an electrode formed from a membrane containing silver chloride did not cause the decomposition of chloride-containing aqueous hydrogen peroxide solutions, whereas a thermal electrolytic silver–silver chloride electrode promoted rapid catalytic decomposition and oxygen evolution.

### Electrodes Based on Membranes Prepared from Organic Ion Exchangers and Neutral Carriers

Membranes from which it has been possible to develop a wide range of electrodes suitable for use in aqueous solutions have been reported (see Table 3.5). As with the membranes based upon insoluble inorganic salts, the ion of interest is present in the membrane as an insoluble species. However, in these types of membranes this insolubility is conferred by combination with an organic reagent which, in turn, must be dispersed and supported in an inert matrix that forms the physical basis for the membrane. Once the ion has been immobilized in the membrane, the potential that develops across the membrane when it separates two solutions containing different concentrations of the ion depends primarily on an ion-exchange process that is similar to the type of process upon which the operation of the glass and the insoluble inorganic salt membrane electrodes depends.

Matrices in which to disperse the electroactive ion-organic moiety combination are water insoluble, porous (pores $ca.$ 0.1 $\mu$m dia.) and hydrophobic materials such as cellulose acetate, Teflon, or thin ($ca.$ 2 mm) hydrophobized glass and ceramic sinters (frits). The use of the latter materials is curtailed by the fragility of the thin discs, and materials such as the Millipore® or Nuclepore® filters are more suitable. Since the electroactive organic material is frequently a solid, it must be dispersed into and held by capillary action within the pores of the matrix as a solution. The solvent used in the preparation of this solution should ideally be totally immiscible with water, be of low volatility, and yet be characterized by a dielectric constant in excess of about 15. Solvents include halogenated aliphatic and aromatic hydrocarbons (bromoheptane, $o$-dichlorobenzene), alcohols (decanol), ethers (diphenyl ether, $o$-nitrophenyl octyl ether), esters of carboxylic acids (phthalates, sebacates, dioctyl adipate), phosphonates [bis($n$-octylphenyl)phosphonate], and phosphates [tris(2-ethylhexyl)phosphate, tritolyl phosphate]. Electrodes formed from membranes in which the electroactive substance is present in solution are often called *liquid-state electrodes*. In this type of electrode, once the determinand ion has entered the membrane its mobility is ensured since it is free to move by diffusion. The selectivity of the membrane, however, is based upon the process by which the ion enters the membrane, and ideally the entry of all ions other than the determinand should be totally inhibited by the unique avidity of the electroactive solute for the determinand. This exclusive avidity is an ideal never attained in practice and in many cases the nature of the solvent used in the preparation of the membrane also affects the selectivity of the membrane.

The design of liquid membrane electrodes is more intricate than electrodes based upon the solid inorganic salt type of membranes. Figure 3.7a shows that the basic design requirement consists of an internal reference half-cell that makes contact with the internal surface of the liquid membrane. In addition to this, it is advisable to provide a reservoir of electroactive solution to replenish that which is lost from the membrane through leaching. Figure 3.7b shows one ingenious design that meets these requirements. In this new type of Orion liquid membrane electrode the membrane assembly, containing the electroactive nonaqueous solution held in a plastic foam, is simply screwed onto the electrode body. The latter arrangement obviates the necessity to handle solutions that are often noxious, and, at the same time, dispenses with the requirement inherent in earlier designs of correctly positioning tiny pieces of membrane material. It should also be noted that the junction between the reference electrode and the wire of the connecting cable is made so that it will be below the level of immersion of the electrode in the test solution when in normal use. This is an important feature if the thermal effects are to be avoided.

The ion–organic moiety combination is not confined to use in liquid membranes, however, since the same electroactive species can sometimes be incorporated into a solid membrane formed with such materials as PVC,

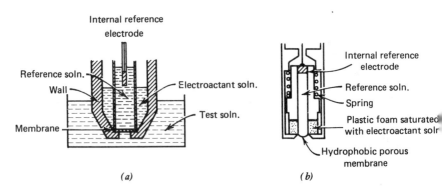

**Figure 3.7.** Liquid membrane electrodes: (a) Diagrammatic representation; (b) The new Orion type.

polythene, and silicone rubber. The preparation of the membrane material is often accomplished by dissolving both the organic electroactant and the polymer in a volatile solvent, and then allowing the solvent to evaporate slowly. A disc cut from this membrane material can be sealed to the body of the electrode with a suitable cement in the manner outlined elsewhere. Electrodes of the type shown in Figure 3.8 are easily constructed (185) from a membrane prepared by the dissolution of PVC powder and the electroactive material in a solvent such as tetrahydrofuran or cyclohexanone. A disc of the membrane material is then sealed to the flush-end of a piece of thick-walled PVC tube with PVC glue. The latter is prepared by dissolving PVC alone in the solvent. Some of the coated wire type of electrodes mentioned earlier and shown in Figure 3.5c are the solid-state analogues of these electrodes and are even more convenient to prepare. It is important to ensure that all the wire extending from the body of the electrode is coated by total immersion of the wire in the organic electroactant–PVC–cyclohexanone (100) or tetrahydrofuran (186) solution. After evaporation of the solvent, the electrodes are usually conditioned by a 0.25–2 h soaking in a solution containing 0.1 mole $L^{-1}$ of determinand ion. Although platinum has been found to be the best conducting base for the organic coating, effective electrodes have also been fabricated using copper and silver wires (187). Graphite has also been used (188).

It is difficult to make a comparison between the relative merits of the liquid membrane, solid membrane, and coated wire types of electrode derived from the same organic electroactant. The liquid membrane electrodes in common use tend to be purchased since these are more difficult to fabricate than the other two types. These commercially available electrodes appear to offer some slight advantage with respect to their durability under conditions of continuous use, but their susceptibility to exhibit a drift in potential ( $ca.$ 1 mV day$^{-1}$) seems somewhat greater than that quoted (190) for the equivalent solid membrane type ( $ca.$ 1 mV week$^{-1}$). These drifts are, of course, negligible when compared with those of the coated wire electrodes for which values of 5 mV day$^{-1}$ and upwards have been reported (191). Even such large drifts need not necessarily inhibit the use of a given type of electrode, provided that it is possible to make frequent standardizations (189) during the course of analyzing solutions of unknown concentrations. It is also recorded (189) that coated wire electrodes can be stored in the dry state in air for periods in excess of six months.

The electroactive ion–organic moiety combinations present in the types of membrane electrodes mentioned earlier all have the common feature of selectively binding small ions either at charges sites of opposite sign to the ion or at a neutral site provided by an organic molecule that has the ability

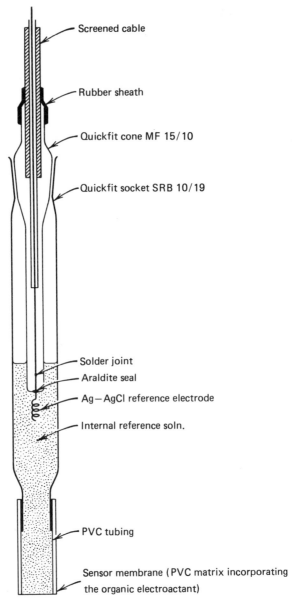

**Figure 3.8.** A PVC matrix membrane ion selective electrode of a type that can be assembled by students (185).

CATEGORIES AND DESCRIPTIONS OF MEMBRANE ELECTRODES   143

to wrap around the ion during complex formation. Compounds that possess one or other of these characteristics have become defined by the following categories:

Organic ion exchangers
Neutral carriers

Regardless of category, the species formed by combination between a given compound and the appropriate determinand must be ideally insoluble and of an adequate conductance to ensure that the electrical resistance of the membrane will not be excessively high. The selectivity of the resulting electrode will depend upon the ability of the compound to combine only with the determinand ion so that all other types of ions are excluded from the membrane.

### *Organic Ion Exchangers*

Anion-responsive membranes are prepared from oil-soluble bulky cations that form salts with determinand anions, or from neutral compounds of low aqueous solubility such as the *o*-phenanthroline complexes of iron(II) and nickel(II). The latter complexes form the electroactants for the perchlorate and nitrate responsive membranes, respectively. Compounds yielding bulky cations are either alkyl/aryl substituted ammonium and phosphonium ions or positively charged dye salts such as gentian violet [*CA* Registry No. *548-62-9*]. Featuring prominently among the ammonium salts used in the preparation of anion responsive membranes is the salt marketed as Aliquat 336S. This product is predominantly *NNN*-trioctyl *N*-methylammonium chloride. Other ammonium salts used for this purpose and marketed under synonyms are Zephiramine (*N*-benzyl *NN*-dimethyl *N*-tetradecylammonium chloride) and Hyamine 1622 (Benzethonium chloride, *CA* Registry No. *121-54-0*). Practical instructions pertaining to the preparation of membrane electrodes based on these types of compounds are given in many of the references quoted in Table 3.5*a*. The corresponding instructions for the preparation of cation-responsive electrodes based on the combination between a determinand cation and a bulky organic anion is found in some of the references given in Table 3.5*b*. Compounds included in this table that can be classified as organic ion exchangers are bulky anions like tetraphenylborate and its derivatives, and the mono and di-alkyl/aryl esters of phosphoric acid. Also included in this category are compounds that form oil-soluble complexes with metal ions of which dithizone (*CA* Registry No. *60-10-6*) is an example.

Table 3.5a. Some Anion-Responsive Membranes Based on Organic Ion Exchangers

| Ion | Membrane Components | Membrane Type | Ref. |
|---|---|---|---|
| *A. Inorganic* | | | |
| Bromide | Aliquat 336S in decanol shaken with sodium bromide (aq) | PVC coated wire | (192) |
| | Aliquat 336S in chloroform shaken with sodium bromide (aq) (solvent removed by distillation) | Solid PVC | (193) |
| Carbonate | Aliquat 336S + 4-butyl-$\alpha,\alpha,\alpha$-trifluoroacetophenone | Liquid | (194, 195) |
| Bicarbonate | Tridecylammonium bicarbonate salt | Liquid | (196) |
| Chloride | Aliquat 336S in decanol shaken with sodium chloride (aq) | PVC coated wire | (192) |
| | Aliquat 336S in chloroform shaken with sodium chloride (aq) (solvent removed by distillation) | Solid PVC | (193) |
| | $N$-benzyl $NN$-dimethyl $N$-octadecylammonium chloride in decanol or nitrobenzene | PVC coated wire | (197) |
| *Chloro-complexes* | | | |
| Chloroantimonate(V) | Sevron Red L(CI Basic Red 17), Sevron Red GL(CI 11,085) or Flavinduline O(CI 50,000)-hexachloroantimonate(V) + $o$-dichlorobenzene | Liquid (in natural rubber) | (216) |
| Chloroaurate(III) | Safranine O(CI Basic Red 2)-tetrachloroaurate(III) + $o$-dichlorobenzene | Liquid (in natural rubber) | (217) |
| Chlorocuprate(II) | Aliquat 336S–chlorocuprate salt | PVC coated wire | (198) |
| Chloroferrate(III) | Aliquat 336S–tetrachloroferrate salt | PVC coated wire | (199) |
| Chloromercurate(II) | Aliquat 336S–chloromercurate salt (also iodide) | PVC coated wire | (200) |
| | Tetradecylphosphonium trichloromercurate + dibutyl phthalate | Solid PVC | (201) |
| Chlorothallate(III) | Tetradecylphosphonium tetrachlorothallate + dibutyl phthalate | Solid PVC | (201) |
| | Sevron Red L(CI Basic Red 17), Sevron Red GL(CI 11,085), Flavinduline O(CI 50,000), or Phenazinduline $O$-tetrachlorothallate + dichlorobenzene | Liquid (in natural rubber) | (216) |

| | | | |
|---|---|---|---|
| Chlorozincate(II) | Aliquat 336S-chlorozincate salt | PVC coated wire | (202) |
| Iodide | Aliquat 336S in chloroform shaken with sodium or potassium iodide (aq) (solvent removed by distillation) | Solid PVC | (193) |
| | Aliquat 336S in decanol shaken with sodium iodide | PVC coated wire | (192) |
| | Aliquat 336S in chloroform shaken with sodium or potassium nitrate (aq) (solvent removed by distillation) | Solid PVC | (193) |
| Nitrate | Aliquat 336S–nitrate salt | Poly(methacrylate) coated wire | (203) |
| | $N$-benzyl $N$-hexadecyl $NN$-dimethylammonium nitrate + decanol | Liquid | (204, 205) |
| | $N$-hexadecyl $NNN$-trimethyl or $NNNN$-tetra-octylammonium nitrates + octanol | Liquid | |
| | $NNN$-tridodecyl $N$-hexadecylammonium nitrate + $o$-nitrophenyl octyl ether | Solid PVC | (206) |
| | $NNNN$-tetraoctyl or $NNNN$-tetrakis(decyl) ammonium nitrates + chlorobenzene | Liquid (cellophane) | (207) |
| | Crystal Violet (CI Basic Violet 3) in tetrachloroethane or nitrobenzene shaken with 1 $M$ sodium nitrate (aq) | Liquid (hydrophobized G4 glass sinter) | (208) |
| | 1,10-Phenanthroline-nickel(II) nitrate + $p$-nitrocymenetris(Bathophenanthroline)-nickel(II) nitrate + $p$-nitrocymene | Solid PVC | (206) |
| | | Liquid (Orion) | (209) |
| | tris(Bathophenanthroline)-nickel(II) nitrate + 2-nitrophenyl octyl ether(A) or 2-nitrophenyl phenyl ether(B) (Either A or B gives electrodes with longer life than those with $p$-nitrocymene) | Liquid or solid PVC | (210) |

(*continued*)

Table 3.5a. (Continued)

| Ion | Membrane Components | Membrane Type | Ref. |
|---|---|---|---|
| Perchlorate | tris(Substituted 1,10-phenanthroline)iron(II) perchlorate + p-nitrocymene | Liquid (Orion) | (209) |
| | tris(Bathophenanthroline)iron(II) perchlorate + nitrobenzene | Solid PVC | (211) |
| | | Liquid | (212) |
| | Aliquat 336S in decanol shaken with sodium perchlorate (aq) | PVC coated wire | (192) |
| | Aliquat 336S in chloroform shaken with sodium perchlorate (aq) (solvent removed by distillation) | Solid PVC | (193) |
| | $N$-Benzyl $NN$-dimethyl $N$-tetradecylammonium perchlorate + dioctyl-phthalate (1:1) | Solid PVC | (213) |
| Perchlorate | Methylene blue perchlorate + nitrobenzene | Liquid | (214, 215) |
| | Dye (0.1%) in tetrachloroethane (TCE) or nitrobenzene (PhNO$_2$) shaken with 1 $M$ perchlorate (aq). [Dye = Gentian Violet (CI Basic Violet 3) in TCE, or Fuchsine (CI Basic Violet 14) in TCE or PhNO$_2$] | Liquid (hydrophobized G4 glass sinter) | (208) |
| | Tetraoctyl or tetraphenyl phosphonium perchlorate (0.01%) in nitrobenzene, dibutyl phthalate or decanol | Liquid | (218) |
| Perrhenate | Brilliant Green (CI Basic Green 1) perrhenate in o-dichlorobenzene | Liquid (in natural rubber) | (219) |
| | Tetraoctyl or tetraphenyl phosphonium perrhenate (0.01%) in nitrobenzene, dibutyl phthalate, or decanol | Liquid | (218) |
| Sulfate | Aliquat 336S in decanol shaken with sodium sulfate (aq) | PVC coated wire | (192) |
| | Aliquat 336S-sulfate + 4-butyl-$\alpha$, $\alpha$, $\alpha$-trifluoroacetophenone (addition of barium sulfate improves selectivity) | PVC coated wire | (220) |

| Ion | Composition | State | Ref. |
|---|---|---|---|
| Tetrafluoroborate | tris(Substituted 1,10 phenanthroline)nickel(II)-tetrafluoroborate + p-nitro cymene | Liquid | (221) |
| | Malachite Green (CI Basic Green 4) (0.1%) in tetrachloroethane or nitrobenzene shaken with 1 M sodium tetrafluoroborate | Liquid (hydrophobized G4 glass sinter) | (208) |
| Thiocyanate | Aliquat 336S in decanol shaken with sodium thiocyanate (aq) | PVC coated wire | (192) |
| | N-Benzyl NN-dimethyl N-octadecylammonium thiocyanate in decanol or nitrobenzene | PVC coated wire | (197) |
| | N-Benzyl NN-dimethyl N-tetradecylammonium thiocyanate + dioctyl phthalate | Solid PVC | (213) |
| | Dye (0.1%) in tetrachloroethane shaken with 1 M thiocyanate (aq). [Dye = Gentian Violet (CI Basic Violet 3) or Alkali Blue (CI acid Blue 119)] | Liquid (hydrophobized G4 glass sinter) | (208) |
| | Gentian Violet (CI Basic Violet 3) or tetraphenyl-arsonium thiocyanate salts + nitrobenzene | Liquid | (222) |

### B. Organic

#### Carboxylates (monoanions)

| Ion | Composition | State | Ref. |
|---|---|---|---|
| Formate | tetrakis(Decyl)ammonium formate + dibutyl phthalate | Solid PVC | (223) |
| Acetate | Aliquat 336S in decanol shaken with sodium acetate (aq) | PVC coated wire | (192) |
| | tetrakis(Decyl)ammonium acetate + dibutyl phthalate | Solid PVC | (223) |
| Acetate, trifluoro-, or 2,4-dichlorophenoxy- | tris(Bathophenanthroline)iron(II)trifluoroacetate in nitrobenzene | Liquid | (224) |
| | Crystal Violet (CI Basic Violet 3)-trifluoroacetate in nitrobenezene, chloroform, or 1,2-dichloroethane | Liquid | (224) |
| | Tetrazolium salts of phenoxyacetic acid | Liquid | (225) |

*(continued)*

147

**Table 3.5a.** (*Continued*)

| Ion | Membrane Components | Membrane Type | Ref. |
|---|---|---|---|
| | Carboxylates (monoanions) (*Continued*) | | |
| Propanoate | tetrakis(Decyl)ammonium propanoate + dibutyl phthalate | Solid PVC | (223) |
| Propanoate, 2-amino-3-phenyl-(phenylalaninate) | Aliquat 336S in decanol shaken with sodium phenylalaninate (aq) | PVC coated wire | (192) |
| Butanoate | tetrakis(Decyl)ammonium butanoate + dibutyl phthalate | Solid PVC | (223) |
| Pentanoate, 2-amino-4-methyl-(leucinate) | Aliquat 336S in decanol shaken with sodium leucinate (aq) | PVC coated wire | (192) |
| Benzoate or salicylate | Aliquat 336S in decanol shaken with sodium benzoate or salicylate | PVC coated wire | (192) |
| Salicylate or hydrogen phthalate | Aliquat 336S salicylate or hydrogen phthalate + dioctyl phthalate | Solid PVC | (213) |
| | Carboxylates (dianions) | | |
| Oxalate | Aliquat 336S in decanol shaken with sodium oxalate (aq) | PVC coated wire | (192) |
| Maleate or phthalate | Crystal Violet (CI Basic Violet 3)-maleate or phthalatein in nitrobenzene, chloroform, or 1,2-dichloroethane | Liquid | (226) |
| | Enolic hydroxy | | |
| 5-ethyl-5-phenyl-barbiturate (phenobarbital) | Aliquat 336S-barbiturate in decanol | PVC coated wire | (227) |

148

| | Sulfonate | | |
|---|---|---|---|
| Benzenesulfonate, napththalene-1-sulfonate | Crystal Violet (CI Basic Violet 3)-benzene-sulfonate or naphthalene-1-sulfonate + nitrobenzene or 1,2-dichloroethane | Liquid | (228) |
| Alkylbenzenesulfonate (surfactants) | Aliquat 336S in decanol shaken with sodium salt of detergent anion | PVC coated wire | (229) |
| | Hyamine 1622(I) + pentadecylbenzenesulfonate. [I = $N$-benzyl $NN$-dimethyl $N$-[5-(4-(1, 1, 3, 3-tetramethyl)butylphenoxy)-3-oxapentyl]ammonium, ion] | Liquid (nylon membrane) | (230) |
| | Zephiramine(I)-didecylbenzenesulfonate + nitrobenzene solidified in naphthalene. [I = $N$-benzyl $NN$-dimethyl $N$-tetradecylammonium] | Naphthalene coated wire | (231) |
| Bile salt anions | $NNN$-Tributyl $N$-hexadecylammonium bromide + sodium taurocholate or sodium deoxytaurocholate + hexachlorobenzene + bromoacetanilide in $o$-dichlorobenzene | Liquid | (232) |
| | Sulfate | | |
| Dodecylsulfate | bis(Diphenylglyoximato)-1,10-phenanthroline-cobalt(III)dodecylsulfate + $o$-dichlorobenzene | Liquid (graphit membrane) | (233, 234) |
| | Aliquat 336S in decanol shaken with sodium salt of detergent anion | PVC coated wire | (229) |

Table 3.5b. Some Cation-Responsive Membrane Based on Organic Ion Exchangers and Neutral Carriers

| Ion | Membrane Components | Membrane Type | Ref. |
|---|---|---|---|
| | A. Inorganic | | |
| Ammonium | Nonactin(I)$^a$ or monactin(II)$^a$ + tris(2-ethylhexyl)phosphate | Liquid (Philips) | (235) |
| Barium | Tetranactin(III)$^a$ + dibutyl or dioctyl phthalate | Solid PVC | (236) |
| | Antibiotic A23187 in nitrobenzene | Liquid (cellulose ester) | (237) |
| | $NNN'N'$-Tetraphenyl-3,6,9-trioxaundecane-diamide(IV)$^a$ in $o$-nitrophenyl octyl ether | Solid PVC | (238, 239) |
| | Tetraphenylborate salt of the barium complex of 4-nonylphenoxypoly(ethyleneoxy)ethanol(V)$^a$ + nitrobenzene | Liquid | (240) |
| | Tetraphenylborate salt of the barium complex of 4-nonylphenoxypoly(ethyleneoxy)ethanol(V)$^a$ + bis($o$-nitrophenyl)ether in $OO$-dioctyl phenylphosphonate | Solid PVC | (241, 242) (243) |
| Calcium | Calcium bis(decyl)phosphate + $OO$-dioctyl phenylphosphonate | Liquid (Orion) | (244) |
| | | Solid PVC | (245, 246) |
| | | PVC coated wire | (100) |
| | Calcium bis(4-octylphenyl)phosphate + $OO$-dioctyl phenylphosphonate | PVC coated graphite (Selectrode) | (247) |
| | Monoesters of phosphoric acid grafted to PVC + plasticizer | Solid PVC | (248) |
| | Calcium bis[4-(1,1,3,3-tetramethylbutyl) phenyl]phosphate + $OO$-dioctyl phenylphosphonate | Solid PVC | (249, 250) |
| | Antibiotic A23187 in nitrobenzene | Liquid (cellulose ester) | (237) |
| | Ethylenedioxybis($NN$-dibenzylacetamide)(VI)$^a$ + $o$-nitrophenyl octyl ether with added tetraphenylborate | Solid PVC | (251, 252) |
| | Diethyl $r$-13, $t$-17, $c$-18, $c$-22 tetramethyl-14, 21-dioxo-16,19,13,22-dioxadiazatetra-triacontanedioate(VII)$^a$ + | Solid PVC | (252, 253) |

| | | | |
|---|---|---|---|
| Copper | Copper(II)-N-phenylbenzohydroxamic acid + 1,1,2,2-tetrachloroethane | Liquid | (254) |
| Divalent cation (Water hardness) | Calcium bis(decyl)phosphate + decanol | Liquid | (255) |
| | Calcium bis(2-ethylhexyl)phosphate + decanol | Solid PVC | (256) |
| Lithium | 2,2-Dimethylpropane-1,3-diyldioxybis(N-heptyl N-methylacetamide)(VIII)$^a$ + tris(2-ethylhexyl) phosphate with added sodium tetraphenylborate | PVC solid | (257) (258) |
| Magnesium | Antibiotic A23187 in nitrobenzene | Liquid (cellulose ester) | (237) |
| Mercury(I) | Palladium-dithizone complex in chloroform used as an extractive membrane for Hg(I) (compared with corresponding Hg(I)-dithizone complex) | Liquid (graphite) | (259) |
| Potassium | Potassium tetrakis(p-chlorophenyl)borate + 3 nitro-o-xylene | Solid PVC | (260) |
| | Potassium tetrakis(p-chlorophenyl)borate with p-hexylnitrobenzene and 3-nitro-o-xylene | Coated wire | (261) |
| | Potassium tetrakis(p-chlorophenyl)borate + a nitroaromatic compound like p-hexylnitrobenzene | Solid PVC | (262) |
| | Potassium tetraphenylborate + 3-nitro-o-xylene or dibutyl phthalate or 2,2,4-trimethylpentyl phthalate | Coated wire | (263, 264) |
| | Dimethyldibenzo-30-crown-10(IX)$^a$ + dipentylphthalate | | |
| | Dimethyldicyclohexyl-18-crown-6 or dicyclohexyl-18-crown-6 + dipentyl phthalate | Solid PVC | (265, 266) |
| | Valinomycin[2001-95-8]$^b$ + diphenyl ether | Liquid (Philips) | (267) |
| | | Solid silicone rubber | (268) |
| | + dibutyl phthalate | Solid PVC | (269) |

(continued)

**Table 3.5b.** (*Continued*)

| Ion | Membrane Components | Membrane Type | Ref. |
|---|---|---|---|
| Potassium (*cont.*) | Valinomycin + DPE + DNP + TEP (3:4:2) [DPE = diphenyl ether, DNP = dinonylphthalate, TEP = tris(2-ethylhexyl)phosphate] | Solid PVC (suitable for clinical use) | (270) |
| | Valinomycin + dioctyl adipate | PVC (graphite Selectrode) | (271) |
| | + dioctyl phthalate | Solid PVC | (271) |
| | | PVC coated wire | (189) |
| | + dipentylphthalate | PVC coated wire (on Ag; AgCl) | (272) |
| | Valinomycin + poly(bisphenol-A-carbonate) + poly(dimethylsiloxane) | Solid block (long life) | (273) |
| | Valinomycin + potassium tetrakis(4-chlorophenyl)borate in dioctyl phthalate | Liquid (micro) | (274) |
| Silver | Silver nitrate (aq.) shaken with 2-[(3-hydroxy-1-oxoinden-2-yl)imino]-1,3-indandione (Ruhemann's Purple) in chloroform | Liquid (graphite) | (275, 276) |
| Sodium | 1,2-Ethylenedioxybis(*NN*-dibenzylacetamide) (VI)$^a$ + dibenzyl ether | PVC solid | (251) |
| | 1,2-Phenylenedioxybis(*N*-benzyl *N*-phenylacetamide)(X)$^a$ + dibutyl decanedioate | PVC solid | (238) |
| | 1,1,1-Propanetriyltris-4-(*N*-heptyl *N*-methyl 3-oxabutanamide) (XI)$^a$ + dibutyl decanedioate | PVC solid | (277) |
| | Monensin (produced by *Streptomyces cinnamonensis*) in nitrobenzene | Liquid (micro) | (278) |

| | | | |
|---|---|---|---|
| Thallium(I) | 4-Nonylphenoxypoly(ethyleneoxy)ethanol(V)[a]–Sr complex | Liquid (cellulose ester) | (237) |
| Uranyl | OO-didecyl dithiophosphate + chlorocyclohexane bis(2-Ethylhexyl)phosphate + | Liquid | (279) |
| | OO-bis(2-ethylhexyl) ethyl-phosphonate or tris(2-ethylbutyl)phosphate | Liquid Solid PVC | (280) (281) |

### B. Organic

| | | | |
|---|---|---|---|
| Antipyrine[60-80-0][b] | Potassium tetraphenylborate + dibutyl phthalate | PVC coated wire | (282) |
| Atropine[51-55-8][b] Brucine[357-57-3][b] Cinchonine[118-10-5][b] | | | |
| Choline[62-49-7][b] | tetrakis(p-Chlorophenyl)borate-acetylcholine + phthalate ester | Solid PVC | (283) |
| Codeine[76-57-3][b] | Potassium tetraphenylborate + dibutyl phthalate | PVC coated wire | (282) |
| Ephedrine[299-42-3][b] | Precipitated cation-tetraphenylborate treated with PVC in tetrahydrofuran + dibutyl or dioctyl phthalate | Solid PVC | (284) |
| Narcotine[128-62-1][b] | Potassium tetraphenylborate + dibutyl phthalate | PVC coated wire | (282) |
| Onium ions, large (e.g., $(C_2H_5)_3N^+C_4H_9$) | Sodium tetraphenylborate + 4,4,4-trifluoro-1-(2-thienyl) butane-1,3-dione + p-nitrocymol | Solid PVC | (285) |
| Papaverine[58-74-2][b] | Potassium tetraphenylborate + dibutyl phthalate | PVC coated wire | (282) |

(*continued*)

Table 3.5b. (Continued)

| Ion | Membrane Components | Membrane Type | Ref. |
|---|---|---|---|
| α-Phenylethylammonium (α-methylbenzylammonium) | 2,3-Butanediyldioxybis[N-ethyl N-(α-methyl-benzyl)acetamide](XII)[a] + o-nitrophenyl octyl ether with added sodium tetraphenylborate | Solid PVC | (286) |
| Propranolol[525-66-6][b] | Precipitated cation-tetraphenylborate treated with PVC in tetrahydrofuran + dibutyl or dioctyl phthalate | Solid PVC | (284) |
| Pyramidon[58-15-1][b] | Potassium tetraphenylborate + dibutyl phthalate | PVC coated wire | (282) |
| Quinine[130-95-0][b] Strychnine[57-24-9][b] Vitamin B1[59-43-8][b] | Tetraphenylborate-vitamin B1 + 1,2-dichloroethane Dipicrylamine-vitamin B1 + nitrobenzene Potassium tetraphenylborate + dibutyl phthalate | Liquid Liquid PVC coated wire | (287) (282) |
| Vitamin B6[8059-24-3][b] | Tetraphenylborate-vitamin B6 + 1,2-dichloroethane Dipicrylamine-vitamin B6 + nitrobenzene Potassium tetraphenylborate + dibutyl phthalate | Liquid Liquid PVC coated wire | (287) (282) |

[a] See corresponding structure (Figure 3.9).
[b] Chemical Abstracts Registry Number.

## Neutral Carriers

These compounds are also included in Table 3.5*b* because cation-responsive membrane electrodes are prepared from them. In contrast to the organic cation-exchangers, however, neutral carriers are oil-soluble uncharged compounds that can selectively extract by complexation determinand cations from an aqueous solution into a liquid or a semisolid membrane (e.g., a PVC membrane). The complexes formed are characterized by a polar internal cavity in which the cation is "trapped" since it becomes completely enveloped by the lipophilic ligand. The dimensions of this cavity are such that a high degree of specificity is shown for a cation of the correct dimensions with the result that electrodes prepared from neutral carriers are generally more selective than those derived from organic cation exchangers. The original, and still the most extensively used neutral carrier, is Valinomycin (*CA* Registry No. *2001-95-8*) which is the electroactant for the potassium membrane electrode. Valinomycin is a depsipeptide (a type of molecule composed alternatively of α-amino acids and α-hydroxy acids) prepared from *Streptomyces fulvissimus* cultures (288). It is composed of three moles each of L-valine, D-α-hydroxyisovaleric acid, D-valine, and L-lactic acid linked alternatively to form a 36-membered ring (289). Another type of ring system is provided by the macrotetrolides or actines of which nonactinic acid {2-[5-(2-hydroxypropyl)-3-methyloxolan-2-yl]propanoic acid} is the parent compound. This tetrahydrofuran (i.e., oxolane) derivative forms a macrocyclic (more than a 10-membered ring) quadruple lactone which, as shown in Figure 3.9, may be substituted in various ways (structures I–III) to provide the electroactants suitable for use in preparing membrane electrodes that respond to ammonium ion. Other types of macrocyclic compounds that have been investigated are the crown ethers (cyclic polyethers), of which dimethyldibenzo-30-crown-10 (structure IX) is an example. The basic functional entity for the compounds can be represented as

$$\boxed{-[CH_2-CH_2-O]_n-}$$

and this type of structure exhibits a marked complexing ability for alkali metal cations. In the suggested nomenclature (290) for these compounds (e.g., structure IX), the number prefix to "crown" refers to the number of ring members, 30 and the suffix refers to the number of heterocyclic oxygens, 10 which are separated according to the basic structure as given. At least 60 of this type of compound have been prepared (291).

Acyclic compounds that contain a central, but smaller, crown-ether system which is modified by attachment of two large hydrophobic *NN*-

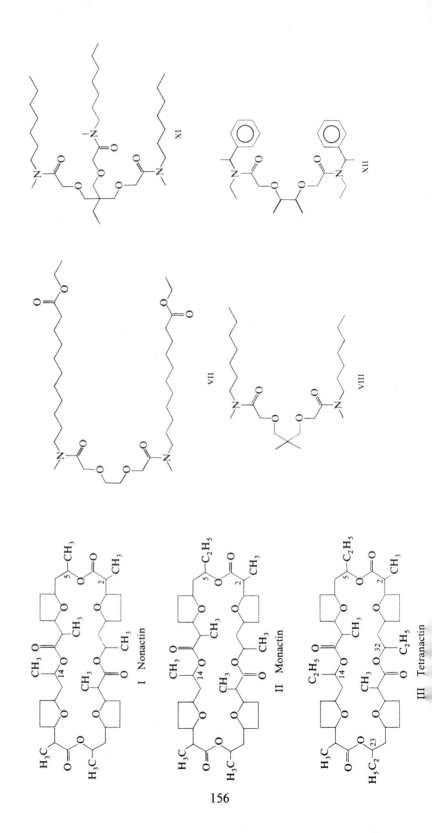

I Nonactin

II Monactin

III Tetranactin

VII

VIII

XI

XII

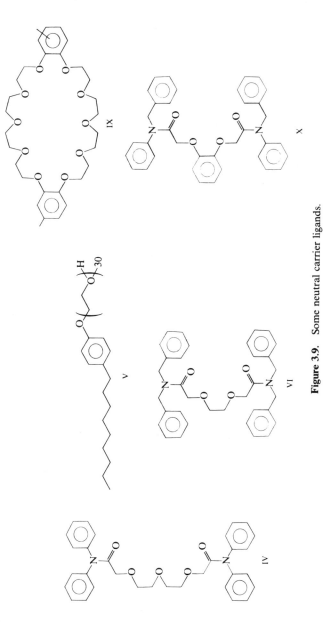

**Figure 3.9.** Some neutral carrier ligands.

disubstituted amide groups have also been investigated. These compounds are diamides of dicarboxylic acids conforming to the basic acyclic skeleton provided by 3, 6-dioxaoctanedioic acid (see structures VI, VII, X, XII) or 3, 6, 9-trioxaundecanedioic acid (see structure IV). The crown ether system has been modified, however, in structures VIII and XI since these compounds contain an extra -$CH_2$ group separating the two ether oxygen atoms. Structure VIII is the diamide of the dicarboxylic acid, 5, 5-dimethyl-3, 7-dioxanonanedioic acid, whereas structure XI is the triamide of the tricarboxylic acid, 1, 1, 1-propanetriyltris-4-(3-oxabutanoic acid). In the presence of the appropriate metal ion these acyclic ligands wrap around the ion to form cyclic complexes.

The research effort to produce compounds suitable as neutral carriers has been intense and at least 200 compounds have been screened as possible carrier molecules (292).

Irrespective of type, organic electroactants are usually solids that must be dispersed within the membrane as solutions. This applies also to the nominally solid membranes such as those based on PVC in which the solvent/plasticizer used in the preparation can affect the subsequent selectivity of the electrode that incorporates the membrane.

In many instances, therefore, it is advantageous to consider the solvent as a composite system, not necessarily confined to a single component, which acts also as a mediator that supplements the role of the organic electroactant. Thus the efficacy of a given organic electroactant can be considerably enhanced by an appropriate selection of the solvent/mediator system since the latter may increase the solubility of the determinand ion in the membrane, and, once in the membrane, the mobility of the ion is affected by the viscosity of the solvent system.

The selectivities of electrodes prepared from solid PVC membranes are greatly influenced by the nature of the plasticizers used in the preparation of the membranes. For example, the selectivities of potassium ion-responsive electrodes prepared from the three different electroactants, potassium tetrakis(*p*-chlorophenyl)borate (261), dimethyldibenzo-30-crown-10 (263), and valinomycin (293), were all found to be dependent upon the type of plasticizer used. Of these three, valinomycin is the most specific toward potassium ion and in this case the most satisfactory plasticizing agent was found to be a mixture of diphenyl ether, dinonyl phthalate, and tris(2-ethylhexyl) phosphate in the ratio 3 : 4 : 2 (270). Generally the selectivities of neutral carrier electroactants depend upon the dielectric constant of the solvent/mediator system, a low dielectric constant favoring selectivity toward monovalent cations. This has been confirmed (294) by comparing the selectivities of a $K^+$-valinomycin electrode and a $Ca^{2+}$-diethyl *r*-13, *t*-17, *c*-18, *c*-22-tetramethyl-14, 21-dioxo-16, 19, 13, 22-dioxadiazatetratriacontanedioate (structure VII) electrode under identical conditions of dielectric constant ($\epsilon$). Each electroactant (1%) was dissolved in two different solvent systems (69%) containing PVC (30%). The first solvent was bis(2-ethylhexyl) hexanedioate ($\epsilon = 4$) and the second was either dibutyl decanedioate or 2-nitrophenyl octyl ether used individually or in mixtures ($\epsilon = 24$). A low $\epsilon$ value for the membrane solvent favored the selectivity of the potassium electrode whereas a high $\epsilon$ value favored the selectivity of the calcium electrode. Even more dependent upon the nature of the solvent/mediator in determining the specificity of the membrane are electrodes based upon organophosphates, in particular, and organic ion exchangers, in general. One of the important applications of ion selective membrane electrodes to analysis is in the routine determination of calcium ion. The most satisfactory electroactants for these are derived from dialkyl hydrogen phosphates, the calcium salt of bis[4-(1, 1, 3, 3-tetramethylbutyl)phenyl]phosphate, in particular, being recommended for this purpose (250). A characteristic of membranes prepared from these types of ester is a gross diminution in specificity when they are prepared using a solvent/mediator of low dielectric constant. Thus a membrane prepared, for example, from bis(2-ethylhexyl)phosphate in decanol senses all divalent cations almost equally, thereby finding application as a "divalent cation" electrode (256). Alternatively, if the phosphate ester is dissolved in the more polar *OO*-dioctyl phenylphosphonate a great enhancement in specificity towards calcium ion occurs and this is the solvent/mediator often used in the preparation of calcium-responsive membranes. *OO*-dioctyl phenylphosphonate can be prepared by the method given in Reference 295.

### Sensors with Modified Membranes.

#### *Gas Sensors*

These sensors or probes are complete electrochemical cells consisting of an indicator electrode and a reference electrode skillfully incorporated into a single unit. This cell is isolated from the gaseous determinand by a gas-permeable, but ion-impermeable, membrane. The cell solution, held as a thin film by this membrane against the surface of the indicator electrode, contains a species to which the indicator electrode is reversible. When the probe is exposed to a medium containing the gaseous determinand, the gas diffuses into this cell solution thereby altering the activity of the indicated ion which, in turn, changes the cell potential. At equilibrium the activity of this ion is related to the partial pressure or concentration of the gas. The gases commonly determined by this method dissolve in water forming species that act either as acids or as bases. If the dissolved gas acts as a weak acid (e.g., $CO_2$), then the $p$H of a solution containing its conjugate base (e.g., $HCO_3^-$) will be decreased when the gas dissolves in it. Conversely, the $p$H of a system will be increased when a gas having basic properties in aqueous solutions (e.g., $NH_3$) enters a solution of its conjugate acid (e.g., $NH_4^+$). These $p$H changes are usually monitored by a glass electrode. A condition which must be fulfilled is that the concentrations of the appropriate conjugate species must be kept sufficiently high in comparison with the gaseous determinand so that it can be assumed that these concentrations are not materially altered when an equilibrium is established across the membrane.

The construction of an ammonia sensor is shown schematically in Figure 3.10. Basically, it is a cell without liquid junction in which the cell solution, containing hydrogen ion and chloride ion, is held within the confines of the membrane. This cell can be regarded as the analogue of cell (3.4) in which the hydrogen electrode is replaced by a flat-ended glass electrode (see Figure 3.10) to yield a cell that can be represented by the notation

$$\text{glass electrode} \underbrace{\left| NH_4^+(0.1\ M), Cl^-(0.1\ M) \right|}_{NH_{3(g)}\ \text{determinand}} AgCl:Ag \qquad (3.20)$$

The double broken lines connecting the vertical line representations of the interfaces made by the glass and silver–silver chloride electrodes with the aqueous cell solution is used to denote the gas-permeable membrane that separates the determinand $NH_{3(g)}$ from the 0.1 $M$ ammonium chloride cell solution. This broken line modification to the abbreviated cell notation is intended to be illustrative rather than a definitive addition. The e.m.f. of cell (3.20) is related to the activity of hydrogen ion present in the aqueous solution and the activity of the chloride ion in this solution by an equation analogous to (3.6),

$$E_{cell} = E_{cell}^\circ - k(\log a_H + \log a_{Cl^-}) \qquad (3.21)$$

As before, $E_{cell}^\circ$ is a calibration factor pertinent to a particular temperature and, in this case, is the difference between the standard potential of the silver–silver chloride electrode and the notional standard potential of the glass electrode (see Chapter 4). Under operational conditions the activity of chloride ion in this cell will be constant, and the hydrogen ion activity will depend upon the equilibrium

$$NH_4^+ \rightleftharpoons NH_3 + H^+$$

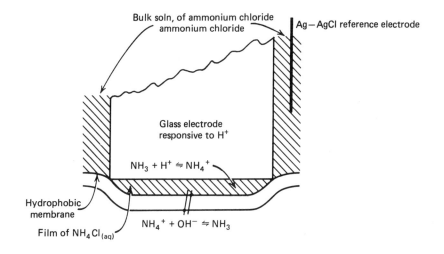

**Figure 3.10.** (*a*) Schematic representation of an ammonia sensor. (*b*) The "Air-gap" electrode (300).

for which the ionization constant is defined as

$$K_a = \frac{a_{NH_3} \cdot a_{H^+}}{a_{NH_4^+}}$$

If it is assumed that the activity coefficient of the nonionic species $NH_3$ is unity and that $a_{NH_4^+}$ in the cell solution is present in a very large excess in comparison with $c_{NH_3}$ emanating from the determinand (296), the expanded form of equation (3.21), that is,

$$E_{cell} = E_{cell}^o - k \log \frac{K_a \cdot a_{NH_4^+}}{c_{NH_3}} \cdot a_{Cl^-}$$

becomes

$$E_{cell} = \text{constant} + k \log c_{NH_3} \quad (3.22)$$

The linear relation between $E_{cell}$ and $\log c_{NH_3}$ forms the basis for a calibration graph from which a measured value of the probe potential may be interpolated in terms of the concentration of ammonia in a sample. It should be noted that the positive sign in equation (3.22) is valid only if the glass electrode is negative with respect to the silver–silver chloride electrode as is implied by cell notation (3.20). This polarity will depend upon the nature of the internal reference solution used in the manufacture of the glass electrode and if this electrode is positive with respect to the silver–silver chloride electrode, the sign in the equation should be negative. Thus the more general form of equation (3.22) is

$$E_{cell} = \text{constant} \pm \text{slope} \log c_{NH_3} \quad (3.23)$$

the polarity of the glass electrode being deducible from the slope of the calibration curve. Both series of calibrant and sample solutions should be adjusted to the same $p$H prior to measurement of the probe potential. If a low limit of concentration is expected, this $p$H should be greater than 12, which is usually obtained by the addition of sodium hydroxide. A lower $p$H is concomitant with a higher limit of detection. Volatile aliphatic and alicyclic amines having $pK_a$ values usually greater than 8 can interfere since, if they pass through the membrane, they will be absorbed in the ammonium chloride solution.

The behavior of sensors that respond to acidic gases are governed by the same basic principles as those described for the ammonia-responsive probe.

Table 3.6. Some Electrochemical Cells Used as Potentiometric Gas Sensors

| Gas | Cell Electrodes Sensing | Reference | Cell Solution (as film) | Membrane | Ref. |
|---|---|---|---|---|---|
| $NH_3$ | $p$H glass | Ag\|AgCl | 0.1–0.01 $M$ $NH_4Cl$ | Microporous PTFE | (298, 297) |
| | $p$H glass | Ag\|AgCl | 0.01 $M$ $NH_4Cl$ + wetting agent[a] | Air gap | (300) |
| | $p$H glass micro (0.01 $M$ HCl filling) | Ag\|AgCl | 0.005 $M$ $NH_4Cl$ + 0.1 $M$ NaCl | Air gap(micro) | (301) |
| | $Ag_2S$ | Ag(?) | 0.1 $M$ $AgNO_3$ + 0.1 $M$ $NH_4NO_3$ | Microporous PTFE | (302) |
| | $Ag_2S$, or $Ag_2S$ + CuS on hydrophobized graphite (Selectrode) | Orion $F^-$ | 0.01 $M$ $AgNO_3$ or $Cu(NO_3)_2$ + 0.1 $M$ $NH_4NO_3$ + 0.01 $M$ NaF | PTFE-polythene | (303) |
| $CO_2$ | $p$H glass | Ag\|AgCl | 0.01–0.005 $M$ $NaHCO_3$ + 0.02–0.1 $M$ NaCl | Microporous PTFE | (297, 304) |
| | $p$H glass | Ag\|AgCl | 0.01 $M$ $NaHCO_3$ + $Cl^-$ + wetting agent[a] | Air gap | (300) |
| | $p$H glass micro (0.01 $M$ HCl filling) | Ag\|AgCl | 0.005 $M$ $NaHCO_3$ + 0.1 $M$ NaCl | Air gap (micro) | (301) |
| | $p$H glass micro (tip dia. 4–8 $\mu$m) | Ag\|AgCl | 0.001 $M$ $NaHCO_3$ + 0.1 $M$ NaCl | Silicone rubber | (305) |
| | Combined $p$H-ref electrode (Metrohm EA-156-x) | | 0.005 $M$ $NaHCO_3$ + 0.05 $M$ NaCl + 0.1% w/v Triton X-100 | Air gap | (306) |
| $CO_2$ and $O_2$[b] (com - $CO_2$(micro) bined) | $p$H glass for | Ag\|AgCl | hydrogel[c] treated electrode + dried $NaHCO_3$ and NaCl[d] | Polystyrene | (307) |
| $Et_2NH$ | $p$H glass | Not stated | 0.1 $M$ $Et_2NH_2Cl$ | Not stated | (297) |
| $NO_x$ | $p$H glass | Ag\|AgCl or Ag\|AgBr | 0.1–0.02 $M$ $NaNO_2$ + NaCl or KBr | Microporous PTFE or Polypropylene | (298, 297) |
| HCN | $Ag_2S$–AgI (?) | Not stated | K Ag(CN)$_2$ | Not stated | (297) |
| $H_2S$ | $Ag_2S$ (?) | Not stated | Citrate Buffer, $p$H 5 | Not stated | (297) |
| $SO_2$ | $p$H glass | Ag\|AgCl | 0.1–0.01 $M$ $K_2S_2O_5$ or $NaHSO_4$ + NaCl | Silicon, rubber, or microporous PTFE | (297, 298) |

[a]Victawet 12 (Stauffer Chemical Co., U.S.A.).
[b]A voltammetric device.
[c]Hydron Copolymer.
[d]Air dried solutions.

Likewise, the responses of the sensors given in Table 3.6 to the appropriate gases are all described by the general form of equation (3.23)

$$E_{cell} = \text{constant} \pm \text{slope} \log c_{gas} \qquad (3.24)$$

This equation is very similar in form to the general equation (3.17) which is applicable to all cells that use ion-selective membrane electrodes. For a given type of glass electrode, the sign of the slope in equation (3.24) is determined by the type of gas. If acidic, the slope will be opposite in sign to that which would be observed for a basic gas. The magnitude of the slope is about 59 mV decade$^{-1}$ of concentration for most gases, the exception being hydrogen sulfide for which the slope is about half this value.

The response time characteristics of these types of gas sensors have been deduced by Ross and co-workers (297) who have proposed a steady-state model as an aid in predicting the effect of membrane properties on the response time. The equation to the model also predicts that for concentration changes of determinand from low to high concentrations, the response time will be almost independent both of the magnitude of the concentration change and the magnitudes of the initial and final concentrations. Conversely, a change from high to low concentration is accompanied by a much longer response time which is calculated to be some 13 times greater for a determinand concentration change of from $10^{-1}$ to $10^{-5}$ $M$ than for the reverse direction. Experimental verification of the model as it applies to the response time of an ammonia sensor has been obtained (298). For ammonia concentrations in excess of $10^{-4}$ $M$, the response times for tenfold concentration increases were found to be independent of the final concentration of ammonia.

The rate of diffusion of the gas through the membrane establishes the rate of response, and although many more highly volatile compounds than those given in Table 3.6 could, in principle, be determined, some would diffuse so slowly that the response of the sensor would be impracticably sluggish. The volume of electrolyte solution containing the conjugate species with which the diffused gas has to equilibrate also influences the rate of response. Thus a small volume of this solution offering a large surface area to the determinand will expedite the response. To these ends a thin film of the electrolyte solution is squeezed against the flat-ended sensing electrode by the tightly fitting membrane, and sometimes a spacer of tissue or cellophane is interposed between the two surfaces. This arrangement also ensures that the film of the electrolyte solution is kept separate from the bulk of this solution, and yet allows a conducting pathway to be maintained between the two portions.

Two types of membrane are used to separate the thin film of electrolyte solution from the determinand gas: microporous and homogeneous. The microporous type consists of a material, usually PTFE, that provides an inert hydrophobic matrix containing very small pores, normally filled with air, through which the gas diffuses. The rate of diffusion of a gas through this material is several orders of magnitude greater than through homogeneous membranes (297). This is to be expected because homogeneous membranes are solid films of plastics, like polypropylene or silicone rubber, and transport of the determinand through these occurs first by dissolution of the gas in the membrane followed by diffusion of the solute through the organic matrix itself. The responses of probes fitted with homogeneous membranes are, therefore, much slower than those fitted with microporous membranes. However, in contrast to homogeneous membranes which are impermeable to water, microporous membranes become permeable to water when they separate two aqueous solutions of different concentrations. This flow of solvent, caused by a difference in osmotic pressures across the membrane, will result in a drift in sensor e.m.f. unless these osmotic pressures are balanced by the addition of an inert electrolyte either to the sample or to the solution contained in the sensor (299). Osmotic drift becomes particularly pronounced if the total concentration of all species in the sample solution is greater than the corresponding osmotic strength of the sensor solution. In some instances it may be possible to counter this by diluting the sample solution with water, but if this is not possible it could be advantageous to tolerate the slower response given by a probe fitted with an homogeneous membrane.

Another model of gas sensor dispenses with the solid membrane entirely replacing it by a layer of air several millimeters thick which separates the determinand solution from the sensing surface (see Figure 3.10). In these "air gap electrodes" (300, 306) the electrolyte solution containing the species conjugate to the determinand is usually applied to the electroactive surface by means of a sponge before every measurement. A wetting agent added to the electrolyte solution ensures the stability of the resulting liquid film. In behavior they conform to the general theoretical model (297) since they can be regarded as the air membrane analogues of the solid membranes. The air gap probes have a high speed of response in comparison with the solid membrane types and also have the advantage that the sensor does not come into direct physical contact with the sample solution. Thus they are free of the deleterious effects that detergent solutions, particulate matter, and organic solvents can have upon a solid membrane. Furthermore, the electrolyte layer can be renewed or even changed according to the requirements of a particular analysis so that the same electrode can be used for measurements of a variety of gases. In the air gap sensors, however, the thickness of the electrolyte film cannot be controlled, and, because it is in direct contact with atmospheric oxygen, there is a possibility that a species in this film could be oxidized or that the film itself could dry out. Susceptibility to such effects are greatly diminished in the solid membrane sensors which also have the advantage that some control of the liquid film thickness is possible by adjusting the tension of the membrane on the surface of the sensing electrode. The liquid film is susceptible to attack by oxygen in the sulfur dioxide probe and hence a solid membrane sensor should be used in this case.

The type of cell used as a basis for the operation of a potentiometric gas probe is not necessarily confined to a cell reversible to hydrogen ion (see Table 3.6). Ion selective membrane electrodes reversible to silver ion, cyanide ion, and hydrogen sulfide ion have provided the sensing electrode for the ammonia, hydrogen cyanide, and hydrogen sulfide probes, respectively. In these instances, however, the relation between the cell e.m.f. and concentration of the gas is also described by equation (3.24). For example, the equilibrium pertinent to the operation of an ammonia probe when a silver sulfide membrane was used as the sensing surface (303) can be represented as

$$2NH_4^+ + Ag^+ \rightleftharpoons Ag(NH_3)_2^+ + 2H^+$$

When a fluoride ion selective electrode was used as the reference electrode in this probe, the slope was found to be $-100$ mV per decade of ammonia concentration.

### *Enzyme and Biological Sensors*

Enzymes are proteins produced by living cells that can catalyze chemical conversions often with a remarkable degree of specificity even when present at very low concentrations. They act as regulators in many biological systems and participate in a reaction by forming an initial transitory complex with the reacting substrate. This activated complex then decomposes liberating the product and the enzyme reverts to its original form. The nature of a particular enzyme determines whether it has the ability to catalyze the synthesis or breakdown of chemical compounds or, alternatively, whether it can promote the transformation of a given functional group into another. Most enzymes are highly specific in that they will catalyze only one specific reaction or will act upon only one type of isomer of a particular class of compound. For example, the enzyme urease is uniquely specific to the decomposition of urea into ammonia and carbon dioxide, whereas some proteolytic enzymes are exclusively sterospecific in that they will promote the hydrolysis of only those peptide bonds that link L-aminoacids. Some enzymes are somewhat less specific with respect to compound or to a particular optical configuration, but are able to catalyze attack upon a particular linkage. These linkages can be of a particular type or, alternatively, can be present at a particular point in a molecule. Thus lipases will hydrolyze a wide range of organic esters, and some hydrolytic enzymes will promote attack at the center of a molecule and others at the ends. In analytical chemistry there has been a steady growth in the utilization of these unique and powerful catalytic properties in such fields as industrial, medical, pharmaceutical, clinical, and food analysis. Enzyme catalyzed reactions can be used to determine the enzyme itself, the substrate, or any activator or inhibitor of the enzymatic reaction. These analytical applications are reviewed regularly (308, 309).

A limitation to the use of enzymes in analysis has been a loss of catalytic activity that arises as a result of their instability in aqueous solutions. To counter this instability, water soluble enzymes can be made insoluble or *immobilized* by combination with some inert matrix or by chemical reaction. Immobilization of enzymes by chemical reaction involves the chemical modification of the molecules by the introduction of groups that drastically reduce the solubility of the enzyme without adversely affecting its catalytic activity. This is difficult to achieve in practice and it is often faster and simpler to physically entrap the enzyme in the type of inert matrix provided by such materials as starch or polyacrylamide. Either one of these methods results in the retention of catalytic activity for long periods of time. Thus it has been noted that a single sample of immobilized glucose oxidase has been used for several thousand determinations (310).

Methods for the successful immobilization of enzymes has made possible the development of electrochemical sensors that can take advantage of the unique catalytic properties of a particular enzyme to produce sensors that possess a high degree of selectivity. However, not all these devices are potentiometric, and the first to be reported (311) was an amperometric glucose sensor in which the enzyme glucose oxidase was used to catalyze the reaction

$$\beta\text{-D-glucose} + 2H_2O + O_2 \xrightarrow{\text{glucose oxidase}} \text{D-gluconic acid} + 2H_2O_2 \quad (3.25)$$

The change in the magnitude of the current produced when a fixed potential was applied between platinum and silver–silver chloride electrodes was proportional to the oxygen consumed in the reaction and thus related to the concentration of glucose. This type of sensor has important applications since there are at least 47 known oxidases acting, with consumption of oxygen, on a wide range of fatty acids, hydroxy acids, amino acids, purines, pyrimidines, aldehydes, carbohydrates, thiols, phenols, and steroids (312).

Although amperometric, they are included in some reviews dealing with ion selective electrodes (313), but are not discussed further here. However, reactions similar to (3.25) can be followed by an indirect potentiometric method. Thus an iodide-selective membrane electrode has been used to monitor reactions such as

$$\text{L-amino acid} + O_2 \xrightarrow[\text{oxidase}]{\text{L-amino acid}} H_2O_2 + \text{ketoacid} + NH_3$$

$$\text{alcohol} + O_2 \xrightarrow[\text{oxidase}]{\text{alcohol}} H_2O_2 + \text{aldehyde}$$

The hydrogen peroxide formed is determined indirectly by measuring the diminution of a known concentration of iodide in the molybdate-catalyzed reaction:

$$3I^- + H_2O_2 + 2H^+ \xrightarrow{MoO_4^{2-}} I_3^- + 2H_2O$$

This sequence of reactions has been used to determine L-amino acids and alcohols by a continuous flow method in which the iodide sensing surface is incorporated into a flow-through cell (314). In such a reaction sequence, an electrode can be used without modification to indicate the activity of the ion in the bulk solution, and the fact that the activity of that ion is influenced indirectly by the products derived from an enzymatic reaction is only incidental insofar as the potentiometric measurement is concerned. A true enzyme sensor, on the other hand, requires some direct modification to the membrane either of a gas probe or to the membrane of an existing type of ion selective electrode. This modification is accomplished by attaching to the sensing surface an additional coating. If the sensor is to be used for the determination of a substrate, this coating will contain the enzyme and, conversely, if it is to be used to measure the activity of an enzyme, the coating will contain the substrate. In either case, reaction between the enzyme and the substrate yields the species normally sensed by the ion selective electrode or the gas sensor. There is no restriction as to the type or configuration of ion selective membrane electrodes that can be used for this application. Thus electrodes fitted with sensing membranes of glass, insoluble inorganic salts, organic ion exchangers, or organic neutral carriers have been used either in the solid-state configuration or as electrodes fitted with an internal reference half-cell. Similarly, there is no restriction as to the type of gas sensor; both the membrane (microporous or solid) and the air gap types have been successfully converted into enzyme sensors. In the air gap version the enzyme can be immobilized either as a coating on the internal surface of the sample cup (315), or as a coating on the bar of the magnetic stirrer (316).

The immobilization of the enzyme against the sensing surface was achieved in some of the earlier enzyme sensors merely by holding a film of a $p$H-stabilized enzyme solution in position by means of a dialysis membrane (317). A more recent method, however, aims at producing a thin cross-linked film containing the enzyme *in situ* upon the sensing surface of a probe (318). This is achieved by spreading a small volume of a solution containing bovine serum albumin over the Teflon membrane of a gas sensing probe and dissolving *in situ* a few milligrams of the enzyme preparation in this solution. Addition of a solution of pentanedial (glutaraldehyde) produces a cross-linked solid film within about 10 min. Adhesion of the film directly to the Teflon membrane, however, was found to be only temporary. Hence it is preferable to position a rubber ring around the circumference of the membrane but flush with the plane of the membrane. During the course of the enzyme film-forming process, the solutions wet this ring and the subsequent adhesion of the film to the ring greatly reinforces the attachment of the film to the Teflon. Enzymes can also be entrapped in polyacrylamide gel (319, 320). Thus a silicone membrane containing nonactin (structure I, Figure 3.9), which formed the sensing surface of an ammonium ion selective electrode, was modified by the attachment of a thin disc of the enzyme urease immobilized in polyacrylamide to produce an electrode sensitive to urea (321). The thin discs of urease-polyacrylamide gel were cut from a gel bar housed in a glass tube (5 mm i.d.) which was fitted with a plunger. By means of the plunger a small portion of the gel could be extruded from which a thin slice could be cut by means of a wet razor blade. The gel was prepared by the light-initiated polymerization of the enzyme-monomer solution (319, 320) using $N$, $N'$-methylenebisacrylamide (0.58 g), acrylamide (5 g), potassium peroxydisulfate (3 mg $K_2S_2O_8$) and riboflavin (3 mg) in $p$H 7 buffer solution (prepared from 25 mL 0.1 M Tris*). A portion of this solution (1 mL) was used to treat 175 mg urease. Once the thin disk of the immobilized urease has been applied to the sensing surface of the electrode, it is held in place by a dialysis membrane. When the enzyme electrode dips into a solution containing urea, ammonium ion is liberated in the disk containing the immobilized enzyme according to the reaction

$$CO(NH_2)_2 + H_2O \xrightarrow{\text{urease}} CO_3^{2-} + 2NH_4^+ \qquad (3.26)$$

*Tris is the synonym for 2-amino-2-hydroxymethyl-1,3-propanediol.

An equilibrium potential with respect to the ammonium ion formed is recorded when the rates of diffusion of urea through the membrane and its reaction in the enzyme layer reaches a steady state. Enzyme sensors based on reaction (3.26) have also been prepared by immobilizing urease on the membranes of carbon dioxide or ammonia gas sensors. The equilibration time when these are incorporated into enzyme probes is slow because it is governed by the rates of diffusion of both products and substrate through the gel and then by the gas itself through the membrane attached to the sensing electrode of the gas probe. This time is much reduced in the air gap sensors when the enzyme can be immobilized on a nylon net affixed to the bar stirrer. The sensitivities of most types of gas sensors depend, however, upon the $pH$ of the solution from which the volatile determinands are to be evolved, maximum sensitivity of the ammonia probe, for example, being observed when the $pH$ of the substrate is greater than 12. The activity of enzymes is also highly dependent upon the $pH$ of the substrate, which generally falls within the range $pH$ 6–8. Outside this range the activities of enzymes are usually totally inhibited. A balance must, therefore, be achieved between the $pH$ for maximum enzyme activity and the $pH$ for the maximum sensitivity of the gas sensor, and the resulting compromise has to favor the activity of the enzyme over the sensitivity of the sensor. In these instances the sensitivity of the enzyme sensor itself is determined by the sensitivity of the gas sensor at the chosen $pH$. The effect of $pH$ upon the sensitivity of an ion selective electrode, with the exception of a glass electrode, is much less marked than with a gas sensor. However, ions that cause interference in the normal operation of the electrode will also interfere in the operation of the electrode when an enzyme containing layer is fitted to its sensing surface. Thus the early urea electrodes, made by coating ammonium ion sensitive glass electrodes (322), were not only sensitive to $pH$ changes but also were susceptible to interferences from sodium and potassium ions which, of course, are present in clinical samples. All these factors should be borne in mind when selecting either an ion selective electrode or a gas sensor as the basis for a potentiometric enzyme sensor. A less common effect that has been reported is the deterioration of the sensing surface of the electrode itself caused by the presence of the enzyme containing layer. Thus a drastic decrease in the hydrogen ion sensitivity of a $pH$ glass electrode was noted when the sensing surface was coated with penicillinase immobilized in polyacrylamide (323). The resultant electrode responded to other cations, particularly sodium ion, probably as a result of penicillinase adsorption on the glass surface during polymerization or by leaching during electrode equilibration. Thus the electrode response to the sodium salt of ampicillin [*CA* Registry No. *69-53-4*] was largely due to the sodium ion rather than the enzymatic hydrolysis of this penicillin to form a penicilloic acid. It is the decrease of $p$H caused by the release of this acid that the glass electrode is expected to monitor. This effect was overcome by interposing a fritted glass disk, upon which the penicillinase was adsorbed, between the sensing surface of the glass electrode and the sample solution (323). Following the publication of a method for the immobilization of penicillinase on glass beads by the covalent linkage of the enzyme to the glass (324), a modified penicillin electrode has been reported which utilizes this method for the immobilization of the enzyme directly on the sensing surface of a glass electrode (325). The potential changes caused by potassium and sodium chlorides on this electrode are mainly accounted for by ionic strength effects.

An extension of the concept of enzyme sensors is the development in recent years of *bioselective sensors*, or *biological electrodes* as they are sometimes called. In these sensors living bacterial cells which, of course, secrete enzymes are held against the sensing surface of a gas probe or an ion selective electrode usually as a moist paste secured by a dialysis membrane. For example L-aspartate can be determined using the enzyme L-aspartase [E.C.* 4.3.1.1.] held against the surface of an ammonia probe (317) or, alternatively, it may be determined using cells of *Bacterium cadaveris* (ATCC† No. 9760) also held against the surface of an ammonia probe (326). In both cases the coating on the probe reacts with the substrate containing L-aspartate to produce ammonia. However, the enzyme coating will lose its activity in time, whereas the coating of living bacteria can be replenished by storing the probe in a suitable growth medium. A disadvantage in the use of living cells is that they can contain enzymes that produce the sensed species, ammonia in this case, from substrates other than the determinand substrate. Thus in the determination of L-arginine using an ammonia probe coated with an air-dried paste containing *Streptococcus faecium* (ATCC No. 9790), the selectivity is inferior to that observed when the probe is coated with the sole enzyme, arginine deaminase [E.C. 3.5.3.6] (327). This is because the bacteria cells, in addition to containing arginine deaminase, also contain other enzymes that produce ammonia from L-asparagine and L-glutamine, making these substances interferences. Sensitivity of the probe toward L-arginine is increased, however, because *Streptococcus faecium* is also a source of ornithine trans carbamylase and carbamate kinase. These too can enter into a reaction sequence initiated by arginine deaminase producing

---

*Enzyme Commission.
†American Tissue Culture Collection.

ammonia. Thus

$$\text{L-arginine} \xrightarrow{\text{arginine deaminase}} \text{citrulline} + NH_3$$

$$\text{citrulline} + H_3PO_4 \xrightarrow{\text{ornithine trans carbamylase}} \text{ornithine} + \text{carbamoyl phosphate}$$

$$\text{carbamoyl phosphate} + ADP \xrightarrow{\text{carbamate kinase}} \text{carbamic acid} + ATP$$

$$\text{carbamic acid} \rightarrow CO_2 + NH_3$$

This bonus yield of ammonia acts, in this case, to enhance the sensitivity with respect to the sensor based on arginine deaminase alone.

The comments made about the effects of $pH$ and interferences upon the base sensors (i.e., gas probe or ion selective electrode) to which enzyme layers are attached apply equally to sensors utilizing living cells. Substrates are usually buffered with respect to $pH$ to ensure the maximum catalytic activity and both pure enzymes and, to a greater extent, bacteria can be adversely affected by the type of the substances used in the preparation of a particular buffer. If a rapid loss in the sensitivity of a sensor is encountered, the incompatibility of a buffer component could be a cause.

Zeroth electrodes either of the conventional platinum type (328), or based upon a solid membrane containing dibutylferrocene (329) have been used as enzyme sensors. Both these sensors depend upon a change in the ratio of activities for the oxidation-reduction couple hexacyanoferrate(III)-hexacyanoferrate(II) as being indirectly indicative of the activity of a determinand substrate. Thus a platinum net electrode coated with the enzyme tyrosinase [E.C. 1.14.18.1] entrapped in a polyacrylamide gel was used to determine phenol according to the reaction sequence

$$\text{phenol} \xrightarrow{\text{tyrosinase}} 1,2\text{-benzenediol} \xrightarrow{O_2} 3,5\text{-cyclohexadiene-1,2-dione}$$

The product, formed in the presence of a saturating level of oxygen, is chemically reduced by hexacyanoferrate(II) to 1,2-benzenediol thus causing a change in the activity ratio of the oxidation-reduction couple. A change in potential emanating from a change in this activity ratio also formed the basis for the determination of lactate (329) using the reaction

$$\text{lactate} + 2Fe(CN)_6^{3-} \xrightarrow[\text{[E.C. 1.1.1.27]}]{\text{lactate dehydrogenase}} \text{pyruvate} + 2Fe(CN)_6^{4-} + 2H^+$$

In this case, however, the enzyme preparation (see Table 3.7) was applied to a PVC membrane containing dibutylferrocene prepared by dissolving PVC (250 mg), dioctylphthalate (500 mg), and dibutylferrocene (300 mg) in tetrahydrofuran (10 ml). Evaporation of this solution contained in a Petri dish (60 cm$^2$) at room temperature produced a thin film (0.1–0.15 mm thick) of membrane material.

Ion selective electrodes have also provided the sensing surfaces which have been used in the development of other types of biological sensors operating on different phenomena. For example, a toad bladder has been used to prepare a sensor, based on a sodium ion-responsive glass electrode, with which to determine the antidiuretic hormone ADH (330). The bladder separates a mucosal solution having an osmolality of about 380 m osmal L$^{-1}$, held against the sensing surface of the electrode, from a serosal solution containing the same concentration of sodium ion but of lower osmolality (~ 250 m. osmol L$^{-1}$). In the absence of the hormone no water can flow across the bladder. However, a property of ADH is that it allows a flow of water in either direction across the bladder, but, in this instance, the difference in osmolality ensures that this flow occurs unidirectionally from the serosal to the mucosal solution. This flow reinforces a diminution in sodium ion activity in the mucosal solution that also occurs in the presence of ADH. This diminution of sodium ion activity around the sensing electrode can be related to the concentration of the hormone.

A recent innovation to the field of clinical chemistry is the use of ion selective electrodes and gas sensors in the rapidly growing field of enzyme immunoassays (331, 332). Preliminary work suggests (333) that these methods could be more sensitive and convenient than radioimmunoassays in the accurate determination of antigens and antibodies present in trace amounts in biological fluids. In the determination of an antigen an appropriate enzyme is used as a label, being linked to the corresponding antibody as an enzyme conjugate. When the antigen is added to a solution of the enzyme substrate in the presence of the enzyme–antibody conjugate, the enzyme-catalyzed reaction of the substrate occurs. If, however, the enzyme–antibody conjugate is incorporated into a membrane attached to the sensing surface of an electrode and if the substrate releases a species to which the electrode responds, the change in concentration of the sensed-species on addition of the antigen produces a change in electrode potential proportional to the concentration of antigen. For example, the hepatitis B surface antigen was determined (333) using an iodide ion selective membrane electrode as the base sensor. The enzyme, horseradish peroxidase [E.C. 1.11.1.7], was linked to rabbit anti-hepatitis B surface antigen to yield the enzyme–antibody conjugate which was incorporated into a proteic

Table 3.7. Some Enzymes and Biological Sensors

| Base Electrode/ Sensor | Substrate | Enzyme/ Bacteria | Buffer | pH | Remarks | Ref. |
|---|---|---|---|---|---|---|
| Ammonia (membrane type) | 5'-Adenosine monophosphate | 5'-Adenylic acid deaminase [EC 3.5.4.6] | Tris$^a$ (0.05 $M$) | 7.5 | Enzyme solution (10 μL) held in place against sensor membrane by a cellophane dialysis membrane | (336) |
| | L-Asparagine | L-Asparaginase [EC 3.5.1.1] | Tris | 7.0 | Enzyme immobilized on 210-mesh nylon net | (337) |
| | Creatinine | Creatininase [EC 3.5.4.21] | Tris(0.15 $M$)-phosphate (0.025 $M$) | 8.5 | Enzyme solution (15 μL) placed between cellophane dialysis membrane and the membrane of the sensor | (338) |
| | L-Methionine | Methioninelyase [EC 4.4.1.11] | Pyrophosphate (0.05 $M$) | 8.7 | Enzyme immobilized by dissolving in a 15% solution of bovine serum albumin followed by the addition of 12.5% pentanedial. 22 other common amino acids, including D-methionine, did not interfere | (339) |
| | Primary amine hydrochlorides$^b$ | Monoamine oxidase [EC 1.4.3.4] | Phosphate | 8.2 | Enzyme activity determined. The amount of NH$_3$ generated is obtained by calculating the difference in NH$_3$ concentration in control and test incubation mixtures. NH$_3$ determined at $p$H 12. | (340) |
| | Urea | Urease [EC 3.5.1.5] | Tris(0.1 $M$) + EDTA (10$^{-3}$ $M$) | 7.0 | Enzyme solution held in place against sensor membrane by a cellophane dialysis membrane | (317) |
| | L-Glutamine | Porcine kidney tissue contains glutaminase [EC 3.5.1.2] | Phosphate (0.1 $M$) | 7.8 | Thin slice of tissue (0.05 mm) held between a monofilament nylon mesh and a dialysis membrane against the sensor surface | (341) |
| | Nitrate | *Azotobacter vinelandii* | Phosphate (0.2 $M$) + sucrose (1%) + INH$^c$ (0.1 $M$) | 7.4 | Nitrate reduced via nitrite to NH$_3$. Wet bacterial cells (5–10 mg) spread onto sensor membrane and held in place by dialysis membrane (0.1 μm pore size) | (342) |
| | L-Glutamine | Bacterium *Sarcina flava* | Tris + MnCl$_2$ (0.01 $M$) | 7.5 | Cells held in position by dialysis membrane | (343) |

| Analyte | Enzyme | Buffer | pH | Notes | Ref. |
|---|---|---|---|---|---|
| | (ATCC 9760) Also used L-Aspartase [EC 4.3.1.1] | $(0.002\ M)$ Tris $(0.1\ M)$ + EDTA $(0.001\ M)$ | | membrane by a dialysis membrane. Correct storage can materially increase life of biological sensor. Storage in the growth medium replenishes supply of active cells *in situ* | |
| L-Arginine | *Streptococcus faecium* (ATCC 9790) | Phosphate $(0.1\ M)$ | 7.4 (28°) | Washed bacterial cells applied to sensor membrane as a paste which, after drying in air, was held in place by a dialysis membrane. Selectivity is not as good as that obtained with arginine deaminase [EC 3.5.3.6] alone | (327) |
| Protein antigen (BSA)[d] Nucleotide (cAMP)[e] | Urease as an enzyme label | Tris + EDTA $(I = 0.1)$ | 7.5 | An enzyme immunoassay for BSA[d] and cAMP[e] in which sensor is used to monitor urease activity | (334) |
| Ammonia (air gap type) | Nitrite reductase [EC 1.6.6.4] | Tris + methyl viologen[f] | 7.0 | Air gap sensor based on glass electrode-silver|silver chloride cell | (344) |
| | L-phenyl-alanine ammonia lyase | Phosphate $(0.1\ M)$ | 7.0 | As above. The solution applied to bulb of electrode was $0.005\ M\ NH_4Cl$ + Victawet 12 | (345) |
| L-phenyl-alanine | Urease [EC 3.5.1.5] | Tris | 8.5 | As above. Immobilized enzyme held by nylon netting in the cup | (315) |
| Urea | | | | Enzyme held by net on magnetic stirring bar | (316) |
| Urea | Urease [EC 3.5.1.5] | Tris | 7.0 | A thin film disc of enzyme entrapped in polyacrylamide gel attached to sensing surface of the electrode | (321) |
| Ammonium ion (monactin type) Carbon dioxide | Gluconate kinase [EC 2.7.1.12] 6-Phospho-D-gluconate dehydrogenase [EC 1.1.99.3] | Glycylglycine $(0.1\ M)$ $MgCl_2 (0.015\ M)$, ATP[g], NADP[h] (both $0.0085\ M$) | 7.0 (37°) | Two-step enzyme sequence. Mixture of enzymes absorbed on a fibre spacer which was held against sensor membrane by a dialysis membrane. Immobilization of the two enzymes by cross-linking with BSA[d] and pentanedial was also used | (346) |
| D-Gluconate | | | | | |
| L-Lysine | L-Lysine decarboxylase [EC 4.1.1.18] | Acetate $(0.5\ M)$ | 5.8 | Enzyme immobilized by cross-linking with BSA[d] and pentanedial | (318) |
| Uric acid | Uricase [EC 1.7.3.3] | Phosphate $(0.1\ M)$ | 6.5 | Sensor membrane coated with enzyme slurry which was held in place with moistened cellophane | (355) |

(*continued*)

Table 3.7. (Continued)

| Base Electrode/ Sensor | Substrate | Enzyme/ Bacteria | Buffer | pH | Remarks | Ref. |
|---|---|---|---|---|---|---|
| | Urea with and without $F^-$ determinand | Urease [EC 3.5.1.5] | Phosphate (0.25 M) | 7.0 | Enzyme immobilized by BSA$^d$-pentanedial method (see above) on silicone membrane of sensor. Urea (0.01 M) in the buffer solution was monitored to a constant value of $E_{cell}$. Departure of cell e.m.f. on addition of $F^-$ inhibitor related to inhibitor concentration | (347) |
| Glass (pH type) | Acetylcholine | Acetylcholine-sterase [EC 3.1.1.7] | Medium of 0.01 M NaCl | 8.0 | Coating of enzyme (50 μm thick) applied to electrode as a film in gelatin. After air drying (25°) this coating was immersed in 2.5% pentanedial (pH 6.8) for 15 min. Enzymatic reaction produces $H^+$ in an amount related to substrate concentration | (348) |
| | Penicillins | Penicillin β-lactamase [EC 3.5.2.6] | $K_2HPO_4$ (0.0004 M) in KCl (0.1 M) | 7.0 | Enzyme immobilized by covalent linkage on to glass beads contained in a column. Passage of solution containing determinand through column causes an increase in $H^+$. Used as a basis for a flowing stream analyzer | (324) |
| | | As above | | | Enzyme immobilized directly on glass electrode by the method given in the above reference | (325) |
| | Benzyl penicillin | As above | Sodium phosphate (0.01 M) | 7.0 | Enzyme solution (100 μL of 50 mg mL$^{-1}$) held against glass electrode by dialysis membrane. Linear relation between $\Delta pH$ and penicillin concentration | (349) |
| | Cephalosporins | *Citrobacter freundii* | Phosphate (0.0005 M) | 7.2 | Cells immobilized in a collagen membrane which was rolled up in a plastic net and contained in a cylindrical reactor through which buffer was pumped. Addition of determinand into flowing stream caused a diminution of pH. Maximum $\Delta pH$ proportional to concentration | (350) |
| | Nicotinic acid | *Lactobacillus arabinosus* (ATCC 8014) | 24-component assay medium | 6.8 | Combined pH-reference electrode used to measure difference in potential between assay medium containing the immobilized bacteria with and without nicotinic acid | (351) |

| Electrode | Analyte | Biological component | Buffer | pH | Notes | Ref. |
|---|---|---|---|---|---|---|
| Glass ($p$Na type) | Antidiuretic hormone (ADH) | Toad bladder | Serosal and mucosal solutions | 8.0 | ADH reduces $Na^+$ concentration in mucosal solution held against electrode surface | (330) |
| Hydrogen sulfide | L-Cysteine | *Proteus morganii* (ATCC 8019), source of cysteine desulfhydrase [EC 4.4.1.1.] | Phosphate (0.1 $M$) | 6.75 (37°) | Bacteria held against the sensor membrane using a dialysis membrane. Poor discrimination between $H_2S$ and $CO_2$ is a limitation | (352) |
| Iodide ion | L-Aminoacids Alcohols | L-Aminoacid oxidase [EC 1.4.3.2] Alcohol oxidase [EC 1.1.3.13] | Phosphate Phosphate | 6.5– 8.4[i] 7.5 | Flow through cell containing iodide ion sensing surface used to monitor changes in iodide ion concentration due to production of $H_2O_2$ which reacts with $I^-$ in the presence of a molybdate catalyst | (315) |
| | Thyroxine | Rat liver microsomes | Citrate (0.05) containing 0.0002 $M$ $Fe(NH_4)_2(SO_4)_2$ | 3.0 | Electrode placed in buffer solution containing thyroxine. Microsomes added and cell e.m.f. measured after 30 min incubation. $E_{cell}$ proportional to $-\log c$ | (353) |
| | Hepatitis B surface antigen/antibody | Horseradish peroxidase [EC 1.11.1.7] used as an enzyme label | Sodium phosphate (0.1 $M$) | 5.0 | An enzyme immunoassay for hepatitis B surface antigen | (333) |
| Zeroth (Platinum) | Phenol | Tyrosinase [EC 1.14.18.1] | $KH_2PO_4$ (0.25 $M$), $Fe(CN)_6^{4-}$ (0.02 $M$) | 6.5 | Enzyme entrapped in acrylamide gel onto a Pt screen electrode. Ratio Fe(III):Fe(II) altered by phenol | (328) |
| (Dibutylferrocene (in PVC membrane) | Lactate | Lactate dehydrogenase [EC 1.1.1.27] | Phosphate (0.1 $M$), EDTA (0.001 $M$), $MgSO_4$ (0.0001 $M$), $K_3FE(CN)_6$ – $K_4FE(CN)_6$ (4:1) | 6.0 | Enzyme + catalase [EC 1.11.1.6] were dissolved in buffer, mixed with gelatin, and applied to PVC membrane containing dibutylferrocene. The enzyme coating was held in place by a dialysis membrane. Lactate alters ratio Fe(III):Fe(II) and changes the cell potential measured with respect to a Ag-AgCl reference electrode | (329) |

[a]Tris: 2-Amino-2-hydroxymethyl-1,3-propanediol. [b]For example, the hydrochlorides of 4-(2-aminoethyl)phenol (tyramine), 3-(2-aminoethyl)-1H-indol-5-ol(serotonin), 2-(4-imidazolyl)ethylamine (histamine), and 1,5-pentanediamine (cadaverine). [c]TNH: Isonicotinic hydrazide (*CA* Registry No. 54-85-3). [d]BSA: Bovine serum albumin. [e]cAMP: Cyclic adenosine-3′,5′-monophosphoric acid (*CA* Registry No. 60-92-4). [f]Methyl viologen: 1,1′-Dimethyl-4,4′-dipyridinium dichloride. [g]ATP: Adenosine-5′-triphosphate (*CA* Registry No. 56-65-6). [h]NADP: Nicotinamide adenine dinucleotide phosphate (*CA* Registry No. 53-59-8). [i]Depending upon the aminoacid being determined.

membrane by a "sandwich principle." First, a membrane was prepared from the unlabeled antibody (i.e., anti-hepatitis B surface antigen) fixed on gelatin. A 10 mm dia. disc of this membrane (about 0.05 mm thick) was soaked initially in a solution of the unlabeled antibody (1 mg $L^{-1}$), then immersed in a very dilute solution of antigen, and finally incubated with a solution of the enzyme-labeled antibody (100 $\mu$g $mL^{-1}$) being washed well with water after each of the first two soakings. After incubation, the membrane was washed with $p$H 5.0 phosphate buffer (0.1 $M$) and then used to cover the sensing surface of the electrode being held in position by a rubber ring. The enzyme activity in the presence of a substrate containing hydrogen peroxide (0.0044 $M$), potassium iodide (0.0001 $M$), and the biological fluid at $p$H 5 was measured utilizing the reaction

$$H_2O_2 + 2I^- + 2H^+ \xrightarrow[\text{peroxidase}]{\text{horseradish}} I_2 + 2H_2O$$

The enzyme activity is indicated by the diminution in the concentration of iodide, and this activity is a function of the antigen present in the biological fluid. Reproducible variations of the electrode potential changes, measured with respect to an Orion 9001 reference electrode, were produced by antigen concentrations in the range 0.5–50 $\mu$g $L^{-1}$. An enzyme immunoassay using an ammonia gas probe as the base sensor has also been described (334).

The enzyme sensors so far discussed in this section have been used to relate an observed equilibrium potential to the concentration of a determinand. In the assay of enzyme activities an alternative method has been reported (335) that relates the observed initial rate of change in the potential of an enzyme electrode directly to the activity of the enzyme. Thus the rate of change that occurs immediately after the addition of the enzyme to the substrate has been found to be directly proportional to the activity of the enzyme. The enzyme sensors used in these kinetically based methods are prepared by one of the methods already outlined in this section. The method is particularly advantageous when coupled enzyme reactions are necessary because it obviates the necessity for the expensive second or coupling enzyme.

Table 3.7 contains a summary of some enzyme and biological sensors that have been developed since 1975 arranged in alphabetical order on the base electrode or sensor. Although a distinction has been made between an enzyme sensor and an unmodified ion selective electrode or gas probe used to monitor reactions catalyzed by enzymes, some of the latter applications are included in the table. A tabulated summary of sensors developed before 1975 is contained in Reference 354.

## Chemically Sensitive Semiconductor Devices (CSSDs)

The ion selective field effect transistor (ISFET), a modified metal oxide semiconductor field effect transistor (MOSFET), is the major type of CSSD which is usually included in classifications of ion selective electrodes (313). In these devices the metal gate of a conventional MOSFET is replaced by an ion selective membrane and the potential developed across this membrane regulates indirectly the current flowing through the device. The magnitude of this current is proportional to the logarithm of the activity of the ion sensed by the membrane. However, since the potential of the cell containing the device is not the measured quantity, it must be excluded from further discussion here on the same grounds that prevented description of the amperometric glucose sensor. A good entry point into the literature of CSSDs is provided by Reference 313.

### REFERENCES

1. V. E. Petrakovich, *Zh. Anal. Khim.* **18**, 1161 (1963).
2. V. P. Apte and R. G. Dhaneshiwar, *Talanta* **13**, 1595 (1966).
3. D. J. G. Ives and G. J. Janz, Eds., *Reference Electrodes*, Academic, New York, 1961; see also G. J. Janz and D. J. G. Ives, *Ann. N.Y. Acad. Sci.* **148**, 210 (1968).
4. A. K. Covington, in *Ion Selective Electrodes*, R. A. Durst, Ed., Chapter 4, N.B.S. Spec. Publ. No. 314, Washington, D.C., 1969.
5. R. G. Bates and J. B. Macaskill, *Pure Appl. Chem.* **50**, 1701 (1978).
6. T. Biegler and R. Woods, *J. Chem. Educ.* **50**, 604 (1973).
7. R. G. Bates, *Determination of pH Theory and Practice*, 2nd ed., Wiley, New York, 1973, p. 289.
8. G. J. Hills and D. J. G. Ives, in *Reference Electrodes*, D. J. G. Ives and G. J. Janz, Eds., Chapter 2, Academic, New York, 1961, p. 102.
9. J. H. Hildebrand, *J. Am. Chem. Soc.* **35**, 847 (1913).
10. R. G. Bates, *Determination of pH Theory and Practice*, 2nd ed., Wiley, New York, 1973, p. 294.
11. R. G. Bates, *Determination of pH Theory and Practice*, 2nd ed., Wiley, New York, 1973, p. 282.
12. J. Caudle, *Ph.D. Thesis*, University of Durham, 1964; quoted in Ref. 4, p. 114.
13. R. G. Bates, *Determination of pH Theory and Practice*, 2nd ed., Wiley, New York, 1973, p. 293.
14. M. H. Peterson and R. E. Groover, *Mater. Prot. Performance* **11**, 19 (1972).
15. G. D. Pinching and R. G. Bates, *J. Res. Natl. Bur. Stand. (U.S.)*, **37**, 311 (1946).

16. R. G. Bates, *Determination of pH Theory and Practice*, 2nd ed., Wiley, New York, 1973, p. 453.
17. Ref. 3, p. 206.
18. H. L. Koppel and A. E. Gealt, *Advan. Instrum.* **27** (Pt. 3), 719 (1972).
19. R. G. Bates, in *Ion Selective Electrodes*, R. A. Durst, Ed., N. B. S. Spec. Publ. No. 314, Washington, D.C., 1969, p. 421.
20. E. A. Ostrovidov and E. A. Bardin, *Zavod. Lab.* **38**, 1327 (1972).
21. J. Keraudy, *C. R. Hebd. Seances Acad. Sci.*, Ser. C **269**, 1241 (1969).
22. T. S. Light, *Anal. Instrum.* **8**, VIII-1, 1 (1970).
23. H. K. Fricke, *Zucker* **14**, No. 7 (1961).
24. Ref. 4, p. 123.
25. G. Mattock, *pH Measurement and Titration*, Heywood, London, 1961, p. 153.
26. P. L. Bailey, *Analysis with Ion Selective Electrodes*, Heyden, London, 1976, p. 22.
27. C. R. Merril, *Nature (London)* **192**, 1087 (1961).
28. T. P. Dirkse, *J. Electrochem. Soc.* **109**, 173 (1962).
29. L. V. Gregor and K. S. Pitzer, *J. Am. Chem. Soc.* **84**, 2671 (1962).
30. Ref. 3, p. 335.
31. G. J. Samuelson and D. J. Brown, *J. Am. Chem. Soc.* **57**, 2711 (1935).
32. Ref. 4, p. 123.
33. A. K. Covington and K. V. Srinivasan, *J. Chem. Thermodyn.* **3**, 795 (1971).
34. R. G. Bates, *Determination of pH Theory and Practice*, 2nd ed., Wiley, New York, 1973, p. 36.
35. P. L. Bailey, *Analysis with Ion Selective Electrodes*, Heyden, London, 1976, p. 28.
36. R. G. Picknett, *Trans. Faraday Soc.* **64**, 1059 (1968).
37. J. V. Leyendekkers, *Anal. Chem.* **43**, 1835 (1971).
38. R. G. Bates, *Determination of pH Theory and Practice*, 2nd ed., Wiley, New York, 1973, p. 54.
39. H. Pallmann, *Koll.-Chem. Beih.* **30**, 334 (1930).
40. R. G. Bates, *Determination of pH Theory and Practice*, 2nd ed., Wiley, New York, 1973, p. 322.
41. J. A. R. Kater, J. E. Leonard, and G. Matsuyama, *Ann. N.Y. Acad. Sci.* **148**, 54 (1968).
42. S. McLeod, H. C. T. Stace, B. M. Tucker, and P. Bakker, *Analyst (London)* **99**, 193 (1974).
43. N. P. Finkelstein and E. T. Verdier, *Trans. Faraday Soc.* **53**, 1618 (1957).
44. D. J. Alner and J. J. Greczek, *Lab. Practice* **14**, 721 (1965).
45. D. J. Alner, J. J. Greczek, and A. G. Smeeth, *J. Chem. Soc. (A)* **1967**, 1205.
46. G. Mattock, *pH Measurement and Titration*, Chapter 8, Heywood, London, 1961.
47. P. L. Bailey, *Analysis with Ion Selective Electrodes*, Heyden, London, 1976, p. 24.
48. J. Jackson, *Chem. Ind. (London)* **1969**, 272.
49. R. G. Bates, *Determination of pH Theory and Practice*, 2nd ed., Wiley, New York, 1973, p. 318.

## REFERENCES

50. A. Albert and E. P. Serjeant, *The Determination of Ionization Constant*, 3rd ed., Chapman and Hall, London, 1984, p. 43.
51. F. J. C. Rossotti and H. Rossotti, *The Determination of Stability Constants*, McGraw-Hill, New York, 1961, p. 146.
52. N. C. Cahoon, *Electrochem. Technol.* **3**, 3 (1965).
53. B. Kratochvil, *Anal. Chem.* **50**, 153R (1978).
54. J. Koryta, *Anal. Chim. Acta* **91**, 1 (1977).
55. J. S. Fritz, *Acid-base Titrations in Non Aqueous Solvents*, Allyn & Bacon, Boston, 1973, p. 62.
56. B. Gutbezahl and E. Grunwald, *J. Am. Chem. Soc.* **75**, 565 (1953).
57. R. C. Larson, R. T. Iwamoto, and R. N. Adams, *Anal. Chim. Acta* **25**, 371 (1961).
58. J. F. Coetzee and J. R. Padmanabhan, *J. Phys. Chem.* **66**, 1708 (1962).
59. L. W. Marple and J. S. Fritz, *Anal. Chem.* **34**, 796 (1962).
60. R. H. Cundiff and P. C. Markunas, *Anal. Chem.* **28**, 792 (1956).
61. S. Bruckenstein and I. M. Kolthoff, *J. Am. Chem. Soc.* **78**, 2974 (1956).
62. I. M. Kolthoff and T. B. Reddy, *Inorg. Chem.* **1**, 189 (1962).
63. C. K. Mann, *Electroanal. Chem.* **3**, 57 (1969).
64. J. Cihalik and J. Simek, *Collect. Czech. Chem. Commun.* **23**, 615 (1958)
65. W. B. Mather, Jr. and F. C. Anson, *Anal. Chim. Acta* **21**, 468 (1959).
66. R. A. Glenn, *Anal. Chem.* **25**, 1916 (1953).
67. H. W. Salzberg and M. Leung, *J. Org. Chem.* **30**, 2873 (1965).
68. C. A. Streuli, *Anal. Chem.* **30**, 997 (1958).
69. V. Plichon, *Bull. Soc. Chim. Fr.* **1964**, 282.
70. P. Arthur and H. Lyons, *Anal Chem.* **24**, 1422 (1952).
71. A. P. Kreshkov, A. N. Yarovenko, and E. M. Etingova, *Zavod. Lab.* **40**, 1068 (1974).
72. B. Kratochvil, E. Lorah, and C. Garber, *Anal. Chem.* **41**, 1793 (1969).
73. O. Bravo and R. T. Iwamota, *J. Electroanal. Chem. Interfacial Electrochem.* **23**, 419 (1969).
74. J. C. Synnott and J. N. Butler, *Anal. Chem.* **41**, 1890 (1969).
75. C. W. Manning and W. C. Purdy, *Anal. Chim. Acta* **51**, 124 (1970).
76. E. L. Johnson, K. H. Pool, and R. E. Hamm, *Anal. Chem.* **38**, 183 (1966).
77. J. Courtot-Coupez and M. Le Demezet, *Bull. Soc. Chim. Fr.* **1967**, 4744.
78. Ref. 3, p. 193.
79. G. Gran and B. Althin, *Acta. Chem. Scand.* **4**, 967 (1950).
80. W. B. Schaap, R. E. Bayer, J. R. Siefker, J-Y. Kim, P. W. Brewster, and F. C. Smith, *Record Chem. Prog.* **22**, 197 (1961).
81. L. A. Knecht and I. M. Kolthoff, *Inorg. Chem.* **1**, 195 (1962).
82. G. Cauquis and D. Serve, *Bull. Soc. Chim. Fr.* **1966**, 302.
83. E. Sutzkover, Y. Nemirovsky, and M. Ariel, *J. Electroanal. Chem. Interfacial Electrochem.* **38**, 107 (1972).
84. B. Burrows and R. Jasinski, *J. Electrochem. Soc.* **115**, 365 (1968).
85. E. Kirowa-Eisner and E. Gileadi, *J. Electroanal. Chem. Interfacial Electrochem.* **25**, 481 (1970).

86. I. Fried and H. Barak, *J. Electroanal. Chem. Interfacial Electrochem.* **27**, 167 (1970).
87. I. Piljac and R. I. Iwamoto, *J. Electroanal. Chem. Interfacial Electrochem.* **23**, 484 (1969).
88. R. C. Murray, Jr. and D. A. Aikens, *Electrochim. Acta* **20**, 259 (1975).
89. A. Cisak and P. J. Elving, *J. Electrochem. Soc.* **110**, 160 (1963).
90. J. Broadhead and P. J. Elving, *Anal. Chim. Acta* **48**, 433 (1969).
91. J. Perichon and R. Buvet, *Electrochim. Acta* **9**, 567 (1964).
92. M. Cremer, *Z. Biol.* **47**, 562 (1906).
93. G. Karreman and G. Eisenman, *Bull. Math. Biophys.* **24**, 413 (1962).
94. O. K. Stephanova, M. M. Shultz, E. A. Materova, and B. P. Nicolsky, *Vestn. Leningr. Univ.* **4**, 93 (1963).
95. I. M. Kolthoff and H. L. Sanders, *J. Am. Chem. Soc.* **59**, 416 (1937).
96. M. S. Frant and J. W. Ross, *Science* **154**, 1553 (1966).
97. A. K. Covington, *Crit. Rev. Anal. Chem.* **3**, 355 (1974).
98. T. Fujinaga, S. Okazaki, and H. Freiser, *Anal. Chem.* **46**, 1842 (1974).
99. J. Ruzicka, C. G. Lamm, and J. C. Tjell, *Anal. Chim. Acta* **62**, 15 (1972).
100. R. W. Cattrall and H. Freiser, *Anal. Chem.* **43**, 1905 (1971).
101. G. J. Moody and J. D. R. Thomas, *Lab. Pract.* **27**, 285 (1978).
102. R. P. Buck, *Anal. Chem.* **46**, 28R (1974).
103. R. P. Buck, *Anal. Chem.* **48**, 23R (1976).
104. R. W. Cattrall, D. M. Drew, and I. C. Hamilton, *Anal. Chim. Acta* **76**, 269 (1975).
105. International Union of Pure and Applied Chemistry, *Recommendations for Nomenclature of Ion-Selective Electrodess* (*Recommendations, 1975*), Pergamon, Oxford, 1976.
106. L. P. Dorsett and D. E. Mulcahy, *Chemistry in Australia* **44**, 241 (1977).
107. G. Eisenman, Ed., *Glass Electrodes for Hydrogen and Other Cations*, Marcel Dekker, New York, 1966.
108. G. Eisenman, R. G. Bates, G. Mattock, and S. M. Friedman, *The Glass Electrode*, Interscience, New York, 1966.
109. L. Van den Berg, *Anal. Chem.* **32**, 628 (1960).
110. R. P. Buck, *Anal. Chem.* **50**, 17R (1978).
111. A. Wikby and B. Karlberg, *Electrochim. Acta* **19**, 323 (1974).
112. A. Wikby, *Talanta* **22**, 663 (1975).
113. R. H. Doremus, Chapter 4 in Ref. 107.
114. E. P. Serjeant and A. G. Warner, *Anal. Chem.* **50**, 1724 (1978).
115. B. Karlberg, *Talanta* **22**, 1023 (1975).
116. C. T. Baker and I. Trachtenberg, *J. Electrochem. Soc.* **118**, 571 (1971).
117. R. Jasinski and I. Trachtenberg, *J. Electrochem. Soc.* **120**, 1169 (1973).
118. R. Jasinski and I. Trachtenberg, *Anal. Chem.* **44**, 2373 (1972).
119. R. Jasinski, I. Trachtenberg, and G. Rice, *J. Electrochem. Soc.* **121**, 363 (1974).
120. J. Vesely, *Chem. Listy.* **65**, 86 (1971); CA **74**, 70836d.
121. H. Hirata and M. Ayuzawa, *Chem. Lett.* **1974**, 1451.

# REFERENCES

122. J. W. Ross, in *Ion Selective Electrodes*, R. A. Durst, Ed., N.B.S. Spec. Publ. No. 314, Washington, D.C., 1969, p. 83.
123. D. J. Crombie, G. J. Moody, and J. D. R. Thomas, *Talanta* **21**, 1094 (1974).
124. W. E. Morf, G. Kahr, and W. Simon, *Anal. Chem.* **46**, 1538 (1974).
125. A. V. Gordievskii, A. F. Zhukov, V. S. Shterman, N. I. Savvin, and Yu. I. Urusov, *Zh. Anal. Khim.* **29**, 1414 (1974).
126. E. H. Hansen, C. G. Lamm, and J. Ruzicka, *Anal. Chim. Acta* **59**, 403 (1972).
127. M. Mascini and A. Liberti, *Anal. Chim. Acta* **64**, 63 (1973).
128. M. S. Mohan and G. A. Rechnitz, *Anal. Chem.* **45**, 1323 (1973).
129. R. P. Buck, in *Ion Selective Electrodes in Analytical Chemistry*, Vol. *1*, H. Freiser, Ed., Chapter 1, Plenum, New York, 1978.
130. E. Pungor and E. Hallos-Rokosinyi, *Acta Chim. Acad. Sci. Hung.* **27**, 63 (1961).
131. E. Pungor, *Anal. Chem.* **39**, 28A (1967); E. Pungor and K. Toth, *Analyst (London)* **95**, 625 (1970).
132. M. Mascini and A. Liberti, *Anal. Chim. Acta* **51**, 231 (1970).
133. G. J. Moody and J. D. R. Thomas, *Selected Ann. Rev. Anal. Sci.* **3**, 59 (1973), quoted in Ref. 129.
134. R. E. Van de Leest, *Analyst (London)* **101**, 433 (1976).
135. T. S. Light and I. L. Swartz, *Anal. Lett.* **1**, 825 (1968).
136. J. Vesely, O. J. Jensen, and B. Nicolaisen, *Anal. Chim. Acta* **62**, 1 (1972)
137. I. C. Popescu, C. Liteanu, and V. Ciovirnache, *Rev. Roum. Chim.* **18**, 145 (1973).
138. J. D. Czaban and G. A. Rechnitz, *Anal. Chem.* **45**, 471 (1973).
139. C. Liteanu, I. C. Popescu, and A. Mocanu, *Rev. Roum. Chim.* **18**, 1467 (1973).
140. J. Vesely, *Collect. Czech. Chem. Commun.* **39**, 710 (1974).
141. G. Papeschi, S. Bordi, and M. Carla, *J. Electrochem. Soc.* **125**, 1807 (1978).
142. M. J. D. Brand, J. J. Militello, and G. A. Rechnitz, *Anal. Lett.* **2**, 253 (1969).
143. A. V. Gordievskii, V. S. Shterman, A. Ya. Syrchenko, N. I. Savvin, A. F. Zhukov, and Yu. I. Urusov, *Zh. Anal. Khim.* **27**, 2170 (1972).
144. A. F. Zhukov, A. V. Vishnyakov, Y. L. Kharif, Y. I. Urusov, F. K. Volynets, E. I. Ryzhikov, and A. V. Gordievskii, *Zh. Anal. Khim.* **30**, 1761 (1975).
145. R. E. Van de Leest, *Analyst (London)* **102**, 509 (1977).
146. G. J. M. Heijne, W. E. Van der Linden, and G. Den Boef, *Anal. Chim. Acta* **89**, 287 (1977).
147. G. J. M. Heijne and W. E. Van der Linden, *Anal. Chim. Acta* **93**, 99 (1977).
148. Y. Umezawa, Y. Imanishi, K. Sawatari, and S. Fujiwara, *Bull. Chem. Soc. Jpn.* **52**, 945 (1979).
149. G. J. M. Heijne, W. E. Van der Linden, and G. Den Boef, *Anal. Chim. Acta* **100**, 193 (1978).
150. H. Hirata and K. Higashiyama, *Anal. Chim. Acta* **57**, 476 (1971).
151. T. Anfalt and D. Jagner, *Anal. Chim. Acta* **55**, 49 (1971).
152. H. Hirata and K. Higashiyama, *Fresenius' Z. Anal. Chem.* **257**, 104 (1971).
153. H. Hirata and K. Higashiyama, *Anal. Chim. Acta* **54**, 415 (1971); *Bull. Chem. Soc. Jpn.* **44**, 2420 (1971).

154. G. A. Rechnitz, G. H. Fricke, and M. S. Mohan, *Anal. Chem.* **44**, 1098 (1972).
155. H. Hirata, K. Higashiyama, and K. Date, *Anal. Chim. Acta* **51**, 209 (1970).
156. J. Vesely, *Collect. Czech. Chem. Commun.* **36**, 3364 (1971).
157. J. Ruzicka and C. G. Lamm, *Anal. Chem. Acta* **54**, 1 (1971); **53**, 206 (1971).
158. L. P. Dorsett and D. E. Mulcahy, *Anal. Lett.* **A11**, 53 (1978).
159. P. K. C. Tseng and W. F. Gutknecht, *Anal. Lett.* **9**, 795 (1976).
160. G. B. Marshall and D. Midgley, *Analyst (London)* **103**, 438 (1978).
161. I. Sekerka and J. F. Lechner, *Anal. Lett.* **9**, 1099 (1976).
162. J. F. Lechner and I. Sekerka, *J. Electroanal. Chem. Interfacial Electrochem.* **57**, 317 (1974).
163. P. K. C. Tseng and W. F. Gutknecht, *Anal. Chem.* **48**, 1996 (1976).
164. O. O. Lyalin and M. S. Turaeva, *Zh. Anal. Khim* **31**, 1879 (1976).
165. M. Mascini and A. Liberti, *Anal. Chim. Acta* **47**, 291 (1969).
166. M. Mascini and A. Liberti, *Anal. Chim. Acta* **53**, 202 (1971).
167. M. Mascini, *Anal. Chim. Acta* **62**, 29 (1974).
168. O. G. Takaishvili, E. P. Motsonelidze, Yu. M. Karachentseva, and P. I. Lavitaya, *Zh. Anal. Khim.* **30**, 1629 (1975).
169. W. D'Olieslager and L. Heerman, *J. Electrochem. Soc.* **126**, 347 (1979).
170. K. Toth, *Ph.D. Thesis*, Veszprem (1964).
171. E. Pungor, J. Havas, and K. Toth, *Acta Chim. Hung.* **41**, 239 (1964).
172. H. Hirata and K. Date, *Bull. Chem. Soc. Jpn.* **46**, 1468 (1973).
173. A. M. G. McDonald and K. Toth, *Anal. Chim. Acta* **41**, 99 (1968).
174. C. J. Coetzee and A. J. Basson, *Anal. Chim. Acta* **57**, 478 (1971); **56**, 321 (1971).
175. H. Hirata and K. Data, *Talanta* **17**, 883 (1970).
176. H. Hirata and K. Higashiyama, *Talanta* **19**, 391 (1972).
177. P. A. Rock, T. L. Eyrich, and S. Styer, *J. Electrochem. Soc.* **124**, 530 (1977).
178. H. Hirata and K. Date, *Anal. Chem.* **43**, 297 (1971).
179. G. G. Guilbault and P. J. Brignac, *Anal. Chem.* **41**, 1136 (1969).
180. I. Sekerka and J. F. Lechner, *Anal. Lett.* **9**, 1099 (1976).
181. I. Sekerka and J. F. Lechner, *J. Electroanal. Chem. Interfacial Electrochem.* **69**, 339 (1976).
182. G. W. S. Van Osch and B. Griepink, *Fresenius' Z. Anal. Chem.* **273**, 271 (1975).
183. G. W. S. Van Osch and B. Griepink, *Fresenius' Z. Anal. Chem.* **283**, 29 (1977).
184. G. W. S. Van Osch and B. Griepink, *Fresenius' Z. Anal. Chem.* **284**, 267 (1977).
185. A. Craggs, G. J. Moody, and J. D. R. Thomas, *J. Chem. Educ.* **51**, 541 (1974).
186. J. E. W. Davies, G. J. Moody, W. M. Price, and J. D. R. Thomas, *Lab. Pract.* **22**, 20 (1973).
187. C. Furnival, G. J. Moody, N. S. Nassory, and J. D. R. Thomas, quoted in Ref 101.
188. A. Ansaldi and S. I. Epstein, *Anal. Chem.* **45**, 595 (1973).
189. R. W. Cattrall, S. Tribuzio and H. Freiser, *Anal. Chem.* **46**, 2223 (1974).

H. Freiser, *Research Development*, December 1976, 28.
190. P. L. Bailey, *Analysis with Ion Selective Electrodes*, Heyden, London, 1976, p. 118.
191. R. W. Cattrall and C-P. Pui, *Anal. Chem.* **47**, 93 (1975).
192. H. J. James, G. Carmack, and H. Freiser, *Anal. Chem.* **44**, 856 (1972).
193. A. Hulanicki and R. Lewandowski, *Chem. Anal. (Warsaw)* **19**, 53 (1974).
194. H. B. Herman and G. A. Rechnitz, *Science* **184**, 1074 (1974).
195. H. B. Herman and G. A. Rechnitz, *Anal. Chim. Acta* **76**, 155 (1975).
196. A. L. Grekovich, E. A. Materova, and N. V. Garbuzova, *Zh. Anal. Khim.* **28**, 1206 (1973).
197. T. Stworzewicz, M. Leszko, and J. Czapkiewicz in *Ion Selective Electrodes*, E. Pungor, Ed., Akademia Kiado, Budapest, 1972, p. 259.
198. R. W. Cattrall and C-P. Pui, *Anal. Chim. Acta* **83**, 355 (1976).
199. R. W. Cattrall and C-P. Pui, *Anal. Chem.* **47**, 93 (1975).
200. R. W. Cattrall and C-P. Pui, *Anal. Chem.* **48**, 552 (1973).
201. A. V. Kopytin, A. F. Zhukov, Yu. I. Urusov, L. A. Kopytina, and A. V. Gordievskii, *Zh. Anal. Khim.* **34**, 465 (1979).
202. R. W. Cattrall and C-P. Pui, *Anal. Chim. Acta* **87**, 419 (1976).
203. B. M. Kneebone and H. Freiser, *Anal. Chem.* **45**, 449 (1973).
204. A. V. Gordievskii, A. Ya. Syrchenkov, V. V. Sergievskii, and N. I. Savvin, *Elecktrokhimiya* **8**, 520 (1972).
205. A. V. Gordievskii, V. S. Shterman, A. Ya. Syrchenkov, N. I. Savvin, and A. F. Zhukov, *Zh. Anal. Khim.* **27**, 772 (1972).
206. J. E. W. Davies, G. J. Moody, and J. D. R. Thomas, *Analyst (London)* **97**, 87 (1972).
207. A. L. Grekovich, E. A. Materova, and V. E. Yurinskaya, *Zh. Anal. Khim.* **27**, 1218 (1972).
208. E. Hopirtean and E. Stefaniga, *Rev. Roum. Chim.* **23**, 137 (1978).
209. J. W. Ross, *U.S. Patent No. 3,483,112* (9 December 1969).
210. A. Hulanicki, M. Maj-Zurawska, and R. Lewandowski, *Anal. Chim. Acta* **98**, 151 (1978).
211. T. J. Rohm and G. G. Guilbault, *Anal. Chem.* **46**, 590 (1974).
212. N. Ishibashi and H. Kohara, *Anal. Lett.* **4**, 785 (1971).
213. N. Ishibashi, A. Jyo, and K. Matsumoto, *Chem. Lett.* **1973**, 1297.
214. M. Kataoka and T. Kambara, *J. Electroanal. Chem. Interfacial Electrochem.* **73**, 279 (1976).
215. E. Hopirtean, M. Preda, and C. Liteanu, *Chem. Anal. (Warsaw)* **21**, 861 (1976).
216. A. G. Fogg, A. A. Al-Sibaak, and C. Burgess, *Anal. Lett.* **8**, 129 (1975).
217. A. G. Fogg and A. A. Al-Sibaak, *Anal. Lett.* **9**, 33 (1976).
218. Y. I. Urusov, V. V. Sergievskii, A. Y. Syrchenkov, A. F. Zhukov, and A. V. Gordievskii, *Zh. Anal. Khim.* **30**, 1757 (1975).
219. A. G. Fogg and A. A. Al-Sibaak, *Anal. Lett.* **9**, 39 (1976).
220. E. W. Baumann, *Anal. Chim. Acta* **99**, 247 (1978).
221. R. M. Carlson and J. L. Paul, *Anal. Chem.* **40**, 1292 (1968).

222. N. Ishibashi and K. Kina, *Bull. Chem. Soc. Jpn.* **46**, 2454 (1973).
223. E. A. Materova and S. A. Ovchinnikova, *Zh. Anal. Khim.* **32**, 331 (1977).
224. N. Ishibashi and A. Jyo, *Microchem. J.* **18**, 220 (1973).
225. N. Hazamoto, N. Kamo, and Y. Kobatake, *J. Assoc. Offic. Anal. Chem.* **59**, 1097 (1976).
226. N. Ishibashi, A. Jyo, and M. Yonemitsu, *Chem. Lett.* **1973**, 483.
227. G. D. Carmack and H. Freiser, *Anal. Chem.* **49**, 1577 (1977).
228. N. Ishibashi, H. Kohara, and K. Horinouchi, *Talanta* **20**, 867 (1973).
229. T. Fujinaga, T. S. Okazaki, and H. Freiser, *Anal. Chem.* **46**, 1842 (1974).
230. S. H. Hoke, A. G. Collins, and C. A. Reynolds, *Anal. Chem.* **51**, 859 (1979).
231. M. Kataoka and T. Kambara, *Bunseki Kagaku* **23**, 1081 (1974); CA **82**, 60427x.
232. T. J. Gilligan, E. L. Cussler, and D. F. Evans, *Biochim. Biophys. Acta* **497**, 627 (1977).
233. N. Ciocan and D. F. Anghel, *Fresenius' Z. Anal. Chem.* **290**, 237 (1978).
234. D. F. Anghel, G. Popescu, and N. Ciocan, *Mikrochim. Acta* **2**, 639 (1977).
235. R. P. Scholer and W. Simon, *Chimia* **24**, 372 (1970).
236. O. K. Stefanova and I. V. Rusina, *Elektrokhimiya* **14**, 882 (1978).
237. A. K. Covington and N. Karesh, *Anal. Chim. Acta* **85**, 175 (1976).
238. W. Simon, E. Pretsch, D. Ammann, W. E. Morf, M. Gueggi, R. Bissig, and M. Kessler, *Pure Appl. Chem.* **44**, 613 (1975).
239. M. Gueggi, E. Pretsch, and W. Simon, *Anal. Chim. Acta* **91**, 107 (1977).
240. R. J. Levins, *Anal. Chem.* **43**, 1045 (1971); **44**, 1544 (1972).
241. A. M. Y. Jaber, G. J. Moody, and J. D. R. Thomas, *Analyst (London)* **101**, 179 (1976).
242. A. M. Y. Jaber, G. J. Moody, and J. D. R. Thomas, *Proc. Anal. Div. Chem. Soc.* **11**, 328 (1976).
243. A. M. Y. Jaber, G. J. Moody, and J. D. R. Thomas, *J. Inorg. Nucl. Chem.* **39**, 1689 (1977).
244. J. W. Ross, *Science* **156**, 1378 (1967).
245. G. J. Moody, R. B. Oke, and J. D. R. Thomas, *Analyst (London)* **95**, 910 (1970).
246. G. H. Griffith, G. J. Moody, and J. D. R. Thomas, *Analyst (London)* **97**, 420 (1972).
247. J. Ruzicka, E. H. Hansen, and J. C. Tjell, *Anal. Chim. Acta* **67**, 155 (1973).
248. L. Keil, G. J. Moody, and J. D. R. Thomas, *Analyst (London)* **102**, 274 (1977).
249. G. J. Moody, N. S. Nassory, and J. D. R. Thomas, *Analyst (London)* **103**, 68 (1978).
250. J. D. R. Thomas, *Lab. Pract.* **27**, 857 (1978).
251. W. E. Morf, D. Ammann, and W. Simon, *Chimia* **28**, 65 (1974).
252. D. Ammann, R. Bissig, M. Gueggi, E. Pretsch, and W. Simon, *Helv. Chim. Acta* **58**, 1535 (1975).
253. D. Ammann, M. Gueggi, E. Pretsch, and W. Simon, *Anal. Lett.* **8**, 709 (1975).
254. W. Szczepaniak, M. Ren, and K. Ren, *Chem. Anal. (Warsaw)* **24**, 51 (1979).
255. K. Sollner, *Ann. N.Y. Acad. Sci.* **148**, 154 (1968).

256. A. Craggs, L. Keil, G. J. Moody, and J. D. R. Thomas, *Talanta* **22**, 907 (1975).
257. M. Gueggi, U. Fiedler, E. Pretsch, and W. Simon, *Anal. Lett.* **8**, 857 (1975).
258. R. C. Thomas, W. Simon, and M. Oehme, *Nature (London)* **258**, 754 (1975).
259. G. E. Baiulescu and N. Ciocan, *Talanta* **24**, 37 (1977).
260. J. E. W. Davies, G. J. Moody, W. M. Price, and J. D. R. Thomas, *Lab. Pract.* **22**, 20 (1973).
261. G. Baum and M. Lynn, *Anal. Chim. Acta* **65**, 393 (1973).
262. E. Hopirtean, C. Liteanu, and E. Stefaniga, *Rev. Roum. Chim.* **19**, 1651 (1974).
263. O. Ryba and J. Petranek, *J. Electroanal. Chem. Interfacial Electrochem.* **67**, 321 (1976).
264. O. Ryba, E. Knizabkova, and J. Petranek, *Collect. Czech. Chem. Commun.* **38**, 497 (1973).
265. O. Ryba and J. Petranek, *Anal. Chim. Acta*, **64**, 425 (1973).
266. J. Petranek and O. Ryba, *Anal. Chim. Acta* **72**, 375 (1974).
267. L. A. R. Pioda, V. Stankova, and W. Simon, *Anal. Lett.* **2**, 665 (1969).
268. J. Pick, K. Toth, E. Pungor, M. Vasak, and W. Simon, *Anal. Chim. Acta* **64**, 477 (1973).
269. B. P. Nikolskii, E. A. Materova, A. L. Grekovich, and V. E. Yurinskaya, *Zh. Anal. Khim.* **29**, 205 (1974).
270. J. G. Schindler, G. Stork, and H. Strueh, *Fresenius' Z. Anal. Chem.* **292**, 396 (1978).
271. U. Fiedler and J. Ruzicka, *Anal. Chim. Acta* **67**, 179 (1973).
272. M. D. Smith, M. A. Genshaw, and J. Greyson, *Anal. Chem.* **45**, 1782 (1973).
273. O. H. Le Blanc and W. T. Grubb, *Anal. Chem.* **48**, 1658 (1976).
274. M. Oehme and W. Simon, *Anal. Chim. Acta* **86**, 21 (1976).
275. G. Baiulescu, V. V. Cosofret, and C. Cristescu, *Revta Chim.* **26**, 429 (1975); *Anal. Abstr.* **29**, 6J74 (1975).
276. V. V. Cosofret, *Rev. Roum. Chim.* **23**, 1489 (1978).
277. M. Geuggi, M. Oehme, E. Pretsch, and W. Simon, *Helv. Chim. Acta* **59**, 2417 (1976).
278. R. P. Kraig and C. Nicholson, *Science* **194**, 725 (1976).
279. E. W. Baumann, *Anal. Chem.* **47**, 959 (1975).
280. W. Szczpaniak, K. Ren, and A. Mickiewicz, *Anal. Chim. Acta* **82**, 37 (1976).
281. D. L. Manning, J. R. Stokely, and D. W. Magowyrk, *Anal. Chem.* **46**, 1116 (1974).
282. E. Hopirtean and E. Stefaniga, *Rev. Roum. Chim.* **21**, 305 (1976).
283. G. Baum, M. Lynn, and F. B. Ward, *Anal. Chim. Acta* **65**, 385 (1973).
284. K. Selinger and R. Staroscik, *Pharmazie* **33**, 208 (1978).
285. R. Scholer and W. Simon, *Helv. Chim. Acta* **55**, 1801 (1972).
286. A. P. Thoma, Z. Cimerman, U. Fiedler, D. Bedekovic, M. Geugi, P. Jordon, K. May, E. Pretsch, V. Prelog, and W. Simon, *Chimia* **29**, 344 (1975).
287. N. Ishibashi, K. Kina, and N. Maekawa, *Chem. Lett.* **1973**, 119.
288. H. Brockmann and G. Schmidt-Kastner, *Chem. Ber.* **88**, 57 (1955).
289. H. Brockmann, M. Springorum, G. Traxler, and I. Hofer, *Naturwissenschaften* **50**, 689 (1963).

290. C. J. Pedersen, *J. Am. Chem. Soc.* **89**, 7017 (1967).
291. C. J. Pedersen and H. K. Frensdorff, *Angew. Chem. Int.* **11**, 16 (1972).
292. W. E. Morf and W. Simon, *Hung. Sci. Instrum.* **41**, 1 (1977).
293. J. G. Schindler, G. Stork and H. J. Strueh, *Fresenius' Z. Anal. Chem.* **292**, 319 (1978).
294. U. Fiedler, *Anal. Chim. Acta* **89**, 111 (1977).
295. A. Craggs, L. Keil, G. J. Moody, and J. D. R. Thomas, *J. Inorg. Nucl. Chem.* **37**, 577 (1975).
296. D. Midgley and K. Torrance, *Analyst (London)* **97**, 626 (1972).
297. J. W. Ross, J. H. Riseman, and J. A. Krueger, *Pure Appl. Chem.* **36**, 473 (1973).
298. P. L. Bailey and M. Riley, *Analyst (London)* **100**, 145 (1975).
299. L. R. McKenzie and P. N. W. Young, *Analyst (London)* **100**, 620 (1975).
300. J. Ruzicka and E. H. Hansen, *Anal. Chim. Acta* **69**, 129 (1974); **72**, 353 (1974).
301. C. P. Pui, G. A. Rechnitz, and R. F. Miller, *Anal. Chem.* **50**, 330 (1978).
302. K. Selinger, R. Staroscik, and F. Malecki, *Chem. Anal. (Warsaw)* **21**, 1153 (1976); *Anal. Abstr.* **32**, 5J106 (1977).
303. T. Anfalt, A. Graneli, and D. Jagner, *Anal. Chim. Acta* **76**, 253 (1975).
304. D. Midgley, *Analyst (London)* **100**, 386 (1975).
305. L. R. Pucacco and N. W. Carter, *Anal. Biochem.* **90**, 427 (1978).
306. F. G. Roemer, L. Puyker, G. B. H. Brinkman, and B. Griepink, *Fresenius' Z. Anal. Chem.* **289**, 35 (1978).
307. D. Parker, D. Delpy, and M. Lewis, *Med. Biol. Eng. Comput.* **16**, 599 (1978).
308. M. M. Fishman and H. F. Schiff, *Anal. Chem.* **44**, 543R (1972); **46**, 367R (1974); **48**, 322R (1976).
309. M. M. Fishman, *Anal. Chem.* **50**, 261R (1978); **52**, 185R (1980).
310. J. Everse, C. L. Ginsburgh, and N. O. Kaplan, *Methods Biochem. Anal.* **25**, 135 (1979).
311. L. C. Clark, Jr. and C. Lyons, *Ann. N.Y. Acad. Sci.* **102**, 29 (1962).
312. A. S. Barker and P. J. Somers, *Top. Enzyme Ferment. Biotechnol.* **2**, 120 (1978).
313. G. H. Fricke, *Anal. Chem.* **52**, 259R (1980).
314. M. Mascini and G. Palleschi, *Anal. Chim. Acta* **100**, 215 (1978).
315. G. G. Guilbault and M. Tarp, *Anal. Chim. Acta* **73**, 355 (1974).
316. G. G. Guilbault and W. Stokbro, *Anal. Chim. Acta* **76**, 237 (1975).
317. D. S. Papastathopoulos and G. A. Rechnitz, *Anal. Chim. Acta* **79**, 17 (1975).
318. W. C. White and G. G. Guilbault, *Anal. Chem.* **50**, 1481 (1978).
319. G. G. Guilbault and J. G. Montalvo, *J. Am. Chem. Soc.* **91**, 2164 (1969).
320. J. G. Montalvo and G. G. Guilbault, *Anal. Chem.* **41**, 1897 (1969).
321. I. Fritz, G. Nagy, L. Fodor, and E. Pungor, *Analyst (London)* **101**, 439 (1976).
322. G. G. Guilbault and J. G. Montalvo, *Anal. Lett.* **2**, 283 (1969).
323. L. F. Cullen, J. F. Rusling, A. Schleifer, and G. J. Papariello, *Anal. Chem.* **46**, 1955 (1974).
324. J. F. Rusling, G. H. Luttrell, L. F. Cullen and G. J. Papariello, *Anal. Chem.* **48**, 1211 (1976).

# REFERENCES

325. C. J. Olliff and J. M. Wright, *J. Pharm. Pharmacol.* **30** (Suppl.), 45P (1978).
326. R. K. Kobos and G. A. Rechnitz, *Anal. Lett.* **10**, 751 (1977).
327. G. A. Rechnitz, R. K. Kobos, S. J. Riechel, and C. R. Gebauer, *Anal. Chim. Acta* **94**, 357 (1977).
328. J. G. Schiller, A. K. Chen, and C. C. Liu, *Anal. Biochem.* **85**, 25 (1978).
329. T. Shinbo and M. Sugiura, *Anal. Chem.* **51**, 100 (1979).
330. S. Updike and T. Treichel, *Anal. Chem.* **51**, 1643 (1979).
331. G. B. Wisdom, *Clin. Chem. (Winston-Salem, N.C.)*, **22**, 1243 (1976).
332. A. H. W. M. Schuurs and B. K. Van Weeman, *Clin. Chim. Acta* **81**, 1 (1977).
333. J-L. Boitieux, G. Desmet, and D. Thomas, *Clin. Chem. (Winston-Salem, N.C.)* **25**, 318 (1979).
334. M. E. Meyerhoff and G. A. Rechnitz, *Anal. Biochem.* **95**, 483 (1979).
335. C. R. Gebauer, M. E. Meyerhoff, and G. A. Rechnitz, *Anal. Biochem.* **95**, 479 (1979).
336. D. S. Papastathopoulos and G. A. Rechnitz, *Anal. Chem.* **48**, 862 (1976).
337. R. Wawro and G. A. Rechnitz, *J. Membr. Sci.* **1**, 143 (1976).
338. M. Meyerhoff and G. A. Rechnitz, *Anal. Chim. Acta* **85**, 277 (1976).
339. K. M. Fung, S. S. Kuan, H. Y. Sung, and G. G. Guilbault, *Anal. Chem.* **51**, 2319 (1979).
340. L. R. Meyerson, K. D. McMurtrey, and V. E. Davis, *Anal. Biochem.* **86**, 287 (1978).
341. G. A. Rechnitz, M. A. Arnold, and M. E. Meyerhoff, *Nature (London)* **278**, 466 (1979).
342. R. K. Kobos, D. J. Rice, and D. S. Flournay, *Anal. Chem.* **51**, 1122 (1979).
343. G. A. Rechnitz, T. L. Riechel, R. K. Kobos, and M. E. Meyerhoff, *Science* **199**, 440 (1978).
344. C-H. Kiang, S. S. Kuan, and G. G. Guilbault, *Anal. Chim. Acta* **80**, 209 (1975).
345. C. P. Hsiung, S. S. Kuan, and G. G. Guilbault, *Anal. Chim. Acta* **90**, 45 (1977).
346. M. A. Jensen and G. A. Rechnitz, *J. Membr. Sci.* **5**, 117 (1979).
347. Tran-Minh Canh and J. Beaux, *Anal. Chem.* **51**, 91 (1979).
348. P. Durand, A. David, and D. Thomas, *Biochim. Biophys. Acta* **527**, 277 (1978).
349. H. Nilsson, K. Mosbach, S. O. Enfors, and N. Molin, *Biotechnol. Bioeng.* **20**, 527 (1978).
350. K. Matsumoto, H. Seijo, T. Watanabe, I. Karube, I. Satoh, and S. Suzuki, *Anal. Chim. Acta* **105**, 429 (1979).
351. T. Matsunaga, I. Karube, and S. Suzuki, *Anal. Chim. Acta* **99**, 233 (1978).
352. M. A. Jensen and G. A. Rechnitz, *Anal. Chim. Acta* **101**, 125 (1978).
353. M. E. Meyerhoff and G. A. Rechnitz, *Anal. Lett.* **12**, 1339 (1979).
354. G. J. Moody and J. D. R. Thomas, *Analyst (London)* **100**, 609 (1975).
355. T. Kawashima and G. A. Rechnitz, *Anal. Chim. Acta* **83**, 9 (1976).

CHAPTER

4

# PROCEDURES OF ANALYTICAL POTENTIOMETRY

The salient features of procedures used in analytical potentiometry have already been alluded to. For example, the need for *calibration standards* when using chemical cells with diffusion was mentioned in Chapter 1. Again in Chapter 1, mention was made of the finite solubilities of electroactive materials used in the construction of electrodes of the second kind and how this property imposes a limit of concentration below which the Nernstian response of even a cell without liquid junction becomes suspect. These remarks apply equally to the quasi-thermodynamic cells with liquid junction in which the ion selective membrane electrodes, described in Chapter 3, are used extensively as indicator electrodes for the analytical applications of potentiometry. Cells used for these applications need to be calibrated under conditions that relate closely to those existing in the analyte solutions. Furthermore, each type of ion selective electrode can be used only within a certain concentration range, the lower limit of which is often determined by the solubility of the membrane components, whereas the upper limit sometimes depends upon complexation effects that may occur at high concentrations of determinand ion. These limits establish the *response range* of the indicator electrode in a cell with liquid junction. Ideally, of course, this electrode would provide the Nernstian response from which it would be possible to specifically and quantitatively determine only the ion of interest. This, in turn, would imply that the electrode responds to only one species among the components of a system. Although it has been shown that ion selective electrodes are not so susceptible to chemical attack as oxidation-reduction electrodes, a property of membrane electrodes is their susceptibility to physical interferences caused by the inability of the membrane material to quantitatively distinguish between two species in the same solution. Thus a given ion selective membrane electrode is never wholly specific to one particular ionic species, but the *selectivity* of its response favors that ionic species over all other types. Important considerations in the establishment of a method utilizing analytical potentiometry include the salient features mentioned which can be summarized as follows:

1. Cell calibration.
2. Response range of the cell.
3. Selectivity of the electrode(s).

Other factors of relevance are the time-dependent factors such as the response time necessary to record a cell potential of suitable stability and the reproducibility of this potential once it is attained. Extraneous effects such as the sensitivity of the constituent electrodes to environmental effects such as temperature, light, and pressure also need to be considered.

## 4.1 METHODS OF CALIBRATION

### Cells without Diffusion

It was shown earlier that the general form of the Nernst equation pertaining to cells without liquid junction can be written as

$$E_{\text{cell}} = E_{\text{cell}}^{\circ} - \frac{k}{n} \log a_2 \qquad (4.1)$$

in which $E_{\text{cell}}^{\circ}$ is the required calibration factor and $a_2$ is the activity of the electrolyte in the cell solution. This electrolyte is the source of cations to which one electrode is reversible and also the source of anions to which the other electrode responds. Provided that the response of each of these electrodes is reproducible, this equation is applicable irrespective of the type of electrode, whether exclusively of the oxidation-reduction type, the ion selective membrane type, or for cells containing one type of each. These cells are not often used in routine analysis because the activity of the species to which one electrode responds must be known in all the standard and analyte solutions containing the determinand ion. The major application of cells without diffusion is, therefore, the determination of thermodynamic properties in solutions, the composition and stoichiometry of which are known. As was pointed out in Chapter 1, the calibration of this type of cell often represents a compromise between the desire to work at the comparatively low ionic strengths for which a form of the Debye–Huckel equation is applicable, and a limit of dilution which can be imposed by the solubility of the electroactants used in the construction of the electrodes.

The general method for the calibration of this type of cell is given elsewhere in this book together with a specific example. These examples are applicable not only for the determination of mean activity coefficients but also for subsequent use in any expression that contains terms in $E^\circ_{\text{cell}}$ and the measured potential across the calibrated electrodes. Although the day-to-day constancy of $E^\circ_{\text{cell}}$ is a very reassuring characteristic since this would confirm that each electrode attains a reproducibly stable potential, this is by no means an essential property of the cell. For example, if the cell

$$\text{Zn(Hg)}|\text{ZnSO}_4|\text{PbSO}_4; \text{Pb(Hg)} \tag{4.2}$$

is calibrated daily at a given temperature, some slight variations in $E^\circ_{\text{cell}}$ as determined by means of the previously given method might be noticed. If, however, after each of these determinations the cell is used to determine the mean activity coefficient of a dilute solution of zinc sulfate always of the same concentration and at the same temperature, this value should remain invariant. If this is indeed the case, then the full calibration procedure can be dispensed with since in this instance, if $E_{\text{cell}}$ is measured for a solution of known molality having a known mean activity coefficient, equation (4.1) can be written as

$$E_{\text{cell}} = E^\circ_{\text{cell}} - \frac{k}{2} \log m^2_{\text{ZnSO}_4} \cdot \gamma^2_{\pm \text{ZnSO}_4}$$

Thus the only unknown will be the calibration factor $E^\circ_{\text{cell}}$. Effectively, by means of such repetitive types of measurements, standards with which to calibrate cells without liquid junction can be established. It would appear that the reproducibility of the mean activity coefficients of standard solutions obtained throughout the scientific world are likely to be superior to the reproducibility of values based on standard reduction potentials.

Oxidation-reduction electrodes are more commonly used in cells without liquid junction than ion selective membrane electrodes because the stability and reproducibility of the latter are often inferior to the oxidation-reduction type. Furthermore, the nature of the application of cells without diffusion makes possible the elimination of the types of interferences that would totally exclude these oxidation-reduction electrodes from consideration in analysis. Nonetheless, ion selective electrodes are used in cells without diffusion for the determination of ionization constants (see Chapter 6) and for the determination of mean activity coefficients. Thus, for example, the measurement of the mean activity coefficients for sodium fluoride solutions and for mixtures of sodium chloride and sodium fluoride have been carried out using the cells (1)

$$\text{Na}^+ \text{ selective glass electrode}|\text{NaF solutions}|\text{F}^-(\text{LaF}_3)\text{selective electrode} \tag{4.3}$$

and

Ag; AgCl|NaCl − NaF solutions|F⁻ (LaF$_3$) selective electrode (4.4)

A sodium amalgam electrode was also used in cell (4.3) in place of the sodium ion selective glass membrane electrode. The fluoride ion selective electrode has also been used in cells similar to cell (4.4) employed for the estimation of the activity coefficients of fluoride ion in mixtures of trace concentrations of sodium fluoride and potassium fluoride in the presence of high concentrations (up to 4 $m$) of sodium chloride, potassium chloride, potassium bromide, and potassium iodide (2). Estimations of individual ion activity coefficients have also been made based upon measurements with a cell similar to cell (4.4) but with a liquid membrane electrode reversible to chloride ion (3). Analogous systems containing ions of alkaline earths have also been studied (4). The interest upon these and similar measurements is centered on the establishment of activity standards for single ions with which to calibrate the cells that contain liquid junctions commonly used in analytical potentiometry. An activity standard is a solution of known composition for which there is good reason to believe that the assigned value for the activity of the determinand ion is very close to its actual value. The first series of activity standards to be established were based upon solutions used to define the N.B.S. scale of $p$H (5). These have been followed more recently (6) by the $p$X ($-\log a_X$) scales where X represents the cations and anions commonly determined with cells containing ion selective membrane electrodes. Data on $p$H standards are also included in Reference 6.

### Activity Standards, $p$H

These standards are prepared from solutions that can contain either separately weighed quantities of a weak acid and its conjugate base in the mole ratio 1 : 1, or a weighed quantity of a weak base partially neutralized by a known quantity of strong acid, or, alternatively, a weighed quantity of the monosodium or monopotassium salt of a weak dicarboxylic acid alone. A characteristic of these buffer solutions is that the hydrogen ion activity changes little when small quantities of strongly acidic or basic contaminants are added to them. A further property of the buffer solutions used in the establishment of a scale of $p$H is that their ionic strength is usually less than 0.1 mole L$^{-1}$. The establishment of this scale is based upon measurements with the following cell for which the diffusion potential is negligible

Pt; H$_2$ (1 atm)|H$^+$ ($p$H buffer solution), Cl$^-$ (known molality)|AgCl; Ag

(4.5)

The electrodes used in this cell can be calibrated at a given temperature using the mean activity coefficient for 0.01 $m$ hydrochloric acid solution* in the equation

$$E^\circ_{cell} = E_{cell} + 2k \log 0.01 \gamma_\pm \qquad (4.6)$$

For solutions contained in cells of the type shown by cell (4.5), this calibration factor $E^\circ_{cell}$ can be combined with the measured e.m.f. of the cell in order to define an acidity function for the solution. Thus from equation (4.1),

$$E_{cell} = E^\circ_{cell} - \frac{k}{2} \log a^2_{H^+} m^2_{Cl^-} \gamma^2_{Cl^-}$$

which, for the known molality of chloride ion contained in the cell, can be written as

$$\frac{E_{cell} - E^\circ_{cell}}{k} + \log m_{Cl^-} = -\log a_H \gamma_{Cl} \qquad (4.7)$$

All the terms on the left-hand side of this equation are known and the thermodynamically valid term $-\log a_H \gamma_{Cl}$ is the acidity function which is usually written as $p(a_H \gamma_{Cl})$. This equation is valid for all calibrated cells containing an electrode reversible to hydrogen ion which is negative with respect to the silver–silver chloride electrode. The corresponding equation for the converse case is

$$p(a_H \gamma_{Cl}) = \frac{E^\circ_{cell} - E_{cell}}{k} + \log m_{Cl^-} \qquad (4.8)$$

The acidity function $p(a_H \gamma_{Cl})$ can be regarded as the thermodynamic precursor to the quasi-thermodynamic quantity $p$H. In the case of the National Bureau of Standards (USA) scale, the $p$H of a given buffer solution is derived by measuring values of $p(a_H \gamma_{Cl})$ for a series of buffer solutions each of which contains the same concentration of buffer component(s) and different but known concentrations of chloride ion. A plot of the experimentally derived values of $p(a_H \gamma_{Cl})$ against the corresponding chloride ion concentrations often gives a straight line relation (see Figure 4.1) which is extrapolated to $m_{Cl^-} = 0$ to yield a value designated as $p(a_H \gamma_{Cl})^0$. The conversion of the latter quantity into the $p$H of the buffer solution without added chloride ion involves an extrathermodynamic as-

*Values of the mean activity coefficient of 0.01$m$ HCl are given on page 94.

sumption as to the magnitude of the trace activity coefficient, $\gamma_{Cl}^0$. This term is the activity coefficient that the chloride ion would have possessed if it were present at the ionic strength produced by the buffer component(s) alone. The procedure adopted for the assignation of a $p$H value from the corresponding $p(a_H\gamma_{Cl})^0$ value can be exemplified by reference to the data presented graphically in Figure 4.1. These data were obtained at 25°C using solutions containing 0.02 $m$ potassium dihydrogen phosphate, 0.02 $m$ disodium hydrogen phosphate, and sodium chloride at the concentrations shown. The extrapolated value of $p(a_H\gamma_{Cl})^0$ was found to be 6.992 at zero chloride ion concentration, and for this condition the ionic strength of the solution is given by

$$I = 0.5\left(m_{K^+} + 2m_{Na^+} + m_{H_2PO_4^-} + 4m_{HPO_4^{2-}}\right)$$

$$= 0.5(0.02 + 0.04 + 0.02 + 0.08) = 0.08 \ m$$

The term $p(a_H\gamma_{Cl})^0$ can be expanded as

$$p(a_H\gamma_{Cl})^0 = -\log a_{H^+} - \log \gamma_{Cl^-}^0 \quad \text{or} \quad pa_H = p(a_H\gamma_{Cl})^0 + \log \gamma_{Cl^-}^0$$

To calculate $pa_H$ from this equation an extrathermodynamic assumption about the magnitude of the activity coefficient for the chloride ion must now be made, and for this purpose the Debye–Huckel equation written in the form of the Bates–Guggenheim convention (see Table 1.2) is used. Thus at 25°C,

$$-\log \gamma_{Cl}^0 = \frac{0.511 I^{1/2}}{1 + 1.5 I^{1/2}} \tag{4.9}$$

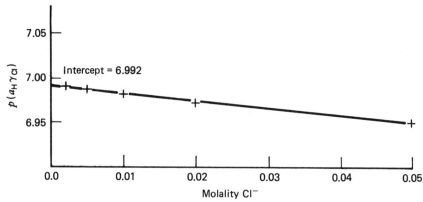

**Figure 4.1.** Extrapolation of the function $p(a_H\gamma_{Cl})$ for 0.02 m equimolal phosphate solutions containing chloride ion.

and therefore the $pa_H$ value of the 0.02 $m$ equimolal phosphate buffer solution is

$$pa_H = 6.992 - \frac{0.511(0.08)^{1/2}}{1 + 1.5(0.08)^{1/2}}$$

$$= 6.891$$

This type of determination has been used (7) for the assignation of the $p$H(S) values to the selected reference solutions given in Table 4.1. The term $p$H(S) is used conventionally as referring specifically to the $pa_H$ values of these selected reference solutions that form the basis for the NBS scale of $p$H. The term $p$H is reserved for an experimentally determined value such as would be determined with the cell

$$\text{Hg; Hg}_2\text{Cl}_2 | 3.5 \ M \ \text{KCl} \| \text{H}^+ \text{ in solution} | \text{glass electrode} \quad (4.10)$$

and the term $pa_H$ remains appropriate for the example given previously.

The operational definition of $p$H is made with reference to the two cells

$$\text{Pt; H}_2 \ (1 \text{ atm}) | \text{solution(S)} \| \text{KCl}( \geq 3.5 \ M) | \text{Hg}_2\text{Cl}_2 ; \text{Hg}$$

$$\text{Pt; H}_2 \ (1 \text{ atm}) | \text{solution(X)} \| \text{KCl}( \geq 3.5 \ M) | \text{Hg}_2\text{Cl}_2 ; \text{Hg}$$

Table 4.1. U.S. National Bureau of Standards $p$H Reference Solutions

| Solution Composition (molality) | $p$H(S) at 25°C |
|---|---|
| Primary reference solutions | |
| KH tartrate (satd at 25°C) | 3.557 |
| 0.5 $m$ KH$_2$ citrate | 3.776 |
| 0.05 $m$ KH phthalate | 4.004 |
| 0.025 $m$ KH$_2$PO$_4$ + 0.025 $m$ Na$_2$HPO$_4$ | 6.863 |
| 0.008695 $m$ KH$_2$PO$_4$ + 0.03043 $m$ Na$_2$HPO$_4$ | 7.415 |
| 0.01 $m$ Na$_2$B$_4$O$_7$ · 10 H$_2$O | 9.183 |
| 0.025 $m$ NaHCO$_3$ + 0.025 $m$ Na$_2$CO$_3$ | 10.014 |
| Secondary reference solutions | |
| 0.05 $m$ K tetroxalate · 2H$_2$O | 1.679 |
| 0.01667 $m$ Tris + 0.05 $m$ Tris · HCl | 7.699 |
| Ca(OH)$_2$ (satd at 25°C) | 12.454 |

*Source*: Reference 6.

in which the same two electrodes are transferred between the reference solution (S) of known pH(S) and solution (X) of unknown $p$H(X). For the cell containing solution (S), the measured e.m.f, $E_s$ is related to $p$H(S) by

$$E_s = E_{cal} + E_j - \left[ E^0_{H^+; H_2} - k \cdot p\text{H}(S) \right]$$

$$E_s = E_{cal} + E_j + k \cdot p\text{H}(S) \tag{4.11}$$

in which $E_{cal}$ is the potential of the calomel reference half-cell. Similarly, the equation relating the measured e.m.f, $E_x$ to $p$H(X) for the other cell is

$$E_x = E_{cal} + E_j + kp\text{H}(X) \tag{4.12}$$

If it is assumed that the junction potential $E_j$ is the same in both cells, then $p$H(X) can be defined by combination of equations (4.11) and (4.12) as

$$p\text{H}(X) = p\text{H}(S) + \frac{E_x - E_s}{k} \tag{4.13}$$

in which $k$ is the Nernst factor at the given temperature (see appendix). Once the cell e.m.f. has been measured for one of the $p$H(S) solutions given in Table 4.1, the operational definition of $p$H given by equation (4.13) allows assignation of a value of $p$H to any solution (X) for which $E_x$ is measured. In practice, however, $p$H is not commonly determined with cells that contain a hydrogen electrode, and electrodes such as the glass, quinhydrone, or antimony electrodes with which the hydrogen electrode is usually replaced often have an imperfect response. It is advisable, therefore, in these instances to use two $p$H standards for the calibration of the cell. If, for example, cell (4.10) is to be used at 25°C to measure the $p$H of samples suspected of having $p$H values in the range 4–7, it would be advisable to select the phosphate and citrate buffer solutions as the calibrating standards for the glass electrode in this cell. At 25° the two standard $p$H values are 6.863 for the phosphate solution and 3.776 for the citrate solution.

The method of standardization involves adjusting the ion activity meter so that the scale records the value of $p$H(S1) when the electrodes are immersed in the phosphate buffer solution, and then to measure the $p$H of the citrate buffer treating this as if it were an unknown $p$H(X). If the glass electrode has a perfect response, then $p$H(X) will be the same as $p$H(S2) and no further action is required. It is more likely, however, that there will be a slight difference between $p$H(S2) of the citrate buffer and the measured value $p$H(X). Under these circumstances certain alternatives are available depending upon the type of ion activity meter used:

1. If the meter is supplied with a slope control, $p$H(X) can be adjusted so that it is coincident with $p$H(S2) of 3.776. If it is found that this adjustment alters the original $p$H setting when the electrodes are reimmersed in the phosphate buffer of $p$H(S1) = 6.863, then the standardization is repeated.

2. The alternative procedure that can be used if the ion activity meter is not fitted with a slope adjustment is to determine the electromotive efficiency of the cell $\beta$. This quantity is defined as

$$\beta = \frac{p\text{H}(\text{S1}) - p\text{H}(\text{X})}{p\text{H}(\text{S1}) - p\text{H}(\text{S2})} \qquad (4.14)$$

in which $p$H(X) is the $p$H recorded when the electrodes are immersed in buffer solution having the published value $p$H(S2). The value of $\beta$ is usually less than unity and, once determined, can be used to correct the recorded $p$H of any solution into a $p$H(X) value (see equation 4.13) provided that this $p$H is within the range encompassed by the two standardizing solutions $p$H(S1) and $p$H(S2). Since $(1 - \beta)$ is the error in $p$H recorded for a change of 1 $p$H unit, then $p$H(X) can be obtained from the recorded $p$H, $p$H(rec) by the equation

$$p\text{H}(\text{X}) = p\text{H}(\text{rec}) + (1 - \beta)[p\text{H}(\text{rec}) - p\text{H}(\text{S1})] \qquad (4.15a)$$

in which $p$H(S1) is the $p$H of the prime standardizing buffer, 6.863 in the example. If $\beta$, as determined by equation (4.14), was found to be 0.9975 $\pm$ 0.0004 and the recorded $p$H of a solution was determined as 5.250, then

$$p\text{H}(\text{X}) = 5.250 + 0.0025(5.250 - 6.863)$$
$$= 5.246$$

The assumption inherent in all $p$H measurements is that the junction potential remains invariant between the standardizing solutions and the sample solution. In absolute terms, therefore, the third decimal place has no significance in $p$H measurements, although for closely matched solutions the error can in some cases be less than 0.007 $p$H unit. This was found (8) for some solutions using cells with liquid junction of the type

Pt; $H_2$|solution X|satd KCl|phosphate buffer ($p$H 6.863)|$H_2$; Pt

$$(4.15b)$$

in which three of the standard buffers were compared with the fourth. The

$p$H of saturated potassium hydrogen tartrate as determined from cell (4.15b) was 3.564 as compared with its standard value 3.557 determined using the $p(a_H \gamma_{Cl})$ method based on cell (4.5). Even closer agreement was obtained for potassium hydrogen phthalate and borax buffers.

In the practical applications of $p$H measurements it has been verified that provided the ionic strength does not exceed 0.1 in a sample solution containing a low concentration of nonelectrolytes, and the $p$H of this solution is between about 2 and 12, then the relation $p$H(X) = $-\log a_{H^+}$ holds with an accuracy of $\pm 0.02$ or better. Insofar as a majority of practical $p$H measurements are made with cells that contain a glass electrode, a method of testing the performance of these electrodes over the range $p$H 0–14 has been developed (9). This method is based on indirect comparisons of glass electrodes with the hydrogen electrode using silver–silver chloride electrodes in cells without liquid junction. A series of specially developed chloride containing buffers provide the cell solutions for these tests that allow the errors of glass electrodes to be determined over the range $p$H 0–14 with an accuracy claimed to be $\pm 0.01$ $p$H.

The effect of temperature upon the values of $p$H(S) for some of the standard buffer solutions given in Table 4.1 can be expressed by the equation

$$p\text{H(S)} = \frac{A}{T} + B + CT + DT^2 \qquad (4.16)$$

where $T$ is the temperature in degrees kelvin. The values of the constants $A$, $B$, $C$, and $D$, given in Table 4.2 are found to express satisfactorily the observed variation with a precision of 0.001–0.003 in $p$H over the tempera-

Table 4.2. Constants of Equation (4.16) for Seven $p$H Standard Solutions

| Solution | Temperature Range (°C) | $A$ | $B$ | $C$ | $10^5 D$ |
|---|---|---|---|---|---|
| KH tartrate (satd at 25°C) | 25–95 | −1727.96 | 23.7406 | −0.075947 | 9.2 |
| KH$_2$ citrate ($m = 0.05$) | 0–50 | 1280.40 | −4.1650 | 0.012230 | 0 |
| KH phthalate ($m = 0.05$) | 0–95 | 1678.30 | −9.8357 | 0.034946 | −2.4 |
| KH$_2$PO$_4$ ($m = 0.025$) Na$_2$HPO$_4$ ($m = 0.025$) | 0–95 | 3459.39 | −21.0574 | 0.073301 | −6.2 |
| KH$_2$PO$_4$ ($m = 0.008695$) Na$_2$HPO$_4$ ($m = 0.03043$) | 0–50 | 5706.61 | −43.9428 | 0.154785 | −15.6 |
| Na$_2$B$_4$O$_7$ ($m = 0.01$) | 0–95 | 5259.02 | −33.1064 | 0.114826 | −10.7 |
| NaHCO$_3$ ($m = 0.025$), Na$_2$CO$_3$ ($m = 0.025$) | 0–50 | 2557.10 | −4.2846 | 0.019185 | 0 |

*Source*: Reference 7.

ture ranges quoted. Application of the equation to the data for five buffers gives the smoothed $p$H(S) values at the temperatures given in Table 4.3. Any slight discrepancy between these values at 25° and the experimentally derived values of Table 4.1 can be attributed to the smoothing effect given by equation (4.16).

The $p$H scale discussed so far is that of the United States of America adopted by the National Bureau of Standards, Washington, D.C. Other scales do exist, but there are only minor differences between the data presented in Tables 4.1 and 4.3 and the data adopted for $p$H standardization in Germany, Hungary, Japan, Poland, Rumania, the Union of Soviet Socialist Republics, and the United Kingdom. These data are all summarized in Reference 6. Many of the experimental aspects of $p$H measurements are described in the books given as references 7 and 10–13. Data for many $p$H buffers are contained in Reference 14 and details of $p$H standards pertinent to the physiological range of $p$H are given in References 15–19.

Table 4.3. Values of $p$H(S) for Five Reference Solutions[a]

| $T$°C | I | II | III | IV | V |
|---|---|---|---|---|---|
| 0 | — | 4.003 | 6.984 | 7.534 | 9.464 |
| 5 | — | 3.999 | 6.951 | 7.500 | 9.395 |
| 10 | — | 3.998 | 6.923 | 7.472 | 9.332 |
| 15 | — | 3.999 | 6.900 | 7.448 | 9.276 |
| 20 | — | 4.002 | 6.881 | 7.429 | 9.225 |
| 25 | 3.557 | 4.008 | 6.865 | 7.413 | 9.180 |
| 30 | 3.552 | 4.015 | 6.853 | 7.400 | 9.139 |
| 35 | 3.549 | 4.024 | 6.844 | 7.389 | 9.102 |
| 38 | 3.548 | 4.030 | 6.840 | 7.384 | 9.081 |
| 40 | 3.547 | 4.035 | 6.838 | 7.380 | 9.068 |
| 45 | 3.547 | 4.047 | 6.834 | 7.373 | 9.038 |
| 50 | 3.549 | 4.060 | 6.833 | 7.367 | 9.011 |
| 55 | 3.554 | 4.075 | 6.834 | — | 8.985 |
| 60 | 3.560 | 4.091 | 6.836 | — | 8.962 |
| 70 | 3.580 | 4.126 | 6.845 | — | 8.921 |
| 80 | 3.609 | 4.164 | 6.859 | — | 8.885 |
| 90 | 3.650 | 4.205 | 6.877 | — | 8.850 |
| 95 | 3.674 | 4.227 | 6.886 | — | 8.833 |

*Source*: Reference 6.
[a] The compositions of these solutions are I: KH tartrate (saturated at 25°); II: KH phthalate, $m = 0.05$ mol kg$^{-1}$; III: KH$_2$PO$_4$, $m = 0.025$ mol kg$^{-1}$, Na$_2$HPO$_4$ $m = 0.025$ mol kg$^{-1}$; IV: KH$_2$PO$_4$, $m = 0.008695$ mol kg$^{-1}$, Na$_2$HPO$_4$ $m = 0.03043$ mol kg$^{-1}$; V: Na$_2$B$_4$O$_7$, $m = 0.01$ mol kg$^{-1}$

## Activity Standards, $p$D

An activity scale $p$D(S), or $-\log a_{D^+}$, the deuterium ion analogue of $p$H(S) has been established using the cell (20): Pt; $D_2$ (1 atm)|$D^+$ ($p$D buffer solution), $Cl^-$ (known molality), $D_2O$|AgCl; Ag for which the values of $E°_{cell}$ are known over a range of temperatures (21). The limiting acidity function, $p(a_D\gamma_{Cl})^0$ for three buffer solutions without added chloride was obtained from which values of $p$D(S) were derived for each of these three buffers by an equation similar to equation (4.9). When deuterium oxide is used as the solvent, the Bates–Guggenheim form of the Debye–Huckel equation equivalent to equation (4.9) at 25°C is

$$-\log \gamma_{Cl}^0 = \frac{0.541 I^{1/2}}{1 + 1.58 I^{1/2}}$$

This equation again assumes that the ion size parameter is 4.56 Å. The values of the Debye–Huckel constants $A$ and $B$ applicable to deuterium oxide are (21):

| $T$°C | 5 | 15 | 25 | 35 | 45 | 50 | 55 |
|---|---|---|---|---|---|---|---|
| $A$ | 0.5232 | 0.5319 | 0.5413 | 0.5513 | 0.5622 | 0.5678 | 0.5736 |
| $B$ | 0.3429 | 0.3488 | 0.3467 | 0.3485 | 0.3504 | 0.3513 | 0.3523 |

The reference values of $p$D(S) for citrate, phosphate, and carbonate buffers are given in Table 4.4.

The behavior of the hydrogen ion-responsive glass electrode in solutions prepared in deuterium oxide has been investigated primarily as a basis for the study of the ionization constants of weak acids (22–24), and it has been established that the potential of this electrode measured with respect to a calomel reference half-cell exhibits a theoretical response in this solvent (25, 26). A comparison between $p$H as determined with a glass electrode for an aqueous solution of buffer, and $p$D determined with the same electrode for the identical concentration of buffer in deuterium oxide has confirmed (26) that the empirical relations (25°C)

$$p\text{D} = p\text{H} + 0.41 \text{ (molar scale)} \quad \text{and} \quad p\text{D} = p\text{H} + 0.45 \text{ (molal scale)}$$

hold for $p$D values greater than 2 but less than 9.

Table 4.4. Reference Values of $p\text{D(S)}$ in Deuterium Oxide from 5–50°C[a]

| $T$°C | Citrate Buffer[a] | Phosphate Buffer[a] | Carbonate Buffer[a] |
|---|---|---|---|
| 5  | 4.378 | 7.539 | 10.998 |
| 10 | 4.352 | 7.504 | 10.924 |
| 15 | 4.329 | 7.475 | 10.855 |
| 20 | 4.310 | 7.449 | 10.793 |
| 25 | 4.293 | 7.428 | 10.736 |
| 30 | 4.279 | 7.411 | 10.685 |
| 35 | 4.268 | 7.397 | 10.638 |
| 40 | 4.260 | 7.387 | 10.597 |
| 45 | 4.253 | 7.381 | 10.560 |
| 50 | 4.250 | 7.377 | 10.527 |

Source: Reference 20.
[a] Compositions (in mol kg$^{-1}$): citrate buffer: $KD_2C_6H_5O_7$ (0.05); phosphate buffer: $KD_2PO_4$ (0.025), $Na_2DPO_4$ (0.025); carbonate buffer: $NaDCO_3$ (0.025), $Na_2CO_3$ (0.025).

## Activity Standards, $p$X

When the hydrogen ion selective glass electrode in the cell (4.10) is replaced by an ion selective electrode that responds reproducibly to changes of an ionic species X contained in the cell solution then the measurement of $p$X ($-\log a_X$) becomes analogous to the measurement of $p$H. The range of values of the cell potentials, particularly for divalent ions, however, is usually smaller than those encountered in $p$H measurements and the ionic strengths of the samples are often greater. Consequently, the magnitude of the liquid junction potential can be large and this effect can only be minimized if the sample and reference solutions have similar compositions. In the establishment of activity standards for solutions, the ionic strengths of which are less than 0.1, recourse can be made to the same assumption as that used in the establishment of the $p$H scale (see equation 4.9); that is,

$$-\log \gamma_{Cl} = \frac{0.511 I^{1/2}}{1 + 1.5 I^{1/2}} \quad (4.17)$$

This approach was used in the calculation of individual ion activities in solutions suitable for the calibration of ion selective electrodes (27), and has as its basis the relation applicable to a solution of an electrolyte $A_r B_s$

$$\gamma_\pm = \left(\gamma_+^r \gamma_-^s\right)^{1/\nu}$$

or

$$\nu \log \gamma_\pm = r \log \gamma_+ + s \log \gamma_- \qquad (4.18)$$

With a knowledge of the experimentally determined mean activity coefficients it is therefore possible to obtain an estimate of the single ion activity coefficients of cations in solutions of their chlorides provided the electrolyte is wholly dissociated. At ionic strengths less than 0.1, combination of equations (4.17) and (4.18) yields

$$r \log \gamma_+ = \nu \log \gamma_\pm + \frac{0.511 s I^{1/2}}{1 + 1.5 I^{1/2}}$$

which would yield for a solution of $CaCl_2$, for example,

$$\log \gamma_{Ca} = 3 \log \gamma_\pm + \frac{1.022 I^{1/2}}{1 + 1.5 I^{1/2}}$$

If a series of cationic activity coefficients, for example, $Na^+$, $K^+$, and $Ca^{2+}$, are thus calculated from the values of the appropriate mean activity coefficients of their chlorides, and it is assumed that these individual ion activity coefficients are identical in solutions of their bromide and nitrate salts, then three pathways leading to the values of $\gamma_{Br^-}$ and $\gamma_{NO_3^-}$ are available. By this type of procedure it has been possible to establish a series of values for the activity coefficient of a given ion at a given concentration. At ionic strengths below 0.1 there is reasonable agreement between these values (27).

This approach fails at higher ionic strengths, however, and the more precise procedure based upon the Stokes–Robinson equation for the mean activity coefficient is used (28, 29). In this equation

$$\ln \gamma_\pm = |z_+ z_-| \ln f_{DH} - \frac{h}{\nu} \ln a_w - \ln[1 + 0.018(\nu - h)] \qquad (4.19)$$

$z_+$ and $z_-$ are the charges of the cation and anion, respectively, $h$ is the hydration number of the total electrolyte, $\nu$ is the number of ions emanating from one electrolyte molecule, $a_w$ is the water activity and $\ln f_{DH}$ is a Debye–Huckel term defined as

$$-\ln f_{DH} = \frac{A I^{1/2}}{\left(1 + B a_i I^{1/2}\right)}$$

in which $A$, $B$, and $a_i$ are the usual terms in the Debye–Huckel equation. Equation (4.19) applies up to molalities of $12/h$, and has been extensively tested for 36 nonassociated electrolytes (29, 30). The application of this Stokes–Robinson hydration theory approach to the establishment of single ion activity coefficients is reviewed in Reference 31. The only extrathermodynamic assumption made in the derivation of equation (4.19) is that the hydration number, $h$ for an electrolyte, does not vary with concentration. Values of $h$ (32) for unassociated electrolytes (see Table 4.5) can therefore be assigned by fitting the experimental data in terms of the parameters $a_i$ and $h$ to the equation. Just as in the establishment of the $p$H scale it was necessary to adopt an extrathermodynamic assumption as to the activity coefficient of the chloride ion, so in the case of $p$X scales it is necessary to adopt a convention as to how the hydration number for an electrolyte can be apportioned between cation and anion. For univalent electrolytes the defining equations are

$$\log \gamma_+ = \log \gamma_\pm + 0.00782(h_+ - h_-)m\phi$$

$$\log \gamma_- = \log \gamma_\pm + 0.00782(h_- + h_+)m\phi$$

where $m$ is the molality, $\phi$ is the osmotic coefficient for the electrolyte, and $h_+$ and $h_-$ are the hydration numbers of cation and anion, respectively. The convention adopted by the National Bureau of Standards, U.S.A. is to assume that the chloride ion remains unhydrated in solution so that its

Table 4.5. Hydration Numbers for Some Halides

| Cation | Chloride | Bromide | Iodide |
|---|---|---|---|
| $H^+$ | 5.5 | 5.8 | 5.7 |
| $Li^+$ | 5.3 | 5.4 | 5.3 |
| $Na^+$ | 3.6 | 3.9 | 3.9 |
| $K^+$ [a] | 2.5 | 2.4 | 2.2 |
| $Rb^+$ | 2.2 | 1.8 | 1.5 |
| $Cs^+$ | 1.9 | 1.5 | 0.7 |
| $NH_4^+$ | 2.2 | — | — |
| $Mg^{2+}$ | 11.8 | — | — |
| $Ca^{2+}$ | 10.8 | — | — |
| $Sr^{2+}$ | 10.4 | — | — |
| $Ba^{2+}$ | 9.5 | — | — |

*Source*: Reference 32.
[a] $KF = 4.1$.

Table 4.6. NBS Values of $p\mathrm{X(S)}$ at 25°C[a]

$X = K^+$, $X = Cl^-$. Solutions prepared from dried reagent grade potassium chloride[a]

| Molality | mol L$^{-1}$ | $\gamma_{K^+}$ | $\gamma_{Cl^-}$ | $pK^+$ | $pCl$ |
|---|---|---|---|---|---|
| 0.001 | 0.000997 | 0.965 | 0.965 | 3.016 | 3.016 |
| 0.01  | 0.00997  | 0.901 | 0.901 | 2.045 | 2.045 |
| 0.1   | 0.0994   | 0.772 | 0.768 | 1.112 | 1.115 |
| 0.2   | 0.1983   | 0.723 | 0.713 | 0.841 | 0.846 |
| 0.3   | 0.2967   | 0.693 | 0.680 | 0.682 | 0.690 |
| 0.5   | 0.4916   | 0.659 | 0.639 | 0.482 | 0.495 |
| 1.0   | 0.9692   | 0.623 | 0.586 | 0.206 | 0.232 |
| 1.5   | 1.4329   | 0.611 | 0.558 | 0.0376 | 0.078 |
| 2.0   | 1.8827   | 0.609 | 0.539 | — | — |

$X = Na^+$, $X = Cl^-$. Solutions prepared from dried reagent grade sodium chloride[b]

| Molality | mol L$^{-1}$ | $\gamma_{Na^+}$ | $\gamma_{Cl^-}$ | $pNa$ | $pCl$ |
|---|---|---|---|---|---|
| 0.001 | 0.000997 | 0.965 | 0.965 | 0.3015 | 3.015 |
| 0.01  | 0.00997  | 0.903 | 0.902 | 2.044 | 2.045 |
| 0.1   | 0.0995   | 0.783 | 0.773 | 1.106 | 1.112 |
| 0.2   | 0.1987   | 0.744 | 0.727 | 0.828 | 0.838 |
| 0.3   | 0.2975   | 0.721 | 0.697 | 0.664 | 0.680 |
| 0.5   | 0.4941   | 0.701 | 0.662 | 0.455 | 0.480 |
| 1.0   | 0.9789   | 0.696 | 0.620 | 0.157 | 0.208 |
| 1.5   | 1.4543   | 0.718 | 0.602 | — | — |
| 2.0   | 1.9200   | 0.752 | 0.593 | — | — |

$X = F^-$ Solutions prepared from dried reagent grade potassium fluoride[c]

| Molality | mol L$^{-1}$ | Activity coefficient ($\gamma_{F^-} = \gamma_{K^+} = \gamma_\pm$) | $pF$ |
|---|---|---|---|
| 0.0001 | 0.0000997 | 0.988 | 4.00$_5$ |
| 0.0005 | 0.000499  | 0.975 | 3.31$_2$ |
| 0.001  | 0.000997  | 0.965 | 3.01$_6$ |
| 0.005  | 0.00498   | 0.927 | 2.33$_4$ |
| 0.01   | 0.00997   | 0.902 | 2.04$_5$ |
| 0.05   | 0.0498    | 0.818 | 1.38$_8$ |
| 0.10   | 0.0996    | 0.773 | 1.11$_2$ |
| 0.20   | 0.1990    | 0.726 | 0.83$_8$ |
| 0.30   | 0.2982    | 0.699 | 0.67$_8$ |
| 0.50   | 0.4961    | 0.670 | 0.47$_5$ |
| 0.75   | 0.7424    | 0.652 | 0.31$_1$ |
| 1.00   | 0.9873    | 0.645 | 0.19$_0$ |
| 1.50   | 1.4729    | 0.646 | $+0.01_4$ |
| 2.00   | 1.9523    | 0.658 | $-0.11_9$ |

Source: Reference 6.
[a] Uncertainties in $pK^+$ and $pCl$ are estimated not to exceed $\pm 0.01$.
[b] Uncertainties in $pNa$ and $pCl$ are estimated not to exceed $\pm 0.01$.
[c] Uncertainty in $pF$ is estimated not to exceed $\pm 0.01$.

## METHODS OF CALIBRATION

Table 4.7. Other $p\mathrm{X(S)}$ Values for Monovalent Cations at 25°C

| Reference Material Molality (mol kg$^{-1}$) | LiCl $p$Li | RbCl $p$Rb | CsCl $p$Cs | NH$_4$Cl $p$NH$_4^+$ |
|---|---|---|---|---|
| 0.01 | 2.033 | 2.044 | 2.045 | 2.043 |
| 0.10 | 1.097 | 1.116 | 1.121 | 1.112 |
| 0.20 | 0.810 | 0.846 | 0.858 | 0.840 |
| 0.50 | 0.406 | 0.495 | 0.518 | 0.483 |
| 1.00 | 0.054 | 0.226 | 0.264 | 0.208 |
| 2.00 | −0.392 | −0.055 | 0.003 | −0.080 |
| 3.00 | −0.754 | −0.232 | −0.157 | −0.261 |
| 4.00 | — | −0.368 | −0.278 | −0.397 |
| 5.00 | — | −0.481 | −0.376 | −0.509 |
| 6.00 | — | — | — | −0.602 |

*Source*: Reference 28.

hydration number is taken as zero. This approach has enabled the establishment of the NBS ionic activity scales which are based upon a series of solutions prepared from potassium chloride, sodium chloride, and potassium fluoride for which values of $p\mathrm{K}^+(\mathrm{S})$, $p\mathrm{Cl(S)}$, $p\mathrm{Na(S)}$, and $p\mathrm{F(S)}$ are calculated (see Table 4.6). These values have been found to be consistent with the observed responses of ion selective electrodes. At ionic strengths below 0.1 the agreement between the values calculated using the earlier approach of Bates and Alfenaar (27) and those quoted in Table 4.6 is within $\pm 0.01$. Other data from which to prepare reference $p\mathrm{X(S)}$ values for some monovalent and divalent cations are given in Tables 4.7 and 4.8, respectively. These data are also based upon the Stokes and Robinson hydration theory approach (33).

Table 4.8. $p\mathrm{X(S)}$ Values for Divalent Cations at 25°C

| Reference Material Molality (mol kg$^{-1}$) | MgCl$_2$ $p$Mg | CaCl$_2$ $p$Ca | SrCl$_2$ $p$Sr | BaCl$_2$ $p$Ba |
|---|---|---|---|---|
| 0.0333 | 1.884 | 1.901 | 1.901 | 1.905 |
| 0.10 | 1.554 | 1.570 | 1.575 | 1.587 |
| 0.20 | 1.321 | 1.349 | 1.360 | 1.389 |
| 0.333 | 1.123 | 1.165 | 1.184 | 1.232 |
| 0.50 | 0.932 | 0.991 | 1.021 | 1.083 |
| 1.00 | 0.463 | 0.580 | 0.646 | 0.777 |
| 2.00 | −0.459 | −0.186 | −0.035 | 0.385 |

The method of using $p\text{X}(S)$ values for the standardization of the X ion selective electrode is analogous to the method outlined for $p$H measurements in which two standards that bracket the activity of the sample are used. Alternatively, a full calibration curve can be constructed by plotting the measured values of $E_{\text{cell}}$ versus the $p\text{X}(S)$ values for a series of standard solutions. The measured value of $E_{\text{cell}}$ for the sample solutions can be interpolated from the graph in terms of the activity of X in the solution. It must be emphasized, however, that such interpolations are not valid if the ionic environment present in the sample solution does not match closely that present in the standard solutions, and are totally inapplicable to solutions of mixed electrolytes at ionic strengths greater than about 0.02.

Although the concentration of an ion rather than its activity is often the required quantity in routine analysis, there are circumstances under which the activity of an ion is of more fundamental interest not only in the applications of analytical chemistry, but also in studies of solution chemistry. This usage of potentiometry is important, therefore, because it provides one of the very few methods available for ion activity determinations. For example, in the analysis of blood serum it is the 'ionized calcium" that is thought to be greater physiological significance than the total calcium as derived from photometric and other methods. If the latter methods are used, then the fraction of the total concentration that is present as free calcium ion (i.e., "ionized calcium") must be computed by means of empirical nomograms. By contrast, utilization of the data (34) given in Table 4.9 permits the single point calibrations (see equation 4.13) of ion selective electrodes at an ionic strength of 0.15 from which it is possible to determine the activity of calcium ion, for example, by direct measurement.

### Concentration Standards

In analytical chemistry the quantity of interest is more likely to be in terms of concentration rather than activity and perhaps the measurement of the

Table 4.9. Three Reference Standards for $p\text{K}^+$ and $p\text{Ca}$ in Blood Serum

| Solution Composition ($M$) | | | Standard Values | | |
|---|---|---|---|---|---|
| NaCl | KCl | CaCl$_2$ | $p$Na | $p\text{K}^+$ | $p$Ca |
| 0.1450 | 0.0042 | — | | 2.504 | — |
| 0.1454 | — | 0.00126 | | — | 3.360 |
| 0.1414 | 0.0041 | 0.00121 | 0.966 | 2.514 | 3.373 |

*Source*: Reference 34.

activity of hydrogen ion rather than its concentration is the exception to this statement. Generally, however, it is much easier to prepare by direct measurement (i.e., weight, volume) a series of concentration standards for any ion than it is to establish a scale of activity standards for that ion since the latter requires the adoption of some theoretical convention as was shown in the previous section. The fundamental requirement when preparing the concentration standard solutions is that these must replicate as closely as possible the conditions that exist in, or that it is possible to create in, the sample solutions whereby it can be assumed that the activity coefficient of the determinand ion will be identical in both sets of solutions. Ideally, therefore, the solvent system must be the same for each in that the background compositions of samples and standards are identical both with respect to the quantity and nature of the neutral electrolytes present. Under these conditions of constant ionic medium, ion selective electrodes may be calibrated directly in terms of concentration. Isotonic saline, a 0.16 $M$ solution of sodium chloride, and many biological media meet the requirements of a constant ionic medium (35) and the establishment of concentration standards for these is of importance in clinical analysis since concentration is the medically significant quantity (36). Just as it was convenient to consider the activity scales $p$H and $p$X separately, so it is also convenient to subdivide the corresponding concentration scales into $pc_H$ and $pc_x$.

## Ionic Concentration, $pc_H$

The concept of a constant ionic medium has been applied to the determination by $p$H titration of the concentration ionization constants (Chapter 6) of weak acids and bases, and the stability constants of metal–ligand complexes (Chapter 8). These types of equilibria all have a term for the hydrogen ion included in the expression relating either the activities or concentrations of the participating species to the appropriate form of the equilibrium constant. If the activity of the hydrogen ion is measured, as in $p$H(X) measurements, and if the concentrations of the other participating species are converted to activities, then the equilibrium constant will be a close approximation to the true thermodynamic constant. This, of course, involves the calculation of the individual activity coefficients for the ionic species which can be done with a degree of certainty provided the ionic strength is low and the system being studied is confined to fairly simple acid-base equilibria. However, because of the variety of different ionic species present in the more complicated metal–ligand equilibria this assignation of individual activity coefficients to the calculated concentrations of the various species is not attempted. Instead, it is more realistic to determine the relevant stability constants under conditions where it can be assumed that the ionic strength is constant, the presumption being that the activity coefficients of these species will also remain constant. If this is so, then the stability constants for a given metal–ligand complex can be expressed wholly in terms of concentrations provided that the measured $p$H(X) can be converted into the corresponding ionic concentration term $pc_H$. This requirement has provided the impetus for the development of ion concentration scales, a typical example of which is credited to McBryde (37).

The cell without liquid junction

$$\text{glass electrode} | H^+ \text{(aqueous solution)}, Cl^- \ (c = 0.001 \ M) | AgCl; Ag \quad (4.20)$$

was used to measure $p(a_H\gamma_{Cl})$ values (see equation 4.7) of aqueous solutions by a method analogous to the measurement of $p$H. Data exists for the $p(a_H\gamma_{Cl})$ values of many buffer solutions (see Chapter 6) and, provided the chloride ion concentration remains the same in both the standard and unknown solutions, the equation analogous to equation (4.13) can be written as

$$p(a_H\gamma_{Cl})_X = p(a_H\gamma_{Cl})_S + \frac{E_S - E_X}{k} \quad (4.21)$$

for a cell in which the response of the glass electrode is perfect. The term $E_S$ is the cell e.m.f when the electrodes are immersed in a solution of known acidity function $p(a_H\gamma_{Cl})_S$ and $E_X$ is the potential when the same electrodes are placed into a solution of unknown acidity function $p(a_H\gamma_{Cl})_X$. (These types of measurements are dealt with in more detail elsewhere in this book.) The ionic concentration of hydrogen ion was assigned to solutions prepared from nitric and perchloric acids in the concentration range 0.1–0.001 $M$ all containing a constant chloride ion concentration of 0.001 $M$ (as sodium chloride). Into those solutions were placed sets of solutions the total anionic concentrations of which with respect to either nitrate (added as the potassium salt) or perchlorate (added as the sodium salt) were, in turn, 0.05, 0.1, 0.2, 0.5, and 1.0 $M$. A further set of perchlorate solutions was prepared with a total anion concentration of 3.0 $M$. Values of $p(a_H\gamma_{Cl})_X$ for each set of solutions were measured and, from the stoichiometry of the solutions, $pc_H$ was calculated. For each electrolyte at a particular value of the ionic strength a linear relationship of the form

$$p(a_H\gamma_{Cl})_X = A + Spc_H \quad (4.22)$$

was obtained from which values of the intercept $A$ and slope $S$ were evaluated. The values of the slopes were close to unity for each set of results allowing this equation to be written in the form

$$pc_H = p(a_H\gamma_{Cl})_X - A \qquad (4.23)$$

The values of $A$, given in Table 4.10, allow the conversion of a measured $p(a_H\gamma_{Cl})_X$ to the corresponding $pc_H$ value if the chloride ion concentration is maintained at 0.001 $M$ in all solutions. They are experimentally confirmed only in the range of $pc_H$ from 1 to 4 although, by analogy with the corresponding $p$H(X) measurements, it appears probable that the relations would apply also to the range over which the glass electrode has a linear response.

Similar equations to (4.22) and (4.23) were obtained when the $p$H(X) values for the preceding series of solutions were measured by McBryde (37) in the absence of added chloride. Thus the relation between $pc_H$ and $p$H(X) was found to be

$$pc_H = p\text{H(X)} - A \qquad (4.24)$$

and these experimentally determined values of $A$ are also included in Table 4.10. The liquid junction potential, introduced as a result of using the glass electrode–calomel reference half-cell combination, [see cell (4.10)] makes the values of $A$ less certain than those derived from the $p(a_H\gamma_{Cl})$ method since the liquid junction potential is particularly pronounced in solutions of acids. Nonetheless, the values of $A$ derived from these acid solutions were confirmed subsequently when the method was tested in solutions of alkali (37)

Table 4.10. Values of $A$ in Equations (4.23) and (4.24) Derived from $p(a_H\gamma_{Cl})_X$ Measurements [Cell (4.20)] and $p$H(X) Measurements at 25°C

| | $p(a_H\gamma_{Cl})_X$ | (Cl$^-$ = 0.001 $M$) | $p$H(X) | | |
|---|---|---|---|---|---|
| $I$ | KNO$_3$ | NaClO$_4$ | KNO$_3$ | NaClO$_4$ | KCl |
| 0.05 | 0.169 | 0.169 | 0.079 | 0.072 | 0.076 |
| 0.10 | 0.207 | 0.197 | 0.097 | 0.082 | 0.104 |
| 0.20 | 0.262 | 0.229 | 0.097 | 0.086 | 0.096 |
| 0.50 | 0.301 | 0.240 | 0.096 | 0.029 | 0.101 |
| 1.00 | 0.297 | 0.174 | 0.058 | −0.065 | 0.047 |
| 3.00 | — | −0.306 | — | −0.493 | — |

from which it was concluded that these values can be used to convert $p\mathrm{H}(\mathrm{X})$ into $pc_\mathrm{H}$ over the useful range of $p\mathrm{H}$. When cells with liquid junction are used, the higher concentrations of added electrolyte, particularly of potassium chloride, offer some advantage over the more dilute solutions. The enhanced precision that was observed when potassium chloride solution was used in the $p\mathrm{H}$ cell was attributed (37) to the effectiveness of this salt in reducing and stabilizing the liquid junction potential. Sodium perchlorate solutions gave $p\mathrm{H}$ readings of excellent stability when the salt bridge of potassium chloride was separated from the solution containing perchlorate by a tube filled with 3.5 $M$ ammonium chloride set in 3% agar. The precision of the results, however, was inferior to that found when potassium chloride was the cell electrolyte.

The usual method of determining the term $A$ in equation (4.24) is by the $p\mathrm{H}$ titration of a known concentration of acid, prepared in an ionic medium appropriate to the subsequent use of the $p\mathrm{H}$ cell, with a standardized solution of strong base. The $p\mathrm{H}(\mathrm{X})$ value, as defined by equation (4.15) for example, is measured after each incremental addition of base and the $pc_\mathrm{H}$ values at these points calculated from the known stoichiometry. The magnitude of each incremental addition of strong base is selected so that a number of values of $A$ may be obtained from readings taken within the range $pc_\mathrm{H}$ 1–3. Provided the ionic environment in the solution for which $pc_\mathrm{H}$ is to be determined resembles closely that used for the determination of $A$, then the error due to the liquid junction potential is likely to be less than that inherent in the routine measurement of $p\mathrm{H}(\mathrm{X})$. At higher $p\mathrm{H}(\mathrm{X})$ values, from pH 3 to 11, it seems probable that the constant ionic medium is effective in assuring the constancy of the liquid junction potential and that $A$ also remains constant within this range.

The constancy of $A$ in the range $p\mathrm{H}$ 3–11 can be checked provided the ionization constants of a series of weak acids or bases are known in terms of concentrations for the ionic medium of interest. The buffered region for a given acid covers the range of $pc_\mathrm{H}$ encompassed by $p\mathrm{K}_c \pm 0.6$ where $p\mathrm{K}_c$ is the ionization constant exponent expressed wholly in terms of concentrations. The stepwise neutralization of a known concentration of weak acid by standardized strong base within this buffered zone thus permits calculation of $pc_\mathrm{H}$, and if $p\mathrm{H}(\mathrm{X})$ is measured after each addition of alkali, then $A$ may be calculated. This approach has been used (38) to determine the relationship between $p\mathrm{H}(\mathrm{X})$ and $pc_\mathrm{H}$ for sodium chloride solutions of ionic strengths 0.04, 0.1, 0.15, and 0.20 by titrating acetic acid and 1,2-ethanediammonium dichloride, in addition to hydrochloric acid, with a standard solution of sodium hydroxide. The values of $p\mathrm{K}_c$ for acetic acid in sodium chloride solutions were taken from the literature (39) as also were the two $p\mathrm{K}_c$ values for the diamine (40). All of the data for the three acid-base systems at the four ionic strengths were colinear, within experimental error, being expressed by

$$p\mathrm{H}(X) = 0.9951\, pc_\mathrm{H} + 0.088 \qquad (4.25)$$

This equation, valid over the range $p\mathrm{H}$ 2–10.5, was written subsequently (41) as

$$pc_\mathrm{H} = p\mathrm{H}(X) - 0.062 \qquad (4.26)$$

confirming the constancy of $A$ over the $p\mathrm{H}$ range suggested by the work of McBryde (37). However, the colinearity of all the data implicit in equation (4.25) is surprising in view of the fivefold change in the ionic strength ($I = 0.04$–$0.20$) of the sodium chloride solutions. The constancy of $A$, therefore, in equation (4.26) can only be as the result of the cancellation of two quantities. The first of these is the liquid junction potential and, of course, the basic assumption in all $p\mathrm{H}$ measurements is that this potential remains invariant between the $p\mathrm{H}$ standards of $p\mathrm{H}(S)$ and the solution of unknown $p\mathrm{H}(X)$. If, as is likely, there is a small difference in the values of the junction potential $\Delta E_j$ between the two solutions, then the true $p\mathrm{H}(X)$ can be defined as

$$p\mathrm{H}(X) = pa_\mathrm{H} + \Delta E_j$$

$pa_\mathrm{H}$ being the measured $p\mathrm{H}$ of solution X. From the definition of an activity term this equation can be written as

$$p\mathrm{H}(X) = pc_\mathrm{H} + \Delta E_j - \log y_\mathrm{H} \qquad (4.27)$$

where $y_\mathrm{H}$ is the activity coefficient of hydrogen ion on the molar scale. The second quantity, therefore, is the activity coefficient of the hydrogen ion which may be constant for a set of data derived at one ionic strength, but will certainly assume a different value at another ionic strength. Since $A = \Delta E_j - \log y_\mathrm{H}$, the constancy of $A$ can only be assured if the increase in $\log y_\mathrm{H}$ as the ionic strength is increased is counterbalanced by a corresponding decrease in $\Delta E_j$. Generally, the liquid junction potential does decrease as the ionic strength of the medium is increased and the ionic environment of this medium tends more towards the ionic environment that exists in the potassium chloride solution of the calomel reference half-cell.

In later work (42), although the constancy of $A$ was found to be valid for the data derived at one ionic strength, the cancellation of the two effects observed in 0.04–0.2 $M$ sodium chloride solutions (38) was not observed for solutions of potassium chloride having similar ionic strengths ($I = 0.05$–$0.20$) covering a similar $p$H range ($p$H 2–12). Thus separate forms of an equation similar to (4.25) were needed for each ionic strength. The analysis by computer of 19 sets of titration data for universal buffer solutions variably composed of 0.003–0.005 $M$ potassium acetate, potassium dihydrogen phosphate, 1,2-ethanediammonium dichloride and borax resulted in the general equation at 25°C

$$pH(X) = A + Spc_H \qquad (4.28)$$

in which both the constants $A$ and $S$ vary with ionic strength. The variation of $A$ was found to be

$$A = 0.2269 - 1.949I + 7.7561I^2$$

and the corresponding variation of $S$ was

$$S = 0.9657 + 0.4752I - 1.532I^2$$

The standard deviation of the slope $\sigma(S)$ was 0.0022, while that of the intercept $\sigma(A)$ was 0.019 $p$H.

The magnitude of the corrections required to convert $p$H(X) to $pc_H$ are apparently greater in potassium chloride solutions than they are in sodium chloride solutions. This is surprising because potassium chloride is almost equitransferent and the liquid junction potential is therefore less in this media than it is in sodium chloride solutions (43). Furthermore, the activity coefficient of hydrogen ion is not likely to be substantially different in these two ionic media, particularly at the lower ionic strengths.

Finally, it should be noted that tables of the hydrogen ion concentration for some buffer solutions prepared in various ionic media are given in Reference 35 and data for $p$H buffers in seawater are contained in References 59 and 61.

### *Ionic Concentrations, $pc_X$*

Just as in the higher range of hydrogen ion concentrations (0.001–0.1) $pc_H$ standards can be prepared by dilution of solutions of strong acids with an electrolyte to yield a condition of constant ionic strength, so $pc_x$ standards can be prepared similarly. For example, the dilution of M(II) nitrates with a potassium nitrate solution to yield a total ionic strength of 0.1 can be used to prepare $pc_{Pb}$ and $pc_{Cd}$ standards covering the range $pc_M$ 2–4. At lower concentrations than this, extraneous effects such as ion adsorption on the walls of the vessel, and, in the preceding example, the complexation of the ion by the chloride ion emanating from the liquid junction of the calomel reference half-cell, can combine to cause appreciable errors. Thus during the calibration of a silver sulfide ion selective electrode with silver nitrate solutions, the lowest practical concentration was found to be $4 \times 10^{-7}$ $M$ when a Teflon® vessel was used. Ion adsorption on the vessel walls was found to be lowest for this material and increased in the order Teflon < Vycor® < polyethylene < hydrophobized Pyrex® < Pyrex (44). In the presence of a complexing agent, however, the response of the silver sulfide electrode can be followed down to a concentration of $10^{-20}$ (45), and silver complexes have been used to calibrate the electrode down to $10^{-24}$ $M$ (46).

Complexing agents are sometimes added, usually in conjunction with a $p$H buffer, to the electrolyte solution chosen to ensure that a condition of constant ionic strength exists in the standard and sample solutions. The resulting solution is often referred to as the *background solution* and its purpose is to mask, by complexation or by precipitation, species that may interfere with the primary response of the indicator electrode. Since the total concentration of electrolyte, $p$H buffer, and the complexant in the background solution is large, it is assumed that when the same quantity of this solution is added to the sample and standard solutions, the ionic medium in all the solutions will be identical not only with respect to ionic strength, but also with respect to $p$H and the degree to which the determinand ion is complexed. It is, of course, the free determinand ion to which the indicator electrode responds.

A masking reagent is one that decreases the concentration of a free metal ion or a ligand to a level where certain of its chemical reactions are prevented (47). There are thus two categories of background solutions depending upon whether the masking agent is selected to suppress interference caused by extraneous metal ions present in the sample solutions or whether its purpose is to suppress interference caused by the presence of an unwanted complexant (i.e., ligand) in the samples. General usage of terms, not entirely appropriate to distinguish between the two functions, has resulted in the categories being described as the following:

1. Total ionic strength adjustment buffers, abbreviated to TISAB.
2. Ligand interference preventive buffers, abbreviated to LIPB.

**Total Ionic Strength Adjustment Buffers.** TISABs were developed originally to diminish the concentrations of ions that interfere in the determination of fluoride ion, and then subsequently applied to the determination of calcium ion. They are now used also to stabilize the valency state of a determinand ion by suppressing the action of any oxidants that may be present in the sample.

Several TISABs, formulated specifically for fluoride ion determinations, have been developed. The first (48), designed to reduce interference caused when calcium, magnesium, chloride, nitrate, phosphate, and sulfate are present at high concentrations, has the following composition:

> Water (500 mL), glacial acetic acid (57 mL), sodium chloride (58 g), and sodium citrate dihydrate (0.3 g) diluted to 1 L.

If the sample solutions contain high concentrations of iron(III) and silicate, then the following TISAB is appropriate (49):

> Water (500 mL), sodium nitrate (85 or 170 g), sodium acetate trihydrate (68 g), and sodium citrate dihydrate (92.4 g) diluted to 1 L.

It should be noted, however, that fluoride ion forms an association complex with sodium ion (1, 50, 51) and it has been reported (52) that it is advantageous to use the corresponding potassium salts particularly when using CDTA (*trans*-1, 2-cyclohexanediyldinitrilotetraacetic acid) as masking agent for metal ions that complex fluoride ion. For example, the following TISAB is effective when aluminium ion is present at high concentrations:

> CDTA (0.011 $M$), potassium acetate (0.75 $M$), acetic acid (0.25 $M$), and potassium chloride (1.0 $M$).

Solutions used for the measurement of $E_{cell}$ are usually prepared by the 1:1 dilution of both samples and standards with these TISABs.

In the determination of calcium ion a TISAB that fulfills the dual function of masking magnesium ion by complexation with 2, 4-pentanedione and also reproducibly complexes calcium ion with iminodiacetic acid has been used (53). The composition of this aqueous solution is:

> Potassium nitrate (40.4 g), disodium iminodiacetate (3.6 g), 2, 4-pentanedione solution (160 mL, 0.5 $M$), ammonia (2 mL, 10 $M$), and ammonium chloride (1.07 g) diluted to 1 L.

Triethanolamine (1 $M$) provides the background solution for the determination of calcium ion in some biological samples.

Antioxidants are added to TISABs used for the determination of species susceptible to oxidation, and also provide a moderately reducing medium for certain other determinations. Thus in the determination of copper(II), formaldehyde or ascorbic acid is added to the TISAB (54) the composition of which is:

Sodium acetate (0.05 $M$), acetic acid (0.05 $M$), sodium fluoride (0.02 $M$), and formaldehyde (0.002 $M$).

The buffer is added to the sample in the ratio 1 : 1 prior to the potentiometric measurement. The acetate $p$H-buffering system ensures the optimal $p$H of 4.8 and the acetate ions themselves also ensure that a fixed ratio of complexes to free copper(II) ions exist both in the sample and standard solutions. Acetate ions also inhibit the adsorption of determinand ions on the walls of the vessel. Furthermore, this TISAB also ensures that there is no loss of copper(II) if iron(III) is present in the solution. Ascorbic acid also fulfills the role of antioxidant in a buffer (55) used in the determination of sulfide. The ionic strength adjustment solution also contains glycine and sufficient sodium hydroxide to yield a $p$H of 13.5.

**Ligand Interference Preventive Buffers.** There are two ways by which a ligand, when present in the sample solutions, can interfere with the correct functioning of the indicator electrode. The ligand can attack the materials used in the construction of the electroactive membrane of the electrode, and/or it can cause a diminution in the concentration of the free determinand ion that is impossible to predict or to allow for unless the identity, property, and concentration of this complexant is known. In these cases an LIPB can be used to mask either or both of these effects. For example (56), in the determination of copper(II) the problems encountered are variations of activities caused by complexation of copper(II) with ligands in the samples, and the abnormal response of the electrode in the presence of some ligands such as nitrilotriacetic acid (NTA) and ethylenediaminetetraacetic acid (EDTA). These types of interferents, including also the chloride ion, can be masked by adding an excess of a complexant that forms a more stable complex with copper(II) than any other ligand present originally in the samples. This is the purpose of an LIPB, which, in the case of the copper(II) determination, is fulfilled by polyamine ligands of which triethylenetetramine (1, 2 ethylenediiminobis [2-ethylamine]) is an example. Thus the avidity of triethylenetetramine is sufficient to displace any other bound ligand in the sample solutions when the tetramine is present at concentrations greater than that of copper(II). The formulation and use of LIPBs is very similar to those of the TISABs already described. They are prepared in a concentrated solution of electrolyte that provides the major contribution to the ionic strength, and, since the complexants are usually derived from weakly acidic and basic compounds, the excess of complexant is used also to provide the $p$H-buffering system. Like the TISABS, the use of

LIPBs is based upon the 1 : 1 dilution of both standards and samples with the buffered solution. The LIPB for the determination of copper(II) has the following composition (56):

Triethylenetetramine (0.40 $M$), nitric acid (0.20 $M$), and potassium nitrate (2.00 $M$).

This composition was found suitable for concentrations of copper(II) in the range $10^{-6}$ to $10^{-2}$ $M$. The LIPB was also effective in reducing the deleterious effects of halides upon the silver sulfide of the membrane. These kinds of anions convert silver sulfide at the membrane surface to the silver salts of the corresponding anions. Such conversions of the membrane are suppressed by the buffer because the solubilities of the silver salts increase in the presence of the tetramine.

Although the presence of triethylenetetramine reduces the concentration of the free copper(II) actually present in the solution to a very low level indeed, it has been shown (57) that the copper ion selective electrode gives a theoretical response to free copper(II) concentrations as low as $10^{-20}$ $M$ in the presence of polyamine buffers.

It is axiomatic that when a calibration is to be performed that the stock solution should be prepared using a salt of high and defined purity and that the water used for this and subsequent dilutions should be of the highest quality. The solutions to be used for the calibration can be prepared by serial dilution from this stock solution. However, if it is necessary to add a background solution in the analysis it is advisable to prepare the calibration solutions at, say, twice the required concentrations. The final twofold dilution is then achieved by adding together equal volumes of a standard solution and the background solution. The concentrations of these solutions should encompass the range likely to be encountered in similarly diluted sample solutions. When obtaining data from which to prepare a calibration curve, $E_{cell}$ is usually measured a minimum of three times at the same temperature for the whole of each series of calibrating solutions. In the first series of measurements the order in which the solutions are presented to the electrodes is from the most concentrated to the most dilute, the second series takes the solutions in the reverse order, and the third repeats the order used in the first series. The calibration curve is prepared by averaging and plotting the mean values of $E_{cell}$ versus $pc_x$. Alternatively, the parameters of this relationship can be obtained by regression analysis to reduce inaccuracy due to drifts in values of $E_{cell}$ or to hysteresis in these values. Interpolation of the concentration of determinand from the measured values of $E_{cell}$ for the sample solutions are valid only if the temperature of these solutions is within 1°C of the temperature used for the calibration, and each sample solution contains the same concentration of background solution as was used for the calibration. For samples of inherently low ionic strength, it can be expected that if identical and swamping amounts of the background components are present throughout, then conditions in the sample solutions will closely resemble those of the standards.

When the ionic strengths of the samples are inherently high, however, as in biological fluids ($I \sim 0.16$) and in seawater ($I \sim 0.7$), the technique of adding a background solution to achieve a condition of a swamping ionic strength becomes ineffective. In the case of body fluids, the ionic strength is sufficiently low that the actual single ion activities are not likely to be too far removed from those predicted by the adopted conventions described earlier in this chapter. Thus activity standards can be formulated for this ionic strength as is evident from the data that were given in Table 4.9. Such calibration data are applicable only if the conditions in the prepared calibration solutions are very similar to those normally encountered in the sample solutions, not only with respect to the activity range of the determinand ion, but also with respect to the composition, $p$H, and range of activities of the other species normally encountered in the samples. This is because these other species can, in some instances, modify the response of the indicator electrode. Moreover, as was pointed out in another section, in biological samples it is probable that concentration, rather than activity, is the medically significant quantity. Taking these factors into consideration, a more suitable method of calibration is to use a set of standard solutions the compositions of which closely parallel the medium of interest as to the composition and range of concentrations of the important inorganic constituents of serum normally encountered in clinical analysis (58). Solutions suggested as being suitable calibrating solution for sodium, potassium, and calcium ion selective electrodes (58) in this application are given in Table 4.11. As originally formulated, these solutions all contained sodium fluoride at a concentration of 0.001 mole L$^{-1}$ which enabled the fluoride ion selective electrode to be used as the reference electrode in the cells without liquid junction

$$\text{Na}^+, \text{K}^+, \text{ or } \text{Ca}^{2+}\text{ISE}|\text{standard} + \text{F}^-|\text{F}^-\text{ISE} \qquad (4.29)$$

This enabled the data obtained with this cell to be compared with the more usual type of cell

$$\text{Na}^+, \text{K}^+, \text{ or } \text{Ca}^{2+}\text{ISE}|\text{standard}\|\text{KCl}|\text{Hg}_2\text{Cl}_2; \text{Hg} \qquad (4.30)$$

Table 4.11. Composition of Calibration Standards[a] Used for Biological Fluids at Constant Ionic Strength of 0.16 mole $L^-$

| $p$H (calcd.)[b] | Total Na$^+$ | K$^+$ | Ca$^{2+}$ | Cl$^-$ | NaNO$_3$ | NaCl | HNO$_3$ | Tris | R[c] |
|---|---|---|---|---|---|---|---|---|---|
| 7.900 | 0.100 | 0.0010 | 0.00030 | 0.0666 | 0.03500 | 0.06500 | 0.0581 | 0.1153 | 1.016 |
| 7.800 | 0.105 | 0.0014 | 0.00038 | 0.0700 | 0.03716 | 0.06784 | 0.05246 | 0.09346 | 1.279 |
| 7.700 | 0.110 | 0.0018 | 0.00046 | 0.0750 | 0.03772 | 0.07222 | 0.04682 | 0.07589 | 1.610 |
| 7.600 | 0.115 | 0.0022 | 0.00054 | 0.0800 | 0.03828 | 0.07672 | 0.04118 | 0.06149 | 2.02. |
| 7.550 | 0.120 | 0.0026 | 0.00062 | 0.0850 | 0.03884 | 0.08116 | 0.03554 | 0.05116 | 2.275 |
| 7.500 | 0.125 | 0.0030 | 0.00070 | 0.0900 | 0.03940 | 0.08560 | 0.02990 | 0.04161 | 2.55. |
| 7.450 | 0.130 | 0.0034 | 0.00078 | 0.0950 | 0.03996 | 0.09000 | 0.02426 | 0.03273 | 2.86 |
| 7.400 | 0.135 | 0.0038 | 0.00090 | 0.1000 | 0.04060 | 0.09440 | 0.01850 | 0.02426 | 3.21. |
| 7.360 | 0.140 | 0.0040 | 0.00100 | 0.1060 | 0.04000 | 0.1000 | 0.01300 | 0.01669 | 3.52. |

*Source*: Reference 58.
[a] Concentrations in mole $L^{-1}$
[b] $p$H calculated at 37°C.
[c] R is the ratio Tris cation/Tris at the calculated $p$H.

that contains a liquid junction. The study revealed that when the ionic strength was varied over the range 0.12–0.20, the slight differences in the values of $E_{cell}$ recorded for cell (4.29) over this range could be accounted for almost wholly by alteration in the activity coefficients. In the case of cell (4.30), however, the junction potential changes with changing ionic strength in such a way as to counteract these activity effects. In practice this fortuitous cancellation means that changes in the ionic strength of samples would produce a smaller error in the cell with liquid junction than in cell (4.29). It will be recalled that a similar cancellation was postulated in the case of some hydrogen ion concentration measurements performed in solutions that also contained sodium chloride.

The comparatively high ionic strength of seawater further stabilizes the activity coefficients of solutes present in small concentrations and has been found to be effective in reducing or eliminating the differences in liquid junction potential in seawater systems (59). Thus it can be regarded as a "constant ionic medium." A panel on conventions for seawater equilibria (60) has found it permissible that artificial seawater solvents used for standard reference solutions be prepared from the three salts sodium chloride, magnesium chloride, and sodium sulfate. A closer approximation to the composition of natural seawater can be achieved, however, by the inclusion of both calcium and potassium ions (59). A typical composition having a salinity of 35% and a formal ionic strength of 0.7229 expressed as molality in water is:

| | | |
|---|---|---|
| NaCl | 0.42664 mole kg$^{-1}$ | 24.061 g kg$^{-1}$ |
| Na$_2$SO$_4$ | 0.02926 | 4.011 |
| KCl | 0.01058 | 0.761 |
| CaCl$_2$ | 0.01077 | 1.153 |
| MgCl$_2$ | 0.05518 | 5.069 |

There is a difference between the formal and the actual ionic strength when synthetic seawaters contain sulfate ion as a result of the ion pairing that occurs when this ion is present. If this effect is allowed for, then the effective ionic strength is 0.6729. Scales of hydrogen ion concentrations have been devised for synthetic seawaters with and without added sulfate. Details of these molality based scales (i.e., $pm_H$ rather than $pc_H$) are given in References 59 and 61. The compositions of other synthetic seawaters are also given or are deducible from the detail given in these references, and the development of a chemical model for seawater is discussed in Reference 62. In some instances synthetic seawaters can be used as the background solution in which to prepare standard solutions of species not contained in them. The alternative, used in the determination of copper(II) at the parts per billion level (63), is to use a standard addition method.

A method of checking the response of electrodes at low concentrations of an electroactive ion is to stabilize the ion by complex formation. Not only does this expedient diminish certain effects, but it can also increase the stability of the system as a whole toward other types of changes caused by contaminants such as those that would alter its $pH$. An example of this application is to be found in the use of metal ion buffers, colloquially called $pM$ *buffers*. In this context the term $pM$ is used as an abbreviation of $pc_M$ and thus denotes $-\log c_M$. The use of $pM$ buffers stems from an idea suggested by Schmid and Reilley (64) who were interested in mercury(II)–EDTA complexes. Subsequently, $pM$ buffers based on EDTA or nitrilotriacetic acid complexes with copper(II) were used (65) for the calibration of the copper(II) ion selective electrode. These two ligands have also formed the basis for $pCa$, $pCd$, $pCu$, and $pPb$ buffers at various $pH$ (66–68) each covering a wide $pM$ range. It must be emphasized, however, that the reliability of these $pM$ values depends wholly upon the reliability of the data from which they are calculated. Serious errors will arise, therefore, unless the temperature and ionic strength in the calibrating solutions, as presented to the indicator electrode, closely duplicate those under which the equilibrium constant data was obtained originally. In some published $pM$ values this basic information is missing and thus it is difficult to check the calculations. It is preferable to avoid uncertainty and to produce one's own $p$M standards based on stability constant data (69–72) using the methods of calculation given by Perrin and Dempsey (14). Note also that methods for producing suitable metal ion and ligand buffers is the subject of a patent held by Radelkis in Budapest, Hungary (73).

### 4.2 STANDARD ADDITION / SUBTRACTION METHODS

These methods are applied advantageously when there is uncertainty as to the background compositions and concentrations of the electrolytes present in the samples, or if the samples contain high concentrations of complexing agents. When the determinand ion concentration is deduced by measuring the change in $E_{cell}$ that results from the addition of a known concentration of determinand ion to the sample solution, the method is referred to as the *standard addition method*. Conversely, the *standard subtraction method* relates the original concentration of determinand ion to the change in $E_{cell}$ when the concentration of that ion is quantitatively diminished by the addition of a known amount of complexant or precipitant to the sample solution. In both methods the background solution is effectively provided by each of the sample solutions and this background may be different, therefore, for each sample within a batch. An important preliminary usually required is that the response of the electrode to the determinand ion must be known by a prior calibration to determine the slope of the $E_{cell}$ versus log $c$ relation. Although background solutions would need to be used for this calibration, they need not bear a close resemblance to the sample solutions for either of the two methods. Once determined, the slope is usually quite stable and needs only periodic checking. However, the concentration range of each sample solution with and without the known addition should lie within that portion of the calibration data for which a constant value of the slope is obtained.

The basic assumptions for both the standard addition and the standard subtraction methods are that the magnitudes of the liquid junction potentials in the respective cells do not change as a result of the "spike" made to the sample solution in the addition method or as a result of the addition of a reagent that causes the calculable diminution of the determinand concentration in the subtraction method. Neither must the activity coefficient of the determinand ion in the sample solution be altered significantly by these additions. Furthermore, if a fraction of the determinand ion is complexed either deliberately in the pretreatment of the sample or as a result of the inherent condition in the sample, then this fraction must also remain constant throughout.

Standard addition/subtraction methods are conveniently subdivided according to the number of known volumes of the appropriate standard solution added to the sample. If, as is common, only one aliquot is added, then the method is classified as a single standard addition or subtraction method depending upon whether the concentration of determinand is increased or diminished by the addition. Of the corresponding multiple standard addition/subtraction methods, the multiple standard addition is the more commonly used. At least five additions of the standard solution are made to the sample solution in these methods. The methods are occasionally referred to as titrations, the single standard addition/subtraction methods sometimes being called *single point titrations*. For the reasons given in Chapter 2, these names are not used here.

### Single Standard Addition / Subtraction Methods

The equation upon which the single standard addition method depends is deducible from the basic Nernst relation written in a form that embraces the foregoing assumptions. Thus when the electrodes are dipped in the sample solution initially, the potential $E_{cell_1}$, is related to the concentration of determinand ion by

$$E_{cell_1} = constant + slope \log c_x f_x \qquad (4.31)$$

The term *constant* is a composite of the ideal potential of the cell when the activity of the determinand ion is unity (i.e., $E_{cell}^\circ$), the contribution of the liquid junction potential to $E_{cell}$, and the constant activity coefficient term. In this equation the value of *slope* has been predetermined as outlined previously and can have either a positive or a negative value depending upon the polarity of the indicator electrode with respect to the reference half-cell. The term $f_x$ is the fraction of determinand ion present in the cell solution that is not complexed and $c_x$ is the unknown concentration in the sample solution. If the volume of the cell solution is $V_x$ and a known volume $V_s$ of a standard solution having a concentration $c_s$ is added, then the total determinand ion concentration will be $(V_x c_x + V_s c_s)/(V_x + V_s)$ and the potential of the cell will now be given by

$$E_{cell_2} = constant + slope \log f_x \left( \frac{V_x c_x + V_s c_s}{V_x + V_s} \right) \qquad (4.32)$$

If $E_{\text{cell}_1}$, is measured before the addition and $E_{\text{cell}_2}$ measured after the addition, then the change in cell potential $\Delta E$ defined as $E_{\text{cell}_2} - E_{\text{cell}_1}$ becomes related to the unknown concentration $c_x$ by

$$\Delta E = slope \log \frac{V_x c_x + V_s c_s}{c_x(V_x + V_s)} \qquad (4.33)$$

provided $f_x$ remains the same throughout. Thus

$$c_x = \frac{c_s V_s}{10^{\Delta E/S}(V_x + V_s) - V_x} \qquad (4.34)$$

where $S$ is the slope. From this equation it is seen that the precision to which $c_x$ can be determined depends upon the reliability to which $c_s$, $V_s$, $V_x$, $S$, and $\Delta E$ are known. The relevant concentration and volume terms can, of course, be prepared and/or measured quite accurately, and the slope, $S$ can usually be determined with a relatively high degree of precision. The principal cause of concern is the uncertainty in the measurement of $E_{\text{cell}}$ values and this uncertainty will become more significant the smaller the value of $\Delta E$. Once again, two opposing effects must be reconciled to ensure the highest precision. Thus the aim must be to make $\Delta E$ as large as possible without changing either $f_x$, the liquid junction potential, or the activity coefficient of the determinand ion as a result of a concomitant large aliquot of added standard solution $V_s$. Thus $V_s$ should be minimized by maximizing $c_s$ and a micrometer syringe is useful in dispensing this solution. The usual technique for making measurements is to add 1 mL of a solution about 100 times more concentrated than the expected concentration of the sample to 100 mL of the sample. It is advisable that $\Delta E$ should not be less than 5 mV, the preferable range being between 10 and 30 mV.

An alternative method not frequently employed because complexants must be absent from the samples is to reverse the procedure by placing the electrodes initially in a known volume of standard solution of the determinand and measuring the change in cell e.m.f. when a known volume of the sample solution is added (74). The equation analogous to (4.34) is

$$c_x = \frac{10^{\Delta E/S}(V_s + V_x)c_s - c_s V_s}{V_x} \qquad (4.35)$$

where $c_x$ and $c_s$ are the sample and standard concentrations, and $V_x$ and $V_s$ are the respective volumes.

The assumptions on which the standard addition method is based apply also to the standard subtraction method, the equation relating the measured change in $E_{cell}$ to the unknown concentration $c_x$ being

$$\Delta E = slope \log \frac{V_x c_x - nV_s c_s}{c_x(V_x + V_s)} \quad (4.36)$$

in which, as in equation (4.33), the subscripts $x$ denote the sample concentration and volume. In this case, however, $V_s$ is the volume of a reagent that reacts quantitatively and exclusively with the determinand ion and $c_s$ is the concentration of this reagent. The reaction is such that $n$ moles of the added reagent cause the removal of one mole of determinant ion. This equation rearranges to:

$$c_x = \frac{nV_s c_s}{V_x - 10^{\Delta E/S}(V_s + V_x)} \quad (4.37)$$

where $S$ is the slope. The same remarks pertain to the precision of this method as were made in the case of the standard addition method. The extent of the concentration change brought about by the addition of the reagent should be such that the original concentration in the sample is approximately halved.

### The Gran and Related Methods of Multiple Standard Additions

In general, the principles of the foregoing single standard addition/subtraction techniques may be extended to include the corresponding multiple additions of the standard solution. A simple treatment of the technique which dispenses with the need to know the value of slope is given in Reference 75, and a novel computer evaluation of this method of multiple standard additions is given in Reference 76. However, the most convenient of the methods of calculation available for the multiple standard addition/subtraction techniques is based upon an adaptation of the method of Gran (77) that was developed originally to linearize potentiotitrimetric curves. In its original form the titration data are transformed into functions which, when plotted against the volume of titrant added, give straight lines that intersect the volume axis at the equivalence point. The Gran transformation rearranges the Nernst equation

$$E_{cell} = E^{\circ}_{cell} + S \log a_x$$

where $S$ is the slope, into a form

$$10^{E/S} = \text{constant} \cdot a_x \quad (4.38)$$

in which the ideal value of the constant would be $10^{E^{\circ}/S}$. The subscript *cell* has been omitted from $E$ and $E°$ in these equations for convenience. Thus $10^{E/S}$ is proportional to $a_x$ and may be plotted, as a measure of $a_x$, against the volume of titrant to yield a straight line relation. The experimentally derived line is then extrapolated back to an intercept on the volume axis at $a_x = 0$, and this volume is the equivalence volume of the titration.

In practice, the conditions that must be observed when using the single addition/subtraction methods apply also to the Gran multistandard addition method. Thus the nature of the background solution must not alter significantly as a result of the additions so that equation (4.38) becomes analogous to that obtained by rearrangement of equation (4.32), for example. That is, for a precalibrated value of the slope,

$$10^{E/S} = \text{constant}\frac{(V_x c_x + V_s c_s)}{(V_x + V_s)}$$

or

$$(V_x + V_s)10^{E/S} = \text{constant}(V_x c_x + V_s c_s) \qquad (4.39)$$

The method of obtaining the unknown concentration $c_x$ is based upon equation (4.39) and is applied as follows. The sample solution is usually prepared in an appropriate background solution (TISAB, LIPB) and a known aliquot volume of this solution $V_x$ is transferred into the measuring cell and $E_{\text{cell}}$ is measured. Known volumes $V_s$ of a standard solution having a known concentration $c_s$ of the determinand are then added sequentially to the cell solution, the value of $E_{\text{cell}}$ being measured after each addition. At least five such standard additions should be made and it is advisable that the total quantity of determinand ion added to the sample solution is such that an overall change of 20–50 mV (for a monovalent ion) occurs in the value of $E_{\text{cell}}$. If the left-hand side of equation (4.39) is plotted against each corresponding value of $V_s$, a straight line plot should be obtained. Extrapolation of this line back to $(V_x + V_s)10^{E/S} = 0$ will yield a negative volume intercept $V_s'$ from which the concentration of the sample solution placed originally in the cell may be calculated since $c_x = V_s' c_s / V_x$. An example of this method is given in Table 4.12.

For routine analysis it is convenient to use special Gran plot paper (10% volume corrected) that is available from Orion Research Inc., among others. The ordinate of the paper is antilogarithmic so that values of $E_{\text{cell}}$ plotted on it become converted to $10^{E/S}$ since it is assumed that the value of $S$ is theoretical. Thus readings of $E_{\text{cell}}$ are plotted on the ordinate against the volume along the abscissa, the scale of the latter being linear. Standardization on procedure is necessary when using this paper in order to ensure that

## STANDARD ADDITION/SUBTRACTION METHODS

**Table 4.12. An Example of Gran's Method**[a]

| Electrode: F$^-$ | | | Sample volume = 100 mL seawater ($V_x$) | | | |
|---|---|---|---|---|---|---|
| Slope: $-59$ mV decade$^{-1}$ | | | Standard NaF soln = 100 ppm F$^-$ ($c_s$) | | | |
| $V_s$ mL | 0.5 | 1.0 | 2.0 | 4.0 | 6.0 | 8.0 | 10.0 |
| $E_{cell}$ (mV) | 72.0 | 66.0 | 58.0 | 46.0 | 39.0 | 33.0 | 29.0 |
| $(V_x + V_s)10^{-E/59}$ | 6.1 | 7.7 | 10.6 | 17.3 | 23.1 | 29.8 | 35.5 |

$$\text{Intercept} = -1.5 \text{ ml} = V_s'$$
$$c_x = c_s V_s'/V_x = 1.5 \text{ ppm}$$
$$\text{Sample contains 1.5 ppm F}^-$$

[a]See equation (4.39).

the volume change as a result of all the additions is exactly 10%. For example, if the initial sample volume in the cell is 100 mL then the total volume of added standard solution must finally be 10 mL. A special Gran ruler has also been prepared that has $10^{E/S}$ scales for monovalent and divalent ions with volumetric correction (78). With the help of this ruler, Gran plots can be done on ordinary millimeter paper. However, electronic calculators have made the full calculations based on equations of the type (4.39) less burdensome than they once were and it is preferable to use these calculations of the value if the slope is not theoretical as is so often the case.

One advantage of the Gran multistandard addition method is that the final result is less dependent upon random errors in individual measurements of $E_{cell}$ since this result is derived from several readings. Another advantage is that the effects of interferents upon the performance of the electrode can be diminished considerably because the final concentration of determinand can be made sufficiently high so that any interference is effectively swamped. The result is then calculated by extrapolation from these interference-free readings.

A variant of the Gran method has been described (79) and is best explained by the following example, which is based upon the determination of fluoride ion in seawater (see also Table 4.12).

**Predetermined Slope (F$^-$ selective electrode) = $-59$ mV decade$^{-1}$**

| Reading Number | 1 | 2 | 3 | 4 | 5 | 6 | 7 | 8 | 9 | 10 | 11 |
|---|---|---|---|---|---|---|---|---|---|---|---|
| Added NaF (ppm) | 0 | 0.01 | 0.05 | 0.5 | 1.0 | 2.0 | 3.8 | 5.7 | 7.4 | 9.1 | 16.7 |
| $E_{cell}$ (mV) | 80 | 80 | 79 | 72 | 66 | 58 | 46 | 39 | 33 | 29 | 14 |

The Nernst equation for these results can be written as

$$E_{cell} = \text{constant} - S\log(x + b)$$

where $x$ is the unknown concentration of fluoride ion in the sample and the values of $b$ are the known concentrations of the added sodium fluoride. From an inspection of the given data it can be concluded that there should be two linear regions in the graph obtained by plotting $E_{cell}$ versus log $b$, as follows:

1. When $b$ is small so that $x \gg b$. Thus

$$E_{cell} = \text{constant} - S \log x$$

which, for a given sample, means that $E_{cell}$ remains substantially constant when small quantities of standard sodium fluoride are added (see Reading numbers 2 and 3).

2. When $b$ is large, such that $b \gg x$, a linear relation of $E_{cell}$ versus log $b$ will be observed for which the slope should be the same as the predetermined slope of the electrode (i.e., $-59$ mV decade$^{-1}$ in the example). Reading numbers 10 and 11 are close to fulfilling this condition within the experimental error of $\pm 1$ mV.

The line connecting these two linear regions will be curved for the readings numbered 4 through 8. If the two straight lines are extrapolated so that they intersect, then the value on the log $b$ axis corresponding to this point of intersection is equal to the logarithm of the concentration of fluoride in the sample, that is, log $x$.

A better method of evaluating $x$, however, is to process the data so that only one linear extrapolation is necessary. To this end a reference reading is chosen arbitrarily from the curved portion of the graph and the changes in the measured values of $E_{cell}$ from this reference can be related to the sample concentration as follows. If the reference reading is chosen so that a known concentration $b_R$ of fluoride has been added to the sample and the corresponding value of $E_{cell}$ is $E_R$, then

$$E_R = \text{constant} - S \log(x + b_R)$$

The corresponding equation for any other standard addition of concentration $b_N$ is

$$E_N = \text{constant} - S \log(x + b_N)$$

and the difference between these equations is

$$E_N - E_R = -S \log \frac{x + b_N}{x + b_R} = \Delta E$$

For the given data $S$ is $-59$ mV decade$^{-1}$; hence

$$10^{\Delta E/59} = \frac{x + b_N}{x + b_R}$$

Letting

$$y = 10^{\Delta E/59}$$

$$b_N = y(x + b_R) - x$$

Hence a plot of the values of $b_N$ versus the corresponding values of $y$ should be linear with an intercept of $-x$ at $y = 0$ and a slope of $x + b_R$. The example data yields the following results by this method.

Let the reference reading be number 4 for which $b_R = 0.5$ ppm and $E_R = 72$ mV.

| Reading Number | 4 | 5 | 6 | 7 | 8 | 9 | 10 |
|---|---|---|---|---|---|---|---|
| $b_N$ | — | 1.0 | 2.0 | 3.8 | 5.7 | 7.4 | 9.1 |
| $\Delta E$ | 0 | 6 | 14 | 26 | 33 | 39 | 43 |
| $y$ | — | 1.3 | 1.7 | 2.8 | 3.6 | 4.6 | 5.4 |

Intercept $= -1.5$; $x = 1.5$ ppm F$^-$
Slope $= 1.9$; $x = 1.4$ ppm F$^-$

An advantage of this method is that it allows an internal check to be made for the consistency of the data since if there are any interferences present, the values of $x$ calculated from the slope and the intercept will be significantly different. In the example these values were obtained by a linear regression analysis of the data, and agree with the sample concentration (1.5 ppm F$^-$) calculated from the same data by Gran's method (see Table 4.12).

Comparative studies have been made on the precision of measurement techniques used with ion selective electrodes (80). Some 11 techniques are reviewed which include the direct measurement of sample concentration after multipoint calibration as well as the standard addition/subtraction methods outlined in the foregoing section. Variants of these methods are also considered in Reference 80 in which the formulae developed make it possible to calculate the expected error for any strategy of standard addition or subtraction and to optimize precision by proper selection of the concentration and volumes of standard solutions.

### Differential Potentiometry

The analytical application of differential potentiometry outlined in Chapter 2 can also be regarded as a standard addition method. Thus it was shown in the discussion of differential potentiometry that the concentration of determinand in the sample half-cell of a concentration cell with liquid junction can be deduced from the volume of a standard solution of determinand which must be added to the reference half-cell in order to obtain a cell e.m.f. of zero. In practice, however, the graphical or least squares interpolation of the results yields more reliable results since the internal consistency of the data can be checked by this method. The method, as modified by R. A. Durst and co-workers (81–83), can also be used with minute amounts of the sample solution ($< 0.1$ mL). For example, the concentration cell with diffusion used for the microanalysis of fluoride ion (83) can be represented as

$$\mathrm{LaF_3 | F^-}(c_x), \mathrm{KNO_3}(0.1\ M) | \mathrm{KNO_3}(0.1\ M) | \mathrm{KNO_3}(0.1\ M), \mathrm{F^-}(c_s) | \mathrm{LaF_3}$$
$$\text{sample} \qquad\qquad \text{salt bridge} \qquad\qquad \text{known concn.}$$

(4.40)

In this cell the junction potential was negligible because the same concentration of background electrolyte (0.1 $M$ $KNO_3$) was used throughout. The sample half-cell in this notation contained 0.01 mL of the sample solution applied to the external surface of the lanthanum fluoride membrane of a fluoride ion selective electrode which was used in the inverted position. In order to use the electrode in this position, electrical contact between the internal surface of the membrane and the internal reference solution was maintained by setting this solution (0.1 $M$ NaF, 0.1 $M$ KCl) as a gel in 4% agar. The geometry of the area circumjacent to the external surface of the membrane was also modified so that the sample solution was confined to the desired area of the membrane. An epoxy dike formed around the membrane was found suitable for this purpose and evaporation of the sample was minimized by placing a polyethylene cap over the entire end of the electrode. A hole drilled in this cap allowed the insertion of the polyethylene capillary end of a salt bridge containing 0.1 $M$ potassium nitrate in 4% agar into the sample solution. The other end of this salt bridge made contact with a solution (100 mL) of sodium fluoride in 0.1 $M$ potassium nitrate contained in a tall form beaker. The half-cell was formed by immersing an unmodified fluoride ion selective electrode into this solution. The addition of known volumes from a microburette of a standard solution of sodium fluoride, also prepared in 0.1 $M$ potassium nitrate, ensured that the concentration of fluoride ion $c_s$ in the half-cell was calculable.

In the two half-cells of cell (4.40) not only does the high concentration of background electrolyte ensure that the junction potential of the cell is negligible, but it also ensures that for practical purposes the activity coefficients of fluoride ion will be identical in each. Thus if there is no bias potential between the electrodes, then equation (2.3) will be applicable. However, in this case the internal reference systems of the two electrodes were different since the solution in the electrode of the sample half-cell had been immobilized in 4% agar. The large bias potential, $\Delta E$ of 116 mV was attributed to this difference. Equation (2.3) must be modified to take account of this and can therefore be written as

$$E_{cell} = \Delta E - S \log \frac{c_x}{c_s} \qquad (4.41)$$

For this equation to be valid the slopes $S$ of the two electrodes must also be identical. An important preliminary to methods based on differential potentiometry is to determine the bias potential between the electrodes and their individual slopes. This is done by calibrating them in a chemical cell with liquid junction against the same reference half-cell (e.g., a calomel reference electrode) using the same series of calibrating solutions. Once obtained, the bias potential is used to correct the observed values of $E_{cell}$ [cell (4.40)] so that when these corrected values are plotted against $\log c_s$, the concentration of fluoride ion in the sample $c_x$ is deducible from the value of $\log c_s$ when the corrected value of $E_{cell}$ is zero. Assuming now that $E_{cell}$ automatically denotes the corrected cell e.m.f., it is best in the early stages to add the incremental volumes of standard solution until $E_{cell}$ is about 30 mV before recording $E_{cell}$ values. Thereafter, the absolute values of $E_{cell}$ will decrease with each incremental addition and then start to increase again as the cell changes polarity after $E_{cell} = 0$. The incremental additions should be continued until $E_{cell}$ is once again about 30 mV (opposite polarity). The concentration of fluoride ($c_s$) added to the 100 mL 0.1 $M$ potassium nitrate ($V_{KNO_3}$) transferred originally to the tall form beaker is calculated for each incremental addition allowing for the volume change by means of the equation

$$c_s = \frac{c_a V_a}{V_{KNO_3} + V_a}$$

where $c_a$ is the concentration of the standard solution and $V_a$ is the volume of this solution added from the microburette. A straight line graph is obtained when the values of $E_{cell}$ are plotted against the corresponding values of $\log c_s$ and this line cuts the $\log c_s$ axis at $E_{cell} = 0$. This point corresponds to the logarithm of the concentration in the sample half-cell, that is, $\log c_x$.

Noting that the volume in the half-cell to which the standard sodium fluoride was added was 100 mL whereas the sample volume in the other half-cell was only 0.01 mL, leads to the conclusion that the number of moles of fluoride ion that need to be added to the 100 mL must be $10^4$ times greater than the amount present in the micro-half-cell to achieve the same concentration in each half-cell. Herein lies the advantage of this type of differential potentiometry which is that it permits the determination of amounts too small to be otherwise added with any degree of precision or accuracy. The method allowed fluoride solution $10^{-3}$ to $2 \times 10^{-6}$ $M$ (containing 190–0.38 nanograms in 0.01 mL) to be determined with an accuracy of 1% over this range of concentrations. To achieve this it was necessary to control the temperature of the solutions by installing the apparatus in a temperature controlled room and to further insulate the concentration cell against local temperature fluctuations by surrounding it with a 2.5–5 cm thickness of polyurethane foam.

Differential potentiometry has also been applied to the determination of parts per billion chloride ion in high purity waters such as boiler feed water or nuclear reactor water (84). In this instance a pair of chloride ion selective electrodes formed part of a concentration cell with liquid junction. The detection limit was found to be 6 ppb chloride ion and the relative standard deviation was 3% at 100 ppb.

## 4.3 SOME RELEVANT PROPERTIES OF ION SELECTIVE ELECTRODES

In assessing the potential utility of a particular type of ion selective electrode to a given application, it is important to know how its response is likely to vary with the $p$H of the medium, the concentration limits between which the electrode is likely to exhibit a linear response, and the degree to which ions other than the determinand ion will affect this response. If the electrode was purchased from a commercial supplier, then the literature provided with it will include this information in the specifications. When the electrode is calibrated by the triplicated method, it is likely that the user will be able to assess the degree to which other extraneous effects, if any, are interfering with the correct functioning of the electrode. The acquisition of these data will allow an assessment to be made of the response time of the electrode, the stability and reproducibility of its potential, and whether it appears sensitive to slight changes in the temperature of the surroundings or whether it is affected by light. It is again emphasized that electrodes with intermetallic connecting junctions above the practical immersion level of the electrode are likely to respond to changes in room temperature. Exclusion from draughts is important in these instances.

When comparing the respective merits of two or more different types of electrodes that respond to the same electroactant, some of the references cited in Tables 3.4, 3.5 and 3.6 contain useful information.

## Influence of $p$H on the Responses of Membrane Electrodes

With the obvious exception of the hydrogen ion responsive glass electrode, hydrogen and hydroxyl ions can be regarded as ubiquitous interferents for all other membrane electrodes. Thus the gas sensors must be used at a particular $p$H (see Table 3.6), whereas for the unmodified membrane electrodes there is a range of $p$H values within which the response of the electrode is not susceptible to changes in $p$H. Generally, the further the $p$H deviates from the lower and/or upper extremities of this range, so the response of the electrode becomes increasingly dependent upon the $p$H. It is for this reason that measurements are usually made in solutions that are buffered with respect to $p$H.

The sodium ion responsive glass electrode is also very responsive to changes in $p$H. These effects have been investigated for various types of glasses (85–87) from which it can be deduced that samples containing sodium ion must be buffered so that $p$H $>$ $p$Na $+$ 3 or 4. The lower limit of sodium ion activity is set, therefore, by the upper $p$H limit of a buffering system which must be as alkaline as is practicable yet must also be free of sodium ion. For example, when the sodium ion concentration in a sample is $< 10^{-7}$ $M$, the $p$H of the buffer should be greater than 11. Measurements have been made at a concentration of $2 \times 10^{-9}$ $M$ using diisopropylamine as buffer (88). The latter caused less subsidiary interference than ammonia, which was used originally.

Membrane electrodes based on inorganic salts, although comparatively insensitive to $p$H change, should be used only within certain $p$H limits. Thus the lanthanum fluoride membrane electrode is best used within the $p$H range 5–8, although anionic responsive electrodes based on the silver salts AgX, where X $=$ Cl$^-$, Br$^-$, SCN$^-$, and I$^-$, can all be used within the $p$H range 3–11. The corresponding mercurous salt membrane electrodes responsive to Cl$^-$, Br$^-$, and SCN$^-$ can be used in the range $p$H 1–6. However, anions derived from weak acids must be determined at $p$H values that ensure that the determinand in the sample is quantitatively in the conjugate base form. This $p$H must be greater than $pK_a + 2$, where $pK_a$ is the ionization constant exponent of the weak acid. Thus the silver iodide membrane can be used for the determination of cyanide ion in the $p$H range 11–13, whereas sulfide ion must be determined at $p$H 13–14 using the corresponding silver sulfide membrane. The responses of cationic responsive membranes derived from silver sulfide–metal sulfide or selenide mixtures also depend on $p$H, the optimum $p$H range tending to be somewhat more restricted than is the case for the anion responsive membranes. At low $p$H these types of membranes are attacked by hydrogen ion, and at higher $p$H the metal ions in the sample are complexed by hydroxyl ion. For membranes that respond to cadmium(II), copper(II), lead(II), mercury(II), and silver(I) these effects are minimized in the $p$H range 4–5. However, the presence of ligands such as ethylenedinitrilotetraacetic acid (EDTA) and nitrilotriacetic acid (NTA) can be used not only to further lower the limit of detection for the electrode, but also to extend this range to higher $p$H. With the exception of the cyanide and sulfide ion-responsive membranes, the $p$H ranges quoted for all these inorganic salt membranes will depend on the concentration of determined ion, the permissible $p$H range being extended considerably at higher concentrations. The ranges quoted herein are, therefore, "safe" $p$H ranges for concentrations of unbuffered determinand ion close to the lower limit of detection for the electrode.

Of the ion selective membranes derived from large organic molecules, the responses of those based upon organic ion exchangers are more influenced by $p$H than those derived from the neutral carriers. This is because the organic ion exchangers tend to be compounds containing functional groups that can take part in acid-base reactions. Inspection of the names of many of these compounds listed in Table 3.5 is sufficient to confirm this, and for the less common types it is advisable to consult the cited reference in order to establish the $p$H that is recommended for best performance. For the commercially available electrodes based on organic ion exchangers that respond to carbonate, chloride, fluoroborate, nitrate, and perchlorate a "safe" $p$H range is 6.5–7.5. This range is suitable also for the calcium and the divalent cation electrodes. As is the case also for anion responsive inorganic salt membrane electrodes, hydroxyl ion interference imposes an upper limit of $p$H upon the use of anion responsive electrodes derived from organic ion exchangers. A similar analogy can be drawn between cation responsive electrodes of the inorganic salt and the organic ion exchanger types; the $p$H at which complexation of the determinand cation by hydroxyl ion starts to become significant being the upper limit of the $p$H range. This effect, of course, is independent of the nature of the membrane electroactant and will thus equally affect the performance of electrodes based on a neutral carrier that responds to calcium ion, for example. Similarly, the performance of the macrotetrolide containing membranes (e.g., monactin) that respond to the conjugate acid $NH_4^+$ will depend upon the $p$H of the sample solution. This should be such that the concentration of ammonia is negligible and this electrode is used, therefore, within the $p$H range 5–7.5. The calcium–neutral carrier type ion selective electrode can be used within the range $p$H 4–10 and valinomycin-based potassium ion selective electrodes within the range $p$H 3–10. As before, the lower limit of $p$H is dictated by the stability of the membrane material toward the hydrogen ion, and in these cases will also include the stability of the accompanying solvent/mediator system.

## Response and Detection Limits

The range of concentrations for which a graph of $E_{cell}$ versus $\log c_A$ is linear describes the response range of an indicator electrode which is selective for species A. Implicit in this definition are the fulfillment of the conditions mentioned earlier as to the composition of the background solution. At high concentrations of the determinand ion some electrodes show deviations from this straight line relationship, whereas at low concentrations all electrodes show some deviation from linearity. As the concentration is decreased below a certain level, the slope of the electrode response becomes progressively smaller than that determined from the linear portion of the graph and the recorded values of $E_{cell}$ become progressively more irreproducible. The causes of the deviations, manifest by the resulting curved region of the graph, are the solubility equilibria of the membrane materials of the electrode, the presence of determinand in reagents added to the solution, and the presence of interfering species. Provided that sufficient calibration points are obtained to define this curve, it can be used to interpolate an unknown concentration of determinand from a measured value of $E_{cell}$. Recourse to this procedure may be inevitable when the concentration of determinand in a sample is very low. An interpretation of these nonideal calibration curves in terms of the previously stated three primary contributing causes has been discussed by D. Midgley (89). This study also considers two secondary effects due to the action of complexing agents and the establishment of steady-state rather than equilibrium conditions.

The lower limit of the response range of an electrode can be defined as that concentration at which the plot of $E_{cell}$ versus $\log c_A$ begins to depart from linearity. Whenever possible the conditions for the analysis should be chosen such that the concentration of determinand is above the lower limit of the response range. The lower limit of detection for an electrode, on the other hand, is more difficult to define unambiguously. In the IUPAC definition (90), this limit of detection is taken as the concentration at the point of intersection of the extrapolated linear segments of the graph of $E_{cell}$ versus $\log c_A$. One of these segments relates to concentrations above the lower limit of the response range, and the other relates to that portion of the curve when $E_{cell}$ changes little as the concentration is further diminished. The earlier IUPAC definition (91), however, will still pertain to some quoted values in the literature. This is defined as the concentration of determinand A at which $E_{cell}$ deviates by $18/z_A$ mV ($z$ = valency of ion A) from the extrapolation of the linear portion of the calibration graph (see Figure 4.2). Both of these arbitrary definitions have been criticized on the grounds that they are not related to analytical performance, in contrast to the statistically based definitions of limit of detection that have been applied to other techniques of chemical analysis (92). It is maintained that the limit of detection should be decided on a statistical basis that allows a solution containing a given concentration of determinand to be discriminated, with a specified degree of confidence, from a blank solution. There are large discrepancies between the IUPAC limit and the statistical limit. An extreme example is the limit of detection for the silver–silver chloride electrode which is 530 $\mu$g L$^{-1}$ of chloride ion by the IUPAC method, whereas the same data yielded a limit of 15 $\mu$g L$^{-1}$ by the statistical method (92). For other electrodes the discrepancy was less, but the IUPAC limit was at least twice that of the statistical limit. A statistical procedure has also been described to establish the lower limits of the response ranges for electrodes (93).

Values for the limits of detection quoted in the literature and in the specifications of commercially available electrodes have usually been determined by some type of graphical interpolation method, generally on solutions of determinand that are unbuffered with respect to the determinand. The effect of these buffers is to lower the limit of detection, and the limit is also lowered if the measurement is carried out in mixed or flowing solutions. Thus data for the lower limit of detection should be treated merely as a guideline. Furthermore, it should be noted that for most routine applications conditions can be created whereby an ion selective electrode will be operating in solutions the concentrations of which will lie within the range of response for the electrode.

## The Selectivity Coefficient

Thus far it has been convenient to assume that by a suitable choice of experimental conditions the effects of ions other than the determinand upon the response of an ion selective electrode can be made negligible. For example, it has been mentioned earlier in this section that it is important to use an electrode within the optimum $p$H range for that electrode to reduce interference by hydrogen or hydroxyl ion. Interferences by other species can sometimes be reduced by the addition of a masking agent and the inclusion of these in TISABs has already been discussed. Selective precipitation can be used in some instances to remove an interferent, an example being the removal of chloride as silver chloride in the potentiometric determination of nitrate ion. Volatile interferents may be removed by heating or by passage of an unreactive gas provided the determinand is not volatile. Ideally, of course, the membrane of a given electrode should respond exclusively to the determinand which it is designed to measure. In practice, however, when an electrode is dipped into a solution containing the determinand ion in the presence of a certain concentration of free interferent, the measured potential will contain contributions from both the determinand and the interferent. Thus in the presence of an interferent B the Nernst equation derived solely in terms of the activity of the determinand A will be inadequate to predict how $E_{cell}$ will vary as the activity of A (i.e., $a_A$) is changed unless allowance is made for the activity of the interferent $a_B$. The appropriate adaptation of the Nernst equation, sometimes referred to as the *Nikolsky equation*, can be written as (91)

$$E_{cell} = \text{constant} + \frac{k}{z_A} \log\left[a_A + k_{A,B}(a_B)^{z_A/z_B}\right] \quad (4.42)$$

The terms of this equation in need of definition are $z_A$ and $z_B$, which are integers with sign and magnitude corresponding to the charges on the determinand ion A and the interferent ion B, respectively, and $k_{A,B}$ which is the *selectivity coefficient* of the electrode. As before, $k$ is the Nernst factor at a particular temperature and the "constant" term is a composite of the standard potential of the indicator electrode, the potential of the reference half-cell, and the junction potential. The value of the selectivity coefficient, $k_{A,B}$ reflects the degree of selectivity of the electrode for the determinand ion A with respect to the interfering ion B. When $k_{A,B} \ll 1$, then the electrode is very much more selective toward the determinand than it is toward the interferent; when $k_{A,B} = 1$, the electrode is equally responsive to the ions A and B; when $k_{A,B} > 1$, the electrode responds preferentially to B rather than to A.

An extensive list of selectivity coefficients for many of the membrane electrodes listed in Tables 3.4 and 3.5 has been prepared for the Commission on Electrochemistry, IUPAC (94). These data are always determined within the concentration exponent range for which the electrode is known to have a linear response when calibrated with the determinand species alone. Two methods have been used to determine selectivity coefficients.

### *The Separate Solution Method*

Measured values of $E_{cell}$ are obtained first for the determinand ion solution of activity $a_A$ in the absence of interferent B, and then for interferent B at the same activity as $a_A$ in a solution that contains no determinand. If these measured values are $E_1$ and $E_2'$, respectively, then equation (4.42) can be rearranged to yield

$$\log k_{A,B} = \frac{z_A(E_2 - E_1)}{k} + \log a_A\left(1 - \frac{z_A}{z_B}\right)$$

For a calibrated electrode of slope $S$ used in a suitable background solution

$$\log k_{A,B} = \frac{E_2 - E_1}{S} + \log c_A\left(1 - \frac{z_A}{z_B}\right)$$

Although many values of selectivity coefficients determined by this method have been reported, they cannot be regarded as reliable. This is because the response of a given electrode in each of the standard solutions A and B is different from its response in a solution containing a mixture of the two at the same concentration. Often the response of the electrode when dipped into a solution containing the interferent alone becomes sluggish and its potential is liable to drift.

### The Mixed Solution Methods

These methods yield more reliable values of $k_{A,B}$ than the separate solution method. The fixed interference method consists of the measurement of $E_{cell}$ for solutions containing the same activity of interferent $a_B$ and varying the activity of the determinand ion $a_A$. This method is preferred over the converse method of holding $a_A$ constant and varying $a_B$ although the latter has been used when hydrogen ion is the interfering ion. The fixed interference method, however, more closely resembles the situation in the samples. The method depends upon the interpolation from the graph of $E_{cell}$ versus $\log c_A$, that concentration of A at which the electrode is responding equally to both ions. If this interpolated concentration of A is $c_A$, and assuming constancy for the activity coefficients under conditions for which $E_{cell}$ is measured, then equation (4.42) becomes

$$c_A = k_{A,B}(c_B)^{z_A/a_B} \qquad (4.43)$$

In the presence of an interfering ion the graph of $E_{cell}$ versus $\log c_A$ ideally has two linear segments interconnected by a curved section. The first of these linear segments is obtained when the range of concentrations of determinand A is such that the electrode yields a Nernstian response, indicating that the concentration of the interferent B is too low to alter this response. The second linear portion is caused when the concentration of A is low so that the electrode responds only to B, and, since the concentration of the latter has been kept constant throughout, the cell potential will also remain constant. The point of intersection when these two linear points are extrapolated yields log $c_A$ from which $k_{A,B}$ may be calculated by equation (4.43). It should again be emphasized that the range of concentrations of the determinand should lie within the range of Nernstian response for the electrode when B is absent.

An alternative method of locating the midpoint of the curved section corresponding to a condition of equal response of the electrode toward A and B must be used if the expected zero electrode response due to the invariant concentration of B cannot be established experimentally. Such a case is represented in Figure 4.2 on which the point Y represents the lower limit of response when the electrode is calibrated with solutions containing the determinand alone. At point X the interferent B starts to cause the electrode to deviate from the linearity observed along the line TX. The concentration at point P is below that at which the response of the electrode departs from linearity at point Y and thus the linear segment of zero electrode response cannot be established. That point R yields the concentra-

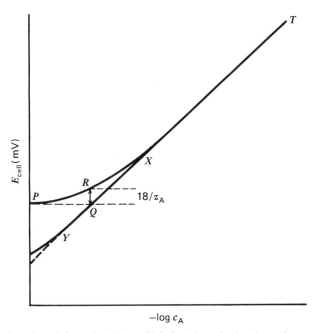

**Figure 4.2.** Location of that point (R) at which the A ion selective electrode responds equally to species A and interferent B.

tion at which the electrode responds equally to the determinand and the interferent is deducible as follows: The concentration corresponding to $R$ is given by equation (4.43) which, when substituted in a modified equation (4.42) yields

$$E_{cell} = \text{constant} + \frac{k}{z_A} \log 2c_A$$

On the other hand, point $Q$ represents the response of the electrode to a solution of concentration $c_A$ when the interferent is absent. The change in $E_{cell}$ between $R$ and $Q$ is given by

$$\Delta E = \frac{k}{z_A} (\log 2c_A - \log c_A)$$

$$= \frac{k}{z_A} \log 2$$

$$\simeq \frac{18}{z_A} \text{ mV at } 25\,°C$$

Thus the midpoint of the curved section at $R$ can be found from a plot of the data by extrapolating the linear portion $TX$ toward $Y$ and then selecting the value of $\log c_A$ at which the line $XP$ differs from $XY$ by $18/z_A$ mV. The antilogarithm of this value is $c_A$ from whence the selectivity coefficient can be calculated from

$$k_{A,B} = \frac{(c_A)^{z_B}}{(c_B)^{z_A}}$$

The values of $k_{A,B}$ given in Table 4.13 are included merely to be indicative of the orders of magnitude likely to be encountered. The values *per se* are of little significance because a given value for a selectivity coefficient is never constant for all concentrations of A and B although it is sometimes constant for a given ratio of the concentration of A to B (95). Unless these data are accompanied by details of the conditions under which they are determined (i.e., $p$H, concentrations of A and B, background solution, temperature, etc.) they can be used only as guidelines as to what species are the principal interferents. Such data are useful when designing a potentiometric method, but the values can never be used to correct an observed $E_{cell}$ for the presence of an interferent. Widely divergent values for a given determinand ion can suggest the use of different types of membrane materials. However, lesser differences for the same membrane materials can never be used as indicating the superiority of one type of electrode over another, unless the two sets of selectivity data were determined under the same conditions.

### Other Characteristics

It has been convenient to assume hitherto that the equilibrium value of the cell e.m.f. at constant temperature becomes available immediately once electrical contact is established between the electrodes and the solution on the one hand, and between the electrodes and the ion activity meter or potentiometer on the other. This presupposes that both electrodes are in such a condition, that their optimum performance is assured, that they respond to a new condition instantaneously on immersion, and that the resulting value of $E_{cell}$ will remain stable indefinitely. Such an assumption also neglects the effects of environmental changes that can occur in the

Table 4.13. Selectivity Coefficients of Some Commercially Available Ion Selective Electrodes

| Electrode | Interferences[a] |
|---|---|
| $Ca^{2+}$ (Beckman) | $Fe^{2+}$ 5, $Cu^{2+}$ 0.33, $Mg^{2+}$ 0.9, $Ba^{2+}$ 0.1 |
| $Ca^{2+}$ (Corning) | $Mg^{2+}$ 0.01, $Ba^{2+}$ 0.01, $Ni^{2+}$ 0.01, $Na^+$ $10^{-3}$ |
| $Ca^{2+}$ (Orion) | $Zn^{2+}$ 3.2, $Fe^{2+}$ 0.80, $Pb^{2+}$ 0.63, $Cu^{2+}$ 0.27, $Mg^{2+}$ 0.01, $Ba^{2+}$ 0.01, $Na^+$ $1 \cdot 6 \times 10^{-3}$ |
| $Ca^{2+} + Mg^{2+}$ (Beckman) | $Zn^{2+}$ 1.0, $Ba^{2+}$ 1.0, $Na^+$ 0.01, $K^+$ 0.01 |
| $Ca^{2+} + Mg^{2+}$ (Orion) | $Zn^{2+}$ 3.5, $Fe^{2+}$ 3.5, $Cu^{2+}$ 3.1, $Ba^{2+}$ 0.94, $Na^+$ 0.01 |
| $NO_3^-$ (Beckman) | $ClO_4^-$ 100, $NO_2^-$ 0.045, $Cl^-$ 0.01, $SO_4^{2-}$ $10^{-5}$ |
| $NO_3^-$ (Corning) | $ClO_4^-$ 1000, $Cl^-$ $4 \times 10^{-3}$, $SO_4^{2-}$ $10^{-3}$ |
| $NO_3^-$ (Orion) | $ClO_4^-$ 1000, $NO_2^-$ 0.06, $Cl^-$ $6 \times 10^{-3}$, $SO_4^{2-}$ $6 \times 10^{-4}$ |
| $ClO_4^-$ (Beckman) | $I^-$ 0.033, $NO_3$ 0.005, $SO_4^{2-}$ $10^{-6}$ |
| $ClO_4^-$ (Corning) | $I^-$ 0.001, $NO_3^-$ 0.001, $SO_4^{2-}$ 0.001 |
| $ClO_4^-$ (Orion) | $OH^-$ 1.0, $I^-$ 0.012, $NO_3^-$ 0.0015, $SO_4^{2-}$ $1.6 \times 10^{-4}$ |
| $K^+$ (Beckman) | $Rb^+$ 2.2, $Cs^+$ 0.5, $NH_4^+$ 0.014, $Na^+$ $2 \times 10^{-4}$ |
| $K^+$ (Corning) | $Rb^+$ 10, $Cs^+$ 20, $NH_4^+$ 0.023, $Na^+$ 0.012 |
| $K^+$ (Phillips) | $Rb^+$ 1.9, $Cs^+$ 0.38, $NH_4^+$ 0.012, $Na^+$ $2.6 \times 10^{-4}$, $Li^+$ $2.1 \times 10^{-4}$ |
| $K^+$ (Orion) | $Rb^+$ 2.2, $Cs^+$ 0.5, $NH_4^+$ 0.05, $Na^+$ 0.09, $Li^+$ 0.03 |
| $NH_4^+$ (Phillips) | $K^+$ 0.12, $Rb^+$ 0.043, $Cs^+$ 0.0048, $Li^+$ 0.0042, $Na^+$ 0.002 |
| $F^-$ (Beckman) | $OH^-$ 0.1 |
| $F^-$ (Orion) | $OH^-$ 0.1 |

[a] Manufacturer's data, see also Reference 94.

solution circumjacent to the electrodes and in the area immediately surrounding the cell. Some factors which must be taken into account in order to ensure that the best performance is obtained from electrodes are summarized below.

### *Conditioning and Storage*

A freshly prepared electrode or a new electrode received from a manufacturer often needs conditioning before use and thereafter it should be stored under suitable conditions when not in use. Mention has already been made of the storage of the hydrogen electrode in water, and the silver–silver chloride electrode which is conditioned in 0.05 $M$ hydrochloric acid for about 16 h and is stored subsequently in 0.01 $M$ hydrochloric acid. The importance of the correct conditioning of glass electrodes was mentioned earlier and the correct storage of these fragile electrodes can substantially reduce the breakage rate.

Electrodes that need to be stored in a wet condition are conveniently positioned in flat-bottomed specimen tubes held in a suitable rack. A rubber bung predrilled with a hole some 2–3 mm greater than the diameter of the stem of the electrode allows the electrode to be slid in and out of the tube without contacting the electroactive part of the electrode. A neoprene "O" ring fitted around the stem of the electrode ensures that the bottom of the electrode does not hit the bottom of the specimen tube and also minimizes the evaporation rate of the solution in the tube when the electrode is positioned in it. When in use the "O" ring also allows the electrode to be positioned correctly in a tall form beaker also fitted with a bung that has a hole of the appropriate diameter drilled through it. Such an arrangement considerably decreases the breakage rate, particularly by students.

It is essential that glass electrodes are always kept in a wet condition once the hydrated layer has been fully developed. The length of time required for this conditioning process depends upon their ultimate use. The 24 h usually recommended by manufacturers is sufficient for results of medium precision although the hydrated layer of the glass is not fully developed after this time. These electrodes are best stored in water. The membranes of the potentiometric gas sensors also need to be kept in a wet condition and the tips of these should be immersed in a portion of the internal filling solution or in deionized water plus any reagents added during the appropriate analytical procedure. Reference half-cells usually require no conditioning and are best stored with their tips immersed in a solution of the same composition as the bridging solution. The internal filling solution should always be kept topped-up and the filling port should be sealed at all other times.

When used regularly, electrodes fitted with membranes of inorganic salts can be stored with the membranes immersed in deionized water. If, however, they are required to be stored for weeks, it is better that they be stored in a dry state with the membrane protected by a plastic cap. In some instances the manufacturers recommend the conditioning of a new electrode by immersing the tip in a 0.001 $M$ solution of the electroactive ion. Usually, however, a rinse in water is all that is required to service an electrode if it has been stored in a dry state for a prolonged period. Solid-state electrodes based on hydrophobized graphite are exceptions in that these need to be conditioned, as previously outlined, and those reversible to divalent ions should not be stored dry.

Electrodes based on organic ion exchangers and neutral carriers generally need to be conditioned immediately before use by a 10 min immersion in a 0.001 $M$ solution of determinand ion. The membranes of these electrodes fail, on an average, after about one month's use. This lifetime is very short in comparison with the glass and inorganic membrane electrodes that usually have lifetimes of two to three years. The evaporation of the solvent/mediator is a problem in the storage of electrodes fitted with membranes containing an organic electroactant. In general, these may be stored for a number of days in a small volume of deionized water or conditioning solution. The Orion series 93 electrodes are of a convenient design that permits the removal of the membrane as part of a module which can be stored dry in a capsule when not in use. This prolongs the life of the membrane considerably.

*Response Time*

According to the IUPAC recommendation (91), the response time is the length of time that elapses between the instant a potentiometric cell is formed by the immersion of its constituent electrodes in a solution, and the first instant at which the cell potential deviates by exactly 1 mV from the final steady-state value of $E_{cell}$. Variations of 30% are not uncommon in the duplicate determination of response time and, at a given temperature, this parameter can depend upon the concentration of the solution previously analyzed, particularly when the concentration changes from high to low. The responses of many electrodes become slower in the presence of an interferent and it is possible for such a substance to have a prolonged effect. An extreme example (96) was produced when a nitrate electrode was transferred from a 0.01 $M$ nitrate solution into a 0.01 $M$ hexafluorophosphate solution. In the latter, a transitory potential of 150 mV was produced and when the electrode was returned to the nitrate solution it took four days to reproduce its original steady-state value. A less drastic "memory" effect (less than 9 min) has also been observed in the dynamic calibration of the fluoride electrode (97) particularly at low concentrations. Transitory "peaks" in the values of $E_{cell}$ have been evident when hydrogen ion responsive glass electrodes have been transferred from solutions of higher $p$H to solutions of lower $p$H (98).

The response time also depends upon the temperature, the higher temperature tending to produce a faster response for identical electrodes transferred between the same two solutions. This is particularly noticeable for the response times of glass electrodes, the resistances of which increase markedly as the temperature is lowered.

Stirring the solution to ensure a linear flow rate of the sample across the membrane decreases the response time for an electrode in comparison to using the electrode in an unstirred solution. In the latter, local conditions may develop in the immediate vicinity of the electrodes whereby the determinand concentration is not representative of the bulk of the solution. A slow drift in potential is sometimes indicative of this, particularly when the electrode is operating near to the lower limit of its response range. With the exception of the hydrogen ion responsive glass electrode, all ion selective electrodes are used in stirred or in flowing solutions. A magnetic stirrer that gives a reasonably constant stirring rate at a given setting is the most widely used device for this purpose. This setting should be chosen to ensure an adequate flow-rate around the electrodes, yet the speed at which the stirrer is operating should not be the cause of electrical noise. A diagnostic that a suitable speed has been chosen is the constancy of the cell potential over a small range of stirrer settings. The concentration of the solution chosen for this test should be near to the lowest concentration expected in the samples. If this is unknown, then the concentration chosen should tend towards the lower limit of response for the electrode. A slow unidirectional drift observed when using magnetically induced stirring is indicative that the heat generated by the stirrer is causing a temperature change in the sample solution. This effect can be eliminated by forming the cell in a beaker surrounded by a water jacket through which water from a thermostatically controlled bath is circulated. An alternative which reduces the rate of temperature change is to insulate the potentiometric cell from the stirrer platform by a 1 cm layer of plastic foam.

The response time for the electrodes of a cell with liquid junction can be reduced by at least tenfold if it is possible to change the concentration of determinand by an amount $\Delta c$ while the electrodes remain in the solution, rather than to record the change in $E_{cell}$ attributable to $\Delta c$ by removing the electrodes from one solution and reimmersing them in the other. This is because the latter transfer causes the cell to be made open-circuit which upsets the equilibria on the membrane surface and destroys the liquid junction of the reference electrode.

These factors affecting the response rate need to be identified in the establishment of a potentiometric method particularly if it is to be used routinely. It is recommended, therefore, that in the development stage the output from the ion activity meter be connected to a chart recorder which greatly facilitates the identification of these effects.

## *Stability and Reproductivity*

The assumption fundamental to analysis by direct potentiometry is that the electrode system is completely stable between the times of calibration and measurement. In general, the maximum time that can be allowed to elapse between successive calibrations depends upon the type of electrode, those based on organic ion exchangers requiring the greatest frequency of calibration checks. Systematic inconsistencies in the triplicated method of calibration are diagnostic in that the stability is such as to require frequent recalibration. In these cases the electrode slope is often reproducible and the inconsistency is due to changes in the value of the term "constant" in equation (4.42), for example. The performance of all electrodes and probes tends to be the most stable at high ionic strengths and at the upper concentration levels within the range of response for the electrodes. So far as it is practicable, therefore, it is advisable to operate the cell under these conditions.

The stability of a cell with liquid junction depends upon the stability of both the indicator electrode and the reference half-cell. This needs to be stated because, in the event of the cell potential being unstable, there is a great tendency to focus attention wholly upon the indicator electrode and to overlook the possibility that the fault might be attributable to the reference half-cell. In fact, a substantial proportion of such problems may be traced to this half-cell and to the liquid junction associated with it. Fouling of the device through which this junction is made is a common cause of instability in the cell potential.

Electrodes of the glass and inorganic salt membrane types are generally quite stable. The hydrogen ion selective glass electrode has been found to be very stable when used in a cell with a silver–silver chloride electrode to measure $p(a_H \gamma_{Cl})$, the acidity function (99). The accuracy of these measurements were comparable to that of the hydrogen electrode, and the stability was well within 0.1 mV over a 24-h period at buffer concentrations of 0.02 mole $kg^{-1}$. The sodium ion selective glass electrode has a short-term reproducibility of $\pm 0.3$ mV within the sodium ion concentration range

0.1–0.001 $M$ and a comparative study has been made of how the apparent standard potential [i.e., the "constant" of equation (4.22)] for this type of electrode varies with age (87). Of the inorganic salt membrane electrodes, the fluoride, silver, halide, and sulfide types can be used to give a stability and reproducibility of ±0.1 mV when used with care as to temperature control, stirring, and so on. For example, the fluoride electrode (100) gave readings that did not change more than 0.1 mV in an hour and its apparent standard potential varied only by 7 mV in a month without change of slope. Deterioration in the stability of response for inorganic salt membrane electrodes occurs, however, after some months of continuous use. The membrane becomes pitted as a result of the dissolution of the membrane material, with the result that small volumes of solution became trapped in these crevices and thus create a "memory" effect when the electrode is transferred to another solution. For example, the stability and response of a copper ion selective electrode deteriorated so that after two months its slope was 16.2 mV decade$^{-1}$ and when it was immersed in a 0.01 $M$ copper nitrate solution the potential drifted at the rate of about 1 mV min$^{-1}$ (101). The correct functioning of this electrode was restored by polishing the membrane with diamond paste followed by treatment with silicone oil. Thereafter, the stability of the reading became ±0.01 mV after 2 min immersion in the 0.01 $M$ copper nitrate solution and the slope was restored to 29.2 mV decade$^{-1}$ at 23°C. The deterioration effects are similarly produced in the cadmium and lead electrodes, but are aggravated in these compared with the copper sulfide membrane as a result of the greater solubilities of cadmium sulfide and lead sulfide.

The stability of electrodes based upon organic electroactants is poor in comparison with the glass and inorganic membrane types. Some guidelines as to the stability of the liquid membrane, solid membrane, and coated wire types of electrode are given elsewhere in this book.

*Temperature Effects*

The potentials of all types of electrodes depend upon the temperature of the medium in which they are immersed. The extent to which temperature control is necessary depends, therefore, upon the precision with which it is intended to measure $E_{cell}$. The use of a thermostatically controlled water bath is recommended if it is intended to measure $E_{cell}$ with a precision better than ±1 mV. It is generally more convenient to pump water from this bath through a jacket surrounding the potentiometric cell than it is to actually mount the cell in the bath. Even for less precise work the temperature change over the period required for the measurement of the potentials for calibrating solutions and the sample solutions should not exceed 1°C. In areas where the ambient temperature is subject to large changes during the day it may be necessary, therefore, to recalibrate the electrodes fairly frequently to correct for any changes in the apparent standard potential and slope.

When the temperature is changed in a cell comprised of an ion selective electrode and a reference half-cell, there can be considerable delay before even a thermal equilibrium is reestablished. This is because the bodies of these electrodes are constructed from materials of low thermal conductivity which inhibits heat transfer to or from the electroactive solution of the reference half-cell and the internal filling solution of the ion selective electrode. Additional time is required, therefore, for the appropriate chemical equilibria to be reestablished. The internal reference electrodes of both the reference half-cell and the ion selective electrode are usually of the metal–insoluble metal salt type and solubility equilibria must be reestablished at the interfaces of these with the internal reference solutions. Additionally, of course, equilibria must be established on both sides of the membrane of the ion selective electrode. All these factors combine to make the reestablishment of an equilibrium potential after a sudden temperature change rather slow. For example, an ion selective electrode can require 2 h to reestablish a steady potential following an abrupt temperature change of $10\,°C$.

The electrode temperature coefficient $dE/dT$ describes the magnitude of the change in the equilibrium potential of the electrode when the temperature of the system is changed by $1\,°C$. These values are usually quoted as nonisothermal temperature coefficients. This means that they were determined using a cell with liquid junction in which the constituent half-cells were identical with respect to electrodes and solutions, but each half-cell was maintained at a different temperature. Isothermal electrode temperature coefficients, on the other hand, are derived from measurements in which the potential of an electrode is measured as a function of temperature against an hydrogen reference half-cell when each half-cell is operated at the same temperature. Values for the nonisothermal temperature coefficients of ion selective electrodes range from $-0.2$ to $+1$ $mV\,°C^{-1}$ (102, 103), whereas values of 1.5 $mV\,°C^{-1}$ and 1.0 $mV\,°C^{-1}$ have been reported, respectively, for the ammonia (104) and carbon dioxide (105) gas probes. The nonisothermal temperature coefficient of a reference half-cell often depends upon the concentration of the standard electroactant contained in it. Thus this temperature coefficient for the calomel half-cell is 0.19 $mV\,°C^{-1}$ when the electrolyte is a saturated solution of potassium chloride, and 0.44 $mV\,°C^{-1}$ when this electrolyte concentration is 3.8 $M$. The corresponding value for a silver–silver chloride reference half-cell containing 3.5 $M$ potassium chloride is about 0.6–0.7 $mV\,°C^{-1}$. The foregoing summary serves to emphasize the need for temperature control when using potentiometric cells even when the application is of comparatively low precision.

The lower limits of the response range and of detection for an electrode are also dependent upon temperature, particularly in those cases where the temperature coefficient of solubility for the membrane material is appreciable. For example, the lower limit of the response range for a silver chloride membrane is about $2 \times 10^{-5}$ $M$ at 5°C and about $5 \times 10^{-5}$ $M$ at 25°C. Temperature changes also affect other equilibria and so may cause changes in the actual concentration of the free determinand in solution. This is an important factor when the determinand ion may take part in acid-base or complex equilibria, brought about, for example, by the use of a metal-ion buffer.

Other factors not considered in the preceding four subsections are hysteresis effects and the sensitivities of electrodes to light. Hysteresis is said to occur in an electrode system when its rate of response on transfer between two solutions, one at state 1 and the other at state 2, is grossly different to that observed when the electrode system is returned from state 2 to state 1. The two forms of hysteresis are concentration-hysteresis and temperature-hysteresis.

Concentration-hysteresis can occur when state 1 represents a high concentration and state 2 is a lower concentration or vice versa. A common example of this is the memory effect induced by the entrapment of the solution within the deep pores or crevices in the membrane. Another cause is when solution 1 contains no interferent, but solution 2 contains an appreciable concentration of interferent. An extreme example of this type is given elsewhere for the nitrate electrode. Under normal operational conditions, however, the overall effect of concentration-hysteresis is to give a spuriously low result, for, although the initial rate of response on transfer of the electrode system may be high, it is followed by a much slower change of potential in the same direction.

Temperature-hysteresis effects are sometimes observed when electrode systems are transferred between two solutions of the same concentration but maintained at different temperatures. This effect has been mentioned in connection with calomel reference half-cells, being attributed in this case to the disproportionation of mercurous chloride. Glass $p$H electrodes also show a temperature-hysteresis, the equilibrium potential being established more rapidly on transfer from lower to higher temperature than in the reverse direction. A study of the effects of sudden temperature changes upon commercially available sodium glass electrodes has revealed a temperature-hysteresis effect in some cases (106).

It is good practice to place all potentiometric cells in such positions that they are protected from the direct incursion of sunlight. Changing intensities of sunlight can cause drifts of 1–5 mV in the potentials of ion selective electrodes that utilize silver salts either in the construction of their membranes or as part of their internal reference electrodes. For example, the reported effects of light on glass electrodes (107, 108) are considered to be the result of either temperature changes or the photosensitivity of the silver–silver chloride internal reference electrode (109). Similar remarks pertain also to reference half-cells which contain silver–silver chloride electrodes (110, 111) that dip into solutions of potassium chloride saturated with respect to silver chloride. The results suggest that the effect of illumination upon these is less marked in the more dilute solutions of potassium chloride than when concentrated solutions of potassium chloride are used.

## REFERENCES

1. J. N. Butler and R. Huston, *Anal. Chem.* **42**, 1308 (1970).
2. J. Bagg and G. A. Rechnitz, *Anal. Chem.* **45**, 1069 (1973).
3. J. V. Leyendekkers, *Anal. Chem.* **43**, 1835 (1971).
4. J. V. Leyendekkers and M. Whitfield, *Anal. Chem.* **43**, 322 (1971); *J. Phys. Chem.* **75**, 957 (1971).
5. R. G. Bates and R. Gary, *J. Res. Natl. Bur. Stand. (U.S.)* **65A**, 495 (1961).
6. International Union of Pure and Applied Chemistry, *Pure Appl. Chem.* **50**, 1485 (1978).
7. R. G. Bates, *Determination of pH Theory and Practice*, 2nd ed., Chapter 4, Wiley, New York, 1973.
8. R. G. Bates, G. D. Pinching, and E. R. Smith, *J. Res. Natl. Bur. Stand. (U.S.)* **45**, 418 (1950).
9. A. K. Covington and M. I. A. Ferra, *Anal. Chem.* **49**, 1363 (1977).
10. L. Linnet, *pH Measurement in Theory and Practice*, Radiometer, Copenhagen (1970).
11. F. G. Shinskey, *pH and pIon Control in Process and Waste Streams*, Wiley, New York, 1973.
12. K. Schwabe, in *Electroanalytical Chemistry*, H. W. Nurnberg, Ed., Chapter VII, Wiley, New York, 1974.
13. C. C. Westcott, *pH Measurements*, Academic, New York, 1978.
14. D. D. Perrin and B. Dempsey, *Buffers for pH and Metal Ion Control*, Chapman and Hall, London, 1974.
15. R. G. Bates, R. N. Roy, and R. A. Robinson, *Anal. Chem.* **45**, 1663 (1973).
16. R. N. Roy, E. E. Swensson, G. LaCross, Jr., and C. W. Kruger, *Anal. Chem.* **47**, 1407 (1975).
17. R. A. Durst, *Clin. Chem. (Winston-Salem, N.C.)* **23**, 298 (1977).
18. R. G. Bates, C. A. Vega, and D. R. White, Jr., *Anal. Chem.* **50**, 1295 (1978).
19. M. Sankar and R. G. Bates, *Anal. Chem.* **50**, 1922 (1978).
20. M. Paabo and R. G. Bates, *Anal. Chem.* **41**, 283 (1969).

21. R. Gary, R. G. Bates, and R. A. Robinson, *J. Phys. Chem.* **68**, 1186 (1964).
22. P. K. Glasoe and F. A. Long, *J. Phys. Chem.* **64**, 188 (1960).
23. K. Mikkelsen and S. O. Nielsen, *J. Phys. Chem.* **64**, 632 (1960).
24. B. M. Lowe and D. G. Smith, *J. Chem. Soc., Faraday Trans. 1* **69**, 1934 (1973).
25. V. Gold and B. M. Lowe, *Proc. Chem. Soc.* **1963**, 140.
26. A. K. Covington, M. Paabo, R. A. Robinson, and R. G. Bates, *Anal. Chem.* **40**, 700 (1968).
27. R. G. Bates and M. Alfenaar, in *Ion Selective Electrodes*, R. A. Durst, Ed., Chapter 6, N.B.S. Spec. Publ. No. 314, Washington, D.C., 1969.
28. R. G. Bates, B. R. Staples, and R. A. Robinson, *Anal. Chem.* **42**, 867 (1970).
29. R. H. Stokes and R. A. Robinson, *J. Am. Chem. Soc.* **70**, 1870 (1948).
30. R. H. Stokes and R. A. Robinson, *Trans. Faraday Soc.* **53**, 301 (1957).
31. R. G. Bates, *Pure Appl. Chem.* **36**, 407 (1973).
32. R. Bates and R. A. Robinson, in *Ion Selective Electrodes*, Conference 1977, E. Pungor and I. Buzas, Eds., Akademia Kiado, Budapest, 1978, p. 3.
33. A. K. Covington, *Crit. Rev. Anal. Chem.* **3**, 355 (1974).
34. A. K. Covington and R. A. Robinson, *Anal. Chim. Acta* **78**, 219 (1975).
35. R. G. Bates, *Determination of pH Theory and Practice*, 2nd ed., Chapter 9, Wiley, New York, 1973.
36. G. W. Neff, W. A. Radke, C. J. Sambucetti, and G. M. Widdowson, *Clin. Chem. (Winston-Salem, N.C.)* **16**, 566 (1970).
37. W. A. E. McBryde, *Analyst (London)* **94**, 337 (1969); **96**, 739 (1971).
38. G. R. Hedwig and H. K. J. Powell, *Anal. Chem.* **43**, 1206 (1971).
39. H. S. Harned and R. A. Robinson, *J. Am. Chem. Soc.* **50**, 3157 (1928); H. S. Harned and F. C. Hickley, *J. Am. Chem. Soc.* **59**, 1284 (1937).
40. D. H. Everett and B. R. W. Pinsent, *Proc. R. Soc. London, Ser. A.* **A125**, 416 (1952).
41. G. R. Hedwig, J. R. Liddle, and R. D. Reeves, *Aust. J. Chem.* **33**, 1685 (1980).
42. A. Avdeef and J. J. Bucher, *Anal. Chem.* **50**, 2137 (1978).
43. J. V. Leyendekkers, *Anal. Chem.* **43**, 1835 (1971).
44. R. A. Durst and B. T. Duhart, *Anal. Chem.* **42**, 1002 (1970).
45. J. W. Ross, in *Ion Selective Electrodes*, R. A. Durst, Ed., Chapter 2, N.B.S. Spec. Publ. No. 314, Washington, D.C., 1969.
46. J. Vesely, O. J. Jensen, and B. Nicolaisen, *Anal. Chim. Acta* **62**, 1 (1972).
47. D. D. Perrin, *Masking and Demasking of Chemical Reactions*, Wiley, New York, 1970, p. 1.
48. M. S. Frant and J. W. Ross, *Anal. Chem.* **40**, 1169 (1968).
49. A. Liberti and M. Mascini, *Anal. Chem.* **41**, 676 (1969).
50. R. A. Robinson, W. C. Duer, and R. G. Bates, *Anal. Chem.* **43**, 1862 (1971).
51. W. C. Duer, R. A. Robinson, and R. G. Bates, *J. Chem. Soc. Faraday Trans. 1*, **68**, 716 (1972).
52. J. Bagg, *Anal. Chem.* **48**, 1811 (1976).
53. A. Hulanicki and M. Trojanowicz, *Anal. Chim. Acta* **68**, 155 (1974).
54. M. J. Smith and S. E. Manahan, *Anal. Chem.* **45**, 836 (1973).
55. R. Boch and H. J. Puff, *Fresenius' Z. Anal. Chem.* **240**, 381 (1968).

56. A. Jyo, T. Hashizume, and N. Ishibashi, *Anal. Chem.* **49**, 1869 (1977).
57. G. Nakagawa, H. Wada, and T. Hayakawa, *Bull. Chem. Soc. Jpn.* **49**, 424 (1975).
58. M. S. Mohan and R. G. Bates, *Clin. Chem. (Winston-Salem, N.C.)* **21**, 864 (1975).
59. K. H. Khoo, R. W. Ramette, C. H. Culberson, and R. G. Bates, *Anal. Chem.* **49**, 29 (1977).
60. I. Hanson, S. Ahrland, R. G. Bates, G. Biederman, D. Dyrssen, E. Hogfeldt, A. E. Martell, J. J. Morgan, P. W. Schindler, T. B. Warner, and M. Whitfield in *The Nature of Seawater*, E. D. Goldberg, Ed., *Physical and Chemical Research Rpt. 1*, Dahleen Konferenzen, Berlin, 1975, p. 263.
61. R. G. Bates and J. B. Macaskill, in *Analytical Methods in Oceanography*, T. R. P. Gibb, Ed., *Adv. Chem. Ser.* **147**, 110 (1975).
62. M. Whitfield, *Proc. Anal. Div. Chem. Soc.* **12** 56 (1975).
63. R. Jansinski, I. Trachtenberg, and D. Andrychuk, *Anal. Chem.* **46**, 364 (1974).
64. R. W. Schmid and C. N. Reilley, *J. Am. Chem. Soc.* **78**, 5513 (1956).
65. R. Blum and H. M. Fog, *J. Electroanal. Chem. Interfacial Electrochem.* **34**, 485 (1972).
66. E. H. Hansen, C. G. Lamm, and J. Ruzicka, *Anal. Chim. Acta* **59**, 403 (1972).
67. E. H. Hansen and J. Ruzicka, *Anal. Chim. Acta* **63**, 115 (1973); **72**, 365 (1974).
68. J. Ruzicka, E. H. Hansen, and J. C. Tjell, *Anal. Chim. Acta* **67**, 155 (1973).
69. D. D. Perrin, *Pure Appl. Chem.* **20**, 133 (1969).
70. L. G. Sillen and A. E. Martell, *Stability Constants of Metal-Ion Complexes*, Spec. Publ. No. 17, Chem. Soc., London, 1964.
71. L. G. Sillen and A. E. Martell, *Stability Constants of Metal-Ion Complexes*, Supplement, Spec. Publ. No. 25, Chem. Soc., London, 1971.
72. D. D. Perrin, *Stability Constants of Metal-Ion Complexes, Part B-Organic Ligands*, Pergamon, Oxford, 1979.
73. J. Havas, M. Kaszas, and M. Patko, *Hungarian Patent* 8509, July 27, 1974.
74. B. Karlberg, *Anal. Chem.* **43**, 1911 (1971).
75. M. Brand and G. A. Rechnitz, *Anal. Chem.* **42**, 1172 (1970).
76. G. Horvai, L. Domokos, and E. Pungor, *Fresenius' Z. Anal. Chem.* **292**, 132 (1979).
77. G. Gran, *Acta Chim. Scand.* **4**, 559 (1950); *Analyst (London)* **77**, 661 (1952).
78. C. C. Westcott, *Anal. Chim. Acta* **86**, 269 (1976).
79. C. J. Rix, A. M. Bond, and J. D. Smith, *Anal. Chem.* **48**, 1236 (1976).
80. G. Horvai and E. Pungor, *Anal. Chim. Acta*, **113**, 287 (1980); **113**, 295 (1980).
81. R. A. Durst and J. K. Taylor, *Anal. Chem.* **39**, 1374 (1967).
82. R. A. Durst, E. L. May, and J. K. Taylor, *Anal. Chem.* **40**, 977 (1968).
83. R. A. Durst, *Anal. Chem.* **40**, 931 (1968).
84. T. M. Florence, *J. Electroanal. Chem. Interfacial Electrochem.* **31**, 77 (1971).
85. G. Mattock, *Analyst (London)* **87**, 930 (1962).
86. S. Phang and B. J. Steel, *Anal. Chem.* **44**, 2230 (1972).
87. M. F. Wilson, E. Haikala, and P. Klvalo, *Anal. Chim. Acta* **74**, 395, 411 (1975).

88. E. L. Eckfeldt and W. E. Proctor, *Anal. Chem.* **43**, 332 (1971).
89. D. Midgley, *Anal. Chem.* **49**, 1211 (1977).
90. International Union of Pure and Applied Chemistry, *Pure Appl. Chem.* **48**, 127 (1976).
91. International Union of Pure and Applied Chemistry, *Recommendations for Nomenclature of Ion-Selective Electrodes (Recommendations, 1975)*, Pergamon, Oxford, 1976.
92. D. Midgley, *Analyst (London)* **104**, 248 (1979).
93. C. Liteanu, I. C. Popescu, and E. Hopirtean, *Anal. Chem.* **48**, 2010 (1976).
94. International Union of Pure and Applied Chemistry, *Pure Appl. Chem.* **51**, 1913 (1979).
95. R. P. Buck, *Anal. Chim. Acta* **73**, 321 (1974).
96. R. E. Reinsfelder and F. A. Schultz, *Anal. Chim. Acta* **65**, 425 (1973).
97. Y. Umezawa, M. Nagata, K. Sawatari, and S. Fujiwara, *Bull. Chem. Soc. Jpn.* **52**, 241 (1979).
98. B. Karlberg, *Talanta* **22**, 1023 (1975).
99. E. P. Serjeant and A. G. Warner, *Anal. Chem.* **50**, 1724 (1978).
100. K. Srinivasan and G. A. Rechnitz, *Anal. Chem.* **40**, 509 (1968).
101. G. Johansson and K. Edstrom, *Talanta* **19**, 1623 (1972).
102. L. E. Negus and T. S. Light, *Instrum. Technol.* **23** (December 1972).
103. E. Lindner, K. Toth, and E. Pungor, in *Ion Selective Electrodes*, E. Pungor, Ed., Akademiai Kiado, Budapest, 1973, p. 205.
104. D. Midgley and K. Torrance, *Analyst (London)* **97**, 626 (1972).
105. D. Midgley, *Analyst (London)* **100**, 386 (1975).
106. K. Bergner, *Anal. Chim. Acta* **87**, 1 (1976).
107. D. D. Perrin and I. G. Sayce, *Chem. Ind. (London)* **1966**, 661.
108. A. F. Milward, *Analyst (London)* **94**, 154 (1969).
109. R. G. Bates, *Determination of pH Theory and Practice*, 2nd ed., Wiley, New York, 1973, p. 390.
110. G. J. Moody, R. B. Oke, and J. D. R. Thomas, *Analyst (London)* **94**, 803 (1969).
111. R. A. McAllister and R. Cambell, *Anal. Biochem.* **33**, 200 (1970).

CHAPTER

5

# ELECTRODES FOR POTENTIOTITRIMETRY

## 5.1 ELECTRODE REQUIREMENTS AND ADVANTAGES OF POTENTIOTITRIMETRY

The performance requirements of electrodes that can be used to detect the equivalence points in titrimetry are much less stringent than those required for electrodes used in direct potentiometry. For potentiotitrimetry it is sufficient that the indicator electrode yield a large change in potential in the vicinity of the stoichiometric equivalence point for the titration reaction, and, as was pointed out in Chapter 2, it is not essential that the monitored value of $E_{cell}$ be capable of interpretation by means of the Nernst equation. Thus electrode systems that would be totally unsuitable for use in direct potentiometry can be applied usefully in potentiotitrimetry. This adds considerably to the range of electrodes that are available for this purpose since all the electrodes suitable for direct potentiometry can also be used in titrimetric applications.

The requirement for the indicator electrode is that it must respond to changes in the concentration of either a species generated by or consumed in the titration reaction, and classically this species is present either as the determinand or as the titrant. For example, if the indicator electrode responds to the titrant, then a sudden change in potential can be expected when the determinand ion has quantitatively reacted. Similarly, if the electrode is responsive to the determinand ion, then the potential will change little once the equivalence point has been reached. In some types of titrations, however, the electrode may respond to both the determinand and the titrant. Titrations with silver salts (argentimetric determinations) using silver-silver halide electrodes are examples of this type, and so is the precipitation titration of fluoride with lanthanum nitrate titrant when a fluoride ion selective electrode is used as the indicator electrode. In these examples the electrode becomes sensitive to the ion activity of the titrant after the equivalence point which, nonetheless, can be detected from the comparatively large change in potential that occurs in its vicinity.

It is not always possible, however, to select or devise an electrode that will respond directly to the determinand and/or the titrant species, and in these instances it is sometimes possible to add an indicating ion for which an electrode is available. This indicating ion may be a reagent that reacts quantitatively with the determinand ion. A known volume of a standard solution of this reagent is added to the determinand ion such that a significant excess of reagent is present. The excess is then determined by the addition of a standard solution of titrant that reacts quantitatively with the reagent. Thus the difference between the number of moles of reagent added originally and that determined by the titration is related to the number of moles of determinand. In these instances the electrode selected should respond either to the reagent added in excess or to the titrant that is used to determine this excess. For example, a platinum indicator electrode can be used in a potentiotitrimetric determination of dyes containing azo or nitro groups (1). These groups are reduced with a standard solution of titanium(III) sulfate in glycerol which is added in excess. This excess is then determined by titration with a standard solution of ammonium iron(III) sulfate, the platinum electrode being used to detect the equivalence point of the reaction $Fe^{3+} + Ti^{3+} \rightarrow Fe^{2+} + Ti^{4+}$. It should be noted that if the glycerol–titanium(III) ratio is greater than 4:1, then the usual interference due to oxidation by air is not observed. A variant of this *indirect titration method*, as it is called, is to use a reagent ion to quantitatively precipitate the determinand. The amount of the ion present in the precipitate after it has been collected, washed, and redissolved is then determined by titration with a standard solution that reacts with the indicating ion. As before, the electrode chosen can respond either to the indicating ion or to the titrant. Thus phenoxyalkylcarboxylic acids in pesticides can be determined (2) by their transformation into insoluble mercury(I) salts. The precipitates are dissolved in $1-2\ M$ nitric acid and the indicating mercury(I) titrated with a standard solution of potassium bromide using a liquid membrane electrode that responds to mercury(I) as the indicator electrode.

In contrast to these indirect methods in which the concentration of indicating ion must equal or exceed that of the determinand ion, it is sometimes possible to employ an indicating species at very low concentration ($10^{-4}$–$10^{-6}\ M$) with respect to the concentrations of determinand and titrant. The use of the mercury electrode in complexometric titrations particularly with ethylenedinitrilotetraacetic acid (EDTA) is an example of this application. These titrations are carried out in the presence of approximately $10^{-6}\ M$ mercury(II)–EDTA complex which is added as the indicating species. The potential of the system $Hg|Hg^{2+}; HgY^{2-}$, in which $Y^{-2}$ is the dianion of EDTA, responds in a Nernstian fashion to $pM^{n+}$ (i.e., $-\log[M^{n+}]$) where $[M^{n+}]$ is the concentration of determinand metal ion.

There is thus a sharp change in potential of the indicating system in the vicinity of the equivalence point when the determinand ion is titrated with EDTA, and this method has been applied to the determination of 29 different metal ions either by direct or indirect titration procedures (3, 4). These *pM electrode systems*, as they are called, are described in more detail elsewhere in this book. A different type of principle governs the employment of iron(II) as an indicator ion in the potentiotitration of iron(III) with EDTA which is monitored with a platinum electrode. In this instance it is sometimes necessary to add one or two drops of a dilute solution of iron(II) to the titrand solution in order to establish the indicating half-cell $Pt|Fe^{3+}; Fe^{2+}$. If this titration is carried out at $p$H 3, the iron(II) is not complexed by the EDTA titrant which reacts selectively with iron(III) causing a large change of potential when this species is fully complexed.

Potentiotitrimetry, therefore, offers a number of advantages over direct potentiometric methods. The method is more versatile, for not only is there a greater variety of electrodes available, but, through the application of the type of indirect and indicating methods outlined, there are a greater number of ways these electrodes can be used. Furthermore, irrespective of the type of indicating system used to detect the equivalence point—whether, for example, colorimetric, conductimetric, photometric, potentiometric, or thermometric—titrimetry is inherently a more accurate method of analysis than direct potentiometry. The constraints upon the application of a stoichiometric chemical reaction to titrimetry are imposed generally by the indicating system used and when titrimetry is applied within these constraints a precision of 1 part in 1000 parts can be obtained. By comparison it would be exceptional to obtain a precision of better than about 10 parts per 1000 by the direct potentiometric method using a cell with liquid junction that consists of an ion selective electrode and a reference half-cell. However, a direct potentiometric method has some advantages over a potentiotitrimetric method in that it is a more rapid method, it is more applicable to trace analysis, and it is more amenable to automation for operation in continuous monitoring.

## 5.2 ELECTRODES FOR ACID–BASE TITRIMETRY

### The Glass Electrode

The cell

glass electrode$|H^+(aq)||3.8\ M\ KCl|Hg_2Cl_2;\ Hg$

commonly used for measurement of $p$H, is also the most widely used to detect the equivalence points of the acid–base titrations that can be performed in water. However, the number of different types of acid–base titrations that are analytically feasible in aqueous solution is very small in comparison to the number that are possible in nonaqueous solvents. This arises from the fact that the autoprotolysis constants for nonaqueous solvents are smaller than that of water, which allows a far greater range of potentials to be observed during a titration than is possible in water. For example, the equivalence point in the titration of the weak base aniline ($pK_a$ in water 4.7) with a standard solution of strong acid cannot be detected from the potential change when the titration is performed in water (autoprotolysis constant = $10^{-14}$), but there is a large change in potential when the titration is carried out in 2-propanol (autoprotolysis constant = $10^{-21}$). A further advantage of using nonaqueous systems for acid–base titrimetry is the increased solubility of many organic compounds in these systems in comparison with their solubilities in water. It is also fortunate that a glass electrode, usually preconditioned and stored in water, can be used as the indicator electrode for titrations performed in many solvent systems. These have included acetic acid (5), acetic anhydride (6), acetonitrile (7), alcohols (8), amides [e.g., dimethylformamide (9), tetramethylurea (10)], ketones [e.g., acetone (11), 2-butanone (12)], pyridine (13), tetrahydrofuran (14), and 1,1,3,3-tetramethylguanidine (15). The glass electrode has also been used in many binary and tertiary mixtures of these solvents that in some instances have contained a significant proportion of such nonpolar solvents as benzene, chloroform, and dioxan. In these applications, however, it is often necessary to either modify the calomel reference half-cell or to replace it by one of the other reference half-cells given in Table 3.3 when using a cell that contains a glass electrode.

### Metal and Metalloid Electrodes

It was discovered in 1925 (16) that a stick of antimony when immersed in an aqueous solution formed a half-cell that responded to changes of $pH$. The potential probably develops as a result of the oxidation of the surface layer of the metal forming antimony trioxide, $Sb_2O_3$, the potential of the couple $Sb|Sb_2O_3$ being then dependent upon hydrogen ion as is shown by the half reaction which is usually written as

$$Sb_2O_{3(s)} + 6H^+ + 6e \rightleftharpoons 2Sb_{(s)} + 3H_2O$$

Nowadays the electrode is seldom used in $pH$ measurements because it is susceptible to the presence of oxidants and reductants, and its performance is also severely impaired by the presence of anions with which antimony forms complexes. The latter include oxalate, tartrate, citrate, and metaphosphate. The presence of a small quantity (0.5 ppm) of copper(II) in the solution also adversely affects the response of the electrode. Despite these limitations, the antimony electrode is still usefully applied in nonaqueous acid–base potentiotitrimetry. Indeed, for some applications the performance of an antimony electrode was found to be equal or superior to that of a glass electrode for equivalence point detection (17). In that study the performances of a variety of metal and metalloid indicator electrodes for acid–base titrations, carried out in dimethylformamide and 4-methyl-2-pentanone under nitrogen, were compared with that of the glass electrode. The titrant was tetrabutylammonium hydroxide (0.1 $M$) prepared in toluene-methanol (3:1 v/v), which also provided the reference half-cell solution, the reference electrode being a platinum wire which was sealed in such a position that it made electrical contact with this titrant solution throughout the titration. In addition to the antimony electrode, the performance of electrodes fabricated from aluminum, gallium, indium, silicon, germanium, and bismuth were found to be as good as or better than the glass electrode giving sharper equivalence point inflections. The aluminum electrode was made from 1 mm diameter wire (99.99% purity) which was mounted in a PTFE holder shaped to fit a B14/23 socket. The antimony and bismuth electrodes (diameter 6 mm, length 15 mm) were formed by pouring the molten metals into Pyrex glass tubes sealed at one end. A copper lead was inserted into the molten metal, and after cooling, the glass was broken away from the lower 10 mm of the rod. Germanium (m.p. 937.4°C) electrodes were made in a similar manner except that the electrical contact was made through a mercury pool on top of the cooled rod. The molten elements were used also to prepare the gallium (m.p. 29.8°C) and the indium (m.p. 156.6°C) electrodes, but in these cases they were poured into J-shaped glass tubes forming electrodes similar in appearance to the mercury amalgam electrodes shown in Figure 1.3. The silicon electrode was a fragment of silicon, about 5 mm wide, fused onto the flat ground end of a length of silica tubing. A mercury pool contact was used.

A fresh length of aluminum wire was used for each determination, and the surfaces of the antimony, bismuth, and germanium electrodes were renewed by slicing a thin layer from the ends of the rods. The surface of the silicon electrode was cleaned by immersing it in hydrofluoric acid and the surfaces of the gallium and indium electrodes were cleaned by melting the elements and then wiping off any dross with a tissue. The cleaned electrodes were exposed to the air for 2 h before being used again.

The gallium electrode was found to be the most sensitive of those investigated in both DMF and 4-methyl-2-pentanone, but the aluminum, antimony, and bismuth electrodes were similar to it in sensitivity. Aluminum and gallium electrodes have the advantage of convenience in use because the gallium electrode can be resurfaced by melting it in hot water and the aluminum electrode is readily made in rod or wire form. Aluminum electrodes have also been used in acid–base titrations in aqueous solution, and an electrode with a carefully polished and degreased trowel-shaped surface (20 mm × 2 mm) was found to yield a total potential change in the titration of potassium hydrogen phthalate with sodium hydroxide some 3.3 times that of an antimony electrode (18). In 99.9% ethanol or in 2:1 benzene–alcohol the potential change was 1.8 times as high, and at the equivalence point the slope was about twice that obtained with the antimony electrode. The preferred form of the electrode, however, appears to be aluminum wire which is clipped immediately before use (19). It has been reported (20) that an aluminum electrode can be used in the following cases:

1. When corrosion-active ions or molecules appear or disappear.
2. When ions or molecules affecting the formation of oxides or the stability of films rather than absorption on the oxide appear or disappear.
3. When, at the equivalence point, ions or molecules affecting the formation of oxides are exchanged by those delaying it.
4. When, at the equivalence point, dissolution of the film surface occurs.

Oxide formation has also been shown to be responsible for the functioning of electrodes fabricated from molybdenum, gold, platinum, and tungsten as indicators in acid–base titrimetry. In the case of the molybdenum electrode the formation of a lower valency oxide was shown to yield an electrode with a good response (21) and this was found to be the case for a gold electrode (22). After irradiation under an infrared lamp (2.25 h at a distance of 17 cm) a gold electrode, formed by sealing gold wire (1 mm dia.) into soft glass tubing (5 mm dia.), gave a change in potential at the equivalence point some 200 mV greater than an electrode that was not irradiated. The improved performance was attributed to the interconversion of gold(III) oxide to gold(II) oxide under the influence of this radiation since it was shown that heat alone did not yield any improvement. A similar type of improvement resulted when a gold electrode was laser-irradiated (23).

The "oxidized" platinum electrode has also been used as an indicator electrode in aqueous and nonqueous media, and has been found reversible to hydrogen ion in the $p$H range 4–13 (24). At $p$H 6.5–9.5 the potentials are said to be determined by the reaction

$$PtO_2 + 2H^+ + 2e \rightleftharpoons Pt(OH)_2$$

when the electrodes are prepared by the following method. They are first boiled in 1 $M$ sodium hydroxide for 30 min and washed with water. Each electrode is then put through 10 cycles of alternately heating it in an ethanol flame for 2 min followed by electrolysis as a cathode in 0.05 $M$ sufuric acid with a current of 10 $\mu$A for 40 min. Thereafter, the electrodes are stored in air and before an electrode is used it is heated in an ethanol flame for 10 sec. The electrode was found suitable as an indicator electrode for titrations of benzoic acids and dinitrophenols in 70% acetone–water with potassium hydroxide prepared in 2-propanol. Titrations of some carboxylates were also performed in this solvent using aqueous hydrochloric acid as titrant. The electrode has also been prepared by cleaning the platinum surface with hot chromic acid followed by heating it in a methanol flame (25).

It has been known for a long time that the potential of a tungsten wire depends upon the acidity of the solution in which it is immersed (26). The $p$H response of untreated tungsten electrodes, however, is not smooth but has inflections which have been explained in terms of a change in the composition of the oxide layer on their surfaces which also causes shifts in the standard electrode potential. Removal of these layers either by chemical reduction with hydrogen at high temperature or by electrolytic reduction in the absence of oxygen yielded an electrode the potential of which was Nernstian over a wide range of $p$H (27). More recently (28), a cylindrical electrode of tungsten 5 mm long and 3 mm in diameter sealed into glass tubing has been used in conjunction with a mercury–mercurous acetate reference half-cell (see Table 3.3) for the titration of alkaloid bases in the nonaqueous system acetic acid–acetic anhydride (1 : 6 v/v). The titrant was perchloric acid (0.1 or 0.01 $M$) in acetic acid, and the results obtained with the tungsten electrode agreed with those obtained when a glass electrode was used as the indicator electrode for these determinations.

The equivalence points of acid–base titrations have also been detected with carbon electrodes. Pyrolytic graphite, a polycrystalline form of carbon that has a high degree of orientation produced by vapor phase deposition, is a material possessing a high degree of impermeability, and freedom from trapped gases and metallic contaminants. The plane that is perpendicular to the plane of deposition has ceramic type behavior, whereas metalloid behavior is observed in the plane of deposition. An electrode produced by sealing a 6-mm cube of this material into the flared end of a borosilicate tube with epoxy cement such that only the metalloid surface was exposed to the solution has been used (29) as indicator for the titration of 2-phenylphenol in pyridine containing about 4% 2-propanol. The titrant was tetrabutylammonium hydroxide in benzene. These titrations were carried out using an H-shaped cell which had a sintered plate sealed into the cross-piece. One arm contained the indicator electrode and the other contained a potassium chloride solution and a calomel electrode. On the basis of the properties of pyrolytic carbon electrodes it was concluded that these function like an oxygen electrode (29). Electrodes prepared from carbon fiber, however, are thought (30) to be dependent upon the formation of carboxylic acid groups on their surfaces by reaction with atmospheric oxygen rather than upon oxygen present in the solution. A carbon fiber electrode is prepared by soldering a 1.5 cm length of 8 $\mu$m diameter fiber to a tin-coated single-strand copper wire. The solder joint is coated with epoxy resin (e.g., Araldite) and sealed into borosilicate glass tubing with the same resin such that about 1 cm of the carbon fiber is exposed. The cross-sectional surface of the carbon fiber is also structurally very different from the longitudinal section (cf. pyrolytic graphite), but in this case both planes yielded almost identical potentials against a saturated calomel electrode when immersed in various aqueous $p$H buffer solutions. In these aqueous solutions the $p$H response of the fiber electrode was linear over the $p$H range 1-13 and was used to detect the equivalent points of acid-base titrations performed in aqueous solution.

Carbon electrodes that are impregnated with epoxide resin, when activated by immersion in 0.2 $M$ potassium permanganate in 0.5 $M$ sulfuric acid, have been used for acid-base titrations in acetic anhydride, acetonitrile (31), anhydrous dimethylformamide (32), and in dimethylsulfoxide (33). In the latter solvent the change in potential in the vicinity of the equivalence point was about twice that obtained with glass, gold, and antimony electrodes. The electrodes were reduced after each titration by immersion in 0.05 $M$ tin(II) chloride in 1 $M$ hydrochloric acid and then oxidized with the permanganate solution before the next titration. The purpose of the epoxy resin is to decrease the surface porosity which, in effect, is similar to hydrophobizing graphite with Teflon. Low porosity graphite electrodes have also been prepared from spectrographically pure graphite powder incorporated into a silicone rubber-based membrane (34), and these have been used as indicator electrodes for acid-base titrations in water and in acetone (35, 36). A great improvement in the magnitude of the inflection about the equivalence point was also observed for these membrane electrodes when they were pretreated with an acid solution of potassium permanganate. More recently (37), it has been shown that pretreatment of graphite membrane electrodes with the aqueous oxidants 0.2 $M$ potassium dichromate, 0.5 $M$ cerium (IV) sulfate, or with a saturated solution of lead(IV) acetate in acetic acid is less effective than treatment with acidified aqueous 0.2 $M$ potassium permanganate in enhancing the potential jumps observed in acid-base titrations performed in acetic acid. The magnitude of these inflections were only reproducible if the electrode was reduced with tin(II) chloride and reactivated with acidified potassium permanganate immediately before the next titration. If this procedure is not adopted between successive titrations, then the potential jump observed in the second titration will be about one-half of that observed using a freshly activated electrode for the first titration. After immersion in acidified permanganate solution for 3–5 min, the activated electrode requires a 20-min period of equilibrium in the test solution before an almost steady potential is attained. Thereafter, stable potentials are attained quite rapidly after the addition of titrant, but it is advisable to allow 2–3 min for stabilization between successive additions of titrant in the vicinity of the equivalence point. The effect of the activation process is attributed to the oxidation of the surface layer of graphite and by the separation of manganese compounds of higher oxidation state onto its surface. Thus a manganese dioxide electrode, formed by the electrolysis of a solution containing manganese(II) sulfate, sodium formate, and formic acid at a platinum anode (area 0.5 cm$^2$) showed stable potentials in the titration of sodium acetate with perchloric acid in acetic acid and a larger potential jump (about 300 mV/0.1 mL HClO$_4$) versus a mercury–mercurous acetate reference half-cell than the carbon-based electrodes. The equivalence points of acid–base titrations among others have also been detected with a carbon electrode obtained from a discharged dry cell (38).

The palladium electrode is in a different category from the foregoing metal and metalloid electrodes in that it operates as an hydrogen electrode by virtue of the capacity of this metal to take up hydrogen gas spontaneously in an amount required to saturate the α-phase of the metal. Once this phase is saturated, the α-palladium hydrogen electrode can be used in solution free of hydrogen gas (39). The electrode has been used in acid–base potentiotitrimetry and is not affected by the presence of oxidizing and reducing agents (40). The response of this electrode has been found superior to the response of the glass electrode for the titrations of solutions of phenols in 2-butanone with a titrant solution of potassium hydroxide in 2-propanol (41). The electrode can be prepared by fusing palladium wire onto a platinum wire which is then flame-sealed through glass. It is charged with hydrogen electrolytically in a 1% aqueous solution of sulfuric acid using a platinum anode, the two electrodes being connected to a 1.5–3 V source. Evolution of gas is observed immediately at the anode but not at the cathode until the palladium is charged with hydrogen. Substantial evolution of hydrogen at this electrode, therefore, signifies that the electrolysis can be terminated. In the titrations performed in 2-butanone, the potential of the palladium electrode was monitored against a reference half-cell prepared by immersing an identically prepared electrode in a solution the composition of which approximated that expected at the endpoint of the titration. As an alternative, the potential was monitored against that of a half-cell formed when a platinum electrode was sealed so that it made contact with the titrant solution. Equilibrium of potential after an addition of titrant was established within a few seconds provided the solution was stirred, and in this respect the performance of the Pd/$H_2$ system was considered superior to that of the glass electrode. The electrode, however, can become poisoned after prolonged use, but can be restored to its original condition by immersing it briefly in hot nitric acid, rinsing, and then gently heating it in an hydrogen flame.

The $p$H response of electrodes prepared from nonstoichiometric bronzes having the formula $M_xWO_3$ has been reported (42). Although this response was not linear, it sufficed for the employment of these electrodes as indicators for acid–base titrations in water and in acetic acid. The change of potential circumjacent to the equivalence point when these titrations were monitored with the tungsten–bronze electrode appeared to be 10–50% larger than were observed when a glass electrode was used in conjunction with the same saturated calomel reference half-cell. The electrodes were prepared from single crystals having the formula $Na_{0.71}WO_3$ formed as a result of the electrolysis of tungsten trioxide (about 36 mole %) and sodium tungstate between a No. 22 chromel wire cathode encased in an alundum tube and a graphite anode at a temperature just above the mp (*ca.* 700°C) of the charge (43). Individual pieces, single crystals, of bronze were cut from larger crystals with a diamond saw. The crystal was sealed to the end of a piece of glass tubing with epoxy cement and connection to the electrometer was made by means of a copper wire immersed in a mercury contact held in the tube.

Other potentiometric indicators that have been used in acid–base titrimetry are the pyrite electrode (44), and the potentials generated across a frit when it separates an immiscible solvent from the aqueous solution in which the titration is performed (45). These potentials were measured between two saturated calomel electrodes.

### Electrodes for Electrolytic Potentiotitrimetry

The passage of a small heavily stabilized current through a pair of indicator electrodes dipping into a stirred solution of titrand and measuring the potential that is generated between them forms the basis of this technique. Usually the two electrodes are fabricated of identical materials although this need not be a requirement. The application of this technique to acid–base titrimetry is not widespread but, nonetheless, it has been used for equivalence point detection in both aqueous and nonaqueous solutions. A pair for platinum electrodes is the system most favored for nonaqueous solvents and when a current, usually in the range of 1–10 $\mu$A, is impressed between them, the shape of the titration curve shows a peak in the values of the observed potentials that is coincident with the volume of titrant required to reach the equivalence point. Thus two platinum electrodes, polarized by passage of a constant current of 1 µA, have been used as the indicating system for the titration of acids dissolved in acetone with tetrabutylammonium hydroxide (46), and also for the titration of organic amines with perchloric acid in a solvent system composed of $m$-cresol and acetonitrile (47). A vacuum tube voltmeter was used to detect the potential peak at the equivalence point. Lithium chloride (0.1 g 30 ml$^{-1}$) was used to increase the conductivity of a 4 : 1 benzene–methanol solvent when long chain carboxylic acids were titrated with sodium methoxide using two platinum electrodes, polarized by a current of 10 µA, as the indicating system (48).

More recent examples of the application of polarized electrodes to acid–base titrimetry in nonaqueous solution are provided by the use of platinum electrodes in the determination of the sulfo and phenol groups in lignosulfonic acid (49), and the use of polarized quinhydrone electrodes in the determination of organic bases in a medium of acetic acid and acetic anhydride (50). The lignosulfonic acid (Registry Number *8062-15-5*) was determined in dimethylsulfoxide by titration with $NNN$-trimethyl-1-hexadecylammonium hydroxide (cetyltrimethylammonium hydroxide) solution prepared in a mixture of methanol and 2-propanol.

In aqueous media the application of differential electrolytic potentiometry to acid–base titrations appears to offer no particular advantage over conventional potentiotitrimetry. However, when dual polarized antimony electrodes were used (51), the method yielded results of high precision and excellent accuracy in situations where the "zero" current method is capable of yielding an acceptable result. In this instance the antimony electrodes were prepared by pouring molten antimony into 5-mm Pyrex tubes, the lower ends of which were drawn out into a 1-mm jet, whereas the upper ends were sealed to a wider diameter (11 mm) tube reamed out to form a small funnel. Precautions were taken to avoid the entrapment of air while the molten metal was being poured into the preheated tubes and contact to the antimony was made by means of a length of 1.6 mm dia. tinned copper wire which was inserted just before the antimony solidified. The 5-mm bore tubing was cut off near the lower end with a diamond wheel and, whenever a fresh antimony surface was required, a further slice was sawn off to expose a known and reproducible area of antimony metal. Potentiometric tests on these electrodes gave a reproducibility between diverse electrodes of 1 mV, the day-to-day reproducibility for a single electrode being better than 1 mV. When used in the $p$H mode they showed a linear response over the range 3–12. When used routinely as indicator electrodes in titrimetry, the two differential electrodes were connected in series to a battery and a single high value resistor such that the required current density was obtained. The product of the voltage and resistance should exceed $5 \times 10^9$ volt-ohms and the optimum current densities depend upon the titration conditions, being in the range 0.1–12 $\mu A\ cm^{-2}$. The potential across the electrodes was monitored with a $p$H meter, and the titrand solution contained a supporting electrolyte of 0.01 $M$ potassium chloride or sulfate. The latter diminished the electromigration of the electroactive species at high dilution and shortened the response time of the electrodes.

## 5.3  ELECTRODES FOR COMPLEXOMETRIC TITRATIONS

These titrations are usually performed to determine the concentration of a metal ion in solution, and the most widely used titrant for this purpose is ethylenedinitrilotetraacetic acid, EDTA. The determinand ion reacts quantitatively with the complexant (e.g., EDTA) to form a soluble complex which reduces the concentration of free determinand ion to a negligible value at the equivalence point. A distinction is made between these titrations in which the determinand ion is effectively removed from solution as a soluble complex, and those in which an insoluble complex is formed. The latter are usually categorized as precipitation titrations.

Electrodes suitable for equivalence point detection in complexometric titrations usually respond to the metal ion being titrated, although electrodes have been developed that respond to some of the visual colorimetric indicators used routinely for these titrations. Ion selective electrodes that respond either directly or indirectly to changes in concentrations of the appropriate determinands are suitable indicator electrodes. However, the most versatile electrode developed is a simple one based on metallic mercury which predates the development of suitable ion selective electrodes by almost a decade.

### The $p$M Electrode Based on Mercury

This electrode (3, 4) is conveniently fabricated in the form of a J-shaped electrode similiar in appearance to the amalgam electrodes shown in Figure 1.3. The cup of the electrode contains a small quantity of pure mercury and when it is immersed in a solution that contains mercury(II), an equilibrium is set up according to the half reaction

$$Hg^{2+} + 2e \rightleftharpoons Hg_{(l)} \quad \text{for which } E° = 0.85\ V$$

When EDTA (represented as the tetraanion $Y^{-4}$) is the titrant, and a small quantity of mercury–EDTA complex ($HgY^{-2}$) is added to the solution containing the determinand metal $M^{n+}$, the resulting half-cell can be represented as an electrode of the third kind, namely,

$$Hg\,|\,Hg^{2+};\,HgY^{2-};\,MY^{(n-4)+};\,M^{n+} \qquad (5.1)$$

The potential of this half-cell is given by the equation

$$E = E^{o\prime} - \frac{k}{2}\log\frac{1}{[Hg^{2+}]} \qquad (5.2)$$

where $E^{o\prime}$ is the formal potential under the experimental conditions and $[Hg^{2+}]$ is the concentration of the uncomplexed mercury(II) in the solution. The term $[Hg^{2+}]$ can be deduced from the stability constant of the reaction $Hg^{2+} + Y^{4-} \rightleftharpoons HgY^{2-}$ which is defined as

$$K_{HgY} = \frac{[HgY^{2-}]}{[Hg^{2+}][Y^{4-}]} \sim 10^{22} \qquad (5.3)$$

Using an analogous definition for the stability constant, $K_{MY}$ of the complex formed between $M^{n+}$ and $Y^{-4}$ allows the term $[Y^{4-}]$ in equation (5.3) to be replaced by the expression

$$[Y^{4-}] = \frac{[MY^{(n-4)+}]}{K_{MY}[M^{n+}]}$$

Thus equation (5.2) can be expanded and written as

$$E = E^{o\prime} - \frac{k}{2}\log\frac{K_{HgY}}{[HgY^{2-}]K_{MY}} - \frac{k}{2}\log\frac{[MY^{(n-4)+}]}{[M^{n+}]} \qquad (5.4)$$

An essential condition for the application of the half-cell is that the stability constant for the mercury complex ($\sim 10^{22}$) is several orders of magnitude greater than the stability constant of the metal determinand complex, otherwise an appreciable amount of $Y^{4-}$ would be removed from the added $HgY^{2-}$ complex. If this condition is met, then $[HgY^{2-}]$ will remain essentially constant throughout the titration, and thus the first two terms of the right-hand side of equation (5.4) will also be constant. Furthermore, in the region of the equivalent point $[MY^{(n-4)+}]$ changes little so that the measured potential of the mercury electrode becomes a linear function of $pM$ (i.e., $\log(1/[M^{n+}])$).

Any substance that reacts extensively with mercury(I) or mercury(II) will cause an interference to the correct functioning of the mercury electrode.

Thus halides, sulfide, and cyanide should be absent from the titrant solution and in some instances, for example, in the titration of barium in ammoniacal solution (3), dissolved oxygen should be removed from the solution. If it is intended, therefore, to measure the potential of the mercury electrode against that of the calomel reference half-cell, it is advisable to use a double junction type of reference electrode in order to avoid contaminating the solution with chloride. The method is applicable over the range of $p$H from 2 to 11 and, as shown in Table 5.1, it is advantageous to subdivide this range into four: $p$H 2, $p$H 4–5.5, $p$H 8–10, $p$H 10. In general, the metal ions that can be titrated at 4–5.5 or at $p$H 8–10 will not interfere with titration at $p$H 2 (52), and alkaline earth ions will not interfere with titrations at $p$H 4–5.5.

Some metal ions, such as aluminum and chromium, form EDTA complexes only slowly at room temperature and in these cases the solution is heated in the presence of an added excess of a standard solution of EDTA. The excess of EDTA is titrated at room temperature with a standard solution of a metal ion that reacts rapidly with EDTA. Zinc(II) is often used for these indirect titrations. A summary of the experimental conditions required for the determination of the individual metal ions listed in Table 5.1 are included in Table 11.1. Mixtures of these metal ions in the different $p$H groups may also be titrated in the same solution. For example, bismuth, cadmium, and calcium have been sequentially titrated (4). At $p$H 2 the first potential jump corresponded to the volume of titrant required to react with bismuth. The $p$H was then changed to 4 by adding sodium acetate–acetic acid buffer and the titration continued until a second inflection in the potential was observed. The amount of EDTA used between the first and second equivalence points corresponded to the cadmium. Finally, the $p$H was adjusted to 8 with ammonium buffer and the calcium was titrated. For all these EDTA titrations the $HgY^{2-}$ complex can be made by dissolving AR mercuric nitrate (0.343 g) in water (100 mL) and adding 0.01 $M$ EDTA (100 mL), or the equivalent volume of EDTA of any other strength, followed by 2 $M$ sodium hydroxide (0.5 ml). Prior to the titration, 1–2 drops of this solution is added to the metal determinand solution which must be buffered with respect to $p$H by the appropriate buffering system given in Table 5.1.

The mercury electrode also functions well when other chelon titrants are used. Thus TRIEN (1,4,7,10-tetrazadecane) and TETREN (1,4,7,10,13-pentaazatridecane) have been used (53), and so has EGTA [3,12-bis(carboxymethyl)-6,9-dioxa-3,12-diazatetradecanedioic acid] (54). The latter is useful because calcium(II) can be titrated in the presence of magnesium(II) and TETREN has the advantage over EDTA in that it is generally more selective. When using these titrants, the mercury(II)–chelon complex is

**Table 5.1. Conditions for Potentiotitrimetry with EDTA Using a Mercury $p$M Electrode**[a]

| $p$H | Buffer Systems | Procedure | Metals Titrated |
|---|---|---|---|
| 2 | $HNO_3$ or $CH_2ClCOOH$ | Direct titrn. with EDTA | Bi, Hg(II), Th |
| 4–5.5 | HOAc or hexamethylenetetramine[b] | Direct titrn. with EDTA | Bi, Cd, Cu, Hg(II), La, Mn(II) Pb, R.E.(III), Sc, V(VI), Y Zn |
| 8–10 | $NH_3$, 2-aminoethanol or $2,2',2''$-nitrilotriethanol[c] | Direct titrn. with EDTA | Ca, Cd, Co, Cu, In, Mg, Ni, Pb, Zn |
|  |  | Addn. of excess EDTA, back-titrn. with std. Ca, Cd, Cu, Mg or Zn soln. | Bi, Ca, Cd, Co, Cr(III), Cu, Hg(II), La, Mg, Ni, Pb, R.E.(III) Sc, Tl(III), Y, Zn |
| 10 | $NH_3$ or 2-aminoethanol | Direct titrn. with EDTA | Ba, Sr, metals titrd. at $p$H 8–10 |
|  |  | Addn. of excess EDTA, back-titrn. with std. Ca, Cd, Cu, Mg or Zn soln. | Ba, Sr, metals titrd. at $p$H 8–10 |

[a]From Reference 52.
[b]1,3,5,7-tetraazatricyclo[3.3.1.1(3,7)]decane.
[c]Triethanolamine.

prepared by a method analogous to that described for the EDTA–mercury(II) complex.

The design of the mercury electrode is not confined to the mercury pool type, and electrodes constructed of mercury coated gold, silver, or platinum are also effective as $p$M electrodes. A mercury coated gold electrode can be easily prepared by dipping a commercial gold plate electrode for about a minute into pure mercury. The electrode is rinsed in water and can be used for only a few titrations before it is necessary to recoat it with mercury. A mercury coated silver electrode can be prepared (55) by fusing a silver wire to form a ball of silver about 5 mm dia. at the end of the wire. After cleaning the wire, it is mounted in a polythene sleeve with the ball extending about 12 mm from the end. The coating is formed by immersing the clean electrode in pure mercury for 3 h after which it is washed with water. Mercury coated platinum electrodes of any geometry are conveniently and rapidly prepared as follows (56).

> The electrode to be coated is first degreased in ethanolic potassium hydroxide and thoroughly cleaned in boiling nitric acid. After rinsing with distilled water the electrode is placed in an electrolytic cell containing 1 $M$ perchloric acid and a mercury pool deep enough to cover the electrode completely when immersed. This mercury is not electrically connected to any part of the system, and is used as a dip for the electrode after the surface platinum oxides have been electrolytically reduced. To this end the potential of the platinum electrode is held at or near the potential of a hydrogen electrode for 10 min in the perchloric acid solution. The electrode then is dipped momentarily into the mercury pool without disconnecting the current and is removed with its coating of mercury. A gold plated electrode coated with mercury has also been prepared electrolytically (57). In this instance the gold was plated onto platinum wire and the mercury deposited onto the gold by electrolytic reduction from a solution of mercury(II) perchlorate. The electrodes with the best performances contained between 20 and 25 weight percent mercury.

The most commonly used of the mercury coated electrodes appears to be that based on silver and these electrodes have been used in the $p$M mode for titrations with EDTA and also for indirect EDTA determinations using mercury(II) as the titrant. It has also been used as a $p$M electrode for the consecutive titration of calcium(II) and magnesium(II) using EGTA as titrant (58). A fiber tipped mercury–mercury(I) sulfate reference half-cell was used in this instance, and the EGTA–mercury(II) complex was prepared as a 0.002 $M$ solution by mixing equivalent amounts of mercury(II) acetate and EGTA (only 1–2 drops of this solution is added). The calcium(II) was titrated at $p$H 8.5–9.0 in an ammonium nitrate–ammonium buffer (*ca.* 1 $M$). The solution was then made 70–80 volume percent in ethanol, adjusted to $p$H 10 with concentrated ammonia, and magnesium(II) titrated with EGTA.

In addition to its use as a mercury electrode of the third kind [see cell (5.1)], the mercury coated silver electrode has also been used extensively as an electrode of the first kind [Hg|Hg(II)] for the indirect titration of many metal–EDTA complexes with mercury(II) titrant. A known volume of 0.05 $M$ EDTA is added to the appropriately buffered metal determinand solution so that an amount in excess of that required to react with the metal ion has been added. The concentration of metal ion is deduced from the volume of a standard solution of mercury(II) nitrate required to react with this excess of EDTA. The method has been applied to the determination of the following metal ions, alone or in mixtures: Bi (59); Hg (60); Ca, Sr, Ba, Cd, Pb, Zn (61); V(62); Cr (63); Rh, Ru, Pd—EDTA or CDTA used (64); Rh in binary mixtures—EDTA or CDTA used (65). It has also been applied to the analysis of lead (66), zinc (67), cobalt (68), pigments, and to some ores (69).

### The Silver Electrode

The half-cell Ag|Ag$^+$ can be used as an indicator electrode of the first kind for the titration of a variety of metal ions with EDTA (70). The stability constant of the silver(I)–EDTA complex is small ($10^{7.3}$) in comparison with that of the mercury complex ($\sim 10^{22}$), and thus the mechanism of operation for this electrode is the antithesis of that for the mercury electrode. In this instance, therefore, the metal determinand is complexed preferentially to silver(I), and when a trace of silver(I) is added to the titrand solution, the potential of the Ag|Ag$^+$ couple remains essentially constant for most of the titration, but in the vicinity of the equivalence point the concentration of silver(I) is sharply diminished. This diminution is reflected in a marked decrease in the potential of the silver electrode from which the equivalence point can be located. The titrand solution must be buffered at a comparatively high $p$H to ensure that the silver(I)–EDTA is formed, and a 0.1 $M$ borate buffer adjusted in the $p$H range 9.0–10.0 suffices for this purpose.

The method has been used (71) for the direct titrations of barium(II), calcium(II), magnesium(II), and strontium(II), and also for the determination of water hardness. These titrations were performed in the presence of $4 \times 10^{-6}$ $M$ silver (I). The method can also be used for the indirect titrations of $Al^{3+}$, $Bi^{3+}$, $Cd^{2+}$, $Co^{2+}$, $Cu^{2+}$, $Fe^{3+}$, $Ga^{3+}$, $In^{3+}$, $La^{3+}$, $Mn^{2+}$, $Nd^{2+}$, $Ni^{2+}$, $Pb^{2+}$, $Sm^{3+}$, $Th^{4+}$, $VO^{2+}$, $Y^{3+}$, $Yb^{3+}$, $Zn^{2+}$. For these indirect titrations, an accurately measured volume of 0.05 $M$ EDTA is added which is in excess of the amount needed to react with the metal ion. This titrand solution, buffered at $p$H 9.0–9.2 with 0.1 $M$ borate and containing 0.05 mL of 0.008 $M$ silver nitrate, is titrated with standardized 0.05 $M$ calcium nitrate.

Because of the necessity of using the silver electrode at high $p$H, it is not so versatile as the mercury electrode in discriminative titrations with EDTA that enable two or more metal ions to be determined in the same solution. However, the silver electrode attains equilibrium more rapidly in alkaline solution than does the mercury electrode. In some cases, particularly in the titration of barium(II) with EDTA, the change in potential circumjacent to the equivalence point is sharper than with the mercury electrode. The silver electrode is also less susceptible to interference from dissolved oxygen at high $p$H. Commercially available silver billet electrodes have been found suitable for these titrations, and these are generally used in conjunction with a saturated calomel half-cell. Electrodes prepared by the electrodeposition of silver onto platinum wire are satisfactory, and one of this type has been used with a glass reference electrode (72) in titrations of barium(II), calcium(II), copper(II), zinc(II), or iron(III) with EDTA. In this application masking agents that did not significantly complex silver(I) were used. Thus in the determination of calcium(II), the masking of iron(III) and aluminum(III) was accomplished with 2,4-pentanedione, triethanolamine, salicylate, or sulfosalicylate. Sulfate was used for the masking of barium in the determination of calcium(II) and magnesium(II), and fluoride for the masking of calcium(II), aluminum(III), and iron(III) in the determination of copper(II).

The silver electrode has been used as a $p$M electrode for the automated titrations of barium(II) and magnesium(II) with DPTA [3,6,9-tris(carboxymethyl)-3,6,9-triazaundecanedioic acid] as the titrant (73). In the case of this titrant, the stability constant for the silver(I)–DPTA complex ($10^{8.6}$) is more than twentyfold stronger than the silver(I)–EDTA complex and in the same order of magnitude as the barium(II) and magnesium(II) complexes permitting the silver electrode to function as a $p$M electrode in a way analogous to the mercury(II)–EDTA electrode. The titration curves for these metals are, therefore, symmetrical, an advantage for automated titrations as opposed to those obtained when EDTA is the titrant in which the concentration of silver(I) remains approximately constant until very close to the equivalence point. Only in the case of calcium(II), which forms a DTPA complex about one hundredfold stronger than the silver(I)–DTPA complex, is a slight asymmetry in the titration curve noticeable, but the deviation from symmetry is too small to cause significant error. In the case of heavy metals which have larger stability constants than calcium(II), direct titrations are also impractical at high $p$H with this titrant but indirect titrations using standard magnesium(II) nitrate solution will give symmetrical curves characteristic of magnesium(II). Titrations with DTPA are also carried out in borate buffer at $p$H 9 and when the sample contains 0.25–0.5 mmol determinand, 0.05 mL of 0.0015 $M$ silver(I) nitrate was added to the titrand solution. The indirect titrations were analogous to those already described for EDTA, except that 0.01 $M$ magnesium(II) nitrate was used as titrant.

The equivalence points of titrations performed with EGTA (74), TRIEN, and TETREN (75) have also been detected with the silver electrode used in conjunction with a saturated calomel reference half-cell. EGTA was used for the selective titrations of calcium(II) in the presence of magnesium(II), aluminum(III), chromium(III), or iron(III) at $p$H 10 (borate buffer). TRIEN and TETREN were used for the titrations of copper(II), cadmium(II), or zinc(II) each in the presence of calcium(II) and magnesium(II), aluminum(III), or iron(III) in the $p$H range 8.0–9.2 (borate buffer). With EGTA the optimum concentration of silver(I) indicator added was $10^{-5}$ $M$, whereas for TRIEN and TETREN a concentration of about $10^{-6}$ $M$ was found satisfactory. The changes in potential in the vicinity of the equivalent points for the TRIEN and TETREN titrants were found to be greater than those obtained with EDTA, but were slightly smaller than were observed for DTPA.

## The Platinum Electrode

This zeroth electrode has been used for EDTA titrations of those metals that exhibit two different oxidation states. Generally, the stability constant of the higher oxidation state complex is much larger than that of the lower oxidation state and thus at a suitably chosen $p$H the lower oxidation state remains uncomplexed throughout the titration. A large potential change, therefore, accompanies the abrupt change in the concentration of the higher oxidation state which occurs in the vicinity of the equivalence point. Other metals that form complexes with EDTA at the given $p$H interfere. Such a procedure has been used (76, 77) to follow the titration of iron(III)–iron(II) mixtures with EDTA, measuring the potential of the iron(III)–iron(II) couple. At $p$H 3, iron(II) remains uncomplexed by EDTA during the titration and the quantitative complexation of iron(III) is accompanied by a large change in potential at the equivalence point. In the determination of iron(III) it is sometimes necessary to add 2–4 drops of $10^{-3}$ $M$ iron(II) solution from a Jones reductor to the iron(III) solution prior to the titration (18).

An adaption of the method is to use a small amount either of the iron(III)–EDTA complex or of the iron(III)–iron(II) couple as the indicating species in the EDTA titrations of other metal ions. Lead(II), zinc(II), cadmium(II), nickel(II), mercury(II), and scandium(III) can be directly titrated by this method (79) and so can bismuth, indium, thorium, and copper (80).

### The Silver–Silver Sulfide–Metal Sulfide Electrodes

The response of this third kind of electrode to the divalent ions of copper, cadmium, mercury, nickel, cobalt, and zinc has been investigated (81). The type of ion to which this type of electrode responds is determined by the incorporation of the appropriate metal sulfide into the electroactive surface of the electrode. Of the divalent metal ions tested, only those of copper, cadmium, and lead showed a rapid and Nernstian response to changes in the free metal ion concentrations within the $p$M range of 1.7–5. The copper sulfide and cadmium sulfide electrodes were found suitable for use in the detection of equivalence points in titrations with EDTA, but the lead sulfide electrode was less satisfactory. These electrodes were made by anodically oxidizing a clean silver rod in a dilute solution of sodium sulfide. The silver–silver sulfide electrode so formed was made sensitive to copper(II), cadmium(II), or lead(II) by immersing it in a dilute solution of sodium sulfide, and then adding a solution of the metal nitrate until the free metal ion concentration was in slight excess at about 0.01 $M$. After rinsing the electrode with water, the electrodes were stored in a saturated solution of the appropriate metal sulfide. The titrations of copper(II), lead(II), and cadmium(II) were performed in solutions buffered at about $p$H 4.7 with 0.005 $M$ acetic acid and sodium acetate.

### Ion Selective Electrodes

In a majority of the applications of these electrodes to complexometric titrations they function directly as $p$M electrodes, just as they do in analysis by direct potentiometry. Thus like the mercury $p$M electrode they may be used to monitor the concentration of the determinand ion during the course of the titration or, alternatively, they can be applied to the indirect titration method by responding to the titrant that is used to react with the excess of complexon. An impediment to the use of some ion selective electrodes, however, is their response time and this applies particularly to electrodes derived from large organic ion exchangers or from neutral carriers. For direct titrations, the response times of these may vary from about 15 sec when the determinand concentration is initially high to several minutes when this concentration is low near the equivalence point. The electrodes based on membranes fabricated from inorganic salts have much faster response times and are thus more suitable for this titrimetric application.

Many of the cation responsive membrane electrodes listed in Tables 3.4 and 3.5 have been used as indicator electrodes in titrations with EDTA and with other complexons and some of these applications are given in Chapter 11. In the discussion here, emphasis is placed on some of the less obvious types of electrodes mentioned in Chapter 3, and also upon those designed specifically for complexometric titrations.

A disadvantage to the use of the mercury and silver electrodes in these titrations is that chloride ions must be removed from the titrand solution. This can be achieved in some instances by passage of the sample solution through a column of an anion exchange resin such as the nitrate form of Dowex 1X-8 or Amberlite IR400 resin (71). The responses of cation selective membrane electrodes on the other hand are not greatly affected by chloride ion concentrations of less than 0.01 $M$, and this ion is not often included in selectivity coefficient data for these electrodes (82). In solutions of high concentrations of chloride ion, however, doubtless the performance of these electrodes would be impaired, and it is interesting to note that coated wire anion responsive electrodes based on the chloro complexes of some transition metals have been used to monitor EDTA titrations in solutions that contain chloride ion concentrations greater than 1 $M$. For example, electrodes based upon the Aliquat 336S-chlorometallate complex (70 parts) in PVC (30 parts), prepared as outlined earlier, have been used in the titration of copper (83), cadmium (84), and zinc (85) with EDTA. A series of ion selective electrodes has also been based on metal complexes of the type $M_x[N(II)L_4]$ where M = Ag, Cu, Pb, Hg, or malachite green (Registry No. *569-64-2*); N(II) = Hg, Zn, Co, or Ni, and L = $I^-$ or $SCN^-$ (86). Among other applications of these types of electrodes, the $PbHg(SCN)_4$ electrode was used in the potentiotitration of lead with EDTA at $p$H 4.0.

The titrations for which the mercury $p$M and the silver(I) oxidation-reduction electrodes function successfully as indicator electrodes have also been monitored by the corresponding mercury(II) (87) and the silver(I) (88) ion selective electrodes. An interesting departure, however, from the equivalent usage of oxidation-reduction and membrane electrodes in these applications is the development of two electrodes that respond to two of the visual indicators commonly used in complexometric titrations. The first is the Eriochrome Black T liquid membrane electrode, and the other is the Chromazurol S electrode.

Eriochrome Black T, 3-hydroxy-4-(1-hydroxy-2-naphthylazo)-7-nitro-naphthalene-1-sulfonic acid, was one of the first and most widely used of the metal indicators (89) employed in the EDTA titrations. The sulfonic acid group is wholly ionized at $p$H 1, and the $pK_a$ values of the two phenolic groups are 6.3 for HIn$^-$ and 11.6 for the dianionic species HIn$^{2-}$. In the $p$H range 7–11 the addition of metallic salts produces a change in color from blue to red which can be attributed to the reaction

$$M^{2+} + HIn^{2-}(\text{blue}) \rightarrow MIn^- + H^+$$

The following cations can be titrated directly with EDTA in the presence of Eriochrome Black T: $Ba^{2+}$, $Cd^{2+}$, $In^{3+}$, $Mg^{2+}$, $Mn^{2+}$, $Pb^{2+}$, rare earths, $Sc^{3+}$, $Sr^{2+}$, $Zn^{2+}$, and $Zr^{4+}$. The electrode is easily made (90) by coating a platinum electrode with an ion pair formed between Eriochrome Black T and Zephiramine (*NN*-dimethyl *N*-tetradecylbenzylammonium chloride). This ion pair, formed in water, is extracted into nitrobenzene which is then dried over anhydrous sodium sulfate. A 1 mL portion of this extract is added to molten naphthalene (3.5 g) and the platinum electrode is dipped into this mixture. The optimum concentration of the ion pair in nitrobenzene was found to be $10^{-4}$ $M$ and its composition can be represented as $Z^+ - H_2In^-$. The electrode was used in the titrations of magnesium(II), cadmium(II), zinc(II), calcium(II), and lead(II) with EDTA in the presence of 2–3 drops of $10^{-3}$ $M$ Eriochrome Black T indicator solution at $p$H 9–10.

Chromazurol S, 3-[(3-carboxy-4-hydroxy-5-methylphenyl)(2,6-dichloro-3-sulfophenyl)methylene]-5-methyl-6-oxo-1,4-cyclohexadiene-1-carboxylic acid, is also a useful visual metal indicator for complexometric titrations (91, 92). It has been used for the titrations of the following metal ions with EDTA: $Al^{3+}$, $Ca^{2+}$, $Cu^{2+}$, $Fe^{3+}$, $Mg^{2+}$, $Ni^{2+}$, rare earths, $Th^{4+}$, and $V^{4+}$. Zephiramine was also the cation of choice from which to form an ion pair with Chromazurol S as a basis for a liquid membrane electrode that responded to this indicator (93). This ion pair was extracted from an aqueous buffer solution of $p$H 3.5 into nitrobenzene, the optimum concentration in this solvent for good electrode response being $10^{-4}$ $M$. This solution was applied to the membrane material, and the internal reference half-cell was formed by dipping a silver–silver chloride electrode into a $10^{-3}$ $M$ aqueous solution of Chromazurol S buffered at $p$H 3.5 that was saturated with respect to silver chloride. The range of linear response of the electrode at $p$H 3.5 (acetate buffer) was from $10^{-2}$ to $10^{-6}$ M with respect to the dianion $H_2In^{2-}$ of Chromazurol S and the potential of the electrode was found to be almost constant over the $p$H range 2–4.8. It was used to detect the equivalence points of the titrations of iron(III), aluminum(III), copper(II), and lead(II) at $p$H 2 (Britton–Robinson buffer) with EDTA. This type of electrode can be used under conditions in which the color change of the visual indicator itself may be obscured.

Zephiramine has also been used (94) to form an ion-pair with the sulfonate anion of 8-hydroxyquinoline-5-sulfonic acid at $p$H 6.5. A liquid membrane electrode that responded in a reproducible fashion to changes in the concentration of sodium 8-hydroxyquinoline-5-sulfonate within the range $10^{-2}$ to $10^{-4.5}$ $M$ was prepared from a solution of this ion-pair in nitrobenzene. The electrode was used to monitor the titration of copper(II) at $p$H 5.4 (acetate buffer) with a standard solution of 8-hydroxyquinoline-5-sulfonic acid. At this $p$H a 2:1 ligand-copper(II) complex is formed.

### The Tungsten Bronze Electrode

Mention has already been made of the application of the tungsten bronze electrode in acid-base potentiotitrimetry. An electrode prepared similarly, but with the formula $Na_xWO_3$ ($x = 0.65$), has been used in the direct titrations of the divalent ions of copper, nickel, cobalt, manganese, lead, zinc, calcium, and magnesium with EDTA (95) at $p$H 10 ($NH_3/NH_4^+$ buffer). Aluminum(III) was determined by an indirect titration of excess EDTA with a standard solution of calcium(II) at this $p$H, and iron(III) was titrated directly in an acidic but unbuffered solution. It was suggested that only one of the following criteria need be met in order that the electrode be used for this purpose:

1. That the titration of the metal ion be performed, directly or indirectly at high $p$H.
2. That the metal ion be easily reducible directly at the electrode surface.
3. That a metal oxidation-reduction couple be present that undergoes a change in concentration ratio at the equivalence point.

### Electrodes for Electrolytic Potentiotitrimetry

Two polarized mercury coated gold electrodes each having an area of about 0.5 cm² have been used in the titrations of calcium(II) and copper(II) with EDTA and other complexons (96). The potential established between these electrodes when a constant current regulated at between 5 and 10 $\mu$A is passed through them is determined by the mercury(II) which is added prior to the titration as the mercury(II)-complexon. Thus, although neither the titrant nor the determinand metal ion are directly electroactive in that they undergo reactions at the polarized electrodes, the concentration of mercury(II) bears a relationship to the determinand ion concentration which, of course, changes rapidly near the equivalence point. When EDTA is used as the titrant, a quantity of the mercury(II)–EDTA complex is added to an appropriately buffered solution of determinand ion. Before the titration begins, some of the mercury(II) is displaced from the complex by the metal ion, the extent of the displacement being a function of solution conditions ($pH$, buffer type and its concentration, and the concentration of the reacting species). A simplified equation representing the displacement is

$$M(X)_p + HgY + (y - p)X \rightleftharpoons Hg(X)_y + MY \tag{5.5}$$

where M is the metal determinand ion, X is a complexing agent added in great excess (usually the buffer), and Y is the tetraanion of EDTA. The charges have been omitted from this equation. The species actually reacting at the electrode is $Hg(X)_y$ and it is the decrease in the concentration of this species that is actually observed as the titration proceeds. The potential difference between the electrodes tends toward zero before the equivalence point is reached because the cathodic and anodic processes are simply the reverse of one another and proceed with negligible overpotential at each electrode. In the vicinity of the equivalence point, however, the potential increases rapidly to a peak, the maximum of which is indicative of the equivalence point. This peak in potential arises from the difference between the cathodic and anodic overpotentials for the two electrode reactions. In contrast to the "zero current" method using a mercury $pM$ electrode in which only a trace of the mercury(II)–EDTA complex is added to the titrand solution, a substantial quantity of this complex is added in the differential electrolytic potentiotitrimetric method. For example, in the titration of calcium(II), 0.6 mmol of mercury(II)–EDTA (i.e., 30 mL of 0.02 $M$) was added to the titrand solution which also contained a $pH$ buffer of 2 $M$ ammonia/ammonium nitrate (8.5 mL), and a 10 mL aliquot of the sample solution containing calcium(II) plus water (120 mL). Before the equivalence point, an excess of free calcium(II) over the concentration of EDTA added is present in the solution and this excess tends to displace mercury(II) from its EDTA complex according to the reaction

$$Ca^{2+} + Hg(NH_3)Y^{2-} + (y - 1)NH_3 \rightleftharpoons CaY^{2-} + Hg(NH_3)_y^{2+}$$

in which $y = 2$ or 3. The mercury(II)–ammine complex undergoes reduction at the cathode, the corresponding anode reaction being the oxidation of mercury metal to reform the ammine complex

$$Hg(NH_3)_y^{2+} + 2e \rightarrow Hg_{(l)} + yNH_3 \qquad \text{(at cathode)}$$

$$Hg_{(l)} + yNH_3 \rightarrow Hg(NH_3)_y^{2+} + 2e \qquad \text{(at anode)}$$

Circumjacent to the equivalence point, however, the reaction occurring at the cathode becomes quite different involving either the reduction of dissolved oxygen or, if the solution is deoxygenated, it becomes

$$NH_4^+ + Hg(NH_3)Y^{2-} + 2e \rightarrow Hg_{(l)} + 2NH_3 + HY^{3-}$$

The anodic reaction does not change, and thus a large potential difference develops across the two electrodes at the equivalence point. In general, sharper inflections are obtained if the concentration of the ammonia buffer added is kept to a minimum to prevent extensive formation of metal determinand–buffer complexes.

Mercury electrodes have also been used in the determination of cadmium with EDTA (97), and in the coulometric titration of cyanide ion with electrogenerated mercury(I) or (II) (98).

Following upon the successful voltammetric and amperometric titrations of various metal ions with EDTA using metal oxide electrodes as anodes (99), it has been shown that these electrodes are also suitable for differential electrolytic potentiotitrations (100). The indicating electrodes were prepared from platinum wire 1 mm in diameter and 3–4 mm in length (area about 10 mm$^2$). The cathode used throughout was an untreated bright platinum wire electrode, whereas the various anodes used were formed from the oxides of bismuth(V), thallium(III), lead(IV), or manganese(IV) which were supported on platinum wires. All these oxides were deposited on a platinum anode electrolytically—the bismuth(V) from a 0.01 $M$ solution of bismuth in nitric acid at a current of 20 $\mu$A for 15 min, thallium(III) from an ammoniacal solution of thallium(I) chloride (1 g L$^{-1}$) for 15 min at 5 $\mu$A, lead(IV) from a solution (50 mL) of 0.01 $M$ lead solution containing $NH_3/NH_4^+$ buffer (5 mL) and sodium tartrate (1 g) at a current of 20 $\mu$A for 30 min, and manganese(IV) at a current of 5 $\mu$A for 15 min from a solution (50 mL) at $p$H 6.5 that contained manganese(II)–EDTA (2 mL containing 2 mg Mn) and 1 mL of 0.025 $M$ zinc(II) solution. Hexamethylenetetramine was used to adjust the $p$H of the manganese solution to $p$H 6.5. Each of these metal oxide electrodes was made the anode in a cell that contained a platinum cathode and through which a controlled current of 2 $\mu$A was passed. The potentials were measured in 25 mL of a nonreducing buffer of $p$H 1–9 with and without added EDTA. Without EDTA, the potential was almost independent of the electrode combination and increased from 1595 to 1880 mV over the $p$H range, oxygen being formed at the anode. Upon the addition of 0.1 mL of 0.01 $M$ EDTA to a solution at a given $p$H, a change in cell potential was observed which was attributed to a change of the anodic reaction from oxygen evolution to oxidation of EDTA. The latter process depended upon the anode oxide type and the change of potential upon the addition of EDTA was greatest for the bismuth(V) oxide anode at $p$H 1 (300 mV), for the manganese(IV) oxide anode at $p$H 3 (480 mV), and for the thallium(III) oxide at $p$H 9 (260 mV).

## 5.4 ELECTRODES FOR PRECIPITATION TITRATIONS

Of the many titrants used in this application, those based on lead(II), mercury(I) and (II), and particularly silver salts are the most versatile. In addition to their obvious use in the analysis of inorganic anions, these titrants also react quantitatively with many organic compounds containing a variety of functional groups and form very sparingly soluble precipitates in aqueous and/or nonaqueous solutions. Potentiotitrimetry is the most frequently used method for these precipitation titrations because, in comparison with the other categories of titrimetry, not many visual indicators are available for equivalence point detection. Equipped, therefore, with electrodes that respond to changes in the concentration of lead(II), mercury(II), and silver(I), plus anion responsive electrodes for sulfide, chloride, and iodide, enable the equivalence points for the direct and indirect titrations of many compounds to be detected. These applications are surveyed in Chapter 13, and thus there is no need here to consider further the types of suitable ion selective membrane electrodes mentioned in Chapter 3.

### Oxidation-Reduction Electrodes Based on Silver

It was reported in 1893 (101) that a silver electrode can be used for the sequential titrations of iodide and bromide. In the form of a commercially available silver billet, a silver wire, or silver plated platinum electrode, this electrode remains the traditional electrode of choice for precipitation titrations. Its use is often associated predominantly with the determination of halides by titration with silver nitrate, and, indeed, the sequential titrations of iodide, bromide, and chloride in a mixture by Martin's method (102) is sometimes a requirement in undergraduate laboratory courses. However, there are many other types of species, both inorganic and organic, that form insoluble silver salts and these too may be determined using a standard solution of silver nitrate as titrant and a silver electrode as indicator electrode. Often the potential of this electrode is monitored against a saturated calomel reference half-cell separated from the titrand solution by a 3 $M$ potassium nitrate salt bridge. Some methods based upon the use of this cell are given in Chapter 13.

A silver electrode is readily converted into a silver–silver sulfide electrode, and as such has been used for the simultaneous determination of hydrogen sulfide and alkyl thiols with silver nitrate as titrant (103). The electrode was prepared from 2 mm dia. silver rod which was polished with fine emery cloth and then further cleaned by soaking in potassium cyanide solution. Although a deposit of silver sulfide can be formed on the silver by anodic deposition, it was found sufficient to form a coating merely by performing a preliminary titration of sodium sulfide in a solution 1 $M$ with respect to sodium hydroxide and 0.05 $M$ with respect to ammonia using standard silver nitrate as titrant. Thereafter the electrode was soaked in alcohol containing sodium acetate and 0.5% silver nitrate for 5 min before use and it was also stored in this solution. The coating must be removed completely and the coating procedure repeated if anomalous titration curves are obtained. When titrating the mixtures, the first inflection point corresponded to the quantitative precipitation of sulfide and the second to the alkyl thiol. A glass electrode served as the reference electrode in these titrations.

In the determination of mercury(II) in alkaline solution with thioacetamide, a silver electrode was made responsive to sulfide, formed by the hydrolysis of thioacetamide, by immersing a commercial billet electrode in an alkaline solution of thioacetamide for 20 min (104). The mercury(II) determinand (5–15 mg), complexed with EDTA (30 mL, 0.1 $M$ disodium salt) in an alkaline solution (20 mL, 2 $M$ NaOH) containing gelatin (20 mL, 1.2%) was titrated with 0.005 $M$ thioacetamide. Gelatin was added to prevent the agglomeration of mercury(II) sulfide on the electrode since this causes spurious results. This addition is said to be unnecessary if a mercury–mercury(II) sulfide electrode is used. It should be noted that the titration of mercury(II) with thioacetamide (i.e., sulfide) is a rapid method offering the additional advantage that halides do not interfere.

A silver electrode has also been made responsive to thiols by its immersion for a short time in a solution of a thiol, for example, cysteine at $p$H 7 (105). The half-reaction

$$AgRS_{(s)} + e \rightleftharpoons Ag_{(s)} + RS^- \tag{5.6}$$

in which the species $RS^-$ arises from the ionization of the thiol RSH, forms the basis for the Nernstian response of this electrode which was prepared for titrimetric applications as follows:

A length (0.15 m) of silver wire (1.27 mm dia.) was coiled around the outside of a 4 mm dia. tube which formed part of a reference half-cell consisting of a silver–silver chloride electrode immersed in saturated potassium chloride solution. Before use, the electrode was immersed briefly in dilute ammonia (1 + 3), rinsed with demineralized water, immersed in dilute nitric acid (1 + 3), for 2 min, and then rinsed again with water. After each titration the electrode must be cleaned by immersion in dilute ammonia (1 + 3). An immersion time of a minute was found to be sufficient when simple thiols were titrated, but 3–5 min was allowed after the titration of a protein (e.g., albumin). The change in potential in the vicinity of the equivalence points were found to be larger when mercury(II) chloride was used as titrant at $p$H 4.5–9.5 than were obtained with either silver nitrate or $p$-chloromercuribenzene sulfonate at $p$H 9.5 or at $p$H 7–9.5, respectively. It must be emphasized that in this application activation of the silver electrode towards the thiol was accomplished merely by immersing the electrode in the titrand solution. Silver electrodes can also be activated towards other sulfur-containing titrants in a similar fashion. These include ethanedithioamide (dithiooxamide), used for the determination of silver, silver–copper mixtures (106), and mercury (107), and also dithiocarbamic acid derivatives used for the titration of copper (108) and indium (109). A silver or a mercury electrode was found to be suitable as an indicator in the titrations of a variety of metal ions with some potassium alkoxymethanedithioates (i.e., potassium alkyl xanthates) prepared as 0.05 $M$ solutions in ethanol (110).

Titrations with sodium tetraphenylborate(III) can also be monitored with a silver electrode. This versatile titrant (111) has been applied to the determination of potassium, rubidium, cesium, and thallium(I), as well as to the determination of ammonium salts, alkylammonium salts, and other quaternary nitrogen compounds. It is used as an aqueous solution adjusted to $p$H 8–9 by the addition of sodium hydroxide (112). The electrode can be prepared from 30 cm of 1 mm dia. silver wire wound in a spiral (113). This is dipped for 20 sec in 6 $M$ nitric acid containing a small amount of sodium nitrite, after which it is rinsed with distilled water, soaked for 2 min in a stirred of 1 $M$ potassium cyanide and finally washed thoroughly with distilled water. This treatment was repeated before each titration. It was found unnecessary to add even a small amount of silver(I) (0.00005 mmol) to the titrand solution in order to establish a reversible potential, since the silver wire soaked for a few minutes in a 2% W/V solution of sodium tetraphenylborate(III) showed a Nernstian response to the anion at concentrations in the range $10^{-4}$–$10^{-6}$ $M$. The electrode was used in conjunction with a mercury–mercury(I) sulfate reference half-cell to monitor the titrations of surfactant quaternary nitrogen salts either alone or in pharmaceutical preparations at $p$H 3.5 (0.2 M acetic acid–acetate buffer). The potential of the electrode can probably be defined by a half-reaction similar to (5.6):

$$AgB(C_6H_5)_{4(s)} + e \rightleftharpoons Ag_{(s)} + B(C_6H_5)_4^- \qquad (5.7)$$

Among other electrodes tested, the silver electrode was also used to indicate the equivalence points of titrations of sodium tetraphenylborate(III) with silver nitrate (114).

## Mercury and Mercury Coated Oxidation-Reduction Electrodes

The J-shaped mercury pool electrode and the mercury-coated silver electrode, already described, can also be employed as indicator electrodes in precipitation titrations. Like the silver electrode, these also respond to sulfur containing titrants, and for some of these applications their performance has been claimed to be superior to the silver electrode. A mercury pool electrode has been used in conjunction with a fiber junction saturated calomel reference half-cell to monitor the titrations of thiols with mercury(II) perchlorate (115). It was also used in the titrations of copper and lead in the presence of each other. The lead(II) was first titrated at $p$H 3–4 with standard potassium dichromate, and the copper(II) remaining in this solution was then titrated with a standardized solution of benzothiazol-2(3$H$)-thione (116). The titration of copper(II) alone at $p$H 3.5–5.8 with benzothiazole-2(3$H$)-thione can also be monitored with a platinum electrode. A mercury electrode has also been found reversible to the cupferronate ion (the anion of $N$-nitroso $N$-phenylhydroxylamine), and was used for the titration of thorium(IV) and cerium(III) with cupferron (the ammonium salt of the hydroxylamine derivative). The addition of cerium cupferronate and fresh mercury cupferronate to the titration cell, forming a mercury-based electrode of the third kind, enhanced the potential change at the equivalence point of the titration of cerium(III) by at least threefold (117).

The use of a mercury-coated silver electrode in the indirect micro- and semimicrodeterminations of silver by the titration of an added excess of standard iodide (118) was found to be more selective and sensitive than the direct traditional method using a silver–silver iodide electrode (119). The trithiocarbonate ion in viscose has also been determined without interference from other sulfidic anions by titration in an air-free solution (100 mL 0.1 $M$ in sodium hydroxide, 0.9 $M$ in sodium acetate) with mercury(II) chloride using a mercury-coated silver electrode (120).

A mercury-coated platinum foil electrode, prepared by a method similar to that given earlier, was used to form the base for a mercury–mercury(II) sulfide electrode (121). The film of mercury(II) sulfide was anodically deposited on the mercury-coated platinum foil from a solution of 0.1 $M$ sodium sulfide using an inert electrode as cathode, and passing the current for 30 sec or for such a period that the film became visible. This electrode was used to monitor the titrations of silver(I), cadmium(II), mercury(II), and lead(II) with thioacetamide. It was found to be more stable than the silver sulfide coated silver wire electrode, the observed reproducibility being within ±2 mV for a series of titrations of the same solution. Furthermore, in contrast to the use of the silver-based electrodes, it was not found necessary to add gelatin to the titrand solution to prevent excessive coating on the electrode surface.

### The Platinum Electrode

The traditional role of a platinum indicator electrode in precipitation titrations is as a zeroth electrode to monitor the changing ratio of an oxidation-reduction couple such as the hexacyanoferrate(III)–hexacyanoferrate(II) couple (122). The titration of zinc with hexacyanoferrate(II) at $p$H 2–3 (123) is an example of this application in which 3–4 drops of a 1% solution of hexacyanoferrate(III) is added to the titrand solution in order to ensure that the couple is coexistent at the electrode surface during the titration. A variant of this method is to use sodium pentacyanonitrosylferrate(III) as titrant (124), which permits both copper(II) and zinc(II) to be determined in the same solution. The optimum $p$H for the titration is 3.0–5.0 for copper and 3.0–5.8 or 12 for zinc. Alternatively, the hexacyanoferrate(III)–hexacyanoferrate(II) couple can be added as an indicating system for a titrant that reacts only with one species of this couple. The titration of sulfate with lead nitrate was an early example of this application (125), and this method has been used more recently in ethanolic solution for the determination of inorganic sulfates in detergents of the alkylnaphthalene type (126). Following the discovery that a platinum wire electrode (0.1–0.2 mm dia.) responds to changes in the ratio hexacyanoferrate(III)/hexacyanoferrate(II) four times faster than a silver wire electrode (0.1–0.2 mm dia), an ultramicrodifferential potentiometric method has been developed for the determination of sulfate (127) using the Pt–Ag pair. This method, however, employs the same principle: When a sulfate solution containing this hexacyanoferrate(III)–hexacyanoferrate(II) couple in the ratio trivalent : divalent of at least 100 : 1 is titrated with a lead salt, the potential of the cell undergoes little change until almost all the sulfate has been precipitated. The first portion of lead(II) in excess sharply alters the hexacyanoferrate(III) to hexacyanoferrate(II) ratio as a result of the precipitation of lead(II) hexacyanoferrate(II), and this results in a large change in potential in the vicinity of the equivalence point. The preparation of the platinum and silver electrodes has been reported in another application of the method (128). In this instance the platinum electrode (12 × 12 mm foil) was washed in concentrated nitric acid and then heated to whiteness with a blow torch before use. The silver wire electrode (1.7 mm dia.) was cleaned by abrasion, washed with distilled water, and then immersed for 4 h in a freshly prepared solution containing potassium hexacyanoferrate(II) trihydrate (126 mg) and potassium hexacyanoferrate(III) (82.2 mg) in water (100 mL). This electrode was washed with water and could be stored in a dry state. It functioned satisfactorily for 15 titrations performed within 3 days of its preparation.

The titration of barium(II) with sulfate has also been monitored by a platinum electrode in the presence of the oxidation-reduction couples iron(III)–iron(II)-$o$-phenanthroline or thallium(III)–thallium(I). In this case the sharp potential change at the equivalence point was attributed to the complexation of iron(III) or thallium(III) with sulfate (129). Similarly, sulfate was titrated with barium chloride in the presence of the same oxidation-reduction systems; the chloro complexes of iron or thallium were formed after the equivalence point was reached.

A departure from these traditional applications of the platinum electrode is the role of this electrode in monitoring the sequential titrations of halides with a mixed titrant 0.1 $M$ in silver nitrate and 0.1 $M$ in thorium(IV) nitrate. It was reported initially (130) that at $p$H 7.2 and at 0° the equivalence points of sequential titrations of the four halides could be located in the order iodide, fluoride, bromide, and chloride. Subsequent work (131), however, could not verify that it was possible to titrate fluoride at 0° but confirmed that bromide and chloride could be titrated at room temperature with a mixed titrant of 0.025 $M$ silver nitrate and 0.05 $M$ thorium(IV) nitrate. The two schools each used a platinum billet electrode to monitor these titrations. The tendency of silver halides to form mixed crystals and solid solutions leading to the coprecipitation of the more soluble halide, for example, chloride when bromide is titrated, is a problem in these sequential titrations. This coprecipitation phenomenon is reduced if the titrand solution is prepared in 0.1 $M$ potassium nitrate (131). The adsorption of bromide and chloride ions on the surface of the platinum was suggested as a possible explanation for the response of the electrode to bromide and chloride. A platinum electrode has been similarly used in a medium of 90% acetone for the determination of traces of chloride and bromide (132).

The platinum electrode appears to respond to changes in the concentration of ethanedithioamide (133). Thus palladium(II) could be determined by titrating a standard solution of ethanedithioamide at $p$H 4.5 (acetate buffer) with the solution of palladium(II) chloride. The titration of palladium(II) with ethanedithioamide, however, was not reproducible.

## Other Oxidation–Reduction Electrodes

Other electrodes used in recent years have included the tungsten, and the zinc and lead amalgam electrodes. In a systematic study of the relative merits of the silver, tungsten, molybdenum, copper, carbon, cadmium, nickel, iron, lead, and tin electrodes as indicator electrodes for the titration of chloride with mercury(II) nitrate, the tungsten electrode was found to give the biggest potential jump at the equivalence point (134). In addition to its use in a cell containing a saturated calomel reference half-cell, it was found also to be effective when used as part of a bimetallic pair with a carbon electrode for the differential potentiotitrimetric indication of this equivalence point. Mercury(II) and mercury(I) nitrates were the titrants used in the determination of tungstate ions in ammonium, potassium, and sodium paratungstates for which the tungsten electrode was found to be suitable as indicator (135). The tungsten electrode was also used in the determination of solutizene (Registry No. *959-79-5*) using mercury(II) nitrate as titrant (136).

Of the amalgam electrodes, the zinc amalgam electrode was the indicator for the titration of the anions of benzimidazole-2(3*H*)-thione, benzothiazol-2(3*H*)-thione and 1,2,3-benzotriazole with zinc(II) sulfate (137) and the lead amalgam (70% lead–30% mercury) electrode for the titration of sulfate in water and soil extracts (138).

## Glass Electrodes

The transient response of a sodium ion-responsive glass electrode toward barium ion was found to be sufficiently persistent for the equivalence point of the titration of sulfate with barium chloride to be recorded as a sharp point of inflection (139). This glass electrode was used in conjunction with a double junction reference half-cell, the junction with the titrand solution being made through 0.2 $M$ hexamethylenetetramine (Hexamine). A portion of the aqueous sulfate solution (5 mL) was adjusted to $p$H 5–6 by adding 0.2 $M$ Hexamine (1 mL), and the acetone and water were added so that a total volume of 50 mL is contained in the titration vessel and the solution is about 70 vol% in acetone. The electrodes were rinsed well with 0.01 $M$ hydrochloric acid before immersion in the solution and the titrant was barium chloride prepared in a 70 vol% solution in 2-propanol.

Mention has already been made of the glass–FeII73 chalcogenide electrode responsive to iron(III). This electrode responds selectively to uncomplexed iron(III) in the presence of iron(III) sulfate and iron(III) hydroxide complexes and can be used for monitoring the titration of sulfate with barium chloride. Within the $p$H range 1.7–2.0 and in the presence of

$10^{-4}$ $M$ iron(III), the addition of the titrant precipitates sulfate liberating uncomplexed iron(III) formerly bound in the sulfate complex, and thus the potential of the electrode changes. Immediately after the equivalence point, however, the potential remains constant.

### Membrane Electrodes Based on Insoluble Inorganic Salts

Less obvious applications of the commercially available fluoride ion selective electrode based on lanthanum fluoride include the monitoring of the titration of lithium in 95% ethanol with ammonium fluoride (140). This method was used for the analysis of concentrated solutions of lithium chloride and lithium nitrate required in the Tramex process for the separation of lanthanides from actinides. Titrations of species that form precipitates on titration with lanthanum(III) salts have also been monitored with a fluoride ion selective electrode. Thus oxalate (0.01 $M$) in a 3:1 ethanol–water medium has been determined by this precipitation titration (141) which is usable within the $p$H range 3–5.5. At these high concentrations of ethanol, however, the equilibrium at the electrode tended to be slower before the equivalence point. The procedure used was to read the potential at a fixed time (3 min) after each addition of lanthanum(III) titrant.

As is to be expected in view of the applications of the silver–silver sulfide oxidation-reduction electrode, the use of ion selective electrodes based upon the silver sulfide membrane has been investigated for similiar applications. Generally, the membrane electrode can be operated more conveniently than the oxidation-reduction type, since it needs neither maintenance nor pretreatment between successive titrations (142). This was found for the titrations of hydrogen sulfide and thiol–hydrogen sulfide mixtures with silver nitrate. Similarly, the sulfide-responsive electrode can be used to monitor the titration of thioacetamide in distilled water, or in the presence of 0.1 or 1 $M$ sodium hydroxide, or in 0.01 $M$ ammonia with standard silver nitrate (143).

The titration of thiols with mercury(II) perchlorate have also been investigated (144). Of the silver sulfide-based electrodes tested, the potential breaks at the equivalence points were largest for the bromide ion selective electrode followed by the iodide-responsive electrode. Acetone was the preferred solvent for many thiols, whereas dioxane or ethanol had limited applicability. The method, however, could not be applied to those heterocyclic compounds which exist predominantly as the thione tautomer. Once again, the response of the electrodes was slower than is normally associated with these types of electrodes. The bromide ion selective electrode was not affected by mercury(II) during several hundred titrations, and only required a gentle polish with number 3/0 emery paper at the commencement of each day's work. It was found advantageous to wipe the membrane of the electrode with tissue paper after each titration.

An iodide-responsive electrode, also based on a silver sulfide membrane, was found suitable for the sequential titrations of iodide, bromide, and chloride (145). These titrations were carried out at $p$H 3.9 with silver nitrate in the presence of aluminum(III) nitrate ($10^{-3}$ $M$) which acted as a coagulant. Accurate analysis of the three was possible under these conditions.

A phosphate-responsive electrode with a membrane mole–ratio composition of $Ag_2S:PbS:PbHPO_4$ of 1:1:1 was based on a hydrophobized graphite Selectrode® (146). The lead hydrogen phosphate was made by heating a solution containing lead nitrate (30 g) in water (70 mL) to boiling and slowly adding to it a dilute solution of phosphoric acid [phosphoric acid (15 mL) mixed with water (20 mL) and diluted to about 50 mL with water]. The mixture was digested at 80–90° for 30 min and allowed to cool. The precipitate was collected, washed with water, and dried at 50°. This solid was mixed with an equimolar mixture of lead sulfide and silver sulfide (supplied with the electrode) in an agate mortar and rubbed into the surface of the electrode with a glass rod. Although the response time of the electrode toward lead(II) was faster than it was toward phosphate, the electrode was used for the titration of phosphate at $p$H 8.0–8.5 with lanthanum(III) chloride. The buffer solution was prepared by dissolving ammonium acetate (38.54 g) in water (about 400 mL), adjusting the $p$H to 8.8 by dropwise addition of concentrated ammonia and diluting to 500 mL with water. These titrations were not affected by some of the interferences such as those by sulfate or chloride found with other titrants and their corresponding indicator electrodes.

### Membrane Electrodes Based on Organic Ion Exchangers

Of the commercially available liquid membrane electrodes, the perchlorate electrode is one of the more versatile in titrimetric applications. It has been used to monitor the titration of large inorganic anions such as perchlorate, perrhenate, periodate, permangante, hexafluorophosphate, and hexachloroplatinate with the precipitant 1,2,4,6-tetraphenylpyridinium acetate, TPPA (147). This titrant (0.05 $M$ in 0.3 $M$ aqueous sodium acetate), standardized against ammonium perchlorate, was used also for the determination of the anions of trinitromethane (nitroform) and some nitroform yielding compounds, phenylborates, 2,4,6-trinitro compounds, some nitro, dinitro, and halogenated phenols and related compounds (148). The perchlorate ion selective electrode was also used to monitor these titrations, and it was found that the nitrate and fluoroborate electrodes also responded to the various anions.

The barrel of a commercially available liquid membrane electrode and the blank membranes supplied with it provided the hardware required for the development of a cesium-sensitive liquid membrane electrode (149) based on cesium tetraphenylborate(III). This salt, prepared by precipitation from 0.02 $M$ cesium nitrate with 0.02 $M$ sodium tetraphenylborate(III), was dried at 120° for 3 h, and applied as a saturated solution in nitrobenzene to the membrane. The internal reference solution was 0.01 $M$ cesium chloride and presumably a silver–silver chloride electrode was immersed in this. The electrode gave a linear response for pure cesium nitrate solutions in the concentration range $10^{-1}$–$10^{-4}$ $M$ and was used for the titration of cesium alone and in mixtures with potassium and rubidium with sodium tetraphenylborate(III). Calcium tetraphenylborate(III) was the titrant of choice for the determination of ammonium ion (150) using liquid membrane electrodes based on tetraphenylborate(III). The membranes of these coated wire electrodes contained 3% sodium tetraphenylborate(III), or 3% ammonium tetraphenylborate(III) or potassium tetraphenylborate(III), 35% poly(vinylchloride), and 62% dibutylphthalate, and each type of electrode had a linear response to ammonium ion. In the titration of ammonium ion with calcium tetraphenylborate(III) all three electrodes gave well-developed titration curves, although when sodium tetraphenylborate(III) was used as titrant the electrode with the sodium tetraphenylborate(III) membrane could not be used.

A picrate-responsive electrode can be used as an indicator electrode in the titration of cations forming insoluble picrates. It has been used, for example, in the determinations of thiourea and silver which are based upon the formation of the insoluble picrate of the silver–thiourea complex (151). This reaction can be represented as

$$2(NH_2)_2C=S + Ag^+ + C_6H_2N_3O_7^- \rightarrow [Ag(SCN_2H_4)_2][C_6H_2N_3O_7]$$

In the titration of thiourea with standard silver nitrate, therefore, the picrate concentration decreased, resulting in a continuous increase of the potential of the picrate-responsive indicator electrode with respect to a double junction silver–silver chloride reference half-cell. After the equivalence point the picrate concentration remained essentially constant, and thus at the equivalence point there was a sharp decrease in the slope of the titration curve. The electrode was based on *NNNN*-tetrapentylammonium picrate and was prepared as follows (152):

> This picrate, prepared by mixing equimolar (0.1 $M$) aqueous solutions of *NNNN*-tetrapentylammonium bromide and sodium

picrate, was extracted into 2-nitrotoluene. The extract was washed with water and then dried over anhydrous sodium sulfate. A 0.01 $M$ solution of $NNNN$-tetrapentylammonium picrate in 2-nitrotoluene was applied to the membrane material supplied with a commercial liquid membrane electrode. The membrane was then positioned in the body of this electrode. The internal reference half-cell was a silver–silver chloride electrode dipping into a 0.01 $M$ solution of sodium picrate prepared in 0.01 $M$ sodium chloride. The picrate electrode was conditioned in 0.01 $M$ sodium picrate for 48 h before use, and it was also stored in this solution.

Diphenylthiocarbazone (Registry No. *60-10-6*), or *dithizon* as it is often called, can be used to extract a number of metal ions into chloroform. The palladium dithizonate complex in this solvent was the electroactant solution which when applied to a hydrophobized graphite membrane formed the basis of a solid state electrode sensitive to mercury(I) (153, 154). The response of this electrode to mercury(I) was explained by the equilibrium

$$2Pd(HDz)_2 + Hg_2^{2+} \rightleftharpoons 2PdDz + Hg_2(HDz)_2 + 2H^+$$

in which Dz represents dithizone. The electrode was used in the titrimetric analysis of inorganic and organic mercury compounds. It was also applied to the indirect titration of organic compounds that form insoluble mercury(I) salts. Thus phenoxyalkylcarboxylic acids in pesticides were determined by the precipitation of the mercury(I) salts followed by the dissolution of the washed precipitate in 2 $M$ nitric acid, and titration of mercury(I) with a standard solution of potassium bromide (155).

Hydrophobized graphite (15 mm long, 6.5 mm dia.) also formed the support for a liquid membrane electrode sensitive to mercury(II) (156). The mercury(II) complex of 1-(2-pyridinylazo)-2-naphthol (*o*-PAN) was the electroactant for this electrode which was prepared as follows:

Mercury(II) nitrate (20 mL 0.01 $M$) at $p$H 6.2 (adjusted with Hexamine) was shaken in a separating funnel with chloroform (20 mL) containing *o*-PAN (0.0249 g) for 3 min. The organic phase was separated and then dried by passage through anhydrous sodium sulfate. The solution was diluted to 100 mL with chloroform, which yielded a solution $5 \times 10^{-4}$ $M$ with respect to the chelate. The hydrophobized graphite membrane was impregnated with a portion of this solution. The response of the electrode to mercury(II) was linear over the $p$M range 1–5 and it was used in the titrations of iodide, bromide, thiocyanate, and chloride with mercury(II) nitrate. It could be applied also to the titrations of the hydrochlorides and hydrobomides of some organic bases. A similarly constructed electrode (157), but based upon the mercury(II) complex with 2-(3-hydroxy-1-oxoinden-2-yl)imino-1,3-indandione (Ruhemann's Purple), has also been used as an indicator electrode for titrations with mercury(II).

Soaps and other anionic surfactants can be titrated with $N$-benzyl $NN$-dimethyl 1-hexadecylammonium chloride or with $NNN$-trimethyl 1-hexadecylammonium bromide (CTAB). An electrode that responds to these surfactants was based on bis(dimethlyglyoxine)-1,10-phenanthroline cobaltate(III)-dodecylsulfate (Registry No. *65436-92-2*), (158). This substance was prepared from cobalt acetate, dimethyglyoxine, and 1,10-phenanthroline in the respective mole ratio 1 : 2 : 3. The acetate complex obtained was converted into the dodecylsulfate salt which was then extracted with a 5 : 1 v/v mixture of $o$-dichlorobenzene and decanol to obtain a 0.005 $M$ solution of the complex in this solvent. The hydrophobized graphite membrane of a solid-state electrode was impregnated with this solution to form an indicator electrode with which it was possible to monitor the titration of soaps and anionic surfactants with $N$-benzyl $NN$-dimethyl 1-hexadecylammonium chloride. These classes of compounds could also be titrated with CTAB (159) using an indicator electrode based on bis(dimethylglyoxine)-1,10-phenanthroline cobaltate(III)-dodecylbenzenesulfonate (Registry No. *63542-03-0*).

### Electrodes for Electrolytic Potentiotitrimetry

The equivalence points of titrations of chloride, thiocyanate, or iodide with silver nitrate, using a pair of polarized silver electrodes (current density 0.8–0.95 $\mu$A cm$^{-2}$), appear as sharp peaks on the titration curve of cell potential versus volume of titrant. These electrodes can be prepared from 0.7 mm dia. (22 SWG) mint silver wire sealed through glass (160) so that 1 cm of this wire is available for immersion in the titrand solution. Before use the surface oxide layer must be removed with boiling 6 $M$ nitric acid and, after washing with distilled water, they can be stored either in distilled water or in dilute silver nitrate solution. When used in the titration of chloride with silver nitrate, for example, the effect of the anodic reaction may be considered as increasing by a minute amount the concentration of silver ion at the electrode surface due to the corrosion of the electrode material. The potential of the electrode, therefore, becomes advanced with respect to the corresponding point in the titration if a "zero" current electrode were being used. Conversely, the deposition of silver on the cathode causes a corresponding retardation of the potential with respect to an unpolarized electrode. For symmetrical reactions and electrodes of identical area, the charge on the two electrodes is equal and opposite, and the potential across them gives the true instantaneous first differential of the titration curve at the operating point. However, with colloidal precipitates such as silver halides adsorption effects can cause the asymmetric differential titration curves which are obtained in practice. This method has been adapted to the semimicro scale titration of nanogram amounts of halides at extreme dilution (161). In this instance the volume of titrand solution was less than 2 mL and quantities of $10^{-5}$ mole of chloride, $10^{-7}$ mole of bromide, and $10^{-8}$ mole of iodide were readily determined. The method was further refined (162) by generating the titrant coulometrically in a 0.5 mL demountable cell. Acceptable results were obtained down to $10^{-11}$ mole of chloride or bromide at $2 \times 10^{-8}$ $M$ in a medium of 0.01 $M$ nitric acid in 80 : 20 methanol/water.

In addition to a pair of silver electrodes, chloridized silver electrodes (i.e., Ag–AgCl) were used in some of the work previously referred to, and this type was also the electrode pair of choice for the argentometric microdetermination of chloride in aluminium metal (163). The silver electrodes were chloridized anodically in 0.1 $M$ hydrochloric acid for 2 min (current density 0.6 mA cm$^{-2}$) and then washed thoroughly with water until the "zero" current bias potential between them in a stirred solution was less than 10 mV. This treatment was necessary before each titration. At a constant current of 1 $\mu$A the potential between the electrodes was measured one minute after each addition of titrant, and this procedure yielded well-defined titration curves.

The use of a polarized pair of silver electrodes was included among other electrode combinations in an investigation of suitable indicator electrodes with which to monitor the titration of sodium tetraphenylborate(III) with silver nitrate (164). A satisfactory response was also obtained using a single silver needle electrode (surface area 10 mm$^2$) cathodically polarized with respect to a calomel reference half-cell (0.1 $M$ NaCl filling solution) by a current of 0.5 $\mu$A. A potassium nitrate salt bridge was used to separate the reference half-cell from the titrand solution. The following additional electrode pairs gave adequate responses: silver–platinum, platinum–silver, graphite–silver, graphite–calomel, graphite–graphite. The graphite electrodes (surface area 5–10 mm$^2$) were prepared by sticking a graphite plate onto the end of a capillary tube of appropriate diameter. Contact between the plate and the connecting wire to the instrument was made through mercury. The current passed through a graphite electrode of about 8 mm² surface area to yield a satisfactory response was 0.2 µA.

Polarized platinum electrodes can also be used to monitor some precipitation titrations. Thus in a study of the titration of organic bases with potassium hexacyanoferrate(II) two platinum wire electrodes (55 mm long, 0.5 mm dia.), placed 3 mm apart and polarized by a current of a few microamps, were used to detect the equivalence point (165). Similarly, a pair of platinum electrodes (area 0.25 cm²) polarized by a current of 2 µA was found suitable for the titration of antimony(III) in 0.5 $M$ sulfuric acid solution with 2-mercapto-$N$-(2-naphthyl)acetamide (thionalide). In this instance the titrant was prepared as a 0.005 $M$ solution in acetic acid (166). A platinum needle electrode (area 15 mm²), cathodically polarized with respect to a calomel reference half-cell (0.1 $M$ NaCl), gave a larger potential change at the equivalence point of a titration of sodium pentacyanonitrosylferrate(III) with 0.05 $M$ mercury(II) nitrate than was observed for a similarly polarized silver electrode (167). However, when 0.1 $M$ silver nitrate was used as titrant, the response of a cathodically polarized silver electrode was superior to that of the platinum electrode.

For some applications the passage of a small polarizing current through ion selective electrodes enhances the changes of potentials at equivalence points with respect to the changes observed when the current is not applied (168). The effect is manifested by a marked increase of the electrode slope accompanied by a marked increase in the sensitivity of the electrode (169). For example, the passage of a 0.5–1.5 µA current through a lead-responsive ion selective electrode was found to increase the sensitivity in the titrations of sulfate (170) and phosphate (171) with lead(II). For the latter application, the determination limit was about 10 µg phosphorus when the titrant was 0.001 $M$ lead(II) perchlorate. The polarity of the electrode is selected on the basis of experimental observation and was negative with respect to a double junction reference half-cell for the titration of phosphate, but positive for the sulfate titration. Equivalence point inflections in precipitation titrations can also be enhanced by reducing the solubility of the precipitated species either by lowering the temperature or by changing the solvent. In potentiotitrimetry the selection of solvent is influenced not only by the solubility of the precipitate in the solvent, but also by the dielectric properties of these media. Thus in the determination of chloride (172) the solvent chosen was either acetone or a mixture of acetic anhydride and acetone in the volume ratio 4:1. Well-defined titration curves were obtained when a silver sulfide ion selective electrode was polarized by a current of $-0.4$ µA, with respect to a double junction reference half-cell, down to a chloride concentration of $4 \times 10^{-6}$ $M$ (7 µg 50 mL⁻¹). In methanol, however, the lower practical limit of these titrations with 0.001 $M$ silver perchlorate was found to be $1.26 \times 10^{-5}$ $M$.

## 5.5 ELECTRODES FOR OXIDATION–REDUCTION TITRATIONS

The platinum electrode is, without doubt, the most versatile and the most widely used of the zeroth electrodes available as indicators in oxidation–reduction titrations. Its use is traditionally associated with titration reactions performed in aqueous sulfuric acid solutions, but it has also proved satisfactory as an indicator electrode for titrations in 10 $M$ phosphoric acid (173), 6 $M$ hydrochloric acid (174), aqueous carbonate solutions (175), in an alkaline medium of mannitol (176), and in strongly alkaline solutions (177). It has also been used for titration reactions performed in acetic acid (178), 1:1 acetone–water (179), acetonitrile (180), dimethylformamide and pyridine (181), ethanol, methanol and 2-propanol (182), 4-methyl-1,3-dioxolan-2-one (proplyene carbonate) (183), 1:4 mixtures of tributylphosphate and carbon tetrachloride (184), and in complex solvent systems such as glycerol, potassium hydroxide, triethanolamine, and dimethylformamide (185). In these applications the platinum electrode is often used in conjunction with a calomel reference half-call, but, as was pointed out in Chapter 2, this electrode may also be used in conjunction with another zeroth electrode having a slower rate of response. Some 30 of these couples in which platinum featured as the indicator electrode have been investigated in the titration of millimolar concentrations of copper(II) in dimethylformamide with titanium(III) prepared in the same solvent (186). Stainless steel, molybdenum, or titanium were found to be satisfactory and so were tungsten, talantum, or gold provided they were preconditioned by soaking in a solution of the titrant. The platinum–graphite couple is sometimes used (187) in aqueous systems, and a wax-impregnated graphite electrode also finds application as an indicator electrode in its own right for some oxidation-reduction titrations (188). Ion selective electrodes, however, are not frequently applied as indicator electrodes in these types of titrations.

### The Platinum Electrode

Unfortunately, there appears to be no method of preparing this electrode that is applicable universally to all types of oxidation-reduction titrations. In much of the recent literature authors neglect to mention what pretreatment, if any, was required in order to obtain the results they report. Platinum wire is the form of the electrode most frequently chosen although foil is sometimes used. Gauze electrodes are reported as being inferior to wire electrodes in the attainment of stable potentials (189). Pretreatment of platinum electrodes is in some applications detrimental to their performance. For example, in the titration of ascorbic acid with 2,6-dichloro-4-(4-hydroxyphenyl)imino-2,5-cyclohexadien-1-one (2,6-dichlorophenol-indophenol), it was found advantageous to use the electrode in the state that it was received from the manufacturer and to store it in water (190). On cleaning these electrodes with acid they required 1 or 2 weeks ageing before they could be used again in the titrations.

When other treatments failed to reactivate an electrode of slow response, heating the platinum wire in an oxygen–gas or oxygen–hydrogen flame to as high a temperature as possible, without actually melting it, was found to be effective (191). This treatment is reminiscent of one of the early methods (192) of heating electrodes at the highest temperature of a Mekker burner, but in this instance they were also cleaned frequently with fused potassium bisulfite. Electrodes have also been prepared as follows (193):

> They were immersed in potassium dichromate–sulfuric acid mixture for 5 min, then thoroughly rinsed with distilled deoxygenated water and allowed to stand for 12 h in 2 $M$ sulfuric acid–0.15 $M$ ferrous sulfate. Immediately before use the electrodes were heated to a dull red heat in a burner flame then plunged into a crystal of sodium bisulfite, care being taken not to amass any bisulfite on the glass around the platinum wire. The electrodes were then heated again to dull red heat, allowed to cool for 90–120 sec and used immediately for the titration of compounds containing a functional carbonyl group in a solution of 1,2-dimethoxyethane with the sodium derivative of fluorene (Registry No. *86-73-7*). It should be noted that this titrant is particularly prone to oxidation.

The correct choice of electrode combinations and the appropriate pretreatment of each electrode can considerably influence the sensitivity of a potentiotitrimetric method. This was shown to be the case in the titration of cobalt with potassium hexacyanoferrate(III). Using a conventional platinum electrode, the minimum amount of cobalt which was determined was 2 mg, but cobalt in the range 1–500 $\mu$g could be determined when a platinum–antimony pair was used (194). The two electrodes were prepared simultaneously by using them to electrolyze a solution of 5% sulfuric acid for 2–3 min at 3 V, the platinum electrode being the cathode. After washing with water, the electrodes remained sufficiently responsive to monitor these titrations for the remainder of the day.

## Other Electrodes

Reference has been made to wax-impregnated graphite electrodes (188) and this type has been found satisfactory in the monitoring of some titrations. Medium density graphite (Grade U-1 in about 30 cm lengths) was impregnated with ceresin wax by immersion in a tube containing the molten wax which was alternately evacuated to 5–10 mm and then exposed to the atmosphere several times until no bubbles were detected arising in the molten wax when the tube was reevacuated (195). The electrodes were then coated with a thin layer of insulating material (Seal-All cement was used) and broken into appropriately sized lengths. The bottom electroactive surface of the electrode was immersed in 0.003% Triton X-100 solution for a minute to ensure that it would be completely wetted by the titrand solution, and electrical connection to the top of the electrode was made by means of an alligator clip which was attached to a section from which the insulating cement had been removed. A fresh area of electroactive surface was exposed when required using a lathe, care being taken to remove any graphite particles from the surface with compressed air.

The central carbon rod taken from a dead dry cell was found to be satisfactory as an indicator electrode for some oxidation-reduction titrations (38). The electrode, used as part of a carbon–tungsten or a carbon–silver pair, was prepared by immersing the rod in concentrated hydrochloric acid for 15 min after which it was washed well with water and dried. Carbon fiber electrodes (30) and hydrophobized graphite electrodes of the Selectrode type (196) have also been used for the detection of equivalence points in oxidation-reduction titrations. A freshly prepared silicone rubber-based graphite gave regularly shaped titration curves similar to those obtained with a platinum electrode when reductants such as 1,4-benzenediol and 2-mercaptoethanol were titrated in acetic acid with lead(IV) tetraacetate (37).

Ion selective electrodes have been developed that respond to titrants commonly used in oxidation-reduction titrimetry, and two examples of these are the Chloramine-T (197) and the perbromate (198) responsive electrodes. The Chloramine-T electrode is based upon the nitrate liquid ion exchanger, tris(bathophenanthroline)-nickel(II) nitrate in $p$-nitrocymene, supplied with an Orion electrode. This was converted to the Chloramine-T salt by diluting the exchanger (2 mL) with dichloromethane (10 mL) and using this to extract Chloramine-T from an aqueous solution (50 mL 0.1 $M$). After evaporating the dichloromethane, the extract, dried over sodium sulfate, was applied to a cellulose acetate membrane which was positioned in the electrode body. The internal reference solution was 0.1 M Chloramine-T in 0.1 $M$ sodium chloride and the half-cell was formed by immersion of a silver–silver chloride electrode in this solution. The silver–silver chloride external reference half-cell was equipped with a double junction, contact with the determinand solution being made through 10% w/v potassium nitrate solution. The Chloramine-T electrode was conditioned in a stirred solution of 0.1 $M$ Chloramine-T for 12 h, and when not in use it was stored in 0.01 $M$ Chloramine-T. The electrode must be protected from direct sunlight and beakers in which the electrode is to be used should be surrounded by black nylon. Although the electrode can be used in monitoring titrations performed with Chloramine-T, for rapid determinations it is more conveniently applied by reacting the determinand with an accurately known amount of Chloramine-T added in excess, and determining the concentration of unreacted Chloramine-T with the electrode (199).

The perbromate ion selective electrode is based upon the perbromate salt of Crystal Violet (Registry No. *548-62-9*) which was precipitated by mixing equimolar (0.05 M) solutions of Crystal Violet and potassium perbromate at $p$H 5.8 (adjusted with 1 $M$ potassium dihydrogen phosphate solution). The perbromate salt was extracted into chorobenzene, and after washing this extract with water, it was dried over anhydrous sodium sulfate. This procedure resulted in the formation of a 0.05 $M$ solution of Crystal Violet perbromate in chlorobenzene. A portion of this solution was applied to the membrane material supplied with a commercially available perchlorate electrode, the internal reference half-cell being a silver–silver chloride electrode dipping into a solution 0.05 $M$ in potassium perbromate and 0.05 $M$ in potassium chloride. The electrode exhibits a Nernstian response to perbromate ion at concentrations $10^{-5}$–$10^{-2}$ $M$ in the $p$H range 2–10, and can be used as an indicator electrode in the titration of iron(II) with potassium perbromate at $p$H 4.

## Electrodes for Electrolytic Potentiotitrimetry

The application of this method to oxidation–reduction titrations has been outlined in Chapter 2. In a great majority of applications platinum is the electrode material of choice. Single platinum wire or foil electrodes, polarized with respect to a reference half-cell are sometimes used, but the configuration most favored is a pair of platinum electrodes through which the current is passed. Gold electrodes are used much less frequently than platinum and are usually prepared from gold foil. In a typical method of preparation (200) two identically sized pieces (13 × 8 mm) were cut from 0.25 mm thick gold sheet (99.6% pure), and a piece of gold wire flame welded to each as an electrical connection. The polarizing circuit consisted of a 50–60 V dry battery connected through a 22 M$\Omega$ resistor across the electrodes allowing a current of 2–3 μA to pass through the titration cell. The electrodes were washed with 1 M sulfuric acid and then with water before each titration. When not in use they were stored in air. Other electrodes that have been used include glassy carbon (201), a mercury drop (202), and tungsten with anodic polarization (203).

## REFERENCES

1. G. N. Nikolaev, V. G. Stavinchuk, and M. K. Saikina, *Zh. Anal. Khim.* **31**, 2038 (1976).
2. N. Ciocan and G. E. Baiulescu, *Anal. Chem.* **50**, 1407 (1978).
3. C. N. Reilley and R. W. Schmid, *Anal. Chem.* **30**, 947 (1958).
4. C. N. Reilley, R. W. Schmid, and D. W. Lamson, *Anal. Chem.* **30**, 953 (1958).
5. M. L. Richardson, *Anal. Chim. Acta* **24**, 46 (1961).
6. E. G. Novikov and V. K. Kondratov, *Zh. Anal. Khim.* **23**, 955 (1968).
7. R. H. Loeppert, Jr., L. W. Zelazny, and B. G. Volk, *Soil Sci. Soc. Am. J.* **41**, 1101 (1977).
8. J. S. Fritz and L. W. Marple, *Anal. Chim.* **34**, 921 (1962).
9. F. C. Trussell and R. E. Lewis, *Anal. Chim. Acta* **34**, 243 (1966).
10. M. S. Greenberg, B. J. Barker, and J. A. Caruso, *Anal. Chim. Acta* **54**, 159 (1971).
11. B. N. Utkin, *Zavod. Lab.* **33** 1381 (1967).
12. J. F. C. Boodts and J. R. Kroll, *Fresenius' Z. Anal. Chem.* **289**, 207 (1978).
13. W. Selig, *Mikrochim. Acta* **1978**(2), 169.
14. L. J. Lohr, *Anal. Chem.* **32**, 1167 (1960).
15. T. R. Williams and M. Lautenschleger, *Talanta* **10**, 804 (1963).
16. I. M. Kolthoff and B. D. Hartong, *Recl. Trav. Chim. Pays-Bas* **44**, 113 (1925).
17. E. J. Greenhow and B. F. Al-Mudarris, *Talanta* **22**, 417 (1975).
18. I. Prazak and J. Grimmer, *Collect. Czech. Chem. Commun.* **24**, 2049 (1959).
19. E. Scarano and M. Mascini, *Talanta* **16**, 707 (1969).
20. L. I. Chekmarev, V. B. Avilov, Yu. F. Klimov, and T. A. Kolmogorova, *Zh. Anal. Khim.* **27**, 1401 (1972); *J. Anal. Chem. USSR* **27**, 1261 (1972).
21. V. T. Athavale, V. P. Apte, M. R. Dhaneshwar, and R. G. Dhaneshwar, *Indian J. Chem.* **6**, 656 (1968).
22. M. R. Dhaneshwar and R. G. Dhaneshwar, *Analyst (London)* **97**, 620 (1972).
23. M. R. Dhaneshwar, R. G. Dhaneshwar, and L. Pandit, *Trans. Soc. Adv. Electrochem. Sci. Technol.* **9**, 105 (1974).
24. V. E. Petrakovich, *Zh. Anal. Khim.* **18**, 1161 (1963); *J. Anal. Chem. USSR* **18**, 1004 (1963).
25. B. P. Gyani and R. Kishore, *Proc. Natl. Acad. Sci. India, Sect. A* **29**, 49 (1960).
26. J. R. Baylis, *Ind. Eng. Chem.* **15**, 852 (1923).
27. S. E. S. El Wakkad, T. M. Salem, H. A. Rizk, and J. G. Ebaid, *J. Chem. Soc.* **1957**, 3776.

# REFERENCES

28. T. J. Pastor and V. J. Vajgand, *Mikrochim. Acta* **1976**(2), 85.
29. F. J. Miller, *Anal. Chem.* **35**, 929 (1963).
30. V. J. Jennings and P. J. Pearson, *Anal. Chim. Acta* **82**, 223 (1976).
31. J. Bercik, M. Cakrt, and Z. Hladky, *Chem. Zvesti* **24**, 290 (1970).
32. J. Bercik, Z. Hladky, and M. Cakrt, *Chem. Zvesti* **24**, 298 (1970).
33. J. Bercik, Z. Hladky, and M. Cakrt, *Fresenius' Z. Anal. Chem.* **261**, 113 (1972).
34. E. Pungor and E. Szepesvary, *Anal. Chim. Acta* **43**, 289 (1968).
35. E. Szepesvary and E. Pungor, *Anal. Chim. Acta* **54**, 199 (1971).
36. E. Szepesvaryne-Rath and E. Pungor, *Magy. Kem. Foly* **77**, 502 (1971).
37. T. J. Pastor, V. J. Vajgand, Z. Simonovic, and E. Szepesvary, *Anal. Chim. Acta* **98**, 233 (1978).
38. M. Natarajan and A. Ramasubramanian, *J. Chem. Educ.* **53**, 663 (1976).
39. S. Schuldiner and J. P. Hoare, *Can. J. Chem.* **37**, 228 (1959).
40. C. Liteanu and I. Haiduc, *Rev. Roum. Chim.* **15**, 1555 (1970).
41. S. Kaufman, *Anal. Chem.* **47**, 494 (1975).
42. M. A. Wechter, H. R. Shanks, G. Carter, G. M. Ebert, R. Guglielmino, and A. F. Voigt, *Anal. Chem.* **44**, 850 (1972).
43. H. R. Shanks, *J. Cryst. Growth* **13 / 14**, 433 (1972).
44. G. V. Makarov, E. A. Buketov, M. A. Edrisova, and L. G. Kozorin, *Zh. Anal. Khim.* **29**, 801 (1974).
45. M. Mioscu and M. Grebla, *Rev. Roum. Chim.* **23**, 1345 (1978).
46. I. Shain and G. R. Svoboda, *Anal. Chem.* **31**, 1857 (1959).
47. G. R. Svoboda, *Anal. Chem.* **33**, 1638 (1961).
48. M. Lugowska and M. Machtinger, *Bull. Soc. Chim. Fr.* **1968**, 5084.
49. M. Forss and K. E. Fremer, *Pap. Puu* **60**, 121 (1978).
50. V. J. Vajgand, T. J. Pastor, and L. J. Bjelica, *Mikrochim. Acta* **1975**, 485.
51. E. Bishop and G. D. Short, *Analyst (London)* **87**, 467 (1962).
52. C. N. Reilley in *Handbook of Analytical Chemistry*, L. Meites, Ed., McGraw-Hill, New York, 1963, pp. 5–26.
53. C. N. Reilley and A. Vavoulis, *Anal. Chem.* **31**, 243 (1959).
54. R. W. Schmid and C. N. Reilley, *Anal. Chem.* **29**, 264 (1957).
55. J. Haslam, D. C. M. Squirrell, and I. G. Blackwell, *Analyst (London)* **85**, 27 (1960).
56. L. Ramaley, R. L. Brubaker, and C. G. Enke, *Anal. Chem.* **35**, 1088 (1963).
57. A. R. Gallego and C. Verejo Morales, *An. Quim.* **73**, 415 (1977); *CA* **88**, 130256r.
58. B. Wallen, *Anal. Chem.* **46**, 304 (1974).
59. H. Khalifa and A. Soliman, *Fresenius' Z. Anal. Chem.* **169**, 81 (1959).
60. H. Khalifa and F. A. Osman, *Fresenius'. Z. Anal. Chem.* **178**, 116 (1960).
61. H. Khalifa and M. M. Khater, *J. Chem. U.A.R.* **10**, 123 (1967); *CA* **70**, 53679e.
62. H. Khalifa and A. El-Sirafy, *Fresenius' Z. Anal. Chem.* **227**, 109 (1967).
63. H. Khalifa and B. Ateya, *Microchem. J.* **13**, 247 (1968).
64. H. Khalifa, A. T. Haj-Hussein, and Y. M. Issa, *Egypt. J. Chem.* **17**, 329 (1974); *CA* **85**, 86777y.

65. H. Khalifa, Y. M. Issa, and A. T. Haj-Hussein, *Fresenius' Z. Anal. Chem.* **282**, 223 (1976).
66. H. Khalifa and A. M. Abdallah, *Microchem. J.* **13**, 726 (1968).
67. H. Khalifa and A. M. Abdallah, *Microchem. J.* **14**, 399 (1969).
68. H. Khalifa and A. I. Atalla, *Egypt. J. Chem.* **17**, 181 (1974); *CA* **85**, 110143m.
69. H. Khalifa and A. I. Atalla, *Egypt. J. Chem.* **17**, 173 (1974); *CA* **85**, 86776x.
70. F. Strafelda, *Collect. Czech. Chem. Commun.* **27**, 343 (1962); **28**, 3345 (1963); **30**, 2320 (1965).
71. J. S. Fritz and B. B. Garralda, *Anal. Chem.* **36**, 737 (1964).
72. W. Kemula, A. Hulanicki, and M. Trojanowicz, *Chem. Anal. (Warsaw)* **14**, 481 (1969).
73. E. D. Olsen and F. S. Adamo, *Anal. Chem.* **39**, 81 (1967).
74. I. E. Lichtenstein, E. Coppola, and D. A. Aikens, *Anal. Chem.* **44**, 1681 (1972).
75. A. Hulanicki, A. Trojanowicz, and J. Domanska, *Talanta* **20**, 1117 (1973).
76. R. Pribil, Z. Koudela, and B. Matyska, *Collect. Czech. Chem. Commun.* **16**, 80 (1951).
77. A. Hulanicki and R. Karwowska, *Talanta* **18**, 239 (1971).
78. W. E. Van der Linden and S. Beijer, *Anal. Chim. Acta* **58**, 472 (1972).
79. F. Strafelda and J. Matousek, *Collect. Czech. Chem. Commun.* **30**, 2334 (1965).
80. T. Nomura, T. Dono, and G. Nakagawa, *Bunseki Kagaku* **14**, 197 (1965); *CA* **62**, 15411f.
81. T. Anfalt and D. Jagner, *Anal. Chim. Acta* **56**, 477 (1971).
82. International Union of Pure and Applied Chemistry, *Pure Applied Chem.* **51**, 1913 (1979).
83. R. W. Cattrall and C-H. Pui, *Anal. Chim. Acta* **83**, 355 (1976).
84. R. W. Cattrall and C-H. Pui, *Anal. Chim. Acta* **88**, 185 (1977).
85. R. W. Cattrall and C-H. Pui, *Anal. Chim. Acta* **87**, 419 (1976).
86. D. E. Ryan and M. T. Cheung, *Anal. Chim. Acta* **82**, 409 (1976).
87. A. Vanni and P. G. Daniele, *Ann. Chim. (Rome)* **65**, 55 (1975).
88. E. A. Kosyuga, A. A. Kalugin, and I. A. Gur'ev, *Zavod. Lab.* **45**, 206 (1979).
89. G. Schwarzenbach and W. Biedermann, *Helv. Chim. Acta* **31**, 678 (1948).
90. A. R. Rajput, M. Kataoka, and T. Kambara, *J. Electroanal. Chem. Interfacial Electrochem.* **66**, 67 (1975).
91. M. Thies, *Fresenius' Z. Anal. Chem.* **144**, 275 (1955).
92. M. Malat and M. Tenorova, *Chem. Listy* **51**, 2135 (1957).
93. M. Kataoka, M. Shin, and T. Kambara, *Talanta* **24**, 261 (1977).
94. M. Sugawara, T. Nakajima, and T. Kambara, *J. Electroanal. Chem. Interfacial Electrochem.* **67**, 315 (1976).
95. M. A. Wechter, P. B. Hahn, G. M. Ebert, P. R. Montoya, and A. F. Voigt, *Anal. Chem.* **45**, 1267 (1973).
96. A. E. Martin and C. N. Reilley, *Anal. Chem.* **31**, 992 (1959).
97. N. Tanaka, I. T. Oiwa, and M. Kodama, *Anal. Chem.* **28**, 1555 (1956).
98. E. P. Przybylowicz and L. B. Rogers, *Anal. Chem.* **30**, 65 (1958).
99. G. Kainz and G. Sontag, *Fresenius' Z. Anal. Chem.* **269**, 267 (1974);

*Mikrochim. Acta* **1975**(1), 171.
100. G. Kraft and H. Dosch, *Fresenius' Z. Anal. Chem.* **281**, 97 (1976).
101. R. Behrend, *Z. Physik. Chem.* **11**, 466 (1893).
102. A. J. Martin, *Anal. Chem.* **30**, 233 (1958).
103. M. W. Tamele, L. B. Ryland, and R. N. McCoy, *Anal. Chem.* **32**, 1007 (1960).
104. B. Coulter and D. G. Bush, *Anal. Chim. Acta* **51**, 431 (1970).
105. T. Y. Toribra and L. Koval, *Talanta* **17**, 1003 (1970).
106. L. H. Kalbus and G. E. Kalbus, *Anal. Chim. Acta* **39**, 335 (1967); **53** 225 (1971).
107. G. E. Kalbus, R. D. Wesley and L. H. Kalbus, *Analyst (London)* **96**, 488 (1971).
108. A. Hulanicki, *Chem. Anal. (Warsaw)* **5**, 881 (1960).
109. Z. I. Ivanova, T. V. Sazhneva, V. K. Chebotarev, and O. E. Shelepin, *Zh. Anal. Khim.* **29**, 466 (1974); *J. Anal. Chem. USSR* **29**, 466 (1974).
110. P. N. Kovalenko, Z. I. Ivanova, V. K. Chebotarev, and V. D. Dionis'ev, *Zh. Anal. Khim.* **24**, 1810 (1969); *J. Anal. Chem. USSR* **24**, 1468 (1969).
111. H. Flaschka and A. J. Barnard in *Advances in Analytical Chemistry and Instrumentation*, C. N. Reilley, Ed., Interscience, New York, 1960, pp. 1-117.
112. S. S. Cooper, *Anal. Chem.* **29**, 446 (1957).
113. S. Pinzauti and E. LaPorta, *Analyst (London)* **102**, 938 (1977).
114. E. Siska and E. Pungor, *Fresenius' Z. Anal. Chem.* **257**, 12 (1971).
115. J. S. Fritz and T. A. Palmer, *Anal. Chem.* **33**, 98 (1961).
116. O. N. Rusina, P. N. Kovalenko, and Z. I. Ivanova, *Zh. Anal. Khim.* **20**, 44 (1965); *J. Anal. Chem. USSR* **20**, 38 (1965).
117. P. N. Kovalenko, Z. I. Ivanova, and I. F. Poyarkova, *Tr. Kom. Anal. Khim. Akad. Nauk SSSR* **17**, 381 (1969); *CA* **72**, 117410m.
118. H. Khalifa and B. Ateya, *Microchem. J.* **12**, 440 (1967).
119. I. M. Kolthoff and J. J. Lingane, *J. Am. Chem. Soc.* **58**, 2457 (1936).
120. H. Dautzenberg, B. Philipp, and J. Schumann, *Faserforsch. Textiltech.* **23**, 372 (1972).
121. T. R. Williams, S. Piekarski, and C. Manning, *Talanta* **18**, 951 (1971).
122. J. A. Antanasiu, *J. Chim. Phys.* **23**, 501 (1926).
123. J. J. Lingane and A. M. Hartley, *Anal. Chim. Acta* **11**, 475 (1954).
124. O. N. Rusina, P. N. Kovalenko, and Z. I. Ivanova, *Zh. Anal. Khim.* **21**, 257 (1966); *J. Anal. Chem. USSR* **21**, 227 (1966).
125. I. M. Kolthoff and N. H. Furman, *Potentiometric Titrations*, 2nd ed., Wiley, New York, 1931, p. 331.
126. J. Alchimowicz, K. Siutowicz, and K. Zwierzak, *Chem. Anal. (Warsaw)* **13**, 871 (1968).
127. T. Kokina, M. N. Petrikova, and I. P. Alimarin, *Zh. Anal. Khim.* **26**, 2237 (1971); *J. Anal. Chem. USSR* **26**, 2003 (1971).
128. G. C. Cortellessa and C. A. Napoli, *Analyst (London)* **93**, 546 (1968).
129. E. M. Zingel, V. G. Korsakov, and I. A. Kedrinskii, *Fiz. Khim.* **1974**(1), 44; *CA* **84**, 83695 pp.
130. D. H. Chou and L. C. Sams, Jr., *Microchem. J.*, **14**, 507 (1969).
131. W. Selig, *Microchem. J.*, **20**, 388 (1975).

132. T. S. Prokopov, *Mikrochim. Acta* **1968**, 401.
133. B. Stankovic, A. Stefanovic, and M. Dugandzic, *Mikrochim. Acta* **1977(1)**, 395.
134. V. A. Kimstach, P. N. Kovalenko, and Z. I. Ivanova, *Zh. Anal. Khim* **25**, 588 (1970); *J. Anal. Chem. USSR* **25**, 508 (1970).
135. O. N. Rusina and R. I. Baglei, *Mater. Nauch. Tekh. Konf. Sev.-Kavkaz. Gornomet. Inst.* **1970**, 68; *CA* **78**, 53179v.
136. V. G. Belikov, K. N. Bagdasarov, V. A. Kimstach, and N. A. Sorokoumova, *Farmatsiya (Moscow)* **25**, 62 (1976); *CA* **85**, 10484e.
137. S. M. Prajapati, S. N. Bhatt, and K. P. Soni, *J. Electrochem. Soc. India* **25**, 133 (1976).
138. C. W. Robbins, D. L. Carter, and D. W. James, *Soil Sci. Soc. Amer., Proc.* **37**, 212 (1973).
139. N. Akimoto and K. Hozumi, *Anal. Chem.* **46**, 766 (1974).
140. E. W. Baumann, *Anal. Chem.* **40**, 1731 (1968).
141. A. Cedergren and G. Sundin, *Anal. Chim. Acta* **94**, 467 (1977).
142. F. Peter and R. Rosset, *Anal. Chim. Acta* **64**, 397 (1973).
143. M. K. Papay, K. Toth, V. Izvekov, and E. Pungor, *Anal. Chim. Acta* **64**, 409 (1973).
144. W. Selig, *Mikrochim. Acta* **1973**, 453.
145. J. Motonaka, S. Ikeda, and N. Tanaka, *Anal. Chim. Acta* **105**, 417 (1979).
146. D. Midgley, *Talanta* **26**, 261 (1979).
147. W. Selig, *Mikrochim. Acta* **1978(2)**, 75.
148. W. Selig, *Mikrochim. Acta* **1978(2)**, 359.
149. C. J. Coetzee and A. J. Basson, *Anal. Chim. Acta* **83**, 361 (1976).
150. E. Hopirtean and E. Stefaniga, *Rev. Roum. Chim.* **20**, 863 (1975).
151. E. Diamandis and T. P. Hadjiioannou, *Mikrochim. Acta* **1977(2)**, 255.
152. T. P. Hadjiioannou and E. P. Diamandis, *Anal. Chim. Acta* **94**, 443 (1977).
153. N. Ciocan, *Rom. Patent* 81211 (1975).
154. G. E. Baiulescu and N. Ciocan, *Talanta* **24**, 37 (1977).
155. N. Ciocan and G. E. Baiulescu, *Anal. Chem.* **50**, 1407 (1978).
156. V. Cosofret, P. G. Zugravescu, and G. E. Baiulescu, *Talanta* **24**, 461 (1977).
157. G. E. Baiulescu and V. V. Cosofret, *Talanta* **23**, 677 (1976).
158. D. F. Anghel, G. Popescu, and N. Ciocan, *Mikrochim. Acta* **1977(2)**, 639.
159. D. F. Anghel and N. Ciocan, *Anal. Lett.* **10**, 423 (1977).
160. E. Bishop and R. G. Dhaneshwar, *Analyst (London)* **87**, 207 (1962).
161. E. Bishop and R. G. Dhaneshwar, *Analyst (London)* **87**, 845 (1962).
162. E. Bishop and R. G. Dhaneshwar, *Anal. Chem.* **36**, 726 (1964).
163. T. Yoshimori, S. Umezaki, and Y. Tanaka, *Bunseki Kagaku* **21**, 1381 (1972); *CA* **78**, 52147h.
164. E. Siska and E. Pungor, *Fresenius' Z. Anal. Chem.* **257**, 12 (1971).
165. M. R. F. Ashworth, H. Gottel, and J. Schneider, *Anal. Chim. Acta* **31**, 17 (1964).
166. B. Kh. Gorbatkova and O. N. Rusina, *Zh. Anal. Khim.* **31**, 2147 (1976); *J. Anal. Chem. USSR* **31**, 1569 (1976).
167. E. Siska and E. Pungor, *Talanta* **19**, 715 (1972).

168. C. Liteanu, I. C. Popescu, and E. Hopirtean in *Ion Selective Electrodes*, E. Pungor, Ed., Akademaiai Kaido, Budapest, 1973, p. 51.
169. I. C. Popescu, C. Liteanu, and L. Savici, *Rev. Roum. Chim.* **18**, 1983 (1973).
170. R. N. Heistand and C. T. Blake, *Mikrochim. Acta* **1972**, 212.
171. W. Selig, *Mikrochim. Acta* **1976(2)**, 9.
172. W. Selig, *Microchem. J.* **21**, 291 (1976).
173. G. G. Rao and N. K. Murty, *Fresenius' Z. Anal. Chem.* **208**, 97 (1965).
174. A. I. Busev and L. Gyn, *Zh. Anal. Khim.* **15**, 191 (1960); *J. Anal. Chem. USSR* **15**, 217 (1960).
175. N. H. Furman and A. J. Fenton, *Anal. Chem.* **32**, 745 (1960).
176. N. Chughtai and J. Dolezal, *Microchem. J.* **20**, 363 (1975).
177. P. Norkus and G. Simkeviciute, *Zh. Anal. Khim.* **26**, 39 (1971); *J. Anal. Chem. USSR* **26**, 30 (1971).
178. K. G. Haeusler, R. Geyer, and S. Rennhak, *Z. Chem.* **17**, 146 (1977).
179. V. I. Ignatov, V. T. Solomatin, and A. A. Nemodruk, *Zh. Anal. Khim* **33**, 2328 (1978); *J. Anal. Chem. USSR* **33**, 1785 (1978).
180. B. C. Verma and S. Kumar, *J. Indian Chem. Soc.* **52**, 528 (1975).
181. Z. Hladky and J. Vrestal, *Collect. Czech. Chem. Commun.* **34**, 984 (1969).
182. A. Ramadan, P. K. Agasyan, and S. I. Petrov, *Zh. Anal. Khim.* **29**, 1144 (1974); *J. Anal. Chem. USSR* **29**, 977 (1974).
183. R. D. Krause and B. Kratochvil, *Anal. Chem.* **45**, 844 (1973).
184. H. C. Mruthyunjaya and A. R. Vasudeva, *Indian J. Chem.* **5**, 430 (1967).
185. G. N. Nikolaev, R. S. Tsekhanskii, and Yu. A. Fedorov, *Zh. Anal. Khim.* **33**, 999 (1978); *J. Anal. Chem. USSR* **33**, 771 (1978).
186. J. T. Stock and R. D. Braun, *Microchem. J.* **15**, 519 (1970).
187. P. Norkus and R. Markeviciene, *Zh. Anal. Khim.* **22**, 1527 (1967); *J. Anal. Chem. USSR* **22**, 1282 (1967).
188. D. P. Smith and M. T. Pope, *Anal. Chem.* **40**, 1906 (1968).
189. G. G. Rao and S. R. Sagi, *Talanta* **9**, 715 (1962).
190. E. E. Spaeth, V. H. Baptist, and M. Roberts, *Anal. Chem.* **34**, 1342 (1962).
191. A. Claasen and L. Bastings, *Fresenius' Z. Anal. Chem.* **202**, 241 (1964).
192. R. C. Van Name and F. Fenwick, *J. Am. Chem. Soc.* **47**, 9 (1925).
193. J. F. Remark and C. A. Reynolds, *Talanta* **23**, 687 (1976).
194. D. A. Lee, *Microchem. J.* **20**, 62 (1975).
195. P. J. Elving and D. L. Smith, *Anal. Chem.* **32**, 1849 (1960).
196. J. Ruzicka, C. G. Lamm, and J. C. Tjell, *Anal. Chim. Acta* **62**, 15 (1972).
197. M. A. Koupparis and T. P. Hadjiioannou, *Anal. Chim. Acta* **94**, 367 (1977).
198. L. A. Lazarou and T. P. Hadjiioannou, *Anal. Lett.* **11**, 779 (1978).
199. M. A. Koupparis and T. P. Hadjiioannou, *Talanta*, **25**, 477 (1978).
200. J. L. Drummond and R. A. Grant, *Talanta*, **13**, 477 (1966).
201. R. Nakashima, M. Furukawa, S. Sasaki, and S. Shibata, *Bunseki Kagaku* **17**, 1066 (1968).
202. W. Buechler, *Helv. Chim. Acta* **47**, 639 (1964).
203. P. K. Agasyn, E. R. Nikolaeva, and R. M. Ryskulbekova, *Zh. Anal. Khim.* **19**, 1219 (1964).

# PART 2
# APPLICATIONS TO THE DETERMINATION OF SOLUTION EQUILIBRIUM DATA

CHAPTER
6

# THE DETERMINATION OF IONIZATION CONSTANTS IN AQUEOUS SOLUTION

## 6.1 INTRODUCTION

**The Definition of Acids and Bases**

According to the Bronsted definition (1), an acid is "a species having a tendency to lose a proton," whereas a base is "a species having a tendency to add on a proton." Hence for every acid HA there is a conjugate base $A^-$:

$$HA \rightleftharpoons H^+ + A^-$$

and for every base B there is a conjugate acid, $BH^+$:

$$BH^+ \rightleftharpoons H^+ + B$$

Thus acetic acid–acetate ion and ammonium ion–ammonia are examples of conjugate acid-base pairs. If HA (or $BH^+$) has a great tendency to lose protons, it follows that its conjugate species $A^-$ (or B) has only a small tendency to accept protons. In other words, if HA (or $BH^+$) is a *strong acid*, $A^-$ (or B) is a *weak base*, and the converse is also true. Acids and bases so defined can only manifest their properties by reacting with bases and acids, respectively. In aqueous solution, the acid species HA or $BH^+$ react with water acting as a base:

$$HA + H_2O(\text{base}) \rightleftharpoons H_3O^+ + A^- \qquad (6.1a)$$

$$BH^+ + H_2O(\text{base}) \rightleftharpoons H_3O^+ + B \qquad (6.1b)$$

and, conversely, the base species $A^-$ or B react with water acting as an acid:

$$A^- + H_2O(\text{acid}) \rightleftharpoons HA + OH^-$$

$$B + H_2O(\text{acid}) \rightleftharpoons BH^+ + OH^-$$

Quantitatively, the acid strength of HA or $BH^+$ relative to the base strength of water is given by the equilibrium constant expressions for equations (6.1a, b) which can be represented as

$$K = \frac{a_{H_3O^+} \cdot a_B}{a_{H_2O} \cdot a_A} \qquad (6.2)$$

where $a_B$ is the activity of the base species and $a_A$ is the activity of the acid species. The usual standard state for water is defined as pure water having unit activity, but the activity of water when solutes are present is less than that of pure water. However, for all practical purposes the activity of water in dilute aqueous solutions may still be regarded as unity, because even in solutions of electrolytes as concentrated as 1 $M$ the vapor pressure is diminished by only 2–4% (2). Letting $H^+$ represent the solvated proton, equation (6.2) can be written, therefore, as

$$K_a = \frac{a_{H^+} \cdot a_B}{a_A} \qquad (6.3)$$

in which $K_a$ is the acidic dissociation constant, more commonly termed the *ionization constant*. This equation can be written in the form

$$pK_a = pH + \log \frac{a_A}{a_B} \qquad (6.4a)$$

where $pK_a$ is the negative logarithm of $K_a$, and is equal to the $pH$ at which the activities of A and B are equal.

Equation (6.4a) not only expresses the strength of acids (i.e., HA), but it also expresses the strength of conjugate acids (i.e., $BH^+$) of bases. The $pK_a$ values of these conjugate acids are commonly spoken of as "the $pK_a$ of the base," "the basic $pK_a$," or "the $pK_a$ for proton addition." Thus $pK_a = 9.25$ for the ammonium ion lies on the same scale as $pK_a = 4.76$ for acetic acid and 10.00 for phenol. The greater the $pK_a$ value, the weaker the substance as an acid or, conversely, the stronger is its conjugate base. For any given solvent, the $pK_a$ scale is convenient for expressing the strengths of both acids and bases. The earlier practice of defining the "basic $pK$," $pK_b$, from the relation:

$$K_b = \frac{a_{BH^+} \cdot a_{OH^-}}{a_B}$$

is thus unnecessary. The $pK_a$ and $pK_b$ scales are interrelated by the equation

$$pK_a + pK_b = pK_w$$

# INTRODUCTION

where $pK_w$ is the negative logarithm of the ionic product of water, $K_w$ ($pK_w \sim 14$), the latter being defined as

$$K_w = a_{H^+} \cdot a_{OH^-}$$

The $pK_a$ values for some 8000 acids, including values for the conjugate species $BH^+$, are given in the compilations listed as References 3–7, and Table 6.1 lists typical ranges of $pK_a$ values for some of the more common types of acids and bases, and Table 6.2 gives specific examples. The reason why so many $pK_a$ values are determined is that a larger number of organic compounds, both natural and synthetic, contain acidic and/or basic groups

Table 6.1. Typical Ranges of $pK_a$ Values for Organic Acids, Including Conjugate Acids of Organic Bases

| Classes of Compounds | Typical $pK_a$ Ranges |
| --- | --- |
| Ethers (as bases) | $-2$ – $-4$ |
| Pyrimidinium ions | 1–2 |
| Aliphatic dicarboxylic acids | 1–4.5 (First dissociation) |
| Anilinium ions | 1–5[a] |
| α-Amino acids | 2–3 (COOH) |
| Monocarboxylic acids | 3–5[b] |
| Hydroxylamine and its esters | 4–5 |
| Pyridinium ions | 4–6 |
| Aliphatic dicarboxylic acids | 5–7 (Second dissociation) |
| Thiophenols | 5–7 |
| α, β-Unsaturated aliphatic aminium ions | 6–9 |
| Imidazolinium ions | 7 |
| Hydroxyheteroaromatics | 7–11 |
| Phenols | 8–10[c] |
| Purines, acid functions | 8–10 |
| α-Amino acids | 9–10.5 ($NH_3^+$) |
| Saturated nitrogen heterocycles | 9–11 ($NH_3^+$) |
| Aliphatic and alicyclic aminium ions | 9–11 |
| Thiols, aliphatic | 9–11 |
| Oximes | 10–12 |
| Guanidinium ions | 11–14 |
| Aldehydes | 11–14 |
| Azoles | 13–16 |
| Alcohols and sugars | 13–16 |

[a] But note 2-nitroanilinium ion, $pK_a = -0.3$.
[b] Trifluoroacetic acid, $pK_a = -0.26$, and 2,4,6-trinitrobenzoic acid, $pK_a = 0.65$.
[c] 2,4,6-Trinitrophenol, $pK_a = 0.22$.

**Table 6.2. Individual $pK_a$ Values of Organic Acids and Bases in Water at 20–25 °C**

| Bases | $pK_a$ | Acids | $pK_a$ |
|---|---|---|---|
| Pyrrole | −3.80 | Methane sulphonic acid | −6.0 |
| Indole | −2.30 | Aminoacetic acid | 2.35 |
| Tetrahydrofuran | −2.10 | 2-Furoic acid | 3.16 |
| Urea | 0.10 | Formic acid | 3.75 |
| 4-Pyrone | 0.10 | Benzoic acid | 4.21 |
| Diphenylamine | 0.77 | Succinic acid | 4.22, 5.64 |
| Oxazole | 0.80 | Acetic acid | 4.76 |
| Pyrimidine | 1.23 | Cyclohexanecarboxylic acid | 4.90 |
| Thiazole | 2.44 | Uric acid | 5.83 |
| Aniline | 4.69 | Thiophenol | 6.52 |
| Quinoline | 4.92 | $p$-Nitrophenol | 7.15 |
| Pyridine | 5.23 | Acetylacetone (enolic form) | 8.13 |
| Isoquinoline | 5.42 | Purine | 8.93 |
| Aminoacetic acid | 9.78 | Phenol | 10.00 |
| Triethylamine | 10.78 | Ethanethiol | 10.54 |
| Ethylamine | 10.81 | 2-Pyridone | 11.65 |
| Diethylamine | 11.09 | Formaldehyde | 13.29 |
| Piperidine | 11.28 | Methanol | 15.5 |
| Acetamidine | ~ 12.40 | Pyrrole | ~ 16.5 |

which govern many of their chemical, physical, and biological properties. For such compounds, the proportion of the species (neutral molecule, anion, cation) that are present at a particular $p$H is determined by the $pK_a$ value and can be calculated from equation (6.4a) conveniently rearranged in the form

$$\frac{a_B}{a_A} = 10^{pH - pK_a} \tag{6.4b}$$

In analytical chemistry, $pK_a$ values assist in the interpretation of acid-base titration curves where multiple acidic or basic sites are present. The optimum conditions for the extraction and isolation of acids and bases from aqueous media can be predicted if the $pK_a$ values are known since the uncharged species is generally less soluble in the aqueous phase. Buffer capacities can be calculated at known $p$H values and buffer concentrations, provided the $pK_a$ of the buffer acid is known. $pK_a$ values are also required in the estimation of microconstants and in the calculation of tautomeric equilibria.

## General Methods for the Determination of $pK_a$ Values

These usually depend upon a quantitative assessment of the ratio of deprotonated/protonated forms for a compound under known conditions of acidity for the solvent medium. Although it is often possible to determine this ratio with high precision, the reliability of any of these methods depends ultimately upon the accuracy with which the appropriate acidity function can either be measured or assigned. The only acidity function of thermodynamic significance is the quantity $p(a_H\gamma_{Cl})$ since this is derived from measurements using a cell without diffusion [e.g., cell (4.5)]. According to equation (6.4), however, the $pK_a$ is directly related to the quasi-thermodynamic quantity $pH$ which is very conveniently measured with a glass electrode typically incorporated into cell (4.10), for example. Thus $pH$ is the acidity function most frequently measured in the determination of ionization constants and is used within the range $pH$ 2–12. Outside this range, the accuracy of $pH$ measurements becomes uncertain as a result of variability both in the magnitude of the liquid junction potential and also in the response of the glass electrode. Acidity scales are available for extending into the strongly acid or alkaline regions the range of ionization constants that can be determined, but these acidity function values are assigned from indirect measurements rather than from the direct measurements associated with, for example, $pH$. A given acidity scale is only appropriate for a given class of compound (8) and failure to select the correct type of scale can lead to large errors in the derived value of the ionization constant. The magnitude of these errors will be independent of the precision of the method used for the determination of the ratio deprotonated form/protonated form. General methods are available for the determination of this ratio. These are based upon the following:

1. A knowledge of the stoichiometry of the solution for which either $p(a_H\gamma_{Cl})$ or $pH$ is the measured quantity.
2. A quantitative analysis of each of the two components in a solution of known acidity function.

The first method is described in this chapter since it is traditionally termed *the potentiometric method*.

The analytical methods (method 2) for the determination of the ratio deprotonated form/protonated form depend upon some difference in property or response that allows a quantitative discrimination to be made between the two forms of the compound. For example, if a study of the spectrum (absorbance versus wavelength) of the pure deprotonated form at high $p$H reveals significantly different features from the spectrum of the pure protonated form obtained at a lower $p$H, then it is possible to deduce the ratio by measuring the absorbances of solutions at values of $p$H that are intermediate between the upper and lower limits. The method is particularly applicable to sparingly soluble aromatic and heteroaromatic acids and bases having solubilities less than $10^{-3}$ $M$. Accurate values of ionization constants can be obtained if solutions are prepared in buffers having known $p(a_H\gamma_{Cl})$ values (9); less accurate values are obtained with $p$H buffer solutions but only approximate values if other types of acidity functions are used (10). The other methods that come into this general category have been reviewed (11).

## 6.2 METHODS BASED UPON MEASUREMENT OF $p(a_H\gamma_{Cl})$

The acidity function $p(a_H\gamma_{Cl})$ for an aqueous solution containing a known molality of chloride ion can be measured in any calibrated cell that contains an electrode reversible to hydrogen ion and an electrode reversible to chloride ion. Often it is sufficient to calibrate the cell by measuring the cell e.m.f. when it contains a 0.01 $m$ solution of hydrochloric acid for which the mean activity coefficients are known. Use of an equation similar to equation (3.13) then allows calculation of $E°_{cell}$. Alternatively, the cell can be calibrated in an analogous manner to that used in $p$H determinations except that two or more solutions of known $p(a_H\gamma_{Cl})$ are used. Accurate measurements of the ionization constants of weak acids and bases are usually associated with the use of the cell in which a hydrogen electrode is used in conjunction with a silver–silver chloride electrode. For example, if the $pK_a$ of a weak acid is to be determined, such a cell can be represented as

$$\text{Pt; } H_2|\text{HA}(m_1), \text{NaA}(m_2), \text{NaCl}(m_3)|\text{AgCl; Ag} \qquad (6.5)$$

In this cell the stoichiometric molal concentrations of the acid HA, its sodium salt NaA, and sodium chloride are all known and are represented in the notation as $m_1$, $m_2$, and $m_3$, respectively. However, accurate values of $pK_a$ can also be obtained more conveniently by replacing the hydrogen electrode of cell (6.5) with a glass electrode provided that care is exercised in the cell calibration. Furthermore, the inconvenience of preparing a pure solid sample of the sodium salt of the acid can be eliminated with slight loss in accuracy by titrating a solution of the acid with a standard solution of sodium hydroxide.

## General Method of Calculation

A measured value of the e.m.f. of the calibrated cell (6.5) is related to the activities of the electroactants $a_{H^+}$ and $a_{Cl^-}$ by the equation

$$E_{cell} = E°_{cell} - k \log a_{H^+} a_{Cl^-}$$

which, provided the chloride ion concentration $m_3$ is known, can be rearranged to yield

$$\frac{E_{cell} - E°_{cell}}{k} + \log m_3 = -\log a_H + \gamma_{Cl^-} \quad (6.6a)$$

Thus under the experimental conditions all the terms of the left-hand side of this equation are known, and the resulting quantity is defined as the acidity function $p(a_H \gamma_{Cl})$. For the ionization of the weak acid, HA equation (6.3) can be written as

$$K_a = \frac{a_{H^+} \cdot a_{A^-}}{a_{HA}}$$

or

$$a_{H^+} = K_a \frac{m_{HA} \cdot \gamma_{HA}}{m_{A^-} \cdot \gamma_{A^-}}$$

and substitution for the quantity $a_{H^+}$ in equation (6.6a) yields on rearrangement:

$$p(a_H \gamma_{Cl}) = \log \frac{m_{HA}}{m_{A^-}} = -\log K_a \frac{\gamma_{HA} \gamma_{Cl^-}}{\gamma_{A^-}} \quad (6.6b)$$

In dilute solution it is assumed that the activity coefficient of the acid HA, being uncharged, is unity which allows this equation to be written as

$$p(a_H \gamma_{Cl}) + \log \frac{m_{HA}}{m_{A^-}} = pK_a - \log \frac{\gamma_{Cl^-}}{\gamma_{A^-}} \quad (6.7)$$

At ionic strengths of less than 0.01 $m$, for which the Bates–Guggenheim convention is valid, the ratio of the activity coefficients $\gamma_{Cl^-}/\gamma_{A^-}$ can be assumed to be equal to unity, and thus a simplified version of equation (6.7) is

$$p(a_H \gamma_{Cl}) + \log \frac{m_{HA}}{m_{A^-}} = pK_a \quad (6.8)$$

At low ionic strengths, therefore, the measured and calculable terms on the left-hand side of this equation yield the thermodynamic $pK_a$ value of a weak monoprotic acid directly. This fortuitous cancellation effect does not occur in the case of weak monofunctional bases, however, as is evident from the equation analogous to (6.7) derived for the process $BH^+ \rightleftharpoons B + H^+$. That is,

$$p(a_H \gamma_{Cl}) + \log \frac{m_{BH^+}}{m_B} = pK_a - \log \gamma_{BH^+} \gamma_{Cl^-} \qquad (6.9)$$

The logarithm of the product of the activity terms may be evaluated by means of the Bates–Guggenheim convention as

$$-\log \gamma_{BH^+} \cdot \gamma_{Cl^-} = \frac{2 A I^{1/2}}{1 + 4.56 B I^{1/2}}$$

and for ionic strengths less than 0.02 $m$ equation (6.9) can be written as

$$p(a_H \gamma_{Cl}) + \log \frac{m_{BH^+}}{m_B} - \frac{2 A I^{1/2}}{1 + 4.56 B I^{1/2}} = pK_a \qquad (6.10a)$$

in which A and B are the constants for the Debye–Huckel equation at the appropriate temperature given in the Appendix. At ionic strengths greater than 0.02 $m$ or for results of the highest accuracy, the validity of the assumptions made in the derivation of equations (6.8) for acids and (6.10a) for bases should be checked. This is done by defining the calculable terms on the left-hand side of these equations as $pK'$ and to plot these values against the corresponding values of the ionic strength. These graphs can be expressed by an equation of the form

$$pK'_a = pK_a \pm \beta I \qquad (6.10b)$$

in which $\beta$ is a constant. Extrapolation of the straight line to $I = 0$ yields an intercept equal to the value of the thermodynamic $pK_a$.

The ratio $m_{HA}/m_{A^-}$ in equation (6.8) should never be regarded as being equal to the stoichiometric ratio $m_1/m_2$ in which $m_1$ and $m_2$ are the respective molalities of the acid and its sodium salt (see cell 6.5). Hydrolysis of these species is compensated for as follows. Material balance of the species demands that

$$m_1 + m_2 = m_{HA} + m_{A^-} \qquad (6.11$$

and electroneutrality of the solutions containing these species demands that the total cations (+ve) must equal the total anions (−ve). That is,

$$m_2 + m_3 + m_{H^+} = m_{A^-} + m_3 + m_{OH^-}$$

In this equation the terms $m_{H^+}$ and $m_{OH^-}$ arise from the ionization interaction with the solvent, water. Hence

$$m_{A^-} = m_2 + m_{H^+} - m_{OH^-} \tag{6.12}$$

Combining this equation with equation (6.11) and rearranging the results yields

$$m_{HA} = m_1 - m_{H^+} + m_{OH^-} \tag{6.13}$$

Thus in terms of the stoichiometric concentration of acid and its sodium salt taken, equation (6.8) can be written as

$$p(a_H \gamma_{Cl}) + \log \frac{m_1 - m_{H^+} + m_{OH^-}}{m_2 + m_{H^+} - m_{OH^-}} = pK_a \tag{6.14}$$

Similarly, if $m_1$ is the stoichiometric concentration of the hydrochloride of a base and $m_2$ is the stoichiometric concentration of free base in the cell solution then equation (6.10a) becomes

$$p(a_H \gamma_{Cl}) + \log \frac{m_1 - m_{H^+} + m_{OH^-}}{m_2 + m_{H^+} - m_{OH^-}} - \frac{2AI^{1/2}}{1 + 4.56 BI^{1/2}} = pK_a \tag{6.15}$$

If $m_1$ approximates $m_2$ and the $pK_a$ is between 5 and 9, both $m_{H^+}$ and $m_{OH^-}$ become negligible. If the $pK_a$ value is less than 5, $m_{H^+}$ must be calculated, and if it is greater than 9, the hydroxyl ion concentration must be included in the calculations. The calculation procedure is as follows:

1. Calculation of the hydrogen ion concentration, $m_{H^+}$. This term must be calculated from the measured value of $p(a_H \gamma_{Cl})$ which requires the Debye–Huckel equation and thus a knowledge of the ionic strength. The latter term is given by

$$\text{either:} \quad I = m_2 + m_3 + m_{H^+} \text{ for an acid HA} \tag{6.16a}$$

$$\text{or:} \quad I = m_1 + m_3 + m_{OH^-} \text{ for a conjugate acid BH}^+ \tag{6.16b}$$

By definition:

$$p(a_H \gamma_{Cl}) = -\log m_{H^+} - \log \gamma_{H^+} \cdot \gamma_{Cl^-}$$

and using the Bates–Gugggenheim form of the Debye–Huckel equation yields

$$-\log m_{H^+} = p(a_H \gamma_{Cl}) - \frac{2AI^{1/2}}{1 + 4.56 BI^{1/2}} \quad (6.17a)$$

or for acids:

$$-\log m_{H^+} = p(a_H \gamma_{Cl}) - \frac{2A\sqrt{m_2 + m_3 + m_{H^+}}}{1 + 4.56 B\sqrt{m_2 + m_3 + m_{H^+}}} \quad (6.17b)$$

Thus in order to calculate the term $m_{H^+}$ required in the expression for the ionic strength of an acid, the contribution of this term to the ionic strength must be known. Equation (6.17) must be solved, therefore, by an iterative process using as a first approximation in the ionic strength terms that $m_{H^+} = 0$. The steps are outlined as follows:

Step (a). Put $I(1) = m_2 + m_3$ and use 6.17 to find $m_{H^+}(1)$

Step (b). Put $I(2) = m_2 + m_3 + m_{H^+}(1)$ and calculate $m_{H^+}(2)$

Step (c). Put $I(3) = m_2 + m_3 + m_{H^+}(2)$ and calculate $m_{H^+}(3)$ \quad (6.18)

The iteration is continued until successive values of $m_{H^+}$ agree.

2. Calculation of the hydroxyl ion concentration, $m_{OH^+}$. If it is assumed that the magnitude of the ionic strength is within the limit for which the Bates–Guggenheim form of the Debye–Huckel equation is applicable, the calculation of $m_{OH^-}$ is comparatively straightforward. Combining the definition for the ionic product of water, $K_w$ (values in the Appendix) that

$$K_w = a_{H^+} \cdot a_{OH^-}$$

with the definition that

$$p(a_H \gamma_{Cl}) = -\log a_{H^+} - \log \gamma_{Cl^-}$$

yields

$$p(a_H \gamma_{Cl}) = -\log \frac{K_w}{m_{OH^-} \gamma_{OH^-}} - \log \gamma_{Cl^-}$$

$$= pK_w + \log m_{OH^-} + \log \frac{\gamma_{OH^-}}{\gamma_{Cl^-}}$$

On the basis of the foregoing assumption, therefore, the activity coefficient terms cancel, and thus

$$-\log m_{OH^-} = pK_w - p(a_H\gamma_{Cl}) \qquad (6.19)$$

The values of the ionic strengths to be used in the determination of the $pK_a$ of a conjugate acid when $m_{OH^-}$ becomes significant is given by

$$I = m_1 + m_3 + m_{OH^-}$$

These values are required for substitution in the Debye–Huckel term of equation (6.15). A specimen calculation of the $pK_a$ for acetic acid is given in Table 6.3.

The limitation of this method of calculation is imposed by the validity of the procedure used to calculate the term $m_{H^+}$ for use in equation (6.14). It has been assumed in equation (6.17) that it is valid to use the Bates–Guggenheim convention which assigns 4.56Å as the value of $a_i$ to be used in the Debye–Huckel equation. In the calculation of $pK_a$, the magnitude of

Table 6.3. Calculation of the Thermodynamic $pK_a$ Value for Acetic Acid (25 °C)[a]

| $p(a_H\gamma_{Cl})$ | $m_1$ | $m_2$ | $m_3$ | $I(1)$[b] | $m_{H^+}(1)$[b] | $I(2)$[b] | $\dfrac{m_1 - m_{H^+}}{m_2 + m_{H^+}}$ | $pK'_a$[c] |
|---|---|---|---|---|---|---|---|---|
| 4.746 | 0.005034 | 0.004845 | 0.005156 | 0.01000 | 0.00002 | 0.01002 | 1.0306 | 4.759 |
| 4.747 | 0.010007 | 0.009688 | 0.010312 | 0.02000 | 0.00002 | 0.02002 | 1.0349 | 4.762 |
| 4.747 | 0.01510 | 0.01453 | 0.01547 | 0.03000 | 0.00002 | 0.03002 | 1.0364 | 4.763 |
| 4.747 | 0.02013 | 0.01937 | 0.02062 | 0.03999 | 0.00003 | 0.04002 | 1.0361 | 4.762 |
| 4.747 | 0.02517 | 0.02422 | 0.02578 | 0.05000 | 0.00003 | 0.05003 | 1.0367 | 4.763 |
| 4.747 | 0.03020 | 0.02906 | 0.03093 | 0.05999 | 0.00003 | 0.06002 | 1.0371 | 4.763 |
| 4.748 | 0.03523 | 0.03391 | 0.03609 | 0.07000 | 0.00003 | 0.07003 | 1.0371 | 4.764 |
| 4.748 | 0.04027 | 0.03873 | 0.04125 | 0.07998 | 0.00003 | 0.08001 | 1.0382 | 4.764 |
| 4.748 | 0.04530 | 0.04360 | 0.04640 | 0.9000 | 0.00003 | 0.09003 | 1.0376 | 4.764 |
| 4.748 | 0.05034 | 0.04845 | 0.05156 | 0.10001 | 0.00003 | 0.10004 | 1.0378 | 4.764 |

[a] Data obtained with the cell Pt; $H_2$|acetic acid ($m_1$), sodium acetate ($m_2$), NaCl ($m_3$)|AgCl; Ag, and recorded in Reference 14. Values of $pK'$ and $I$ were submitted to a linear regression analysis which yielded as intercept the *thermodynamic* $pK_a = 4.760 \pm 0.002$. The relation between $pK'_a$ and $I$, described by equation (6.10b), is $pK' = 4.760 + 0.044I$.
[b] Values of $I(1)$, $m_{H^+}(1)$, and $I(2)$ were calculated using equations (6.18) followed by equation (6.17b). In this example the values of $m_{H^+}$ are small, and hence one round of calculations sufficed to obtain invariant values of $I$ and $m_{H^+}$. At 25° the appropriate form of the Debye–Huckel equation based on the Bates–Guggenheim convention is $-\log \gamma_\pm = 0.5108 I^{1/2}/(1 + 1.5 I^{1/2})$.
[c] The term $pK'_a$ in this example is defined by equation (6.8).

$m_{H^+}$ and hence the ratio of $m_{HA}/m_{A^-}$ in equation (6.8) depends upon the value of $a_i$ used. The stronger the acid, the greater the correction required for the magnitude of $m_{H^+}$ and, therefore, the more dependent the calculated value of $pK_a$ will be upon this indeterminate quantity $a_i$. This effect was very evident in the calculation of the $pK_a$ for 2,6-dihydroxybenzoic acid at 25° (12):

| $a_i =$ | 1 | 2 | 3 | 4 | 5 | 6 | 7 Å |
|---|---|---|---|---|---|---|---|
| $pK_a =$ | 1.077 | 1.114 | 1.147 | 1.178 | 1.205 | 1.228 | 1.252 |

However, for acids weaker than this the effect becomes less pronounced as can be seen in the calculated $pK_a$ values for formic acid at 25° (13):

| $a_i =$ | 0 | 3 | 4.5 Å |
|---|---|---|---|
| $pK_a =$ | 3.749 | 3.750 | 3.750 |

It is, therefore, the method of calculation that imposes the limit of utility for this potentiometric method rather than the accuracy with which the acidity function $p(a_H\gamma_{Cl})$ can be measured. Furthermore, the effect will not be limited to acids of the type HA, but will also affect the accuracy of the determination of ionization constants for the conjugate acid species $BH^+$. For a given acid, the effect upon the calculated $pK_a$ value will be more pronounced when using dilute solutions in cell (6.5), for example, than it will be when the stoichiometric concentrations of $m_1$ and $m_2$ are larger [see equations (6.12) and (6.13)]. It is difficult, therefore, to make explicit statements as to the value of $pK_a$ below which this effect will become significant, but it is advisable to check the magnitude of this effect when calculating the ionization constants of acidic species stronger than $pK_a$ 3.

## Measurement of $p(a_H\gamma_{Cl})$

The methods of preparation of the hydrogen and silver–silver chloride electrodes used in cell (6.5) for the measurement of the acidity function $p(a_H\gamma_{Cl})$ have been outlined in Chapter 3. The basic design of the all-glass cells has not changed substantially from that used in the 1930s (15), although the use of the interchangeable ground-glass joints has facilitated the assembly of these cells (16). The cell assembly is mounted in a constant temperature bath controlled to 0.01°, and nowadays the cell e.m.f. can be conveniently measured to the nearest 0.01 mV by means of a digital voltmeter. Alternatively, values of $E_{cell}$ can be measured using a potentiometer that is standardized with a calibrated Weston cell maintained at constant temperature. In both cases it is advisable to couple a strip chart recorder either directly to the digital voltmeter, or to the detector of the potentiometer to establish that the recorded value of the cell potential is, in fact, an equilibrium value. These values must be corrected for the barometric pressure and the vapor pressure of water at the given temperature by adding the corrections given in Table 3.1. It is also essential to determine the value of the calibration factor $E_{cell}^{\circ}$ at a given temperature by obtaining the corrected value of $E_{cell}$ when the cell contains 0.01 $m$ hydrochloric acid. Equation (3.13) and the mean activity coefficient data for this solution allow $E_{cell}^{\circ}$ to be calculated. At 25° this should be within the range 0.22230–0.22272 V (17).

It has been known for a number of years that precise measurements were also possible by replacing the cumbersome hydrogen electrode of cell (6.5) with the more convenient glass electrode (18), and that the resulting cells could be used for the accurate measurement of the $pK_a$ values of weak acids (19). The glass electrode used was of such a low resistance that the e.m.f. of the cell could be measured with a potentiometer coupled to a sensitive galvanometer (5600 mm $\mu A^{-1}$). To obtain the necessary low resistance, the bulb of the glass electrode had to be large (30 mm dia.), and this electrode was conditioned by immersion in 0.1 $M$ hydrochloric acid for several weeks (19). Conventional glass electrodes have also been used in conjunction with a silver–silver chloride electrode to form a cell without diffusion, the e.m.f. of which was measured with a potentiometer (0.01 mV) that was equipped with an electrometer as detector (12). A similar arrangement was used in assessing the accuracy of the hydrogen ion selective glass electrode (20). Provided that care was taken in screening the electrode connections to the potentiometer and that the glass electrode was conditioned by soaking for a minimum of two weeks in water, measurements of the acidity function $p(a_H\gamma_{Cl})$ at 25° were reproducible and agreed with the published values for the buffer solutions used. In these instances the time taken for a stable value of e.m.f. to be attained after immersion of the electrodes in a solution was 10–15 min. Once a stable potential was established it remained at the same value for three hours which was the maximum duration that the pen recorder coupled to the output of the electrometer detector was allowed to run.

A cell containing a glass electrode can be standardized to measure the acidity function $p(a_H\gamma_{Cl})$ in an analogous manner to the standardization of a cell to be used for the measurement of $pH$. Quite frequently the necessary basic data pertaining to cell (6.5), from which the acidity functions of various buffer solutions can be calculated, is included with the derived $pH$ buffer solutions in the original publications. For example, the $p(a_H\gamma_{Cl})$ values given in Table 6.4 were selected and were calculated from the basic

**Table 6.4.** $p(a_H\gamma_{Cl})$ Values for Some Solutions From 0–50 °C

| Solution Numbers | $p(a_H\gamma_{Cl})$ Values at $T\,°C$ | | | | | | | | | |
|---|---|---|---|---|---|---|---|---|---|---|
| | 0 | 5 | 10 | 15 | 20 | 25 | 30 | 35 | 40 | 45 |
| $1^a$ | 2.084 | — | 2.085 | — | 2.086 | 2.087 | 2.088 | — | 2.089 | — |
| $2^b$ | 3.941 | 3.916 | 3.897 | 3.880 | 3.868 | 3.856 | 3.849 | 3.841 | 3.837 | 3.834 |
| $3^c$ | 6.972 | 6.939 | 6.913 | 6.889 | 6.870 | 6.857 | 6.846 | 6.836 | 6.831 | 6.827 |
| $4^d$ | 10.411 | 10.339 | 10.276 | 10.218 | 10.164 | 10.115 | 10.071 | 10.031 | 9.995 | 9.960 |

$^a$Solution 1: 0.01 m HCl (23).
$^b$Solution 2: 0.05 m potassium dihydrogen citrate + 0.01 m KCl (24).
$^c$Solution 3: 0.0825 m $KH_2PO_4$, 0.0825 m $Na_2HPO_4$ + 0.01 m NaCl calculated from data of Reference by means of the equation

$$p(a_H\gamma_{Cl}) = \frac{1.00033\,E_{cell} + 0.001128k - E^\circ_{AgCl;Ag}}{k} + \log 0.01$$

where $k$ is the Nernst factor at the appropriate temperature with values of $E^\circ_{AgCl;Ag}$ taken from Reference 23.
$^d$Solution 4: 0.025 m $NaHCO_3$, 0.025 m $Na_2CO_3$ + 0.01 m KCl (24).

e.m.f data originally published. If the e.m.f. data were published before 1 January 1948, it is necessary to convert these values in international volts to absolute volts by multiplying them by the factor 1.00033 (21). In the standardization of the cell

$$\text{Ag; AgCl}|H^+(a_{H^+}), Cl^- \text{ (known molality)}|\text{glass}|pH7\text{ buffer}$$
$$+ Cl^-|\text{AgCl; Ag} \quad (6.20)$$

two $p(a_H\gamma_{Cl})$ standards are required, and are selected so that the values of solutions of unknown $p(a_H\gamma_{Cl})$ will be within the range encompassed by these two standards. The polarity implied by the notation of cell (6.20) is that often observed when the commercially supplied glass electrode contains an internal filling solution of about $pH$ 7. Glass electrodes with an internal filling solution of hydrochloric acid (0.01–0.1 M) are less commonly supplied nowadays, but if one of these types is to be used, then the polarity of the resulting cell will be reversed with respect to cell (6.20). The Nernst expression for cell (6.20) is

$$E_{cell} = \left(E^\circ_G - k\log\frac{1}{a_{H^+}}\right) - \left(E^\circ_{AgCl;Ag} - k\log a_{Cl^-}\right)$$
$$= E^\circ_{cell} + k\log a_{H^+} \cdot a_{Cl^-} \quad (6.21)$$

The positive term in this equation becomes negative if a cell with a reversed polarity is used. The acidity function, deducible from equation (6.21), is therefore defined as

$$p(a_H\gamma_{Cl}) = \frac{E°_{cell} - E_{cell}}{k} + \log m_{Cl^-} \qquad (6.22)$$

If the electrodes are immersed in a solution of known acidity function $p(a_H\gamma_{Cl})_s$ containing a chloride ion molality $m_1$ and the e.m.f. of this cell is measured as $E_{cell1}$, then

$$p(a_H\gamma_{Cl})_s = \frac{E°_{cell} - E_{cell1}}{k} + \log m_1$$

On transferring the same electrodes to a solution of unknown acidity function $p(a_H\gamma_{Cl})_x$ but of known chloride ion concentration $m_2$, the value of $p(a_H\gamma_{Cl})_x$ can be calculated if the e.m.f. of this cell $E_{cell2}$ is measured from

$$p(a_H\gamma_{Cl})_x = p(a_H\gamma_{Cl})_{s1} + \log\frac{m_2}{m_1} - \frac{\Delta E}{k} \qquad (6.23)$$

in which $\Delta E$ is the difference between $E_{cell2}$ and $E_{cell1}$. This equation, of course, presupposes that the slopes of the electrodes are exactly equal to $k$, the Nernst factor at the given temperature. It is necessary, therefore, to check this by measuring the acidity function of a second standard solution $p(a_H\gamma_{Cl})_{s2}$, treating this as if it were an unknown. This procedure then becomes analogous to a $p$H standardization in which the electromotive efficiency, as defined by equation (4.14), is determined. The corresponding equation becomes

$$\beta = \frac{p(a_H\gamma_{Cl})_{s1} - p(a_H\gamma_{Cl})_x}{p(a_H\gamma_{Cl})_{s1} - p(a_H\gamma_{Cl})_{s2}} \qquad (6.24)$$

in which $p(a_H\gamma_{Cl})_x$ is the $p(a_H\gamma_{Cl})$ value calculated from equation (6.23) when the electrodes are immersed in the solution having the published value $p(a_H\gamma_{Cl})_{s2}$. For a theoretical response $p(a_H\gamma_{Cl})_x$ should equal $p(a_H\gamma_{Cl})_{s2}$ and thus the ratio $\beta$ would be unity for this case. If it is found that $\beta < 1$, then $1 - \beta$ is the correction which must be applied to compensate for the actual response of the electrodes. Under this condition the observed value, $p(a_H\gamma_{Cl})_{obs}$ for a solution whose acidity function lies between those of the two standards can be corrected to a true value by applying the expression

$$p(a_H\gamma_{Cl})_{true} = p(a_H\gamma_{Cl})_{obs} + [p(a_H\gamma_{Cl})_{obs} - p(a_H\gamma_{Cl})_{s1}](1 - \beta)$$
$$(6.25)$$

in which $p(a_H\gamma_{Cl})_{s1}$ is the $p$H of the prime standardizing buffer solution. When the internal filling solution of the glass electrode is $p$H 7 buffer, it is expedient to use an equimolal phosphate buffer solution containing chloride having a $p(a_H\gamma_{Cl})$ value of around 7 as this prime standardizing buffer (see Table 6.4).

When the chloride ion concentration is maintained identical in all the standardizing and test solutions, then the term $\log m_2/m_1$ in equation (6.23) disappears and $p(a_H\gamma_{Cl})_x$ becomes defined as

$$p(a_H\gamma_{Cl})_x = p(a_H\gamma_{Cl})_s + \frac{(E_{cell1} - E_{cell2})}{k} \qquad (6.26)$$

This equation is the analogue of equation (4.13) which was used to define $p$H($X$), and $p(a_H\gamma_{Cl})_x$ may be determined similarly to $p$H($x$) by using an ion activity meter as a "$p(a_H\gamma_{Cl})$ meter." Suitable solutions with which to calibrate this meter and to check the response of the electrodes at 25° are found in Reference 22.

0.02 $M$ KH$_2$PO$_4$, 0.02 $M$ Na$_2$HPO$_4$, 0.01 $M$ NaCl     $p(a_H\gamma_{Cl}) = 6.983$ (20)
0.01 $m$ hydrochloric acid                                       $p(a_H\gamma_{Cl}) = 2.087$ (23)
0.04958 $M$ potassium dihydrogen citrate                         $p(a_H\gamma_{Cl}) = 3.856$ (24)
 +0.01 $M$ KCL

The solids used in the preparation of these solutions can be prepared from the recrystallized AR grade reagents dried at 120° for one hour. Potassium dihydrogen citrate can be obtained (Merck) in a condition suitable for the direct preparation of $p$H standard buffer solutions according to DIN 19266. This solid can be used as supplied, but once the bottle has been opened it is best stored over silica gel. Solutions of hydrochloric acid are usually prepared from the constant boiling acid (25), but it has been reported that these solutions can also be made up from the acid supplied in ampoules for the preparation of 1 $M$ hydrochloric acid (26). Although these solutions contained 0.001% w/v mercury (II) chloride, this did not affect the measured e.m.f. values of cell (6.5) with respect to the values obtained when the solutions were prepared from constant boiling acid.

The calibration of the ion activity meter to register $p(a_H\gamma_{Cl})$ for aqueous solutions which must contain 0.01 $M$ sodium or potassium chloride is similar to the calibration for $p$H. The electrodes of cell (6.20) are immersed in the phosphate buffer and, using the asymmetry potential control, the recorded value is adjusted so that the meter indicates 6.98$_3$. The electrodes are then each washed three times with the 0.01 $m$ hydrochloric acid solution, then immersed in this solution, and the meter adjusted to read $2.08_7$ using the slope control. If large adjustments of the asymmetry potential and slope control are necessary, it is advisable to repeat the procedure until reproducible values are obtained. The linearity of the electrodes are then checked using the potassium dihydrogen citrate solution, and the recorded value is usually within 0.003 of the published $p(a_H\gamma_{Cl})$ value of $3.85_6$. If the glass electrode is supplied with an internal reference solution of hydrochloric acid then the meter should first be standardized with 0.01 $m$ hydrochloric acid thereby reversing the order given previously. The time taken to record an invariant value of $p(a_H\gamma_{Cl})$ is usually within 10 min, and the detection of this equilibrium value is made easier by coupling a strip chart recorder to the output of the ion activity meter.

Values of $p(a_H\gamma_{Cl})$ of some solutions over the temperature range 0–50° are given in Table 6.4. The carbonate–bicarbonate solution (Solution 4) is prepared from "primary standard" grade sodium carbonate dried at 250° for 90 min, and from AR grade sodium bicarbonate dried over a mixture of Drierite and molecular sieves for about 2 days at room temperature. The sodium carbonate is stored over calcium chloride and Drierite. It should be noted that although the data for Solution 3 are old, some of the $p(a_H\gamma_{Cl})$ values calculated from the e.m.f. values given in Reference 27 have been verified in the author's laboratory. The factor $0.001128k$ in the footnote equation of Table 6.4 converts the given chloride ion concentration of 0.010026 $m$ to exactly 0.01 $m$, and the constant 1.00033 in this equation converts the observed data in international volts to absolute volts (21). Thus all the data in the table pertain only to chloride ion concentrations of 0.01 $m$, and this concentration has been chosen because either sodium chloride or potassium chloride can be used interchangeably as the source of chloride ion. At concentrations greater than this there is a discrepancy between $p(a_H\gamma_{Cl})$ values obtained in identical concentrations of the two salts as a result of the difference between the mean activity coefficients of these solutes which increases as the concentration increases.

Amine and zwitterionic buffers are not included in Table 6.4 because the acidity function relevant to these is $p(a_H\gamma_{Br})$ rather than $p(a_H\gamma_{Cl})$. It is usual to employ a silver–silver bromide electrode in conjunction with a hydrogen electrode to establish the $p$H of these buffers because the lower solubility of silver bromide, as compared with silver chloride, makes the silver–silver bromide electrode more suitable for use in these solutions (28). However, satisfactory results were obtained for the $pK_a$ value of the ammonium ion using a cell containing a glass electrode and a silver–silver chloride electrode (29), and the latter electrode has also been used in a variety of amine-type buffers (26).

## The Potentiometric Titration Method

The ionization constants of monoprotic acids of the types HA and $BH^+$ are conveniently determined by a titration in which the acidity function $p(a_H\gamma_{Cl})$ is measured by means of an ion activity meter at various stages of neutralization during the titration. Although the values obtained by this method are not so accurate as those obtained by the method given earlier and exemplified by the value for acetic acid in Table 6.3, they are capable of results within 0.01 $pK_a$ unit of the value obtained using every refinement of technique and apparatus.

When a solution of weak acid of known initial molar concentration $C_t$, prepared in 0.01 $M$ sodium chloride, is being titrated with a standard solution of sodium hydroxide, also prepared in 0.01 $M$ sodium chloride, the cell in which the species HA and NaA are coexistent can be represented as

$$\text{Ag; AgCl}|\text{Cl}^-(0.01\ M), \text{HA}(c_1), \text{NaA}(c_2)|\text{glass electrode} \quad (6.27)$$

Under these circumstances it is convenient to write equation (6.14) in terms of molar concentrations as

$$pK_a = p(a_H\gamma_{Cl}) + \log\frac{c_1 - c_{H^+} + c_{OH^-}}{c_2 + c_{H^+} - c_{OH^-}} \quad (6.28)$$

At a given point in the buffered region of the titration curve, therefore, the concentration of acid remaining after the addition of a known concentration of sodium hydroxide is given by

$$c_1 = C_t - c_{NaOH}$$

and, similarly, the concentration of sodium salt formed is

$$c_2 = c_{NaOH}$$

Substitution for these terms in equation (6.28) yields the general equation to the titration curve

$$pK_a = p(a_H\gamma_{Cl}) + \log\frac{C_t - c_{NaOH} - c_{H^+} + c_{OH^-}}{c_{NaOH} + c_{H^+} - c_{OH^-}} \quad (6.29)$$

in which $C_t$ is the total concentration of acid originally present corrected for the dilution of the titrant solution as a result of the addition of titrant.

The cell corresponding to the notation (6.27) when a base B prepared in 0.01 $M$ sodium chloride is titrated with standard perchloric acid solution, also prepared in 0.01 $M$ sodium chloride, is

$$\text{Ag; AgCl}|\text{Cl}^-(0.01\ M), \text{B}(c_2), \text{BHClO}_4(c_1)|\text{glass electrode} \quad (6.30)$$

In terms of the total concentration of base initial taken $C_t$ and the concentration of titrant added $c_{\text{HClO}_4}$, the concentrations $c_1$ and $c_2$ are given by

$$c_2 = C_t - c_{\text{HClO}_4} \quad \text{and} \quad c_1 = c_{\text{HClO}_4}$$

At suitable points on the titration curve, therefore, the $pK_a$ values of the species $BH^+$ may be calculated by incorporation of these terms in equation (6.15) which then becomes

$$pK_a = p(a_H\gamma_{\text{Cl}}) + \log\frac{c_{\text{HClO}_4} - c_{H^+} + c_{OH^-}}{C_t - c_{\text{HClO}_4} + c_{H^+} - c_{OH^-}} - \frac{2AI^{1/2}}{1 + 4.56\,BI^{1/2}}$$

(6.31)

Once again, $C_t$ must be corrected for volume changes as a result of the addition of titrant. Since the unit of concentration in this equation is mole per liter, rather than the corresponding unit of mole per kilogram used in equation (6.15), the values of the constants A and B must be selected from the unit volume section of Table A.2. It should also be noted that the measured quantity written as $p(a_H\gamma_{\text{Cl}})$ in these equations should more correctly be written as $p(a_H y_{\text{Cl}})$, where $y_{\text{Cl}}$ is the activity coefficient of chloride ion on the molar scale. At the concentrations used in these titrations, however, the two terms may be used interchangeably with negligible error.

The methods of calculation of $c_{H^+}$ and $c_{OH^-}$ in equations (6.29) and (6.31) are as described earlier, except when calculating $c_{H^+}$ the appropriate unit volume values of the constants A and B must be used in equation (6.17a). As a rough guide, it is advisable to test whether it is necessary to correct for $c_{H^+}$ when the measured value of $p(a_H\gamma_{\text{Cl}})$ is less than 5, and for $c_{OH^-}$ when $p(a_H\gamma_{\text{Cl}}) > 9$.

The results obtained in the determination of the $pK_a$ values for benzoic acid and for 2-amino-2-hydroxymethyl-1,3-propanediol (Tris) are given in Tables 6.5 and 6.6, respectively. The $p(a_H\gamma_{\text{Cl}})$ values at 25° were measured by an ion activity meter supplied with asymmetry and slope controls which was calibrated to read $p(a_H\gamma_{\text{Cl}})$ directly. To this end, the calibrating solutions were contained in stoppered 150 mL tall form beakers that were suspended in a thermostatically controlled water bath maintained at 25°C. When transferring the glass and the silver–silver chloride electrodes between

Table 6.5. Determination of the $pK_a$ for Benzoic Acid (25 °C)[a]

| 1 M NaOH mL | $p(a_H \gamma_{Cl})$ | $C_t$ | $c_2 = c_{NaOH}$ | $I$[c] | $c_{H^+}$[c] | $c_1 = C_t - c_{NaOH}$ | $\log\dfrac{(c_1 - c_{H^+})}{(c_2 + c_{H^+})}$ | $pK_a$[b] |
|---|---|---|---|---|---|---|---|---|
| 0.10 | 3.440 | 0.00999 | 0.00100 | 0.01145 | 0.00045 | 0.00899 | 0.770 | 4.210 |
| 0.20 | 3.673 | 0.00998 | 0.00200 | 0.01265 | 0.00027 | 0.00798 | 0.531 | 4.204 |
| 0.25 | 3.778 | 0.00998 | 0.00249 | 0.01270 | 0.00021 | 0.00749 | 0.431 | 4.209 |
| 0.30 | 3.871 | 0.00997 | 0.00299 | 0.01316 | 0.00017 | 0.00698 | 0.333 | 4.204 |
| 0.35 | 3.961 | 0.00997 | 0.00349 | 0.01363 | 0.00014 | 0.00648 | 0.242 | 4.203 |
| 0.40 | 4.048 | 0.00996 | 0.00398 | 0.01409 | 0.00011 | 0.00598 | 0.157 | 4.205 |
| 0.45 | 4.132 | 0.00996 | 0.00448 | 0.01457 | 0.00009 | 0.00548 | 0.072 | 4.204 |
| 0.50 | 4.219 | 0.00995 | 0.00498 | 0.01506 | 0.00008 | 0.00497 | −0.015 | 4.204 |
| 0.55 | 4.303 | 0.00995 | 0.00547 | 0.01553 | 0.00006 | 0.00448 | −0.097 | 4.206 |
| 0.60 | 4.391 | 0.00994 | 0.00596 | 0.01601 | 0.00005 | 0.00398 | −0.184 | 4.207 |
| 0.65 | 4.484 | 0.00994 | 0.00646 | 0.01650 | 0.00004 | 0.00348 | −0.276 | 4.208 |
| 0.70 | 4.584 | 0.00993 | 0.00695 | 0.01698 | 0.00003 | 0.00298 | −0.374 | 4.210 |
| 0.75 | 4.694 | 0.00993 | 0.00744 | 0.01747 | 0.00003 | 0.00249 | −0.482 | 4.212 |
| 0.80 | 4.823 | 0.00992 | 0.00794 | 0.01796 | 0.00002 | 0.00198 | −0.609 | 4.214 |

[a] Benzoic acid (0.1221 g, dried over conc. $H_2SO_4$ *in vacuo* overnight) was dissolved in 0.01 M sodium chloride (100 mL) at 40°. This solution at 25° was titrated with 1 M sodium hydroxide prepared in 0.01 M sodium chloride.
[b] Mean $pK_a$ = 4.207 ± 0.007 (standard deviation 0.003).
[c] Values obtained from the second round of calculations [see equations (6.18)].

Table 6.6. Determination of the $pK_a$ for 2-Amino-2-hydroxymethyl-1,3-propanediol (Tris) at 25 °C[a]

| 1 M HClO (mL) | $p(a_H\gamma_{Cl})$ | $C_t$ | $c_1$ ($c_1 = c_{HClO_4}$) | $c_2$ ($c_2 = C_t - c_1$) | $pK'_a$ | $I$ | $\dfrac{1.023\sqrt{I}}{1+1.5\sqrt{I}}$ | $pK_a$[b] |
|---|---|---|---|---|---|---|---|---|
| 0.10 | 9.111 | 0.00999 | 0.00100 | 0.00899 | 8.157 | 0.01100 | 0.093 | 8.064 |
| 0.20 | 8.770 | 0.00998 | 0.00200 | 0.00798 | 8.169 | 0.01200 | 0.096 | 8.073 |
| 0.25 | 8.650 | 0.00998 | 0.00249 | 0.00749 | 8.171 | 0.01249 | 0.098 | 8.073 |
| 0.30 | 8.541 | 0.00997 | 0.00299 | 0.00698 | 8.173 | 0.01299 | 0.100 | 8.073 |
| 0.35 | 8.444 | 0.00997 | 0.00349 | 0.00648 | 8.175 | 0.01349 | 0.101 | 8.074 |
| 0.40 | 8.352 | 0.00996 | 0.00398 | 0.00598 | 8.175 | 0.01398 | 0.103 | 8.072 |
| 0.45 | 8.265 | 0.00996 | 0.00448 | 0.00548 | 8.178 | 0.01448 | 0.104 | 8.074 |
| 0.50 | 8.181 | 0.00995 | 0.00498 | 0.00497 | 8.182 | 0.01498 | 0.106 | 8.076 |
| 0.55 | 8.093 | 0.00995 | 0.00547 | 0.00448 | 8.180 | 0.01547 | 0.107 | 8.073 |
| 0.60 | 8.000 | 0.00994 | 0.00596 | 0.00398 | 8.175 | 0.01596 | 0.109 | 8.066 |
| 0.65 | 7.908 | 0.00994 | 0.00646 | 0.00348 | 8.177 | 0.01646 | 0.110 | 8.067 |
| 0.70 | 7.810 | 0.00993 | 0.00695 | 0.00298 | 8.178 | 0.01695 | 0.111 | 8.067 |
| 0.75 | 7.700 | 0.00993 | 0.00744 | 0.00249 | 8.175 | 0.01744 | 0.113 | 8.062 |
| 0.80 | 7.575 | 0.00992 | 0.00794 | 0.00198 | 8.178 | 0.01794 | 0.114 | 8.064 |

[a] Tris (0.1211 g dried *in vacuo* at 40° for 16 h) dissolved in 0.01 M sodium chloride (100 mL). This solution was titrated with 1 M perchloric acid prepared in 0.01 M sodium chloride.
[b] Mean $pK_a = 8.070 \pm 0.008$ (standard deviation 0.005).
[c] $pK'_a = p(a_H\gamma_{Cl}) + \log c_2/c_1$ [see equation (6.31)]. Corrections for $c_{H^+}$ or $c_{OH^-}$ are not required in this example.

these solutions each electrode was washed three times with the solution into which it was going to be immersed, but before they were immersed in the titrand solution they were each washed with 0.01 $M$ sodium chloride solution maintained at 25° in the bath. No attempt was made to mop or dry the electrodes with tissue paper or filter paper during these transfers. The titrand solution was contained in a double-walled beaker that was mounted on the platform of a magnetic stirrer, and the temperature of this solution was kept at 25° by pumping water from the bath through the jacket of the beaker. The titrant, prepared in 0.01 $M$ sodium chloride solution, was added from a micrometer syringe fitted with a Luer needle to which was attached a length of polythene canula tubing. During an addition of titrant, the tip of the canula tubing was immersed in the titrand solution which was stirred by a magnetized bar during the addition and for about 30 sec after. However, when a value of $p(a_H\gamma_{Cl})$ was being recorded, the tip of the tubing was lifted out of the solution and the magnetic stirring was switched off. A chart recorder connected to the output of the ion activity meter was used to detect the stable reading of $p(a_H\gamma_{Cl})$ which usually occurs within 2–3 min after the addition. The meter, as supplied, had a scale that could be expanded so that each scale division corresponded to 0.01 $p$H unit, or, in this case, 0.01 unit of $p(a_H\gamma_{Cl})$. Thus although the values of $p(a_H\gamma_{Cl})$ given in Tables 6.5 and 6.6 are estimated to 0.001 unit, the uncertainty in these readings is likely to be 0.003 unit, but the trace on the recorder chart helped in the assignation of the guessed third decimal place. In view of this, the $pK_a$ value of 4.207 ± 0.007 for benzoic acid is in good agreement with the values 4.2050 ± 0.0015 (19), 4.204 ± 0.005 (30) and 4.208 ± 0.002 (20). Similarly, the value of 8.070 ± 0.008 obtained for Tris agrees, within the experimental error, with the values 8.075 (4) and 8.069 (5) obtained with more elaborate equipment.

The advantage in measuring $p(a_H\gamma_{Cl})$ rather than $p$H (see Section 6.3) is that the acidity function is a thermodynamically valid quantity whereas $p$H, being related to the activity of a single ion, is not. Furthermore, when $p$H is being measured during a titration it is assumed that the diffusion potential across the junction of the reference half-cell and the titrand solution is the same as it was in the standardizing buffer solutions, and that this liquid junction potential remains constant throughout the titration.

## 6.3 THE $p$H TITRATION METHOD

The determination of ionization constants by potentiometric titration in which $p$H is measured is the method most frequently used in practice. Despite the objections concerning the quasi-thermodynamic nature of $p$H and the uncertainty of the constancy of the liquid junction potential, the $pK_a$ values obtained by this method are often in good agreement with those derived from $p(a_H\gamma_{Cl})$ measurements using cell (6.5), for example, with every refinement of technique. It must be borne in mind, however, that although $p$H meters designed to measure $p$H with a precision of $\pm 0.001\, p$H unit are often used in these determinations, the relation $p\text{H}(X) = -\log a_{H^+}$ holds with an accuracy that can only be stated to be better than $\pm 0.02$ unit for buffer solutions of similar ionic strengths. In potentiometric titrations the ionic strength may differ significantly from that of the standardizing buffers and, at best, the error in the derived $pK_a$ value is likely to deviate from the true value by 0.02 $pK_a$ unit. It is for this reason that in assessing the accuracy of $pK_a$ values, the $p$H titration method is regarded as an approximate method rather than as an accurate method.

## Method of Calculation

When a solution of a weak monoprotic acid of known initial molar concentration $C_t$ is being titrated with a standard solution of sodium hydroxide, the cell used for the measurement of $p$H during the titration can be represented as

$$\text{Hg; Hg}_2\text{Cl}_2 | 3.5\, M\, \text{KCl} || \text{HA}(c_1), \text{NaA}(c_2) | \text{glass electrode} \quad (6.32)$$

In this cell the measured $p$H is related to $pK_a$ by equation (6.4) which can be written as the analogue of equation (6.28). Thus

$$pK_a = p\text{H} + \log\frac{(c_1 - c_{H^+} + c_{OH^-})\, y_{HA}}{(c_2 + c_{H^+} - c_{OH^-})\, y_{A^-}} \quad (6.33)$$

where $y_{HA}$ and $y_{A^-}$ are the molar activity coefficients of the nonionized portion of the weak acid and the ionized portion, respectively. Assuming, as before, that the activity coefficient of the species HA, being nonionized, is unity leads to

$$pK_a = p\text{H} + \log\frac{(c_1 - c_{H^+} + c_{OH^-})}{(c_2 + c_{H^+} - c_{OH^-})} - \log y_{A^-} \quad (6.34)$$

Substituting the relationships that $c_1 = C_t - c_{NaOH}$ and $c_2 = c_{NaOH}$, together with the Debye–Huckel equation for $-\log y_{A^-}$ written in the Bates–Guggenheim convention, yields the analogue of equation (6.29) as

$$pK_a = p\text{H} + \log\frac{C_t - c_{NaOH} - c_{H^+} + c_{OH^-}}{c_{NaOH} + c_{H^+} - c_{OH^-}} + \frac{AI^{1/2}}{1 + 4.56\, BI^{1/2}} \quad (6.35)$$

The equation relating $pK_a$ to the measured $pH$ for the titration of a base of initial concentration $C_t$ with a standard solution of hydrochloric acid can be similarly derived, the analogue of equation (6.31) being

$$pK_a = pH + \log\frac{c_{HCl} - c_{H^+} + c_{OH^-}}{C_t - c_{HCl} + c_{H^+} - c_{OH^-}} - \frac{AI^{1/2}}{1 + 4.56\,BI^{1/2}} \quad (6.36)$$

The values of A and B to be used in equation (6.35) and (6.36) should be in terms of unit volume (see Table A.2), and it should be anticipated that it may be necessary to use the term $c_{H^+}$ if the $pH$ is below 5 and $c_{OH^-}$ if the $pH$ is above 9. The ionic strength during the titration of a monoprotic acid is given by an equation similar to (6.16a) as

$$I = c_{Na^+} + c_{H^+} \quad (6.37a)$$

in which $c_{Na^+}$ is equal to the concentration of sodium hydroxide added. It can also be deduced that the ionic strength during the titration of a base is

$$I = c_{Cl^-} + c_{OH^-} \quad (6.37b)$$

where $c_{Cl^-}$ is equal to the concentration of hydrochloric acid titrant added.

When the $pK_a$ of a weak acid is being calculated and it is found that the contribution of the term $c_{Na^+}$ to the ionic strength is much greater than the contribution of $c_{H^+}$, it is sufficient to assume that $c_{H^+} = a_{H^+}$. Thus equation (6.37a) becomes

$$I = c_{Na^+} + a_{H^+} \quad (6.38)$$

in which $a_{H^+}$ is calculated directly from the measured $pH$. However, when it is not valid to make this assumption, equation (6.38) becomes the first step in a series of successive approximations. The initial value of the ionic strength is then used in the equation:

$$-\log c_{H^+} = pH - \frac{AI^{1/2}}{1 + 4.56\,BI^{1/2}} \quad (6.39)$$

in order to calculate $c_{H^+}$ and hence a new value of the ionic strength. The iterative process is continued until successive values of both $I$ and $c_{H^+}$ agree.

The procedure for calculating the ionic strength during the titration of a base is very similar to the method described. The first approximation is that

$$I = c_{Cl^-} + \frac{K_w}{a_{H^+}}$$

in which $K_w$ is the ionic product of water at the given temperature (see Table A.4) and $a_{H^+}$ is calculated from the measured $p$H. Having calculated the initial values for the ionic strength, $c_{OH^-}$ can be calculated from

$$-\log c_{OH^-} = pK_w - pH + \frac{AI^{1/2}}{1 + 4.56\,BI^{1/2}} \qquad (6.40)$$

These values are then substituted in equation (6.37b) and the sequence by means of equation (6.40) repeated until successive values of $I$ and $c_{OH^-}$ become invariant. When it is necessary to calculate values for $c_{H^+}$ or $c_{OH^-}$ for inclusion in equation (6.35) for acids or equation (6.36) for conjugate acids, these values are calculated by equations (6.39) or (6.40) as applicable.

The experimental technique used in the $p$H titration method is very similar to that employed in the potentiometric titrations in which the acidity function $p(a_H \gamma_{Cl})$ is measured. There is, of course, no necessity to prepare the titrand and titrant solutions in 0.01 $M$ sodium chloride when $p$H is being measured, and with this exception the titration procedure is identical. The standardization of the electrodes are as described in Chapter 4 and the washing of them during transfer between the standardizing buffer solutions is as described previously. If, as is usual, the titrand solution is prepared in water, then the electrodes should be washed with water that has been maintained at the temperature of the standardizing buffers before immersing them in the titrand solution.

The results for benzoic acid given in Table 6.7 were obtained using a $p$H meter calibrated in 0.02 unit divisions. The $pK_a$ value of 4.19 ± 0.02 must be regarded as a pseudo-thermodynamic $pK_a$ since it is derived from $p$H measurements, but nevertheless it is in fair agreement with the value 4.205 obtained at 20° (3) with cell (6.5).

### Practical or Concentration-Dependent $pK_a$ Values

Thus far in this chapter the methods of calculations described yield the thermodynamic $pK_a$ value or a close approximation to it. Because these calculations have incorporated activity coefficients, the values are really extrapolated values that pertain only to a condition of zero ionic strength at a given temperature. As such they are useful in correlating the effects of substituents upon the $pK_a$ value of the unsubstituted acid or conjugate acid in terms of the free energy changes involved. Since the basis for the relationships is the equation

$$\Delta G° = 2.306\,RT\,pK_a$$

Table 6.7. Results of a $pH$ Titration of Benzoic Acid (20 °C)[a]

| 0.1 M KOH (mL) | $pH$ | $c_2$ | $a_{H^+}$ | $I$ | $c_{H^+}$ | $c_1$ | $\log\dfrac{c_1 - c_{H^+}}{c_2 + c_{H^+}}$ | $pK_a'$[b] | $f(I)$[c] | $pK_a$[d] |
|---|---|---|---|---|---|---|---|---|---|---|
| 0.5 | 3.38 | 0.00104 | 0.00042 | 0.00146 | 0.00043 | 0.00938 | 0.78 | 4.16 | 0.02 | 4.18 |
| 1.0 | 3.63 | 0.00206 | 0.00023 | 0.00229 | 0.00025 | 0.00825 | 0.54 | 4.17 | 0.02 | 4.19 |
| 1.5 | 3.83 | 0.00306 | 0.00015 | 0.00321 | 0.00016 | 0.00714 | 0.34 | 4.17 | 0.03 | 4.20 |
| 2.0 | 3.99 | 0.00404 | 0.00010 | 0.00414 | 0.00011 | 0.00596 | 0.15 | 4.14 | 0.03 | 4.17 |
| 2.5 | 4.16 | 0.00500 | 0.00007 | 0.00507 | 0.00007 | 0.00500 | −0.01 | 4.15 | 0.03 | 4.18 |
| 3.0 | 4.34 | 0.00594 | 0.00005 | 0.00599 | 0.00005 | 0.00396 | −0.19 | 4.15 | 0.04 | 4.19 |
| 3.5 | 4.53 | 0.00686 | 0.00003 | 0.00689 | 0.00003 | 0.00294 | −0.37 | 4.16 | 0.04 | 4.20 |
| 4.0 | 4.76 | 0.00777 | 0.00002 | 0.00779 | 0.00002 | 0.00194 | −0.61 | 4.15 | 0.04 | 4.19 |
| 4.5 | 5.12 | 0.00865 | — | 0.00865 | — | 0.00097 | −0.95 | 4.17 | 0.04 | 4.21 |

[a]Benzoic acid (0.1221 g, dried over conc. $H_2SO_4$ *in vacuo* overnight) was dissolved in 47.5 mL water at 40 °C.
[b]$pK_a' = pH + \log(c_1 - c_{H^+})/(c_2 + c_{H^+})$.
[c]The following function of the ionic strength is defined as $f(I)$ at 20°: $f(I) = 0.507 I^{1/2}/(1 + 1.5 I^{1/2})$.
[d]$pK_a = pK' + f(I)$ [see equations (6.34) and (6.35)]. Mean $pK_a = 4.19 \pm 0.02$ (standard deviation 0.01).

it follows that for these to be valid, the $pK_a$ values upon which the correlations are based must also have thermodynamic validity. It is only under these conditions that the effects of substituents can be discussed in terms of factors which affect free energies. Such factors form the basis for the Hammett and Taft equations which are the most widely used of the methods for the prediction of $pK_a$ values (31). However, a thermodynamic value of the $pK_a$ has little relevance when confronted with a problem, for example, that requires the calculation of the fractions of acidic and conjugate base components at a given $p$H in a medium of relatively high ionic strength. For this purpose a $pK_a$ value that has been determined in the medium of interest is required. These values are sometimes referred to as *practical* or *concentration-dependent* $pK_a$ values, and they may be derived wholly in term of concentration or they may be *mixed constants*. The values given in the columns marked "$pK'_a$" in Table 6.6 and 6.7 are examples of mixed constants since they have been calculated from equations in which $p(a_H\gamma_{Cl})$ or $p$H are activity-derived terms, whereas the ratios of ionized/nonionized species have been calculated in terms of concentrations. Mixed constants can, therefore, be defined in terms of the stoichiometric concentration $c_1$ of the acid (HA or BH$^+$) and the stoichiometric concentration $c_2$ of the base (A$^-$ or B) by

$$pK_a^M = p\text{H} + \log \frac{c_1 - c_{\text{H}^+} + c_{\text{OH}^-}}{c_2 + c_{\text{H}^+} - c_{\text{OH}^-}} \quad (6.41)$$

When applied specifically to the titrations of acids, the log term of this equation becomes identical to the log term of equation (6.35), and, similarly, it becomes equal to the log term of equation (6.36) when bases are titrated. Unless the ionic strength of the solution is such that $c_{\text{H}^+}$ or $c_{\text{OH}^-}$ can be calculated from the measured $p$H value as described, equation (6.41) is valid only when $c_{\text{H}^+}$ or $c_{\text{OH}^-}$ are small in comparison with the stoichiometric concentration $c_1$ and $c_2$. Provided that the ionic strength of a solution is the same as that for which the $pK_a^M$ value was determined, the ratio ionized species/nonionized species may be calculated for any measured value of $p$H using the relationship corresponding to equation (6.4b). For acids this relationship is

$$\frac{c_{\text{A}^-}}{c_{\text{HA}}} = 10^{p\text{H} - pK_a^M}$$

and for bases it is

$$\frac{c_{\text{BH}^+}}{c_{\text{B}}} = 10^{pK_a^M - p\text{H}}$$

In Chapter 4 methods for using cells such as (6.32) to measure hydrogen ion concentration were discussed. These methods enable another form of a practical ionization constant to be defined wholly in terms of concentrations by measurement of $pc_H$ in media of constant ionic strength. The equation corresponding to (6.41) can be written as

$$pK_a^c = pc_H + \log\frac{c_1 - c_{H^+} + c_{OH^-}}{c_2 + c_{H^+} - c_{OH^-}} \qquad (6.42)$$

in which $K_a^c$ is referred to as *the concentration ionization constant*. There is no difficulty in solving this equation when it is necessary to allow for the magnitude of $c_{H^+}$ because this quantity can be calculated directly from the measured $pc_H$. Difficulty arises, however, when it is necessary to correct for $c_{OH^-}$, since the calculation of this term requires a value for the ionic product of water in the ionic medium of interest. At ionic strengths of 0.1 and less, the magnitude of this concentration dependent value of the ionic product of water $K_w^c$ can be deduced from the thermodynamic value at the given temperature $K_w$ using the definition that

$$K_w = c_{H^+} c_{OH^-} y_{OH^-} y_{H^+}$$

$$= K_w^c y_{OH^-} y_{H^+}$$

Hence

$$pK_w^c = pK_w - \frac{2A I^{1/2}}{1 + 4.56 \, B I^{1/2}} \qquad (6.43)$$

in which the values of $pK_w$ to use in this equation are given in the Appendix. Provided that $c_1$ and $c_2$ exceed 0.01 $M$ and the values of $pc_H$ are less than 11.8, the error in the value of $pK_a^c$ will not exceed 0.02 as can be seen from the calculated values of $pK_w^c$ given in parenthesis in Table 6.9. At ionic strengths greater than 0.1 the value of $pK_w^c$ must be determined as outlined in Section 6.4.

## 6.4 THE DETERMINATION OF THE IONIC PRODUCT OF WATER

The ionization of pure water, following the definition for ionization processes in water given by equations (6.1), can be represented by the autoionization

$$H_2O + H_2O \rightleftharpoons H_3O^+ + OH^-$$

for which the equilibrium constant expression is

$$K = \frac{a_{H_3O^+} \cdot a_{OH^-}}{a_{H_2O}^2}$$

As with other ionization processes, the standard state of pure water is defined as unit activity which leads to the expression

$$K_a = a_{H_3O^+} \cdot a_{OH^-}$$

as being the equivalent of equation (6.3) in dilute solution. This constant is called the *ionic product of water* and is usually written in the form

$$K_w = a_{H^+} \cdot a_{OH^-} \qquad (6.44)$$

with the understanding that the symbol $H^+$ represents the hydrated proton. As with other ionization constants, there is a thermodynamic value which in this case is defined by equation (6.44), and practical or concentration-dependent values which are defined in terms of concentration (molal, $m$ or molar, $c$) by

$$K_w^c = m_{H^+} \cdot m_{OH^-}$$

This section aims to outline how these quantities can be determined.

### The Thermodynamic Value of $pK_w$

The method depends upon the measurement of the acidity function $p(a_H \gamma_{Cl})$ in a series of solutions that all contain the same concentration of a strong alkali, but each solution contains a different, and known, concentration of chloride ion prepared from a salt that contains the same cation as the strong alkali. A cell typically used for this determination (32) can be represented as

$$Pt; H_2 | KOH(0.01\ m), KCl(m_3) | AgCl; Ag \qquad (6.45)$$

in which the electrodes had been calibrated previously in order to obtain a value for $E_{cell}^\circ$. Provided that the concentration $m_3$ is known, the acidity function can be defined by equation (6.6) as

$$p(a_H \gamma_{Cl}) = \frac{E_{cell} - E_{cell}^\circ}{k} + \log m_3$$

Recognizing that

$$p(a_H\gamma_{Cl}) = -\log(m_{H^+}\gamma_{H^+}\gamma_{Cl^-}) \tag{6.46}$$

and that $K_w$ can be written as

$$K_w = \frac{\gamma_{H^+}\gamma_{OH^-}}{a_{H_2O}} \cdot m_{H^+}m_{OH^-}$$

leads, on elimination of $m_{H^+}$ between these two equations, to

$$p(a_H\gamma_{Cl}) = -\log\left[\frac{K_w a_{H_2O}\gamma_{H^+}\gamma_{Cl^-}}{\gamma_{OH^-}\gamma_{H^+}m_{OH^-}}\right]$$

Combining the measured and known terms on the left-hand side yields

$$p(a_H\gamma_{Cl}) + \log\frac{1}{m_{OH^-}} = pK_w + \log\frac{\gamma_{OH^-}\gamma_{H^+}}{a_{H_2O}} - \log\gamma_{H^+}\gamma_{Cl^-}$$

Thus for cell (6.45), a plot of $[p(a_H\gamma_{Cl}) + 2.000]$ versus ionic strength will yield the value of $pK_w$ at $I = 0$ since $a_{H_2O}$ is unity under this condition, and by reason of the definition of activity coefficients, the other terms on the right of the equation vanish.

### Practical or Concentration-Dependent $pK_w^c$ Values

Combining the definition that $K_w^c = m_{H^+}m_{OH^-}$ with the definition of $p(a_H\gamma_{Cl})$ given by equation (6.46) yields

$$p(a_H\gamma_{Cl}) = -\log\frac{K_w^c}{m_{OH^-}}\gamma_{H^+}\gamma_{Cl^-}$$

from whence

$$pK_w^c = p(a_H\gamma_{Cl}) + \log\frac{1}{m_{OH^-}} + 2\log\gamma_\pm \tag{6.47}$$

The term $\gamma_\pm$ in this equation is the mean activity coefficient of hydrochloric acid which must be determined in the medium of interest by independent measurements. Thus the e.m.f. values of two cells must be measured separately. These cells can be represented as

$$\text{Pt; H}_2|\text{MOH}(0.01\ m), \text{MA}(m_2); \text{MCl}(m_3)|\text{AgCl}; \text{Ag} \tag{6.48}$$

and

$$Pt; H_2|HCl(0.01\ m), MA(m_2); MCl(m_3)|AgCl; Ag \qquad (6.49)$$

It is assumed in the notations that the medium of interest, MA is a strong electrolyte in which $A^-$ is not chloride. To determine $pK_w^c$ for a solution of MA of known concentration, $m_2$, the $p(a_H\gamma_{Cl})$ values of each cell are measured for at least three values of $m_3$ ranging from 0.01–0.05 $m$. These values are then plotted against $m_3$ to yield $p(a_H\gamma_{Cl})°$ values at zero chloride ion concentration. The mean activity coefficient for 0.01 $m$ hydrochloric acid in a solution of MA containing no chloride is obtained from the relevant $p(a_H\gamma_{Cl})°$ value using the equation

$$\log \gamma_{H^+}\gamma_{Cl^-} = 2\log \gamma_\pm = -\log m_{H^+} - p(a_H\gamma_{Cl})° \qquad (6.50)$$

However, under the conditions specified in the two cells, and assuming the mean activity coefficient of hydrochloric acid is the same in both cells, this equation can be combined with equation (6.47) to yield

$$pK_w^c = p(a_H\gamma_{Cl})°_{KOH} + 4.000 - p(a_H\gamma_{Cl})°_{HCl}$$

Table 6.8. Values of $p(a_H\gamma_{Cl})$ in 0.01 $M$ Hydrochloric Acid and in 0.01 $M$ Potassium Hydroxide Solutions Containing Potassium Chloride[a]

| Solution KOH ($m$) | KCl ($m$) | \multicolumn{8}{c}{$p(a_H\gamma_{Cl})$ Values} |||||||||
|---|---|---|---|---|---|---|---|---|---|
| | | 0° | 10° | 20° | 25° | 30° | 40° | 50° | 60° |
| .01 | 0.01 | 12.947 | 12.539 | 12.171 | 12.000 | 11.838 | 11.540 | 11.268 | 11.023 |
| – | 0.02 | 12.948 | 12.539 | 12.172 | 12.001 | 11.840 | 11.542 | 11.270 | 11.025 |
| – | 0.03 | 12.949 | 12.540 | 12.172 | 12.002 | 11.840 | 11.543 | 11.271 | 11.025 |
| – | 0.04 | 12.949 | 12.541 | 12.173 | 12.003 | 11.841 | 11.543 | 11.271 | 11.025 |
| – | 0.05 | 12.949 | 12.541 | 12.174 | 12.004 | 11.842 | 11.545 | 11.272 | 11.025 |
| HCl | KCl | | | | | | | | |
| .01 | — | 2.089 | 2.088 | 2.088 | 2.089 | 2.089 | 2.093 | 2.093 | 2.093 |
| – | 0.01 | 2.119 | 2.118 | 2.119 | 2.120 | 2.120 | 2.125 | 2.127 | 2.129 |
| – | 0.02 | 2.139 | 2.139 | 2.140 | 2.141 | 2.142 | 2.147 | 2.148 | 2.150 |
| – | 0.03 | 2.154 | 2.154 | 2.155 | 2.156 | 2.158 | 2.165 | 2.166 | 2.169 |
| – | 0.05 | 2.176 | 2.177 | 2.178 | 2.180 | 2.181 | 2.186 | 2.189 | 2.192 |

[a] Calculated from the data given in Reference 32.

Table 6.9. Values of $pK_w^c$ for Solutions of Sodium Chloride and Potassium Chloride at Various Temperatures[a]

| I | 0° NaCl | 0° KCl | 10° NaCl | 10° KCl | 20° NaCl | 20° KCl | 25° NaCl | 25° KCl | 30° NaCl | 30° KCl | 40° NaCl | 40° KCl | 50° NaCl | 50° KCl | 60° NaCl | 60° KCl |
|---|---|---|---|---|---|---|---|---|---|---|---|---|---|---|---|---|
| 0.02 | 14.824 | 14.827 | 14.418 | 14.418 | 14.048 | 14.047 | 13.876 | 13.877 | 13.713 | 13.714 | 13.411 | 13.412 | 13.138 | 13.137 | 12.893 | 12.889 |
|  | (14.828)[b] |  | (14.418) |  | (14.049) |  | (13.877) |  | (13.713) |  | (13.413) |  | (13.138) |  | (12.890) |  |
| 0.03 | 14.802 | 14.086 | 14.400 | 14.396 | 14.028 | 14.028 | 13.856 | 13.857 | 13.693 | 13.693 | 13.391 | 13.391 | 13.116 | 13.115 | 12.868 | 12.869 |
|  | (14.808) |  | (14.398) |  | (14.028) |  | (13.856) |  | (13.691) |  | (13.391) |  | (13.116) |  | (12.868) |  |
| 0.06 | 14.776 | 14.770 | 14.358 | 14.360 | 13.989 | 13.991 | 13.818 | 13.821 | 13.655 | 13.657 | 13.350 | 13.355 | 13.074 | 13.078 | 12.825 | 12.830 |
|  | (14.766) |  | (14.356) |  | (13.986) |  | (13.813) |  | (13.648) |  | (13.348) |  | (13.071) |  | (12.823) |  |
| 0.11 | 14.728 | 14.736 | 14.322 | 14.328 | 13.951 | 13.959 | 13.779 | 13.787 | 13.615 | 13.623 | 13.310 | 13.321 | 13.032 | 13.042 | 12.783 | 12.792 |
|  | (14.724) |  | (14.314) |  | (13.943) |  | (13.770) |  | (13.605) |  | (13.303) |  | (13.026) |  | (12.777) |  |
| 0.21 | 14.691 | 14.702 | 14.282 | 14.295 | 13.913 | 13.924 | 13.741 | 13.755 | 13.576 | 13.591 | 13.271 | 13.287 | 12.993 | 13.009 | 12.743 | 12.757 |
| 0.51 | 14.656 | 14.671 | 14.251 | 14.265 | 13.880 | 13.895 | 13.707 | 13.724 | 13.543 | 13.561 | 13.236 | 13.254 | 12.958 | 12.973 | 12.705 | 12.719 |
| 1.01 | 14.675 | 14.721 | 14.269 | 14.314 | 13.897 | 13.942 | 13.724 | 13.768 | 13.558 | 13.602 | 13.248 | 13.294 | 12.967 | 13.009 | 12.712 | 12.749 |
| 2.01 | 14.782 | 14.855 | 14.379 | 14.445 | 14.006 | 14.067 | 13.822 | 13.892 | 13.666 | 13.722 | 13.351 | 13.407 | 13.061 | 13.113 | 12.795 | 12.846 |
| 3.01 | 14.930 | 14.995 | 14.528 | 14.580 | 14.155 | 14.198 | 13.979 | 14.020 | 13.810 | 13.849 | 13.488 | 13.526 | 13.194 | 13.228 | 12.915 | 12.954 |

[a] Calculated from the data in References 32 and 33 in terms of molal concentrations.
[b] Values in parenthesis are calculated from $pK_w^c = pK_w - 2AI^{1/2}/(1 + 4.56\, BI^{1/2})$ [see equation (6.43)].

in which $p(a_H\gamma_{Cl})^\circ_{KOH}$ is the extrapolated value obtained from cell (6.48) and $p(a_H\gamma_{Cl})^\circ_{HCl}$ is the corresponding value for cell (6.49).

Some $p(a_H\gamma_{Cl})$ values for solutions of 0.01 $m$ hydrochloric acid and 0.01 $m$ potassium hydroxide with which to calibrate cells containing a glass electrode in place of the hydrogen electrode in cells (6.48) and (6.49) are given in Table 6.8. It can be anticipated that the glass electrode will be in error in solutions of 0.01 $m$ potassium hydroxide, but at a given temperature the variation of $p(a_H\gamma_{Cl})$ with ionic strength for these solutions is not great. For example, on changing the ionic strength from 0.02 to 3 $m$ with potassium chloride, the change in $p(a_H\gamma_{Cl})$ for 0.01 $m$ potassium hydroxide was only 0.15 unit at 25°. Over such a small range, therefore, the error in the electrode is likely to remain constant and will be compensated for in the calibration. It is important, however, that carbon dioxide be excluded from all solutions when taking these measurements.

Values of $pK_w^c$ in solutions of potassium chloride and sodium chloride are given in Table 6.9, and values in potassium nitrate and in seawater are contained in References 34 and 35, respectively. It must be noted, however, that hydrogen ion in seawater is not measured on the same concentration scale, but rather on one in which

$$[H^+]_{seawater} = [H^+] + [HSO_4^-]$$

## 6.5 POLYFUNCTIONAL ACIDS AND BASES

These are compounds that have more than one ionizable group per molecule and range from simple diprotic acids such as 4-hydroxybenzoic acid [$pK_1$ = 4.67; $pK_2$ = 9.37 (7)] through compounds such as EDTA with six ionizable groups [$pK_1 \sim 0.3, pK_2 \sim 1, pK_3 \sim 2, pK_4 \sim 2.7, pK_5 \sim 6$, and $pK_6 \sim 10$ (7)] to polypeptides and proteins with many ionizable groups. The basic techniques already described in this chapter can be used to investigate the ionization properties for these compounds and the aim here is to outline some of the quantitative aspects by which these properties are expressed and calculated from potentiometric data.

### Bifunctional Compounds with Widely Separated Ionization Steps

When a compound has two ionizable groups that are separated by at least three units of $pK_a$, then the $pK_a$ of one group may be determined without interference from the simultaneous ionization of the other. It is convenient for the purposes of the determination and the subsequent calculations to divide these bifunctional compounds into categories according to the nature of the ionizable groups as follows:

1. Diprotic acids (dibasic acids).
2. Diprotic conjugate acids (diacidic bases).
3. Ampholytes

### Diprotic acids

The method of determination and calculation of the $pK_a$ of the stronger acid group ($pK_1$) is exactly as described in Sections 6.2 and 6.3. Although the techniques used to determine the ionization constant exponent of the weaker acidic group $pK_2$ do not differ from those that are used to determine $pK_1$, the methods of calculation must be modified to take account that the ionization in this case involves two charged species, a monoanion being in equilibrium with a dianion:

$$HA^- \rightleftharpoons A^{2-} + H^+$$

If, for example, an accurate thermodynamic value of the ionization constant for this process is required, the cell

$$Pt; H_2|MHA(m_1), M_2A(m_2); MCl(m_3)|AgCl; Ag \qquad (6.51)$$

would be assembled in which M represents either the potassium or sodium salt of each species. The method of determination is very similar to the example given in Table 6.3 in that the acidity function values $p(a_H\gamma_{Cl})$ are measured for a series of carefully prepared solutions in which $m_1$, $m_2$, and $m_3$ are known. In this case, however, the equation equivalent to 6.6a is

$$p(a_H\gamma_{Cl}) + \log\frac{m_{HA^-}}{m_{A^{2-}}} = -\log K_2 \cdot \frac{\gamma_{HA^-}\gamma_{Cl^-}}{\gamma_{A^{2-}}} \qquad (6.52)$$

Bearing in mind that

$$-\log \gamma_{A^{2-}} = \frac{2^2 A I^{1/2}}{1 + 4.56 B I^{1/2}}$$

yields

$$p(a_H\gamma_{Cl}) + \log\frac{m_{HA^-}}{m_{A^{2-}}} = pK_2 - \frac{2A I^{1/2}}{1 + 4.56 B I^{1/2}} \qquad (6.53)$$

which assumes that the Debye–Huckel equation written as the Bates–Guggenheim convention can be applied equally for all the activity coefficient terms in equation (6.52). At low values of ionic strength this assumption appears valid, but for accurate results it is better to define a term $pK'_2$ as

$$pK'_2 = p(a_H\gamma_{Cl}) + \log\frac{m_{HA^-}}{m_{A^{2-}}} + \frac{2AI^{1/2}}{1 + 4.56\,BI^{1/2}} \quad (6.54)$$

These values of $pK'_2$ which are close approximations to the thermodynamic value of $pK_2$ are then plotted against $I$ and extrapolated to zero ionic strength to yield the true thermodynamic value.

If the measured values of $p(a_H\gamma_{Cl})$ are outside the range 5–9 then, by the same arguments put forth in the derivation of equation (6.14), it can be shown that equation (6.54) must be written as

$$pK'_2 = p(a_H\gamma_{Cl}) + \log\frac{(m_1 - m_{H^+} + m_{OH^-})}{(m_2 + m_{H^+} - m_{OH^-})} + \frac{2AI^{1/2}}{1 + 4.56\,BI^{1/2}}$$

$$(6.55)$$

Under these conditions the complete expression for the ionic strength is given by

$$I = m_1 + 3m_2 + m_3 + 2m_{H^+} - 2m_{OH^-} \quad (6.56a)$$

in which $m_{H^+}$ is obtained if $p(a_H\gamma_{Cl}) < 5$ by a series of successive approximations similar to those described earlier and $m_{OH^-}$ is calculated from equation (6.19) when $p(a_H\gamma_{Cl}) > 9$.

For the routine determination of the $pK$ values of diprotic acids it is usual to employ the potentiometric titration technique, the values of $pK_1$ and $pK_2$ being obtained from a single titration with two equivalents of standard carbonate-free alkali. As with monoprotic acids, the potentiometrically determined quantity may be either $p(a_H\gamma_{Cl})$ or $pH$ which allow the values of $pK_2$ to be calculated from suitable additions of titrant covering the range 1.2–1.8 equivalents of alkali. If $p(a_H\gamma_{Cl})$ is measured after each addition of standard alkali covering the range 0.2–1.8 mol equivalent, then $pK_1$ can be calculated from the first equivalent by equation (6.29), and for the second equivalent (i.e., 1.2–1.8 equiv.) a close approximation to the thermodynamic value of $pK_2$ will be given by

$$pK_2 = p(a_H\gamma_{Cl}) + \log\left(\frac{2C_t - c_{NaOH} - c_{H^+} + c_{OH^-}}{c_{NaOH} - C_t + c_{H^+} - c_{OH^-}}\right) + \frac{2AI^{1/2}}{1 + 4.56\,BI^{1/2}}$$

$$(6.56b)$$

In this equation $C_t$ is the stoichiometric concentration of the acid $H_2A$ originally taken allowing for volume changes as a result of the addition of titrant, and $c_{NaOH}$ is the total concentration of standard alkali added, *including* that used in the neutralization of the stronger group. The values of ionic strength for use in equation (6.56b) are given for each of the second equivalent additions of titrant by

$$I = 2c_{NaOH} - C_t + 2c_{H^+} - c_{OH^-} + c_{NaCl} \qquad (6.57)$$

where $c_{NaCl}$ is the concentration of sodium or potassium chloride present in the titrand and titrant solutions (see Table 6.5). The equations corresponding to (6.56b) and (6.57) if $pH$ is the measured quantity are

$$pK_2 = pH + \log\left(\frac{2C_t - c_{NaOH} - c_{H^+} + c_{OH^-}}{c_{NaOH} - C_t + c_{H^+} - c_{OH^-}}\right) + \frac{3AI^{1/2}}{1 + 4.56\,BI^{1/2}} \qquad (6.58)$$

and

$$I = 2c_{NaOH} - C_t + 2c_{H^+} - c_{OH^-} \qquad (6.59)$$

The terms in these equations are as previously defined.

### *Diprotic Conjugate Acids (Diacidic Bases)*

The ionization of these bifunctional compounds are described by the equilibria

$$BH^{2+} \stackrel{K_1}{\rightleftharpoons} BH^+ + H^+$$

$$BH^+ \stackrel{K_2}{\rightleftharpoons} B + H^+$$

which is analogous to the system used for acids. The ionization constant exponent of the stronger base group is, therefore, given by $pK_2$ which is the first group to be titrated if, as is usual, the free base is dissolved in water and titrated with a standard solution of strong acid. The quantity pertinent to this discussion is, therefore, $pK_1$, the value of $pK_2$ being determined as described in Sections 6.2 and 6.3.

A suitable cell with which to measure the acidity function for solutions containing the dication $BH_2^{2+}$, the monocation $BH^+$, and chloride can be represented as

$$Pt; H_2 | BH_2X_2(m_1), BHX(m_2), NaCl(m_3) | AgCl; Ag$$

in which $X$ represents a suitable anion (e.g., $ClO_4^-$, $NO_3^-$, $Cl^-$). The general equation relating $p(a_H \gamma_{Cl})$ and the concentrations $m_1$ and $m_2$ to $pK_1$ can be written similarly to equation (6.55) as

$$pK_1' = p(a_H \gamma_{Cl}) + \log\left(\frac{m_1 - m_{H^+} + m_{OH^-}}{m_2 + m_{H^+} - m_{OH^-}}\right) - \frac{4 A I^{1/2}}{1 + 4.56 \, B I^{1/2}} \quad (6.60)$$

the ionic strength in this case being given by

$$I = 3m_1 + m_2 + 2m_{OH^-} - m_{H^+} + m_3$$

If the values of $pK_1'$ are plotted against the corresponding values of ionic strength and the resulting line extrapolated to $I = 0$, the intercept yields the thermodynamic value of $pK_1$.

The technique for the $p(a_H \gamma_{Cl})$ titration of a free diacidic base is similar to that employed in obtaining the results for Table 6.6. The equation to obtain an estimate of the thermodynamic $pK_1$ value from the second equivalent of standard perchloric acid is

$$pK_1 = p(a_H \gamma_{Cl}) + \log\left(\frac{c_{HClO_4} - C_t + c_{OH^-} - c_{H^+}}{2C_t - c_{HClO_4} - c_{OH^-} + c_{H^+}}\right) - \frac{4 A I^{1/2}}{1 + 4.56 \, B I^{1/2}} \quad (6.61)$$

In this equation the ionic strength can be calculated from

$$I = 2c_{HClO_4} - C_t + 2c_{OH^-} - c_{H^+} + c_{NaCl} \quad (6.62)$$

The term $c_{HClO_4}$ is the total concentration of perchloric acid added *including* that used in the neutralization of the stronger basic group, and volume corrections should be applied to this term and also to the term $C_t$. The equations to be used when a $pH$ titration is performed with standard hydrochloric acid titrant are

$$pK_1 = pH + \log\left(\frac{c_{HCl} - C_t + c_{OH^-} - c_{H^+}}{2C_t - c_{HCl} - c_{OH^-} + c_{H^+}}\right) - \frac{3 A I^{1/2}}{1 + 4.56 \, B I^{1/2}}$$

and

$$I = 2c_{HCl} - C_t + 2c_{OH^-} - c_{H^+}$$

which correspond to equations (6.61) and (6.62), respectively.

### Ampholytes

For the purpose of this discussion these are compounds like 3-aminophenol [$pK_1 = 4.31(-NH_2)$ (6) $pK_2 = 9.86$ ($-OH$) (7)] that contain an acidic and a basic group. Since these two ionizations give rise only to singly charged ions, the $pK$ values of the groups may be determined separately using one of the procedures described for monofunctional compounds in Sections 6.2 and 6.3.

### Overlapping $pK_a$ Values

The ionization processes of a bifunctional compound are said to overlap when the two $pK_a$ values are separated by less than three $pK_a$ units. If such a compound is titrated, the ionization of the second group commences before the neutralization of the first group is complete. The equation used to resolve the ionization constants for these two processes is often referred to as the Speakman equation (36) which can be written in the form:

$$a_{H^+}\frac{(1-\bar{h})}{2-\bar{h}}K_1^M + K_1^M K_2^M = \frac{a_{H^+}^2 \bar{h}}{2-\bar{h}} \qquad (6.63a)$$

The term $\bar{h}$ in this equation is defined as

$$\bar{h} = \frac{c_{NaOH} + c_{H^+} - c_{OH^-}}{C_t} \qquad (6.63b)$$

and is the average number of protons removed from the weak acid molecule by the addition of base. If, as is usual, the values of $a_{H^+}$ are derived from $pH$ measurements and these are substituted directly in equation (6.63), then the constants derived from the least squares treatment of the data will yield the values of the mixed constants $K_1^M$ and $K_2^M$. This is because $\bar{h}$ has been derived in terms of concentrations whereas the activity of the hydrogen ion is assumed to be the measured quantity. A computer program which resolves the two constants as the corresponding thermodynamic values $K_1^T$ and $K_2^T$ has been published for diprotic acids, diprotic conjugate acids, and ampholytes (37). Inasmuch as $pH$ is a pseudo-thermodynamic quantity, the values of $pK_1^T$ and $pK_2^T$ obtained by this method must also be regarded as pseudo-thermodynamic, but they are often found to be in good agreement with values determined under more rigorous conditions.

The Speakman equation has also been applied to the determination of the thermodynamic ionization constants of diprotic acids when the quantity

$p(a_H\gamma_{Cl})$ has been measured accurately in cells equipped with hydrogen and silver–silver chloride electrodes (38). If a term $P$ is defined as

$$P = a_H\gamma_{Cl}$$

then equation (6.63) can be written

$$\frac{P}{\gamma_{Cl^-}}\left(\frac{1-\bar{h}}{2-\bar{h}}\right)K_1^M + K_1^M K_2^M = \frac{P^2\bar{h}}{\gamma_{Cl^-}^2(2-\bar{h})} \quad (6.64)$$

For a diprotic acid the thermodynamic constants $K_1^T$ and $K_2^T$ are related to the corresponding mixed constants by

$$K_1^T = K_1^M \gamma_{HA^-} \quad \text{and} \quad K_2^T = \frac{K_2^M \gamma_{A^{2-}}}{\gamma_{HA^-}}$$

and thus equation (6.64) becomes

$$\frac{P(1-\bar{h})}{2-\bar{h}}\frac{K_1^T}{\gamma_{Cl^-}\gamma_{HA^-}} + \frac{K_1^T K_2^T}{\gamma_{A^{2-}}} = \frac{P^2\bar{h}1}{(2-\bar{h})\gamma_{Cl^-}^2} \quad (6.65)$$

At ionic strengths of less than $0.05\ m$ it can be assumed that the activity coefficients are independent of the nature of the ions, and thus the product $\gamma_{Cl^-}\gamma_{HA^-}$ can be equated to $\gamma_{Cl^-}^2$, which allows equation (6.65) to be formulated as

$$XK_1^T + \frac{K_1^T \cdot K_2^T \gamma_{Cl^-}^2}{\gamma_{A^{2-}}} = Y \quad (6.66)$$

A method of solving an equation of this type for conditions under which it is possible to neglect the contributions of $m_{H^+}$ or $m_{OH^-}$ in the calculation of $\bar{h}$, is described in Reference 38. Equation (6.66) also forms the basis for the calculation of $K_1^T$ and $K_2^T$ from titrations in which $p(a_H\gamma_{Cl})$ is the measured quantity. Provided that the $p(a_H\gamma_{Cl})$ value after the addition of 0.1 equivalent of alkali is found to be between 3 and 4, and the initial concentration of acid taken exceeds 0.005 $M$, then the values of $c_{H^+}$ to use in the calculation of $\bar{h}$ by equation (6.63a) may be obtained from

$$-\log c_{H^+} \cong -\log a_{H^+} = p(a_H\gamma_{Cl}) - \frac{AI^{1/2}}{1 + 4.56\ BI^{1/2}}$$

in which the ionic strength is approximated as

$$I \cong c_{NaCl} + c_{NaOH}$$

These approximations make it feasible to use a hand calculator to perform a least squares solution of equation (6.66) to obtain a good estimate of $K_1$ as the slope. The intercept $Z$ is defined as

$$Z = K_1^T K_2^T \frac{\gamma_{Cl^-}^2}{\gamma_{A^{2-}}}$$

which, when combined with the value of $K_1^T$, makes it possible to calculate $c_{HA^-}$ and $c_{A^{2-}}$ and hence the ionic strength for each addition of alkali. Thus it can be shown that

$$c_{HA^-} = \frac{K_1^T \cdot P \cdot C_t}{(P^2 + K_1^T P + Z)}$$

$$c_{A^{2-}} = \frac{Z \cdot C_t}{(P^2 + K_1^T P + Z)} \quad (6.67)$$

and

$$I = 0.5(c_{NaOH} + c_{HA^-} + 4c_{A^{2-}} + a_{H^+}) \quad (6.68)$$

Defining a function of the ionic strength as

$$FS = \frac{I^{1/2}}{1 + 4.56\, BI^{1/2}}$$

and assuming that the Bates–Guggenheim convention is valid at the ionic strengths involved, leads to the formulation of equation (6.66) as

$$\frac{K_1^T X}{10^{2A \cdot FS}} + K_1^T K_2^T = \frac{Y}{10^{2A \cdot FS}} \quad (6.69)$$

At 25° this equation becomes

$$\frac{K_1^T X}{10^{1.023\, FS}} + K_1^T K_2^T = \frac{Y}{10^{1.023\, FS}}$$

Application of the least squares procedure to values of $Y/10^{1.023\, FS}$ and $X/10^{1.023\, FS}$ allows a better estimate of $K_2^T$ and usually confirms the value of $K_1^T$ obtained from the initial least squares treatment. The $pK_a$ values obtained for butanedioic acid (succinic acid) by this method were 4.21 and

5.65 which are in agreement with the accepted values of 4.209 and 5.638 at 25° (39). The same titration results, when substituted in a computer program that will resolve up to six overlapping $pK_a$ values, yielded $pK_1 = 4.198 \pm 0.001$ and $pK_2 = 5.640 \pm 0.003$ (40).

## Polyelectrolytes

The general form of the Speakman equation is retained when the number of overlapping $pK_a$ values exceeds two, as can be seen from the equation derived for a tetraprotic acid with four overlapping $pK_a$ values:

$$K_1^M a_{H^+}^3 \frac{(1-\bar{h})}{(4-\bar{h})} + K_1^M K_2^M a_{H^+}^2 \frac{(2-\bar{h})}{(4-\bar{h})}$$

$$+ K_1^M K_2^M K_3^M a_{H^+} \frac{(3-\bar{h})}{(4-\bar{h})} + K_1^M K_2^M K_3^M K_4^M = \frac{a_{H^+}^4 \cdot \bar{h}}{(4-\bar{h})}$$

Equations of this type can also be solved by the method of least squares for which computer subroutines are available. In the determination of the ionization constants of polyprotic acids with more than three ionizable groups, the inherently more accurate $p(a_H \gamma_{Cl})$ titration method offers no advantage over the usual $p$H titration method because the error introduced by the method of calculation exceeds the likely experimental error during the measurement of $p$H. For example, the thermodynamic $pK_a$ values calculated for benzenehexacarboxylic acid (mellitic acid) were $pK_1$ approximately 0.8, $pK_2$ 2.28, $pK_3$ 3.52, $pK_4$ 5.15, $pK_5$ 6.52, $pK_6$ 7.71 (41). The error in calculating the ionic strength from the calculated ionic compositions of the solution, by equations similar in type to equations (6.67) and (6.68), becomes progressively greater as the number of charges on the anionic species increase. When the penta-anion is the predominant species the error is multiplied fifteenfold and during the formation of the hexa-anion the error is multiplied by a factor of 21. These errors are, of course, reflected in the values of the anionic activity coefficients. These must also be calculated on the assumption that the Debye–Huckel equation, which also contains a $z^2$ term, can still be validly applied to these multicharged species even if precautions are taken to ensure that the ionic strength does not exceed 0.1 $M$. Thus the errors in converting the mixed constants to thermodynamic values by equations similar in type to (6.69) become progressively greater as the charges on the anionic species increase, and although a test of convergence can be incorporated into the iterative process of the computer program, the method of calculation can, at best, yield only an estimate of the thermodynamic constants. These constants are, in themselves, of little significance because they are "macroscopic constants" that cannot be related to the ionizing property of a particular group without further experimental work. By contrast, concentration-dependent constants determined under a given set of conditions can enable the ionic composition to be calculated at any $p$H measured under those conditions. Such information is essential in the calculation of stability constants of metal complexes and in the formulation of buffer solutions. However, in order to calculate these concentration-dependent constants for compounds such as EDTA or benzenehexacarboxylic acid that start to undergo ionization at low $p$H, it is essential to measure $pc_H$ during the titration rather than $p$H because of the need to know $c_{H^+}$ in the calculation of $\bar{h}$ (see equation 6.63b). It is necessary, therefore, to carry out these titrations in media of very high ionic strength so that multicharged anions (e.g., $A^{6-}$; $I = 21c_{A^{6-}}$) make a negligible contribution to the ionic strength.

Each value of $p\mathrm{K}_1$, $p\mathrm{K}_2$... determined by the preceding methods is, in fact, a composite of the ionization constants for at least two other ionization processes that cannot be detected during the determination. When, for example, a diprotic acid represented as $HX—YH$ is titrated with one equivalent of alkali, the first proton may be removed to yield two possible monoanions $^-X—YH$ or $HX—Y^-$. On addition of a further equivalent of alkali each of these monoanions form the same dianion $^-X—Y^-$ so that the overall ionization of the diprotic acid must be represented as

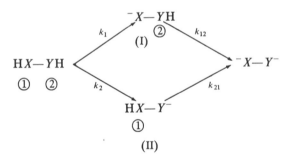

If the protons of the diprotic acid be arbitrarily labeled ① and ②, then one pathway of its ionization involves loss of proton ① to form anion (I) for which the *microscopic ionization constant* is $k_1$. Anion (I) then loses proton ② for which the microscopic constant is $k_{12}$, the subscripts indicating the order by which the protons have been removed from the diprotic acid. The other pathway by means of anion (II) involves loss of proton ② for which the constant is $k_2$, followed then by loss of proton ① for which the constant is $k_{21}$. The experimentally determined macroscopic constants $\mathrm{K}_1$ and $\mathrm{K}_2$ are written in terms of the concentration of an undefined monoanion $c_{HA^-}$ which is really the sum of the concentrations of the two species $^-X-YH$ and $HX-Y^-$. Thus the macroscopic constants are related to the microscopic constants by

$$K_1 = k_1 + k_2 \quad \text{and} \quad K_2 = \frac{k_{12} \cdot k_{21}}{k_{12} + k_{21}}$$

The four microscopic constants are not independent because both pathways yield the same dianion $^-X-Y^-$ hence

$$k_1 k_{12} = k_2 k_{21}$$

The values of these microscopic constants cannot be calculated from $K_1$ and $K_2$ unless one of the values can be assigned from an independent measurement. In the case of dicarboxylic acids the value of $k_1$ or $k_2$ ($k_1 = k_2$) may be assigned by assuming that its value will be the same as the ionization constant of the corresponding monomethyl or monoethyl ester of the acid. A similar assumption is often made for the zwitterionic aminocarboxylic acids, except in these cases $k_1$ is not equal to $k_2$ and the value of the ionization constant of the ester is assigned to the process involving the nonzwitterionic form

$$\overset{+}{N}H_3-R-COOH \overset{k_2}{\rightleftharpoons} NH_2-R-COOH + H^+$$

Much of the work upon these types of zwitterionic equilibria can be traced through Reference 42. The assignation of microscopic constants for compounds more complex than these becomes rapidly more difficult as the number of acidic groups increase. The maximum number that has been determined is 12 for the tyrosine molecule that contains one amino, one carboxy, and one phenolic group (43). A compound such as benzenehexacarboxylic acid has a total of $6 \times 2^{6-1} = 192$ microscopic constants of which $2^6 - 1 = 63$ are independent. An additional 57 relationships are necessary before these values can be determined from the 6 experimentally determined ionization constants. These figures have been quoted to emphasize that it is useless to attempt an analysis of the ionization behavior of a polymeric acid such as polymethacrylic acid even if it were possible to determine the macroscopic $pK_a$ values accurately. Yet it is possible using the potentiometric titration technique to obtain much information about how these acids and other polyelectrolytes behave in aqueous solution.

Potentiometric titrations of polymeric acids (polyacids), polyampholytes, and polybases in which $pH$ is the measured quantity are performed on solutions of the polyelectrolyte prepared in dilute solutions of sodium or potassium chloride. Before the titration, the titrand concentration is often determined by some other technique. For example, in the case of proteins the determination of nitrogen by the Pregl micro-Kjeldahl method (44) is recommended (45) and usually the protein content of the titrand solution is within the range 0.5–1%. It is possible to deduce from the titrations of proteins much information about the number and types of ionizable groups such as $\alpha$-COOH, side chain —COOH, $\alpha$-$\overset{+}{\text{N}}\text{H}_3$, side chain —$\overset{+}{\text{N}}\text{H}_3$, —OH and —SH groups (46–48). Detailed descriptions of the experimental procedures used are available (49). Thermodynamic characterization of biopolymeric phase transitions in the $p$H range where polyelectrolytes and ionizable polypeptides undergo alteration in the macromolecular conformation is also possible from a study of $p$H-titration curves obtained at various temperatures (50).

A basic assumption in these titrations is that a single type of ionizable group, many of which may be present in the polymer, can be expressed in terms of a single microscopic ionization constant. For example, in a polymeric carboxylic acid that contains $n$ identical and independent acid groups if, for a single group in a large number of identical molecules, $\alpha$ is the chance that the group be found in the deprotonated state, then $1 - \alpha$ is the chance it is found in the protonated state. Thus the microscopic constant for that group can be expressed as

$$k = a_{\text{H}^+} \cdot \frac{\alpha}{(1-\alpha)}$$

This equation can be expressed in terms of $\bar{h}$ (see equation 6.63b) since the number of protons removed from all $n$ groups (i.e., $\bar{h}$) will equal $n\alpha$. Similarly, the number of protonated groups will be $n - \bar{h}$ or $n(1 - \alpha)$. Therefore,

$$pk = p\text{H} - \log\frac{\bar{h}}{(n-\bar{h})} = p\text{H} - \log\frac{\alpha}{(1-\alpha)} \qquad (6.70)$$

If the basic assumption is valid, then at constant ionic strength $pk$ should be independent of the degree of neutralization of the acid as is the case for the $pK_a$ value for a simple acid like benzoic acid. In fact, a linear relation between $pk$ and $\alpha$ or $\bar{h}$ is observed for many polyacids indicating that as the negative charges are progressively increased during the titration the loss of protons becomes progressively more difficult. Despite the fact that these observations show that the $n$ groups are neither identical nor independent, much information has been deduced from graphs of $pk$ versus $\alpha$ or $\bar{h}$, particularly insofar as the elucidation of conformational changes of polyelectrolytes are concerned (51, 52). Some of the more recent work in this important application of potentiometry is summarized in Table 6.10.

Table 6.10. Properties of Some Polyelectrolytes That Have Been Investigated by Potentiometric Titrations

| Polyelectrolyte | Property Investigated | Ref. |
|---|---|---|
| | A. Polyacids | |
| Poly(acrylic acid) [9003-01-4] | The effects of Ni(II) and Cu(II) complexing on the titration curves were studied. Different $pK_a$ values were obtained at various dilutions of the polymer which indicated that the metal ion formed complexes with two —COOH groups at moderate concentrations and with more than two at low concentrations. | (53) |
| | Variation of apparent $pK$ with degree of dissociation $\alpha$ was represented by a second degree polynomial in $\alpha$. | (54) |
| | The concentration dependence of the titration curves were explained in term of the radius of the rodlike polyelectrolyte. | (55) |
| | Methods described for extending potentiometric investigations to low degrees of ionization. | (56) |
| | Conformational changes studied in aqueous urea and xylose solutions. | (57) |
| Stereoregular | Isotactic poly(acrylic acid) [25584-52-5] with helical configuration shown to have a different titration curve from the syndiotactic acid which had a zigzag conformation. | (58) |
| | Isotactic acid retained its configuration after ionization, but syndiotactic poly(acrylic acid) [25568-87-0] was extended by ionization. | (59) |
| Chloroacetic acids–poly(vinyl alcohol) | The apparent $pK$ values of these carboxyalkylated poly(vinyl alcohol) derivatives differed little in acid strength and ranged from 4.0–4.4 in $H_2O$ and 2.95–3.45 in 2 $M$ KCl for the trichloro, dichloro, and monochloro derivatives. | (60) |

(*continued*)

**Table 6.10.** *(Continued)*

| Polyelectrolyte | Property Investigated | Ref. |
|---|---|---|
| Fulvic acids | Equivalence point calculations and acidic functional groups. | (61) |
| | Two acids, extracted from sludge–soil mixtures were shown to contain functional groups ranging in acidity from very strong acid groups, probably $SO_3H$ ($pH < 2$), through numerous —COOH groups to $N$-containing groups and —OH and —SH groups; some ionized at $pH > 10$. | (62) |
| Poly(guluronic acid) [*36562-70-6*] | Titrations with KOH and $Ca(OH)_2$ compared. | (63) |
| Humic acids | Titration curves of acids from four Indian soils. Acids were purified with HCl + HF, dialysis, and ion exchange. | (64) |
| | Titration revealed that at $pH$ 7 one —COOH group had been titrated, whereas at $pH$ 9.5 two phenolic groups and three—COOH groups had been titrated. Metal complexes were also studied. | (65) See also (66, 67) |
| | Combined titration with ISE responding to Cu(II), Pb(II), and Cd(II) used to determine the stability constants for these metals. | (68) |
| Poly(maleic acid) [*26099-09-2*] | Acidity of the primary —COOH groups increased through the counterion series $Me_4N^+$, $Li^+$, $K^+$, whereas that of secondary —COOH groups increase with decrease in crystallographic radius of the counterion. | (69) |
| | Titration curves indicated that the acid in the absence of salts behaved as a poly(tetracarboxylic acid) due to interaction between two adjacent groups and their next door neighbors. No interaction was observed in the presence of salts due to their shielding effects and acid behaved as a poly(dicarboxylic acid). | (70) |

Table 6.10. (Continued)

| Polyelectrolyte | Property Investigated | Ref. |
|---|---|---|
| Poly(mannuronic acid) [29894-36-8] | Titrations with KOH and $Ca(OH)_2$ compared. | (63) |
| Poly(methacrylic acid) [25087-26-7] | Relationship between titration curves and conformational changes in isotactic poly(methacrylic acid) [25068-55-7] and syndiotactic poly(methacrylic acid) [25750-36-1] discussed. Technique enables calculation of electrostatic free energy. | (71) |
| | Evidence for the existence of thermodynamically irreversible conformation changes in isotactic poly(methacrylic acid) was obtained. | (72) |
| | Deformational swelling and potentiometric behavior of ionized poly(methacrylic acid) gels discussed. | (73, 74) |
| | Study of conformational changes in aqueous urea and xylose solutions. | (57) |
| | Titrations with KOH and $Ca(OH)_2$ compared. | (63) |
| 2-Hydroxyethyl methacrylate–methacrylic acid copolymer [31693-08-0] | Investigation of the deformational swelling, and potentiometric behavior of these ionized water-swollen gels, used in biomedical and separation applications, revealed that with increasing degrees of neutralization the swelling degree increased strongly. | (75) |
| | Conformational transition of these copolymers, indicated as a result of titration, was distinct only with copolymers containing $\geq$ 48 mole % methacrylic acid. Transition is probably not of the expansion of the coiled form of the macromolecule into the expanded form, but rather is related to the local configurational and conformation structure. | (76) |
| Methacrylic acid type cation exchangers | Electrochemical properties determined by titration. | (77) |
| Pectic acid [9046-40-6] | Titrations with KOH and $Ca(OH)_2$ compared. | (63) |

(continued)

**Table 6.10.** *(Continued)*

| Polyelectrolyte | Property Investigated | Ref. |
|---|---|---|
| | *B. Polyampholytes* | |
| Poly(adenylic acid) | CD spectrum and titration data indicate that at high ionic strength the double-helical conformation of the individual strands of the aggregate is preserved. The single strand—double helix transition region of the titration curve shifts to lower $pH$ values as the ionic strength is increased. | (78) |
| Poly(L-alanine, glutamic acid) | Titration data for this copolymer, prepared by thermal condensation, showed that the intrinsic $pK$ of the —COOH groups is 3.92 which is different from that of the $\alpha$-linked poly(glutamic acids) but similar to the $\gamma$-linked acids. | (79) |
| Poly(p-amino-L-phenylalanine) | Titration data and CD spectra showed that the nonionized polymer can assume two ordered conformations depending on the temperature. A right-handed helical form is present at room temperature, but at temperatures higher than 40°, a new ordered conformation corresponding to a $\beta$-structure is found as is typified by its slow rate of formation and infrared absorption. | (80) |
| Poly(aspartic acid) | Ratio of $\alpha$ to $\beta$ linkages of the polymer, prepared by thermal condensation of L or DL-aspartic acid followed by alkaline hydrolysis, was determined to be 7:3 by titration. | (81) |
| 1,3-Benzenedisulfonyl chloride-L-lysine copolymer | Titration in salt-free solutions showed that the transition from a compact form to an extended form was shifted towards higher values of neutralization with increasing length of tetraalkylammonium counterions and disappeared gradually as acetone was added to the titration medium. | (82) |

**Table 6.10.** *(Continued)*

| Polyelectrolyte | Property Investigated | Ref. |
|---|---|---|
| Poly(S-carboxyethyl-L-cysteine) [*34306-63-3*] | Reversibility and time dependence of the titration curves were examined by different methods to establish conditions for obtaining equilibrium curves. The Henderson–Hasselbach plot of the titration curve yields a clear distinstion between the $\beta$-form and random coil and permits estimation of the content of $\beta$-form at a given $p\mathrm{H}$. | (83) |
| Poly(S-carboxymethyl-L-cysteine) | For titrations in 0.005–0.2 $M$ NaCl, ORD reveals that a $\beta$-structure to random coil transition occurs. The standard free energy change per amino acid residue for the transition from nonionized random coil to nonionized $\beta$-form was determined to be $ca.$-750 cal mole-residue$^{-1}$. | (84) |
|  | For polymers of high molecular weight the change of the titration curve with time indicated that the $\beta$-coil transition became sharper and the transition free energy increased as equilibrium was approached. This suggested that equilibrium data were not obtained previously and that reversibility is not necessarily sufficient to confirm the equilibrium. | (85) |
| Poly(L-glutamic acid) [*24938-00-9*] | Helix-coil transitions in 0.2 $M$ NaCl and in its mixtures with dioxan were studied using spectropolarimetry and viscometry in addition to potentiometric titration. By means of the latter the enthalpy and entropy differences between the helical and coillike states of the uncharged molecules were determined from titrations performed in the temperature range 8–50°. | (86) |
|  | The helix-coil transition in 0.05 and 0.005 $M$ aqueous KCl showed that the transition depended upon the polymer concentration in the range 0.006–0.04 monomole$^{-1}$ in 0.005 $M$ KCl. | (87, 88) |

*(continued)*

**Table 6.10.** *(Continued)*

| Polyelectrolyte | Property Investigated | Ref. |
|---|---|---|
| | Transitions in aqueous dioxane mixtures studied. | (89) |
| | The concentration dependence of the titration curves were explained. | (55) |
| Poly($\alpha$-DLl-glutamic acid) [*25513-46-6*] | Titrations reveal that a conformational transition from a contracted, regularly ordered H-bonded structure to an expanded random chain occurs at $p$H 3–5. | (90) |
| Poly(L-glutamic acid) sodium salt [*26247-79-0*] | The effects of neutral salts on the conformational change were studied by titration and ORD, and showed that the free energy of formation of the nonionized $\alpha$-helix from the nonionized random coil decreased very rapidly in the limit of zero added salt. Above 0.1 $M$ salt, where the free energy determination becomes meaningful, the anions and cations investigated do not affect the change in free energy with the exception of $Li^+$. | (91) |
| L-Glutamic acid derivativesPoly(L-histidyl-L-alanyl-$\alpha$-L-glutamic acid) Poly(L-lysyl-L-alanyl-$\alpha$-L-glutamic acid) | Potentiometric titration and viscosity measurements were interpreted in terms of the hydrodynamic size of these polypeptides. | (92) see also (93) |
| Poly(L-histidine) [*26062-48-6*] | Titration and microcalorimetric measurements were used to characterize conformational changes with varying $p$H in the solution in terms of the enthalpy and free energy changes accompanying the order–disorder transitions. | (94) |
| Poly(L-lysine) [*25104-18-1*] | Enthalpy and entropy differences between the helical and coil-like states of the uncharged molecule were determined from the titration curve obtained in 0.2 $M$ NaCl. | (95) |
| Poly(L-lysine hydrochloride) [*28826-16-6*] | CD measurements during the titration in 0.1 $M$ KCl at 40–80° showed a transition from a random coil to | (96) |

Table 6.10. (Continued)

| Polyelectrolyte | Property Investigated | Ref. |
|---|---|---|
| Poly(L-lysine$^x$, L-isoleucine$^y$) | $\alpha$-helical structure (at 40–50°) which subsequently transformed into a $\beta$-configuration. At temperature > 60° the $\beta$-structure was formed directly. Copolymers of L-lysine and isoleucine containing 4–15% isoleucine were studied by titration and CD spectroscopy. With increasing isoleucine content $\beta$-sheet formation was favored over $\alpha$-helix formation at higher $p$H and room temperature. The fraction of $\beta$-sheet present as a function of $p$H was calculated from the titration data as were also the thermodynamic quantities for the transition. | (97) |
| Poly(L-lysine$^x$, L-leucine$^y$) | Copolymers of L-lysine and L-leucine containing 0–0.41 mole fraction L-leucine were studied by titration and CD spectroscopy. Among other properties, the free energy change for the conversion of 1 mole residue from nonionized helix to nonionized coil was obtained from the titration data for each copolymer up to a leucine mole fraction of 0.16. | (98) |
| Poly(L-lysine$^x$, L-valine$^y$) | Copolymers of L-lysine and L-valine containing 0–13% L-valine were studied in 0.1 $M$ KF. The fraction of $\beta$-sheets versus $p$H was calculated for poly(L-lysine$^{86.8}$, L-valine$^{13.2}$) at 25° from the titration data and so were the thermodynamic quantities for the coil to $\beta$-sheet transition. | (99) |
| A L-proline-poly(styrene)sorbent | The $pK_a$ values of this sorbent, prepared by treating L-proline with a cross-linked chloromethylated styrene polymer, were 2.2 for the —COOH groups and 9.5 for the —NH$_2$ groups. The stability constants of the copper complexes were close to the values for the complexes of $N$–benzyl-L-proline. | (100) |

(continued)

**Table 6.10.** *(Continued)*

| Polyelectrolyte | Property Investigated | Ref. |
|---|---|---|
| Poly(L-tyrosine) [*25619-78-7*] | Titration data was combined with light-scattering measurements and ir spectroscopy to characterize the conformational changes exhibited when the $pH$ was varied. | (101) |
| | Thermodynamic quantities for the folding of the random coil to the $\alpha$-helix in 10% ethanol, and from the random coil to the $\beta$-structure in water and in 10% ethanol were calculated from the titration data. | (102) |
| | CD spectroscopy showed that the conformational transitions occur within the same degrees of ionization over which the titration curve shows a characteristic hump. | (103) |
| Human immuno-gamma globulin | The buoyant density and $pH$ titration curves in 3 $M$ CsCl were recorded. The $pH$ titration established the intrinsic $pK$ values for the acidic and hystidyl residues of the molecule and confirmed the hypothesis that it conformed to the oil-drop model of protein conformation. | (104) |
| $\beta$-Lactoglobulin | The standard free energy change in the dissociation of uncharged (i.e., isoionic) protein into two uncharged (isoionic) subunits was estimated from the titration curves using a method analogous to that applied to the helix-coil transition of poly(L-glutamic acid). | (105) |
| Globular proteins | Discrete charge calculations of the potentiometric titration curves for sperm whale myoglobulin, hemoglobulin $\alpha$-chain, and cytochrome c. | (106) See also (107) |
| Native apolipoproteins | A comparison of the ionization behavior of the human apo A-II, apo C-I, apo C-III proteins, and of their complexes with dimyristoyllecithin was based on the potentiometric titration of the basic and acidic residues and spectrophotometric titration of the phenolic groups. | (108) See also (109) |

**Table 6.10.** *(Continued)*

| Polyelectrolyte | Property Investigated | Ref. |
|---|---|---|
| Ovalbumin | Titration in 2.4 $M$ CsCl covering the $p$H range 2–12 yield the apparent $p$K values for carboxyls, imidazoles, $\epsilon$-amino, and phenolic residues of 4.3, 6.3, 9.6, and 10.9, respectively. Between half and all of the 10 tyrosine residues titrated below $p$H 11.5. | (110) |
| Papain | The ionization behavior of groups at the active site was determined from the $p$H dependence of the difference in protein content of papain and the methylthio derivative of the —SH group at the active site of papain. | (111) |
| Pepsinogen | Conformational transition in this protein was investigated by titration and ORD. The transition occurred in the range of ionization of the three histidines of pepsinogen. | (112) |
| Proteins and nucleoproteins of tobacco mosaic virus strains and mutants | The position of acid, basic, and phenolic amino acid residues was determined by potentiometric titration and spectrophotometry. | (113) |
| Ribonuclease A and its complex with 3′-cytosine monophosphate | The ionization behavior of the enzyme and its complex was characterized by potentiometry and calorimetry. | (114) |

*C. Polybases*

| Polyelectrolyte | Property Investigated | Ref. |
|---|---|---|
| Poly(ethylenimine) [*9002-98-6*] | Protonation did not exceed ~70% even at low $p$H on titration, and the resulting curve was not of the monomeric type even at high ionic strengths. | (115) |
|  | Titration data was evaluated subsequently taking account of nearest-neighbor interaction. | (116) |
| Poly(vinylamine) [*26336-38-9*] | Found to be a stronger base than poly(ethylenimine) and, like the latter, yielded a titration curve which differed from the monomeric type. | (115) |
| Atactic poly(2-vinyl pyridine) [*25014-15-7*] | That a conformational change occurred in the titration was deduced from a plateau in the $p$H plot of the stoichiometric proton concentration degree of ionization. | (117) |

## REFERENCES

1. J. N. Bronsted, *Rec. Trav. Chim.* **42**, 718 (1923); *Chem. Rev.* **5**, 231 (1928).
2. H. S. Harned and B. B. Owen, *The Physical Chemistry of Electrolytic Solutions*, 3rd ed., Reinhold, New York, 1958, p. 574.
3. G. Kortum, W. Vogel, and K. Andrussow, *Dissociation Constants of Organic Acids in Aqueous Solution*, Butterworths, London, 1961.
4. D. D. Perrin, *Dissociation Constants of Organic Bases in Aqueous Solution*, Butterworths, London, 1965.
5. D. D. Perrin, *Dissociation Constants of Organic Bases in Aqueous Solution, Supplement 1972*, Butterworths, London, 1972.
6. D. D. Perrin, *Dissociation Constants of Inorganic Acids and Bases*, Butterworths, London, 1969.
7. E. P. Serjeant and B. Dempsey, *Ionisation Constants of Organic Acids in Aqueous Solution*, Pergamon, Oxford, 1979.
8. C. H. Rochester, *Acidity Functions*, Academic, London, 1970.
9. P. D. Bolton, F. M. Hall, and I. H. Reece, *Spectrochim. Acta* **22**, 1149 (1966).
10. A. Albert and E. P. Serjeant, *The Determination of Ionization Constants*, 3rd ed., Chapman and Hall, London, 1984.
11. R. F. Cookson, *Chem. Rev.* **74**, 5 (1974).
12. B. M. Lowe and D. G. Smith, *J. Chem. Soc., Faraday Trans. 1* **69**, 1934 (1973).
13. E. J. King, *Acid-Base Equilibria*, Pergamon, Oxford, 1965, p. 47.
14. R. G. Bates, *Determination of pH Theory and Practice*, 2nd ed., Wiley, New York, 1973, p. 455.
15. H. S. Harned and J. O. Morrison, *Am. J. Sci.* **33**, 161 (1937).
16. R. Gary, R. G. Bates, and R. A. Robinson, *J. Phys. Chem.* **68**, 1186 (1964).
17. R. G. Bates and J. B. Macaskill, *Pure Appl. Chem.* **50**, 1701 (1978).
18. A. K. Covington and J. E. Prue, *J. Chem. Soc.* **1955**, 3693; **1955**, 3701.
19. E. J. King and J. E. Prue, *J. Chem. Soc.* **1961**, 275.
20. E. P. Serjeant and A. G. Warner, *Anal. Chem.* **50**, 1724 (1978).
21. R. G. Bates and R. Gary, *J. Res. Natl. Bur. Stand. (U.S.)* **65A**, 495 (1961).
22. I. R. Davies, E. P. Serjeant, and A. G. Warner, *J. Chem. Educ.* **54**, 649 (1977).
23. R. G. Bates and V. E. Bower, *J. Res. Natl. Bur. Stand. (U.S.)* **53**, 283 (1954).
24. B. R. Staples and R. G. Bates, *J. Res. Natl. Bur. Stand. (U.S.)* **73A**, 37 (1969).
25. G. D. Pinching and R. G. Bates, *J. Res. Natl. Bur. Stand. (U.S.)* **37**, 311 (1946).
26. A. K. Covington and M. I. A. Ferra, *Anal. Chem.* **49**, 1363 (1977).
27. R. G. Bates and S. F. Acree, *J. Res. Natl. Bur. Stand. (U.S.)* **34**, 373 (1945).
28. R. N. Roy, R. A. Robinson, and R. G. Bates, *J. Am. Chem. Soc.* **95**, 8231 (1973).
29. A. K. Covington and J. Caudle, *J. Chem. Educ.* **49**, 552 (1972).
30. J. G. Travers, K. G. McCurdy, D. Dolman, and L. G. Hepler, *J. Solution. Chem.* **4**, 267 (1975).
31. D. D. Perrin, B. Dempsey, and E. P. Serjeant, *$pK_a$ Prediction for Organic Acids and Bases*, Chapman and Hall, London, 1981.

# REFERENCES

32. H. S. Harned and W. J. Hamer, *J. Am. Chem. Soc.* **55**, 2194 (1933).
33. H. S. Harned and G. E. Mannweiler, *J. Am. Chem. Soc.* **57**, 1873 (1935).
34. R. F. Jameson and M. F. Wilson, *J. Chem. Soc., Dalton Trans.* **1972**, 2607.
35. C. H. Culberson and R. M. Pytkowicz, *Mar. Chem.* **1**, 309 (1973).
36. J. C. Speakman, *J. Chem. Soc.* **1940**, 855.
37. Ref. 10 p. 65.
38. M. Eden and R. G. Bates, *J. Res. Natl. Bur. Stand. (U.S.)* **62**, 161 (1959).
39. R. G. Bates and R. Gary, *J. Res. Natl. Bur. Stand. (U.S.)* **65A**, 495 (1961).
40. I. R. Davies and E. P. Serjeant, *unpublished results*.
41. A. G. Warner and E. P. Serjeant, *unpublished results*.
42. B. van de Graf, A. J. Hoefnagel, and B. M. Wepster, *J. Org. Chem.* **46**, 653 (1981).
43. R. B. Martin, J. T. Edsall, D. B. Wetlaufer, and B. R. Hollingworth, *J. Biol. Chem.* **233**, 1429 (1958).
44. A. Hiller, J. Plazin, and D. D. Van Slyke, *J. Biol. Chem.* **176**, 1401 (1948).
45. G. E. Perlmann, *Methods Enzymol.* **26** Part C, 413 (1972).
46. R. K. Cannan, A. H. Palmer, and A. C. Kirkbrick, *J. Biol. Chem.* **142**, 803 (1942).
47. C. Tanford and Y. Nozaki, *J. Biol. Chem.* **234**, 2874 (1959).
48. Y. Nozaki, L. G. Bunville, and C. Tanford, *J. Am. Chem. Soc.* **81**, 5523 (1959).
49. J. Steinhardt and S. Beychok, in *The Proteins*, H. Neurath, Ed., 2nd ed., Vol. 2, Academic, New York, 1964, p. 139.
50. T. M. Birshtein and O. B. Ptitsyn, *Conformation of Macromolecules*, Wiley, New York, 1966, p. 277.
51. M. Nagasawa and A. Holtzer, *J. Am. Chem. Soc.* **86**, 538 (1964).
52. D. S. Olander and A. Holtzer, *J. Am. Chem. Soc.* **90**, 4549 (1968).
53. J. McLaren, J. D. Watts, and A. Gilbert, *J. Polym. Sci., Part C*, No. **16**, 1903 (1967).
54. M. Mandel, *Eur. Polym. J.* **6**, 807 (1970).
55. K. Nitta and S. Sugai, *J. Phys. Chem.* **78**, 1189 (1974).
56. I. Muresan and L. Zador, *Rev. Roum. Chim.* **20**, 1119 (1975).
57. O. F. Schaefer, *Colloid Polym. Sci.* **255**, 550 (1977).
58. Y. Kawaguchi and M. Nagasawa, *J. Phys. Chem.* **73**, 4382 (1969).
59. P. Monjol, *Bull. Soc. Chim. Fr.* **1972**, 1313.
60. E. A. Hassan and M. M. B. El-Sabbuh, *Indian J. Chem.* **11**, 1062 (1973); *Egypt. J. Chem.* **17**, 907 (1974).
61. D. S. Gamble, *Can. J. Chem.* **50**, 2680 (1972).
62. G. Sposito and K. M. Holtzclaw, *Soil Sci. Soc. Am. J.* **41**, 330 (1977).
63. R. Kohn, *Chem. Zvesti* **27**, 218 (1973).
64. R. K. Shah, I. M. Bhatt, and M. R. Choksi, *Chem. Era* **12**, 422 (1976).
65. M. Adhikari, K. K. Chakrabarti, and G. Chakrabarti, *J. Indian Chem. Soc.* **55**, 439 (1978).
66. R. K. Shah, M. R. Chokshi, and K. P. Soni, *J. Indian Chem. Soc.* **54**, 912 (1977).
67. M. Adhikari, J. Roy, and G. C. Hazra, *J. Indian Chem. Soc.* **55**, 332 (1978).

68. T. Takamatsu and T. Yoshida, *Soil Sci.* **125**, 377 (1978).
69. N. Muto, T. Komatsu, and T. Nakagawa, *Bull. Chem. Soc. Jpn.* **46**, 2711 (1973).
70. G. Barone and E. Rizzo, *Gazz. Chim. Ital.* **103**, 401 (1973).
71. M. Nagasawa, *Pure Appl. Chem.* **26**, 519 (1971).
72. J. C. Leyte, H. M. R. Arbouw-van der Veen, and L. H. Zuiderweg, *J. Phys. Chem.* **76**, 2559 (1972).
73. J. Hasa, M. Ilavsky, and K. Dusek, *J. Polym. Sci., Polym. Phys. Ed.* **13**, 253 (1975).
74. J. Hasa and M. Ilavsky, *J. Polym. Sci., Polym. Phys. Ed.* **13**, 263 (1975).
75. M. Ilavsky, K. Dusek, J. Vacik, and J. Kopecek, *J. Appl. Polym. Sci.* **23**, 2073 (1979).
76. E. Prokopova, M. Stol, E. Knizakova, and M. Bohdanecky, *Makromol. Chem.* **180**, 615 (1979).
77. L. K. Shataeva, I. A. Chernova, J. Vacik, and G. V. Samsonov, *Izv. Akad. Nauk SSSR Ser. Khim.* **1977**, 353.
78. V. Vetterl and A. Polaskova, *Stud. Biophys.* **69**, 159 (1978).
79. E. Kokufuta, T. Terada, S. Suzuki, and K. Harada, *BioSystems* **10**, 299 (1978).
80. E. Peggion, A. Cosani, M. Palumbo, M. Terbojevich, and M. Goodman, *Biopolymers* **15**, 2227 (1976).
81. E. Kokufuta, S. Suzuki, and K. Harada, *BioSystems* **9**, 211 (1977).
82. J. C. Fenyo, J. Beaumaris, and E. Selegny, *J. Polym. Sci. Polym. Chem. Ed.* **12**, 2659 (1974).
83. H. Maeda and S. Ikeda, *Biopolymers* **10**, 2525 (1971).
84. S. Makino and S. Sugai, *Biopolymers* **9**, 1049 (1970).
85. H. Maeda and S. Ikeda, *Biopolymers* **14**, 1623 (1975).
86. V. Bychkova, O. B. Ptitsyn, and T. V. Barskaya, *Biopolymers* **10**, 2161 (1971).
87. K. Nitta, M. Yoneyama, and N. Ohno, *Biophys. Chem.* **3**, 323 (1975).
88. N. Kono and A. Ikegami, *Biopolymers* **4**, 823 (1966).
89. M. Morcellet and C. Loucheux, *Polymer* **16**, 785 (1975).
90. E. Homma, K. Ogawa, and T. Watanabe, *Chem. Lett.* **11**, 1285 (1974).
91. G. Conio, V. Terfiletti, F. Bodria, C. Troglia, and E. Patrone, *Biopolymers* **13**, 1483 (1974).
92. H. J. Goren, L. Grandan, A. W. L. Jay, and N. Lotan, *Biopolymers* **16**, 1541 (1977).
93. D. S. Sharp, R. Almassy, L. G. Lum, K. Kinzie, J. S. V. Zil, and J. B. Ifft, *Biopolymers* **15**, 757 (1976).
94. M. Terbojevich, A. Cosani, E. Peggion, F. Quadrifoglio, and V. Crescenzi, *Macromolecules* **5**, 622 (1972).
95. T. V. Barskaya and O. B. Ptitsyn, *Biopolymers* **10**, 2181 (1971).
96. A. Cosani, M. Terbojevich, L. Romanin-Jacur, and E. Peggion, *Pept., Polypeptides, Proteins, Proc. Rehovot. Symp. 2nd.* **1974**, 166; *CA* **83**, 115195y.
97. B. Walter and G. D. Fasman, *Biopolymers* **16**, 17 (1977).
98. C. R. Snell and G. D. Fasman, *Biopolymers* **11**, 1723 (1972).
99. R. Mandel and G. D. Fasman, *Biopolymers* **14**, 1633 (1975).

# REFERENCES

100. Yu. A. Zolotarev, A. A. Kurganov, and V. A. Davankov, *Talanta* **25**, 493 (1978).
101. M. B. Senior, S. L. Gorrell, and E. Hamori, *Biopolymers* **10**, 2389 (1971).
102. D. Pedersen, D. Gabriel, and J. Hermans, Jr., *Biopolymers* **10**, 2133 (1971).
103. A. Cosani, M. Palumbo, M. Terbojevich, and E. Peggion, *Inst. J. Pept. Protein Res.* **6**, 457 (1974).
104. J. E. Ruark and J. B. Ifft, *Biopolymers* **14**, 1161 (1975).
105. A. Holtzer and M. Nagasawa, *J. Am. Chem. Soc.* **93**, 606 (1971).
106. J. B. Matthew, S. H. Friend, L. H. Botelho, L. D. Lehman, G. I. H. Hanania, and F. R. N. Gurd, *Biochem. Biophys. Res. Commun.* **81**, 416 (1978).
107. K. Nitta and S. Sugai, *Biopolymers* **11**, 1893 (1972).
108. F. Soetewey, M. J. Lievens, R. Vercaemst, M. Rosseneu, H. Peeters, and V. Brown, *Eur. J. Biochem.* **79**, 259 (1977).
109. M. Rosseneu, F. Soetewey, M. J. Lievens, R. Vercaemst, and H. Peeters, *Eur. J. Biochem.* **79**, 251 (1977).
110. J. B. Ifft and L. G. Lum, *C. R. Trav. Lab. Carlsberg* **38**, 339 (1971); *CA* **76**, 123085h.
111. S. D. Lewis, F. A. Johnson, and J. A. Shafer, *Biochemistry* **15**, 5009 (1976).
112. G. E. Perlmann, A. Oplatka, and A. Katchalsky, *J. Biol. Chem.* **242**, 5163 (1967).
113. G. Paulsen, *Z. Naturforsch.* **B 27**, 427 (1972).
114. M. Flogel and R. L. Biltonen, *Biochemistry* **14**, 2603 (1975).
115. C. J. Bloys van Treslong and A. J. Staverman, *Recl. Trav. Chim. Pays-Bas* **93**, 171 (1974).
116. C. J. Bloys van Treslong, *Recl. Trav. Chim. Pays-Bas* **97**, 13 (1978).
117. T. Kitajima-Yamashita, *Polym. J.* **4**, 262 (1973).

CHAPTER
7

# ACID–BASE EQUILIBRIA IN NONAQUEOUS SOLVENTS

## 7.1 INTRODUCTION

By contrast to its application in aqueous solutions, potentiometry has only limited application in fundamental investigations of ionization processes that occur in nonaqueous solvents. This is because equilibria additional to that represented by the simple ionization $HA \rightleftharpoons H^+ + A^-$ often occur in nonaqueous solutions, and electrodes that respond to the solvated proton in these media do not sense these extraneous processes. On the other hand, acid–base potentiotitrimetry performed in nonaqueous solutions has a greater application than it has in water because the ranges of potentials it is possible to measure in nonaqueous solvents are often much greater than is possible in aqueous solution. This allows a wider range of weak acids and bases to be titrated to a discernable equivalence point than is possible in water. Thus this chapter emphasizes the role of potentiometry in obtaining data pertinent to these analytical applications and deals, therefore, with methods used for the calibration of cells and the way these are used to obtain values of the ionization constants and related quantities. Some of the data used in the calibration of potentiometric cells have been obtained using other techniques among which conductimetry and spectrophotometry have featured prominently (1). It is beyond the scope of this book to describe the application of these techniques other than to note the reasons why it has been necessary to use them. This is because the nonaqueous solvents that are useful in acid–base potentiotitrimetry have much lower dielectric constants than water, and in solutions of lower dielectric constant effects that are not operative in water must be considered. Prominent among these in acid–base equilibria is the observation that the concentration of the anion of a weak acid ($A^-$), for example, may be considerably smaller than that calculated from the stoichiometric type of approach described in Chapter 6 for aqueous solutions. In solutions of low dielectric constant a small value of the equilibrium constant is associated with the process $MA \rightleftharpoons M^+ + A^-$, in which $A^-$ is the anion of a weak acid and $M^+$ is a soluble cation such as the tetraethylammonium ion. Furthermore, the salt may also take part in other equilibria such as

$$MA + M^+ \rightleftharpoons M_2A^+ \quad \text{and} \quad MA + A^- \rightleftharpoons MA_2^- \qquad (7.1)$$

In some solvents the anion $A^-$ may not be stabilized by solvation, but is stabilized instead by forming a hydrogen bonded complex with the undissociated acid

$$A^- + HA \rightleftharpoons A^-.....HA \qquad (7.2)$$

The *homoconjugation complexes*, as they are called, are also formed by the species $BH^+$ which is stabilized similarly by reaction with the free base

$$BH^+ + B \rightleftharpoons BH^+...B$$

In a given solvent homoconjugation depends greatly upon the structure of the acidic species, and is found to be markedly reduced in those acids that can be stabilized by intramolecular hydrogen bonding (2). Similarly, subtle variations in the polarizability or other properties of the anion can affect the magnitude of the equilibrium constant for the formation of the homoconjugation complex [e.g., for reaction (7.2)]. That these types of effects may be in operation can be shown by an analysis of potentiometric data, but these data in themselves are insufficient to identify and quantitatively deduce the magnitude of the equilibrium constants for some of these processes. A symptom of homoconjugation is that the change in $p$H observed for a change in neutralization of an acid from 10 to 90% neutralization is grossly exaggerated with respect to the change in $p$H when a homoconjugate complex is not formed. In the latter case the $p$H changes by $\log 10/90 = -0.95$ to $\log 90/10 = +0.95$ or by approximately 1.9 units over this range of neutralization for titrations in which $c_{H^+}$ or $c_{OH^-}$ remain negligible in comparison with the concentrations of the species derived from the acid. However, when a stable homoconjugate complex is formed, for example, when the equilibrium constant $K_f$ for reaction (7.2) is $10^4$ (see Table 7.3) and the original concentration of acid taken is 0.1 $M$, the range of $p$H covered for the same change in the degree of neutralization is about 7.6 units (2a). The magnitude of this range for a given acid depends upon the original concentration taken, becoming less as the original concentration of acid taken is reduced. Thus if the titration is performed in very dilute solution (e.g., $10^{-4}$ $M$ when $K_f = 10^4$), the range of $p$H covered from 10 to 90% neutralization becomes only slightly greater than the 1.9 units observed when homoconjugation is not a factor. The $p$H titration method used for the determination of $p\text{K}$ values in aqueous solution cannot be applied, therefore, when homoconjugation is a factor. However, provided that $c_{\text{H}^+}$ is negligible in comparison with $c_{\text{HA}}$ or $c_{\text{A}^-}$, the measured $p\text{H}$ at half-neutralization ($c_{\text{HA}} = c_{\text{A}^-}$) is independent of the magnitude of $K_f$ for the homoconjugate complexes and such measurements afford an estimate of the $p\text{K}$ of the acid.

Despite the complications referred to, the Bronsted definition of acids and bases given earlier is applicable also to acid–base equilibria in nonaqueous solvents, the more general form of equilibrium (6.1) being

$$\text{HA} + \text{SH} \rightleftharpoons \text{SH}_2^+ + \text{A}^-$$

In this equilibrium SH represents the solvent and $\text{SH}_2^+$ represents the solvated proton, these two species being $\text{H}_2\text{O}$ and $\text{H}_3\text{O}^+$, respectively, for the aqueous processes considered in Chapter 6. When the acid HA is dissolved in acetonitrile, for example, the equilibrium can be represented as

$$\text{HA} + \text{CH}_3\text{CN} \rightleftharpoons \text{CH}_3\text{CN}^+\text{H} + \text{A}^-$$

and, similarly, when a base is dissolved in glacial acetic acid the process $\text{B} + \text{SH} \rightleftharpoons \text{BH}^+ + \text{S}^-$ is a generalization for

$$\text{B} + \text{CH}_3\text{COOH} \rightleftharpoons \text{BH}^+ + \text{CH}_3\text{COO}^-$$

In the first example acetonitrile has acted as a base, and in the second the solvent glacial acetic acid has acted as an acid.

### Solvent Classification

Although more descriptive classifications have been proposed (2a), it is sufficient for the purpose of this book to classify solvents according to their proton donor-acceptor properties into two general types:

*Amphiprotic*: Solvents that possess both acidic and basic properties.
*Aprotic*: Solvents that are neither acidic nor basic.

### *Amphiprotic Solvents*

Like water, these solvents undergo self-ionization or *autoprotolysis* that may be represented by the general case

$$\text{SH} + \text{SH} \rightleftharpoons \text{SH}_2^+ + \text{S}^- \qquad (7.3)$$

Again, like water, the extent to which the equilibrium lies on the right-hand side of equation (7.3) is reflected in the magnitude of the ionic product of the solvent which is now generally referred to as the *autoprotolysis constant* and designated as $K_{SH}$:

$$K_{SH} = a_{SH_2^+} \cdot a_{S^-}; \quad \text{compare } K_W = a_{H_3O^+} a_{OH^-} \quad (7.4)$$

Values of the autoprotolysis constants for nonaqueous solvents range from values much larger than the ionic product of water ($K_W = K_{SH} = 10^{-14}$), for example, formic acid,

$$HCOOH + HCOOH \rightleftharpoons HCOOH_2^+ + HCOO^-; \quad K_{SH} = 10^{-6.2}$$

through solvents with about the same $K_{SH}$ as water, for example, glacial acetic acid,

$$CH_3COOH + CH_3COOH \rightleftharpoons CH_3COOH_2^+ + CH_3COO^-;$$

$$K_{SH} = 10^{-14.9}$$

to solvents like 2-methyl-2-propanol (*tert*-butyl alcohol),

$$(CH_3)_3COH + (CH_3)COH \rightleftharpoons (CH_3)_3COH_2^+ + (CH_3)_3CO^-;$$

$$K_{SH} = 10^{-28.5}$$

and acetonitrile,

$$CH_3CN + CH_3CN \rightleftharpoons CH_3CNH^+ + CH_2CN^-; \quad K_{SH} \sim 10^{-33}$$

The cations formed in these processes are called the *lyonium* ions, whereas the anions are called *lyate* ions. The glass electrode is found to respond in a Nernstian fashion to the solvated proton ($SH_2^+$) in these and other solvents, and by analogy with water it follows that in solvents with much smaller autoprotolysis constants than water the glass electrode can assume a much larger range of potentials than it can in water. This, in turn, implies that as a first approximation, it is possible by titration in solvents with small values of the autoprotolysis constant to discriminate or differentiate between a wider range of acids or bases than is possible in solvents with comparatively large values of $K_{SH}$. There are, of course, other factors that must be considered in deciding whether two acids could be titrated sequentially in a given solvent, but the magnitude of the autoprotolysis constant is a fundamental quantity in arriving at such a decision. Water and glacial acetic acid, for example, have autoprotolysis constants of similar magnitude, but water is a much weaker acid than acetic acid. In water strong acids such as perchloric, hydrochloric, and nitric acids, although they have different intrinsic acid strength, all form the same lyonium ion ($H_3O^+$) and thus appear to be of equal acid strength. The reaction of any one of these acids with water reduces their acid strength to form the weaker acid $H_3O^+$, and thus all are *leveled* to the same strength. In glacial acetic acid, however, acids are leveled to the higher ultimate strength of the lyonium ion $CH_3COOH_2^+$, and the extent of the reaction

$$HA + CH_3COOH \rightarrow CH_3COOH_2^+ + A^-; \quad (A^- = ClO_4^-, Cl^-, NO_3^-)$$

is different for each acid, permitting the *differentiation* or *discrimination* of acids that are leveled in water. Similarly, the strongly proton-donating or *protogenic* nature of glacial acetic acid leads to a ready protonation of bases some of which cannot be protonated in aqueous solution as a result of the weaker protogenic nature of the water molecule. Thus weak acids such as anilinium ion ($pK_a = 4.6$ for $BH^+$ in water) that cannot be titrated to a discernible equivalence point in water can be titrated successfully in glacial acetic acid. Other *protogenic amphiprotic* solvents include formic acid, sulfuric acid, and hydrofluoric acid. In contradistinction to these, certain amphiprotic solvents are proton-acceptors or *protophilic*, and this category includes solvents such as liquid ammonia and 1,2-ethanediamine. The autoprotolysis constant for the latter $K_{SH} = 10^{-15.3}$ for the autoionization

$$2NH_2C_2H_4NH_2 \rightleftharpoons NH_2C_2H_4\overset{+}{N}H_3 + NH_2C_2H_4NH^-$$

is similar in magnitude to that of water. However, in this case the solvent is a much weaker acid than water and is, therefore, useful as a medium for the titration of weakly acidic substances. In this solvent acids having ionization constants in water larger than $10^{-5}$ are all leveled to about equal strength. Amphiprotic solvents that can neither be classified as protogenic nor as protophilic can be considered as being *ampholytic* in that they have only weakly acidic or weakly basic properties in comparison with water. Amides (e.g., dimethylformamide, $K_{SH} \sim 10^{-17}$), nitriles (e.g., acetonitrile, $K_{SH} \sim 10^{-33}$), and ketones (e.g., 2-butanone, $K_{SH} = 10^{-25.5}$) are examples within this category, and such solvents are suitable media in which to titrate either weak acids or weak bases.

## Aprotic Solvents

Apart from perhaps a weakly solvating effect, aprotic solvents do not interact with either acids or bases. This category includes solvents such as the aliphatic and aromatic hydrocarbons, for example, petroleum ethers, toluene, and benzene, and chlorinated hydrocarbons among which chloroform is often used. In potentiotitrimetry, aprotic solvents are used to modify the dielectric constants of titration media composed predominantly of amphiprotic solvents. Their addition can result in the enhancement of the potential jump observed in the vicinity of the equivalence point, the suppression of undesirable side reactions, and in the solubilization of the reaction products of the titration. Precipitation of the latter during a titration can adversely affect the response of the indicator electrode.

## 7.2 GENERAL PROPERTIES OF AMPHIPROTIC SOLVENTS

Ideally, in its analytical applications an amphiprotic solvent should possess a high dielectric constant, and also a low value of the autoprotolysis constant since the magnitude of the latter determines the range of potentials that will be available in that solvent. Unfortunately, these two ideals are not mutually compatible as is shown by the data collected in Table 7.1. The influence of these factors in the application of potentiometry to the investigation of acid-base equilibria in nonaqueous solvents is outlined in the following together with an enumeration of other properties.

### The Dielectric Constant

Most organic solvents have dielectric constants lower than that of water $[\epsilon = 78.3 \, (25°)]$ and do not dissolve ionic compounds as readily. In solvents of low dielectric constant the dissociation constants of salts are small, the resistance is high, and the response of electrodes tends to be less reliable than in solvents with higher dielectric constants. It has been shown (19) that for solutions of tetrabutylammonium picrate the formation constants for the process

$$(Bu)_4N^+ + (picrate)^- \rightleftharpoons (Bu)_4N \text{ picrate}$$

obtained by conductimetry in a number of solvents are directly proportional to the inverse of the dielectric constant for the solvent. Thus a large value of the formation constant is associated with a solvent of low dielectric constant indicating a small degree of dissociation in that solvent. Whereas in water

Table 7.1. Autoprotolysis Constants (25°C) and Other Properties of Some Amphiprotic Solvents

| Solvent | mp (°C) | bp (°C) | Refractive Index $n_D$ | Density | Dielectric Constant | Viscosity cP | $pK_{SH}$ | Refs.[a] |
|---|---|---|---|---|---|---|---|---|
| *Acids* | | | | | | | | |
| Acetic acid | 16.5 | 118.0 | 1.36995(25°) | 1.04365(25°) | 6.2(25°) | 1.04(30°) | 14.45 | 3 |
| Formic acid | 8.3 | 100.5 | 1.36938(25°) | 1.214(25°) | 58.5(16°) | 1.97(25°) | 6.2 | 4 |
| *Alcohols* | | | | | | | | |
| Methanol | −97.49 | 64.51 | 1.32855(20°) | 0.78675(25°) | 32.6(25°) | 0.44(25°) | 16.7 | 5 |
| Ethanol | −114.5 | 78.32 | 1.35941(25°) | 0.78506(25°) | 24.3(25°) | 1.08(25°) | 19.5 | 6 |
| Propanol | −126.1 | 97.2 | 1.38343(25°) | 0.79950(25°) | 19.7(25°) | 2.00(25°) | 19.4 | 6 |
| 2-Propanol | −89.5 | 82.3 | 1.37538(25°) | 0.7808(25°) | 18.3(25°) | 1.77(30°) | 20.8 | 6 |
| 2-Methyl-2-propanol | 25.5 | 82.2 | 1.3878(20°) | 0.7887(20°) | 10.9(30°) | — | 28.5 | 7 |
| n-Butanol | −89.3 | 117.2 | 1.39931(20°) | 0.8098(20°) | 17.1(25°) | 2.27(30°) | 21.8 | 8 |
| 1,2-Ethanediol | −13.0 | 197.6 | 1.43063(25°) | 1.1135(20°) | 37.7(25°) | 13.55(30°) | 15.8 | 9 |
| 1,2-Propanediol | −59.0 | 188.2 | 1.4324(20°) | 1.036(25°) | 32.0(20°) | — | 17.2 | 9 |

(*continued*)

Table 7.1. (Continued)

| Solvent | mp (°C) | bp (°C) | Refractive Index $n_D$ | Density | Dielectric Constant | Viscosity cP | $pK_{SH}$ | Refs.[a] |
|---|---|---|---|---|---|---|---|---|
| *Amides* | | | | | | | | |
| Formamide | 2.55 | 210.5 | 1.44682(25°) | 1.12918(25°) | 109.5(25°) | 3.30(25°) | 16.8 | 10 |
| NN-Dimethylformamide | −61.0 | 153.0 | 1.4269(25°) | 0.9445(25°) | 36.7(25°) | 0.80(25°) | 27.0 | 11 |
| N-Methyl-2-pyrrolidinone | −24.7 | 204.0 | 1.4684(20°) | 1.0260(25°) | 32(25°) | 1.69(25°) | 24.3 | 12 |
| *Amines* | | | | | | | | |
| Hydrazine | 2.0 | 113.5 | 1.46979(22.3°) | 1.0036(25°) | 51.7(25°) | 0.97(25°) | 13.0 | 13 |
| 1,2-Ethanediamine | 11.0 | 117.0 | 1.4532(20°) | 0.895(25°) | 12.5(25°) | 1.54(25°) | 15.3 | 14 |
| 2-Aminoethanol | 10.3 | 170.8 | 1.4539(20°) | 1.0117(25°) | 37.7(25°) | 18.95(25°) | 5.1(20°) | 15 |
| *Ketones* | | | | | | | | |
| 2-Butanone | −86.3 | 79.6 | 1.3788(20°) | 0.8054(20°) | 18.5(25°) | — | 25.5 | 16 |
| 2-Hexanone | −57.0 | 128.0 | 1.4007(20°) | 0.81127(20°) | — | — | 25.0 | 16 |
| *Other* | | | | | | | | |
| Acetonitrile | −45.7 | 81.6 | 1.3416(25°) | 0.77683(25°) | 36.0(25°) | 0.325(30°) | ~33.0 | 17 |
| Dimethylsulfoxide | 18.55 | 189.0 | 1.4742(30°) | 1.096(25°) | 46.7(25°) | 2.00(30°) | ~33.0 | 18 |
| | | | | | | | ~35.1 | 77 |

[a] References to values of $pK_{SH}$.

strong 1:1 electrolytes become substantially dissociated at concentrations of less than 1 $M$, in solvents of dielectric constant less than 40 these electrolytes become appreciably dissociated only at concentrations less than $10^{-3}$ $M$. For the concentrations usually encountered, the principal species in solution can be assumed to be an ion pair (20) when the formation constant is greater than $10^3$ (i.e., dissociation constant $< 10^{-3}$). Conductimetric, spectrophotometric, and solubility measurements (21) have provided evidence that other types of association reactions in addition to ion-pair formation occur in solvents of low dielectric constant (see reaction 7.1). The stabilization of an anion through homoconjugation with the undissociated acid was mentioned earlier. The undissociated acid can also participate in self-association reactions such as $2HA \rightleftharpoons (HA)_2$ in solutions of low dielectric constant (22) and these have been studied by differential vapor pressure techniques (23). In analysis by nonaqueous acid–base titrimetry these effects can be made manifest by abnormally shaped titration curves in which an extra inflection is sometimes evident (24).

The activity coefficients of ions in a solution of low dielectric constant are much smaller than they are for a solution of identical concentration in water. For many purposes the activity coefficients for aqueous solutions having an ionic strength of less than $10^{-3}$ $M$ are often assumed to be unity. In solutions of low dielectric constant, however, such an assumption would be erroneous, and despite the need to use solutions more dilute than $10^{-3}$ $M$ to minimize these effects, a check should always be made of the approximate magnitude of the activity coefficient. To this end both the dielectric constant $\epsilon$ and the density of the pure solvent $d_0$ are needed. The magnitude of the constants A and B in the usual form of the Debye–Huckel equation for the activity coefficient of a single ion

$$-\log \gamma_i = \frac{Az^2 I^{1/2}}{1 + Ba_i I^{1/2}} \tag{7.5}$$

are given by

$$A = \frac{1.825 \times 10^6}{(\epsilon T)^{3/2}} d_0^{1/2} \tag{7.6}$$

and

$$B = \frac{50.29}{(\epsilon T)^{1/2}} d_0^{1/2} \tag{7.7}$$

where $T$ is the absolute temperature. At 298°K the values for 2-methyl-2-propanol, that is, *tert*-butylalcohol ($\epsilon \sim 10.9$), 2-propanol ($\epsilon = 18.3$), and acetonitrile ($\epsilon = 36.0$) are, respectively, A = 8.75, B = 0.784, A = 4.00, B = 0.602, and A = 1.45, B = 0.428 in comparison with their values in water of A = 0.511 and B = 0.329. As with the Bates–Guggenheim convention in water, the value of $a_i$ can usually be assumed to be 4.5 Å. Thus the single-ion activity coefficient for a univalent ion at a concentration of $10^{-3}$ M in 2-methyl-2-propanol is 0.564 in comparison to its value in water of 0.965 for the same concentration at 25°.

### The Autoprotolysis Constant and the Relative Scale of Acidity

Values of the autoprotolysis constants are usually expressed in a similar manner to $p\text{K}_w$ as $-\log \text{K}_{\text{SH}}$ (Table 7.1). As a generalization, the values quoted for those solvents of $p\text{K}_{\text{SH}} > 20$ can be regarded as being minimum estimates and the actual values are probably greater than those quoted. These low values stem from the presence of ubiquitous trace contaminants such as water and carbon dioxide, and also from the presence of impurities that arise in some instances from the decomposition of the solvent itself. A further problem in some solvents has been finding a base that is sufficiently strong so that meaningful measurements can be taken. In this regard the use of crown ethers such as the cryptands to promote the ionization of potassium 2-methyl-2-propanolate (potassium *tert*-butoxide) was an innovation in the determination of $p\text{K}_{\text{SH}}$ for 2-methyl-2-propanol (7). The value obtained as a result (28.5) was considerably larger than the value (22.2) reported hitherto (25). Likewise, the $p\text{K}_{\text{SH}}$ values for the ketones may also be low. On the basis of the range of potentials possible in 2-butanone it has been concluded that $p\text{K}_{\text{SH}}$ for this solvent will be about 31 (26). On a similar basis it would appear that $p\text{K}_{\text{SH}}$ for the acetone (not given in Table 7.1) is of at least the same magnitude (27).

In amphiprotic solvents the potential of a cell comprised of a glass electrode and a reference electrode can be related to the activity of the solvated proton ($\text{SH}^+$) by the same type of equation that is applicable to aqueous cell solutions:

$$E_{\text{cell}} = \text{constant} - \text{slope} \cdot pa_{\text{SH}^+} \qquad (7.8)$$

The range of $p$H values in aqueous solution is often considered as covering $p$H 0–14, which on the millivolt scale yields a range of $E_{\text{cell}}$ values covering 826 mV at 25°. The useful analytical range, however, is somewhat less than this and could be arbitrarily fixed by taking the $p$H of a solution $2 \times 10^{-3}$ with respect to perchloric acid ($p$H = 2.7) as the lower limit and the $p$OH of a solution $2 \times 10^{-3}$ $M$ in sodium hydroxide ($p$OH = 2.7) as the upper limit of $p$H ($p$H = 14 − 2.7 = 11.3). This practical range of $E_{cell}$, therefore, covers 59(11.3 − 2.7) ≅ 500 mV, and within this range all analytically useful acid–base titrations are performed. In a hypothetical solvent of $p$K$_{SH}$ = 28 and with favorable dielectric properties that exclude ion-pair and other association effects, the practical range could be similarly defined as covering a millivolt range equivalent to $p$H 2.7–25.3 or about 1500 mV. The practical range available in the hypothetical solvent is about three times greater than in water, and thus this solvent is likely to offer a better titration medium for the sequential determination of each acid in a mixture of weak acids than exists in water. This is found in practice and in 2-methyl-2-propanol ($p$K$_{SH}$ = 28.5), for example, a mixture of five different acids has been determined potentiotitrimetrically from the separate inflection points for each acid (27). 2-Methyl-2-propanol, therefore, has a greater discriminative or differentiating effect than water ($p$K$_{SH}$ = 14.0) which, in turn, is a better discriminative solvent than either formic acid ($p$K$_{SH}$ = 6.2) or 2-aminoethanol ($p$K$_{SH}$ = 5.1). It should be noted that the high viscosity of the latter (18.95 cP) would also exclude the use of the latter as a titration medium at room temperature.

The practical millivolt range of a real solvent, as opposed to the hypothetical case considered, can be defined as

$$E_S = E_{HNP}(HClO_4) - E_{HNP}(Et_4NOH) \quad (7.9)$$

The term $E_S$ is called the *relative scale of acidity* (28) for a given solvent and is the difference in millivolts between the half-neutralization potential of perchloric acid $E_{HNP}(HClO_4)$ and the half-neutralization potential of a strong base $E_{HNP}(Et_4NOH)$ determined with the same pair of electrodes at identical concentrations. The value of $E_{HNP}(HClO_4)$ is determined by measuring the potential between an electrode reversible to the solvated proton and a reference half-cell when the cell solution contains equimolar concentrations of perchloric acid and tetraethylammonium perchlorate. The corresponding value of $E_{HNP}(Et_4NOH)$ is measured similarly, but using the same equimolar concentrations of tetraethylammonium hydroxide and tetraethylammonium perchlorate. Values of $E_S$ for solvents cannot be meaningful unless the concentration at which these half-neutralization potentials were determined is kept constant for the solvents investigated. Russian workers have apparently adopted a concentration of 0.02 $M$ as standard (29) and on this basis have prepared a histogram (28) from which the relative scale of acidity for each of 26 solvents can be conveniently compared with that of water (see Figure 7.1). The advantage of such a practical scale for a given solvent over the scale based upon its value of

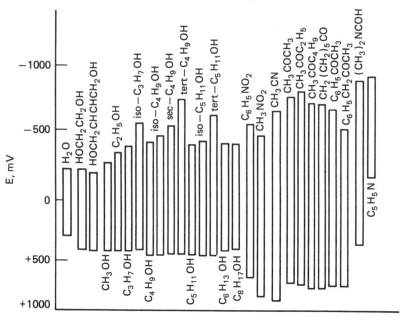

Figure 7.1. Relative scale of acidity of some nonaqueous solvents (28).

$pK_{SH}$ is that it allows for ion-pair formation and the effects of association upon the actual concentrations of the species present in the solutions. In this regard the constituent half-neutralization potentials of equation (7.9) are similar in concept to the formal reduction potentials mentioned in Chapter 1. Furthermore, in defining this scale of relative acidity the value of $pK_{SH}$ for the particular solvent need not be known, but will be related indirectly to the magnitude of $E_S$. The latter is, of course, also related to the range of proton potentials available in the solvent, which is an important consideration in choosing a solvent for nonaqueous acid–base potentiotitrimetry. The actual values of the constituent half-neutralization potentials relative to those determined in water are also indicative of the protophilic, protogenic, or ampholytic nature of the solvent which provides additional qualitative information as to the nature of compounds that might be titrated in the solvent.

The exponent of the autoprotolysis constant $pK_{SH}$ and the relative scale of acidity for a solvent have been combined to define the *relative acidity constant* for the solvent as (28)

$$K_E = \frac{E_S}{pK_{SH}} \qquad (7.10)$$

The magnitude of $K_E$ reflects the differentiating or discriminative capability of the solvent, the solvents with the greatest discriminative capability having the largest values of $K_E$. On this scale, water would have a $K_E$ value of about 36 in comparison with values of 39 for diols, 50 for acetonitrile, and 64 for dimethylformamide (28). This scale, however, offers no apparent advantage over the practically determined relative scale of acidity for a solvent, particularly as the values of the autoprotolysis constant for a number of analytically useful solvents may be either in error by several orders of magnitude or unknown. Even when the $pK_{SH}$ value for a solvent is known relatively accurately, the range of proton potentials deduced from the value of the autoprotolysis constant may not bear any resemblance to the relative scale of acidity actually available in that solvent for the sequential titrations of the components of a mixture of weak acids, for example. The presence of trace amounts of water in the solvent together with the water generated by the titration reaction itself tends to reduce the actual range of potentials available. Sometimes the titrant is prepared in a mixture of solvents one of which may be a stronger acid than the solvent selected as the titration medium. For example, the presence of comparatively small quantities of the stronger acidic solvent methanol in a titration medium composed of 2-methyl-2-propanol will reduce the relative scale of acidity of the latter considerably.

## Other Properties*

The temperature range over which the solvent remains a liquid, the vapor pressure of the solvent at room temperature, and whether the solvent is excessively hygroscopic, toxic, or viscous are factors that must also be considered when choosing an appropriate amphiprotic solvent. A low vapor pressure at room temperature is an advantage in that increases in concentration as a result of solvent evaporation are minimized. When using nonaqueous solvents, however, it is generally good technique to use apparatus that is closed to the atmosphere, not only to counter solvent evaporation, but also to minimize the ingress of atmospheric contaminants. If accurate measurements are planned, it is advisable to design the apparatus so that the transfer of solutions in and out of the apparatus is performed in an atmosphere of nitrogen. It is advisable also to regard all nonaqueous solvents as being hygroscopic and, of course, the precautions described will ensure that the pure solvent remains uncontaminated. Under these conditions the effects of the toxicity and odor of the solvent are also lessened. The Merck Index (30) is a good source of information concerning the toxicity of

---

*Useful information is available in Reference 28a.

substances, and it is good practice to assume all organic solvents to be toxic both in the liquid and vapor phases and to be absorbed readily through the skin.

The density of the solvents must be taken into account when calculating activity coefficients using the constants $A$ and $B$ defined by equations (7.6) and (7.7), respectively. The density must also be taken into account when using volumetric apparatus calibrated for use with aqueous solutions. It has been found that some organic liquids can be measured accurately with such apparatus (31), but it should also be noted that the coefficient of cubic expansion of most organic solvents is much larger than that of water. Significant volumetric errors can only be avoided, therefore, by ensuring that all solutions are brought to the same temperature before they are measured and that the laboratory temperature remains fairly constant. Considerable errors can be expected when highly viscous liquids are used with volumetric apparatus calibrated for water.

## 7.3 POTENTIOMETRIC MEASUREMENTS WITH CELLS CONTAINING A GLASS ELECTRODE IN SELECTED NONAQUEOUS SOLVENTS

The calibration of a cell with liquid junction that contains a glass electrode and a suitable reference electrode dipping into a solution of acid or base prepared in a nonaqueous amphiprotic solvent differs from the type of calibration that is used for aqueous solutions. It is worth recalling that the $p$H scale with its well-defined standardizing aqueous buffer solutions was developed entirely from potentiometric measurements. That this was possible is due fundamentally to the observation that dilute solutions of hydrochloric acid are completely ionized in water and that it was possible, therefore, to calibrate the cell Pt; $H_2 | $HCl (known molality)$|$AgCl; Ag by extrapolation of the measured values of $E_{cell}$ versus a function of the ionic strength in order to obtain the calibration factor $E^{\circ}_{cell}$. Once $E^{\circ}_{cell}$ was known it was possible to measure $p(a_{H^+}\gamma_{Cl^-})$ for buffer solutions and from these to derive the $p$H values. This approach is valid only for those solvents of large dielectric constant in which a strong acid such as hydrochloric acid is wholly ionized. Thus, for example (32), similar types of measurements can be performed in formamide ($\epsilon = 110$) but not in dimethylformamide ($\epsilon = 37$). In the latter solvent the overall dissociation constant exponent for hydrogen chloride referring to the equilibrium constant

$$K_D = \frac{c_{H^+} \cdot c_{Cl^-}}{c_{HCl} + c_{H^+Cl^-}}$$

is $pK_D$ approximates 3.5 (33). This refers to the overall dissociation which includes a concentration term for the ion pair $H^+Cl^-$ that exists in solutions of lower dielectric constant. The glass electrode was, therefore, calibrated in dimethylformamide using buffer solutions prepared from substances the $pK$ values of which were determined both by conductimetry and spectrophotometry. This indirect method of calibration is inherently less accurate than that used in aqueous solutions and the rate of response of the glass electrode in nonaqueous solutions is considerably slower than it is in aqueous solutions. The method of calibrating a cell to measure $pa_H$ values in a particular solvent is often unique to that solvent and this section deals with aspects of cell design, the conditioning, calibration, and use of the glass electrode in obtaining data from which the $pK$ value of an acid or base in the solvent may be derived. To this end, each solvent is considered separately.

## General Cell Design

An H-shaped cell that accommodates the glass electrode in one vertical arm, the reference electrode in the other, with the salt bridge in the horizontal section is the general design most favored. For example, in the cell described for use with acetonitrile (34) the horizontal section was of 12 mm diameter and contained two sintered glass discs of fine porosity sealed 3.5 cm apart. Between these two discs a filling tube of 4 mm diameter was sealed to the upper surface and a drainage tube fitted with a stopcock was sealed to the lower surface. One vertical limb of the H was of 20 mm diameter and provided the compartment for the glass electrode, whereas the other vertical limb was formed into a small compartment of about 10 mm diameter. The solution contained in this compartment and in the horizontal section was a 0.1 $M$ solution of tetraethylammonium perchlorate prepared in acetonitrile. Electrical contact between this solution and a separate reference half-cell containing a 0.01 $M$ solution of silver nitrate in acetonitrile was made through a capillary side arm sealed to the half-cell vessel. This side arm also contained the silver nitrate solution. Through the bottom of this vessel a platinium wire had been sealed and coated electrolytically with silver (ex. 0.05 $M$ argentocyanide solution) to form the reference electrode. All sections of the cell were stoppered to prevent ingress of atmospheric contaminants. The dimensions of the glass electrode compartment were such that the potential of the glass electrode with respect to the reference half-cell could be measured using a 5 mL aliquot of a stock solution of the tetraethylammonium salt of acid under study. This solution was diluted subsequently by the incremental additions of known amounts of the pure acid stock solution, the value of $E_{cell}$ being measured after each addition.

The bridge solution in the horizontal section was changed after each series of measurements.

## Measurement in Glacial Acetic Acid

Glacial acetic acid was the first organic solvent to be used for the determination of bases (35). It is more protogenic than water and is used as a titration medium for bases that are too weak to be titrated in aqueous solution. For samples containing a mixture of bases it is often the solvent of choice in which to determine the total amount of base present. The anhydrous solvent (mp 16.55–16.60°) is prepared (36) by adding the calculated amount of acetic anhydride to reagent grade (99.8%) glacial acetic acid, refluxing for 20 h and distilling. The acid has also been dried by passage through a column of 5A Molecular Sieve (37).

The application of potentiometry to the investigation of acid–base equilibria in anhydrous acetic acid has relied upon data obtained initially by spectrophotometry. For a cell that contains an electrode reversible to the solvated proton and an appropriate reference electrode, the Nernst equation relating $E_{cell}$ to the activity of hydrogen ion in a solution of acetic acid can be written as

$$E_{cell} = \left(E_{cell}^° + E_j\right) + k \log a_{H^+} \qquad (7.11)$$

Although strong acids are almost wholly dissociated in acetic acid, the low dielectric constant of the solvent causes the positive and the negatively charged ions to remain primarily as ion pairs. This excludes the possibility of obtaining the calibration factor ($E_{cell}^° + E_j$) by the measurement of $E_{cell}$ for a solution of a calculated activity of hydrogen ion. If, however, an overall dissociation constant defined as

$$K_{HX} = \frac{a_{H^+} \cdot a_{A^-}}{a_{HX} + a_{H^+X^-}} \cong \frac{a_{H^+}^2}{c_{HX}} \qquad (7.12)$$

is known, then the cell can be calibrated using a solution of known acid concentration $c_{HX}$. Combination of equations (7.11) and (7.12) yields

$$E_{cell} = \left(E^° + E_j\right) + \frac{k}{2} \log K_{HX} + \frac{k}{2} \log c_{HX} \qquad (7.13)$$

from whence measurement of $E_{cell}$ can yield the calibration factor $E^° + E_j$. In the original work (38) the chloranil electrode was calibrated by this method using the $K_{HX}$ value for hydrogen chloride ($pK = 8.55$) determined by spectrophotometry (39).

An analogous procedure was used for bases, the ionization constants of which are customarily reported as $pK_B$ values rather than as $pK_a$ values for the species $BH^+$. Thus the equation analogous to (7.12) is

$$K_B = \frac{a_{BH^+} \cdot a_{Ac^-}}{a_B + a_{BH^+Ac^-}} \cong \frac{a_{Ac^-}^2}{c_B} \qquad (7.14a)$$

The term $a_{Ac^-}^2$ can be eliminated from this equation in favor of the corresponding term $a_{H^+}$ by using the definition of the autoprotolysis constant for glacial acetic acid which is

$$K_{SH} = a_{H^+} \cdot a_{Ac^-}$$

Thus

$$a_{H^+} = \frac{K_{SH}}{(K_B c_B)^{1/2}} \qquad (7.14b)$$

and substitution of this term for $a_{H^+}$ in equation (7.11) yields

$$E_{cell} = \left(E_{cell}^\circ + E_j\right) + k \log K_{SH} - \frac{k}{2} \log K_B - \frac{k}{2} \log c_B \qquad (7.15)$$

This equation was solved for $K_{SH}$ by measuring the potential of the calibrated cell when it contained a solution of known concentration of pyridine, the overall dissociation constant of which had been determined spectrophotometrically (40) as $pK_B = 6.10$. The value obtained for $pK_{SH}$ was 14.45 at 25° (7). This pioneering work of Kolthoff and Bruckenstein (7, 38–40) has formed the basis for the subsequent determinations of $pK_B$ values using the glass electrode to measure the activity of the solvated proton in glacial acetic acid at 25°.

Before use the glass electrode was conditioned in anhydrous glacial acetic acid for two weeks and was stored in this solvent between measurements (41). When immersed in a solution of base, about 30 min was required before a stable potential was obtained (42). The cell, consisting of a glass electrode and a sleeve type calomel electrode that contained a saturated solution of lithium chloride in acetic acid, was calibrated for the determination of $pK_B$ values of bases using the following data:

| | |
|---|---|
| Lithium acetate | $pK_B = 6.79$ |
| Sodium acetate | $pK_B = 6.58$ |
| Urea | $pK_B = 10.18$ |

Measurements of $E_{cell}$ were made separately on 0.004 $M$ solutions of these three bases, a minimum of four e.m.f. readings being taken for each solution (43). A graph was constructed of $E_{cell}$ versus $pK_B$ and the measured value of $E_{cell}$ for 0.004 $M$ solutions of other bases interpolated from the graph in terms of their $pK_B$ values. A variant of this method is to use equation (7.15) and apply it to two bases B1 and B2 one of which is a standard. This yields the equation

$$\frac{2(E_{B1} - E_{B2})}{k} = \log c_{B2} K_{B2} - \log c_{B1} K_{B1} \qquad (7.16)$$

If solutions are prepared so that $c_{B2} = c_{B1}$, then $pK_{B2}$ can be calculated by reference to a standard base such as sodium acetate, the $pK_B$ of which was determined to be 6.58 both by conductimetry and spectrophotometry (38). Equation (7.16) can be written as

$$\frac{2\Delta E}{k} = pK_{B1} - pK_{B2} \qquad (7.17)$$

Two series of solutions of the bases B1 and B2 are prepared separately covering a range of identical concentrations not exceeding 0.004 $M$. The measured value of $E_{cell}$ for each solution is plotted against its concentration for both series on the same graph in order to check that the response of the electrode is linear for both bases. Values of $\Delta E$ for identical concentrations of the bases are interpolated from the graph in order to calculate the unknown $pK_B$ from equation (7.17).

A method that has been used to measure $pa_H$ in solutions of anhydrous acetic acid with a glass electrode also employs sodium acetate as a standard (44). From equation (7.14b) the $pa_H$ of a 0.1 $M$ solution of sodium acetate in acetic acid can be calculated as

$$pa_H = pK_{SH} - \tfrac{1}{2}pK_B + \tfrac{1}{2}\log c_B$$

$$= 14.45 - 3.29 - 0.5 = 10.66 \qquad (7.18)$$

If a $p$H meter fitted with a wide ranging asymmetry potential control is operated in the mV mode, and the potential adjusted to read 1066 mV when the electrodes are immersed in 0.1 $M$ sodium acetate, then $pa_H$ for any other solution for which $E_{cell}$ is measured is given by

$$pa_H = 10.66 - \frac{(1066 - E)}{59.1} = \frac{E - 436}{59.1}$$

where $E$ is the observed potential in millivolts. Other standards with which to check that this equation validly represents the response of the glass electrode can be calculated for solutions of 0.1 $M$ perchloric acid and 0.1 $M$ sodium perchlorate using the known $p$K value of 4.87 (38) for perchloric acid. Thus the $pa_H$ for 0.1 $M$ perchloric acid calculated from equation (7.12) is 2.94 and the $pa_H$ for 0.1 $M$ sodium perchlorate calculated from equation (7.18) is 6.42. When the cell was standardized with 0.1 $M$ sodium acetate, $pa_H$ 10.66, the actual values calculated from equation (7.19) for 0.1 $M$ sodium perchlorate and 0.1 $M$ perchloric acid were 6.42 and 2.98, respectively, indicating that the liquid junction potential remained constant over this range and that the response of the glass electrode was satisfactory. The materials were prepared by the following methods:

*Sodium acetate*: This was recrystallized twice from acetic acid and dried at 200° for 24 h.

*Sodium perchlorate*: The anhydrous reagent was recrystallized from acetic acid, and dried at 180° for 24 h.

*Perchloric acid* (39): A 1 $M$ stock solution was prepared by adding a known amount of analyzed reagent grade 72% perchloric acid to chilled anhydrous acetic acid (750 mL). The mixture was frozen and a calculated amount of acetic anhydride added cautiously. The solution was found to contain 0.22% water (Karl Fischer method) and to be 0.975 $M$ in perchloric acid at 25.5°. It was standardized titrimetrically using pure potassium hydrogen phthalate.

*Lithium acetate* (38): Lithium acetate dihydrate (30 g) was added cautiously to boiling acetic anhydride (75 mL), and acetic acid (25 mL) was added after the reaction ceased. After cooling, benzene (50 mL) was added. The precipitate was filtered, washed thoroughly with benzene, and dried at 50° *in vacuo* for one week.

*Urea*: The reagent grade material was recrystallized from acetic acid, washed with ethanol, and dried to constant weight in a vacuum oven at 90°.

The ion-pair dissociation constants $K_{BHClO_4}$ pertaining to reactions of the type

$$BH^+ClO_4^- \rightleftharpoons BH^+ + ClO_4^-$$

have also been calculated from measurements of the e.m.f. of cells in which the glass electrode was the indicator electrode. This method is also a comparative one that utilizes the values of $K_B = 10^{-6.58}$ for sodium acetate and $K_{BHClO_4} = 10^{-5.85}$ (38) as the basic standards for the method (43). These values are included in Table 7.2.

Table 7.2. Overall Dissociation Constants for Bases and Salts in Glacial Acetic Acid (25°)

| Compound | $pK_B$ | $pK_{BHX}$ | $pK_B(H_2O)^a$ |
|---|---|---|---|
| Acetamide | $10.50^c$ | $5.25^{c,e}$ | 14.60 |
| Acetanilide | $10.96^c$ | $4.53^{c,e}$ | ~15.00 |
| Benzylamine | $6.12^c$ | $5.18^{c,e}$ | 4.62 |
| Butylamine (butanamine) | $5.86^c$ | $4.93^{c,e}$ | 3.34 |
| 2-Butanamine | $5.68^c$ | $4.89^{c,e}$ | 3.44 |
| NN-Dibutyl-1-butanamine | $5.15^c$ | $4.14^{c,e}$ | 5.40 |
| 2,5-Dichloroaniline | $9.48,^b 10.25^c$ | $5.73^{c,e}$ | 12.47 |
| NN-Diethylaniline | $5.78^b$ | $5.82,^{b,e} 6.84^{b,f}$ | 7.43 |
|  | $5.20^c$ | $4.28^{c,e}$ |  |
| NN-Diethylethanamine (triethylamine) | $5.00^c$ | $3.81^{c,e}$ | 3.21 |
| NN-Dimethylaniline | $5.55^c$ | $4.44^{c,e}$ | 8.85 |
| 1,1-Dimethylhydrazine | $6.10^d$ |  | $6.82^d$ |
| Diphenylamine | $10.20^c$ | $5.85^{c,e}$ | 13.21 |
| 1,3-Diphenylguanidine | $5.15^c$ | $3.92^{c,e}$ | 3.88 |
| N-(4-Ethoxyphenyl)acetamide (phenacetin) | $10.90^c$ | $4.73^{c,e}$ | 11.80 |
| N-Ethylaniline | $9.10^c$ |  | 8.88 |
| N-Ethylethanamine (diethylamine) | $5.55^c$ | $4.59^{c,e}$ | 2.96 |
| Hydrazine | $5.82^d$ |  | $6.03^d$ |
| 2-Methoxyaniline | $10.25^c$ | $5.79^{c,e}$ | 9.47 |
| 4-Methoxyaniline | $10.60^c$ | $5.14^{c,e}$ | 8.64 |
| 2-Methyl-2-propanamine | $5.70^c$ | $4.89^{c,e}$ | 3.31 |
| o-Nitroaniline | $11.05^c$ | $4.51^{c,e}$ | 14.26 |
| N-Phenylbenzylamine | $9.73^c$ | $6.38^{c,e}$ | — |
| Pyridine | $6.10,^b 5.97^c$ | — | 8.77 |
| Thiourea | $9.14^g$ | $6.26^{c,e}$ | 15.20 |
| m-Toluidine | $10.50^c$ | $5.31^{c,e}$ | 9.29 |
| Tribenzylamine | $5.36^c$ | $6.71^{c,e}$ | — |
| Urea | $10.24^b$ | $6.96^{b,f}$ | 13.90 |
|  | $10.18^c$ | $5.90^{c,e}$ |  |
| Lithium acetate | $6.79^b$ | — | — |
| Lithium perchlorate | — | $5.31^{c,e}$ | — |
| Potassium acetate | $6.15^b$ | — | — |
| Sodium acetate | $6.58^b$ | — | — |
| Sodium perchlorate | — | $5.48^{b,e}$ | — |

[a] Data from Reference 45.
[b] Values from Reference 38.
[c] Values from Reference 43.
[d] Values from Reference 44.
[e] $X = ClO_4^-$.
[f] $X = Cl^-$.
[g] Values from Reference 46.

## Measurements in Acetonitrile

Acetonitrile is a weaker base and also a weaker acid than water with a much smaller value of the autoprotolysis constant ($\sim 10^{-33}$) than water ($10^{-14}$). Thus a much larger range of $pa_H$ values is available in acetonitrile which makes feasible the sequential titrations of the components of mixtures of acids or bases that would be impossible in water. The autoprotolysis constant for acetonitrile, defined as

$$K_{SH} = a_{CH_3CN^+H} \cdot a_{CH_2CN^-}$$

was determined (17) by measuring the overall ionization constant for the strong base tetramethylguanidine conductimetrically taking stringent precautions to prevent contamination of the solutions by carbon dioxide. The equilibrium can be represented as

$$B + CH_3CN \rightleftharpoons BH^+ + CH_2CN^-$$

for which the overall ionization constant $K_B$ was found to be $1.26 \times 10^{-9}$ at 25°. The value of $K_{BH^+}$, the ionization constant of the conjugate acid of this base, defined as

$$K_{BH^+} = \frac{a_B \cdot a_{H^+}}{a_{BH^+}}$$

was determined potentiometrically as $5.0 \times 10^{-24}$. The product of these two constants yields $K_{SH} = 6.3 \times 10^{-33}$ at 25° or $pK_{SH} = 32.2$. This value of $pK_{SH}$ must be regarded as a minimum value because the conductance of the solutions was found to be very small and was greatly affected by traces of impurities.

Acids that are completely ionized in water ($\epsilon = 78.5$) may also be highly ionized in acetonitrile, but the lower dielectric constant of this solvent ($\epsilon = 36.0$) leads to the formation of ion pairs that are incompletely dissociated. Thus hydrobromic acid ($pK$ 5.5), sulfuric acid ($pK$ 7.25), hydrochloric acid ($pK$ 8.9), and nitric acid ($pK$ 8.9) all appear to be weak acids in acetonitrile, the $pK$ values referring to the overall equilibrium for the processes

$$HA + CH_3CN \rightleftharpoons CH_3CN^+H \cdot A^- \rightleftharpoons CH_3CN^+H + A^-$$

By contrast, perchloric acid appears to be completely dissociated in acetonitrile. However, despite this, solutions of perchloric acid cannot be used to calibrate the glass electrode because the hydrogen ion activity of these solutions decreases on standing (34). The effect becomes noticeable one hour after preparation of the solution, and after a year's aging, the hydrogen ion activity decreases by more than a factor of $10^3$. The total acidity of the solution as determined by titration remains unchanged, however, and the effect has been attributed to the formation of a polymer of acetonitrile which is a stronger base than the monomeric solvent (48). Thus

$$CH_3CN^+H + \text{polymer} \rightleftharpoons \text{polymer } H^+ + CH_3CN$$

The effect on the shape of the titration curve for a base as experimentally determined with a cell containing a glass indicator electrode is most noticeable after the equivalence point, the electrode recording a considerably higher $pa_H$ than corresponds to the concentration of free perchloric acid. A consequence of this is that the curves are not symmetrical about the equivalence point. Nevertheless, the inflections that are observed are sharp and precise.

Values of $pa_H$ in acetonitrile were measured (34) with the glass electrode incorporated into the H-shaped cell configuration which can be represented as

$$Ag \left| \begin{array}{c} 0.01\ M\ AgNO_3\ \text{in} \\ CH_3CN \end{array} \right| \begin{array}{c} TEAP\ \text{in} \\ CH_3CN \end{array} \left| \begin{array}{c} a_{H^+}\ \text{in} \\ CH_3CN \end{array} \right| \text{glass electrode} \quad (7.19)$$

In this notation TEAP is an abbreviation for tetraethylammonium perchlorate. Before being used in this cell, the linearity of the glass electrode was first checked with aqueous $p$H buffers, and when not in use it was stored in water. Immediately before use in acetonitrile the glass bulb was rinsed several times with absolute ethanol and dried in a current of dry nitrogen. The electrode was then soaked in purified acetonitrile for 20 min before being positioned in the cell through a tightly fitting polythene cap. Stable potentials ($\pm 2$ mV) were recorded in 5–10 min.

Cell (7.19) can be calibrated to measure $pa_H$ at $25°$ by using two buffer systems; one based upon picric acid ($pK = 11.0 \pm 0.1$), and the other based upon $o$-nitroaniline ($pK = 4.85$). The $pa_H$ values for solutions containing the picric acid–picrate buffer system, and for solutions containing the $o$-nitroaniline–$o$-nitroanilinium system were calculated from these spectrophotometrically determined (49) $p$K values as described in the following.

1. *The $pa_H$ values of buffer solutions containing picric acid and tetraethylammonium picrate.* It has been shown (50) that picric acid undergoes simple ionization up to relatively high concentrations and that tetraethylam-

monium picrate is virtually completely dissociated even in 0.1 $M$ solution. Thus $pa_H$ values of dilute solutions containing these two species can be calculated using the $pK$ value of picric acid (HPi) provided that the concentrations $c_{HPi}$ and $c_{Pi^-}$ are known

$$pa_H = 11.0 + \log\frac{c_{HPi}}{c_{Pi^-}} - 1.45(I)^{1/2} \quad (7.20a)$$

In this equation the activity coefficients have been approximated by the Debye–Huckel limiting equation

$$-\log y_i = AI^{1/2} \quad (7.20b)$$

where A has the value calculated by equation (7.6) and $I$ is the concentration of tetraethylammonium picrate, $c_{Pi^-}$. In the preparation of the buffer solutions (49), $c_{Pi^-}$ was maintained in the range $1.4–7.1 \times 10^{-3}$, whereas $c_{HPi}$ was varied over the wider range $9.5 \times 10^{-5}–8.6 \times 10^{-2}$. This enabled a range of $pa_H$ values extending from 12.2 to 9.6 to be used for the calibration of the cell. From a graph of $E_{cell}$ versus $pa_H$, the values of the constant and slope could be obtained from which the $pa_H$ of a solution could be calculated by means of the equation

$$pa_H = \frac{\text{constant} - E_{cell}}{\text{slope}} \quad (7.21)$$

2. *The $pa_H$ values of buffer solutions containing o-nitroaniline and o-nitroanilinium perchlorate.* The $pa_H$ values of mixtures of o-nitroaniline, concentration $c_B$, and its conjugate acid, concentration $c_{BH^+}$, were calculated similarly to the picric acid–picrate buffer system. In this case, however, the equation is

$$pa_H = 4.85 + \log\frac{c_B}{c_{BH^+}} + 1.45(I)^{1/2} \quad (7.22)$$

The value of $c_{BH^+}$ was calculated stoichiometrically from the volume of a standard solution of perchloric acid added to a solution containing a known concentration of o-nitroaniline. These buffer solutions were prepared in $2 \times 10^{-3}$ $M$ tetraethylammonium perchlorate, and thus the ionic strength is given by $I = 0.002 + c_{HClO_4}$. Calibrating solutions covering the $pa_H$ range 4.20–6.48 were prepared, keeping the concentrations of tetraethylammonium perchlorate constant at 0.002 $M$ and the total concentration of the cationic and base forms of o-nitroaniline constant at $3.40 \times 10^{-3}$ $M$.

The concentration of perchloric acid added to the solutions were such that the concentration $c_{BH^+} = c_{HClO_4}$ and ranged from $9.1 \times 10^{-5}$ to $2.90 \times 10^{-3}$ M in the calibrating solutions. By plotting the values of $E_{cell}$ versus $pa_H$ on the same graph as the results for the picric acid–picrate buffer system the validity of equation (7.21) can be confirmed.

Other acid–base pairs could, of course, be used simlarly to prepare calibrating solutions for the glass electrode provided the $p$K values are known. Homoconjugation can be suspected, however, if a plot of $E_{cell}$ versus $pa_H$ is not linear and the apparent slope deviates grossly from the expected Nernst factor, 59.2 mV decade$^{-1}$ at 25°. It must be emphasized that the $pa_H$ values obtained are very dependent upon the purity of acetonitrile used in the preparation of the solution. The following methods are used for the purification of this and other substances used in the cell calibration.

1. *Acetonitrile* (51). Acetonitrile (3 L) was shaken for 24 h with activated alumina (200 g previously heated at 250° for 4 h). The solvent was decanted and the treatment repeated with a fresh portion of alumina. Water was removed by shaking the decanted solvent with portions of anhydrous calcium chloride (100–150 g) until the water content was less than 0.2%. This was followed by shaking the solvent for 24 h with anhydrous magnesium sulfate (100–150 g) and then for 1 h with phosphorus pentoxide. The solvent was distilled at atmospheric pressure through a column 1 m long and 2 cm internal dia. packed with stainless steel wool, the whole apparatus being protected against ingress of water with drying tubes filled with anhydrous calcium chloride. The first and last 100 mL portions of distillate were discarded. The water content was determined by Karl Fischer titration performed on a weighed quantity of acetonitrile dissolved in pure methanol. The concentration of water was usually within the range 0.7–2.0 mM. The solvent was contained in 1 L Pyrex bottles fitted with interchangeable stoppers and these were stored in the dark. When required, the solvent was dispensed by means of a glass siphon system fitted with the appropriately sized joints and protected against moisture with a calcium chloride drying tube. Traces of ammonia that formed after standing a few months when the solvent contained 1 to 2 mmole L$^{-1}$ of water was detected from the increase in absorption of a $10^{-4}$ M solution of picric acid at 420 μm. Freshly prepared batches were found to contain less than $10^{-6}$ mole L$^{-1}$ of ammonia. Other methods are given in Reference 52.

2. *Picric acid* (49). The acid (30 g) was dissolved in hot benzene (500 mL) and, after cooling, was transferred to a separating funnel

and shaken with 3 portions of 12 $M$ hydrochloric acid (3 × 10 mL). The benzene layer was shaken with water (10 mL portions) until the aqueous layer gave a negative test for chloride. The benzene solution was then evaporated at 50–60° *in vacuo* until the picric acid crystallized. The product was recrystallized from benzene and dried *in vacuo* at 70°.

3. *Tetraethylammonium picrate* (50). A saturated aqueous solution of picric acid (0.05 $M$) was titrated with a 10% aqueous solution of tetraethylammonium hydroxide to just beyond the equivalence point (detected with a glass electrode). The solution was evaporated until crystallization occurred. This product was recrystallized twice from water and then from 90% ethanol, and finally was dried over phosphorus pentoxide in a vacuum desiccator.

4. *o-Nitroaniline* (51). Before use the pure compound was recrystallized from ethanol and dried for 3 h at 60° at atmospheric pressure.

5. *Perchloric acid* (53). A stock solution of 1 $M$ perchloric acid in anhydrous acetic acid was prepared by a method similar to that described elsewhere in this book. Standard solutions of perchloric acid in acetonitrile were then prepared from this stock solution.

6. *Tetraethylammonium perchlorate* (54). A hot 1 $M$ aqueous solution of tetraethylammonium bromide was added slowly and with stirring to an equivalent amount of a hot 0.1 $M$ aqueous solution of sodium (or hydrogen) perchlorate. After cooling in ice water the precipitate was collected on a Buchner funnel and washed with ice cold water until the wash liquor was free from bromide. The salt was recrystallized from water and dried in a vacuum oven at 60°.

The determination of ionization constants of acids and conjugate acids in acetonitrile is rather more complicated than is implied by equations (7.20) and (7.22). That equation (7.20) could be applied to picric acid stems from the suppression of complex formation due to homoconjugation as a result of stabilization of the acid by intramolecular hydrogen bonding of the hydroxy group through the two ortho nitro groups. Picric acid is, therefore, atypical, and for the general case of the ionization of acids allowance must be made for the stabilization of the anions through combination with the undissociated acid. The equation derived (49) relating the measured hydrogen ion activity to the ionization constant $K_{HA}$ allows for this homoconjugation effect:

$$K_{HA}^2 c_a - a_H y K_{HA}\left[(c_a + c_s) + K_f(c_s - c_a)^2\right] + a_H^2 y^2 c_s = 0 \quad (7.23)$$

In this equation $c_a$ and $c_s$ are the known stoichiometric concentrations of the acid and the salt, respectively, $y$ is the activity coefficient (molar scale) calculated by equation (7.20b), and $K_f$ is the formation constant of the homoconjugation complex referring to the equilibrium

$$HA + A^- \rightleftharpoons A\ldots HA^-$$

This equilibrium constant $K_f$ may be expressed in terms of the ratio of the measured $a_H$ to the activity of the hydrogen ion at the half-neutralization point $a_{HNP}$. If this ratio, $r$ at constant ionic strength, is expressed as

$$r = \frac{a_H}{a_{HNP}}$$

then $a_H = rK_{HA}/y$ and substitution of this term in equation (7.23) yields

$$K_f = \frac{c_s r^2 - r(c_a + c_s) + c_a}{r(c_s - c_a)^2} \quad (7.24)$$

In the determination of the $pK_{HA}$ values of aromatic carboxylic acids, for example (55), $pa_H$, values were measured for solutions containing mixtures of $4 \times 10^{-4}$–$10^{-2}$ $M$ acid, and $10^{-3}$–$6 \times 10^{-3}$ $M$ of its tetraethylammonium salt. The values of the half-neutralization potential at the various concentrations of acid and salt were obtained at $c_a/c_s = 1$ from plots of $E_{cell}$ versus log $(c_a/c_s)$. From the value of the half-neutralization potential ($E_{HNP}$), the value of $a_{HNP}$ may be calculated from the cell calibration equation

$$pa_{HNP} = \frac{(\text{constant} - E_{HNP})}{\text{slope}} \quad (7.25)$$

Thus the values of $K_f$ may be calculated for each of the solutions from equation (7.24), and hence equation (7.23) may be solved for $K_{HA}$. Values of $pK_{HA}$ and $K_f$ determined by this method are included in Table 7.3. The tetraethylammonium carboxylate salts were prepared by dissolving the acid in a minimum volume of absolute alcohol and adding water (100 mL). These solutions were neutralized to the phenolphthalein endpoint by the addition of a 1 $M$ aqueous solution of tetraethylammonium hydroxide after which they were evaporated to dryness. The residues were then recrystallized from an ethyl acetate–ethanol mixture and dried *in vacuo* at 50–70°.

The determination of the ionization constants for bases referring to the process

$$BH^+ + CH_3CN \rightleftharpoons B + CH_3CN^+H$$

is often simpler than for acids since the homoconjugation constants $K_f$ for the equilibrium

$$B + BH^+ \rightleftharpoons BHB^+$$

are much smaller and can be neglected (see Table 7.3). The maximum value of $K_f$ determined spectrophotometrically for 25 bases was 35 (56) which would have negligible effect when substituted in equation (7.23). In such cases, therefore, it is sufficient to measure $pa_H$ and to calculate $pK_{BH^+}$ from the equation

$$pK_{BH^+} = pa_H - \log\frac{c_B}{c_{BH^+}} - 1.45I^{1/2} \qquad (7.26)$$

The method adopted (56) was to prepare four or five buffer solutions containing a constant concentration ($5 \times 10^{-4}$ M) of the perchlorate or picrate salt of the base in a constant ionic medium of 0.01 M tetraethylammonium perchlorate solution. To each of these solutions was added a different amount of base so that the final concentration of the base in the solutions ranged from $2 \times 10^{-4}$ to $2 \times 10^{-3}$ M. The aminium picrates were

Table 7.3. Ionization Constant Exponents and Homoconjugation Formation Constants, $K_f$ for some Acids and Bases in Acetonitrile at 25°

| Acids | $K_f$ | $pK_{HA}$ | Bases | $K_f$ | $pK_{BH^+}$ |
|---|---|---|---|---|---|
| Benzoic acid[a] | $4 \times 10^3$ | 20.7 | Ammonia[d] | 11 | 16.5 |
| m-Bromobenzoic acid[a] | $6 \times 10^3$ | 19.5 | Methylamine[d] | 35 | 18.4 |
| p-Hydroxybenzoic acid[a] | $2 \times 10^3$ | 16.7 | Dimethylamine[d] | 31 | 18.7 |
| o-Hydroxybenzoic acid[a] | $1 \times 10^3$ | 20.8 | Trimethylamine[d] | 6 | 17.6 |
| p-Nitrobenzoic acid[a] | $7 \times 10^3$ | 18.7 | Ethylamine[d] | 25 | 18.4 |
| 3,5-Dinitrobenzoic acid[a] | $1 \times 10^4$ | 16.9 | Triethylamine[d] | Nil | 18.5 |
| Phenol[b] | $6 \times 10^{5c}$ | 26.6 | Piperidine[d] | 26 | 18.9 |
| o-Nitrophenol[b] | 100 | 22.0 | Pyrrolidine[d] | 32 | 19.6 |
| p-Nitrophenol[b] | $8.5 \times 10^{4c}$ | 20.7 | Pyridine[d] | 4 | 12.3 |
| 2,4-Dinitrophenol[b] | 100 | 16.0 | Aniline[d] | Nil | 10.6 |
| 2,4,6-Trinitrophenol[b] | Nil | 11.0 | 1,3-Diphenylguanidine[d] | Nil | 17.9 |
|  |  |  | Tetramethylguanidine[e] | — | 23.3 |

[a] From Reference 55.
[b] From Reference 47.
[c] Value refers to the overall reaction $A^- + 2HA \rightleftharpoons (HA)_2A^-$.
[d] From Reference 56.
[e] From Reference 17.

prepared by titrating an ethanolic solution of the base with a saturated solution of picric acid using a glass electrode to detect the equivalence point. These salts can generally be recrystallized from ethanol, and then dried *in vacuo*. The perchlorate salts were prepared by titration of the base in an appropriate solvent with 70% aqueous perchloric acid.

## Measurements in Dimethylformamide

Dimethylformamide is a protophilic solvent that is more basic than water or acetonitrile and is a useful solvent for the titration of many types of individual acids (57–60). The formation of homoconjugation complexes occurs to a lesser extent in dimethylformamide than in acetonitrile with the result that the shapes of the titration curves are often normal in the buffer region. It is a good solvent for a wide range of polar and nonpolar organic compounds, and it will also dissolve many inorganic perchlorates, especially those of the alkali metals and the alkaline earths. Chlorides, in general, are only sparingly soluble, the exception being lithium chloride (52). Dimethylformamide hydrolyzes slightly, especially when wet, to form formic acid. The solvent has been purified by the following procedure (61) to yield a product that contained less than $5 \times 10^{-6}$ $M$ acidic or basic impurities and less than 10 ppm water:

> Reagent grade dimethylformamide (400 mL) was treated with Linde AW-500 Molecular Sieves (90 g) and then purified by vacuum fractional distillation from phosphorus pentoxide (25–30 g). Before transferring the slurry of phosphorus pentoxide and dimethylformamide to the three-necked distillation flask (600 mL capacity), the apparatus was flushed with nitrogen. During the distillation a capillary bleed attached to the nitrogen supply was of such a diameter as to maintain a pressure of between 2.5 and 8.0 mm of nitrogen in the apparatus. The distillation temperature was within the range 33–49° and the first fraction (75–100 mL) was found to contain about $10^{-5}$ $M$ basic impurities and was discarded. The second fraction (75–200 mL) was collected and stored under nitrogen at $-20°$ in a freezer. The purified solvent was used within 48 h following preparation.

The glass electrode has been proved reliable for use in dimethylformamide without modification (59). However, an electrode modified by replacing the internal reference solution with a dimethylformamide solution containing about $10^{-3}$ $M$ silver perchlorate, picric acid, and tetraethylammonium perchlorate has also been used (61). In this instance the internal reference electrode was a silver wire positioned in the electrode through a rubber serum bottle cap. These modified electrodes were stored in either the pure solvent or in a solution containing about $10^{-3}$ $M$ picric acid. Unmodified glass electrodes have also been stored in the pure solvent when not in use (59), but other workers have preferred to store them in water (62). In the latter case the glass electrode was rinsed with ethanol, dried under a flow of compressed air, and immersed in pure dimethylformamide for at least 30 min before use. Whether stored in water or in the solvent, equilibrium potentials were found to be established within 15 min in acidic solutions. In basic media ($pa_H \sim 20$), however, up to 2 h were needed to reach steady values (62). The glass electrode has been incorporated into the same type of cell as (7.19) in which all solutions were prepared in dimethylformamide (59). It has also been used in conjunction with a calomel reference half-cell, the salt bridge of which was a dimethylformamide solution saturated both in mercury(I) chloride and in tetraethylammonium picrate (62).

The glass electrode can be calibrated for use in dimethylformamide by methods analogous to those used for its calibration in acetonitrile. In this case, however, the buffer systems used have been based on either 4-chloro-2,6-dinitrophenol or 2,6-dinitrophenol (59). The $p$K values of these acids, determined spectrophotometrically at 25°, are 4.70 and 5.77, respectively. If the 2,6-dinitrophenol buffer is used, then solutions are prepared containing tetraethylammonium 2,6-dinitrophenolate (concentration range 5.11–8.45 × $10^{-3}$) and 2,6-dinitrophenol (concentration range 1.47–29.2 × $10^{-3}$ $M$). The $pa_H$ values for these solutions can be calculated by

$$pa_H = 5.77 + \log \frac{c_s}{c_a} - 1.55 I^{1/2} \qquad (7.27)$$

where $c_s$ is the concentration of the tetraethylammonium 2,6-dinitrophenolate, and $c_a$ is the concentration of 2,6-dinitrophenol. Values of $E_{cell}$ are measured for a series of these buffer solutions and plotted against $pa_H$ calculated by equation (7.27) to obtain a cell calibration equation of the same type as equation (7.21). Tetraethylammonium 2,4-dinitrophenolate can be prepared by titrating a solution of 2,6-dinitrophenol in ethanol with a 1 $M$ aqueous solution of tetraethylammonium hydroxide almost to the equivalence point as monitored with a glass electrode, and evaporating the solution to dryness. The excess phenol was removed by washing the residue with anhydrous ether after which the solid was recrystallized from ethyl acetate–petroleum ether and dried *in vacuo* at 70°. 2,6-Dinitrophenol (reagent grade) was recrystallized from water and dried *in vacuo* at 50° (63).

Alternatively, it has been reported that the glass electrode can be calibrated using solutions of known concentrations of picric acid provided that these concentrations do not greatly exceed $10^{-3}$ $M$ (61). At concentra-

Table 7.4. Values of $pc_H$ (Molar Scale) of Acid–Salt Buffer Solutions in Dimethylformamide[a]

| Acid (HA) | Salt | $pc_H$ |
|---|---|---|
| Picric | KA | 2.20 |
| 2,6-Dinitrophenol | KA | 6.05 |
| Salicylic | NaA | 7.94 |
| $p$-Nitrobenzoic | LiA | 10.00 |
| Benzoic | LiA | 11.70 |

[a] Solutions 0.02 $M$ with respect to both acid and salt.

tions of less than $10^{-3}$ $M$, picric acid was shown spectrophotometrically to be completely dissociated, and this was confirmed by the observation that a linear relationship exists between $E_{cell}$ and $pc_H$ over the concentration range $4.4 \times 10^{-5}$–$1.2 \times 10^{-3}$ ($pc_H = 4.36 - 2.92$). The electrode can be calibrated, therefore, by titration of the purified solvent with a dilute solution ($\sim 1.5 \times 10^{-2}$ $M$) of picric acid. A difference of only 0.05 $pc_H$ unit exists between this scale and that based on the more concentrated buffer solution containing picric acid and potassium picrate given in Table 7.4 together with four other reference $pc_H$ solutions (64).

Because homoconjugation constants are at least 10 times smaller in dimethylformamide than in acetonitrile, the ionization constants of many acids may be determined from measurements of $pa_H$ in dilute solutions of the acid and its salt using the simple expression

$$pK = pa_H - \log\frac{c_s}{c_a} + 1.55\sqrt{c_s} \qquad (7.28)$$

A series of solutions are prepared in which the concentrations of the tetraethylammonium salt of the acid ($c_s$) are kept constant and the concentrations of the acid ($c_a$) are varied. A check should be made that when values of $E_{cell}$ for these solutions are plotted against $\log c_a/c_s$, a linear relation is obtained with a slope that agrees with the value of the slope obtained from the calibration of the cell. A gross discrepancy between the two slopes is indicative that homoconjugation of the anion is occurring, invalidating the use of equation (7.28). The method given for acetonitrile should then be followed. A method for the determination of the $pK$ values for the diprotic acids malonic, succinic, and dimethylmalonic acid, and for the determination of the formation constants of the homoconjugation complexes of these acids is given in Reference 62. In contrast to the ionization of these acids in water, there is a great difference between the values of the successive ionization constants $K_1$ and $K_2$, $K_2$ being at least $10^{9.5}$ times smaller than $K_1$.

The ionization constant exponent of the conjugate acid species $BH^+$ in dimethylformamide may be determined by the same method as that described for bases in acetonitrile. The appropriate form of equation (7.26) is

$$p K_{BH^+} = p a_H - \log \frac{c_B}{c_{BH^+}} - 1.55\sqrt{I} \qquad (7.29)$$

Provided that dilute solutions are used in the determination, the extent of homoconjugation in this solvent can be regarded as negligible. It should be noted (see Table 7.5) that there appears to be no strong base in dimethylformamide, the $pK_{BH^+}$ of tetramethylguanidine being 13.6 in comparison to its value of 23.3 in acetonitrile. For this reason it has not been possible to obtain an acceptable value for the autoprotolysis constant of dimethylformamide. However, it appears that the value $pK_{SH} = 27$ (11) is a reasonable estimate (62).

## Measurements in *N*-Methyl-2-Pyrrolidinone

Insofar as it is possible to make comparisons, the acid-base properties of this commercially available solvent appear to be similar to those of dimethylformamide. Thus the published $pK$ values for acids (66) in *N*-methyl-2-pyrrolidinone are similar to their values in dimethylformamide. Benzoic acid, for example, has a $pK$ of 12.3 in both solvents. Other comparisons can be made as follows, the $pK$ in dimethylformamide being given in parentheses for each acid: Acetic acid 13.3 (13.5); 2-hydroxybenzoic acid 8.6 (8.2); 4-nitrobenzoic acid 10.5 (10.6); 3-nitrophenol 14.3 (15.4); 4-nitrophenol 12.5 (12.3); 2,4-dinitrophenol 6.8 (6.3); 4-chlorophenol 16.5 (16.8). It is more difficult to make comparisons between the $pK$ values of bases in the two solvents for only two substances have been determined in both. Furthermore, the values are reported as $pK_{BH^+}$, referring to the cationic acids in Table 7.5, but as $pK_B$ values in Reference 66. However, taking the autoprotolysis constant exponent $pK_{SH}$ for *N*-methyl-2-pyrrolidinone to be 24.3 (12) allows the following comparisons of $pK_{BH^+}$ to be made: triethylamine 8.7 (9.2); diphenylguanidine 8.9 (9.1). Once again, the values in dimethylformamide, given in parentheses, are of the same order of magnitude as those determined in *N*-methyl-2-pyrrolidinone. The solvent combines a high boiling point, low vapor pressure, and good solving power with a lower toxicity than dimethylformamide. It has been used as a titration medium for the determination of sterically hindered phenols, bisphenols, and hydroxyphenyl ketones (67).

Table 7.5. $p$K Values of Acids in Dimethylformamide (25°)

| Acid | $p$K | log $K_f$ | $p$K $(H_2O)^a$ | Ref. |
|---|---|---|---|---|
| Hydrochloric acid | 3.20 | 2.2 | — | 59 |
| *Carboxylic Acids* | | | | |
| Acetic acid | 13.50 | 2.6 | 4.76 | 59 |
| Benzoic acid | 12.27 | 2.4 | 4.21 | 59 |
| 2,6-Dihydroxybenzoic acid | 3.56 | — | 1.05 | 59 |
| 3,5-Dinitrobenzoic acid | 8.95 | 3.2 | 2.79 | 59 |
| 2-Hydroxybenzoic acid | 8.24 | 1.7 | 3.08 | 59 |
| 4-Nitrobenzoic acid | 10.60 | 2.6 | 3.43 | 59 |
| *Phenols* | | | | |
| 3-Chlorophenol | 16.29 | — | 9.12 | 61 |
| 4-Chlorophenol | 16.78 | — | 9.41 | 61 |
| 4-Chloro-2,6-dinitrophenol | 4.70 | — | 2.96 | 59 |
| 2,4-Dinitrophenol | 6.33 | — | 4.07 | 59 |
| 2,6-Dinitrophenol | 5.77 | — | 3.70 | 59 |
| 3,5-Dinitrophenol | 11.30 | 3.2 | 6.68 | 59 |
| 3-Nitrophenol | 15.43 | — | 8.36 | 61 |
| 4-Nitrophenol | 12.34 | — | 7.16 | 61 |
|  | 12.64 | 2.3 | — | 59 |
| 3-Trifluoromethylphenol | 15.70 | — | 8.95 | 61 |
| 2,4,6-Trinitrophenol (Picric acid) | 1.60 | — | 0.33 | 59 |
| *Heteroaromatic Acids* | | | | |
| 4-Amino-5-bromopyrimidin-2(1$H$)-one | 16.20 | — | 10.52 | 60 |
| 4-Amino-5-methylpyrimidin-2(1$H$)-one | 19.00 | — | 12.40 | 60 |
| 5-Aminopyrimidine-2,4(1$H$, 3$H$)-dione | 16.00 | — | 9.52 | 60 |
| 4-Aminopyrimidin-2(1$H$)-one (Cytosine) | 18.90 | — | ~ 12.30 | 60 |
| 5,6-Dihydro-5-methylpyrimidine-2,4 (1$H$, 3$H$)-dione | 18.50 | — | 11.70 | 60 |
| 5,6-Dihydropyrimidine-2,4(1$H$, 3$H$)-dione | 18.30 | — | 11.74 | 60 |
| 2,3-Dihydro-2-thioxopyrimidin-4(1$H$)-one (2-Thiouracil) | 12.10 | — | 7.75 | 60 |
| 3,4-Dihydro-4-thioxopyrimidin-2(1$H$)-one (4-Thiouracil) | 12.20 | — | 7.82$^c$ | 60 |
| 5-Methylpyrimidine-2,4(1$H$, 3$H$)-dione (Thymine) | 15.50 | — | 9.90 | 60 |
| 5-Nitropyrimidine-2,4(1$H$, 3$H$)-dione | 9.10 | — | 5.56 | 60 |
| Pyrimidine-2,4(1$H$, 3$H$)-dione (Uracil) | 15.10 | — | 9.43 | 60 |
| Pyrimidine-2(1$H$)-one | 14.50 | — | 9.17 | 60 |

Table 7.5. (Continued)

| Acid | | $pK$ | $\log K_f$ | $pK\,(H_2O)^a$ | Ref. |
|---|---|---|---|---|---|
| *Diprotic Acids* | | | | | |
| Butanedioic acid (Succinic acid) | I | 10.40 | $5.0^b$ | 4.21 | 62 |
| | II | 19.20 | — | 5.64 | |
| Dimethylpropanedioic acid | | | | | |
| (Dimethylmalonic acid) | I | 8.30 | $7.3^b$ | 3.03 | 62 |
| | II | 21.50 | — | 5.73 | |
| Propanedioic acid (Malonic acid) | I | 7.80 | $6.4^b$ | 2.85 | 62 |
| | II | 20.80 | — | 5.70 | |
| Sulfuric acid | I | 3.10 | — | — | 59 |
| | II | 17.20 | — | — | |
| *Cationic Acids* | | | | | |
| 4-Ammoniomethylpyridinium | | | | | |
| ($4\text{-}N^+H_3\text{—}CH_2\text{—}C_6H_4N^+H$) | | 3.89 | — | — | 61 |
| Ammonium | | 9.45 | — | — | 59 |
| Anilinium | | 4.36 | — | 4.60 | 59 |
| 4-Benzylpyridinium | | 4.86 | — | 5.59 | 61 |
| Butylammonium | | 9.10 | — | 10.65 | 59 |
| 4-Cyanomethylpyridinium | | 3.71 | — | — | 61 |
| Diphenylguanidinium | | 9.10 | — | 10.12 | 59 |
| 4-Ethylpyridinium | | 5.09 | — | 6.03 | 61 |
| 4-Hydroxylmethylpyridinium | | 4.53 | — | 5.33 | 61 |
| 4-Methylpyridinium | | 5.07 | — | 5.99 | 61 |
| Pyridinium | | 5.06 | — | 5.23 | 61 |
| Tetramethylguanidinium | | 13.65 | — | 13.60 | 59 |
| Triethylammonium | | 9.25 | — | 10.82 | 59 |

$^a$ Values from References 46 and 65.
$^b$ For the formation of the species $H_2A_3^{4-}$.
$^c$ Value from Reference 60.

An investigation of the cell

$$\text{Ag; AgCl} \left| \begin{array}{c} \text{Satd. KCl in} \\ \text{methanol} \end{array} \right| \left| \begin{array}{c} \text{HA in } N\text{-methyl} \\ \text{-2-pyrrolidinone} \end{array} \right| \text{glass electrode} \quad (7.30)$$

revealed that a linear relationship existed between $E_{\text{cell}}$ and $\log c_H$ with a slope of 59 mV decade$^{-1}$ for solutions of perchloric, picric, and 4-toluenesulfonic acids over the concentration range approximately 0.01–0.002 $M$. This observation formed the basis for the calibration of the cell using stoichiometric concentrations of 4-toluenesulfonic acid (66).

For the determination of the $pK$ values of acids in $N$-methyl-2-pyrrolidinone, the electrodes of cell (7.30) were transferred between a standard solution (0.005 $M$) of 4-toluenesulfonic acid and a half-neutralized solution of the acid under investigation. The latter solution was prepared by titrating a 0.01 $M$ solution of the weak acid (25 mL) with 0.5 mole equivalent of a 0.1 $M$ solution of tetraethylammonium hydroxide prepared in a 9:1 benzene–methanol medium. The difference in potential, $\Delta E$ volts, between the standardizing strong acid solution of concentration $c_{HA}$ and the half-neutralized solution of weak acid was related to the $pK$ of the latter at 25° by

$$pK = \frac{\Delta E}{0.059} - \log c_{HA} - 2\log y_i \qquad (7.31)$$

in which all the terms on the right-hand side are known. The term $\log y_i$ can be approximated by equation (7.20b) for which the relevant value of A is 1.98 at 25°.

The $pK_B$ values of bases were determined similarly (66), a standard solution of the strong base tetraethylammonium hydroxide replacing 4-toluenesulfonic acid as the reference solution in cell (7.30). A linear relation between $E_{cell}$ and the logarithm of the concentration of tetraethylammonium hydroxide over the range of 0.01–0.002 $M$ was observed in this instance. The difference in potential between a solution of weak base (0.01 $M$) half-neutralized with 4-toluenesulfonic acid titrant (0.1 $M$), and the tetraethylammonium hydroxide reference solution was interpreted in terms of the $pK_B$ for the equilibrium

$$B + HS \rightleftharpoons BH^+ + S^-$$

by an equation analogous to (7.31).

It should be reiterated that the half-neutralization potential is independent of any homoconjugation effects (34), and provided that the glass electrode can be reliably calibrated, this method offers an alternative to those already described for the determination of $pK$ values. In subsequent work (67) the 4-toluenesulfonic acid calibrating solution, referred to previously, was replaced by a half-neutralized solution of benzoic acid ($pK$ 12.3). For solutions of identical ionic strength and concentrations the difference between the half-neutralization potentials of the acid of the unknown $pK_{HA}$ and that of the standard acid having a known $pK_{St}$ is related to $pK_{HA}$ at 25° by

$$pK_{HA} = pK_{St} - \frac{\Delta E}{0.059} \qquad (7.32)$$

where $\Delta E$ is the difference in half-neutralization potential (i.e., $E_{HA} - E_{St}$). The reproducibility of values of $pK_{HA}$ determined by this method is about $\pm 0.2$ $pK$ units.

## Measurements in Dimethylsulfoxide

Dimethylsulfoxide is a versatile solvent for both organic and inorganic compounds. In general, iodides, bromides, chlorides, perchlorates, and nitrates are very soluble, but fluorides, sulfates, and carbonates are not (52). It is an amphiprotic solvent being slightly more protophilic than dimethylformamide, and is a potentially useful titration medium for carboxylic acids, phenols, and carbon acids. The range of $pa_H$ in this solvent has been extended up to $pa_H$ 35 by using a hydrogen electrode to measure these values in solutions of the sodium salt of dimethylsulfoxide in anhydrous dimethylsulfoxide (70). This sodium salt ($Na^+CH_3SOCH_2^-$), commonly called *sodium dimsyl*, has been used as a titrant in dimethylsulfoxide for very weak carbon acids (see Table 7.6) using a visual indicator (71). However, the solvent has not gained wide acceptance as a medium for potentiotitrimetry and this is possible due to the slow rate of response of the glass electrode in the anhydrous solvent ($<$ 0.005% water) at $pa_H$ values greater than 13. For an unmodified glass electrode about 45 min was required to attain an equilibrium potential (72), whereas for an electrode modified by replacing the internal filling solution with one prepared in dimethylsulfoxide or with mercury, a period of at least 15 min was required (73). It has been reported, however, that if a water content of approximately 0.3% can be tolerated in the solvent, an unmodified glass electrode attains reproducible potentials in 1–5 min and has been used up to $pa_H = 28$ (74). In this instance tetrabutylammonium hydroxide was used as titrant.

The cell (72),

Ag|0.01 $M$ AgNO$_3$ in DMSO||HA in DMSO|glass electrode

can be calibrated with the same buffer systems based on 2,6-dinitrophenol or 4-chloro-2,6-dinitrophenol that were used for the calibration of the cell in dimethylformamide. If solutions containing tetraethylammonium 2,6-dinitrophenolate (concentration range $1.3 \times 10^{-2}$–$3.6 \times 10^{-3}$ $M$) and 2,6-dinitrophenol (concentration range $6 \times 10^{-4}$–$5 \times 10^{-2}$) are used, then the equation analogous to (7.27) is

$$pa_H = 3.84 + \log \frac{c_s}{c_a} - 1.16\sqrt{I} \qquad (7.33)$$

Table 7.6. $p$K Values of Acids in dimethylsulfoxide (25°)

| Acid | $p$K (DMSO) | $p$K ($H_2O$) |
|---|---|---|
| *Carboxylic Acids* | | |
| Acetic acid | 12.6,[a] 12.3[b] | 4.76[c] |
| 3-Aminobenzoic acid | 11.6[d] | 4.74[c] |
| 4-Aminobenzoic acid | 12.7[d] | 4.68[c] |
| Benzoic acid | 10.9[b], 11.0[d], 11.1[a], 11.0[e] | 4.21[c] |
| 3-Bromobenzoic acid | 9.68[e] | 3.81[c] |
| 2-Chlorobenzoic acid | 9.29[e] | 2.90[c] |
| 4-Chlorobenzoic acid | 10.0[b], 10.1[d] | 3.99[c] |
| 4-Chloro-3-nitrobenzoic acid | 8.62[e] | — |
| 3,4-Dichlorobenzoic acid | 9.20[e] | 3.64[c] |
| 3,5-Dichlorobenzoic acid | 8.81[e] | 3.54[c] |
| 2,6-Dihydroxybenzoic acid | 3.1[a] | 1.05[c] |
| 3,4-Dimethylbenzoic acid | 11.41[e] | 4.50[c] |
| 2,4-Dinitrobenzoic acid | 6.52[e] | 1.42[c] |
| 3,5-Dinitrobenzoic acid | 7.40[e], 7.4[a] | 2.79[c] |
| 4-Ethoxybenzoic acid | 11.5[d] | 4.80(20°)[c] |
| 2-Hydroxybenzoic acid | 6.8[a], 6.6[e] | 3.08[c] |
| 3-Hydroxybenzoic acid | 11.1[d] | 4.08[c] |
| 4-Hydroxybenzoic acid | 11.8[d] | 4.50[c] |
| 3-Methylbenzoic acid | 11.0[d] | 4.25[c] |
| 4-Methylbenzoic acid | 11.2[b] | 4.37[c] |
| 2-Nitrobenzoic acid | 8.18[e] | 2.21[c] |
| 3-Nitrobenzoic acid | 9.17[e] | 3.46[c] |
| 4-Nitrobenzoic acid | 8.9[b], 8.9[d], 9,04[e] | 3.43[c] |
| *Phenols* | | |
| 4-Bromophenol | 15.5[b] | 9.37[c] |
| 4-Chloro-2,6-dinitrophenol | 3.5[a] | 2.96[c] |
| 2,6-Di-*tert*-butyl-4-nitrophenol | 7.6[a] | 11.70[c] |
| 2,6-Dinitrophenol | 3.84[a] | 3.70[c] |
| 3,5-Dinitrophenol | 10.6[a] | 6.68[c] |
| 4-Methoxyphenol | 18.2[b] | 10.21[c] |
| 4-Methylphenol | 17.9[b] | 10.28[c] |
| 2-Nitrophenol | 11.1[a] | 7.23[c] |
| 4-Nitrophenol | 11.0[a], 10.8[b] | 7.16[c] |
| Phenol | 16.4[a], 17.1[b] | 9.99[c] |
| Picric acid | −1[a] | 0.33[c] |
| *Carbon Acids* | | |
| 9-Carboxymethylfluorene | 10.2[b], 10.3[d] | 12.88[b] |
| Fluoradene | 10.5[d] | — |
| Fluorene | 22.7[b], 20.5[d] | 21.0($H_-$)[b] |

**Table 7.6.** (*Continued*)

| Acid | $pK$ (DMSO) | $pK$ (H$_2$O) |
|---|---|---|
| Fluorene-9-carbonitrile | 8.4[b] | 11.41[b] |
| Indene | 18.5[d] | — |
| 4,5-Methylenephenanthrene | 20.0[d] | 21.2(H$_-$)[d] |
| 9-Methylfluorene | 19.7[d] | 21.8(H$_-$)[d] |
| Nitroethane | 15.7[b], 13.9[d] | 8.9[d] |
| Nitromethane | 16.0[b], 15.9[d] | 10.2[d] |
| 2,4-Pentanedione | 13.4[d] | 9.03[c] |
| 1,12-*o*-Phenylenedihydropleiadene | 14.4[d] | — |
| 1-Phenyl-1,3-butanedione | 12.1[f] | 8.9[c] |
| 9-Phenylfluorene | 16.4[d] | 18.6(H$_-$)[d] |
| Propanedinitrile | 11.0[d] | 11.20[c] |
| Triphenylmethane | 28.8[d] | — |
| *Nitrogen Acids* RNH$_2$ ⇌ R$\overline{\text{N}}$H + H$^+$ | | |
| 4-Chloro-2-nitroaniline | 18.6[b], 15.1[d] | 17.1[d] |
| 2,4-Dinitroaniline | 15.6[b] | 15.0[b] |
| 2,4-Dinitrodiphenylamine | 12.4[d], 12.7[b] | 13.8[d] |
| 4-Nitroaniline | 20.8[b] | 18.9[b] |
| 4-Nitrodiphenylamine | 14.3[d] | 15.7(H$_-$)[d] |
| 2,3,5,6-Tetrachloroaniline | 17.1[d] | 19.2(H$_-$)[d] |
| *Cationic Acids* | | |
| Anilinium | 3.6[b], 3.6[a] | 4.60[g] |
| Butylammonium | 11.1[a] | 10.65[g] |
| 4-Chloroanilinium | 3.0[b] | 3.98[g] |
| Dibutylammonium | 10.0[a] | 11.25[g] |
| Diethylammonium | 10.5[a] | 11.04[g] |
| Dipropylammonium | 10.1[b] | 11.00[g] |
| Ethylammonium | 11.0[a] | 10.70[g] |
| Piperidinium | 10.5[b] | 11.20[g] |
| Pyridinium | 3.4[a] | 5.23[g] |
| Tetramethylguanidinium | 13.2[a] | 13.6[g] |
| Tributylammonium | 8.4[a] | 9.93[g] |
| Triethylammonium | 9.0[a] | 10.82[g] |
| Trimethylammonium | 9.0[b] | 9.75[g] |

[a] Values from Reference 72.
[b] Values in approximately 0.3% H$_2$O from Reference 74.
[c] Values from Reference 65.
[d] Values from Reference 79.
[e] Values from Reference 78.
[f] Values from Reference 73.
[g] Values from Reference 45.

where 3.84 is the mean of the spectrophotometrically and conductimetrically determined $p\mathrm{K}$ values. The equation for solutions of 4-chloro-2,6-idinitrophenol (concentration range $6 \times 10^{-4}$–$5 \times 10^{-2}$ $M$) and tetraethylammonium 4-chloro-2,6-dinitrophenolate (concentration $1 \times 10^{-2}$ $M$) is

$$p\mathrm{a}_\mathrm{H} = 3.50 + \log\frac{c_\mathrm{s}}{c_\mathrm{a}} - 1.16\sqrt{I} \qquad (7.34)$$

Dilute solutions of 4-toluenesulfonic acid have also been used for the calibration of a similar type of cell containing a glass electrode (73). Their use is based upon the assumption that $p\mathrm{a}_\mathrm{H}$ can be calculated directly from the stoichiometric concentration of acid taken. Picric acid solutions, $p\mathrm{K}$ approximately $-0.3$ (75), prepared by titrating a standard solution of the acid into dimethylsulfoxide, have been similarly used to calibrate the glass electrode (74).

The methods available for the potentiometric determination of ionization constants in dimethylsulfoxide are similar to those already described for acetonitrile and the amide solvents. They have been determined by measuring the $p\mathrm{a}_\mathrm{H}$ values of solutions that contain known weights of the acid and its tetraethylammonium salts (72). Potentiometric titrations of a dilute solution ($< 10^{-3}$ $M$) of the acid with a standardized solution of the highly reactive titrant cesium dimsyl (73) have also yielded data from which the ionization constant of the acid could be calculated by the equation

$$p\mathrm{K} = p\mathrm{a}_\mathrm{H} - \log\frac{c_\mathrm{s}}{c_\mathrm{a}} + 1.16 I^{1/2} \qquad (7.35)$$

At these low concentrations homoconjugation was confirmed to be negligible since plots of $E_\mathrm{cell}$ versus $\log c_\mathrm{a}/c_\mathrm{s}$ yielded straight-line graphs that had slopes within $\pm 10\%$ of the theoretical. It should be noted, however, that dimsyls of the alkali metals require special precautions both in their preparation and in their use, and it has been suggested (74) that the $p\mathrm{K}$ values obtained using these as titrants may be low as a result of the ion pair formation that typifies the behavior of the metal salts of weak acids in dimethylsulfoxide. Such impediments were apparently not encountered when tetrabutylammonium hydroxide was used as titrant (74) in the determination of $p\mathrm{K}$ values by the half-neutralization potential method. This titrant (0.5 equivalent) was added to a $2.5 \times 10^{-3}$ $M$ solution of the acid (10 mL) under an atmosphere of argon at 25° and the equilibrium potential measured after 2–5 min with the cell

Ag; AgCl|Satd. Et$_4$NCl in DMSO|0.03 $M$ Et$_4$NCl in DMSO|

half-neutralized acid|glass electrode

When the internal filling solution of the glass electrode was replaced by a solution prepared in dimethylsulfoxide, the equilibration time was reduced to 1–2 min. The cationic acid 2,4-dinitrodiphenylaminium ion, the $pK$ of which was determined to be 12.7, was used as the reference compound. The difference between the half-neutralization potential of this acid and the half-neutralization potential of an acid of unknown $pK_{HA}$ was used to calculate $pK_{HA}$ by means of equation (7.32). Values up to approximately $pK = 23$ were determined by this method for which the solvent was purified as follows:

> Reagent grade solvent was heated with sodium hydroxide solution for four hours on a steam bath after which it was distilled *in vacuo*. The vacuum distillation was then repeated in the presence of calcium hydride using an apparatus fitted with bleed tube connected to an argon supply. The water content, determined by Karl Fischer titration, was 0.3%.

Dimethylsulfoxide containing a lower water content (0.01% w/w) can be obtained following the procedure given in Reference 76, and another method of purification is given in Reference 77. It is apparent that a small quantity of water is advantageous in lowering the equilibration time required for the glass electrode. In the half-neutralization potential method just described the addition of the titrant increases the amount of water present still further. The titrant was prepared by concentrating under vacuum a 10% aqueous solution of tetrabutylammonium hydroxide. To this concentrated solution (50–60%), dimethylsulfoxide was added to yield a 0.4 $M$ stock solution of the titrant. The concentration of the titrant was determined daily by titration in aqueous medium with standard hydrochloric acid. The $pK$ values determined by the half-neutralization potential method in dimethylsulfoxide containing approximately 0.3% water are similar in magnitude to the values obtained by other methods in purer specimens of this solvent (see Table 7.6). A collection of the half-neutralization potentials for 34 substituted benzoic acids measured with respect to benzoic acid is given in Reference 80. These data may be interpolated in terms of an approximate order of magnitude of the ionization constant exponent for each substituted acid using the value of the $pK$ of benzoic acid in Table 7.6 together with equation (7.32).

## Measurements in 2-Methyl-2-Propanol

Of the anhydrous alcohols used as titration media, 2-methyl-2-propanol (*tert*-butylalcohol) has the lowest autoprotolysis constant ($K_{SH} = 10^{-28.5}$)

and, therefore, the largest range of possible potentials. It is particularly useful as a medium for titrations that will discriminate between carboxylic acids and phenols. This is because the solvating ability of 2-methyl-2-propanol is such that the formation of stable homoconjugation complexes is inhibited, and, in consequence, the titration curves are comparatively flat in comparison to those obtained in a solvent like acetonitrile in which stable complexes are formed. At a given concentration, therefore, the range of potentials covered by a titration of an acid in 2-methyl-2-propanol is much less than in acetonitrile, and in the sequential titration of two or more acids the buffer regions are more discrete, leading to clearly defined breaks in the regions of the equivalence points. However, in contrast to solvents with higher dielectric constants, the tetraalkylammonium salt of the acids are poorly dissociated in 2-methyl-2-propanol, the equilibrium constants for reactions of the type

$$R_4NA \rightleftharpoons R_4N^+ + A^- \quad (7.36)$$

often being very small ($\sim 10^{-5}$). In view of this, the values of $pK$ calculated from potentiometric measurements alone are uncertain unless the value of the equilibrium constant for reaction (7.36) has been determined previously by conductimetry. This allows calculation of $a_{A^-}$ by equation (7.38a) and the subsequent use of this quantity in an equation similar to (7.38b) to calculate $pK$ from the measured $pa_H$.

The response of the glass electrode in 2-methyl-2-propanol was found to parallel the response of the hydrogen electrode over the $pa_H$ range 2.8–14.6 when these electrodes were used in conjunction with a 0.1 $M$ AgNO$_3$|Ag reference half-cell fitted with a 0.01 $M$ tetraethylammonium perchlorate salt bridge (7). The glass electrode was stored in water when not in use (21) and required 10–20 min to reach a stable and reproducible potential ($\pm 2$ mV). The cell can be standardized using the data given in Table 7.7 which have been verified by both the glass and the hydrogen electrodes. The $pa_H$ values for solutions containing only pure trifluoromethanesulfonic acid at a given stoichiometric concentration can be calculated by a method similar to that which would pertain to aqueous solutions of an acid of known thermodynamic $pK_a$ value. Thus equation (6.33) can be written as

$$a_H = \frac{K_a(c_1 - c_{H^+})y_{HA}}{(c_2 + c_{H^+})y_{A^-}}$$

in which $c_2$, the concentration of the tetraethylammonium salt of the acid in this case, is zero. The remaining product in the denominator, $c_{H^+}y_{A^-}$, can be

POTENTIOMETRIC MEASUREMENTS WITH CELLS 403

Table 7.7. Some $pa_H$ Values for Solutions Prepared in 2-Methyl-2-Propanol at 25°[a]

| Acid | $c_{HA}$ | $c_{Et_4NA}$ | $a_{A^-}$ | $pa_H$ |
|---|---|---|---|---|
| Trifluoromethanesulfonic[b] | $1.17 \times 10^{-2}$ | 0 | $1.55 \times 10^{-3d}$ | 2.81 |
| | $5.48 \times 10^{-3}$ | 0 | $1.02 \times 10^{-3d}$ | 2.99 |
| | $1.92 \times 10^{-3}$ | 0 | $5.49 \times 10^{-4d}$ | 3.26 |
| 3,4 Dimethylbenzoic[c] | $1.36 \times 10^{-2}$ | $1.75 \times 10^{-3}$ | $1.02 \times 10^{-4e}$ | 13.28[f] |
| | $8.63 \times 10^{-4}$ | $1.75 \times 10^{-3}$ | $1.02 \times 10^{-4e}$ | 14.48[f] |
| | $5.22 \times 10^{-4}$ | $1.056 \times 10^{-3}$ | $7.85 \times 10^{-5e}$ | 14.59[f] |

[a] See Reference 7.
[b] $pK = 3.55$ (see Table 7.8).
[c] $pK = 15.41$ (see Table 7.8).
[d] $a_{A^-} = a_{H^+}$, calculated by equation (7.37).
[e] Calculated from $c_{Et_4NA}$ using $K_d = 6.3 \times 10^{-6}$ (Table 7.8) for reaction (7.36).
[f] $pa_H = 15.41 - \log c_{HA}/a_{A^-}$.

assumed equal to the hydrogen ion activity and thus

$$a_{H^+} = \sqrt{K_a(c_1 - c_{H^+})} \qquad (7.37)$$

in which $c_1$ is the stoichiometric concentration of the acid now termed $c_{HA}$ in Table 7.7. This equation can be solved for $a_{H^+}$ by a method of successive approximations using the following sequence:

STEP 1. Put $a_{H^+} = a_1 \sim \sqrt{K_a c_{HA}}$, calculate the first estimate of the molar activity coefficient from the Debye–Huckel equation adapted to the constants relevant to 2-methyl-2-propanol:

$$-\log y = \frac{8.73 a_1^{1/2}}{1 + 3.39 a_1^{1/2}}$$

and then calculate the first estimate of $c_{H^+}$ as

$$c_{H^+} = c_1 = \frac{a_1}{y}$$

STEP 2. Put $a_{H^+} = a_2 \sim \sqrt{K_a(c_{HA} - c_1)}$, and calculate $y$ and hence a new value of $c_H$.

STEP 3. Repeat Step 2 until successive values of $a_{H^+}$ are constant (i.e., $a_{n-1} = a_n$). Usually about eight iterations are sufficient.

The method that would be used to calculate $p$H values for aqueous solutions cannot be applied, however, to the data given in Table 7.7 for solutions containing known concentrations of 3,4-dimethylbenzoic acid $c_{HA}$ and its tetraethylammonium salt $c_{Et_4NA}$. This is because the latter salt is largely undissociated, and hence the actual activity of the anion $a_{A^-}$ must be calculated from the dissociation constant of the salt $K_d$ (see reaction 7.36) and the value of $c_{Et_4NA}$. The equation for this purpose is similar to equation (7.37)

$$a_{A^-} = \sqrt{K_d(c_{Et_4NA} - c_{A^-})} \qquad (7.38a)$$

and so is its solution. An approximation that dispenses with the need to use the Debye–Huckel equation has been used for calculation of $a_{A^-}$ values given in Table 7.7. Thus equation (7.38a) has been approximated as

$$a_{A^-} \sim \sqrt{K_d(c_{Et_4NA} - a_{A^-})}$$

to yield values of $pa_H$ that agree with those calculated from values of $a_{A^-}$ obtained using equation (7.38a). These $pa_H$ values were calculated using the equation

$$pa_H = 15.41 - \log \frac{c_{HA}}{a_{A^-}} \qquad (7.38b)$$

These methods can, of course, be applied to calculate $pa_H$ values for solutions of the acids and their tetraethylammonium salts from the $pK$ values collected in Table 7.8 for those entries in which $K_d$ values for the salts have also been recorded.

The materials required to prepare the solutions listed in Table 7.7 were purified as follows:

*2-Methyl-2-propanol* (21). A reagent quality sample was shaken with calcium hydride (2 g L$^{-1}$), decanted, and distilled at atmospheric pressure in a 1 m column packed with glass helices. This procedure was repeated, and the alcohol finally distilled by itself to yield a product containing 0.05–0.1% water as determined by Karl Fischer titration.

*Trifluoromethanesulfonic acid (Triflic acid)* (21). A commercial sample (3M Co.) was used without purification.

*3,4-Dimethylbenzoic acid* (81). A commercial sample (K and K Co.), mp 167° was used.

Table 7.8. $pK$ Values of Acids and Dissociation Constants of their Tetraethylammonium Salts $K_d$ in 2-Methyl-2-Propanol at 25°

| Acid | | $pK$ | $K_d^b$ | $pK(H_2O)^a$ | Ref. |
|---|---|---|---|---|---|
| *Sulfonic Acids* | | | | | |
| Trifluoromethanesulfonic acid | | 3.55 | $1.9 \times 10^{-5}$ | — | 21 |
| 2,4,6-Trinitrobenzenesulfonic acid | | 3.85 | — | — | 21 |
| *Carboxylic Acids, Monoprotic* | | | | | |
| Acetic acid | | 14.27 | $3.3 \times 10^{-5}$ | 4.76 | 82 |
| Benzoic acid | | 15.10 | — | 4.20 | 83 |
| 3-Bromobenzoic acid | | 13.48 | $2.2 \times 10^{-5}$ | 3.81 | 82 |
| Chloroacetic acid | | 12.24 | $3.4 \times 10^{-5}$ | 2.87 | 82 |
| 4-Chloro-3-nitrobenzoic acid | | 11.75 | — | $3.29^e$ | 82 |
| Cyanoacetic acid | | 10.68 | — | 2.47 | 82 |
| Cyclohexanecarboxylic acid | | 15.82 | — | 3.89 | 82 |
| 2,3-Dibromopropanoic acid | | 11.71 | — | $2.17^e$ | 82 |
| Dichloroacetic acid | | 10.27 | $3.3 \times 10^{-5}$ | 1.35 | 82 |
| 3,4-Dichlorobenzoic acid | | 12.97 | — | 3.64 | 82 |
| 3,4-Dimethylbenzoic acid | | 15.41 | $6.3 \times 10^{-6}$ | 4.50 | 82 |
| 3,5-Dinitrobenzoic acid | | 10.60 | $4.9 \times 10^{-5}$ | 2.82 | 82 |
| Methoxycarboxylacetic acid (Me.H.malonate) | | 12.52 | — | $3.35^e$ | 84 |
| 2-Methoxycarboxylbenzoic acid (Me.H.phthalate) | | 13.21 | — | — | 82 |
| 3-Methoxycarboxylpropanoic acid (Me.H.succinate) | | 14.22 | — | 4.49 | 82 |
| (Z)-3-Methoxycarboxylpropenoic acid (Me.H.maleate) | | 12.77 | — | $2.94^e$ | 82 |
| 4-Nitrobenzoic acid | | 12.04 | $5.7 \times 10^{-5}$ | 3.43 | 82 |
| *Carboxylic Acids, Diprotic* | | | | | |
| 1,2 Benzenedicarboxylic acid | I | 13.21 | — | 2.95 | 84 |
| | II | $—^d$ | — | 5.41 | |
| Butanedioic acid | I | 12.0 | $5.8 \times 10^{-5}$ | 4.21 | 84 |
| | II | $—^d$ | $8.2 \times 10^{-6}$ | 5.64 | |
| (Z)-Butenedioic acid (Maleic acid) | I | 7.65 | — | 1.91 | 84 |
| | II | $—^d$ | — | 6.33 | |
| Propanedioic acid | I | 9.70 | — | 2.85 | 84 |
| | II | $—^d$ | — | 5.70 | |
| *Phenols* | | | | | |
| 4-Bromophenol | | 19.16 | $4.5 \times 10^{-5}$ | 9.36 | 82 |
| 3,5-Dichlorophenol | | 17.12 | — | 8.18 | 82 |

(*continued*)

## Table 7.8. (Continued)

| Acid | $pK$ | $K_d{}^b$ | $pK(H_2O)^a$ | Ref. |
|---|---|---|---|---|
| 3,5-Dichloro-2,4,6-trinitrophenol | 4.70 | $4.6 \times 10^{-5}$ | −0.70 | 82, 21 |
| 3.5-Dinitrophenol | 13.40 | $3.8 \times 10^{-4}$ | 6.68 | 82 |
| 3-Nitrophenol | 17.17 | $1.0 \times 10^{-4}$ | 8.36 | 82 |
| 4-Nitrophenol | 14.48 | — | 7.16 | 82 |
| 4-Nitro-3-trifluoromethylphenol | 12.77 | — | $6.41^e$ | 82 |
| 3-Trifluoromethylphenol | 18.62 | — | 8.95 | 82 |
| 2,4,6-Trinitrophenol | 4.77 | $7.4 \times 10^{-5}$ | 0.33 | 82 |
|  | 5.06 | — | — | 85 |
|  | 5.46 | $4.7 \times 10^{-5b}$ | — | 21 |
| *Heteroaromatic* | | | | |
| 5,5-Diethylpyrimidine-2,4,6(1$H$, 3$H$, 5$H$)-trione (Veronal, Barbital) | 16.8 | $2.3 \times 10^{-5}$ | 8.02 | 82 |

$^a$ Values from Reference 65.
$^b$ Defined as $K_d = (a_{Et_4N^+})a_{A^-}/c_{Et_4NA}$.
$^c$ For ethoxycarbonyl derivative.
$^d$ Too weak to be determined.
$^e$ Value quoted in Reference 82.
$^f$ For tetrabutylammonium picrate.

*Tetraethylammonium 3,4-dimethylbenzoate* (81). A solution of 3,4-dimethylbenzoic acid in ethanol was neutralized to a potentiotitrimetrically determined equivalence point with a 1 $M$ aqueous solution of tetraethylammonium hydroxide. The resulting solution was evaporated to dryness and the solid washed with anhydrous ether. The product was recrystallized from a mixture of ethyl acetate and ethanol, and was dried at 60° *in vacuo*.

## Measurements in Some Ketones

Although ketones have been used successfully for a number of years as titration media for carboxylic acid, phenols, and nitrogen bases (86–88), there appears to be a paucity of information as to the quantitative aspects of acid-base equilibria in these solvents. The values of the autoprotolysis constant exponents for 2-butanone and 2-hexanone given in Table 7.1 were determined potentiometrically (16) using a glass electrode and a silver–silver chloride electrode but little detail is given as to the actual method used in the cell calibration. These solvents were used subsequently for the titration of mixtures of mono and dicarboxylic acids.

Measurements of $pa_H$ have been made in acetone (dielectric constant = 20) using the cell (89)

Pt; Ag|satd. AgNO$_3$ in Me$_2$CO|Et$_4$NClO$_4$ in Me$_2$CO|

test soln.|glass electrode (7.39)

This can be calibrated with buffer solutions containing picric acid, $c_{HPi}$ ranging from $3.72 \times 10^{-4}$ to $1.873 \times 10^{-2}$, and tetraethylammonium picrate, $c_{Pi^-} = 1.24 - 2.2 \times 10^{-3}$, for which the $pa_H$ values can be calculated from the spectrophotometrically determined $pK$ value of 9.26 by an equation analogous to (7.20):

$$pa_H = 9.26 - \log\frac{c_{HPi}}{c_{Pi^-}} - 3.98\sqrt{I} \qquad (7.40)$$

The ionic strength term in this equation is taken as $c_{Pi^-}$. It has been confirmed that, within the given range of concentrations of tetraethylammonium picrate, this salt is wholly dissociated in acetone. The range of the calibration can be extended using similarly calculated $pa_H$ values for solutions containing 2-nitroaniline and perchloric acid using the reported $pK$ value of 4.82.

A cell similar to (7.39) has been used for $pa_H$ measurements in 4-methyl-2-pentanone (methyl isobutyl ketone), except that the salt bridge contained a 0.1 $M$ solution of tetrabutylammonium perchlorate prepared in this solvent (90). This cell was calibrated initially in buffer solutions containing perchloric acid and tetrabutylammonium perchlorate, the compositions and $pa_H$ values of which are given in Table 7.9. These data are based on the conductimetrically determined values for the $pK$ of perchloric acid and the dissociation constant for the tetrabutylammonium salt. It should be noted that when the $pa_H$ values of such solutions were less than about 4, the glass electrode did not attain a constant potential. Under this condition of $pa_H$, the solutions became red indicating reaction with the solvent. The solutions of perchloric acid were prepared by dilution of a 0.92 $M$ stock solution of perchloric acid in glacial acetic acid with the solvent. The low dielectric constant of 4-methyl-2-pentanone (12.92) is comparable with that of 2-methyl-2-propanol (10.9), and thus in common with the latter solvent the salts of acids are poorly dissociated in this ketone. In comparison with 2-methyl-2-propanol, however, 4-methyl-2-pentanone has a lower anionic solvating power with the consequence that stable homoconjugation complexes can be formed. For example, the equilibrium constant for the

Table 7.9. $pa_H$ Values of Solutions Containing Perchloric Acid and Tetrabutylammonium Perchlorate in 4-Methyl-2-Pentanone (25°)

| $c_a{}^a$ | $c_s{}^b$ | $pa_H{}^c$ |
|---|---|---|
| $10^{-3}$ | $10^{-3}$ | 4.16 |
| $5 \times 10^{-4}$ | $5 \times 10^{-4}$ | 4.70 |
| $10^{-3}$ | $10^{-2}$ | 4.52 |
| $2 \times 10^{-3}$ | $2 \times 10^{-2}$ | 4.37 |
| $5 \times 10^{-4}$ | $10^{-1}$ | 5.13 |
| $10^{-3}$ | $10^{-1}$ | 4.83 |
| $2 \times 10^{-3}$ | $10^{-1}$ | 4.53 |

$^a c_a$ = molar concentration perchloric acid.
$^b c_s$ = molar concentration tetrabutylammonium perchlorate.
$^c$Data from Reference 90 corrected for liquid junction potential and the presence of acetic acid in the perchloric acid solutions.

reaction

$$A^- + HA \rightleftharpoons A^- \ldots HA$$

is $2 \times 10^4$ for the formation of the 3,5-dinitrobenzoate...3,5-dinitrobenzoic acid complex, and the dissociation constant for tetraethylammonium 3,5-dinitrobenzoate is $7 \times 10^{-5}$. In cases like this the $pK$ value cannot be determined by measuring the $pa_H$ of a half-neutralized solution of the acid, and neither can it be obtained by the method outlined for buffer solutions prepared in 2-methyl-2-propanol. However, an adaptation of the $pa_H$ at half-neutralization method can be used in order to obtain an estimate of the $pK$ (90). The values of $pa_H$ are measured for a series of buffer solutions all containing equimolar concentrations of the acid and its tetraalkylammonium "salt" at large dilution. These values are plotted against the concentration and extrapolated to zero concentration at which the anion is completely dissociated and the activity coefficient is unity. The $pK$ is equal therefore, to the $pa_H$ at $c = 0$.

In the picric acid–tetrabutylammonium picrate buffer system, homoconjugation is negligible, and the dissociation constant of tetrabutylammonium picrate is $5 \times 10^{-4}$. It is thus possible to calculate $pa_H$ values for dilute solutions ($\sim 10^{-3}$ M) containing known concentrations of each solid from the value of $pK_{HPi} = 11.0$ using equation (7.38) to calculate $a_{A^-}$ and hence

$$pa_H = 11.0 - \log \frac{c_{HPi}}{a_{A^-}}$$

These buffer solutions were used daily to check the calibration of the glass electrode (90). It was also reported that the glass electrode attained equilibrium within 5 min in relatively acidic solutions and in 30 min in solutions with a $p$H of 12 or greater. The solvent and the salt can be prepared as follows:

*4-Methyl-2-pentanone* (90). Reagent grade solvent was shaken with dry calcium chloride for about 10 h. This operation was repeated three times. After decantation the solvent was passed through a column of chromatographic alumina and then distilled at least three times over phosphorus pentoxide in a dry and carbon dioxide-free stream of nitrogen. The distillation column was packed with steel wool. A red coloration appeared upon the initial addition of phosphorus pentoxide but by the third distillation this coloration was very weak and developed very slowly. A final distillation was carried out without addition of the drying agent and the solvent stored in bottles fitted with a siphon allowing the solvent to be dispensed without contact with the atmosphere. The water content did not exceed $2.2 \times 10^{-3}$ $M$ as determined by Karl Fischer titration, and the conductivity ($1.2 \pm 0.2 \times 10^{-7}$ $\Omega^{-1}$ cm$^{-1}$) did not change on storing the solvent for two months.

*Tetrabutylammonium picrate* (21) can be prepared by neutralizing picric acid in aqueous solution with carbonate-free tetrabutylammonium hydroxide, evaporating to dryness and recrystallizing from ethyl acetate or ethyl acetate–ethanol mixtures. The salt was dried *in vacuo* at 60° for 6 h.

## 7.4 MEASUREMENTS IN MIXED AQUEOUS SOLVENTS

In some instances organic solvents containing up to about 20 wt % water offer advantages over the anhydrous solvent as the solvent system for potentiotitrimetry. Not only does the addition of water increase the dielectric constant of the medium thereby increasing its solvating power with respect to anions, but it also has the effect of substantially increasing the rate of response and the reproducibility of the glass electrode particularly at high $p$H values. At the same time the ability of the solvent to dissolve covalent compounds is preserved when the medium contains such a high proportion of the solvent. However, the consequence of introducing water into the solvent is to substantially decrease the titration range of the medium with respect to that available in the pure solvent. A great diminution in the autoprotolysis constant exponent $pK_{SH}$ is indicative of this. For example, the $pK_{SH}$ value for dimethylsulfoxide is reduced from ~ 33 for the pure solvent to $pK_{SH} = 18.4$ (25°) for a mixture 80% DMSO–20% water (91). A concomitant decrease in the differentiating or discriminative properties of the medium can thus be expected. The determination of ionization constants in such media provide the necessary information to decide whether it is possible to titrate sequentially a mixture of two or more acids to discrete equivalence points.

Interest in acid-base equilibria in mixed aqueous solvents is not confined, however, to media that contain a high proportion of organic solvent. Although media containing 10–50 wt % organic solvent are not frequently used in potentiotitrimetry, they are often used for studying acid-base behavior by the potentiometric method. Thus the aqueous $pK$ values of compounds too insoluble to be determined by the methods described in Chapter 6 have been obtained indirectly by potentiometric measurements in partially aqueous media. In some instances a plot of the $pK$ values so obtained against either the corresponding values of solvent composition or the reciprocal of the dielectric constant $(1/\epsilon)$ can be extrapolated reliably to either 0% organic component or to the value of $1/\epsilon$ for pure water at the given temperature (i.e., 1/78.36 at 25°). Quite frequently, however, the compounds are insoluble even when the organic component is present in the range 10–20% by weight. In these cases the extrapolation can be very uncertain. Interest can also be centered upon the substituent effects that operate within a given series of organic acids or bases. These are sometimes compared by determining the $pK$ value for each individual compound in an aqueous-organic medium of fixed composition, often about 50% w/w. It should be noted, however, that the composition of an aqueous-organic solvent mixture expressed as a weight percentage tends to give a false impression as to the actual molecular composition of the medium. Thus mixtures which are 81.3% dimethylsulfoxide–18.7% water or 72% ethanol–28% water by weight still contain as many molecules of water as of the organic component. Association and other effects noted in Section 7.3 are usually negligible in such mixtures which enables cells containing a glass electrode to be calibrated by methods similar to those used in water.

### Measurements of $p(a_H \gamma_{Cl})$ and $p$H in Water–Methanol and Water–Ethanol Mixtures

These mixtures have been popular media in which to study acid-base behavior for a long time (92), and some of their properties are given in Table 7.10. The measurement of the acidity function $p(a_H \gamma_{Cl})$ for a range of

**Table 7.10. Properties of Water–Alcohol Mixtures at 25°C**

*Water–Methanol*[a]

| MeOH (wt %) | Mole % | $d$ | $\epsilon$ | $A^b$ | $B^b$ | $pK_{SH}$ |
|---|---|---|---|---|---|---|
| 0 | 0 | 0.9971 | 78.4 | 0.511 | 0.329 | 14.00 |
| 20.0 | 12.32 | 0.9645 | 70.0 | 0.595 | 0.342 | 14.04[c] |
| 40.0 | 27.26 | 0.9319 | 60.9 | 0.721 | 0.360 | 14.08[c] |
| 43.3 | 30.04 | 0.9240[c] | 59.4 | 0.745 | 0.363 | 14.09 |
| 60.0 | 45.75 | 0.8914 | 51.7 | 0.901 | 0.383 | 14.15[c] |
| 64.0 | 49.99 | 0.8824 | 49.8 | 0.948 | 0.388 | 14.17 |
| 84.2 | 74.98 | 0.8328[c] | 40.7 | 1.247 | 0.417 | 14.57 |
| 90.0 | 83.50 | 0.8156 | 37.9 | 1.373 | 0.427 | 14.94[c] |
| 94.2 | 90.13 | 0.8033[c] | 35.8 | 1.484 | 0.436 | 15.24 |

*Water–Ethanol*[d]

| EtOH | Mole % | $d$ | $\epsilon$ | $A^b$ | $B^b$ | $pK_{SH}$ |
|---|---|---|---|---|---|---|
| 10.0 | 4.16 | 0.9800 | 72.8 | 0.565 | 0.338 | 14.22 |
| 15.0 | 6.46 | 0.9732 | 70.0 | 0.598 | 0.343 | 14.32(14.33[e]) |
| 20.0 | 8.91 | 0.9660 | 67.1 | 0.634 | 0.350 | 14.39 |
| 25.0 | 11.53 | 0.9587 | 64.2 | 0.675 | 0.356 | 14.46 |
| 30.0 | 14.35 | 0.9511 | 61.2 | 0.723 | 0.363 | 14.54 |
| 35.0 | 17.39 | 0.9419 | 58.1 | 0.777 | 0.371 | 14.61(14.57[e]) |
| 40.0 | 20.68 | 0.9315 | 55.1 | 0.837 | 0.379 | 14.69 |
| 45.0 | 24.24 | 0.9210 | 52.0 | 0.908 | 0.388 | 14.77 |
| 50.0 | 28.11 | 0.9085 | 48.9 | 0.989 | 0.397 | 14.84(14.88[e]) |

[a] Data from References 93 and 94.
[b] Constants of the Debye–Huckel equation.
[c] Graphically interpolated values.
[d] Data interpolated graphically from the values given in Reference 95.
[e] Values from Reference 96.

alcohol–water compositions with the cell

$$\text{glass electrode} | \text{H}^+(a_\text{H}), \text{Cl}^-(\text{known molality}) | \text{AgCl}; \text{Ag} \quad (7.41)$$

can be accomplished using the general procedures described for aqueous solutions in Chapter 6. The values for necessary standardizing solutions, given in Tables 7.11 and 7.12, were obtained using the same type of hydrogen electrode and silver–silver chloride electrode combination as was used to obtain $p(a_\text{H}\gamma_\text{Cl})$ values for the aqueous buffer solutions.

Table 7.11.
$p(a_H\gamma_{Cl})$ and $pa_H$ Values in Methanol–Water Mixtures (Wt %)

*Approximate values ( ± 0.02) of $p(a_H\gamma_{Cl})$ at 25°C[a]*

| Soln. No. | Methanol (%) | | | | | | |
|---|---|---|---|---|---|---|---|
|  | 0 | 20 | 40 | 50 | 70 | 80 | 90 |
| 1 | 2.205 | 2.31 | 2.46 | 2.56 | 2.86 | 3.24 | 3.85 |
| 2 | 4.168 | 4.50 | 4.91 | 5.13 | 5.70 | 6.12 | 6.85 |

*$p(a_H\gamma_{Cl})$ Values in 50% methanol at various temperatures[b]*

| Soln. No. | Temperature (°C) | | | | | | |
|---|---|---|---|---|---|---|---|
|  | 10 | 15 | 20 | 25 | 30 | 35 | 40 |
| 3 | 8.197 | 8.178 | 8.163 | 8.151 | 8.143 | 8.138 | 8.137 |
| 4 | 5.674 | 5.666 | 5.661 | 5.660 | 5.662 | 5.667 | 5.676 |
| 5 | 5.950 | 5.933 | 5.919 | 5.910 | 5.903 | 5.901 | 5.901 |

*$pa_H$ Values at 25°C at round methanol percentages[c]*

| Soln. No. | Methanol (%) | | | | | | |
|---|---|---|---|---|---|---|---|
|  | 10 | 20 | 40 | 50 | 70 | 80 | 90 |
| 6 | 2.19 | 2.25 | 2.38 | 2.47 | 2.76 | 3.13 | 3.73 |
| 7 | 4.30 | 4.48 | 4.87 | 5.07 | 5.57 | 6.01 | 6.73 |

*$pa_H$ Values in 50% methanol at various temperatures[d]*

| Soln. No. | Temperature (°C) | | | | | | |
|---|---|---|---|---|---|---|---|
|  | 10 | 15 | 20 | 25 | 30 | 35 | 40 |
| 8 | 8.072 | 8.051 | 8.034 | 8.021 | 8.011 | 8.004 | 8.001 |
| 9 | 5.586 | 5.577 | 5.571 | 5.568 | 5.569 | 5.573 | 5.580 |
| 10 | 5.863 | 5.844 | 5.829 | 5.818 | 5.811 | 5.806 | 5.806 |

| | |
|---|---|
| Soln. 1 | 0.01 $m$ oxalic acid + 0.01 $m$ ammonium hydrogen oxalate + 0.002 $m$ NaCl. |
| Soln. 2 | 0.01 $m$ succinic acid + 0.01 $m$ lithium hydrogen succinate + 0.002 $m$ NaCl. |
| Soln. 3 | 0.01 $m$ KH$_2$PO$_4$ + 0.01 $m$ Na$_2$HPO$_4$ + 0.01 $m$ NaCl. |
| Soln. 4 | 0.01 $m$ acetic acid + 0.01 $m$ sodium acetate + 0.01 $m$ NaCl. |
| Soln. 5 | 0.01 $m$ sodium hydrogen succinate + 0.01 $m$ NaCl. |
| Soln. 6 and 7 | same buffer concentrations as Soln. 1 and 2, but without chloride. |
| Soln. 8, 9, and 10 | same buffer concentrations as Soln. 3, 4, and 5 but without chloride. |

[a] Interpolated from data in Reference 94 at round methanol percentages.
[b] Data from Reference 97.
[c] Data from Reference 94.
[d] Data from Reference 97.

Table 7.12. $p(a_H\gamma_{Cl})$ Values for Solution Prepared in Three Water–Ethanol Mixtures (wt % EtOH)

| Solutions | $m_3$ | −10° | | | −5° | | | 0° | | | 25° | | |
|---|---|---|---|---|---|---|---|---|---|---|---|---|---|
| | | 10% | 20% | 40% | 10% | 20% | 40% | 10% | 20% | 40% | 10% | 20% | 40% |
| Hydrochloric acid (concn. = $m_3$)[a] | 0.01 | 2.096 | 2.103 | 2.129 | 2.095 | 2.102 | 2.131 | 2.095 | 2.102 | 2.132 | 2.097 | 2.106 | 2.141 |
| | 0.02 | 1.823 | —[c] | 1.871 | 1.823 | —[c] | 1.874 | 1.824 | —[c] | 1.875 | 1.826 | —[c] | 1.888 |
| | 0.03 | —[c] | —[c] | 1.722 | 1.667 | —[c] | 1.726 | 1.668 | —[c] | 1.727 | 1.672 | —[c] | 1.743 |
| 0.05 m HAc + 0.05 m NaAc + NaCl ($m_3$)[b] | 0.01 | —[d] | 5.183 | 5.623 | 4.977 | 5.151 | 5.596 | 4.957 | 5.127 | 5.573 | 4.917 | 5.073 | 5.528 |
| | 0.02 | —[d] | 5.184 | 5.624 | 4.977 | 5.153 | 5.596 | 4.957 | 5.126 | 5.571 | 4.916 | 5.072 | 5.527 |
| | 0.03 | —[d] | 5.185 | 5.621 | 4.978 | 5.152 | 5.593 | 4.957 | 5.126 | 5.570 | 4.916 | 5.072 | 5.524 |
| 0.025 m KH$_2$PO$_4$ + 0.025 m Na$_2$HPO$_4$ + NaCl ($m_3$)[b] | 0.01 | 7.483 | 7.755 | —[e] | 7.423 | 7.685 | —[e] | 7.372 | 7.626 | —[e] | 7.212 | 7.428 | 7.741 |
| | 0.02 | 7.468 | 7.735 | —[e] | 7.408 | 7.668 | —[e] | 7.358 | 7.608 | —[e] | 7.200 | 7.411 | 7.713 |
| | 0.03 | 7.454 | 7.717 | —[e] | 7.394 | 7.650 | —[e] | 7.345 | 7.591 | —[e] | 7.186 | 7.395 | 7.688 |

[a] Values calculated from data in Reference 98. (E° values given in Reference 99 used.)
[b] Values calculated from data in Reference 99.
[c] No data available.
[d] Solutions froze.
[e] Salts incompletely soluble.

The $pa_H$ values recorded in Table 7.11 and 7.13 were derived from the relevant $p(a_H\gamma_{Cl})$ values by means of the extrapolated (i.e., $m_{Cl} = 0$) quantity $p(a_H\gamma_{Cl})°$ as described for aqueous buffer solutions. The appropriate values for the Debye–Huckel equation constants $A$ and $B$, calculated by equations (7.6) and (7.7), respectively, are given in Table 7.10. The convention of setting the value of the parameter $a_i$ to the same value (i.e., 4.56 Å) as that selected by the Bates–Guggenheim convention for aqueous solutions appears to be generally applicable also in water–alcohol mixtures. Exceptions have been noted, however, and in 20% ethanol in water (by weight) the best fit of the data appears when $a_i$ is in the range 6.2–8.5 Å (99). The error in deriving the $pa_H$ values as a result of this does not exceed 0.005 $p$H unit. These derived $pa_H$ values of the buffer solutions when they are used as reference solutions are identified with $p$H(S) in the operational definition of $p$H. The measurement of $p$H(X) in ethanol–water mixtures using the low temperature values of $p$H(S) given in Table 7.13 for calibration requires the use of a modified reference half-cell and a special glass electrode. Although the latter type is commercially available, their resistance is very high at temperatures below 0°C and their response is concomitantly slow. Furthermore, too little is yet known about the residual liquid junction potential in these solvents to permit the assessment of the internal consistency of the $p$H scale defined by these reference points (99).

The use of the hydrogen ion-responsive glass electrode in cells similar to (7.41) for measurements in ethanol–water mixtures was reported some years

Table 7.13. $pa_H$ Values in Three Ethanol–Water Mixtures[a]

| Temperature (°C) | $pa_H$ (% EtOH) | | |
|---|---|---|---|
| | 10 | 20 | 40 |
| 0.05 m HAc + 0.05 m NaAc | | | |
| −10 | — | 5.075 | 5.498 |
| − 5 | 4.881 | 5.044 | 5.470 |
| 0 | 4.861 | 5.021 | 5.445 |
| 25 | 4.822 | 4.967 | 5.395 |
| 0.025 m KH$_2$PO$_4$ + 0.025 m Na$_2$HPO$_4$ | | | |
| −10 | 7.376 | 7.638 | — |
| − 5 | 7.315 | 7.569 | — |
| 0 | 7.263 | 7.508 | — |
| 25 | 7.104 | 7.310 | — |

[a] Data from Reference 99.

ago (100). Results obtained using the glass electrode in methanol–water mixtures have led to the conclusion that the standard potential of the glass electrode is independent of solvent composition in this medium as, indeed, it should be if the electrode is behaving as a reversible hydrogen electrode (101). The use of cell (7.41) to determine the ionization constants of monoprotic acids in ethanol–water mixtures by a comparative method that dispenses with the need to rely upon other standardization data has also been described (102). A more recently developed method also uses cell (7.41) for the determination of ionization constants in organic solvent–water mixtures without formal standardization of the electrodes (103). This method has the advantage that the slope of the cell response is also checked during the determination, and it is claimed that the experimental procedure is simplified with respect to the earlier methods. Cell (7.41) has also been used to obtain the autoprotolysis constant exponents $pK_{SH}$ for alcohol–water mixtures (95) and a host of other aqueous organic mixtures (104). The values of $pK_{SH}$ for the ethanol–water mixtures given in Table 7.10 were derived from these data.

### Determination of Ionization Constants in Aqueous Dimethylsulfoxide

Although the potential of the glass electrode is quickly established and is highly reproducible in mixtures containing up to 80 wt % (48 mole percent) of dimethylsulfoxide (91), the measurement of the acidity function $p(a_H\gamma_{Cl})$ has been inhibited by the significant solubility of silver chloride in dimethylsulfoxide. However, the cells without liquid junction of the type

glass electrode | HCl, $KNO_3$ in aqueous DMSO | AgCl; Ag

glass electrode | KOH, KCl in aqueous DMSO | AgCl; Ag

have been used (105, 106) in aqueous solutions containing up to 94 wt % (78 mole %) dimethylsulfoxide without any reported deleterious effects upon the performance of the silver–silver chloride electrode. There would appear, therefore, to be no impediment to the measurement of $p(a_H\gamma_{Cl})$ in these mixtures leading to the establishment of $pa_H$ scales as have been obtained for alcohol–water mixtures. In the absence of such data, cells with liquid junction represented as

Glass electrode | HCl($c_{HCl}$), $Et_4NOH$($c_{Et_4NOH}$), 0.1 $M$ $Et_4NBr$ |

| 0.001 $M$ KCl in 50% DMSO | AgCl; Ag

have been calibrated to measure $pc_H$ in 80% (w/w) dimethylsulfoxide (91). A 0.01 $M$ solution of hydrochloric acid prepared in this medium was titrated with 0.02 $M$ tetraethylammonium hydroxide also in 80% dimethylsulfoxide. $E_{cell}$ was measured after each addition and the value of $c_{H^+}$ calculated from

$$c_{H^+} = c_{HCl} - c_{Et_4NOH} \tag{7.42}$$

This procedure allowed the value of the $constant_1$ (i.e., $E_{cell}^{\circ} + E_j$) to be calculated at 25°C from

$$E_{cell} = constant_1 + 0.05916\, pc_H \tag{7.43}$$

Similarly, the cell was also calibrated in the alkaline region by titrating a standard solution of tetraethylammonium hydroxide with standard hydrochloric acid which allowed calculation of $c_{OH^-}$. A second constant $constant_2$ was calculated from

$$E_{cell} = constant_2 - 0.05916\, pc_{OH}$$

From these data the exponent of the autoprotolysis constant expressed in terms of concentrations $pK_{SH}^c$ was also calculated from

$$pK_{SH}^c = pc_H + pc_{OH} = \frac{constant_2 - constant_1}{0.05916} \tag{7.44}$$

The concentration ionization constant exponents $pK_a^c$ have been determined at constant ionic strength by potentiometric titration (91) in which the quantity $pc_H$ was measured. A 0.01 $M$ solution of the weak acid also 0.01 $M$ in hydrochloric acid and 0.1 $M$ in tetraethylammonium bromide was prepared in 80 weight percent dimethylsulfoxide in water. This solution was titrated with two equivalents of tetraethylammonium hydroxide which was prepared in the same solvent mixture and presumably contained 0.1 $M$ tetraethylammonium bromide. The values of $pc_H$ calculated by means of equation (7.42) for the $E_{cell}$ readings of the first equivalent allowed $constant_1$ of equation (7.43) to be evaluated, and from the $E_{cell}$ readings recorded during the addition of the second equivalent the values of $pc_H$ pertinent to the stages of neutralization of the weak organic acid could be calculated. The values of $pK_a^c$ corresponding to these degrees of neutralization could be calculated, therefore, by equation (6.42); that is,

$$pK_a^c = pc_H + \log\left(\frac{c_1 - c_{H^+} + c_{OH^-}}{c_2 + c_{H^+} - c_{OH^-}}\right) \tag{7.45}$$

The term $c_2$ is the stoichiometric concentration of the acid converted to its anion, and $c_1$ is the stoichiometric concentration of the nonionized acid remaining. When necessary, the titration can be continued after the equivalence point to obtain $constant_2$. This allows the autoprotolysis constant exponent $pK_{SH}^c$ to be calculated by equation (7.44) which, in turn, permits calculation of the term $c_{OH^-}$ in equation (7.45). Some values of the autoprotolysis constant exponents for dimethylsulfoxide–water mixtures have been included in Table 7.14. These $pK_{SH}$ values are expressed in terms of activities. At ionic strengths of less than 0.1 they may be converted into the corresponding concentration terms $pK_{SH}^c$ by the equation

$$pK_{SH}^c = pK_{SH} - \frac{2AI^{1/2}}{(1 + 4.56\,BI^{1/2})}$$

for which the appropriate values of the constants A and B are also given in Table 7.14.

The 80% dimethylsulfoxide–water mixture is a good medium for the potentiotitrimetry of aliphatic monocarboxylic acids. The change of poten-

Table 7.14. Some Properties of Dimethylsulfoxide–Water Mixtures at 25°C[a]

| (wt %) DMSO | Mole Fraction | $\epsilon$ | $d$ | $A^b$ | $B^b$ | $pK_{SH}^c$ |
|---|---|---|---|---|---|---|
| 0.00 | 0.00 | 78.40 | 0.99704 | 0.510 | 0.328 | 13.99 |
| 11.83 | 0.03 | 78.05 | 1.01173 | 0.517 | 0.332 | 14.24 |
| 24.61 | 0.07 | 77.55 | 1.02945 | 0.527 | 0.336 | 14.57 |
| 30.02 | 0.09 | 77.25 | 1.03739 | 0.532 | 0.338 | 14.73 |
| 41.38 | 0.14 | 75.95 | 1.05756 | 0.551 | 0.344 | 15.28 |
| 53.55 | 0.21 | 74.55 | 1.07219 | 0.570 | 0.349 | 15.85 |
| 59.11 | 0.25 | 73.00 | 1.08060 | 0.591 | 0.354 | 16.34 |
| 72.66 | 0.38 | 68.48 | 1.08858 | 0.653 | 0.367 | 17.48 |
| 80.01 | 0.48 | 64.80 | 1.08926 | 0.709 | 0.378 | 18.58 |
| 85.69 | 0.58 | 60.98 | 1.08583 | 0.776 | 0.389 | 19.69 |
| 90.21 | 0.68 | 57.10 | 1.09888 | 0.861 | 0.404 | 20.74 |
| 93.89 | 0.78 | 53.55 | 1.09786 | 0.948 | 0.417 | 21.71 |

[a] Data from Reference 106.
[b] Constants of the Debye–Huckel equation.
[c] Calculated from the data in Reference 106 assuming an ionic strength of 0.01 M. The values agree with the values interpolated from Reference 105 to within 0.05 unit up to a weight percent of 62% dimethylsulfoxide.

tial in the vicinity of the equivalence point is greater in this medium than it is in water, and its excellent solvating ability makes possible the titrations of acids like pentadecanoic and octadecanoic acids that are too insoluble to be titrated in water (91).

## 7.5 HALF-NEUTRALIZATION POTENTIALS

When a glass electrode and a suitable reference half-cell are dipped into a solution containing equimolar concentrations of nonionized acid and its derived ionized species, the potential that develops is called *the half-neutralization potential*, designated as $E_{HNP}$. Provided that the salt is completely ionized in the particular solvent, this quantity will be related to the $pK$ of the acid or base if the concentration of the lyate or lyonium ion (e.g., $c_{H^+}$ or $c_{OH^-}$ in water) is small in comparison with the stoichiometric concentrations of the two conjugate species. Thus if the cell has been calibrated, $E_{HNP}$ is related to the $pa_H$ at half-neutralization by equation (7.25) which, in turn, is equal to the $pK$ of the acid. Alternatively, if the $pK$ of one acid is known in the solvent of interest, then the difference between the half-neutralization potential of that acid and the half-neutralization potential of an acid of unknown $pK$ can be interpreted directly in terms of the latter by equation (7.32). These $pa_H$ and/or $pK$ data are not necessarily available, however, and this deficiency is of particular relevance to some of the more complex solvent mixtures that are used in analysis by potentiotitrimetry. Furthermore, the measurement and applications of half-neutralization potentials generally predate the systematic types of investigations that yielded the data presented in Section 7.3, and are far easier to obtain. Consequently, there is a greater systemization of data relating to half-neutralization potentials in nonaqueous solvent systems than there is for ionization constant data in these media. This systemization stems from the observation that the relative acidities of a group of structurally related compounds appear to be merely modified rather than fundamentally changed by varying the solvent. Thus to a first approximation, the acid-base behavior of amines in several organic solvents was found to parallel their behavior in aqueous solution as reflected in the magnitude of their $pK$ values in water (107). In a given solvent system, therefore, the value of the half-neutralization potential for each member of a closely related series of compounds when plotted against its corresponding $pK$ value in water is found to yield a straight line graph for the series.

Half-neutralization potentials are, in themselves, useless data that would vary, for example, with each individual glass electrode used in their determination and would also vary from day to day for a given glass electrode.

It has become customary, therefore, to measure half-neutralization potentials with respect to the half-neutralization potential of a standard acid or base and to express the difference as $\Delta$HNP. For acidic compounds $\Delta$HNP is usually defined as the difference between the half-neutralization potential of the compound in question and the half-neutralization potential of benzoic acid. Similarly, for bases $\Delta$HNP is usually defined as the difference between the half-neutralization potential of the given base and the half-neutralization potential of 1,3-diphenylguanidine. That is, for acid $X$:

$$\Delta\text{HNP} = E_{\text{HNP}}(\text{acid } X) - E_{\text{HNP}}(\text{benzoic acid}) \qquad (7.46a)$$

for base $Y$:

$$\Delta\text{HNP} = E_{\text{HNP}}(\text{base } Y) - E_{\text{HNP}}(1,3\text{-diphenylguanidine}) \qquad (7.46b)$$

Negative values of $\Delta$HNP signify that acid $X$ is a weaker acid than benzoic acid, or that base $Y$ is a stronger base than 1,3-diphenylguanidine (i.e., its conjugate acid is a weaker acid than the 1,3-diphenylguanidium ion). Thus for a given solvent system and a given titrant, values of $\Delta$HNP should be reasonably reproducible ($\pm 10$ mV) throughout the scientific world although some further variation can be expected if the values were determined at different concentrations. The two standard substances can be prepared for use as follows:

*1,3-Diphenylguanidine* (mp 148°): Recrystallize from toluene three times followed by drying at 100°.
*Benzoic acid*: Microanalytical grade reagent dried *in vacuo* over sulfuric acid overnight at room temperature.

Half-neutralization potentials are usually determined by the potentiometric titration of a known concentration of the compound with a standardized titrant added in such aliquots that a reliable graphical representation of the titration curve of $E_{\text{cell}}$ versus percent compound neutralized can be obtained. The value of $E_{\text{cell}}$ corresponding to exactly 50% neutralization is interpolated from the graph as $E_{\text{HNP}}$. A suitable concentration of compound for the determination is 0.01 $M$ and it is recommended that the initial volume of the titrand solution be adjusted so that the total concentration of all species is exactly 0.01 $M$ when 0.5 equivalent of titrant has been added. The solutions should be protected against contamination by water or carbon dioxide, and it is often necessary to seal the vessel containing the titrand solution and to blanket the surface of this solution with nitrogen. The half-neutralization potential of the standard substance [see equations

(7.46)] should be determined at least once per day and this determination should be used also as a check of the concentration of titrant. The titrants most commonly used are perchloric acid for bases and a tetraalkylammonium hydroxide for acids.

For a series of closely related acids or bases, the relation between $\Delta$HNP for each individual member and its $pK_a$ value in water ($pK_{H_2O}$) is given by the linear equation

$$\Delta\text{HNP} = a + b \cdot pK_{H_2O} \qquad (7.47)$$

Equations of this type obtained for the same series of compounds in two or more different solvents have been used to predict which of the solvents would have the best discriminative ability for the possible sequential titration of two members of the series. The basis for this prediction is the magnitude of the constant $b$(mV). For example, the following six versions of equation (7.47) were obtained for some *ortho*-substituted phenols in six different alcohols (108):

methanol $\Delta\text{HNP} = 318 - 59\,pK_{H_2O}$   ethanol $\Delta\text{HNP} = 409 - 72\,pK_{H_2O}$

propanol $\Delta\text{HNP} = 412 - 72\,pK_{H_2O}$   2-propanol $\Delta\text{HNP} = 476 - 81\,pK_{H_2O}$

butanol $\Delta\text{HNP} = 404 - 70\,pK_{H_2O}$   2-methyl-2-propanol $\Delta\text{HNP} = 442 - 77\,pK_{H_2O}$

From these results the discriminative or differentiating ability of the solvents were adjudged to be 2-propanol ($b = -81$) > 2-methyl-2-propanol ($b = -77$) > the remainder. In water the theoretical slope is about 59 mV per unit of $pK_{H_2O}$ indicating that these two alcohols are better differentiating solvents than water. Measured values of $\Delta$HNP have also been inserted in equations of this type in order to assess the $pK_{H_2O}$ values for compounds within the series that are either too insoluble or too prone to decomposition to be determined directly in water (108). Furthermore, given the $pK_{H_2O}$ values of two compounds within the same series but of unknown $\Delta$HNP, these equations allow an assessment to be made of the likelihood that the two could be titrated sequentially to discrete equivalence points. Normally, a separation of about 150 mV between the two $\Delta$HNP values is sufficient to achieve this.

It should be noted that $\Delta$HNP values are very dependent upon the nature of the titrant and particularly upon the solvent used for the preparation of the titrant solution. Whenever possible, the same solvent should be used for the preparation of both the titrant and titrand solutions, although in some instances the insolubility of the titrant may render this impracticable.

Table 7.15. Some Applications of Measurements of Half-Neutralization Potentials

| Compound Type | Solvent | Titrant | Remarks[a] | Ref. |
|---|---|---|---|---|
| $m$ and $p$ subs. benzoic acids | Nine different | $(Bu)_4NOH$ | $\Delta$HNP values measured with respect to benzoic acid. For acetonitrile $a = -655$, $b = 157$; for 4-methyl-2-pentanone $a = -616$, $b = 145$; for dimethylformamide $a = -601$, $b = 144$. Equations relating $\Delta$HNP to Hammett sigma constants $\sigma$ also given. | 109 |
| $o$, $m$, and $p$ subs. benzoic acids | Pyridine | $(Bu)_4NOH$ | $\Delta$HNP values measured with respect to benzoic acid. For $m$ and $p$ subs. benzoic acid $a = -646$ and $b = 156$. For $o$ subs. $a = -36$ and $b = 10$. The titration of aliphatic mono and dicarboxylic acids were also investigated. | 110 |
| $o$, $m$, and $p$ subs. benzoic acids | Dimethylsulfoxide | $(Me)_4NOH$ | The $\Delta$HNP values of 34 benzoic acids measured with respect to benzoic acid. For $o$-subs. $a = -463$ and $b = 134$, and for $m$ and $p$ subs. $a = -652$ and $b = 156$. The solutions at $E_{HNP}$ contained < 1.5% water. | 80 |

(*continued*)

Table 7.15. (Continued)

| Compound Type | Solvent | Titrant | Remarks[a] | Ref. |
|---|---|---|---|---|
| Various classes of carboxylic acids | N-Methyl-2-pyrrolidinone | 0.1 $M$ (Et)$_4$NOH in C$_6$H$_6$-MeOH | $\Delta$HNP values measured with respect to benzoic acid. For monocarboxylic acids $a = -366$, $b = 84$; for $o$ subs. benzoic acid $a = -274$, $b = 70$; for $m$ and $p$ subs. benzoic acids $a = -601$, $b = 140$, for phenols $a = -666$, $b = 90$; for dicarboxylic acids $pK_1$ $a = -2424$, $b = 549$; for $pK_2$ $a = -1408$ and $b = 283$. | 111 |
| Benzene-1,2,3 and 1,2,4-tricarboxylic acids and 1,2,3,4 and 1,2,4,5-tetracarboxylic acids. | Various | | A significant discriminating effect was found when dimethyl formamide was the solvent and tetraethylammonium hydroxide the titrant. | 112 |
| Phenols | Nine Different | (Bu)$_4$NOH | 2-Methyl-2-propanol was found to be the best discriminative solvent of those investigated which included CH$_3$CN, CH$_3$COCH$_3$, and C$_5$H$_5$N. | 113 |

| | | | |
|---|---|---|---|
| alcohols | (Bu)₄NOH in C₆H₆–MeOH | $\Delta$HNP values measured with respect to benzoic acid. The best discriminative solvent was found to be 2-propanol for which $a = 476$ and $b = -81$ for $o$-subs. phenols, and $a = 461$ and $b = -76$ for $m$ and $p$ subst. | 108 |
| Phenols | (Bu)₄NOH | $\Delta$HNP values measured with respect to benzoic acid. Constant $a = -1138$ and $b = 151$. Titration behavior of miscellaneous acids also investigated. | 114, 115 |
| Subs. phenylureas | (Bu)₄NOH mixed methoxide in 9:1 C₆H₆–MeOH | $E_{HNP}$ data related to Hammett sigma $\sigma$ values. | 116 |
| Seven acids selected on the basis of $pK_{H_2O}$ values and other properties rather than on structure. | C₆H₆–CH₃CN of various compositions | Sodium methoxide in benzene | The relation between $E_{HNP}$ and $pK_{H_2O}$ was studied in mixtures of the two solvents having dielectric constants in the range $\epsilon = 5$–35. The best discriminative mixture, as reflected in the largest slope for the relation (62.5), had $\epsilon = 30$ which corresponded to 79.5% (w/w) CH₃CN. This composition | 117 |

(*continued*)

Table 7.15. (Continued)

| Compound Type | Solvent | Titrant | Remarks[a] | Ref. |
|---|---|---|---|---|
| Nitrogen bases, substituted phosphines | Nitromethane | 0.05 M HClO$_4$ in CH$_3$NO$_2$ | was used for the discriminative titrations of sulfonamides. $\Delta$HNP values measured with respect to 1,3-diphenylguanidine. For aliphatic and aromatic amines $a = 78.5$ and $b = -77.5$; for amides and ureas $a = 688$ and $b = -83$. | 118,119 |
| Phenothiazines and sympathomimetic amines | Five solvents | HClO$_4$ in each solvent | Correlations were obtained between $E_{HNP}$ and the $pK_b$ values of these bases in water. | 120 |
| Nitrogen compounds in petroleum | 1:2 C$_6$H$_6$–CH$_3$CN, 1:2 C$_6$H$_6$–acetic anhydride | HClO$_4$ in dioxane | Compounds classified according to their $pK_{H_2O}$ values. The types of compounds that may be discriminated by titration in CH$_3$CN and/or acetic anhydride are described. | 121 |
| 34 Nitrogen bases. Nitrogen-containing compounds in asphaltenes | Nitrobenzene and acetophenone | HClO$_4$ in dioxane | The 34 model compounds were resolved into five classes based on $pK_{H_2O}$ and $E_{HNP}$. Nitrogen-containing compounds present in | 122 |

| | | | |
|---|---|---|---|
| Trialkoxy (phenylamino)silanes | 2-Butanone-CHCl$_3$, CH$_3$CN–CHCl$_3$ | HClO$_4$ in 2-butanone | asphaltenes were titrated under similar conditions, and could be categorized into a titratable class (pyridine or aniline types) and a nontitratable class. The latter included compounds such as pyrroles, amides, and phenazines. A correlation was established between ΔHNP (measured with respect to 1,3-diphenylguanidine) and the electronegativity of the substituents. | 123 |
| Dihydro-1,4-benzodiazepines | Glacial acetic acid or 2-butanone | HClO$_4$ in the given solvent | Relation between ΔHNP (measured with respect to 1,3-diphenylguanidine) and the Hammett substituent constant σ established for the 7-substituted-1,3-dihydro-2$H$-1,4-benodiazapin-2-ones and the corresponding 2-thiones. | (124) |

---

[a]Including constants $a$ and $b$ of equation (7.47) where appropriate.

Recourse must be made to a different solvent system in these cases, and the titrant is prepared in this solvent as a comparatively concentrated solution to minimize the effect of the foreign solvent upon the acid-base characteristics of the titration medium. It can be seen, for example, in Table 7.15 that benzene–methanol is a mixture that is sometimes chosen in which to prepare tetraethylammonium hydroxide. The presence of methanol in the titrand solution can, however, exert a greater deleterious effect upon the acid-base properties of the solvent than water (125) in some cases, and its presence can greatly affect the shape of the titration curve and thus the half-neutralization potential. The magnitude of the effect depends upon the nature of the solvent and also upon the type of compound being titrated. Its effect is generally more noticeable in the titration of very weak acids and has been reported (125) in acetone ($\epsilon = 21$) and pyridine ($\epsilon = 13$). The effect has also been noted in the titrations of diprotic acids in acetonitrile and in dimethylsulfoxide, but appears to be minimal in the titration of these acids in 2-methyl-2-propanol (84). The main reason why the curve is little affected by methanol in 2-methyl-2-propanol is that the dianionic salt of the diprotic acid in this solvent behaves as a uni-univalent electrolyte, whereas the dianion is more strongly solvated by methanol (or water) in acetonitrile or dimethylsulfoxide than the monovalent species in 2-methyl-2-propanol (84). A generally more satisfactory titrant is tetrabutylammonium hydroxide prepared in 2-propanol (126).

### REFERENCES

1. I. M. Kolthoff and M. K. Chantooni, Jr., in *Treatise on Analytical Chemistry*, I. M. Kolthoff and P. J. Elving, Eds., 2nd ed., Part 1 (Vol. 2), Wiley, New York, 1979, p. 239.
2. J. F. Coetzee and G. R. Padmanabhan, *J. Phys. Chem.* **69**, 3193 (1965).
2a. I. M. Kolthoff, *Anal. Chem.* **46**, 1992 (1974).
3. S. Bruckenstein and I. M. Kolthoff, *J. Am. Chem. Soc.* **78**, 2974 (1956).
4. L. P. Hammett and A. J. Deyrup, *J. Am. Chem. Soc.* **54**, 4329 (1932).
5. S. Kilpi and H. Warsila, *Z. Phys. Chem.* (*Leipzig*) **A117**, 427 (1934).
6. A. Teze and R. Schaal, *Compt. Rend.* **253**, 114 (1961).
7. I. M. Kolthoff and M. K. Chantooni, Jr., *Anal. Chem.* **51**, 1301 (1979).
8. A. P. Kreshkov, N. Sh. Aldarova, and N. T. Smolova, *Zh. Fiz. Khim.* **43**, 2846 (1969).
9. K. K. Kundu, P. K. Chattopadhyay, D. Jana, and M. N. Das, *J. Phys. Chem.* **74**, 2633 (1970).
10. F. M. Verhoek, *J. Am. Chem. Soc.* **58**, 2577 (1936).
11. M. Breant and G. Demange-Geurin, *Bull. Soc. Chim. Fr.* **1961**, 2935.
12. A. P. Kreshkov, N. Sh. Aldarova, and B. B. Tanganov, *Zh. Fiz. Khim.* **44**, 2089 (1970); *Russian J. Phys. Chem.* **44**, 1186 (1970).

# REFERENCES

13. L. J. Vieland and R. P. Seward, *J. Phys. Chem.* **59**, 466 (1955).
14. S. Bruckenstein and L. M. Mukherjee, *J. Phys. Chem.* **66**, 2228 (1962).
15. C. Jacquinot-Vermesse and R. Schaal, *Compt. Rend.* **254**, 3679 (1962); M. Teze and R. Schaal, *Bull. Soc. Chim. Fr.* **1962**, 1372.
16. A. P. Kreshkov and L. G. Yarmakovskaya, *Zh. Anal. Khim.* **29**, 572 (1974); *J. Anal. Chem. USSR* **29**, 487 (1974).
17. I. M. Kolthoff and M. K. Chantooni, Jr., *J. Phys. Chem.* **72**, 2270 (1968).
18. R. Stewart and J. Jones, *J. Am. Chem. Soc.* **89**, 5069 (1967).
19. Y. H. Inami, H. K. Bodensh, and J. B. Ramsey, *J. Am. Chem. Soc.* **83**, 4745 (1961).
20. E. Grunwald and L. J. Kirshenbaum, *Introduction to Quantitative Analysis*, Prentice-Hall, Engelwood Cliffs, N.J., 1972, p. 109.
21. M. K. Chantooni, Jr. and I. M. Kolthoff, *J. Phys. Chem.* **82**, 994 (1978).
22. G. Allen and E. F. Caldin, *Quart. Rev. (London)* **7**, 255 (1953).
23. J. F. Coetzee and R. M. Lok, *J. Phys. Chem.* **69**, 2690 (1965).
24. D. M. Morman and G. A. Harlow, *Anal. Chem.* **39**, 1869 (1967).
25. A. P. Kreshkov, N. T. Smolova, N. Sh. Aldarova, and N. A. Gabidulina, *Zh. Anal. Khim.* **26**, 2456 (1971).
26. L. N. Bykova and S. I. Petrov, *Zh. Anal. Khim.* **25**, 5 (1970); *J. Anal. Chem. USSR* **25**, 1 (1970).
27. J. S. Fritz, *Acid-Base Titrations in Non Aqueous Solvents*, Allyn and Bacon, Boston, 1973, p. 33.
28. A. P. Kreshkov, *Talanta* **17**, 1029 (1970).
28a. J. F. Coetzee (Ed.), *Recommended Methods for Purification of Solvents and Tests for Impurities*, Pergamon, Oxford, 1982.
29. A. P. Kreshkov, N. Sh. Aldarova, and B. V. Tanganov, *Zh. Anal. Khim.* **25**, 362 (1970); *J. Anal. Chem. USSR* **25**, 307 (1970).
30. M. Windholtz, Ed., *The Merck Index*, 9th ed., Merck, Rahway, N.J., 1976.
31. E. B. Buchanen, Jr., *Talanta* **13**, 1599 (1964).
32. U. N. Dash, *Aust. J. Chem.* **30**, 2621 (1977).
33. I. M. Kolthoff, M. K. Chantooni, Jr., and H. Smagowski, *Anal. Chem.* **42**, 1622 (1970).
34. I. M. Kolthoff and M. K. Chantooni, Jr., *J. Am. Chem. Soc.* **87**, 4428 (1965).
35. J. B. Conant and N. F. Hall, *J. Am. Chem. Soc.* **49**, 3047 (1927); **49**, 3062 (1927); **52**, 4436 (1930).
36. W. P. Tappmeyer and A. W Davidson, *Inorg. Chem.* **2**, 823 (1963).
37. E. A. Burns and E. A. Lawler, *Anal. Chem.* **35**, 803 (1963).
38. S. Bruckenstein and I. M. Kolthoff, *J. Am. Chem. Soc.* **78**, 2474 (1956).
39. I. M. Kolthoff and S. Bruckenstein, *J. Am. Chem. Soc.* **78**, 1 (1956).
40. S. Bruckenstein and I. M. Kolthoff, *J. Am. Chem. Soc.* **78**, 10 (1956).
41. O. W. Kolling and W. L. Cooper, *Anal. Chem.* **42**, 758 (1970).
42. O. W. Kolling and E. A. Mawdsley, *Inorg. Chem.* **9**, 408 (1970).
43. O. W. Kolling, *Anal. Chem.* **40**, 956 (1968).
44. E. A. Burns and E. A. Lawler, *Anal. Chem.* **35**, 803 (1963).
45. D. D. Perrin, *Dissociation Constants of Organic Bases in Aqueous Solution*, Butterworths, London, 1965; Supplement 1972, Butterworths, London, 1972.
46. O. W. Kolling, *Anal. Chem.* **44**, 414 (1972).

47. J. F. Coetzee and G. R. Padmanabhan, *J. Phys. Chem.* **69**, 3193 (1965).
48. I. M. Kolthoff, M. K. Chantooni, and S. Bhowmik, *Anal. Chem.* **39**, 1627 (1967).
49. I. M. Kolthoff and M. K. Chantooni, Jr., *J. Am. Chem. Soc.* **87**, 4428 (1965).
50. J. F. Coetzee and G. R. Padmanabhan, *J. Phys. Chem.* **66**, 1708 (1962).
51. I. M. Kolthoff, S. Bruckenstein, and M. K. Chantooni, Jr., *J. Am. Chem. Soc.* **83**, 3927 (1961).
52. C. K. Mann, *Electroanal. Chem.* **3**, 57 (1969).
53. J. F. Coetzee and I. M. Kolthoff, *J. Am. Chem. Soc.* **79**, 6110 (1957).
54. I. M. Kolthoff and J. F. Coetzee, *J. Am. Chem. Soc.* **79**, 870 (1957).
55. I. M. Kolthoff and M. K. Chantooni, Jr., *J. Phys. Chem.* **70**, 856 (1966).
56. J. F. Coetzee and G. R. Padmanabhan, *J. Am. Chem. Soc.* **87**, 5005 (1965).
57. J. S. Fritz, *Anal. Chem.* **24**, 306 (1952); **24**, 674 (1952).
58. J. S. Fritz and R. T. Keen, *Anal. Chem.* **24**, 308 (1952).
59. I. M. Kolthoff, M. K. Chantooni, Jr., and H. Smagowski, *Anal. Chem.* **42**, 1622 (1970).
60. U. Ya. Mikstais, N. T. Smolova, and A. Ya. Veveris, *Zh. Anal. Khim.* **32**, 362 (1977); *J. Anal. Chem. USSR* **32**, 287 (1977).
61. C. D. Ritchie and G. H. Mergerle, *J. Am. Chem. Soc.* **89**, 1447 (1967).
62. E. Roletto and J. Juillard, *J. Solution Chem.* **3**, 127 (1974).
63. I. M. Kolthoff, M. K. Chantooni, Jr., and S. Bhowmik, *J. Am. Chem. Soc.* **88**, 5430 (1966).
64. R. G. Bates, *Determination of pH Theory and Practice*, 2nd ed., Wiley, New York, 1973, p. 207.
65. E. P. Serjeant and B. Dempsey, *Ionization Constants of Organic Acids in Aqueous Solution*, Pergamon, Oxford, 1979.
66. A. P. Kreshkov, Ya. A. Gurvich, G. M. Gal'pern, and N. F. Kryuchova, *Zh. Anal. Khim.* **27**, 1166 (1972); *J. Anal. Chem. USSR* **27**, 1043 (1972).
67. A. P. Kreshkov, Ya. A. Gurvich, and G. M. Gal'pern, *Zh. Anal. Khim.* **28**, 2440 (1973); *J. Anal. Chem. USSR* **28**, 2158 (1973).
68. E. Fischer, *J. Chem. Soc.* **1955**, 1382.
69. P. G. Sears, W. H. Fortune, and R. G. Blumenshine, *J. Chem. Eng. Data* **11**, 406 (1966).
70. J. Courtot-Coupez and M. L. Demezet, *C.R. Seances. Acad. Sci.* **268**, 1438 (1968).
71. G. C. Price and M. C. Whiting, *Chem. Ind.* (*London*) **1963**, 775.
72. I. M. Kolthoff, M. K. Chantooni, Jr., and S. Bhowmik, *J. Am. Chem. Soc.* **90**, 23 (1968).
73. C. D. Ritchie and R. E. Uschold, *J. Am. Chem. Soc.* **89**, 1721 (1967).
74. B. A. Korolev, T. V. Levandovskaya, and M. V. Gorelik, *Zh. Obshch. Khim.* **48**, 157 (1978); *J. Gen. Chem. USSR* **48**, 135 (1978).
75. R. L. Benoit and C. Buisson, *Electrochim Acta* **18**, 105 (1973).
76. I. M. Kolthoff and T. B. Reddy, *Inorg. Chem.* **1**, 189 (1962).
77. W. S. Matthews, J. E. Bares, J. E. Bartmess, F. G. Bordwell, F. J. Cornforth, G. E. Drucker, Z. Margolin, R. J. McCallum, G. J. McCollum, and N. R. Vanier, *J. Am. Chem. Soc.* **97**, 7006 (1975).

# REFERENCES

78. I. M. Kolthoff and M. K. Chantooni, Jr. *J. Am. Chem. Soc.* **93**, 3843 (1971).
79. C. D. Ritchie and R. E. Uschold, *J. Am. Chem. Soc.* **90**, 2821 (1968).
80. L. N. Bykova, S. I. Petrov, L. A. Karaseva, E. M. Gorynina, and A. D. Galitsyn, *Reakts. Sposobn. Org. Soedin.* **12**, 55 (1975); *Organic Reactivity (USSR)* **12**, 53 (1975).
81. M. K. Chantooni, Jr. and I. M. Kolthoff, *J. Am. Chem. Soc.* **92**, 7025 (1970).
82. M. K. Chantooni, Jr. and I. M. Kolthoff, *Anal. Chem.* **51**, 133 (1979).
83. L. Marple and J. S. Fritz, *Anal. Chem.* **35**, 1223 (1963).
84. I. M. Kolthoff and M. K. Chantooni, Jr., *Anal. Chem.* **50**, 1440 (1978).
85. M. K. Chantooni, Jr. and I. M. Kolthoff, *J. Phys. Chem.* **79**, 1176 (1975).
86. D. B. Bruss and G. E. A. Wyld, *Anal. Chem.* **29**, 232 (1957).
87. J. S. Fritz and S. S. Yamamura, *Anal. Chem.* **29**, 1079 (1957).
88. J. S. Fritz and C. A. Burgett, *Anal. Chem.* **44**, 1673 (1972).
89. M. Foltin and P. Majer, *Collect. Czech. Chem. Commun.* **43**, 95 (1978).
90. J. Juillard and I. M. Kolthoff, *J. Phys. Chem.* **75**, 2496 (1971).
91. M. Georgieva, G. Velinov, and O. Budevsky, *Anal. Chim. Acta* **90**, 83 (1977).
92. M. Mizutani, *Z. Phys. Chem.* **118**, 318, 327 (1925).
93. E. J. King, *Acid–Base Equilibria*, Pergamon, Oxford, 1965, p. 252.
94. C. L. de Ligny, P. F. M. Luykx, M. Rehback, and A. A. Wieneke, *Recl. Trav. Chim. Pays-Bas* **79**, 713 (1960).
95. E. M. Woolley, D. G. Hurkot, and L. G. Hepler, *J. Phys. Chem.* **74**, 3908 (1970).
96. B. Gutbezahl and E. Grunwald, *J. Am. Chem. Soc.* **75**, 565 (1953).
97. M. Paabo, R. A. Robinson, and R. G. Bates, *J. Am. Chem. Soc.* **87**, 415 (1965).
98. M. Sankar, J. B. Macaskill, and R. G. Bates, *J. Solution Chem.* **8**, 887 (1979).
99. R. G. Bates, H. P. Bennetto, and M. Sankar, *Anal. Chem.* **52**, 1598 (1980).
100. E. Grunwald and B. J. Berkowitz, *J. Am. Chem. Soc.* **73**, 4939 (1951).
101. K. C. Ong, R. A. Robinson, and R. G. Bates, *Anal. Chem.* **36**, 1971 (1964).
102. J. O. Frohlinger, R. A. Gartska, H. W. Irwin, and O. W. Steward, *Anal. Chem.* **40**, 1409 (1968).
103. C. C. Panichajakul and E. M. Woolley, *Anal. Chem.* **47**, 1860 (1975).
104. E. M. Woolley and R. E. George, *J. Solution Chem.* **3**, 119 (1974).
105. E. M. Woolley and L. G. Hepler, *Anal. Chem.* **44**, 1520 (1972).
106. P. Fiordiponti, F. Rallo, and F. Rodante, *Z. Phys. Chem. (Frankfurt am Main)* **88**, 149 (1974).
107. H. K. Hall, Jr. *J. Phys. Chem.* **60**, 63 (1956).
108. A. P. Kreshkov, L. N, Bykova, and Z. G. Blagodatskaya, *Zh. Anal. Khim.* **23**, 123 (1968); *J. Anal. Chem. USSR* **23**, 97 (1968).
109. R. R. Miron and D. M. Hercules, *Anal. Chem.* **33**, 1770 (1961).
110. C. A. Streuli and R. R. Miron, *Anal. Chem.* **30**, 1978 (1958).
111. G. M. Gal'pern, Ya. A. Gurvich, and N. F. Kryuchkova, *Zh. Anal. Khim.* **25**, 1819 (1970); *J. Anal. Chem. USSR* **25**, 1561 (1970).
112. A. I. Vasyutinskii and A. A. Tkach, *Zh. Anal. Khim.* **24**, 911 (1969); *J. Anal. Chem. USSR* **24**, 726 (1969).
113. N. T. Crabb and F. E. Critchfield, *Talanta* **10**, 271 (1963).

114. C. A. Streuli, *Anal. Chem.* **32**, 407 (1960).
115. L. N. Bykova, *Zh. Anal. Khim.* **26**, 224 (1971); *J. Anal. Chem. USSR* **26**, 200 (1971).
116. M. L. Cluett, *Anal. Chem.* **34**, 1491 (1962).
117. A. K. Amirjahed and M. I. Blake, *J. Pharm. Sci.* **63**, 696 (1974).
118. C. A. Streuli, *Anal. Chem.* **31**, 1652 (1959).
119. C. A. Streuli, *Anal. Chem.* **32**, 985 (1960).
120. L. G. Chatten and L. E. Harris, *Anal. Chem.* **34**, 1495 (1962).
121. B. E. Buell, *Anal. Chem.* **39**, 756 (1967).
122. L. J. Darlage, H. N. Finkbone, S. J. King, J. Ghosal, and M. E. Bailey, *Fuel*, **57**, 479 (1978).
123. N. A. Makarova, E. A. Kirichenko, and A. I. Ermakov, *Zh. Anal. Khim.* **28**, 972 (1973); *J. Anal. Chem. USSR* **28**, 860 (1973).
124. A. V. Bogat-skii, S. A. Andronati, Z. I. Zhilina, O. P. Rudenko, I. A. Starovoit, and T. K. Chumachenko, *Zh. Obshch. Khim.* **42**, 2571 (1972); *J. Gen. Chem. USSR* **42**, 2562 (1972).
125. G. A. Harlow and G. E. A. Wyld, *Anal. Chem.* **30**, 73 (1958).
126. G. A. Harlow, C. Noble, and G. Wyld, *Anal. Chem.* **28**, 78 (1956).

CHAPTER

8

# THE DETERMINATION OF STABILITY CONSTANTS BY POTENTIOMETRIC TITRATION

## 8.1 INTRODUCTION

This chapter outlines the application of potentiometry to the determination of the equilibrium constants for reactions between metal ions (M) and ligands (L) in aqueous solution. Following the conventional practice of omitting the charges and ignoring the role of the solvent, these equilibria can be represented as

$$M + nL \rightleftharpoons ML_n \tag{8.1}$$

for which the stability constant is defined as

$$\beta_n = \frac{[ML_n]}{[M][L]^n} \tag{8.2}$$

the square brackets representing concentrations in moles per liter. It is important to recognize that all the coordination positions on the metal ion are usually occupied by water so that reaction (8.1) simply involves replacement of one or more molecules of water by the ligand. When the ligand can only donate one pair of electrons to form a coordinate bond it is said to be *unidentate*, and when two or more pairs can be donated the ligand is said to be *multidentate*. A *chelate* is formed when a multidentate ligand forms two or more coordinate bonds with the same metal atom thereby forming a ring structure. When a multidentate ligand coordinates with two or more central metal atoms, such as $Fe_2(OH)_2^{4+}$, the complex is said to be *polynuclear*. Conversely, a *mixed-ligand complex* is formed when a single metal atom coordinates with more than one type of ligand other than water. Generally, the analytically important mixed-ligand complexes contain both a monodentate ligand and a chelating ligand thereby forming a mixed complex of different physical and chemical properties from either of the parent complexes. These differences in properties may sometimes be used for the selective determination of the metal ion or the ligand.

Interest in the determination of stability constants of metal ion complexes was greatly stimulated by the work of J. Bjerrum (1) who elaborated a general method for the determination and calculation of the stability constants for metal ammine complexes. It was shown that unidentate ligands invariably are added in a succession of steps which can be represented generically as

$$ML_{n-1} + L \rightleftharpoons ML_n \tag{8.3}$$

for which each stepwise formation constant could be defined as

$$K_n = \frac{[ML_n]}{[ML_{n-1}][L]} \tag{8.4}$$

The formation of many metal chelates was also found to follow a stepwise pattern as represented by reaction (8.3). This reaction type when applied to the formation of the copper–glycine chelate, for example, can be represented by the equilibria

$$Cu^{2+} + NH_2CH_2COO^- \rightleftharpoons Cu(NH_2CH_2COO)^+$$

$$Cu(NH_2CH_2COO)^+ + NH_2CH_2COO^- \rightleftharpoons Cu(NH_2CH_2COO)_2$$

and the corresponding values of the stepwise formation constants $K_1$ and $K_2$ are

$$K_1 = \frac{[Cu(NH_2CH_2COO)^+]}{[Cu^{2+}][NH_2CH_2COO^-]} \quad \text{and}$$

$$K_2 = \frac{[Cu(NH_2CH_2COO)_2]}{[Cu(NH_2CH_2COO^+)][NH_2CH_2COO^-]} \tag{8.5}$$

These stepwise formation constants $K_1$ and $K_2$ are, of course, related to the stability constant for the overall reaction corresponding to (8.1); that is,

$$Cu^{2+} + 2NH_2CH_2COO^- \rightleftharpoons Cu(NH_2CH_2COO)_2$$

by

$$\beta_2 = K_1 K_2 = \frac{[Cu(NH_2CH_2COO)_2]}{[Cu^{2+}][NH_2CH_2COO^-]^2} \tag{8.6}$$

## INTRODUCTION

Stability constants of these types are often determined by $p$H titration at constant ionic strength. When a solution of a metal ion is added to a solution of the protonated ligand, protons are liberated. For the glycine–copper(II) example this situation can be represented as

$$2\text{NH}_3^+\text{CH}_2\text{COO}^- + \text{Cu}^{2+} \rightleftharpoons \text{Cu}(\text{NH}_2\text{CH}_2\text{COO})_2 + 2\text{H}^+$$

Thus quantitative formation of the $\text{CuL}_2$ complex occurs only if the protons emanating from this equilibrium are removed, usually by reaction with hydroxyl ion. Titration of the chelating agent with a solution of sodium or potassium hydroxide occurs, therefore, at lower $p$H values than those observed when the chelating agent is titrated alone as reference to Figure 8.1 will confirm.

Some thousands of stability constants have been determined since 1945 and these are recorded in the compilations listed as References 2–8. It is

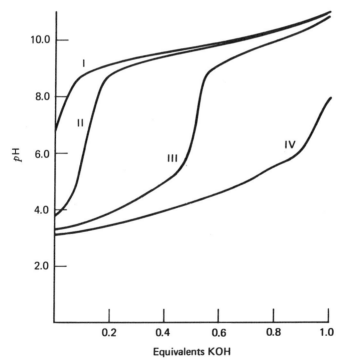

**Figure 8.1.** $p$H Titration curves for 0.02 $M$ glycine in a medium of 0.15 $M$ potassium nitrate: Curve I, 0.02 $M$ glycine; Curve II, 0.02 $M$ glycine + 0.001 $M$ copper(II); Curve III, 0.02 $M$ glycine + 0.005 $M$ copper(II); Curve IV, 0.02 $M$ glycine + 0.01 $M$ copper(II).

quite probable, therefore, that the neophyte, confronted with the problem of determining a stability constant of a newly synthesized chelating agent, would find a system in these compilations that could serve as a model for the determination. It is beyond the scope of this book to describe these determinations in detail and the reader is referred to References 9–14 for the basic principles and for guidance as to the best method of calculation. Reference 14 would serve as an excellent introduction to the topic.

## 8.2 THE $p$H TITRATION METHOD

### Experimental Outline

A majority of the recorded stability constants of metal chelates and the more recently investigated mixed-ligand chelate complexes have been determined by this method in cells with liquid junction that contain a glass indicator electrode. These cells are calibrated so that the measured potential can be interpreted either in terms of $p$H or in terms of $pc_H$, depending upon whether the stability constants are to be recorded as "mixed constants" at "constant ionic strength" or as "concentration constants" at a given ionic strength. Quite frequently the ionic strength used in the determination of concentration stability constants is adjusted so that it is 3 $M$ with respect to sodium perchlorate in which case the "Wilhelm" type of reference half-cell described in Chapter 3 (Figure 3.3) can be used. This, of course, can be modified for other concentrations of sodium perchlorate. If $p$H($X$) is to be measured at constant ionic strength, the latter should not greatly exceed the ionic strengths of the standardizing buffers. A medium of 0.15 $M$ potassium nitrate is often used, and this appears to be a reasonable compromise in this regard since it is also approximately equal to the ionic strength of biological interest. In these circumstances a 3.5 $M$ calomel reference half-cell fitted with a double junction and containing a bridge solution 1.8 $M$ with respect to both potassium chloride and potassium nitrate has been found suitable (15).

Regardless of whether $p$H or $pc_H$ is the potentiometrically derived quantity, it is essential that a constant ionic strength be maintained throughout the titrations needed to determine the stability constant of a metal chelate. In the comparatively simple determination of the overlapping thermodynamic $pK_a$ values for a diprotic acid the procedure used to calculate the ionic strength and hence $c_H$ for insertion in equation (6.63) is quite tedious despite a simplifying assumption. Similarly, the term $c_H$ is also required in the calculation of stability constants [see equation (8.9)], and in these cases the calculation of a "titration-generated" ionic strength is greatly complicated by a much wider diversity of solute species. Not only must the concentrations of the free and bound ligand species be calculated, but the charge and distribution of the free and bound metal ions must be known in order to calculate the ionic strength for each addition of titrant. These calculations would require a value for the formation constant of each complex in the solution, and doubtless an iterative process could be developed for this purpose if reliable values of the constants were obtainable. Frequently, however, the values of the successive formation constants lie closely together and although the value for the overall stability constant can be determined quite reliably, the actual values of its component formation constants are often uncertain. It is far more expedient, therefore, to perform measurements of $p$H upon a solvent of constant ionic strength, and to convert $a_H$ to $c_H$ for use in equation (8.9) by determining an approximate value of the activity coefficient for the hydrogen ion in this medium. To this end, a solution of strong acid of known concentration is titrated alone in the chosen medium (16). From the measured $p$H and the known stoichiometry of the solution, the mean approximate value of the activity coefficient $y$ for the medium may be calculated using the values obtained at each addition by applying the equation $y = a_H/c_H$. This value of $y$ also contains a small correction factor for the residual junction potential, and is used subsequently to calculate $c_H$ in the numerator of equation (8.9). It must be emphasized, however, that $a_H$ calculated directly from the $p$H is used in the denominator of this equation.

The choice of salts from which to prepare the background electrolyte necessary to maintain a constant ionic strength is governed by the nature of both cation and anion. Ideally, neither of these ions should enter into a competitive equilibrium with the system being studied, but in practice these effects can only be minimized by the appropriate choice of electrolyte. The use of large singly charged anions such as perchlorate (radius ~ 0.235 nm) and nitrate (radius ~ 0.189 nm) minimize the electrostatic interactions that exist between anions and metal ions. These anions, therefore, are the most frequently used both as the counterion to the metal ion and in the background electrolyte. The cations chosen for the latter are generally confined to the alkali metal cations lithium, sodium, and potassium since these have a comparatively low tendency to form complexes. Thus sodium perchlorate, prepared by the method of Biedermann and Cavatta (17), is frequently used as the electrolyte for the background solution. In these instances the metal ion to be investigated is used as its perchlorate salt. These can be prepared by dissolving the metal carbonate or oxide in a small excess of perchloric acid and recrystallizing the product from water. Acid-free standard solutions (0.05 $M$) of some divalent metal perchlorates have also been prepared by ion exchange (18), and a variety of transition metal perchlorates have been obtained using only gaseous reagents (19). If the background solution is to be of potassium nitrate, then this solution and the solutions of metal nitrates are prepared from analytical grade reagents recrystallized from water. Accurate standardization of all stock solutions is essential if reliable values of the stability constants are to be obtained.

The actual technique used for the determination of stability constants by the $p$H-titration method is essentially the same as that described for the determination of ionization constants. First, the ionization constants for the fully protonated ligand are determined. If the ligand is either a base or an ampholyte, then it should be dissolved in an appropriate number of equivalents of acid and the protonated compound titrated with the number of equivalents of alkali required to deprotonate it. This alkali should be prepared in the background solution used to maintain a constant ionic strength in the titrand. If $p$H is the measured quantity, then the ionization constants are calculated as mixed constants by the methods described in Chapter 6; if $pc_H$ is measured, then the $pK_a$ values are expressed wholly in terms of concentrations. This titration is then repeated under identical conditions in the presence of a known total concentration of metal ion. Typically, this titrand solution would contain the ligand at a concentration in the range 2–10 m$M$ and the metal ion at a concentration in the range 1–5 m$M$. Because the metal cation displaces the hydrogen cation, the second titration curve lies at lower $p$H values than the first one. If only one proton is liberated from a fully protonated ampholyte as a result of complex formation, then the two curves will be superimposable for the first equivalent of base.

It is strongly recommended that the two titrations be repeated using a different concentration of ligand in the first titration and a different ratio of total ligand : total metal in the second. A gross discrepancy between stability constants calculated from the values obtained in the second titration of each series is indicative that substantial quantities of mixed or polynuclear complexes are interfering. The simple method of calculation outlined in the following would not cope with such a contingency which, fortunately, is quite rare for many of the systems investigated in dilute solutions. Computer programs are available which make allowance for the presence of these other complexes. An initial assumption is made that only the simple complexes are formed, as in the manual method of calculation. However, the programs usually contain a diagnostic that will indicate the presence of other species together with a means of allowing for and estimating the magnitudes of their effects. The use of such programs is also advantageous since it allows the experimenter to employ a much greater degree of flexibility in the range of concentrations used for the titrations than could be contemplated if the results had to be calculated manually. For example, in the investigation of the complex formed between one ligand and one metal, 11 titrations were performed at a single temperature yielding a total of 210 readings to be processed (15). It is not surprising, therefore, that semiautomatic titrators (20) and a fully automatic computer-operated system (21) have been devised and used in acquiring data from which stability constants have been calculated.

A variant of the $p$H-titration method involves the use of an additional indicator electrode reversible to the metal ion under investigation. Cadmium and lead-amalgam electrodes containing 3% of the metal have been used (22) in conjunction with a glass indicator electrode to study the complex formation of cadmium(II) and lead(II) with ethylenedinitrilo-$N$-(2-hydroxyethyl)-$N,N',N'$-triacetic acid (HEDTA). The J-shaped amalgam electrodes (see Figure 1.3) were used also to investigate the complexes of these ions with ethylenedinitrilotetraacetic acid EDTA (23). The complex formation between lead(II) and some dicarboxylic acids have also been studied with a lead amalgam electrode (24, 25).

## Rudimentary Calculations

The basic equation, sometimes referred to as the Irving–Rossotti equation (26), for the calculation of stability constants can be written as

$$\bar{n} + \sum_{1}^{N} (\bar{n} - n)\beta_n [L]^n = 0 \tag{8.7}$$

in which $N$ is the maximum ligand number, and $\bar{n}$ is the average ligand number which gives the mean number of ligands bound to one metal atom. The term $n$ has the same connotation as in equations (8.1) and (8.2), and is an integer having a value $n = 1, \cdots, n = N$. Thus for systems such as the glycine–copper complexes mentioned earlier in which $N = 2$, equation (8.7) can be written as

$$\frac{\bar{n}}{(\bar{n} - 1)[L]} = \frac{(2 - \bar{n})[L]}{(\bar{n} - 1)} \beta_2 - K_1 \tag{8.8}$$

It should be noted that the term $\beta_1$ is conventionally replaced in equations of these types by the first stepwise formation constant $K_1$ [see equations (8.5) and (8.6)]. The $p$H titration method, therefore, requires the calculation of the two functions $\bar{n}$ and $[L]$ for each relevant point on the titration curve. The function $[L]$ is the concentration of *free* chelating species that will be the concentration of dianion if the compound is a diprotic acid, the concentration of monoanion if the compound is an ampholyte such as glycine, or the concentration of uncharged molecular species if the compound is a diacidic base. When $N = 2$, $\bar{n}$ can have any value between 0 and 2.

The function $[L]$ is calculated from the experimentally derived values of $p\mathrm{H}$ or $pc_\mathrm{H}$, the stoichiometry of the titrand solution, and the protonation constants of the ligand. These protonation constants are the reciprocal of the corresponding ionization constants and can be regarded as the stepwise protonation formation constants for the ligand. Thus for the citrate ion ligand ($L^{3-}$) the first protonation constant $K_{\mathrm{H},1}$ refers to the process

$$L^{3-} + \mathrm{H}^+ \rightleftharpoons HL^{2-} \quad \text{for which} \quad K_{\mathrm{H},1} = \frac{[HL^{2-}]}{[L^{3-}][\mathrm{H}^+]} = \frac{1}{K_{a3}}$$

where $K_{a3}$ is the third ionization constant. It is convenient also to define the overall proton formation constants $\beta_{\mathrm{H},2}$ and $\beta_{\mathrm{H},3}$ as

$$\beta_{\mathrm{H},2} = \frac{[\mathrm{H}_2 L^-]}{[L^{3-}][\mathrm{H}^+]^2} = \frac{1}{K_{a3} \cdot K_{a2}} \quad \text{and}$$

$$\beta_{\mathrm{H},3} = \frac{[\mathrm{H}_3 L]}{[L^{3-}][\mathrm{H}^+]^3} = \frac{1}{K_{a3} K_{a2} K_{a1}}$$

For a complexant that can lose $m$ protons on forming the metal complex, the free ligand concentration $[L]$ when $a_{\mathrm{H}^+}$ is assumed to be the measured quantity is given by

$$[L] = \frac{mC_{\mathrm{H}_m L} - c_{\mathrm{NaOH}} - c_{\mathrm{H}^+} + K_w^c/c_{\mathrm{H}^+}}{a_{\mathrm{H}^+} K_{\mathrm{H},1} + 2a_{\mathrm{H}^+}^2 \beta_{\mathrm{H},2} + , \cdots , m a_{\mathrm{H}^+}^m \beta_{\mathrm{H},m}} \tag{8.9}$$

The term $K_w^c$ is the value of the ionic product of water in terms of the concentrations of hydrogen ion and hydroxyl ion at the selected ionic strength for the determination. It is required only when $c_{\mathrm{OH}^-}$ becomes significantly large in comparison with the other terms of the numerator. As mentioned earlier, the term $c_{\mathrm{H}^+}$ must be calculated from the measured $p\mathrm{H}$ using the approximate activity coefficient previously determined for the medium of constant ionic strength. If, however, $pc_\mathrm{H}$ is the measured quantity, $c_{\mathrm{H}^+}$ is used in both the numerator and denominator of equation (8.9). The term $C_{\mathrm{H}_m L}$ in this equation is the concentration of fully protonated complexant present initially and corrected at each point on the titration curve for the volume change caused by the addition of standard alkali. The protonation of any sulfonate group introduced into the complexant molecule as a means of increasing its aqueous solubility is usually not attempted.

The function $\bar{n}$ can be calculated from $[L]$ using the equation

$$\bar{n} = \frac{C_{H_mL} - \alpha_{L(H)}[L]}{C_M} \qquad (8.10)$$

in which $\alpha_{L(H)}$ is defined as

$$\alpha_{L(H)} = 1 + a_{H^+}K_{H,1} + a_{H^+}^2\beta_{H,2} + ,\cdots, a_{H^+}^m\beta_{H,m}$$

The term $C_M$ is the total metal ion concentration originally present in the titrand solution corrected for the added volume of titrant, and $C_{H_mL}$ has already been defined.

For the many complexes in which the total ligand number is two, equation (8.8) can be solved graphically by plotting values of $\bar{n}/(n-1)[L]$ against the corresponding values of $(2-\bar{n})[L]/(\bar{n}-1)$ to yield a straight line of which the slope is $\beta_2$ and the intercept is $-K_1$. It will be noticed in most instances that the points corresponding to $\bar{n}$ values of 0.9–1.1 are excessively sensitive to experimental error, and these values should be excluded if equation (8.8) is to be solved by the method of least squares. Generally, the most reliable values are to be obtained for the results corresponding to the $\bar{n}$ ranges of 0.2–0.8 and 1.2–1.8. A worked example is given in Reference 27. Methods of evaluating the constants of equation (8.7) graphically when the ligand number is greater than two are given in References 11, 12, and 14.

## Computerized Methods of Calculation

When $\bar{n}_i$ and $[L]_i$ have been calculated for a given point $i$ of the titration then, if the following definitions are made,

$$y_i = \bar{n}_i \quad \text{and} \quad x_{i,1} = (1-\bar{n})[L],$$

$$x_{i,2} = (2-\bar{n})[L]^2, \cdots, x_{i,n} = (n-\bar{n})[L]^n$$

equation (8.7) for that point can be written as

$$y_i = x_{i,1}K_i + x_{i,2}\beta_2 + \cdots x_{i,n}\beta_n \qquad (8.11)$$

An analogous equation may be written for each of all the other experimental points, and, in principle, there is little difficulty in solving such "linear" equations by computer. However, since the errors in the $x$ and the $y$ terms of equation (8.11) vary by orders of magnitude throughout the titration and both are derived from the same function $[L]$ [see equation (8.10)], these errors are correlated. Under these conditions the statistics are complex, and their relevance to the computer evaluation of stability constants has been reviewed (28). It is sufficient to state here that the computer programs fit the equilibrium constants to the titration data by using iterative methods that successively refine initial estimates of the constants so that their final values enable a terminating convergence diagnostic to be satisfied. Gross errors in the estimation of the initial values of the constants by the experimenter has been reported to lead to a convergence failure in some circumstances (29). However, in the light of all the stability constant data available there seems little excuse for these excessively erroneous estimates.

Of the many computer programs written for the evaluation of stability constants (28), LETAGROP VRID (20) and SCOGS (31) are the most widely quoted and both have wide application. SCOGS can deal with multireactant systems but is limited to $p$H titration data. LETAGROP (32) can deal additionally with many types of data, but each type requires the use of one or more purpose-written subprograms. Both LETAGROP and SCOGS have been criticized on the grounds that they possess certain mathematical defects that may prevent the attainment of convergence to satisfactory solutions (33). The program MINIQUAD 75 (34) is said to overcome these defects and always lead to convergence of the estimates of the equilibrium constants. PSEUDO PLOT (35) computes and plots titration curves which are compared, by superposition, with experimental ones. By varying the metal : ligand ratios and absolute concentrations it provides a searching check on the adequacy of the types of complexes postulated (36). A more recent program, DALSFEK (37), uses a damped least-square iterative method to fit equilibrium constants to potentiometric and spectrophotometric data.

Although listings for the programs mentioned have either been published or are available from the authors, their adaptation and implementation for one's own use is difficult. A volume currently in press (38) describes in detail computer programs for the computation of stability constants (39). It is understood that this volume deals with the application and implementation of general computer programs for the calculation of stability constants from different types of primary data.

Mention must also be made of programs that utilize the derived values of stability constants to calculate the equilibrium concentrations in solutions containing metal ions and complex-forming species. Such knowledge is of importance not only in the establishment of the most favorable conditions for a given determination, but also in assessing the extent to which other species interfere. Programs such as COMICS (40) and its modifications (41, 42), HALTAFALL (43), and EQUIL (44) have wide applicability for these purposes.

## REFERENCES

1. J. Bjerrum, *Metal Ammine Formation in Aqueous Solution*, P. Haase and Son, Copenhagen, 1941, reprinted 1957.
2. L. G. Sillen and A. E. Martell, *Stability Constants of Metal-Ion Complexes*, Special Publication Nos. 17 and 25, Chemical Society, London, 1964, 1971.
3. E. Hogfeldt, *Stability Constants of Metal-Ion Complexes—Part A—Inorganic Ligands*, in preparation.
4. D. D. Perrin, *Stability Constants of Metal-Ion Complexes—Part B—Organic Ligands*, Pergamon, Oxford, 1979.
5. A. E. Martell and R. M. Smith, *Critical Stability Constants, Vol. 1: Amino Acids*, Plenum, New York, 1974.
6. R. M. Smith and A. E. Martell, *Critical Stability Constants, Vol. 2: Amines*, Plenum, New York, 1975.
7. A. E. Martell and R. M. Smith, *Critical Stability Constants, Vol. 3: Other Organic Ligands*, Plenum, New York, 1977.
8. R. M. Smith and A. E. Martell, *Critical Stability Constants, Vol. 4: Inorganic Ligands*, Plenum, New York, 1976.
9. F. J. C. Rossotti and H. Rossotti, *The Determination of Stability Constants*, McGraw-Hill, New York, 1961.
10. A. Ringbom, *Complexation in Analytical Chemistry*, Interscience, New York, 1963.
11. M. T. Beck, *Chemistry of Complex Equilibria*, Van Nostrand Reinhold, London, 1970.
12. J. Inczedy, *Analytical Applications of Complex Equilibria*, Ellis Horwood (Wiley), Chichester, 1976.
13. E. Wanninen (Ed.), *Essays on Analytical Chemistry*, Pergamon, Oxford, 1977, pp. 51–101.
14. H. Rossotti, *The Study of Ionic Equilibria*, Longman, London, 1978.
15. H. Stunzi, D. D. Perrin, T. Teitei, and R. L. N. Harris, *Aust. J. Chem.* **32**, 21 (1979).
16. C. W. Childs and D. D. Perrin, *J. Chem. Soc. A* **1969**, 1039.
17. G. Biedermann and L. Cavatta, *Acta Chem. Scand.* **15**, 1347 (1961).
18. E. P. Serjeant, *Nature* **186**, 963 (1960).
19. G. Biedermann and D. Ferri, *Chem. Scr.* **2**, 57 (1972).
20. F. L. G. Bouserie, G. Windels, and H. P. Thun, *Chem. Instrum.* **5**, 21 (1973).
21. S. Gobom and J. Kovacs, *Chem. Scr.* **2**, 103 (1972).

22. N. Oyama, T. Shirato, H. Matsuda, and H. Ohtaki, *Bull. Chem. Soc. Jpn.* **49**, 3047 (1976).
23. N. Oyama, H. Matsuda, and H. Ohtaki, *Bull. Chem. Soc. Jpn.* **50**, 406 (1977).
24. A. Olin and P. Svanstrom, *Acta Chem. Scand.*, Ser. A **A29**, 849 (1975).
25. A. Hamman, A. Olin, and P. Svanstrom, *Acta Chem. Scand.*, Ser. A **A31**, 384 (1977).
26. H. Irving and H. S. Rossotti, *J. Chem. Soc.* **1953**, 3397.
27. A. Albert and E. P. Serjeant, *The Determination of Ionization Constants*, Chapter 10, Chapman Hall, London, 1984.
28. F. J. C. Rossotti, H. S. Rossotti, and R. J. Whewell, *J. Inorg. Nucl. Chem.* **33**, 2051 (1971).
29. P. Gans and A. Vacca, *Talanta* **21**, 45 (1974).
30. L. G. Sillen, *Acta Chem. Scand.* **18**, 1085 (1964).
31. I. G. Sayce, *Talanta* **15**, 1397 (1968); **18**, 653 (1971).
32. N. Ingri and L. G. Sillen, *Arkiv. Kemi.* **23**, 47 (1964).
33. A. Sabatini, A. Vacca, and P. Gans, *Talanta* **21**, 53 (1964).
34. P. Gans, A. Sabatini, and A. Vacca, *Inorg. Chim. Acta* **18**, 237 (1976).
35. A. M. Corrie, G. K. R. Makar, M. L. D. Touche, and D. R. Williams, *J. Chem. Soc., Dalton Trans.* **1975**, 105.
36. D. D. Perrin, *Talanta* **24**, 339 (1977).
37. R. M. Alcock, F. R. Hartley, and D. E. Rogers, *J. Chem. Soc., Dalton Trans.* **1978**, 115.
38. D. D. Perrin, *personal communication* (1981).
39. D. J. Leggett (Ed.), *Computational Methods for the Determination of Stability Constants* (tentative title), Plenum, New York, in proof.
40. D. D. Perrin and I. G. Sayce, *Talanta* **14**, 833 (1967).
41. G. A. Cumme, *Talanta* **20**, 1009 (1973).
42. G. Ginsburg, *Talanta* **23**, 149 (1976).
43. N. Ingri, W. Kakolowicz, L. G. Sillen, and B. Warnqvist, *Talanta* **14**, 1261 (1967).
44. Tung-Po I and G. H. Nancollas, *Anal. Chem.* **44**, 1940 (1972).

# PART 3
# APPLICATIONS TO TITRIMETRIC ANALYSIS

CHAPTER

9

# THE BASICS OF TITRIMETRY

The classification of titration reactions and the means of detecting the completion of these reactions by the monitoring of cell potentials have been discussed in Part 1 of this book. Thus in Chapter 2 a description of the types of cells used in potentiotitrimetry was given, and in Chapter 5 some of the electrodes developed specifically for this technique were described. However, the emphasis thus far has been upon the means the equivalence point detection rather than upon the basic principles on which titrimetric methods depend. In this chapter, therefore, some of the salient features pertinent to the titrimetric technique are discussed.

## 9.1  GENERAL PRINCIPLES

### Criteria for Suitable Reactions

It has been shown that standard reduction potentials are useful data in predicting whether a reaction between a given oxidant and a given reductant would occur. Provided that the data were completely appropriate to the course of the reaction, the criterion for a spontaneous reaction was that $E^{\circ}_{cell}$ should be positive, which of course implies [equation 1.3b)] that $\Delta G°$ for the reaction should be negative. The same criterion applies to all reactions and the quantity $\Delta G°$ really defines either the value of $\Delta G$ when the quotient $Q$ of equation (1.1) is unity, or the condition that pertains at the completion of the reaction when the change of free energy will be zero. In the latter case the reaction

$$aA + bB \rightarrow lL + mM \qquad (9.1)$$

reaches a state of equilibrium, and hence equation (1.1) at a particular temperature becomes

$$\Delta G° = -RT \ln Q \qquad (9.2)$$

Since $\Delta G°$, $R$, and $T$ are all constant, it follows also that $Q$ is constant and, for this specified condition, $Q$ now becomes the equilibrium constant for the reaction. This is defined similarly as

$$K = \frac{a_L^l \cdot a_M^m}{a_A^a a_B^b} \tag{9.3}$$

Given, therefore, the values for the standard free energies of formation for the species A, B, L, and M, the overall standard free energy change for reaction (9.1) may be calculated, and hence its equilibrium constant may be evaluated by equation (9.2). For the reaction to be classified as thermodynamically quantitative, the value of the equilibrium constant would have to be infinitely large. Such a condition, however, is not thermodynamically feasible, and the maximum value for $K$ that could be experimentally derived would be about $10^{40}$. By contrast, the values of the equilibrium constants for analytically useful titrimetric reactions are often in the range $10^8$–$10^{15}$. For example, the equilibrium constant of the titration reaction between acetic acid ($pK_a = 4.76$) and sodium hydroxide in dilute solution (25°C), as represented by the equilibrium

$$CH_3COOH + OH^- \rightleftharpoons CH_3COO^- + H_2O$$

is given by

$$K = \frac{a_{CH_3COO^-}}{a_{CH_3COOH} \cdot a_{OH^-}} = \frac{K_a}{K_w} \cong \frac{10^{-4.76}}{10^{-14.00}} = 10^{9.24};$$

$$(\Delta G° = -8.314 \times 298 \ln K = -52.7 kJ).$$

Similarly, for the precipitation titration reaction,

$$Ag^+ + Cl^- \rightleftharpoons AgCl_{(s)}$$

$$K = \frac{a_{AgCl}}{a_{Ag^+} \cdot a_{Cl^-}} = \frac{1}{K_{sp}} \cong \frac{1}{1.8 \times 10^{-10}} = 10^{9.75}$$

At the equivalence point of each of these titrimetric reactions, therefore, if 25 mL of 0.1 $M$ determinand has reacted with 25 mL of 0.1 $M$ titrant, then the amount of determinand not reacted will be less than 0.01% of the original quantity taken. Generally, a reaction is considered suitable for routine titrimetry if 99.90% of the determinand can be titrated leaving only 0.1% untitrated.

The foregoing types of reactions are usually characterized by very low values of the energy of activation which is identifiable with a very fast rate of reaction at room temperature. However, numerous reactions of comparatively high activation energy are used in titrimetric analysis, and in these cases the titration reactions are sometimes carried out at higher temperatures, usually in the range 40–60°C. Another way to increase the rate of a titration reaction is to use a catalyst which has the effect of lowering the activation energy for a particular path along which the reaction then proceeds. In oxidation–reduction titrations, where the reactions are sometimes slow, catalysts are often applied to speed up the reactions.

The value of $\Delta G°$ for the reaction, therefore, and its rate constant are the fundamental physical quantities that determine whether a reaction is suitable for titrimetric analysis. The criteria for a useful reaction stem from these properties and can be summarized as follows:

1. The reaction should be stoichiometric; that is, there should be a definite whole-number ratio between $a$ and $b$ in reaction (9.1).
2. The equilibrium constant for the reaction should be large. For usual analytical accuracy the reaction must be at least 99.9% complete when a stoichiometric amount of titrant has been added.
3. The rate of the chemical reaction should be rapid.
4. There must be a method of determining the point in the titration at which a stoichiometric amount of titrant has been added and the titration reaction is complete. The cells used for this purpose have already been described in Part 1, and the methods used to deduce this equivalence point are discussed in Section 9.2.

Provided these criteria are met, then it is axiomatic that the number of moles of determinand B in reaction (9.1) can be deduced from the number of moles A required to react quantitatively with B. In this case reactant A would have been used in the form of a *standard solution* for the determination of the concentration of B. The concentration of this standard solution of A may have been determined beforehand against another standard solution, or it could have been prepared by dissolving a known weight of pure compound A in a known volume (or weight) of the solvent used for the titration. If the latter procedure had been used, then compound A would be referred to as the *primary standard* for the determination.

Although titrimetric analysis has been associated almost exclusively with volume measurements (i.e., volumetric analysis) employing calibrated measuring vessels such as burettes, pipettes, and measuring flasks, the development of accurate top-pan balances has made measurement by weight almost as convenient as measurement by volume. These weight measurements are inherently more accurate than volume measurements and have been recommended for greater analytical accuracy (1). They have been used for high-precision titrimetry (2), and polythene dispensing bottles of special design (3) have been found to be a convenient alternative to the traditional volumetric ware. The employment of these weight techniques does not, of course, eliminate any error that may be inherent in the detection of the equivalence point.

## Primary Standards

There are very few substances that are available in a sufficiently high grade of purity to be suitable for use as primary standards in titrimetric analysis. Ideally, the substance should be absolutely pure, but this state can never be attained in practice. In this regard it is worth recalling (4) that 1 g of high-quality water with an impurity content of 1 part in $10^9$ would still contain about $10^{13}$ particles of foreign species, with probably every radiochemically stable element present. By contrast, a substance is considered sufficiently pure for use as a standard in titrimetric analysis if the major component present exceeds 99.9% of the total. Ideally, however, the following requirements (5) should be met if the substance is to be classified as a primary standard:

1. The substance should be readily purified and easily preserved in a pure dry state.
2. It should not be hygroscopic or efflorescent, otherwise the substance would be altered during weighings in air.
3. It should have a high molecular weight so that weighing errors may be negligible.
4. It should be readily soluble under the conditions in which it is employed.
5. It should be capable of being tested for impurities by tests of known sensitivity, and the total amount of impurities should not exceed 0.02%.
6. The elements in the substance should be such that disturbance of the natural isotopic abundance would not materially affect the molecular weight.

These requirements are very difficult to satisfy in practice. Under most circumstances recrystallization is generally considered to be a very effective method of purification, and so it is with respect to almost all impurities with one exception—the solvent. In the purification of substances, seemingly suitable for use as primary standards, the difficulty of removing the solvent so that this impurity alone is present at levels below 0.02% often becomes the major task. During the usual process of crystallization the rate of crystal growth is rapid, with the result that the mother liquor is held not only on the surfaces of the crystals, but also in the interstices of masses of crystals which may be in close proximity or partially grown together (4). Normally, the solvent may be easily evaporated from the surfaces of crystals, but once it is held in the interstices it is seldom removable by drying. It can then be removed only by heating the substance to a temperature at which the vapor pressure of the trapped liquid becomes sufficient to rupture the crystals. In some instances the compound may decompose before this temperature is reached (6). The amount of mother liquor entrapped by crystals can be controlled to some degree by varying the rate of crystal growth, for example, but some inclusion of mother liquor can usually be detected by microscopy. In general, the careful control of the conditions whereby the growth of large single crystals is promoted yield products that are relatively free from inclusions. For example, details of the conditions necessary to grow large crystals of borax, potassium hydrogen phthalate, and sulfamic acid free from inclusions and suitable for use as primary standards in acid-base titrimetry are given in Reference 7. These were prepared in the form of large, perfectly clear, single crystals, each crystal weighing between 0.1 and 0.2 g. For some substances, however, the growth of large single crystals is apparently very difficult and in these instances the mother liquor inclusion can sometimes be reduced to a low level (< 0.05%) by drying the finely ground substance (100-mesh sieve) over a suitable desiccant under vacuum for a prolonged period.

A variety of chemical primary standards is available from the U. S. National Bureau of Standards, Washington, D. C., and these are provided with an overall certification of purity. The following substances are often used as primary standards for titrimetry.

### *Acid–Base Titrimetry*

**Potassium Hydrogen Phthalate ($KHC_8H_4O_4$).** This substance was first suggested as a primary standard in 1915 (8), and is used almost exclusively as the primary standard for the standardization of strong bases. It is readily available in pure form (99.99%), it dissolves easily in water [solubility at 25, 35, and 100°C is 10.25%, 12.70%, and 36.10%, respectively (9)], and it is stable on heating up to 135°C. However, the $pK_a$ of the hydrogen phthalate ion is 5.1, and this is its only real disadvantage since the equivalence point occurs in the alkaline region. Consequently, the solutions of bases must be free from carbonate.

**Sulfamic Acid.** When purified according to the recommended method (10), this substance can be used as a primary standard for the standardization of bases. In comparison to potassium hydrogen phthalate it is a strong acid [$pK_a$ = 1.0 (11)] and hence the presence of carbonate in the base will not greatly affect the equivalence point. A disadvantage, however, is its comparatively low molecular weight, and its susceptibility to react with moisture at 100°C producing ammonium hydrogen sulfate by hydrolysis. It must be dried, therefore, *in vacuo* over phosphorous pentoxide. It can be used as a primary standard for titrations in nonaqueous media, and is soluble in dimethylformamide, 1,2-ethanediamine, butylamine, and dimethylsulfoxide. It is only sparingly soluble in acetone and acetic acid.

**Sodium Carbonate.** It has been common practice for many years to use this substance as the primary standard of choice for standardizing solutions of acids. The recommended method of preparation (5) involves the conversion of purified sodium bicarbonate to the carbonate by heating at 270°C to constant weight. The substance is available with a purity within the limits 100 ± 0.02%. It should be noted that the concentration of carbon dioxide at the endpoint is not reproducible because the equilibrium $CO_2 + H_2O \rightleftharpoons H_2CO_3$ is comparatively slow. Elimination of carbon dioxide can be accelerated by use of the enzyme carbonic acid anhydrase (12) or by raising the temperature.

**2-Amino-2-Hydroxy-1,3-Propanediol (THAM or TRIS).** This substance has frequently been proposed or used as a primary standard for the standardization of strongly acidic titrants (13–18) and is sold commercially as such. It appears, however, that crystals of THAM supplied as primary standard grade material may not always be of the requisite quality, and could contain up to 0.7% of included mother liquid (6). A method recommended for the drying of this substance is to carry out an initial vacuum drying of the lightly ground material. The material is then ground again, sifted through a 100-mesh sieve, and finally dried in vacuum at room temperature over anhydrous magnesium perchlorate for 48 h. The preliminary drying of the lightly ground material prevents the formation of a gum during the drying of the finely ground material. The purity of the material was found to be 99.91% using a coulometric standardization procedure. The purity of an NBS Standard Reference Material sample SRM723 was found to be 99.967% by the same procedure when the material, as received, was dried in a vacuum over magnesium perchlorate for two weeks. Prolonged drying of finely ground THAM, this time over phosphorus pentoxide for 40 days, was also used in an investigation of this substance as a standard for solution calorimetry (19). A difference was noted between the results obtained between the ground and the unground material. Evidently, the particle size must be small if the final drying is to be effective, and the grinding process has, in many cases, been overlooked when preparing this substance for use as a titrimetric standard. It is worth recalling that originally it was recommended (13) that the material should be ground to pass a 50-mesh sieve and be dried at 60° and a pressure lower than 10–15 torr for 12 h or, preferably, dried over phosphorus pentoxide at < 5 torr for 24–36 h.

THAM is soluble in most solvents and has been used as a primary standard for titrimetry in solvents such as acetic acid (15) and 2-methoxyethanol. Solvents containing an excess of acetic anhydride should not be used because the amine function may be impaired through acetylation. Solutions of THAM ($pK_a$ = 8.1) absorb carbon dioxide. Thus standardization of acids is best accomplished by weighing accurately the amount of THAM that will be approximately equal to one mole equivalent of acid and dissolving it in the appropriate solvent immediately before titration.

**Other Substances.** The "ionic associate" types of compounds bis[1,2-dihydro-1,5-dimethyl-2-phenylpyrazol-3(3$H$)-one] perchlorate [Reg. No. *18923-31-4*] (I) and bis[1-butyl-2,6-dimethylpyridin-4(1$H$)-one]perchlorate [Reg. No. *51449-19-5*] (II) have been proposed as being suitable for use as primary standards in nonaqueous titrimetry (20–23). Both compounds have excellent solubility characteristics in a variety of solvents and have high molecular weight (i.e., I = 476.92, II = 458.98). Ionic associate (II) exists as a monohydrate under normal room conditions, but drying under vacuum at 60° for 24 h effectively removes the water of hydration. No explosion hazard was experienced with either compound at 110° (23) and both compounds showed excellent stability when their standard solutions in glacial acetic acid were stored for 6 weeks. Both substances are monoacidic bases and can be titrated with perchloric acid solutions according to conventional stoichiometry; they are also monobasic acids and so can be titrated with tetraalkylammonium hydroxide titrants. They have been available commercially (I and II from Eastman, EKX10891 and EK11311, respectively), but they are no longer listed in the recent catalogues available in Australia.

### *Complexometric Titrations (24)*

A large number of substances have been recommended (25) as primary standards and for preparation of standard metal ion solutions with which to standardize solutions of chelons such as ethylenedinitrilotetraacetic acid (EDTA) and ethylenedinitrilo-$N$-(2-hydroxyethyl)-$N,N',N'$-triacetic acid (HEDTA). The commonly supplied salts of these acids cannot be used as primary standards largely due to the uncertainty as to the exact extent to which they are hydrated. For example, although the disodium salt of EDTA, $Na_2H_2C_{10}H_{12}O_8N_2,2H_2O$ (M = 372.24) is available commercially in a high degree of purity, these samples have a moisture content of 0.3–0.5%. This moisture can be eliminated by drying at 80°C and 50% humidity (referred to 25°C) and constant weight is attained only after 4 days.

Calcium carbonate is a substance used frequently as a primary standard for these types of titrations. It is easily obtainable in a high degree of purity, it can be dried without having to fix the temperature within narrow limits, and it is not hygroscopic. Metals such as copper, nickel, zinc, iron, or bismuth are available commercially in purities of 99.99% or better, and consideration can also be given to oxides such as zinc oxide or mercury(II) oxide. These are available in a high degree of purity and are not hygroscopic. Lead(II) chloride has also been found particularly suitable as a standard substance, for it is easily purified by recrystallization from water, contains no water of crystallization, is stable on drying and in storage, absorbs practically no moisture, and has a high equivalent weight. A primary standard of high equivalent weight (388.65) that is stable indefinitely at 110°C has been proposed to fill the dual purpose as a standard for complexometry and for acid-base titrimetry (26). This substance is the 1:1 complex formed between cadium(II) and HEDTA, which can be represented as CdHY (Y = HEDTA). The complex is a comparatively strong monoprotic acid that can be used for the standardization of alkalis. When used for the standardization of EDTA, however, the complex must first be destroyed by boiling with aqueous ammonium peroxydisulfate for a few minutes. Therein lies its chief disadvantage.

### *Oxidation–Reduction Titrations*

**Oxidants.** Potassium dichromate, available as a standard reference material from the National Bureau of Standards (U. S.), is the most widely used and studied (27, 28) oxidant for this purpose. Analytical grade material, thrice recrystallized from water followed by the drying of the pulverized solid at 140–150°C to constant weight before use, is of sufficient purity for most purposes. Potassium iodate and potassium bromate of a sufficiently high purity ($\geq$ 99.8%) to be used as subsidiary standards are commercially available. Both these salts should be ground before drying at 140–150°C to constant weight. However, potassium hydrogeniodate, $KH(IO_3)_2$ (M = 389.91) appears to be a better standard than potassium iodate being available commercially at a purity of > 99.95%. It is dried to constant weight at 100°C (several hours required), and it can also be used as a primary standard for acid-base titrimetry. It has the advantage of being a stronger monoprotic acid than the hydrogen phthalate ion.

**Reductants.** Arsenic(III) oxide and sodium oxalate are the principal reductants available as primary standards. Both are available from the National Bureau of Standards (U. S.) at purities in excess of 99.95%. It has been reported (29), however, that very finely divided arsenic(III) oxide is difficult to transfer from weighing bottles to volumetric flasks even though a 100 mm funnel was used to facilitate this transfer. Some airborne loss of the fluffy powder occurred. This difficulty did not arise when the material became available in a more crystalline form. Although sodium oxalate is commonly used as a primary standard, its use is recommended only when oxalate is to be determined (30). For best absolute accuracy, arsenic(III) oxide is recommended (30). However, sodium oxalate has been used more recently in fairly concentrated solutions of perchloric acid (instead of sulfuric acid) in the presence of manganese(II) sulfate catalyst for the standardization of potassium permanganate solutions (31). Under these conditions the reaction becomes feasible at room temperature although previous warming of the solution is recommended. This is in contrast to the usual Fowler–Bright method (32) which calls for the heating of the solution to 55–60°C in the vicinity of the equivalence point.

Thiourea has been proposed as a reductimetric standard and its solutions have been used for the standardization of iodine-containing oxidants (33). This substance is soluble in water (17.12% w/w at 25°C) and can be recrystallized from ethanol after which it is dried at 100°C. It does not show any tendency to undergo decomposition and its solutions are stable for about 10 weeks. A disadvantage is its low molecular weight (76.12).

*Precipitation Titrations*

Silver metal of high purity (99.999%) is available at this time and solutions prepared from this can be used in the analysis of many inorganic and organic determinands by precipitation titration. Coulometrically generated silver ions provide an accurate standard, and as such have been used for the standardization of solutions of halides to 0.005% (34). It is more usual, however, to standardize solutions of silver nitrate employing either sodium chloride or potassium chloride as the primary standards. Both these salts are available commercially in a high state of purity. They should be finely ground and dried at 130–150°C to constant weight before use. When used as a standard in solution calorimetry (19), potassium chloride (Merck-Suprapur. Art. 4938) was finely ground and dried over phosphorus pentoxide for 40 days. It is also possible to standardize solutions of silver(I) using thiourea as the standard (35).

## Units of Concentration

With the adoption by the International Union of Pure and Applied Chemistry of the mole as the base unit of quantity (36), the mole is no longer a unit of mass, but is one of amount of substance, and terms such as gram-molecule and gram-ion are obsolete. The mole is now defined as:

...the amount of substance of a system which contains as many elementary entities as there are carbon atoms in 0.012 kilogrammes of carbon-12. The elementary entities must be specified and may be atoms, molecules, ions, electrons, other particles or specified groups of such particles.

A consequence of this definition is that there has been a move toward abandoning the use of long-established terms such as *normality*, *normal solution*, and *equivalent*. The objections (37) to these center around the variability in the definitions of these terms which, in comparison with the mole concept, introduce what are regarded as unnecessary complications in the teaching of analytical stoichiometry. For example, there can be only one definition for one mole of a triprotic acid whereas, theoretically, there can be three titrimetric definitions for one equivalent of that acid depending upon the number of moles of solvated protons that are titrated. If the titrimetric reaction is represented as

$$H_3A + NaOH \rightarrow Na_2HA + H_2O$$

then one equivalent of $H_3A$ will be equal to its formula weight. If, however, the reaction being considered is

$$H_3A + 3NaOH \rightarrow Na_3A + 3H_2O$$

then one equivalent is equal to one-third the formula weight of $H_3A$. Similar variability in definition is found also for some oxidants and reductants, a widely quoted example being that for the equivalent of potassium iodate. In the reaction

$$IO_3^- + 6H^+ + 6e \rightarrow I^- + 3H_2O$$

the equivalent is one-sixth the formula weight of potassium iodate, whereas for the Andrews titration in concentrated hydrochloric acid the reaction is

$$IO_3^- + 6H^+ + 4e \rightarrow I^+ + 3H_2O$$

from whence the equivalent is one-fourth the formula weight.

In the recent literature, however, the use of the terms *normality*, *normal solution*, and *equivalent* has persisted, and it is argued (38) that the abandonment of these terms introduces a practical difficulty when reactions that are reproducible but nonstoichiometric are considered. For many reactions it is not always possible to specify the composition of a sample solution in terms of a single reacting component of known stoichiometry. This is particularly the case in the analysis of certain mixtures of polymers, in biochemical determinations, in solutions of mixtures of metal ions of different oxidation states, or in the determination of the degree of unsaturation for a reaction mixture in organic synthesis. Nevertheless, in these cases it is of great value to state explicitly the equivalence between the standardized titrant and the analyte solution in terms of the number of equivalents of titrant reacted per unit volume of sample solution. Accordingly, in view of the continued practical value and widespread use of the concepts of normality and equivalent it has been considered essential that these terms should be retained (38). In order that the definition of an equivalent be compatible with the foregoing definition of a mole, the following definition has been recommended (38):

The equivalent of a substance (in Germany, the Val) is that amount of it which, in a specified reaction, combines with, releases or replaces that amount of hydrogen that is combined with 3 grammes of carbon-12 in methane $^{12}CH_4$. Although the term equivalent now specifies an amount of substance rather than a mass it will be obvious that one equivalent of NaOH will have a mass of 0.03997 kg and one equivalent in $H_2SO_4$ a mass of 0.049039 kg and so on in acid-base reactions in aqueous media.

Following on from this definition a *normal solution* is:

That solution which contains one equivalent (German 1 Val) of a defined species per $dm^3$ according to the specified reaction.

Bearing in mind that one mole contains one Avagadros number of atoms, molecules, ions, electrons,..., the equivalent can also be expressed in terms of this constant which has the numerical value of $6.022045 \times 10^{23}$ (39). Thus in acid–base reactions one equivalent can be defined as the mass of the substance that supplies or combines with $6.022045 \times 10^{23}$ hydrogen ions. In oxidation-reduction reactions it is the amount that supplies or combines with $6.022045 \times 10^{23}$ electrons, and in precipitation and complexometric titrations it is the amount that supplies or reacts with $6.022045 \times 10^{23}$ of monovalent metal ions, or with one-half that number of divalent metal ions, and so on.

## Stoichiometry

If reaction (9.1) fulfills the criteria necessary for a titrimetric reaction, it is sufficient to write it as

$$a\text{A} + b\text{B} \rightarrow \text{products} \tag{9.4}$$

which can be used to imply that $a$ moles of titrant A react with $b$ moles of determinand B. To calculate the amount of B in the titrand solution it is necessary to determine the volume $V_E$ (or sometimes the weight) of the titrant solution A of known concentration $C_A$ (moles L$^{-1}$) required to react quantitatively with B. Ideally, this volume (or weight) $V_E$ should be identical to the exact volume of titrant required to reach the *equivalence point* of the titration. In practice, however, the experimentally determined *endpoint* of the titration, regardless of the technique by which this is determined, will differ to some extent from the ideal equivalence point. Within this constraint, the weight in milligrams of determinand B ($w_B$) in the titrand solution is related to the volume of titrant (mL) to reach the endpoint $V_E$ by

$$w_B = \frac{b}{a} C_A V_E M_B$$

where $M_B$ is the formula weight of B(g). The analogous equation when the concentration of titrant A is expressed in normality $N_A$ (equivalents L$^{-1}$) is

$$w_B = N_A V_E Q_B$$

where $Q_B$ is the equivalent weight of determinand B. The accuracy with which $w_B$ can be determined is predominantly dependent upon the accuracy with which the equivalence volume $V_E$ can be assessed. In potentiotitrimetry the basic quantities required for this purpose are the monitored values of $E_{cell}$, or a function thereof, and the corresponding values of $V_A$, the volume of titrant. The types of cells used are described in Chapter 2 and the electrodes in Chapter 5. An outline of how the basic potentiotitrimetric data is used to evaluate $V_E$ is given in Section 9.2.

## 9.2 METHODS OF ASSESSING THE EQUIVALENCE VOLUME, $V_E$

This discussion is confined to the monitoring of titrations using a chemical cell with diffusion consisting of an indicator electrode dipping into the titrand solution and a reference half-cell making electrical contact with this solution through a salt bridge. As mentioned in Chapter 1, the shape of a plot of $E_{cell}$ versus volume of titrant is sigmoidal, being typified by Figure 1.4. The experimental endpoint volume is deduced from this curve and is usually taken as being the point of maximum slope observed when the amount of titrant added is circumjacent to the stoichiometric completion of the reaction. Mention has already been made that there is a small error between the volume corresponding to this point of maximum slope (i.e., the endpoint) and the true equivalence volume. This is known as the *titration error*, and its magnitude depends upon the equilibrium constant of the titration reaction and the concentration of determinand in the titrand solution. Although in many cases this error can be made $< 0.1\%$ by a judicious choice of experimental conditions, it must be emphasized at this stage that it is always advisable to standardize the titrant with a known amount of determinand using the same procedure, electrodes, and instrument as are to be used in the actual determination. Such a standardization of procedure will effectively cause the cancellation of the titration error. The adoption of this procedure is even more important when using a special reference half-cell, the potential of which has been adjusted to be equal to the potential of the indicator electrode at the equivalence point. In this Pinkhof–Treadwell method the endpoint volume is taken as the volume of titrant required to give a zero deflection on a galvanometer or some other device connected across the cell. It is recommended that the titrant be standardized at least once a day when using this technique, and this recommendation applies also when using the other techniques outlined in Chapter 2. These are the techniques of differential potentiotitrimetry, electrolytic potentiotitrimetry, and differential electrolytic potentiotitrimetry, the titration curves of which were discussed in Chapter 2.

The first titrations to be monitored by potentiometric means were reported in 1893 (40), and in the intervening years much interest has been centered upon the best way of deducing the equivalence volume from the titration curve. This interest seems to have intensified over the last 20 years or so, and since the early 1960s there has been a proliferation of papers dealing with potentiotitration curves and their mathematical interpretation. In a survey of more than 100 of these papers it emerges that there are three types of methods used currently for assessing the equivalence volumes of acid–base, complexometric, oxidation–reduction, and precipitation titrations. These are as follows:

Graphical methods
Methods based on the linearization of the titration data
Curve-fitting methods

The graphical methods are the simplest and most widely used and have as their aim the detection of the point of maximum slope observed when $E_{cell}$ (or $pH$ or $pX$) is plotted ($y$-axis) against volume of titrant ($x$-axis). The advantage of this type of plot is that only simple arithmetical treatment of the data is required and the cell does not have to be calibrated.

Linearization of the titration data requires that the mathematical equation to the titration curve be known and that this equation can be arranged into the linear form $y = mx + c$ ($m$ = slope, $c$ = intercept) yielding the equivalence volume as an intercept or as a point of intersection. Although accurate standardization of the cell is not always necessary for these linearization methods, the slope of the indicator electrode must be known and must remain constant. In some instances it is also advisable to ensure that the ionic strength remain fairly constant throughout the titration. When using this type of method it is usual to obtain at least six data points before and after the equivalence point in the regions of the plot of $E_{cell}$ versus volume titrant where the change of $E_{cell}$ with added titrant is comparatively small. The equivalence volume is obtained by extrapolation of these points. Linear titration plots are advantageously applied to those cases where there is likely to be a considerable difference between the point of maximum slope of a conventional titration curve and the actual equivalence volume. For example, in those precipitation titrations when a precipitate of the form $A_iB_j$ ($i \neq j$) is produced, the point of maximum slope does not necessarily coincide even approximately with the equivalence volume.

Curve-fitting methods usually combine data obtained during the titration with a theoretical equation for the titration curve in such a way as to yield the best estimates of the parameters in that equation. The $E_{cell}$ data must be reliable for this purpose, and hence it is necessary to pay meticulous attention to the calibration of the cell. The application of this type of method, therefore, requires a potentiometric titration rather than a monitoring of potential which is the distinguishing feature of potentiotitrimetry. Nonetheless, it is included in this section because curve-fitting methods can be used to assess the equivalence volume under conditions so unfavorable that its evaluation in any other way would be impossible. The application of these methods does not depend upon the existence of a point of maximum slope on the titration curve and, therefore, yields useful results in titrations of species at concentrations below that at which the point of maximum slope disappears. Access to a digital computer is considered essential to obtain the best estimates of the titration parameters and computer programs are available for this purpose.

## Graphical Methods

Near to the equivalence point the potential of the indicator electrode changes greatly in comparison with the changes observed earlier in the titration when a substantial amount of determinand remained to react with the titrant. In the assessment of the equivalence volume, therefore, interest is traditionally centered upon the shape of the curve in the region where the rate of change of potential for a given incremental volume is at a maximum. The titrant is usually added in equal incremental volumes $\Delta V$ in this region to yield the typical set of results given in Table 9.1.

Titration curves are often categorized as *symmetrical curves* or *asymmetrical curves* depending upon whether they are symmetrical about the point of maximum slope or *inflection point* as it is often called. Symmetrical titration curves are obtained typically in strong acid–strong base titrations and in those precipitation and oxidation–reduction titrations where the stoichiometric coefficients [$a$ and $b$ in reaction (9.4)] for the titration reaction are unity. It is often stated erroneously that the inflection point is coincident with the true equivalence point for these symmetrical or *isovalent* titration reactions. Such a conclusion was reached as a result of ignoring the volume change that occurs during the titration (41, 42). A more rigorous mathematical treatment taking into account this dilution effect reveals that the inflection point and the equivalence point will never coincide for so-called symmetrical titration curves (43, 44). For strong acid–strong base or isovalent ($a:b = 1$) precipitation titrations the point of maximum slope always precedes the equivalence point. Only for heterovalent titration reactions ($a \neq b$) producing unsymmetrical curves may there be conditions under which the inflection point and the equivalence point will coincide. It can be shown also for isovalent complexometric titration curves that there will be a discrepancy between the two points even if volume changes are neglected (45).

For symmetrical titration reactions the difference between the true equivalence volume $V_E$ and the volume corresponding to the inflection point $V_I$ depends upon the concentration of the reactants and the equilibrium constant for the titration reaction. The titration error ($V_I - V_E$) increases as the reactants become more dilute and is greater for a reaction having a small value of the equilibrium constant than for one which has a large equilibrium constant. For acid–base titrations the magnitude of this equilibrium constant is reflected in the value of the ionization constant for the acid. At a given concentration the weaker the acid (i.e., $K_a$ small), the greater will be the titration error. It can be shown (43) that in titrations of weak acids this error is no greater than 0.1% if, at an equivalence point concentration of 0.1 $M$, the $pK_a$ of the acid is less than 9. The error increases with dilution, however, and rises to 1% if an acid having $pK_a = 9$ is titrated so that the concentration of the conjugate base at the equivalence point is 0.01 $M$.

The titration error for isovalent precipitation titrations (i.e., $a = b$) is less than 0.1% if the concentration exceeds $10^3$ $(K_{sp})^{1/2}$ where $K_{sp}$ is the solubility product, and similarly for complexometric titrations it will be < 0.1% if the product of the stability constant and concentration exceeds $5 \times 10^3$. For heterovalent precipitation titrations the relevant information is given in the second paper listed as Reference 43. It is again emphasized that these titration errors will cancel if the standardization of the titrant is performed using a standard sample of the determinand having a concentration close to those expected in the samples for analysis. Under this condition the volume of titrant commensurate with the inflection point suffices for the calculation of the concentration of determinand.

The inflection point can sometimes be found directly by inspection of the curve of $E_{cell}$ versus volume of titrant if, circumjacent to this point, there is a very large change in potential for very small incremental additions of titrant. However, it is more usual to have to resort to a method of geometric construction for the location of this point and a number of such methods are available. Tubb's method of circle fitting (46) is an example of these

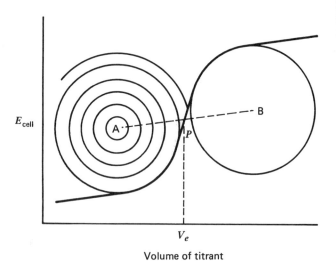

**Figure 9.1.** Tubbs method (46) of locating the inflection point $P$ and thence the corresponding volume of titrant $V_e$.

methods, and depends upon the observation that titration curves have the shape that can be thought of geometrically as being described by the arcs of two impenetrable disks that lie on the same plane in contact with each other (see Figure 9.1). The inflection point lies at the point of contact of the two resulting circles through which the line linking the centers of these circles passes. This is the geometrical basis for the method which can be applied also to asymmetric titration curves. The centers of the two arcs of the circles are located on the graph by means of a template made from a thin rigid transparent plastic sheet upon which is marked a series of concentric circles that increase in diameter in steps of about 1 cm up to a maximum diameter of about 30 cm. The center of these circles is located on the template by a small hole through which the point of a pencil will fit. To locate the inflection point on the titration curve, the template is maneuvered so that the arc of one of the circles on the template is coincident with the arc made by one branch of the titration curve. The center of this circle is marked on the graph. The center of the arc described by the other branch of the titration curve is then located and marked similarly and a line connecting the two centers is drawn. The inflection point is taken as the point where this line intersects the titration curve and the assessed equivalence volume is taken as the volume of titrant that corresponds to this point.

A device that is also used directly on the titration curve of $E_{cell}$ versus volume of titrant (47) depends upon the refraction produced when portions of the curve are viewed through two parallel glass rods (20 cm long, 0.2 cm dia.) mounted on a special ruler (Figure 9.2). This ruler, which is used parallel to the $E_{cell}$ axis, is attached at right angles to another ruler along which it can be moved and which ensures that the two parallel glass rods remain at right angles to the volume axis when the titration curve is viewed through them. As a result of refraction, two points appear on the glass rods which are images of the curve, and at the inflection point the distance between these refracted points is maximal and is proportional to $\Delta E$. This point is located on the curve by marking the graph with a pencil through the two slots marked G on Figure 9.2 and drawing a line between them. It is necessary, therefore, that these slots be located so that they are coincident with an imaginary line drawn between the two glass rods.

A method devised in 1919 (48) is still probably the most widely used of all methods for finding the inflection point of titration curves. In this method the quantity $\Delta E/\Delta V$ is plotted against volume of titrant to yield two curves that, when produced, intersect at the maximum value for $\Delta E/\Delta V$. This represents the point of maximum slope in the original curve, and the volume of titrant corresponding to this maximum is assumed to equal the equivalence volume. An objection to this method is the difficulty of accurately extrapolating the two curves to a precisely defined point of

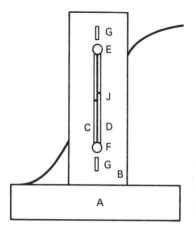

Figure 9.2. Ruler for endpoint location by graphical differentiation. A, B: plastic rulers; C, D: glass rods; E, F: holes to insert rods. G: slots for marking the position of the endpoint; J: refracted images of the curve (47).

intersection even when the $E_{cell}$ data are reliable. However, because drifts in $E_{cell}$ values are often encountered in this critical region of the curve, the values of $\Delta E/\Delta V$ often have scatter which greatly increases the uncertainty of extrapolating the two curves. A modified graphical procedure has been proposed which, it is claimed, can be used to find the inflection point precisely even with moderate experimental scatter (49).

A number of arithmetical methods are available for obtaining the point of maximum slope directly from the titration curve data. The most widely used of these has come to be known as Kolthoff's method (50–52), which is based upon the original work (48). An adaptation of this method is best explained by reference to the data given in Table 9.1.

It should be noted that the volume readings, either taken directly from the burette or interpolated from a graph of $E_{cell}$ versus $V$, should be such that there is a constant difference $\Delta V$ between successive volume readings. From the data three volume readings are selected that straddle the maximum value observed for $\Delta E$. In the example these correspond to reading numbers 3, 4, and 5, and it is clear that the inflection volume $V_I$ lies between 21.10 and 21.20 mL. The first significant volume reading for this method is,

Table 9.1. Some Typical Titration Curve Data Circumjacent to the Equivalence Point

| Reading Number | 1 | 2 | 3 | 4 | 5 | 6 |
|---|---|---|---|---|---|---|
| $V$ (mL) | 20.90 | 21.00 | 21.10 | 21.20 | 21.30 | 21.40 |
| $E_{cell}$ (mV) | 290 | 310 | 340 | 480 | 545 | 575 |
| $\Delta E$ | 20 | 30 | 140 | 65 | 30 | — |

therefore, taken as 21.10 mL, and this is the value assigned to $V'$ in equation (9.5). The maximum potential change is denoted by $\Delta E_m$ (140 mV), the potential change before it as $\Delta E_b$ (30 mV), and the one after it as $\Delta E_a$ (65 mV). The volume $V_I$ is given by

$$V_I = V' + \Delta V \left[ \frac{\Delta E_m - \Delta E_b}{2\Delta E_m - (\Delta E_b + \Delta E_a)} \right] \quad (9.5)$$

from whence $V_I$ for this example is calculated as

$$V_I = 21.10 + 0.1 \left[ \frac{140 - 30}{280 - (30 + 65)} \right] = 21.16 \text{ mL}$$

The other method outlined here is Fortuin's method (53) which is mathematically more rigorous than Kolthoff's method and is claimed to be more accurate. A nomogram (Figure 9.3) is required for the application of this method, which uses the same three data as were used in the previous example. Again, the incremental volumes $\Delta V$ must be equal. However, in this method the values of $\Delta E$ are arranged in decreasing order of magnitude $\Delta_0$, $\Delta_1$, and $\Delta_2$ so that for the example of Table 9.1 $\Delta_0 = 140$, $\Delta_1 = 65$, and $\Delta_2 = 30$. The volume $V'$ in this case corresponds with the potential between $\Delta_0$ and $\Delta_1$ (i.e., 21.20 mL). From these $\Delta$ values two ratios are defined as

$$R_1 = \frac{\Delta_1}{\Delta_0} \quad \text{and} \quad R_2 = \frac{\Delta_2}{\Delta_1}$$

$$= 0.464 \qquad\qquad = 0.462$$

Lining up these values of $R_1$ and $R_2$ on the nomogram allows a value for an inflection point parameter $p_i$ to be selected (i.e., $p_i = 0.19$ for the example). If, as in this example, $\Delta_0$ precedes $\Delta_1$, then $V_I$ is calculated from

$$V_I = V' - p_i \Delta V = 21.20 - 0.1 \times 0.19 = 21.18 \text{ mL}$$

For the converse case of $\Delta_1$ preceding $\Delta_0$, $V_I$ is given by

$$V_I = V' + p_i \Delta V$$

Although these two methods, in common with a third method credited to Hahn (54), require that equal incremental volumes be taken for the calculation of $V_I$, a method is available that interpolates the volumes corresponding to equal values of $\Delta E$ from the titration curve (55). A further method

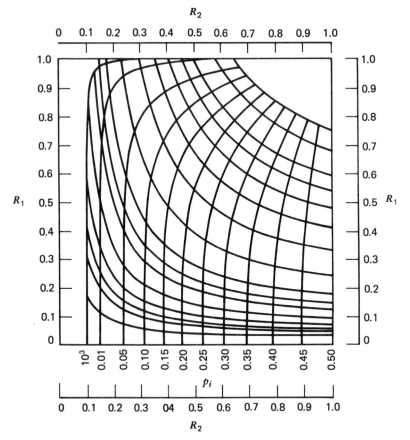

**Figure 9.3.** Nomogram to Fortuin's method (53).

has been described that requires neither constant values of $\Delta V$ nor equal values of $\Delta E$ (56).

The value of $V_I$ is assumed to equal the true equivalence volume $V_E$ in these methods, which have as their objective the determination of the point of maximum slope for the curve. A basic assumption inherent in the application of the methods is that the potentials monitored by the indicator electrode are not affected by the presence of interfering ions, which will limit the magnitude of $\Delta E$ in the region of the equivalence point and thus the precision of the measurement. The concentration of the species to which the indicator electrode responds is usually very small in this region, and thus the recorded potentials will be very susceptive to the presence of interfering ions. Such interferences can seriously influence the position of the inflection point on the titration curve, and, in some instances, serious errors (> 1%) can result (57). These effects have been considered in relation to precipitation titrations (58) and chelometric titrations (59–61). If ions likely to interfere are known to be present in the titrand solution and these cannot be masked, it is better to use a linearization method rather than a method that relies solely upon potentials monitored in the region of the equivalence point. Although the error will not be eliminated entirely by this expedient, it will certainly be reduced.

## Linearization of Titration Curves

The principal method used for the linearization of titration data is the method of Gran (62), the basic equations for which were given earlier. When this method is applied to titrimetric data two functions are derived; one for the readings obtained before the equivalence point and the other for the readings after the equivalence point. Each function $F$ is linearly dependent on the volume of titrant $V$, and both equal zero at the equivalence point. Thus the volume required to reach this point can be located from the intercept of one or both lines with the volume axis. The functions $F$ derived by Gran are summarized later; $V_0$ is the initial volume of the titrand solution including the volume of any reagent solution, such as a buffer solution, a masking solution, an ionic strength adjustment solution added before the titration; $E$ is the measured value of $E_{cell}$; and $S$ is the calibration slope of the electrode. It should be noted that $E/S$ can be written as $-p\text{H}$ if this is the monitored quantity, or as $-p\text{X}$ if the activity of an ion other than the hydrogen ion is monitored. The abbreviation e.p. is used for equivalence point:

1. Strong acid–strong base

    a. Before e.p. $F = (V_0 + V)10^{E/S}$
    b. After e.p. $F = (V_0 + V)10^{-E/S}$

2. Weak monobasic acid–strong base

    a. Before e.p. $F = V \times 10^{E/S}$
    b. After e.p. as 1(b)

3. Weak monoacidic base–strong acid

    a. Before e.p. $F = V \times 10^{-E/S}$
    b. After e.p. as 1(a)

4. Precipitation titration $aA + bB \rightarrow A_aB_b$. Electrode responding to B (titrant).
   a. Before e.p.  $F = (V_0 + V)10^{-bE/aS}$
   b. After e.p.   $F = (V_0 + V)10^{E/S}$

5. Precipitation titration $aA + bB \rightarrow A_aB_b$. Electrode responding to A (determinand)
   a. Before e.p.  $F = (V_0 + V)10^{E/S}$
   b. After e.p.   $F = (V_0 + V)10^{-aE/bS}$

6. Complexometric titration $aA + bB \rightarrow A_aB_b$. Electrode responding to B (titrant)
   a. Before e.p.  $F = (V_0 + V)^{1-(1/a)} \times V^{1/a} \times 10^{-bE/aS}$
   b. After e.p.   as 4(b).

7. Complexometric titration $aA + bB \rightarrow A_aB_b$. Electrode responding to A (determinand)
   a. Before e.p.  as 5(a)
   b. After e.p.   $F = (V_0 + V)^{1-(1/b)} \times 10^{-aE/bS}$

8. Oxidation of determinand $n_B A_{red} + n_A B_{ox} \rightarrow n_B A_{ox} + n_A B_{red}$
   a. Before e.p.  $F = V \times 10^{-n_A E/S}$
   b. After e.p.   $F = 10^{n_B E/S}$

9. Reduction of determinand $n_B A_{ox} + n_A B_{Red} \rightarrow n_B A_{red} + n_A B_{ox}$
   a. Before e.p.  $F = V \times 10^{n_A E/S}$
   b. After e.p.   $F = 10^{-n_B E/S}$

An arbitrary constant $k$ is often included in these Gran functions because, at the time of the original publication, the exponent terms were most conveniently evaluated using tables of antilogarithms. For example, the function for the titration of a strong acid by a strong base is often written as

$$F = (V_0 + V)10^{k-pH} \quad \text{or} \quad F = (V_0 + V)10^{k+E/S}$$

The value assigned to $k$ was such that the antilogarithms would fall in a suitable range such as 0–1. If $pH$ were the measured quantity, $k$ would be given an integer $pH$ value, but if $E_{cell}$ were measured in millivolts, $k$ would be given a convenient value of a whole number of hundreds of millivolts. However, it is now sufficient to set $k = 0$ for all the functions since calculators that include $10^x$ functions are available, greatly facilitating the calculation of these Gran functions. It is advisable to use at least four significant figures for this purpose.

Mention has been made of the special Gran plot paper (10% volume corrected) that is available. This also simplifies the treatment of titrimetric data provided that the following conditions are met:

1. The expected concentration of determinand in the titrand is such that a titrant can be prepared that will yield a dilution close to 10%.
2. The calibration slope of the indicator electrode is close to $\pm 58$ mV per decade for univalent electrodes or $\pm 29$ mV per decade for divalent electrodes.

Curved plots will result if either of these conditions are not satisfied when using this paper. In addition to these factors, curved plots can also result because the derivations of the functions are not completely rigorous. Thus the effect of the variation in the activity coefficients has been neglected but this has been found to be negligible in most practical titrations if the volume of titrants remains less than 1.3 times the equivalence volume (63). The concentration ranges for precipitation titrations within which the Gran functions deviated from linearity have also been calculated (63) and, in general, Gran functions for the precipitation reactions $aA + bB \rightarrow A_aB_b$ are reliable if $-\log K_{sp} \geq 5(a + b)$ where $K_{sp}$ is the solubility product. They are reliable also for strong acid–strong base titrations, but should not be applied to the results of titrations of weak acids in aqueous solution if $pK_a \leq 3$ or $pK_a \geq 8$. Improved functions that allow for hydrolysis and activity coefficients have been proposed for such titrations (64), but these require that the $pK_a$ value(s) of the acid be known and that the cell be carefully calibrated so that $p$H can be measured accurately.

When used within the constraints mentioned, the advantage of the Gran method is that only about six readings of $E_{cell}$ and volume of titrant are needed to establish each branch of the titration curve. Thus the titrant can be added in constant and comparatively large incremental volumes throughout the titration. If six readings are taken before the equivalence point and six after, then extrapolation of the two appropriate functions to $F = 0$ gives two independent values for the equivalence volume. Furthermore, the equilibrium values of $E_{cell}$ are generally more rapidly attained and more stable in the "buffered" regions of the titration curve than they are near to the equivalence point with the result that a reliable extrapolation can be made. The extrapolation of only one branch of the titration curve is useful in locating the equivalence volume if, for some reason, this cannot be reached. It has been recommended (65) that regression analysis be used as a means of deducing the equivalence volume, thus obviating the necessity to draw the graph. This approach has also been used in a computer program that was applied to the titration of iron(II) with cerium(IV) (66). A detailed evaluation of the accuracy of Gran's method when applied to the determination of the total alkalinity and carbonate content in seawater has been made by means of the program HALTAFALL (67). Modified functions were suggested for this application. The application of Gran plots and other logarithmic diagrams to complexometric titrations has been investigated (68). A detailed study has also been made on the application of Gran's method to the titration of fluoride with lanthanum(III) (69).

Other methods for the linearization of titration curves require that the cell be carefully calibrated and assume the constancy of the liquid junction potential. These are, therefore, examples of the application of direct potentiometry to titrimetry. Most attention has been centered upon the linearization of acid–base titration curves, and, as mentioned (64), it is usually necessary to know the $pK_a$ value of the titrand under the conditions chosen for the titration. An exception to this latter requirement is a method that relies upon the choice of suitable chemical conditions in order that linear titration curves be obtained directly (70). In the titration of a weak acid, for example, if a swamping concentration of its sodium salt be added at the start, then it follows from the equation $c_{HA} = c_{H^+} K_a^c c_{A^-}$ that $c_{HA}$ during the titration will be directly proportional to the measured values of $c_{H^+}$. If these values of $c_{H^+}$ are plotted against the volume of titrant added, then it can be shown that a straight line will be obtained that intersects the volume axis at the equivalence volume $V_E$. Volume corrections are unnecessary for this method since the species HA and $A^-$ are diluted equally during the titration. Linear titration plots are obtained when $c_{A^-} \cong 50 c_{HA}$, as exemplified by the titration of 0.01 $M$ acetic acid in the presence of 0.5 $M$ sodium acetate. This salt can also be used in swamping concentrations for the titrations of acids stronger than acetic acid, whether they are monoprotic, diprotic, or in mixtures. In these cases one is merely titrating acetic acid in the present of excess acetate. For acids weaker than acetic acid a swamping concentration of sodium sulfite was used. The method has been applied also to oxidation–reduction titrations and involves keeping one of the components of the couple at a constant concentration. Since the formal potential depends upon the ratio $c_{ox}/c_{red}$, it follows that if the concentration of oxidized species $c_{ox}$ is kept constant, $E_{cell}$ depends only upon the concentration of reduced species, $c_{red}$. Hence when $c_{red}$ is titrated with a suitable oxidant, a plot of $E_{cell}$ versus volume of oxidant will be linear, the line again intersecting the volume axis at $V_E$. This approach was used in the determination of $V_E$ for the titration of iodine with thiosulfate. Gran's method cannot be applied in this instance because iodide must be added from the beginning to keep the iodine in solution, and the Gran functions assume that all the iodide present must be generated by the titration. It can be shown for the titration represented by

$$I_3^- + 2S_2O_3^{2-} \rightarrow 3I^- + S_4O_6^{2-}$$

that if the titrand contains a large excess of iodide (0.8 $M$) with respect to iodine ($\sim$ 0.05 $M$), then the derived linear equation is

$$V_E - V = \text{constant} \times 10^{2E/k}$$

where $E$ is the measured value of $E_{\text{cell}}$ and $k$ is the Nernst factor at the temperature of the titration. To keep the ionic strength constant at approximately 0.8 $M$ throughout the titration, the ionic strength of the titrant solution (0.1 $M$ thiosulfate) was adjusted to 0.8 $M$ with potassium chloride. Under these conditions it was assumed that the liquid junction potential, the activity coefficients, and thus the formal potential of the cell remained constant throughout the titration.

Refinements in the linearization of acid–base $p$H titration curves have been based upon rearrangement of equations similar to 6.33 which are solved by computer programs involving iterative procedures. Thus programs have been written, and are presumably available, for the simultaneous determination of precise equivalence points and $pK_a$ values (71), for the determination of mixtures of strong and weak acids (72), for the resolution of mixtures of weak monoprotic acids in solution (73), and for the determination of partially neutralized weak acids or weak bases (74). These methods can sometimes be used even when the acid is so weak that no inflection is apparent from the titration curve.

**Curve-Fitting Methods**

The application of these methods stem from the use of computer programs like HALTAFALL (75) which have been used to calculate the equilibrium concentrations present in a solution given the appropriate equilibrium constants. With such a program it is possible to generate sets of theoretical $E_{\text{cell}}$ versus volume of titrant curves. By varying all the parameters according to a suitable strategy, a best fit between the experimental and calculated curves is obtained from which the total concentration of the determinand species can then be evaluated. The potentiality of this method, known as *multiparametric curve fitting*, was recognized in 1971 (76) when it was concluded that the accuracy and precision of the method was likely to be very high. A related procedure was used shortly afterwards (77) to analyze binary mixtures of formic, acetic, and propanoic acids for which detection of the endpoints by conventional means is not possible. Since then the methods have been applied predominantly to acid-base titrations in aqueous solution.

Although the programs devised are elegant, the results obtained using them will be only as reliable as the experimental data that are processed by them. It is essential, therefore, that as much care be exercised in the preparation of solutions (carbonate-free) and in the calibration of the cell to measure either $pH$ or $pc_H$ (see Chapter 4) as would be used for a precise determination of a $pK$ value at a given temperature. Generally, the $pK_a$ value of the determinand need not be known since this can usually be evaluated by the program. However, it is essential to ensure that if $pH$ is the measured quantity, the value of $K_a$ and the activity coefficient of the hydrogen ion remain constant throughout the titration. To this end, a condition of constant ionic strength is maintained in both the titrand and titrant solutions by the addition of an inert electrolyte such as potassium chloride. The ionic strengths used range from 0.1 to 3.0 M. If $pH$ is measured, then the activity coefficient of the hydrogen in the titration medium must be calculated to compare the experimental curve with the computer-generated curves. Calibration of the cell directly in terms of $pc_H$ obviates this necessity. Usually about a dozen points which cover the range from 0.1 to 1.05 equivalent of titrant are sufficient for one titration which should be carried out at constant temperature in a sealed beaker under an atmosphere of nitrogen. From these data it is possible to locate the equivalence point with an accuracy and a precision that cannot be approached by other techniques and even to obtain useful and reliable results under conditions so unfavorable that other techniques fail completely. Thus multiparametric curve fitting has been applied in the titrations of weak bases with strong acids at extreme dilutions (78), to titrations with an unstandardized reagent (79), to the titration of sodium dodecanoate (80), and to the determination of strong acids in the presence of weak acids (81). The relevant programs CFT3 and, presumably, CFT4 are said to be available (79). A listing of the program ACBA, which can also be applied to the titrations of mixtures of acids or bases, is attached as an appendix to Reference 82. A computer program that uses a simpler version of multiparameter curve fitting has been adapted to on-line use for the titration of mixtures of a weak acid and a strong acid (83). In addition to the curve-fitting program, the software for the automatic titration system was developed to calibrate the cell, to control the titration, and record the titration curve. A monitor program controlled the three latter real-time tasks and also allowed the curve-fitting program to be run as a background task during the recording of the next titration curve. A complete titration and calculation required about 20 min.

# REFERENCES

1. K. Eckschlager, *Errors Measurements and Results in Chemical Analysis*, Van Nostrand Reinhold, London, 1969, p. 34.
2. R. W. Perry and H. J. Scullion, *Analyst (London)* **94**, 801 (1969).
3. T. D. Rice, *Anal. Chim. Acta* **97**, 213 (1978).
4. E. Wichers, *Anal. Chem.* **33(4)**, 23A (1961).
5. Analytical Standards Sub-Committee, *Analyst (London)* **90**, 251 (1965).
6. W. F. Koch, D. L. Biggs, and H. Diehl, Talanta **22**, 637 (1974).
7. A. Madej and A. Rokosz, *Chemical Anal. (Warsaw)* **18**, 227 (1973); **20**, 1115 (1975); **21**, 31 (1976); **21**, 271 (1976).
8. F. D. Dodge, *Ind. Eng. Chem.* **7**, 29 (1915).
9. W. S. Hendrixson, *J. Am. Chem. Soc.* **42**, 724 (1920).
10. Analytical Standards Sub-Committee, *Analyst (London)* **92**, 587 (1967).
11. E. J. King and G. W. King, *J. Am. Chem. Soc.* **74**, 1212 (1952).
12. A. L. Underwood, *Anal. Chem.* **33**, 955 (1961).
13. J. H. Fossum, P. C. Markunas, and J. A. Riddick, *Anal. Chem.* **23**, 491 (1951).
14. T. H. Whitehead, *J. Chem. Educ.* **36**, 297 (1959).
15. T. R. Williams and J. D. Harley, *Chemist-Analyst* **50**, 114 (1961).
16. J. E. Ruch and F. E. Critchfield, *Anal. Chem.* **33**, 1569 (1961).
17. E. W. Wilson, Jr. and D. F. Smith, *Anal. Chem.* **41**, 1903 (1969).
18. G. Marinenko, *Natl. Bur. Stand. (U. S.) Technical Notes 543*, Nov. 1970, p. 56.
19. R. Rychly and V. Pekarek, *J. Chem. Thermodyn.* **9**, 391 (1977).
20. A. I. Busev, B. E. Zaitsen, V. K. Akimov, Ya. Chelikhovskii, and F. Kopetski, *Zh. Obshch. Khim.* **38**, 534 (1968); *J. Gen. Chem. USSR* **38**, 523 (1968).
21. A. I. Busev, V. K. Akimov and I. A. Emel'yanova *Zh. Anal. Khim.* **23**, 616 (1968); *J. Anal. Chem. USSR* **23**, 525 (1968).
22. J. T. Alessi, D. G. Bush, and J. A. Van Allan, *Anal. Chem.* **46**, 443 (1974).
23. A. K. Mukherji, *Anal. Chim. Acta* **77**, 331 (1975).
24. G. Schwarzenbach and H. Flaschka, *Complexometric Titrations*, Methuen, London, 1969, p. 146.
25. C. N. Reilley, A. J. Barnard, Jr., and R. Puschel, in *Handbook of Analytical Chemistry*, L. Meites, Ed., McGraw-Hill, New York, 1963, p. 3-94.
26. J. E. Powell, J. S. Fritz, and D. B. James, *Anal. Chem.* **32**, 954 (1960).
27. G. Marinenko and J. K. Taylor, *J. Res. Natl. Bur. Stand. (U.S.)* **67A**, 453 (1963).
28. R. Belcher, C. L. Chakrabarti, and W. I. Stephen, *Analyst (London)* **94**, 20 (1969).
29. A. J. Zielen, *Anal. Chem.* **40**, 139 (1968).
30. I. M. Kolthoff, H. A. Laitiner, and J. J. Lingane, *J. Am. Chem. Soc.* **59**, 429 (1937).
31. O. A. Ohlweiler and A. M. H. Schneider, *Anal. Chim. Acta* **58**, 477 (1972).
32. R. M. Fowler and H. A. Bright, *J. Res. Natl. Bur. Stand. (U.S.)* **15**, 493 (1935).
33. B. C. Verma, S. M. Ralhan, and N. K. Ralhan, *Mikrochim Acta* **1976(1)**, 201.
34. G. Marinenko and J. K. Taylor, *J. Res. Natl. Bur. Stand. (U.S.)* **67A**, 31 (1963).

35. R. Soloniewicz, *Chem. Anal. (Warsaw)* **22**, 177 (1977).
36. I.U.P.A.C. *Manual of Symbols and Terminology for Physico-chemical Quantities and Units*, Pure Appl. Chem. **21**, 20 (1970).
37. J. Lee, *Education in Chemistry* **2**, 229 (1965).
38. I.U.P.A.C., *Recommendations of Usage of the Terms 'Equivalent' and 'Normal'*, Information Bulletin No. 36, August 1974.
39. I.U.P.A.C. *Manual of Symbol and Terminology for Physico-chemical Quantities and Units*, Pure Appl. Chem. **51**, 1 (1979).
40. R. Behrend, *Z. Physik. Chem.* **11**, 466 (1893).
41. P. S. Roller, *J. Am. Chem. Soc.* **50**, 1 (1928); **54**, 3485 (1932); **57**, 98 (1935).
42. J. N. Butler, *J. Chem. Educ.* **40**, 66 (1963).
43. L. Meites and J. A. Goldman, *Anal. Chim. Acta* **29**, 472 (1963); **30**, 18 (1964); **30**, 28 (1964); **30**, 200 (1964).
44. L. Meites and T. Meites, *Anal. Chim. Acta* **37**, 1 (1967).
45. W. Lund, *Talanta* **23**, 619 (1976).
46. C. F. Tubbs, *Anal. Chem.* **26**, 1670 (1954).
47. A. M. Wahbi, S. Ebel, and C. P. Christiansen, *Fresenius' Z. Anal. Chem.* **271**, 344 (1974).
48. J. C. Hostetter and H. S. Roberts, *J. Am. Chem. Soc.* **41**, 1337 (1919).
49. S. R. Cohen, *Anal. Chem.* **38**, 158 (1966).
50. I. M. Kolthoff and N. H. Furman, *Potentiometric Titrations*, Wiley, New York, 1931, p. 95.
51. I. M. Kolthoff and H. A. Laitinen, *pH and Electrotitrations*, Wiley, New York, 1944, p. 110.
52. I. M. Kolthoff, E. B. Sandell, E. J. Meehan, and S. Bruckenstein, *Quantitative Chemical Analysis*, Macmillan, London, 1969, p. 946.
53. J. M. H. Fortuin, *Anal. Chim. Acta* **24**, 175 (1961).
54. F. L. Hahn, *Fresenius' Z. Anal. Chem.* **163**, 169 (1958).
55. F. L. Hahn, *Fresenius' Z. Anal. Chem.* **183**, 275 (1961).
56. J. F. Yan, *Anal. Chem.* **37**, 1588 (1965).
57. F. A. Schultz, *Anal. Chem.* **43**, 502 (1971).
58. P. W. Carr, *Anal. Chem.* **43**, 425 (1971).
59. F. A. Schultz, *Anal. Chem.* **43**, 1523 (1971).
60. P. W. Carr, *Anal. Chem.* **44**, 452 (1972).
61. T. Anfalt and D. Jagner, *Anal. Chem.* **45**, 2412 (1973).
62. G. Gran, *Analyst (London)* **77**, 661 (1952).
63. C. McCallum and D. Midgley, *Anal. Chim. Acta* **65**, 155 (1973).
64. D. Midgley and C. McCallum, *Talanta* **21**, 723 (1974); **23**, 320 (1976).
65. C. Liteanu and D. Cormos, *Talanta* **7**, 25 (1960); **7**, 32 (1960).
66. T. J. MacDonald, B. J. Barker, and J. A. Caruson, *J. Chem. Educ.* **49**, 200 (1972).
67. I. Hansson and D. Jagner, *Anal. Chim. Acta* **65**, 363 (1973).
68. A. Johansson, *Talanta* **20**, 89 (1973).
69. T. Eriksson, *Anal. Chim. Acta* **58**, 437 (1972).
70. A. Johansson, *Talanta* **22**, 945 (1975).

71. T. N. Briggs and J. E. Stuehr, *Anal. Chem.* **46**, 1517 (1974).
72. C. McCallum and D. Midgley, *Anal. Chim. Acta* **78**, 171 (1975).
73. C. McCallum and D. Midgley, *Chem. Ind. (London)* **1978**, 844.
74. D. Midgley and C. McCallum, *Fresenius' Z. Anal. Chem.* **290**, 230 (1978).
75. N. Ingri, W. Kakotowicz, L. G. Sillen, and B. Warnqvist, *Talanta* **14**, 1261 (1967).
76. T. Anfalt and D. Jagner, *Anal. Chim, Acta* **57**, 165 (1971).
77. F. Ingman, A. Johansson, S. Johansson, and R. Karlsson, *Anal. Chim. Acta* **64**, 113 (1973).
78. D. M. Barry and L. Meites, *Anal. Chim. Acta* **68**, 435 (1974).
79. D. M. Barry, L. Meites, and B. H. Campbell, *Anal. Chim. Acta* **69**, 143 (1974).
80. S. L. Young, E. Matijevic, and L. Meites, *J. Phys. Chem.* **78**, 2626 (1974).
81. D. Mutlow and L. Meites, *Anal. Chim. Acta* **92**, 285 (1977).
82. G. Arena, E. Rizzarelli, and S. Sammartano, *Talanta*, **26**, 1 (1979).
83. M. Bos, *Anal. Chim. Acta* **90**, 61 (1977).

# CHAPTER 10
# ACID–BASE TITRIMETRY

## 10.1 INTRODUCTION

In Chapter 7 it was mentioned that the range of potentials that can be monitored by one of the hydrogen ion-responsive electrodes, described in Chapter 5, during the course of an acid–base titration depends upon the value of the autoprotolysis constant for the titration medium. Thus the range in water was shown to be small in comparison to the ranges available in some other solvent, and this restricts the applications of potentiotitrimetry in water considerably. If the traditional graphically based methods described in Chapter 9 for the assessment of the equivalence volume are used, aqueous titrations are confined to acids (HA) having $pK_a$ values $\leq 9$ and to conjugate acids (BH$^+$) having $pK_a$ values $\geq 5$ for the error to be $\leq 0.1\%$ at endpoint concentrations of 0.1 $M$. This range depends upon the dilution, and if the dilution is increased tenfold, then the $pK_a$ range is diminished by one $pK_a$ unit at each end of the range if an accuracy of $\leq 0.1\%$ is to be maintained. The use of Gran's method to obtain the equivalence volume can be used to increase this $pK_a$ range by about one unit at either end for a given concentration if only one line of the plot, corresponding to the $pH$ readings before the endpoint, is extrapolated. The limits of application of the more sophisticated methods of linearization and multiparametric curve fitting is established by the limits within which it is possible to measure the $pH$ accurately. This is approximately between $pH$ 2 and 11 if a glass electrode is used in a cell with liquid junction. Even this latter range is small in comparison to the ranges theoretically possible if some of the organic solvents discussed in Chapter 7 are used as titration media. In practice, however, the magnitude of the range available in a given solvent is curtailed considerably by the presence of water, either as an impurity in the solvent or generated by the titration itself. In some instances the presence of small amounts of water is desirable otherwise the response of the indicator electrode would be too slow for the titration to be considered suitable for routine use. Despite this curtailment, the applications of acid–base potentiotitrimetry in nonaqueous media now exceed the applications in water as evidenced by the great increase in the number of methods published during the last 20 years. Quite often binary or even tertiary mixtures of solvents are the media used for these titrations. However, water must remain the solvent of choice if at all possible for reasons of convenience and economy. Methods developed for the enhancement of the acidity or basicity of compounds in aqueous solution are, therefore, important in this respect since they afford the means whereby determinands, which hitherto could not be titrated alone because of unsuitable ionization characteristics in water, can be titrated as a result of a modification to their ionization processes effected by these enhancement methods.

Carbon dioxide should be regarded as the ubiquitous interferent for all acid–base titrations, whether performed in aqueous or nonaqueous media. Absorption of carbon dioxide by alkali titrants can be minimized by storing the solutions in bottles sealed with a bung carrying a soda–lime guard tube and a delivery tube. If a syringe is to be used to dispense the titrant, this can be filled by siphon action through the delivery tube. Alternatively, if a burette is to be used, this should also be protected by a soda–lime guard tube and before filling should be purged with nitrogen. The titrant should then be pumped into the burette by nitrogen presaturated with respect to the solvent vapor. The titrand solution should be contained in a beaker sealed with a bung through which holes have been bored to accommodate the electrodes, the delivery tube from the burette or syringe, and an inlet tube for nitrogen. As mentioned earlier, "O"-ring seals suffice for the location of the electrodes and significantly reduce their breakage rate in comparison to the breakages experienced when they are held tightly in a bung. Although it is sometimes necessary to purge the titrand solution of carbon dioxide by passage of presaturated nitrogen through this solution before the titration, it is better to direct the flow of nitrogen across the surface of the solution rather than through it during the titration when $E_{cell}$ is being monitored. Stirring is best accomplished by magnetic means provided that the titrand solution is insulated against temperature rises which this mode of stirring can induce. For nonaqueous titrations it is worth reiterating that the coefficients of cubic expansion for most organic solvents are high in comparison with water, and hence in order to avoid significant volumetric error the temperature must remain constant throughout the time taken to standardize the titrant and perform the determination. Substances used as primary standards in acid–base titrimetry were discussed earlier and suitable indicator electrodes were described in Section 5.2.

## 10.2 TITRATIONS IN AQUEOUS SOLUTIONS

### Titrants

The titrants commonly used are standard solutions of hydrochloric acid for the titration of bases and carbonate-free sodium or potassium hydroxide for the titrations of acids. Some laboratory supply houses market these as concentrated solutions contained in sealed ampoules. Dilution of these solutions to the specified volume with distilled water yields standard solutions that are satisfactory for most purposes. When preparing solutions of sodium hydroxide, however, it is advisable to open, transfer, wash, and dilute the contents of the ampoules using a stream of nitrogen to minimize the absorption of carbon dioxide. Distilled water, freed of carbon dioxide either by boiling for 10–15 min or by bubbling nitrogen through it, should be used for the dilution. Alternatively, details of a method that can be used for the preparation of carbonate-free potassium hydroxide solution using an ion exchange technique are available (1).

## Enhancement of Acidity

Acids with $pK_a$ values > 9 cannot be determined accurately by conventional acid–base titrimetry unless a neutral substance is added to the titrand solution that will increase the acidity of the determinand. Quite frequently this is achieved by complexation. A well-documented example is the titration of boric acid ($pK \sim 9.2$) in the presence of polyols with which boric acid forms complexes. These act as stronger monoprotic acids than boric acid and can be titrated accurately. Among the complexing agents that have been used, mannitol (2) is by far the most popular, although boric acid reacts with several other polyols to form complexes which are acidic enough to be titrated directly with alkali hydroxide solution. Of these polyols, glucose, fructose, sorbitol, and 1,2-ethanediol have been claimed to be equal or superior to mannitol. A more recent detailed potentiometric study has shown that sorbitol gives slightly sharper endpoints than the more commonly used mannitol, but both are satisfactory for the titration of $5 \times 10^{-2}$–$5 \times 10^{-4}$ $M$ solutions of boric acid (3). A solution containing 5% (w/v) gives a sufficient enhancement of acidity over this concentration range of boric acid. This method of enhancement has been used for the determination of boron in organoboron compounds. The oxidants and oxidation procedures used for the conversion of some of these compounds to boric acid have included trifluoroperoxyacetic acid (4), alkaline peroxydisulfate (5), and a micro Carius procedure (6). Boron and nitrogen in amineboranes have been determined by an acid hydrolysis followed by the consecutive determinations of both the boric acid and the amine salt by titration with alkali (7). In the determination of diborane, boric acid was quantitatively produced by hydrolysis in water (8). The method has also been applied in the determination of boron in samples of titanium boride (9) and in boron silicides (10). Boric acid was produced by oxidation with peroxydisulfate in sulfuric acid in the first method, and by three types of sodium carbonate fusions in the second.

Hexitols like mannitol and sorbitol are themselves very weak acids, the $pK_a$ values for which are about 13.3. However, it has been reported (11) that these compounds can be determined as diacidic bases by titration of their complexes formed with molybdenum(VI) or tungsten(VI). If the concentration ratio of $MO_4^{2-}$ (M = Mo or W) to hexitol is about 4:1, the concentration of hexitol can be derived using Gran's method for the linearization of the titration curve.

The acidity of some phenols has been enhanced by the addition of the nonionic surfactant nonylphenoxypoly(ethylenoxy)ethanol to the solution (12). The phenolic compounds formed 1:3, 1:5, or 1:7 adducts with this ethoxylated nonylphenol. These have $pK_a$ values in water ($pK_{H_2O}$) within the range 6.5–7.0 for the 1:3 adducts, 5.8–6.0 for the 1:5 adducts, and 5.4–5.7 for the 1:7 adducts. Titrations of phenol ($pK_{H_2O}$ 10.0), 1,2-benzenediol ($pK_{H_2O}$ 9.4, ~ 13.0), 1,3-benzenediol ($pK_{H_2O}$ 9.2, 11.3), and 1,4-benzenediol ($pK_{H_2O}$ 9.9, 11.6) with sodium hydroxide at concentrations of $10^{-4}$ $M$ were possible when the concentration of ethoxylated nonylphenol was $7 \times 10^{-4}$ $M$.

Polymeric compounds added to the titrand solution have also been used to sharpen the inflection in the titration curve of weak carboxylic acids in dilute solutions. Titration of dilute solutions ($10^{-4}$–$10^{-5}$ $M$) of weak acids such as acetic and trimethylacetic acids in the presence of poly[dimethylaminoethyl methacrylate] resulted in the polyelectrolyte acting as a buffer up to the equivalence point at which a sharp break in the titration curve occurred (13). Acids of the type HA and $BH^+$ have been titrated in the presence of a nonionic resin [Amberlite XAD-2]. The alterations in the shape of the titration curves obtained when solutions of diphenylamine hydrochloride, benzylamine hydrochloride, and $m$-nitrophenol were titrated were quantitatively accounted for by considering the sorption of both the neutral conjugate species (B or HA) and the ionic conjugate species ($A^-$ or $BH^+$) on the resin (14). It was concluded that titrations in the presence of this resin permit the accurate determination of the $BH^+$ charge-type which is too weak to be titrated alone. The observed opposite sign of the $pH$ shift in the titrations of HA and $BH^+$ charge-type acids should allow the development of discriminative titrations of acids with similar strengths but of different charge type.

The distribution of carboxylic acids between an aqueous phase and an immiscible organic solvent during the course of the titration of the aqueous phase has been investigated as a means of increasing the difference in the apparent strengths of the acids being titrated (15). After the addition of each portion of titrant the two phases were shaken until equilibrium was established and then the $pH$ of the aqueous phase was measured. The solvents used were carbon tetrachloride, ether, and benzene. The shape of the titration curve is influenced by the distribution and dimerization constants of the acid for the given aqueous–nonaqueous system. These vary greatly, and their magnitude depends upon the nature of the substituent attached to the carboxy group. Thus it is reportedly possible to titrate the following binary mixtures of acids in which each component acid has a $pK$ value in the range 4.8–4.9: nonanoic acid–acetic acid, nonanoic acid–pentanoic acid, nonanoic acid–hexanoic acid, octanoic acid–pentanoic acid. Two-phase solvent systems have also been used to determine chelate-forming copper extraction reagents by acid–base titration. The organic acid, liberated into the aqueous phase by the addition of salicylaldoxime or 8-quinolinol samples prepared in chloroform to copper sulfate solution (50 mL ~ 0.05 $M$), was titrated with sodium hydroxide to $p$H 4.7 (16). A large number of organic ammonium ions have been quantitatively titrated in a heterogeneous medium consisting of an aqueous phase and pentachlorophenol dissolved in dichloromethane (17). The pentachlorophenol forms ion pairs with the cation liberating protons which are titrated with sodium hydroxide. The cations determined by this technique included tetrapropylammonium bromide [Reg. No. *1941-30-6*], hyoscyamine sulfate [*620-61-1*], and strychnine nitrate [*66-32-0*].

The liberation of hydrogen ion as a result of complex formation has been used as a basis for the determination of metal ions (18). Thus calcium(II), cadmium(II), cobalt(II), copper(II), mercury(II), manganese(II), nickel(II), lead(II), and zinc(II) have been determined in unbuffered or slightly buffered solutions by titration with a standard EDTA–NaOH solution having a $p$H approximately 10.6. Preliminary neutralization with dilute sodium hydroxide was necessary before titration so that the $p$H ranged between 4.4 and 5.0 for cadmium(II), cobalt(II), copper(II), manganese(II), nickel(II), lead(II), and zinc(II); about $p$H 5.6 for mercury(II); to $p$H 6.5–6.9 for calcium(II) or magnesium(II). Similarly, a complexant can sometimes be determined in a one-phase partially aqueous solution through complexation with a metal ion to yield an amount of hydrogen ion in proportion to the concentration of complexant. For example, bisthiosemicarbazones with alkyl side chains have been determined by this means in the presence of copper(II). In this case the two moles of hydrogen ion produced by this complexation were titrated with standard sodium hydroxide solution in a dioxan–water medium (19).

## Enhancement of Basicity

Although weak bases such as aniline [$pK_a$ = 4.60 (25°)] cannot be titrated in water to a discernible inflection point, bases with a $pK_a$ > 2 can be titrated successfully (20) in concentrated aqueous solutions of neutral salts (6–8 $M$). The method has also been applied to discriminate between the individual nitrogen atoms of polyfunctional amines, and between strong and weak bases in single titrations. The presence of the salt decreases the aqueous solubility of the amines considerably, but this effect can be countered by adding methanol to the titrand solution. Should this be necessary, basicity enhancement can be accomplished by using a neutral electrolyte such as lithium chloride which is appreciably soluble in aqueous methanolic solutions. A method has been developed for the determinations of $m$ and $p$-phenylenediamines, adipic and carbonic acid dihydrazides, diazabicyclooctane, and dihydroxydiethylpiperazine based on their titrations with 1 $M$ hydrochloric acid in a concentrated lithium chloride solution (6-8 $M$) or a 4-5 $M$ solution of calcium chloride (21). It has been proposed (22) that the mode of action of the electrolyte in enhancing the basicity results from the reduction of the water activity in the solution which has the effect of reducing the hydration of the hydrogen ion, the activity of which is thereby increased at a given concentration. Protonation equilibria are then displaced according to the reaction

$$H_3O^+ \cdot nH_2O + B \rightarrow BH^+ + (n+1)H_2O$$

According to this proposal, sufficiently soluble nonelectrolytes should, therefore, have a similar effect provided that they are not comparable to water in basic strength or solvating power. In confirmation of this it was observed that the addition of sugars produce a similar enhancement of the endpoint in the titration of aniline with hydrochloric acid. In the titration of some organic bases used as drugs it has been claimed that aqueous phenol solutions (90:10 v/v) are superior to concentrated salt solutions in that this system affords a greater versatility in the number of bases that can be titrated (23). Enhancement in basicity is also produced in some organic solvents as a result of the addition of neutral salts in high concentrations. Thus very weak bases in water such as urea ($pK_a \sim 0.1$), caffeine ($pK_a \sim -0.1$), and $m$-nitroaniline ($pK_a \sim -0.25$) have been titrated successfully in solutions of acetone and other ketones containing 3 $M$ lithium perchlorate (24). In these cases the titrant was perchloric acid.

The use of heterogeneous titration media, as exemplified by the two-phase aqueous-organic solvent systems applied to the titrations of acids, does not appear to have been investigated to the same extent for bases. However, major alterations in the shape of titration curves have been reported (25) when solutions of weak bases prepared in 0.1 $M$ aqueous solutions of lithium chloride were titrated in the presence of a sulfonic acid cation exchange resin (Dowex 50W-X8, 200-400 mesh, $H^+$ form) converted to its lithium form. These major shifts were found to depend strongly on the charge type of the base and the electrolyte concentration, which facilitates the discriminative titration of bases with similar $pK_a$ values but of different charge-types (e.g., $A^-$ and B).

## 10.3 TITRATIONS IN NONAQUEOUS SOLVENTS

### Solvents and Applications

The general properties of some selected solvents were discussed in Section 7.2, and methods were described for the determination of ionization constants in seven of these solvents (Section 7.3). The number of solvents in which such potentiometric measurements have been made is small in comparison with the number that have been used for potentiotitrimetry, as evidenced by the inclusion of over 50 solvent systems listed in Table 10.1. This list of titration media is by no means exhaustive, and neither are the applications given for each solvent system. It is hoped, however, that such a compilation will provide guidance in the selection of a suitable titration medium for analogous types of compounds not listed therein. In some cases the solvent systems have been selected by the workers on the basis of a systematic study using the type of $pK_{H_2O}$–$\Delta$HNP relations described in Section 7.4 and given in Table 7.15. In other cases $\Delta$HNP values or HNP values alone have been measured in a number of media as the basis for selection of a titration medium that allows sequential titration of two or more compounds. This is often possible if the $\Delta$HNP or HNP values differ by at least 150 mV. However, for the reasons stated earlier, it is preferable to use $\Delta$HNP values for this purpose.

Analytical reagent grade solvents are often used without further purification in the preparation of the titration media. It is usual to check that the solvent system contains no acidic or basic impurities by performing a blank titration on the same volume as is to be used in the presence of the determinand. As was previously mentioned, the presence of water inhibits the range of potentials available in a given solvent which, in turn, reduces the utility of the medium for the sequential titration of two or more components of a mixture. Frequently, however, attempts to remove traces of water from the solvent are regarded as superfluous because the principal cause of contamination from water arises as a result of the addition of titrant. For most purposes a water content of up to 1% at the equivalence point can be tolerated although it is good practice to reduce the extent of the contamination to below this level. Other grades of solvents are purified usually by fractional distillation after pretreatment with either molecular sieves, passage through a column containing activated alumina, or treatment with ion exchange resins to remove either acidic (Amberlyst A-26) or basic impurities (Dowex 50W-X8, 100 mesh).

Table 10.1. Some Applications of Acid–Base Potentiotitrimetry in Various Nonaqueous Media

| Solvent System | Types of Titrations Performed (With References) |
|---|---|
| | *Acetic Acid* |
| Alone | Acetates of 15 metal ions (26). Salts of weak acids such as acetate, phthalate citrate, tartrate (27). Indirect for $SO_4^{2-}$ by means of barium acetate (28). Salts of nitrocompounds (e.g., nitroalkanes; 29). Mixtures of hydrazine and 1,1-dimethylhydrazine (30). Benzodiazepine derivatives (31). Cephalosporin [*11111-12-9*] antibiotics (32). Aminophenols (33). Thioethers such as diethylsulfide and L-methionine [*63-68-3*]; basicity enhanced with $Hg(OAc)_2$ (34). Heterocyclic amines, thioethers, and thioamides; basicity enhanced with $Hg(OAc)_2$ (35). |
| In binary mixtures | |
| $+Ac_2O$ (1:1) | Better discriminating solvent for strong acids than HOAc alone (36). |
| $+CH_3CN$ | Mixtures of aminoacids (37). |
| $+MeCOEt$ (1:1) | Heterocyclic nitrogen bases (38). |
| $+HCOOH$ (15:1) | Pyridine, nitrogen, and phosphoric acid in copolymers (39). |
| In ternary mixtures | |
| $+Ac_2O$ + dioxane | Precision titration of monosodium glutamate [*142-47-2*] (40). |
| $+Ac_2O$ + $CHCl_3$ (1:2:15) | Titration of rare earth acetates of bases (41). |
| $+Ac_2O$ + $CHCl_3$ | Ba laurate and Cd laurate as bases in stabilizer LAST DP-4 [*51731-72-7*] (42). Mixtures of rare earth acetates (43). |
| $+$Dioxan $+$ HCOOH (2:2:1) | Tetracycline [*60-54-8*] and derivatives as bases (44). |
| $+Ac_2O$ + $C_6H_6$ (4:5:10) | Papverine-HCl [*61-25-6*] and theobromine [*83-67-0*] in Theopaverine [*8075-58-9*] tablets (45). |
| | *Acetic Anhydride* |
| Alone | Amine oxides and phosphine oxides (46). Caprolactam [*105-60-2*] (47). Mixtures of |

**Table 10.1.** (*Continued*)

| Solvent System | Types of Titrations Performed (With References) |
|---|---|
| | N-oxides of aliphatic amines, hydrogenated and aromatic heterocyclic bases (48). Pyridyl-containing carbamates, thiocarbamates, dithiocarbamates, dithiocarbazates, and thioureas (49). 1,5-alkylated cyclopenta[b]pyrrole and cyclohexa[b]pyrrole (50). Dithiocarbamates (51). Subs. azoles (e.g., imidazoles; 52). Traces dodecalactam [*947-04-6*] (53). Various tetracycline hydrochlorides (54). Weak bases such as meprobamate [*57-53-4*], carisoprodol [*78-44-4*], and other alkyl esters of carbamic acid (55). |
| In binary mixtures + Me$_2$CO | Mixtures of codeine [*76-57-3*] and some other drugs of basic character (56). |
| | *Acetonitrile* |
| Alone | Fourteen nitrogen bases such as ethanolamines, quinolines, acridines, and *p*-toluidines (57). Sulfanilamides (58). Mixtures primary, secondary, and tertiary amines (59). |
| In binary mixtures + C$_6$H$_6$ (1:4) | Binary mixtures of diamines with diamines and monoamines (60). |
| + CHCl$_3$ (1:4) | Best solvent found for mixtures of diamines (61). |
| + CHCl$_3$ | Alkoxy(phenylamino)silanes (62). |
| In ternary mixtures + Me$_2$CO + CHCl$_3$ (2:1:1) | Mixtures of aminopyrine [*58-15-1*], guaiacyl-*o*-hydroxybenzoate nicotinate, and lidocaine [*137-58-6*] as hydrochlorides (63). |
| | *Alcohols* |
| 1,2-Ethanediol | Determination of the antibiotic Tubocin [*13292-46-1*] (64). |
| Ethanol + C$_6$H$_6$ (3:7) | Mixtures of diorganodithiophosphoric acids and their Zn salts (65). |
| Methanol alone | H$_2$SO$_4$ in presence of other inorganic and |

(*continued*)

**Table 10.1.** (*Continued*)

| Solvent System | Types of Titrations Performed (With References) |
|---|---|
| | organic acids (66). Aromatic polycarboxylic acids (67). Mixtures HOAc-HF-HNO$_3$ (68). Acid enhancement of aromatic carboxylic acids with Ca(NO$_3$)$_2$ (69). Det. of P in poly(styrylphosphonic acid) (70). $\beta$-diketones as acids (71). Organophosphinic acids R$_2$P(:O)OH (72). Analysis of 6-benzylaminopurine (73). |
| In binary mixtures +1,2 Ethanediol (2:1) | Det. of monobasic phosphates of mono and divalent cations (74). |
| 2-Methoxyethanol | Tertiary amines in the presence of primary and secondary amines (75). Aldehydes and ketones indirectly by oximation with excess hydroxylammonium formate (76). |
| 2-Methyl-2-propanol | Mineral, carboxylic, and dicarboxylic acids and phenols (77). Mixtures of phenols (78). Very weak acids such as ethylacetoacetate by enhancement with tetrabutylammonium ion (79). Metal ions as perchlorates in the presence of EDTA (80). Det. of commercial benzophenone (LIX) extractants (81). |
| 2-Propanol alone | Amine oxide–tertiary amine mixtures (82). Phenols (83). Thiols; acidity enhanced by halogenated derivatives of nitrobenzene (84). Carbonyl compounds in the presence of NH$_2$OH · HCl and morpholine (85). Mixtures of dicarboxylic acids and their diphenyl esters (used as plasticized and monomers) (86). |
| In binary mixtures +CHCl$_3$ (1:4) | Mixtures of dicarboxylic acids and their monoesters (87). |
| +Diethyleneglycol (1:2) | Mixture of alkyl sufates and sulfuric acids (88). |
| +DMF (1:1) | La, Y, Sc nitrates, bromides, and perchlorates as acids (89). |
| +1,2-Ethanediol | Titration of benzimidazole and 2-CH$_2$OH, 2-CH$_2$CH$_2$OH, and 2-Ph CH$_2$OH derivatives (90). |

**Table 10.1.** (*Continued*)

| Solvent System | Types of Titrations Performed (With References) |
|---|---|
| +1,2-Ethanediol (1:1) | Det. of 1-amino-4-methylpiperazine and methylpiperazine in mixtures (91). |
| +Dioxane (5:1) | Alkyl and arylthiols (92). |

*Amides*

| | |
|---|---|
| Dimethylformamide (DMF) alone | Bis(aryl and alkylamine)tetrabromotellurium adducts (93). Hydroxamic acids (94). Mixtures p-methylbenzoic acid and 1,4 benzenedicarboxylic acids (95). Mixtures Ga, In, Tl(III) (96). Analysis of aspirin and acetaminophen mixtures (97). Acetaminophen and salicylamide mixtures (98). Microdet. phenols, carboxylic acids, and phenolic acids (56 described; 99). Enhancement of basicity with LiCl and HOAc for det. of various amino groups (100). Metal dithiophosphates (e.g., $[(BuO)_2P(S)S]_2Zn$) as acids (101). Tetracyclines and their hydrochlorides (102). |
| In binary mixtures +Toluene | Det. of terminal-$NH_2$ groups in aromatic polyamides or polysufides (103). |
| N-Methyl-2-pyrrolidinone | Bis(phenols), hydroxyphenylketones, alkyl and arylphenols (104). Aliphatic mono and dicarboxylic acids, benzoic acids, phenols (105). |
| Tetramethylurea | Barbiturates and sulfa drugs as acids (106). Analgesics such as phenylbutazone [*50-33-9*] and indomethacin [*53-81-1*] (107). |

*Dimethylsulfoxide (DMSO)*

| | |
|---|---|
| In binary mixtures +$C_6H_6$ (1:4) | Melamine [*108-78-1*] and cyanuric acid [*108-80-5*] (108). |
| +$C_6H_6$ (9:1) | Alkylphenols (109). |
| +2-PrOH | R $C_6H_4N_2BF_4$ where R = 4-$NO_2$, 3-$NO_2$, 4-CN, 4-Br, 4-Me, or 4-OMe (110). |

*Ketones*

| | |
|---|---|
| Acetone alone | Aniline and diphenylamine mixtures (111). Mixtures of $\beta$-nitroalcohols (112). Polynuclear |

(*continued*)

**Table 10.1.** (*Continued*)

| Solvent System | Types of Titrations Performed (With References) |
|---|---|
| | phenolic compounds (113). Titration Al, Ga, In and Tl(I) as iodides (114). Titration of rare earth iodides (115). Det. As as $AsI_3$ in waste products (116). Det. $H_2SO_4$ in mixtures (117). Det. triphenyltin compounds (118). Enhancement of basicity of weak bases with LiCl (119). Titration nitroguanidine (120). Det. of organogermanium mixtures (121). Det. of polyethylene polyamines (e.g., triethylenetetramine; 122). Det. of La, Mg, and Ca as metal iodides (123). Mixtures of 4-aminobenzoic acid, quinine-HCl, salicylamide, and acetylsalicylic acid (124). Orotic acid [*65-86-1*] in mixtures with its derivatives (125). |
| In binary mixtures | |
| $+C_6H_6$ (1:1) | Solasodine [*126-17-0*] in fruits of *Solanum* species (126). |
| $+C_6H_6$ | $AsI_3$ in presence of Al, Sn(IV), Sb(III), Ge(IV), Se(IV), Hg(II), and Bi (127). |
| +BuOH (25:1) | $BeBr_2$ in presence of Li, Na, K, Rb, Cs, Sr, Ba. Other titrations discussed (128). |
| +MeOH (4:1) | Rare earth nitrates as acids (129). |
| +MeOH (3:1) | Titration of $H_m$ ($MCl_6$); M = Ti, Mo, Pt, Sn (130). |
| +2-PrOH (10:1) | Mixtures of pyridine carboxylic acids (131). |
| $+H_2O$ (9:1) | Organic bases (e.g., $MeNH_2$, $Et_2NH$, codeine [*76-57-3*], pyridine, salts, and mixtures; 132). |
| $+H_2O$ (4:1) | Mixtures of nitric and nitrous acids (133). Polycomponent mixtures of dicarboxylic acids (134). |
| $+H_2O$ | $H_3BO_3$ and alkyl borates. Mixtures $H_3BO_3$ with carboxy and mineral acids (135). Titration of *NN*-disubstituted 1,2-ethanediamines (136). |
| In ternary and other mixtures | |
| $+CHCl_3 + 2,6-Me_2C_6H_3OH$ | Carboxy groups in poly(caprolactam), (137). |
| $+C_6H_6 + CHCl_3$ | |
| +MeOH (10:10:5:1) | Mixtures of ephedrine-HCl [*50-98-1*], theophylline [*58-55-9*], and phenobarbital [*50-6-6*] (138). |

**Table 10.1.** (*Continued*)

| Solvent System | Types of Titrations Performed (With References) |
|---|---|
| 2-Butanone Alone | Mixtures of strong acids (e.g., $H_2SO_4$), weak (carboxylic), and very weak acids (phenols; 139). Mixtures of bases such as piperidine and 3,5-dimethylpyrazole (140). Mixtures of weak acid salts and weak acids (141). Pyridyl-substituted ureas (142). Pyridyl-substituted carbamates, thiocarbamates, etc. (143). Perfluoroalkanedicarboxylic acids (144). Mixtures of $HClO_4$ and $HNO_3$ (145). Mixtures $AsI_3$, $SbI_3$, $BiI_3$ (146). Ternary mixtures of acids (e.g., $HCl$-$HF$-$H_2SO_4$) (147). Triphenylamine dyes (148). |
| In binary mixtures +MeOH (10:1) | Pyridine and quinoline bases in the presence of their oxidation products (149). |
| +$Et_2NH$ (1:2) | Representative mono, di, and trinitro derivatives of benzene as acids (150). |

*Nitromethane*

Titration of dipiperidylbenzaminals (151).

*Phenols*

| (In binary mixtures) *o*-Cresol + $CHCl_3$ (7:3) | Det. of carboxy groups in poly(ethylene terephthalate) (152). |
|---|---|
| Phenol + $H_2O$ (9:1) | Titration of organic bases (23). |

*Pyridine*

Phenolic-OH compounds (153). Other organic hydroxyls with 3,5-dinitrobenzoyl chloride (154). Alkoxy groups by means of alkylpyridinium iodide (155). Acylamidines and related compounds (156). 2,4,7-Trinitrofluoren-9-one molecular complexes with polynuclear aromatics (157). 4-Nitrophenylhydrazones as weak acids (158). Dimethylsulfate indirectly (159). Subs. hexaphenyldisiloxanes (160). Naphthyl esters of carboxylic acids (161). Benzanilides as weak acids (162). 4-Aminobenzoic acid-formaldehye-4-bromophenol copolymers (163). Microtitrn of pentaerythri-

(*continued*)

| Solvent System | Types of Titrations Performed (With References) |
| --- | --- |
| In binary mixtures + THF (2:1) | tol tetranitrate (164). Dicarboxylic acids [$(CH_2)_n(COOH)_2$, $n = 0-8$ and 11] (165). Ammonium arenesulfonates (166). Bemegride [*64-65-3*] (167).<br><br>Phenylureas, anilides, and alkyl and phenylcarbamates (168). |
| | *Tetrahydrofuran (THF)* |
| | Det. of carboxylic acid chlorides (169). |
| | *Tetramethylguanidine* |
| | Titration of substituted phenols and substituted benzoic acids (170). It is less hazardous than pyridine (171, 172). |
| | *Tetraalkoxysilanes* |
| $(MeO)_4Si$ or $(EtO)_4Si$ | Mixtures of benzoic acids (173). Mixtures of monosubs. anilines (174). Aniline derivatives (175). Organic base picrates (176). |

The glass electrode was used in most of the solvent systems listed in Table 10.1, although a Pt–Pt pair was used in dimethylformamide (83), an Sb–Pt pair in pyridine (153), tungsten indicator electrodes in 2-butanone (144) and in pyridine–tetrahydrofuran (168), and a hydrogen electrode in 1,1,3,3-tetramethylguanidine (170). A description of the principal investigations with these types of electrodes is given in Section 5.2.

### Acid Titrants

The principal classes of acids used in nonaqueous acid–base titrimetry are as follows:

1. Mineral acids.
2. Sulfonic acids.
3. Lewis acids.

#### Mineral Acids

Of the acids in these three classes, perchloric acid is the most frequently employed. It is usually prepared as a solution in acetic acid (176) or dioxane. A 0.1 $M$ solution of perchloric acid in dioxane is prepared by diluting 72% $HClO_4$ (8.4 mL) to 1 L (177). Solutions of the acid have been prepared in other solvents such as acetonitrile and 4-methyl-2-pentanone, and have also been found satisfactory as titrants when prepared in nitromethane (178) and 2-propanol (179). Although perchloric acid is stable in acetic acid and can even be distilled (180), alcoholic solutions of perchloric acid should *never be heated* because under anhydrous conditions violently explosive perchlorate esters may be formed. A solution of perchloric acid in acetone is not sufficiently stable to use as a titrant, and for the titration of bases in acetone a perchloric acid titrant prepared in dioxane has been used (179). Perchloric acid has been reported as being unstable in a solvent medium of 9:1 acetic anhydride–acetic acid, its concentration changing by 10–15% in a single day (36). In this instance 2,4,6-trinitrobenzenesulfonic acid was selected as the titrant. When prepared as a titrant in glacial acetic acid, dioxane, or 2-propanol, perchloric acid can be standardized against potassium hydrogen phthalate. However, 1,3-diphenylguanidine is used as a primary standard when perchloric acid is prepared in acetonitrile or nitromethane (179). It is suggested that the titrant should be standardized daily when prepared in these solvents. The purification of 1,3-diphenylguanidine was given earlier.

Hydrochloric acid is not used often as a titrant in nonaqueous media. When it is used it is generally prepared as a solution in an alcohol or a mixture of alcohols (74, 82, 130) for titrations in which the determinand has been dissolved in a similar mixture of alcohols, in acetone, in binary mixtures of acetone and an alcohol (130), or in pyridine (156). For example, in pyridine the titrant was used as a 0.2 $N$ solution prepared by diluting concentrated hydrochloric acid with a 1:1 mixture of 1,2-ethanediol and 2-propanol. Standard solutions of hydrochloric acid have also been prepared in acetone (181) and dioxane (182) for the titrations of solutions of acetates prepared in ternary solvent mixtures that contain acetic acid as a minor component.

### Sulfonic Acids

It has already been mentioned that the stability of the titrant 2,4,6-trinitrobenzenesulfonic acid is superior to that of perchloric acid in a solvent mixture of 9:1 acetic anhydride–acetic acid. Generally, unsubstituted arylsulfonic acids are considerably weaker acids than perchloric acid in acetic acid and in 4-methyl-2-pentanone (183), but substitution of nitro groups considerably increases the strength of the sulfonic acid group with the result that 2,4-dinitrobenzenesulfonic acid is almost as strong as perchloric acid, and 2,4,6-trinitrobenzenesulfonic acid is stronger. In a solvent mixture of 9:1 acetic anhydride–acetic acid, however, 2,4,6-trinitrobenzenesulfonic acid is weaker than perchloric acid. If necessary, arylsulfonic acids can be purified by treating an aqueous solution of the acid with an excess of barium hydroxide. After filtration the solution is passed through a column of Dowex 50W-X8, H-form, 100–200 mesh. The eluate is transferred to a rotary evaporator and water removed *in vacuo* at steam bath temperatures. The moist crystalline mass is then dried in a vacuum desiccator over phosphorus pentoxide.

Other sulfonic acids which have been used as titrants are fluorosulfonic acid (184) and trifluoromethanesulfonic acid (185). Both these acids are available commercially. Fluorosulfonic acid is a slightly stronger acid than perchloric acid in acetic acid and trifluoromethylsulfonic acid (triflic acid) is of similar strength to perchloric acid in this solvent. These acids do not appear to offer any major advantages over perchloric acid other than that potassium salts may be titrated without the risk of the formation of precipitates.

## *Lewis Acids*

Typical titrants in this category are prepared from compounds such as boron(III) bromide, aluminum(III) bromide, gallium(III) bromide, tin(IV) chloride, and titanium(IV) chloride which have unfilled electron shells and react as acids by accepting an electron pair from a base. Their use has been limited to the titration of tertiary aromatic amines with tin(IV) or titanium(IV) chlorides in acetyl chloride (186); the titration of nitrogen bases in acetonitrile with these chlorides (187) and also with boron(III) bromide, gallium(III) bromide, or aluminum(III) bromide (57), and the titration of oxygen bases such as alcohols with tin(IV) chloride (187). Most of these types of bases formed 1:1 adducts with the titrant, commonly referred to as $AB_1$ adducts, whereas some of the oxygen bases formed $AB_2$ adducts. These titrants are susceptible to hydrolysis by moisture, and are, therefore, handled in a dry box.

## Base Titrants

The following are the principal classes of titrants in this category:

1. Hydroxides, alkoxides, and related salts of alkali metals.
2. Hydroxides of tetraalkylammonium cations.
3. Strong nitrogen bases.

Of these classes, the tetraalkylammonium hydroxides are the most widely used. Benzoic acid is often employed as the prime standard against which these titrants are standardized.

### Hydroxides, Alkoxides, and Related Salts of Alkali Metals

Solutions of potassium hydroxide prepared in various alcohols have been used to titrate solutions of acids dissolved in methanol (67), in dimethylformamide (96), in acetone (114–116), in 2-butanone (145) and in media containing binary mixtures of these solvents, for example, 25 : 1 acetone–butanol (128). For these applications the titrant was used as a solution in ethanol or 2-propanol. As a solution in 2-propanol or, less frequently, in butanol, potassium hydroxide has featured prominently in the titration of Group IIA metal iodides, Group IIIA metal bromides, nitrates, and perchlorates, Group IIIB metal chlorides, bromides, and iodides, and group VA metal iodides by the Russian schools. Some of these applications have been listed in Table 10.1 as titrations in 2-propanol-DMF (89), in DMF (96), in acetone (114, 115, 123), in acetone–benzene (127), in acetone–butanol (128), and in 2-butanone (146).

Sodium methoxide has been used for the titrations of weak acids in alcoholic solutions (72), in DMF (99), in acetone (113), and in pyridine (161). It is prepared by dissolving metallic sodium in dry methanol and diluting this solution with benzene so that the benzene content in the titrant solution exceeds 80% v/v. As was mentioned previously, the presence of methanol can adversely affect the acid–base properties of the titrand solution, and the proportion of this solvent must, therefore, be kept small. Very weak acids (e.g., phenylureas, anilides, and alkyl and phenylcarbamates) have been titrated in a water-free medium of 2 : 1 pyridine–tetrahydrofuran under nitrogen with a solution of lithium diisopropylamide prepared in the same solvent system (168). The potentials were monitored with a tungsten indicator electrode which was incorporated into the following cell:

Ag; AgCl|3MKCl|0.1 $M$ LiBr in 2 : 1 pyridine–THF|titrand|W

Within this category of titrants, mention must be made of the use of sodium dimsyl for the titration of the even weaker carbon acids in DMSO (188). In his instance, however, a visual indicator was used.

### Tetraalkylammonium Hydroxides

The two principal titrants in this class are tetraethylammonium hydroxide, favored by the Russian workers (189), and tetrabutylammonium hydroxide (190). Both titrant solutions are often prepared in binary mixtures of benzene and methanol, benzene being the predominant component for the reasons already stated. These solutions have been used for the titrations of weak acids in 2-propanol (83), 2-methyl-2-propanol (77), dimethylformamide (94), tetramethylurea (106), acetone (117), 2-butanone (143), pyridine (162), 1,1,3,3-tetramethylguanidine (171), and in binary mixtures of some of these solvents. The stabilities of tetraalkylammonium hydroxides in 2-propanol have been investigated at 35° and 50° (191). Tetramethylammonium hydroxide was found to be very stable having a half-life of 26 days at 50°, the tetraheptyl and tetrahexyl derivatives had half-lives of 154 and 127 days, respectively, at 35°, the tetrabutyl derivative had a half-life of 123 days at 35°, and the half-life of tetraethylammonium hydroxide was 3 days at 35°. The rate of decomposition is very dependent upon the temperature, and a titrant stored in a refrigerator at $-15°$ is about 16,000 times more stable than when it is stored at room temperature. Although tetramethylammonium hydroxide was found to be the most stable of those tested, it is not widely used because its salts are only sparingly soluble in the commonly used nonaqueous media. The presence of water has a great stabilizing influence upon these quaternary ammonium titrants when they are prepared in 2-propanol. Thus within the concentration range studied, the half-life (in days) of tetraethylammonium hydroxide, for example, was found to be related to the water content $c_{H_2O}$ (in weight percent) of the 2-propanol solvent by the empirical relationship (35°)

$$\log t_{1/2} = 0.643 c_{H_2O} + 0.415$$

The enhanced stability has been attributed to three factors. These are the reduction in basicity of the solution as a result of the relatively high acidity of water, the effect of water in decreasing the ion association, and the reduction of the direct component of the inductive effect of the positively charged nitrogen atom upon the $\beta$-hydrogen atoms as a result of the increase in the dielectric constant of the medium. Presumably the presence of methanol in the titrant solution would also exert some stabilizing influence for similar reasons. The nature of the cation was also found to have an effect upon the shape of the titration curves obtained when diprotic acids, dissolved in 2-propanol, were titrated with solutions of tetraalkylammonium hydroxides also prepared in 2-propanol (192). The effect is sufficiently great to cause some dicarboxylic acids to titrate as monoprotic acid with one titrant and as diprotic acids with another. With uncharged and cationic monoprotic acids, however, the structure of the quaternary ammonium titrant had no effect upon the apparent strength of these acids a reflected by the shapes of their titration curves.

Tetraalkylammonium hydroxide titrants can be prepared by either the ion exchange method (193) or the silver oxide method (194) from the corresponding iodides. Details of these methods as applied to the preparation of tetrabutylammonium hydroxide solutions are given in Reference 195. They can also be prepared more conveniently as titrants in 2-propanol by the potassium hydroxide method (196), which is based upon the reaction between quaternary ammonium chlorides and a standard solution of potassium hydroxide in 2-propanol. A measured volume of 0.2 $M$ potassium hydroxide in 2-propanol is treated with a slight excess of the dry quaternary ammonium chloride, and this mixture is shaken vigorously after which the precipitated potassium chloride is allowed to settle. The supernatant liquor is then decanted into another container which is equipped to permit the delivery of titrant and yet protects it from carbon dioxide. A disadvantage of the method is that the titrant contains 100–200 ppm potassium which can have a deleterious effect upon the titrations of very weak acids. If necessary, this contaminant can be reduced to a negligible concentration by treating the titrant with a strong acid ion exchange resin in its quaternary ammonium form. Finally, concentrated stock solutions of tetrabutylammonium hydroxide are available commercially as 40% aqueous solutions or as 25% methanolic solutions. Solutions of 0.1 $M$ tetrabutylammonium hydroxide prepared, for example, in toluene–methanol or 2-propanol–methanol are also available commercially.

### *Strong Nitrogen Bases*

Although the use of these amine-type titrants is not widespread, bases such as cyclohexylamine (66, 168), 1,3-diphenylguanidine (147), and piperidine (171) have been used principally in the titrations of strong mineral acids in 2-butanone (147), in 2:1 diethyleneglycol–2-propanol (171), in 4:1 acetone–water (197), and in methanol (66). Thus sulfuric acid has been titrated in the presence of alkylsulfates (171) and in the presence of hydrofluoric and hydrochloric acids (147). The titration of the sulfuric acid generated when a mixture of gallium, indium, and thallium sulfates in methanol were passed through a cation exchange resin formed a basis or the analysis of binary mixtures of these salts (197). An exception to these types of application is the titration of aliphatic and aromatic acyl chlorides (R-COCl) in tetrahydrofuran with a standard solution of cyclohexylamine also prepared in tetrahydrofuran (168). Carboxylic acids are not titrated, and the presence of any free hydrochloric acid that may be present in the sample can be corrected for by titration with tripropylamine.

## REFERENCES

1. A. Albert and E. P. Serjeant, *The Determination of Ionization Constants*, 3rd ed., Chapman and Hall, London, 1984, p. 22.
2. J. A. M. Van Liempt, *Z. Anorg. Allg. Chem.* **111**, 151 (1920).
3. R. Belcher, G. W. Tully, and G. Svehla, *Anal. Chim. Acta* **50**, 261 (1970).
4. R. D. Strahm and M. F. Hawthorne, *Anal. Chem.* **32**, 530 (1960).
5. I. Dunstan and J. V. Griffiths, *Anal. Chem.* **33**, 1598 (1961).
6. D. G. Shaheen and R. S. Braman, *Anal. Chem.* **33**, 893 (1961).
7. H. C. Kelly, *Anal. Chem.* **40**, 240 (1968).
8. U. Seidelmann, *Fresenius' Z. Anal. Chem.* **291**, 47 (1978).
9. V. R. Negina, E. A. Kozyreva, A. V. Balakshina, and L. S. Chikisheva, *Zavod. Lab.* **34**, 278 (1968); *CA* **69**, 64316q.
10. A. J. Frank, *Anal. Chem.* **35**, 830 (1963).
11. M. Mikesova and M. Bartusek, *Collect. Czech. Chem. Commun.* **43**, 1867 (1978).
12. K. H. Mohr and F. Wolf, *Fresenius' Z. Anal. Chem.* **233**, 269 (1968).
13. A. Shatkay and S. Ehrlich-Rogozinsky, *Anal. Chem.* **39**, 75 (1967).
14. S. Puon and F. F. Cantwell, *Anal. Chem.* **49**, 1256 (1977).
15. I. M. Korenman and I. A. Gur'ev, *Zh. Anal. Khim.* **30**, 1898 (1975); *J. Anal. Chem. USSR* **30**, 1601 (1975).
16. V. A. Semenov, *Zh. Anal. Khim.* **30**, 1058 (1975); *J. Anal. Chem. USSR* **30**, 893 (1975).
17. K. Gustavii, P. A. Johansson, and A. Brandstrom, *Acta Pharm. Suec.* **13**, 391 (1976).
18. F. Frenzel, *Fresenius' Z. Anal. Chem.* **276**, 189 (1975); **278**, 23 (1976); **283**, 269 (1977).
19. M. J. M. Campbell, R. Grzeskowiak, G. G. Jenkinson, and I. D. M. Turner, *Analyst (London)* **97**, 70 (1972).
20. F. E. Critchfield and J. B. Johnson, *Anal. Chem.* **30**, 1247 (1958).
21. A. P. Kreshkov, L. N. Shvetsova, G. P. Svistunova, and E. A. Emelin, *Zh. Anal. Khim.* **26**, 369 (1971): *J. Anal. Chem. USSR* **26**, 321 (1971).
22. M. Ojeda, R. Perez, and P. A. H. Wyatt, *Anal. Chem.* **38**, 1068 (1966).
23. F. Pellerin and D. Baylocq-Quinault, *Ann. Pharm. Fr.* **33**, 479 (1975); *CA* **85**, 10471y.
24. W. L. Schertz and G. D. Christian, *Anal. Chem.* **44**, 755 (1972).
25. F. F. Cantwell and D. J. Pietrzyk, *Anal. Chem.* **46**, 344 (1974).
26. A. T. Casey and K. Starke, *Anal. Chem.* **31**, 1060 (1959).
27. M. L. Richardson, *Anal. Chim. Acta* **24**, 46 (1961).
28. G. Goldstein, O. Menis, and D. L. Manning, *Anal. Chem.* **33**, 266 (1961).
29. H. Feuer and B. F. Vincent, *Anal. Chem.* **35**, 598 (1963).
30. E. A. Burns and E. A. Lawler, *Anal. Chem.* **35**, 802 (1963).
31. K. H. Beyer and W. Sadee, *Arch. Pharm. (Weinheim)* **300**, 667 (1967); *CA* **68**, 72284x.
32. F. Barbato, A. Bava, C. Grieco, C. Silipo, and A. Vittoria, *Ann. Chim.*

(*Rome*) **68**, 575 (1978).
33. W. Jedrzejewski and J. Badecka-Jedrzejewska, *Chem. Anal.* (*Warsaw*) **14**, 73 (1969); *CA* **70**, 111454r.
34. C. Casalini, G. Cesarano, and G. Mascellani, *Farmaco, Ed. Prat.* **31**, 447 (1976); *CA* **85**, 166699q.
35. G. Mascellani and C. Casalini, *Anal. Chem.* **47**, 2468 (1975).
36. D. J. Pietrzyk, *Anal. Chem.* **39**, 1367 (1967).
37. A. P. Kreshkov, N. Sh. Aldarova, and G. V. Turovtseva, *Dokl. Akad Nauk SSSR* **169**, 1093 (1966); *CA* **65**, 14085a.
38. K. C. M. Rao and P. R. Naidu, *Talanta* **19**, 1465 (1972).
39. E. A. Emelin, V. E. Lozhkin, L. S. Matskevich, and O. A. Markova, *Zavod. Lab.* **37**, 1185 (1971); *CA* **76**, 47529p.
40. N. Adler, *Anal. Chem.* **34**, 1668 (1962).
41. A. P. Kreshkov, A. N. Yarovenko, and L. A. Bondareva, *Zh. Obshch. Khim.* **39**, 2388 (1969); *CA* **72** 74356d.
42. A. P. Kreshkov, A. N. Yarovenko, P. I. Selivokhin, L. A. Bondareva, and L. A. Bernshtein, *Plast. Massy 1974* **69**; *CA* **81**, 25968g.
43. A. N. Yarovenko, L. B. Kuznetsova, L. A. Bondareva, and O. D. Bartikova, *Tr. Vses. Konf. Anal. Khim. Nevodnykh, Rastvorov Ikh Fiz-Khim. Svoistvam, 3rd. 1971* (*Pub. 1974*) **1**, 192–194; *CA* **84**, 25390z.
44. S. H. Hansen, *Arch. Pharm. Chemi, Sci. Ed.* **1**, 89 (1973); *CA* **80**, 74369a.
45. E. Sell, *Pol. J. Pharmacol. Pharm.* **27**, 561 (1975); *CA* **84**, 111728x.
46. D. C. Wimer, *Anal. Chem.* **34**, 873 (1963).
47. V. K. Akimov, B. N. Kolokolov, and S. M. Gel'fer, *Zh. Anal. Khim.* **21**, 729 (1966); *CA* **65**, 9735e.
48. N. N. Bezinger, G. D. Gal'pern, N. G. Ivanova, and G. A. Semeshkina, *Zh. Anal. Khim.* **23**, 1538 (1968); *CA* **70**, 43911k.
49. E. G. Novikov and V. K. Kondratov, *Zh. Anal. Khim.* **23**, 955 (1968); *CA* **69**, 64483s.
50. N. N. Bezinger and G. D. Gal'pern, *Zh. Anal. Khim.* **23**, 635 (1968); *CA* **69**, 16069q.
51. M. A. Bernard and M. M. Borel, *Bull. Soc. Chim. Fr.* **1972**, 1663; *CA* **77** 83311r.
52. S. Veibel and L. B. Kuznetsova, *Anal. Chim. Acta* **65**, 163 (1973).
53. V. N. Kotova, T. V. Platonova, and V. E. Petrakovich, *Zh. Anal. Khim.* **30**, 349 (1975); *CA* **83**, 59319c.
54. H. Ellert, R. Ceglarski, and A. Regosz, *Farm. Polska* **22**, 185 (1966); *CA* **67**, 5733j.
55. A. Zdunska, W. Kwapiszewski, and B. Dulowska, *Chem. Anal.* (*Warsaw*) **21**, 681 (1976); *CA* **86** 21826q.
56. E. Sell and D. Dajzer, *Ann. Acad. Med. Gedanensis* **6**, 127 (1976); *CA* **86**, 127363b.
57. C. B. Riolo and R. Perego, *Ann. Chim.* (*Rome*) **54**, 552 (1964); *CA* **61**, 6990c.
58. A. P. Kreshkov, S. V. Vasileva, and L. N. Balyatinskaya, *Farmatsiya* (*Moscow*) **23**, 53 (1974); *CA* **81**, 126823m.

59. V. Cosofret and C. Stefanescu, *Rev. Chim.* (*Bucharest*) **28**, 677 (1977); *CA* **88**, 44545c.
60. A. P. Kreshkov, L. N. Bykova, and I. P. Pevzner, *Zh. Anal. Khim.* **19**, 890 (1964); *CA* **61**, 12638f.
61. A. P. Kreshkov, L. N. Bykova, and I. D. Pevzner, *Dokl. Akad. Nauk SSSR* **150**, 99 (1963); *CA* 59 8126b.
62. N. A. Makarova, *Zh. Anal. Khim.* **28**, 972 (1973); *CA* **79**, 73356m.
63. G. Tortolani, *Farmaco. Ed. Prat.* **29**, 272 (1974); *CA* **81**, 68638s.
64. P. Zikolov and G. Ovcharova, *Farmatsiya* (*Sofia*) **26**, 20 (1976); *CA* **87** 90765b.
65. M. M. Fialko and A. N. Balashova, *Zh. Anal. Khim.* **30**, 1637 (1975); *CA* **84** 138091w.
66. W. Stuck, *Fresenius' Z. Anal. Chem.* **177**, 338 (1960); *CA* **55**, 16274b.
67. A. I. Vasyutinskii and A. A. Tkach, *Zh. Anal. Khim.* **24**, 911 (1969); *CA* **71**, 77117s.
68. H. G. Griffin and W. E. Sonia, Jr., *Anal. Chem.* **41**, 1488 (1969).
69. A. Blazsek-Bodo and C. Liteanu, *Rev. Roum. Chim.* **17**, 1771 (1972); *CA* **78**, 37751g.
70. R. V. Yarkova, G. F. Atknin, and V. M. Aksenenko, *Zavod. Lab.* **39**, 146 (1973); *CA* **79**, 7156v.
71. Y. K. Agrawal and J. P. Shukla, *Indian J. Chem.* **13**, 94 (1975); *CA* **82**, 164493g.
72. T. Jasinski, A. Modro, and T. Modro, *Chem. Anal.* (*Warsaw*) **10**, 929 (1965); *CA* **64**, 14968b.
73. U. Mikstais, A. P. Kreshkov, A. Veveris, and N. T. Smolova, *Zh. Anal. Khim.* **33**, 605 (1978); *CA* **89**, 1582r.
74. A. P. Kreshkov and L. B. Kuznetsova, *Zh. Anal. Khim.* **22**, 1046 (1967); *CA* **67**, 113483m.
75. J. E. Ruch and F. E. Critchfield, *Anal. Chem.* **33**, 1569 (1961).
76. J. E. Ruch, J. B. Johnson, and F. E. Critchfield, *Anal. Chem.* **33**, 1566 (1961).
77. J. S. Fritz and L. W. Marple, *Anal. Chem.* **34**, 921 (1962).
78. N. T. Crabb and F. E. Critchfield, *Talanta* **10**, 271 (1963).
79. L. W. Marple and J. S. Fritz, *Anal. Chem.* **35**, 1431 (1963).
80. L. W. Marple and G. J. Scheppers, *Anal. Chem.* **38**, 553 (1966).
81. D. J. Barkley, *Anal. Chim. Acta* **105**, 83 (1979).
82. L. D. Metcalfe, *Anal. Chem.* **34**, 1849 (1962).
83. A. P. Kreshkov, L. N. Bykova, and Z. G. Blagodatskaya, *Zh. Anal. Khim.* **23**, 123 (1968); *CA* **68**, 92799x.
84. S. I. Obtemperanskaya and T. A. Egorova, *Zh. Anal. Khim.* **24**, 1439 (1969); *CA* **72**, 28244t.
85. E. A. Gribova and Yu. P. Bavrina, *Zavod. Lab.* **39**, 945 (1973); *CA* **80**, 10127n.
86. A. P. Kreshkov, N. Sh. Aldarova, A. I. Izvneev, K. N. Shulunova, T. Kh. Batorova, and A. D. Markov, *Plast. Massy* **1976**, 69; *CA* **85**, 136970d.
87. A. P. Kreshkov, N. A. Kazaryan, and K. N. Shulunova, *Zh. Anal. Khim.* **23**,

1199 (1968); *CA* **69**, 102870a.
88. G. V. Poltevskii, A. I. Yakushin, and V. P. Savel'yanov. *Zh. Anal. Khim.* **32**, 636 (1977); *CA* **87**, 77973y.
89. A. P. Kreshkov, S. M. Milaev, and V. K. Manzhigeeva, *Tr. Mosk. Khim.-Tekhnol. Inst.* **75**, 169 (1973); *CA* **81**, 145147k.
90. D. P. Dani and K. P. Soni, *J. Inst. Chem. (India)* **49**, 158 (1977); *CA* **87**, 145407u.
91. L. Polaczek and W. Jakubowski, *Chem. Anal. (Warsaw)* **22**, 105 (1977); *CA* **87**, 189522g.
92. S. I. Obtemperanskaya and T. A. Egorova, *Vestn. Mosk. Univ. Khim.* **24**, 115 (1969); *CA* **72**, 50754f.
93. T. Jasinski and Z. Kokot, *Chem. Anal. (Warsaw)* **11**, 967 (1966); *CA* **67** 7751n.
94. T. W. Stamey, Jr. and R. Christian, *Talanta* **13**, 144 (1966).
95. F. C. Trussell and R. E. Lewis, *Anal. Chim. Acta* **34**, 243 (1966).
96. A. P. Kreshkov, S. M. Milaev, and G. A. Khanturgaev, *Zh. Anal. Khim.* **27**, 1503 (1972); *CA* **77**, 172303g.
97. M. I. Walash, S. P. Agarwal, and M. I. Blake, *Can. J. Pharm. Sci.* **7**, 123 (1972); *CA* **78**, 62233x.
98. M. I. Blake and L. B. Shumaker, *J. Assoc. Off. Anal. Chem.* **56**, 653 (1973).
99. S. S. M. Hassan and M. T. M. Zaki, *Talanta* **22**, 843 (1975).
100. E. A. Emelin, *Zh. Anal. Khim.* **30**, 335 (1975); *CA* **82**, 171644k.
101. D. Morel and M. Marichy, *Talanta* **24**, 582 (1977).
102. A. Regosz, *Rozpr. Wydz. 3: NaukMat.-Przyr., Gdansk. Tow. Nauk.* **8**, 217 (1973); *CA* **83**, 152411m.
103. E. A. Emelin, V. Savinov, and L. B. Sokolov, *Zh. Anal. Khim.* **28**, 1188 (1973); *CA* **79**, 146956u.
104. A. P. Kreshkov, Ya. A. Gurevich, and G. M. Gal'pern, *Zh. Anal. Khim.* **28**, 2440 (1973); *CA* **80**, 90983b.
105. G. M. Gal'pern, Ya. A. Gurevich, and N. F. Kryuchkova, *Tr. Vses. Nauch. -Issled. Proekt. Inst. Monomerov* **2**, 166 (1970); *CA* **76**, 104618t.
106. M. Greenberg, B. J. Barker, and J. A. Caruso, *Anal. Chim. Acta* **54**, 159 (1971).
107. M. I. Walash and M. Rizk, *Indian J. Pharm.* **39**, 82 (1977); *CA* **87**, 157253d.
108. R. Morales, *Anal. Chem.* **40**, 1148 (1968).
109. M. M. Fialko, A. N. Balashova, and A. F. Lyashenko, *Zh. Anal. Khim.* **25**, 1648 (1970); *CA* **74**, 27859b.
110. S. I. Petrov, L. M. Buchneva, and L. A. Kazitsyna, *Zh. Anal. Khim.*, **32**, 2436 (1977); *CA* **88**, 182077c.
111. B. N. Utkin, *Zavod. Lab.* **33**, 1381 (1967); *CA* **68**, 74919g.
112. V. M. Aksenenko and E. G. Aksenenko, *Zavod. Lab.* **34**, 535 (1968); *CA* **69**, 56883g.
113. S. K. Chatterjee, *Indian J. Chem.* **7**, 605 (1969); *CA* **71**, 45631p.
114. S. M. Milaev and G. A. Khanturgaev, *Zh. Anal. Khim.* **29**, 472 (1974); *CA* **81**, 32885f.

115. A. P. Kreshkov, S. M. Milaev, and V. K. Manzhigeeva, *Zh. Anal. Khim.* **29**, 914 (1974); *CA* **81**, 85620x.
116. A. P. Kreshkov, S. M. Milaev, and G. I. Khanturgaeva, *Zavod. Lab.* **40**, 926 (1974); *CA* **82**, 50963e.
117. P. K. Agasyn, G. N. Sharapova, and O. V. Katorina, *Zavod. Lab.* **43**, 1046 (1977); *CA* **88**, 181862t.
118. R. Litan, J. Basters, A. Martijn, and T. Van der Molen, *J. Assoc. Off. Anal. Chem.* **61**, 1504 (1978); *CA* **90**, 67507x.
119. W. L. Schertz and G. D. Christian, *Anal. Chem.* **44**, 755 (1972).
120. V. M. Aksenenko, E. G. Aksenenko, and E. N. Gromova, *Zavod. Lab.* **31**, 1191 (1965); *CA* **64**, 18407e.
121. A. P. Kreshkov, I. F. Kolosova, and Z. P. Dobronevskaya, *Fiz-Khim. Metody Anal. Kontr. Proizvod., Mater. Konf. Rab. Vuzov (Vyssh. Uch. Zaved.) Zavod Lab. Yugo-Vostoka SSSR, 4th 1971 (Pub. 1972)* **1**, 1; *CA* **80**, 115822p.
122. M. Zgoda and S. Petri, *Chem. Anal. (Warsaw)* **16**, 175 (1971); *CA* **75**, 14710s.
123. A. P. Kreshkov, S. M. Milaev, and V. K. Manzhigeeva, *Tr. Mosk. Khim-Tekhnol. Inst.* **81**, 92 (1974); *CA* **84**, 11884b.
124. I. Grabowska and K. Weclawska, *Farm. Pol.* **32**, 363 (1976); *CA* **85**, 149186c.
125. A. Veveris, U. Mikstais, and I. Jurgevica, *Latv. PSR Zinat. Akad. Vestis, Kim. Ser.* **1977**, 498; *CA* **88**, 44544b.
126. L. Telek, *J. Pharm. Sci.* **66**, 699 (1977); *CA* **87**, 18363b.
127. G. I. Khanturgaeva, S. M. Milaev, and A. P. Kreshkov, *Zh. Anal. Khim.* **30**, 95 (1975); *CA* **83**, 52751d.
128. A. P. Kreshkov, S. M. Milaev, and M. M. Baldanov, *Zh. Anal. Khim.* **27**, 2349 (1972); *CA* **78**, 118887h.
129. A. P. Kreshkov, S. M. Yarovenko, S. M. Milaev, and N. Sh. Aldarova, *Zh. Anal. Khim.* **21**, 34 (1966); *CA* **64**, 11859g.
130. A. P. Kreshkov, P. Ya. Yakovlev, L. B. Kuznetsova, and N. A. Protsenko, *Zavod. Lab.* **38**, 923 (1972); *CA* **78**, 23547h.
131. V. K. Kondratov and E. G. Novikov, *Zh. Anal. Khim.* **22**, 1881 (1967); *CA* **68**, 74935j.
132. K. I. Evstratova, V. I. Kurov, N. A. Goncharova, A. I. Ivanova, and V. Ya. Solomko, *Zh. Anal. Khim.* **22**, 1160 (1967); *CA* **68**, 56416m.
133. V. K. Polovnyak, K. N. Mochalov, R. Sh. Safin, and A. F. Makhotkin, *Zavod. Lab.* **34**, 1294 (1968); *CA* **70**, 74027f.
134. A. P. Kreshkov, L. N. Bykova, and N. T. Smolova, *Zh. Anal. Khim.* **19**, 156 (1964); *CA* **60**, 15146e.
135. R. T. Savel'yanova and V. P. Savel'yanov, *Zh. Anal. Khim.* **31**, 2056 (1976); *CA* **86**, 182423w.
136. M. Zgoda and J. Kalinka, *Chem. Anal. (Warsaw)* **16**, 1259 (1971); *CA* **76**, 94345k.
137. M. J. Maurice, *Anal. Chim. Acta* **26**, 406 (1962).
138. E. Sell and D. Rajzer, *Chem. Anal. (Warsaw)* **21**, 933 (1976); *CA* **85**, 198219z.
139. A. P. Kreshkov, L. N. Bykova, and N. A. Kazaryan, *Izv. Vyssh. Uchebn. Zaved., Khim. Khim. Tekhnol.* **4**, 20 (1961); *CA* **55**, 14165i.

140. A. P. Kreshkov, L. N. Bykova, and N. Sh. Shemet, *Zh. Anal. Khim.* **16**, 331 (1961); *CA* **56**, 4107a.
141. A. P. Kreshkov, A. N. Yarovenko, and I. Ya. Zel'manova, *Dokl. Akad. Nauk SSSR* **143**, 348 (1962); *CA* **57**, 2823f. See also *CA* **59**, 1088b.
142. V. K. Kondratov and E. G. Novikov, *Zh. Anal. Khim.* **22**, 1245 (1967); *CA* **68**, 9176y.
143. V. K. Kondratov and E. G. Novikov, *Zh. Anal. Khim.* **23**, 631 (1968); *CA* **69**, 32863z.
144. M. N. Chelnokova and L. N. Dubrovina, *Zh. Anal. Khim.* **23**, 1076 (1968); *CA* **69**, 83208x.
145. N. I. Denisova and A. A. Golubeva, *Zh. Anal. Khim.* **27**, 1221 (1972); *CA* **77**, 121797p.
146. A. P. Kreshkov, S. M. Milaev, G. A. Khanturgaev, and G. I. Khanturgaeva, *Zh. Anal. Khim.* **29**, 98 (1974); *CA* **80**, 115704b.
147. P. Ya. Yakovlev, A. P. Kreshkov, L. B. Luznetsova, and N. A. Protsenko, *Zh. Anal. Khim.* **30**, 1544 (1975); *CA* **84**, 25437v.
148. J. F. C. Boodts and J. R. Kroll, *Fresenius' Z. Anal. Chem.* **289**, 207 (1978); *CA* **88**, 154309m.
149. V. K. Kondratov, N. D. Rus'yanova, N. V. Malysheva, and L. P. Yurkina, *Zh. Anal. Khim.* **22**, 1585 (1967); *CA* **68**, 26790m.
150. S. I. Petrov and V. A. Drozdov, *Zh. Khim.* **1969**, Abstr. No. 7G213; *CA* **72**, 18312z.
151. T. U. Urisbaev, A. V. Shchelkunov, and S. I. Petrov, *Zh. Anal. Khim.* **33**, 2235 (1978); *CA* **90**, 132354m.
152. M. J. Maurice and F. Huizinga, *Anal. Chim. Acta* **22**, 363 (1960).
153. E. J. Greenhow and J. W. Smith, *Analyst (London)* **84**, 457 (1959).
154. W. T. Robinson, Jr., R. H. Cundiff, and P. C. Markunas, *Anal. Chem.* **33**, 1030 (1961).
155. R. H. Cundiff and P. C. Markunas, *Anal. Chem.* **33**, 1028 (1961).
156. B. H. Beggs and R. D. Spencer, *Anal. Chem.* **34**, 1590 (1962).
157. R. H. Cundiff and P. C. Markunas, *Anal. Chem.* **35**, 1323 (1963).
158. W. T. Robinson, Jr., A. J. Sensabaugh, and P. C. Markunas, *Anal. Chem.* **35**, 770 (1963).
159. W. M. Banick, Jr., and E. C. Francis, *Talanta* **13**, 979 (1966).
160. G. Schott and E. Popowski, *Z. Chem.* **8**, 265 (1968); *CA* **69**, 56862z.
161. A. Groagova and V. Chromy, *Analyst (London)* **95**, 548 (1970).
162. T. V. Kaskik and G. A. Rassolova, *Zh. Anal. Khim.* **32**, 1600 (1977); *CA* **88**, 58029z.
163. S. K. Chatterjee and L. S. Pachauri, *Polymer* **19**, 596 (1978); *CA* **89**, 180460u.
164. W. Selig, *Microchim Acta* **1978** (2), 169; *CA* **89**, 131855t.
165. T. Jasinski and H. Smagowski, *Chem. Anal. (Warsaw)* **10**, 1321 (1965); *CA* **64**, 14967b.
166. T. Jasinski and R. Korewa, *Chem. Anal. (Warsaw)* **13**, 1319 (1968); *CA* **71**, 9489q.
167. S. Ueoka, S. Okada, S. Iga, H. Isaka, and K. Yoshimura, *Bunseki Kagaku* **20**,

1196 (1971); *CA* **75**, 144042m.
168. W. Hausmesser, W. Gerlach, and C.-H. Roeder, *Fresenius' Z. Anal. Chem.* **292**, 23 (1978); *CA* **89**, 156963q.
169. L. J. Lohr, *Anal. Chem.* **32**, 1167 (1960).
170. J. A. Caruso, G. G. Jones, and A. I. Popov, *Anal. Chim. Acta* **40**, 49 (1968).
171. T. R. Williams and J. Custer, *Talanta* **9**, 175 (1962).
172. T. R. Williams and M. Lautenschleger, *Talanta* **10**, 804 (1963).
173. T. Jasinski and Z. Kokot, *Chem. Anal. (Warsaw)* **12**, 809 (1967); *CA* **68**, 111226b.
174. W. Rodziewicz and Z. Kokot, *Chem. Anal. (Warsaw)* **11**, 961 (1966); *CA* **66**, 121888s.
175. W. Rodziewicz and Z. Kokot, *Chem. Anal. (Warsaw)* **11**, 175 (1966); *CA* **64**, 18386h.
176. D. C. Wimer, *Anal. Chem.* **30**, 77 (1958).
177. J. C. Fritz, *Anal. Chem.* **22**, 578 (1950).
178. C. A. Streuli, *Anal. Chem.* **31**, 1653 (1959).
179. L. G. Chatten and L. E. Harris, *Anal. Chem.* **34**, 1495 (1962).
180. J. S. Fritz, *Acid–Base Titrations in Non Aqueous Solvents*, Allyn and Bacon, Boston, 1973, p. 48.
181. P. K. Agasyan, N. I. Stenina, and G. N. Sharapov, *Zavod. Lab.* **40**, 1069 (1974); *CA* **82**, 67780r.
182. E. A. Gribova and Yu. P. Bavrina, *Zavod. Lab.* **42**, 925 (1976); *CA* **86**, 64994k.
183. D. J. Pietrzyk and J. Belisle, *Anal. Chem.* **38**, 969 (1966).
184. R. C. Paul, S. K. Vasisht, K. C. Malhotra, and S. S. Pahil, *Anal. Chem.* **34**, 820 (1962).
185. E. S. Lane, *Talanta* **8**, 849 (1962).
186. J. Singh, R. C. Paul, and S. S. Sandhu, *J. Chem. Soc.* **1959**, 845.
187. E. T. Hitchcock and P. J. Elving, *Anal. Chim. Acta* **27**, 501 (1962); **28**, 301 (1963).
188. G. C. Price and M. C. Whiting, *Chem. Ind. (London)* **1963**, 775.
189. A. P. Kreshkov, L. N. Bykova, and N. A. Mkhitaryan, *Zh. Anal. Khim.* **14**, 529 (1959).
190. R. H. Cundiff and P. C. Markunas, *Anal. Chem.* **28**, 792 (1956).
191. G. A. Harlow, *Anal. Chem.* **34**, 1487 (1962).
192. G. A. Harlow, *Anal. Chem.* **34**, 1482 (1962).
193. G. A. Harlow, C. M. Noble, and G. E. A. Wyld, *Anal. Chem.* **28**, 787 (1956).
194. R. H. Cundiff and P. C. Markunas, *Anal. Chem.* **28**, 792 (1956).
195. J. S. Fritz, *Acid–Base Titrations in Non Aqueous Solvents*, Allyn and Bacon, Boston, 1973, p. 133.
196. G. A. Harlow and G. E. A. Wyld, *Anal. Chem.* **34**, 172 (1962).
197. A. P. Kreshkov, E. N. Sayushkina, and N. A. Timasheva, *Zavod. Lab.* **37**, 1421 (1971); *CA* **76**, 80620j.

CHAPTER
11

# COMPLEXOMETRIC TITRATIONS

## 11.1 A SURVEY OF METHODS

Although the stability constants of thousands of metal–ligand complexes have been reported (References 2–8, Chapter 8), the number of ligands used as titrants in complexometry is comparatively small. Generally, the ligands suitable for complexometric titrations form water-soluble, stable 1 : 1 complexes with the metal ion determinand at the equivalence point, the reactions being represented by

$$M(H_2O)_n + L \rightarrow M(H_2O)_{(n-1)}L + H_2O$$

The magnitude of the stability constant is such that the concentration of free metal ion at the equivalence point is negligibly small. The generic term *chelon* is used to describe this class of ligand which includes polyaminocarboxylic acids, polyamines, and, more recently, the cryptands.

Of the polyaminocarboxylic acids, ethylenedinitrilotetraacetic acid, EDTA, is the most widely used titrant. Like glycine, EDTA exists as a zwitterion as is represented by Structure I in Figure 11.1. The trivial name EDTA has been used indiscriminately in the literature to represent the free acid, its hydrated disodium salt, and the tetraanion. Among the many synonyms, Chelaton II, Complexone II, Edetic acid, and Sequestrene have been used for the free acid, and Chelaton III, Complexone III, Titriplex III, Trilon B, and Versene have been used for the disodium salt of ethylenedinitrilotetraacetic acid. Generally, in analytical chemistry EDTA is usually taken to mean the disodium salt of the acid since this is the form in which it is most frequently employed as a titrant. Structure II is 1,2-cyclohexanediyldinitrilotetraacetic acid often referred to as CDTA or as DCTA. This reagent often forms stronger complexes than does EDTA although the rate of formation of these can be slower than with EDTA. It is used advantageously for the titration of calcium(II) in the presence of barium(II), a titration not possible with EDTA because the magnitudes of the stability constants for the complexes of these ions do not differ sufficiently. With CDTA, however, there is an appreciable difference between the stability constants of the barium complex (log K = 8.0) and the calcium complex (log K = 12.5) which is sufficient to allow calcium to be titrated independently of barium. EGTA (see structure III) is the trivial name for ethylenebis(oxyethylenenitrilo)tetraacetic acid, a reagent with which it is possible to titrate calcium (log K = 11.0 for Ca complex) in the presence of magnesium (log K = 5.2 for Mg complex). Such selectivity is not possible with EDTA because log K for calcium is only two units larger than that of magnesium with the result that both ions are titrated. TTHA or ethylenedinitrilo-$N, N'$-diacetic-$N, N'$-bis(ethylenenitrilodiacetic acid) (see structure IV) is unique with respect to the others of Figure 11.1 because it is decadentate and can, therefore, form not only 1:1 complexes but also binuclear complexes. For example, thorium, zirconium, indium, and

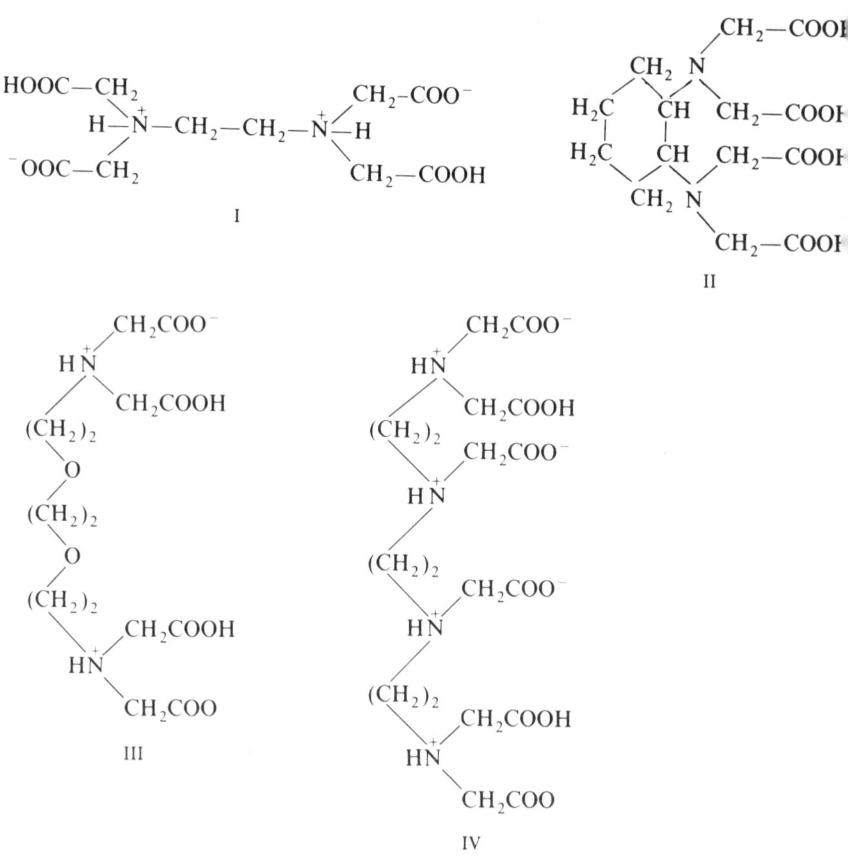

Figure 11.1. Some polyaminocarboxylic acids used as titrants in complexometric titrations: I, ethylenedinitrilotetraacetic acid, EDTA; II, 1,2-cyclohexanediyldinitrilotetraacetic acid, CDTA or DCTA; III, ethylenebis(oxyethylenenitrilo)tetraacetic acid, EGTA; IV, ethylenedinitrilo-$N, N'$-diacetic-$N, N'$-bis(ethylenenitrilodiacetic acid).

manganese are measurable as 1 : 1 ML complexes with this reagent, whereas aluminum, iron, and gallium form $M_2L$ complexes.

Polyamines are not widely used as reagents for complexometric titrations. TRIEN and TETREN, trivial names for 1,4,7,10-tetraazadecane and 1,4,7,10,13-pentaazatridecane, respectively, have been found satisfactory as titrants for mixtures of metal ions that are difficult to analyze using the polyaminocarboxylic acid titrants. It is for this reason that they have been included in Table 11.1. The synonyms for these compounds are triethylenetetramine (TRIEN) and tetraethylenepentamine (TETREN).

Like most of the polyaminocarboxylic acids and the polyamines, the cryptands that have so far found application as complexometric titrants also form 1 : 1 complexes with the metal ion determinand. The cryptands, however, are unique in that they form inclusion compounds called *cryptates* in which the species being complexed is contained (or "buried") inside a molecular cavity (or "crypt"). In this respect they are similar in type to the macropolycyclic compounds classified previously as neutral carriers. The dimensions of the molecular cavity determine the type of species that will fit within it, and for this reason they possess a comparatively high degree of selectivity with respect to the polyaminocarboxylic acids and the polyamines. At the time of the literature survey for this book three cryptands had been described as complexometric titrants for monovalent cations, and the structures of these are given in Figure 11.2. The synonyms for these compounds are Kryptofix 211 (structure V = 4,7,13,18-tetraoxa-1,10-diazabicyclo[8.5.5]tricosane), Kryptofix 221 (structure VI = 4,7,13,16,21-pentaoxa-1,10-diazabicyclo[8.8.5]tricosane), and Kryptofix 222 (structure VII = 4,7,13,16,21,24-hexaoxa-1,10-diazabicyclo[8.8.8]hexacosane), and they have been used for the titrations of lithium, sodium, and potassium ions, respectively (see Serials 60, 65, and 76 in Table 11.1).

In this chapter the emphasis is toward summarizing how these chelons are used in practice, and to this end a number of analytical procedures are given in Table 11.1. The basic theory of complexometric titrations is adequately covered in modern textbooks dealing with quantitative chemical analysis at the senior undergraduate level, and this is not repeated here.

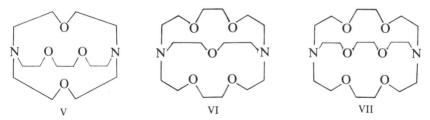

Figure 11.2. Cryptands used as complexometric titrants: V, Kryptofix 211; VI, Kryptofix 221; VII, Kryptofix 222.

Table 11.1. Electrodes and Procedures for some Complexometric Determinations

| Serial Number Determinand | Chelon | Titration Type | Electrode | Buffer | Procedures and Remarks | Ref. |
|---|---|---|---|---|---|---|
| 1. Al | EDTA | Indirect with Ca | Ag|Ag$^+$ | H$_3$BO$_3$(0.1 M) + NaOH(0.1 M) adj. to 9.2 | Add an accurately measured volume of 0.05 M EDTA which is in excess of the amount needed to react with 0.05 to 0.5 mmol Al(III) in 30–50 mL soln. Add the buffer soln. (50 mL), adj. to pH 9.2 and then add 0.008 M AgNO$_3$ (0.05 mL). Titr. with std. 0.05 M Ca(NO$_3$)$_2$. | 1, 2 |
| 2. — | EDTA | Indirect with Zn | Hg|Hg$^{2+}$ | 0.5 M acetate adj. to pH 4.6 | 0.1 M Al(III) (10 mL) was acidified (pH 1–2) and boiled for 1 min. 0.01 M EDTA (15 mL) was added to the hot soln. and, after cooling, the buffer soln. (15–20 mL) and 1–2 drops 10$^{-3}$ M Hg(II)–EDTA were added. The XS EDTA was titr. with std. Zn(II) soln. | 3 |
| 3. Al + Fe | CDTA | Indirect With Fe(III) | Pt | Acetate pH 5. | See Serial 44. | 26 |
| 4. Ba | EDTA | Direct | Ag|Ag$^+$ | H$_3$BO$_3$(0.1 M) + NaOH (0.1 M) adj. to 9.2 | Add the buffer soln. (50 mL) to a soln. containing 0.05–0.5 mmol Ba(II) (~ 50 mL) followed by 0.008 M AgNO$_3$ (0.05 mL). Titr. with 0.05 M EDTA. For samples containing ≧ 0.01 mmol, add 0.01 mL Ag$^+$ soln. and titr. with 0.01 M EDTA. Cl$^-$ must be removed.[a] | 1, 2 |
| 5. — | EDTA | Direct | Hg|Hg$^{2+}$ | 0.5 M NH$_3$ adj. to ~ 10 with HNO$_3$ | Add the buffer soln. (10–25 mL) and Hg(II)–EDTA complex (1 mL, 10$^{-3}$ M) to a known aliquot (15–25 mL) of Ba(II) (≦0.05 M). After deoxygenating the soln. with N$_2$, titr. with std. EDTA (0.005–0.05 M) slowly near endpoint, while excluding O$_2$. | 3 |

| | | | | | | |
|---|---|---|---|---|---|---|
| 6. Ba | CDTA | Direct | $Cu^{2+}$ ISE ($Ag_2S$-CuS) | $NH_3$ | Concd. $NH_3$ (10 mL) is added to a soln. (~40 mL) containing between 0.003 and 0.3 mmol Ba(II), and the soln. titr. with 0.002 or 0.1 M CDTA. | 4 |
| 7. — | CDTA | Indirect with Hg(II) | Hg(Ag)\|$Hg^{2+}$ | 10% Hexamine[i] (200 mL) + 0.5 M NaOH (~16 mL) pH 11. | Add an XS of 0.05 M CDTA to a known volume of soln. containing 0.07–10 mg Ba(II) followed by the buffer soln. (35–40 mL). Titr. XS CDTA with std. Hg(II). | 5 |
| 8. Bi(III) | EDTA | Direct | Hg\|$Hg^{2+}$ | $p$H Adj. to 1.5. | Adjust the $p$H of a soln of Bi(III) (25 mL ~0.005 M) to 1.5 with $NH_3$ or $HNO_3$. Add Hg(II)–EDTA and titr. with std. EDTA. | 3 |
| 9. — | EDTA | Indirect with Ca | Ag\|$Ag^+$ | $H_3BO_3$ (0.1 M) + NaOH (0.1 M) adj. to 9.2 | Proceed as in Serial 1. | |
| 10. — | EDTA | Indirect with $HgY^{2-}$ | $Hg^{2+}$ ISE | Hexamine,[i] $p$H 4. | To a soln. Contain Bi(III) add an XS of std. $HgY^{2-}$ complex (Y = EDTA) and sufficient hexamine (~0.5 g) to raise $p$H to ~4. Heat the soln. to 80–90°, cool, and titr. the liberated Hg(II) with std. EDTA. | 6 |
| 11. Sequential Bi, Cd, Ca | EDTA | Direct | Hg\|$Hg^{2+}$ | Various | Determine Bi(III) at $p$H 1.2–2 (adj. with $HNO_3$), Cd(II) at $p$H 4.0 (HOAc–NaOAc), and Ca(II) at $p$H 8.0 ($NH_3$ buffer) by titrn. in the presence of Hg(II)–EDTA complex (1 drop $10^{-3}$ M) with std EDTA. | 3 |

(*continued*)

Table 11.1. (Continued)

| Serial Number Determinand | Chelon | Titrn. type | Electrode | Buffer | Procedures and Remarks | Ref. |
|---|---|---|---|---|---|---|
| 12. Buformin[b] [692-13-7] | EDTA | Indirect with Cu(II) | $Cu^{2+}$ ISE liq. membrane | | This determination involves the conversion of the alkylbiguanide (L) into the complex $CuL_2SO_4$ by the addn. of XS ammoniacal std. $CuSO_4$ soln. to an aq. soln. containing the biguanide hydrochloride (30–100 mg) and titrating the XS Cu(II) with std. EDTA. | 7 |
| 13. Ca | EDTA | Direct | $Ag|Ag^+$ | $H_3BO_3$ (0.1 $M$) + NaOH (0.1 $M$) adj. to 9.2 | Proceed as in Serial 4. | 1 |
| 14. — | EDTA | Direct | $Hg|Hg^{2+}$ | 0.5 $M$ Ethanolamine adj. to 8.5 with $HNO_3$ | Add the buffer soln. (10 mL) and Hg(II)–EDTA complex (1 drop $10^{-3}$ $M$) to a known aliquot (15–25 mL) of Ca(II) ($\leq$0.05 $M$) and titr. with std. EDTA (0.05 or 0.005 $M$). | 3 |
| 15. — | EDTA | Direct | $Hg|Hg^{2+}$ | 0.05 $M$ $NH_3$ adj. to $p$H 10 with $HNO_3$ | Add the buffer soln. (10–25 mL) and Hg(II)–EDTA complex (1 drop $10^{-3}$ $M$) to a known aliquot (15–25 mL) of Ca(II) ($\leq$0.05 $M$) and titr. with std. EDTA (0.05 or 0.005 $M$). | 3, 8 |
| 16. — | EDTA | Direct | $Ca^{2+}$ ISE liq. membrane | $NH_3 \sim 10$ | Adjust the $p$H of a soln. (10 mL) containing 1–10 mg Ca(II) to $p$H 10 by adding 7 $M$ $NH_3$ dropwise. Titr. with std. EDTA. Precision is improved in the presence of Li, Na, and K. When Ba, Mg, and Zn are present, the single inflection represents the combined titrn. of Ca(II) and these ions. In the presence of Sr(II) the shape of the curve is distorted and the endpoint cannot be detected. $PO_4^{3-}$, $F^-$, $CO_3^{2-}$, and oxalate interfere. | 10, 11, 12 |

| | | | | | |
|---|---|---|---|---|---|
| 17. Ca | EDTA | Direct | $Cd^{2+}$ ISE Selectrode | 0.1 $M$ Borax ~ 9 | Add the buffer soln. (15 mL) to a known aliquot (25 mL) of Ca(II) ($\leqq 0.01$ $M$) followed by 0.01 $M$ Cd(II)–EDTA complex (1 mL). Dilute the soln. to 100 mL and titr. with EDTA. | 13 |
| 18. — | EDTA | Direct | $Cu^{2+}$ ISE | $p$H 10 | Add 0.5 $M$ Cu(II)–EDTA complex (1 mL) to a soln. containing 0.005–0.007 mol Ca(II) (~ 20 mL) and adj. the $p$H to 10 with 0.1 $M$ NaOH. Titr. with std. 0.01 $M$ EDTA added from a 1 mL microburette. Proceed as in Serial 6 except that EDTA is used as titrant. | 14 |
| 19. — | EDTA | Direct | $Cu^{2+}$ ISE | $NH_3$ | | 4 |
| 20. Ca in dolomite and glass | EGTA | Direct | Ag\|$Ag^+$ | Borate $p$H 10 | The soln. containing 0.25–0.35 mmol Ca(II), 2,4 pentanedione at a total concn. of 0.02–0.1 $M$, and $10^{-5}$ $M$ Ag(I) is buffered at $p$H 10. Macro quantities of Al(III), Cr(III), and Fe(III) are effectively masked by the dione and do not interfere in the titrn. of the soln. with std. EGTA. | 15 |
| 21. Ca and Mg sequential | EGTA | Direct | Hg(Ag)\|$Hg^{2+}$ | $NH_3$-$NH_4NO_3$ $p$H 8.8–9 and $p$H 10 | Add 1 $M$ $NH_4NO_3$ (2 mL) to the sample which contains 0.005–0.1 mol of both Ca and Mg, and adjust the $p$H to 8.8–9.0 by the dropwise addition of concd. $NH_3$. Add a few drops of 0.002 $M$ Hg(II)–EGTA complex and titr. with 0.01 $M$ EGTA to the first inflection point corresponding to the amount of Ca(II) present. Stop the titrn. shortly after this point and make the soln. 80% v/v in ethanol. Adj. the $p$H to 10 with concd. $NH_3$ and complete the titrn. of Mg(II) with std. EGTA. (See also water hardness serials 109–111). | 16 |

(*continued*)

Table 11.1. (Continued)

| Serial Number Determinand | Chelon | Titrn. type | Electrode | Buffer | Procedures and Remarks | Ref. |
|---|---|---|---|---|---|---|
| 22. Ca | CDTA | Indirect with Hg(II) | Hg(Ag)|Hg$^{2+}$ | 10% Hexamine$^i$ (200 mL) + 0.5 M NaOH (~16 ML) $p$H 11. | An XS of CDTA (0.05 M) is added to a known volume of soln. containing 0.02–6 mg Ca(II) followed by the buffer soln. (35–40 mL). The XS CDTA is titrated with std. Hg(II). | 5 |
| 23. Cd | EDTA | Direct | Hg|Hg$^{2+}$ | 0.5 M Acetate adj. to $p$H 4.6 | Proceed as in Serial 14. | 3 |
| 24. — | EDTA | Direct | Cd$^{2+}$ ISE Selectrode | 0.5 M Acetate adj. to $p$H 4.8 | Add the buffer soln. (10 mL) to a known aliquot (25 mL) of Cd(II) ($\leq 0.01$ M). Dilute to ~100 mL and titr. with std. EDTA. | 13 |
| 25. — | EDTA | Direct | Cu$^{2+}$ ISE (Ag$_2$S–CuS) | NH$_3$ | Proceed as in Serial 6 except that EDTA is used as titrant. | 4 |
| 26. — | Polyamines TRIEN$^c$ TETREN$^d$ | Direct | Ag|Ag$^+$ | Saturated borax $p$H 8.0–9.2 | To the sample containing 0.002–0.2 mmol Cd(II) add 10$^{-4}$Ag–TRIEN complex (1 mL) and satd. borax (50 mL). Dilute the soln. to ~100 mL, adj. $p$H to 8.0, and titr. with std. 0.001–0.05 M TRIEN. The titrn. can be successfully performed at $p$H 8.0 in the presence of tenfold XS of Ca(II) and Mg(II) and in the presence of equimolar Al(III) at $p$H 9.2. Tenfold XS of Al(III) is masked in 0.05 M 2,4-pentanedione at $p$H 8.7 and equimolar Fe(III) is masked in 0.01 M sulfosalicylic acid at $p$H 9.0. Similar results were obtained with TETREN titrant. | 17 |
| 27. — | EDTA | Indirect with Ca | Ag|Ag$^+$ | H$_3$BO$_3$(0.1 M) + NaOH (0.1 M) adj. to 9.2 | Proceed as in Serial 1. | 1 |

| # | Analyte | Titrant | Method | Electrode | Buffer | Procedure | Ref |
|---|---|---|---|---|---|---|---|
| 28. | Cd | EDTA | Indirect with Cu | $Cu^{2+}$ ISE $[Ag_2S/CuS(1:1)]$ | 0.05 M Acetate pH 4.75 | An XS of $10^{-3}$ M EDTA is added to a $\geq 5 \times 10^{-4}$ M soln. of Cd(II) and the soln. buffered at pH 4.75 (10 mL). The XS is titrated with std. Cu(II) soln. ($SO_4^{2-}$ or $NO_3^-$) added in 0.005 or 0.01 mL increments at 2 min intervals from a microburette. The concn. of Cu(II) should be such that a total of between 0.1 and 0.2 mL are needed to reach the equiv. point. Proceed as in Serial 22. | 18 |
| 29. | — | CDTA | Indirect with Hg(II) | $Hg(Ag)|Hg^{2+}$ | 10% Hexamine$^i$ (200 mL) + 0.5 M NaOH (~ 3 mL) pH 8-9. $NH_3$ pH 9.5–10.0 | | 5 |
| 30. | Ce(III) | EDTA | Indirect with Zn | $Hg|Hg^{2+}$ | $NH_3$ pH 9.5–10.0 | Add XS EDTA (15 mL 0.005 M) to the Ce(III) soln. (10 mL 0.005 M) and adj. the pH to 9.5–10.0 with concd. $NH_3$. Add Hg(II)–EDTA complex (1 drop $10^{-3}$ M) and titr. with std. Zn(II) soln. | 3 |
| 31. | Co(II) | EDTA | Direct | $Hg|Hg^{2+}$ | 0.5 M $NH_3$ adj. to pH 10 ($HNO_3$) | Proceed as in Serial 15. | 3 |
| 32. | — | EDTA | Direct | $Cd^{2+}$ ISE | Acetate pH 5.5 | A 0.02–0.4 M soln. buffered at pH 5.5 and containing 0.01 M Cd(II)–EDTA complex (1 mL) is titrated with std. EDTA. | 19 |
| 33. | — | EDTA | Indirect with Ca | $Ag|Ag^+$ | $H_3BO_3$(0.1 M) + NaOH (0.1 M) adj. to pH 9.2 | Proceed as in Serial 1. | 1 |
| 34. | Cr(III) | EDTA | Indirect with Hg(II) | $Hg^{2+}$ ISE | Hexamine$^i$ pH ~ 4 | Add std. EDTA in XS to the Cr(III) soln., adj. pH to ~ 4 with hexamine (~ 0.5 g), boil the soln. for 2–3 min, and cool and titr. the XS EDTA with std. Hg(II) soln. For analysis of multicomponent Cr(III) mixtures see Reference 20. | 6 |

(*continued*)

Table 11.1. (Continued)

| Serial Number Determinand | Chelon | Titration Type | Electrode | Buffer | Procedures and Remarks | Ref. |
|---|---|---|---|---|---|---|
| 35. Cr(III) | EDTA | Indirect with Zn | Hg\|Hg$^{2+}$ | 0.2 $M$ Acetate adj. to $p$H 3.5 | Add the buffer soln. (10 mL) to a portion (5 mL) of Cr(III) (~ 0.02 $M$) followed by an XS of EDTA (10 mL 0.02 $M$), and boil the soln. for 10 min (violet complex formed). Cool the soln, adj. its $p$H to 4.8, add EDTA–Hg(II) complex (1 drop $10^{-3}$ $M$), and titr. with std. Zn(II) soln. | 3 |
| 36. Cu(II) | EDTA | Direct | Hg\|Hg$^{2+}$ | 0.5 $M$ Acetate adj. to $p$H 4.6 | Proceed as in Serial 15. | 3 |
| 37. Sequential Cu + M(II) | EDTA | Direct | Cu$^{2+}$ ISE [Ag$_2$S/CuS(1:1)] | 0.5 $M$ Acetate $p$H 4.75 | The soln. (10 mL) containing Cu(II) and M(II) (M = Zn, Cd, or Mn) and buffered at $p$H 4.75 is titr. in 0.005 or 0.01 mL increments added at 2 min intervals from a microburette. The concn. of std. EDTA should be such that a total of between 0.1 and 0.2 mL are needed to reach equiv. point. Equimolar concns. of Cu and Zn and Cu and Cd can be discriminated at concn. $> 10^{-4}$ $M$, whereas for Cu and Mn(II) the minimum advisable conc. is $10^{-3}$ $M$. | 21 |
| 38. — | Polyamines TRIEN$^c$ TETREN$^d$ | Direct | Ag\|Ag$^+$ | Saturated borax soln. | To the sample containing 0.002–0.2 mmole Cu(II) add $10^{-4}$ $M$ Ag–TRIEN complex (1 mL) and satd. borax (50 mL). Dilute the soln. to ~ 100 mL, adj. the $p$H to 8.0, and titr. with std. 0.001–0.05 $M$ TRIEN. A tenfold excess of Ca(II) and Mg(II) causes no interference and Al(III) in equimolar quantities can be tolerated when the titr. is performed at $p$H 9.1. Fe(III) is masked in 0.01 $M$ sulfosalicylic acid at $p$H 8.8. Similar results are ob- | 17 |

| # | Analyte | Titrant | Method | Electrode | Buffer/Conditions | Procedure | Ref. |
|---|---|---|---|---|---|---|---|
| 39. | Cu | EDTA | Indirect with Ca | Ag|Ag$^+$ | H$_3$BO$_3$(0.1 $M$) + NaOH (0.1 $M$) adj. to $p$H 9.2 | Proceed as in Serial 1. | 1 |
| 40. | — | CDTA | Indirect with Hg(II) | Hg(Ag)|Hg$^{2+}$ | 10% Hexamine$^i$ (200 mL) + 0.5 $M$ NaOH (~ 3 mL) $p$H 8–9. | Proceed as in Serial 22. | 5 |
| 41. | Chelons NTA$^e$ + EDTA NTA + DTPA | Various | Direct with Fe(III) | Pt|Fe$^{2+}$, Fe$^{3+}$ | NaOAc (83g 3H$_2$O) + HOAc (23 mL) in 1 L $p$H 4.6 | To a soln. containing up to 2 mmol of a mixture of NTA with EDTA or DTPA in the ratio <10:1 add the buffer soln. (20 mL) and ferroin$^j$ soln. (1 mL) and dilute to 100–200 mL with H$_2$O. If the ratio of NTA:EDTA or NTA:DTPA is <1:10 add a known amount of std. 0.05 $M$ Bi(III) soln. to bind the bulk of EDTA or DTPA and bring the ratio in to the range from 10:1 to 1:10 to avoid adding excessive amounts of Fe(III). Titr. the soln. with std. 0.05 $M$ FeCl$_3$. The first potential break gives the Fe(III) consumed by EDTA or DTPA and the second that consumed by NTA. For determination of chelons in detergents and in waters and sewage see References 23 and 24, respectively. | 22 |
| 42. | Fe(III) dil. solns. | EDTA | Direct | Pt|Fe$^{2+}$, Fe$^{3+}$ | $p$H 1.2–2.0 | The $p$H of a soln. containing Fe(III) (10$^{-6}$ to 10$^{-3}$ $M$) was adjusted to 1.5–2.0 (H$_2$SO$_4$ or HClO$_4$) and 10$^{-3}$ $M$ Fe(II) (2–4 drops from Jones reductor) was added. The titrn. with EDTA was carried out at 60–70° under N$_2$, 5 titrn. points being obtained before the equivalence point which was obtained by the linearization of the curve. | 25 |

(*continued*)

Table 11.1. (Continued)

| Serial No. Determinand | Chelon | Titrn. type | Electrode | Buffer | Procedures and Remarks | Ref. |
|---|---|---|---|---|---|---|
| 43. Fe(III) | EDTA | Direct | $Cu^{2+}$ ISE | HOAc(50 mL) + NaOAc $3H_2O$ (270 g) in 1 L $p$H 5 | Add 0.5 $M$ Cu(II)–EDTA complex (1 drop) to a slightly acidic ($p$H ~ 2) soln. (~ 20 mL) containing 0.005–0.007 mmol Fe(III) followed by the buffer soln. (~ 1 mL). Titr. with std. 0.01 $M$ EDTA added from a 1 mL microburette. | 14 |
| 44. Fe(III) ipo 70-fold XS Al(III) | DTPA | Direct | Pt\|$Fe^{2+}$, $Fe^{3+}$ | $ClCH_2COOH$ (94.5 g) + NaOH (3.6 g) in 1 L $p$H 1.6<br><br>NaOAc $3H_2O$ (280 g) + HOAc (50 mL) in 1 L $p$H 5. | To a slightly acidic soln. of Fe(III) (mg amounts) add the buffer soln. (5 mL), 0.05 $M$ Fe(II) (1 drop from Jones reductor), dilute to ~ 100 mL, and titr. with 0.05 $M$ DTPA.<br><br>*Note*: Al(III) may then be determined by adding an XS of std. CDTA, and the buffer soln. (15 mL $p$H 5) and titrating the XS with std. 0.05 $M$ $FeCl_3$ soln. | 26 |
| 45. Fe(III) | DTPA | Direct | Pt\|$PtCl_6^{2-}$ | $p$H 2 | The $p$H of the soln. containing Fe(III) is adj. to $p$H 2, then 0.4% $Na_2PtCl_6$ (2–3 drops) is added and the soln. titr. with std. DTPA. The system is specific for Fe(III) because of the adsorption of this ion on the Pt electrode and the high stability constant of the chelate. | 27 |
| 46. — | EDTA | Indirect with Ca | Ag\|$Ag^+$ | $H_3BO_3$ (0.1 $M$) + NaOH (0.1 $M$) adj. to $p$H 9.2 | Proceed as in Serial 1. | 1 |
| 47. — | EDTA | Indirect with Cu(II) | $Cu^{2+}$ ISE [$Ag_2S$/CuS (1:1)] | 0.05 M Acetate adj. to $p$H 4.75 | Proceed as in Serial 28. | 18 |
| 48. — | EDTA | Indirect with | $Hg^{2+}$ ISE | Hexamine$^i$ $p$H ~ 4 | Proceed as in Serial 10. | 6 |

| | | | | | | |
|---|---|---|---|---|---|---|
| 49. Ga(III) | EDTA | Indirect with Ca | Ag|Ag$^+$ | H$_3$BO$_3$ (0.1 $M$) + NaOH (0.1 $M$) adj. to $p$H 9.2 | Proceed as in Serial 1. | 1 |
| 50. Ga(III) | EDTA | Indirect with Zn | Hg|Hg$^{2+}$ | 0.2 $M$ Acetate adj. to $p$H 4.6 | Add the buffer soln. (10 mL) to a portion (5 mL) of Ga(III) (~ 0.02 $M$), heat the soln. and add an XS of std. EDTA (10 mL 0.02 M). Cool the soln., add Hg(II)–EDTA complex (1 drop 10$^{-3}$ $M$), and titr. XS EDTA with a std. Zn(II) soln. | 3 |
| 51. Hf(IV) | EDTA | Indirect with Cu(II) | Hg|Hg$^{2+}$ | $p$H 4 | Add XS std. EDTA to the Hf(IV) soln. (5 mL 0.05 $M$), add acetate buffer, and adj. to $p$H 4 with NaOH. After adding Hg(II)–EDTA complex (1 drop 10$^{-3}$ $M$) titr. the XS EDTA with std. Cu(II). | 3 |
| 52. Hg(II) | EDTA | Direct | Hg|Hg$^{2+}$ | 0.5 $M$ Acetate adj. to $p$H 4.6 | Add the buffer soln. (15–25 mL) to a known aliquot (15–25 mL) of Hg(II) ($\leq$ 0.05 $M$) and titr. with std. EDTA. | 3 |
| 53. — | EDTA | Direct | Cu$^{2+}$ ISE | HOAc (60 mL) + NaOAc 3H$_2$O (270 g) in 1 L $p$H 5 | Add the buffer soln. (1 mL) to a soln. containing 0.005–0.007 mmol Hg(II) (20 mL) and titr. with std. 0.01 $M$ EDTA added from a microburette (1 mL).$^g$ | 14 |
| 54. — | CDTA | Direct | Hg(Ag)|Hg$^{2+}$ | 10% Hexamine$^i$ $p$H 6–10 | Buffer the Hg(II) soln. (0.01–0.05 $M$) in the $p$H range 6–10 and titr. with std. CDTA. $\Delta E$ in the vicinity of equivalence point is 199–225 mV/0.1 mL of 0.05 $M$ titrant which is greater than that observed in similar titrations with EDTA. | 28, 29 |
| 55. — | CDTA | Indirect with Hg(II) | Hg(Ag)|Hg$^{2+}$ | Hexamine$^i$ (2.5 g) + 0.01 $M$ NaOH (5 mL) | A suitable volume of the buffer is added to a soln. containing $\geq$ 0.02 mg Hg(II) followed by an XS of 0.001 $M$ CDTA. Titr. this XS with std. 0.01 $M$ Hg(II). | 28 |

(*continued*)

Table 11.1. (Continued)

| Serial No. Determinand | Chelon | Titrn. type | Electrode | Buffer | Procedures and Remarks | Ref. |
|---|---|---|---|---|---|---|
| 56. In(III) | EDTA | Direct | $Hg\|Hg^{2+}$ | 0.5 $M$ Acetate adj. to $p$H 4. | Proceed as in Serial 15. | 3 |
| 57. — | EDTA | Indirect with Ca | $Ag\|Ag^+$ | $H_3BO_3$ (0.1 $M$) + NaOH (0.1 $M$) adj. to $p$H 9.2 | Proceed as in Serial 1. | 1 |
| 58. — | EDTA | Indirect | $Hg\|Hg^{2+}$ | 0.2 $M$ Acetate adj. to $p$H 4.6 | Add the buffer soln. (25 mL) to a portion (10 mL) of ~ 0.01 $M$ In(III) followed by an XS of EDTA (10 mL 0.02 $M$) and EDTA–Hg(II) complex (1 drop $10^{-3}$ $M$). Titr. with std. Zn(II) soln. | 3 |
| 59. — | CDTA | Indirect with Hg(II) | $Hg(Ag)\|Hg^{2+}$ | 10% Hexamine[i] (200 mL) + 0.5 $M$ NaOH (~ 3 mL) $p$H 8–9. | Proceed as in Serial 7. | 5 |
| 60. K | Kryptofix 222 | Direct | $K^+$ ISE (glass?) | 1 $M$ Triethanolamine $p$H 9.5–10.0 | The $p$H of a weakly acid soln. of $K^+$ in the concn. range 0.0001–0.1 $M$ is adj. with 1 $M$ triethanolamine to $p$H 9.5. Titr. this soln. with std. Kryptofix 222. $K^+$ (0.78 mg) can be det. in the presence of equal amounts of Li, Cs, and Mg. | 30 |
| 61. La(III) | EDTA | Direct | $Cu^{2+}$ ISE | $p$H 10 | Proceed as in Serial 18. | 14 |
| 62. — | EDTA | Indirect with Ca | $Ag\|Ag^+$ | $H_3BO_3$ (0.1 $M$) + NaOH (0.1 $M$) adj. to $p$H 9.2 | Proceed as in Serial 1. | 1 |
| 63. — | EDTA | Indirect with Zn | $Hg\|Hg^{2+}$ | $NH_3$ $p$H 9.5–10.0 | Proceed as in Serial 30. | 3 |
| 64. — | CDTA | Indirect with Hg(II) | $Hg(Ag)\|Hg^{2+}$ | 10% Hexamine[i] (200 mL) + 0.5 $M$ NaOH (~ 8 mL) $p$H 10 | Proceed as in Serial 7. | 5 |

| # | Reagent | Method | Electrode | Medium/Buffer | Remarks | Ref |
|---|---|---|---|---|---|---|
| 65. Li | Kryptofix 211 | Direct | $Na^+$ glass[h] | 1 M Triethanolamine $pH$ 9.5–10.0 | The $pH$ of a weakly acid soln. of $Li^+$ in the conc. range 0.001–0.1 M is adj. with 1 M triethanolamine to $pH$ 9.5. This soln. is titr. with a std. soln. of Kryptofix 211. $Li^+$, concn. $10^{-3}$ M, can be determined in the presence of equal amounts of K, Rb, Cs, $NH_4^+$, Mg, Ca, Sr, and Ba, but Na interferes. | 30 |
| 66. Mg | EDTA | Direct | $Ag|Ag^+$ | $H_3BO_3$ (0.1 M) + NaOH (0.1 M) adj. to $pH$ 9.2. | Proceed as in Serial 4. | 1 |
| 67. — | EDTA | Direct | $Hg|Hg^{2+}$ | 0.5 M Ethanolamine adj. to $pH$ 8.5 with $HNO_3$ | Proceed as in Serial 14. | 3 |
| 68. — | CDTA | Direct | $Cu^{2+}$ ISE | $NH_3$ | Concd. $NH_3$ (10 mL), sodium citrate (0.5 mmol) is added to a soln. (~ 40 mL) containing between 0.003 and 0.3 mmol Mg(II), and the soln. titr. with 0.002 or 0.1 M CDTA. | 4 |
| 69. — | CDTA | Indirect with Hg(II) | $Hg(Ag)|Hg^{2+}$ | 10% Hexamine[i] (200 mL) + 0.5 NaOH (~ 16 mL) $pH$ 11. | Proceed as in Serial 7. | 5 |
| 70. Mn(II) | EDTA | Direct | $Hg|Hg^{2+}$ | 0.05M Acetate adj. to $pH$ 4.6 | Proceed as in Serial 14. | 3 |
| 71. — | EDTA | Direct | $Cd^{2+}$ ISE | $NH_3$ + ascorbic acid $pH$ 10. | A 0.02–0.4 M soln. buffered at $pH$ 10 and containing 0.01 M Cd–EDTA complex (1 mL) is titr. with std EDTA. | 19 |
| 72. — | EDTA | Direct | $Cu^{2+}$ ISE | $NH_3$ | Concd. $NH_3$ (10 mL), ascorbic acid (0.1 g) is added to a soln. (~ 40 mL) containing between 0.003 and 0.3 mmole Mn(II), and the soln. titr. with 0.002 or 0.1 M EDTA. | 4 |

(*continued*)

Table 11.1. (Continued)

| Serial No. Determinand | Chelon | Titrn. type | Electrode | Buffer | Procedures and Remarks | Ref. |
|---|---|---|---|---|---|---|
| 73. Mn(II) | EDTA | Indirect with Ca | Ag\|Ag$^+$ | H$_3$BO$_3$(0.1 M) + NaOH (0.1M) adj. to pH 9.2 | Proceed as in Serial 1. | 1 |
| 74. Mo(VI) | EDTA | Indirect with Fe(III) | Pt (?) | pH 1.5–2 | Mo(VI) (5–100 mg) was reduced to Mo(V) with hydrazine dihydrochloride at pH 1.5–2, and an XS of std. EDTA added. This XS was titr. with std. FeCl$_3$ soln. | 31 |
| 75. Metals in Oils | EGTA | Direct | Hg\|Hg$^{2+}$ | 8.8% v/v Ethanolamine pH 7–8 | The oil or additive contg Ba, Ca, Zn, and/or Mg is dissolved in a mixed aqueous-organic solvent,$^j$ buffered to pH 7–8 and titr. with std. EDTA to determine total metal content. A second inflection is seen if Mg(II) is present. The titrn. is repeated after addition of SO$_4^{2-}$ to mask Ba$^{2+}$ which is determined by difference. Zn(II) is determined separately at pH 4–4.5 (acetate buffer) by titrn. with std. CDTA using a Ag electrode. Dialkyldithiophosphate interferes and is removed by anion exchange. | 32 |
| 76. Na | Kryptofix 221 | Direct | Na$^+$ glass | 1 M Triethanolamine pH 95–10.0 | The pH of a weakly acid soln. of Na$^+$ in the concn. range 0.001–0.1 M is adj. with 1 M triethanolamine to pH 9.5. This soln. is titr. with a std. soln. of Kryptofix 221. Na$^+$ (0.46 mg) can be determined in the presence of equal amounts of NH$_4^+$, Rb, Cs, and Mg, but Li, K, Ca, Sr, and Ba interfere. | 30 |

| | | | | | |
|---|---|---|---|---|---|
| 77. Nd(III) | EDTA | Indirect with Ca | Ag\|Ag$^+$ | H$_3$BO$_3$(0.1 $M$) + NaOH (0.1 $M$) adj. to $p$H 9.2 | 1 |
| 78. — | EDTA | Indirect with Zn | Hg\|Hg$^{2+}$ | NH$_3$ $p$H 9.5–10.0 | 3 |
| 79. Ni | EDTA | Direct | Hg\|Hg$^{2+}$ | 0.5 M NH$_3$ adj. to $p$H 10 with HNO$_3$ | 3 |
| 80. — | EDTA | Direct | Cd$^{2+}$ ISE | Acetate $p$H 5.5 | 19 |
| 81. — | EDTA | Indirect with Ca | Ag\|Ag$^+$ | H$_3$BO$_3$(0.1 $M$) + NaOH (0.1 $M$) adj. to $p$H 9.2 | 1 |
| 82. — | EDTA | Indirect with Cu(II) | Cu$^{2+}$ ISE [Ag$_2$S/CuS (1:1)] | 0.05 $M$ Acetate $p$H 4.75 | 18 |
| 83. Nitro-glycerin | EDTA | Indirect | Pt\|Fe$^{3+}$, Fe$^{2+}$ | $p$H 2.5 | 33 |

A mixture of nitroglycerin (0.1–0.15 g), HOAc (25 mL) and HCl (25 mL 1:1) is deoxygenated with N$_2$. The nitroglycerin is reduced with Fe(II) (25 mL, 0.070 $M$) by refluxing the mixture for 10–15 min and after cooling the $p$H is adj. to 2.5 with 30% NaOH. The Fe(III) formed is titr. with 0.2 $M$ EDTA. A blank titrn. is performed on the reagents submitted to an identical procedure.

| | | | | | |
|---|---|---|---|---|---|
| 84. Pb(II) | EDTA | Direct | Hg\|Hg$^{2+}$ | 0.5 $M$ Acetate adj. to $p$H 4.6. | 3 |

Proceed as in Serial 14.

(*continued*)

Table 11.1. (Continued)

| Serial No. Determinand | Chelon | Titrn. type | Electrode | Buffer | Procedures and Remarks | Ref. |
|---|---|---|---|---|---|---|
| 85. Pb(II) | EDTA | Direct | $Cu^{2+}$ ISE ($Ag_2S$-CuS) | $NH_3$ | Concd. $NH_3$ (10 mL), sodium tartrate (0.5 mmol) is added to a soln. (~ 40 mL) contg. between 0.003 and 0.3 mmol Pb(II) and the soln. titr. with 0.002 or 0.1 M EDTA. | 4 |
| 86. — | EDTA | Direct | $Pb^{2+}$ ISE | $NH_3$–$NH_4Cl$ $p$H 10 | Pb(II) is complexed by the addition of potassium sodium tartrate (Rochelle salt) in a 10:1 mol ratio [versus Pb(II)]. The soln. is buffered at $p$H 10 and titr. with std. EDTA. The lower limit of concn. of Pb(II) is $3.5 \times 10^{-6}$ M to yield a change of 20 mV around the equiv. point. | 34 |
| 87. — | EDTA | Indirect with Ca | Ag\|$Ag^+$ | $H_3BO_3$ (0.1 M) + NaOH (0.1 M) adj. to $p$H 9.2 | Proceed as in Serial 1. | 1 |
| 88. — | CDTA | Indirect with Hg(II) | Hg(Ag)\|$Ag^+$ | 10% Hexamine$^i$ (200 mL) + 0.5 M NaOH (3 mL) $p$H 8–9. | Proceed as in Serial 7. Pb in pigments has been similarly determined by titrn. with EDTA. | 5, 35 |
| 89. Pb(II) in lead glasses | DTPA | Direct | Ag\|$Ag^+$ (glass reference electrode) | $H_3BO_3$ (61.83g) + NaOH (20 g) in 2 L $H_2O$ $p$H 9.0–9.5 | Dissolve lead glass or lead silicate (0.5 g) in HF and $HClO_4$ (Pt dish) and heat to dryness. Add $H_2O$ (15 mL), concd. $HNO_3$ (2 mL), and evaporate to dryness. Repeat the additions of $H_2O$ and $HNO_3$, and then to the boiling soln. add 30% $H_2O_2$ (3–4 drops). Continue boiling for 15 min then dilute soln. to 100 mL, add Na tartrate (2 g), and adj. $p$H to ~ 9 using phenolphthalein as indicator. Add the buffer soln. (10 mL) and check that $p$H = $9.0 \pm 0.1$ (critical). Add 0.0002 M $AgNO_3$ (1 drop) and titr. with std. DTPA. | 36 |

| No. | Ion | Titrant | Method | Electrode | Medium | Remarks | Ref. |
|---|---|---|---|---|---|---|---|
| 90. | Pr(III) | EDTA | Indirect with Zn | Hg\|Hg$^{2+}$ | NH$_3$ $p$H 9.5–10.0 | Proceed as in Serial 30. | 3 |
| 91. | — | CDTA | Indirect with Hg(II) | Hg(Ag)\|Hg$^{2+}$ | 10% Hexamine$^i$ (200 mL) + 0.5 $M$ NaOH (~ 8 mL) $p$H 10 | Proceed as in Serial 7. | 5 |
| 92. | Rh + M | EDTA or CDTA | Indirect with Hg(II) | Hg(Ag)\|Hg$^{2+}$ | Hexamine$^i$ $p$H 8–11 | M = Pd(II), Ru(III), Ni(II), Fe(III), Co(II), Y(III), La(III), or Th(IV). To determine the total Rh + M add 0.03–0.003 $M$ EDTA or CDTA (4.5–11 mL) in XS to the sample soln. 0.01–0.001 $M$ in Rh and M (1–5 mL). Boil the soln. 5–10 min, cool and buffer the soln. in the $p$H range 8–11. Titr. with std. 0.03–0.003 $M$ Hg(II). To determine M only use the same procedure but omit boiling the soln. Rh does not interfere. CDTA must be used when the mixture contains Pd(II). | 37 |
| 93. | Sc(III) | EDTA | Indirect with Zn | Hg\|Hg$^{2+}$ | NH$_3$ $p$H 9.5–10.0 | Proceed as in Serial 30. | 3 |
| 94. | Sm(III) | EDTA | Direct | Cu$^{2+}$ ISE | HOAc (10 mL) + NaOAc 3H$_2$O (270 g) in 1 L $p$H 5. | Proceed as in Serial 18. | 14 |
| 95. | — | EDTA | Indirect with Ca | Ag\|Ag$^+$ | H$_3$BO$_3$ (0.1 $M$) + NaOH (0.1 $M$) adj. to $p$H 9.2 | Proceed as in Serial 1. | 1 |
| 96. | — | EDTA | Indirect with Zn | Hg\|Hg$^{2+}$ | NH$_3$ $p$H 9.5–10 | Proceed as in Serial 30. | 3 |
| 97. | Sr | EDTA | Direct | Ag\|Ag$^+$ | H$_3$BO$_3$ (0.1 $M$) + NaOH (0.1 $M$) adj. to $p$H 9.2 | Proceed as in Serial 4. | 1 |
| 98. | — | EDTA | Direct | Hg\|Hg$^{2+}$ | 0.5 $M$ NH$_3$ adj. to $p$H 10 with HNO$_3$ | Proceed as in Serial 15. | 3 |

(*continued*)

Table 11.1. (Continued)

| Serial No. Determinand | Chelon | Titrn. type | Electrode | Buffer | Procedures and Remarks | Ref. |
|---|---|---|---|---|---|---|
| 99. Sr | EDTA | Direct | $Cu^{2+}$ ISE | $NH_3$ | Proceed as in Serial 6 except that EDTA is used. | 4 |
| 100. — | CDTA | Indirect with Hg(II) | $Hg(Ag)\|Hg^{2+}$ | 10% Hexamine$^i$ (200 mL) + 0.5 M NaOH (~ 16 mL) pH 11. | Proceed as in Serial 7. | 5 |
| 101. Th(IV) | EDTA | Direct | $Hg\|Hg^{2+}$ | pH 3.2 | An aliquot of Th(IV) soln. is added to 0.001 M $HNO_3$ (90 mL) contg. Hg(II)–EDTA complex (1 drop $10^{-3}$ M). Adj. the pH to 3.2 with 3 M $NH_3$ and titr. with std. EDTA. | 3 |
| 102. — | EDTA | Direct | $Cu^{2+}$ ISE | HOAc (60 mL) + NaOAc $3H_2O$ (270 g) in 1 L pH 5. | Add 0.5 M Cu–EDTA complex (1 drop) to a slightly acid (pH ~ 2) soln. (~ 20 mL) contg. 0.005–0.007 mol Th(IV) followed by the buffer soln. (1 mL). Titr. with std. 0.01 M EDTA added from a 1 mL microburette. | 14 |
| 103. — | EDTA | Indirect with Ca | $Ag\|Ag^+$ | $H_3BO_3$ (0.1M) + NaOH (0.1 M) adj. to pH 9.2 | Proceed as in Serial 1. | 1 |
| 104. Tl(III) | EDTA | Indirect with Zn | $Hg\|Hg^{2+}$ | 0.2 M Acetate adj. to pH 4.0 | Add the buffer soln. (25 mL) to a portion (10 mL) of Tl(III) (~ 0.01 M) followed by an XS of EDTA (10 mL 0.02 M) and EDTA–Hg(II) complex (1 drop $10^{-3}$ M). Titr. rapidly with std. Zn(II). | 3 |
| 105. V(IV) | EDTA | Direct | Pt | 0.5 M Acetate adj. to pH 3.9 | Proceed as in Serial 14. Abnormally shaped titrn. curves are obtained when a Hg electrode is used to monitor the titrn. of vanadyl ion. | 3 |

| | | | | | |
|---|---|---|---|---|---|
| 106. V(IV) | EDTA | Direct | $Cd^{2+}$ ISE | Acetate $p$H 4.5 | The soln. buffered at $p$H 4.5 is titr. with EDTA in the presence of Cd(II)–EDTA complex (1 mL 0.01 $M$). | 38, $\overline{39}$ |
| 107. — | EDTA | Indirect with Ca | Ag|Ag$^+$ | $H_3BO_3$ (0.1 $M$) + NaOH (0.1 $M$) adj. to $p$H 9.2 | Proceed as in Serial 1. | 1 |
| 108. V(V) | EDTA | Indirect with Hg(II) | Hg(Ag)|Ag$^+$ | 10% Hexamine$^j$ (200 mL) + 0.5 $M$ NaOH (~ 2 mL) $p$H 8. | Treat the vanadate soln. ~ 0.025 $M$ (1–4 mL) with 1 : 1 $H_2SO_4$ (6–8 drops). Wait until the yellow color fades (~ 1 min), add $Na_2SO_3$ (0.3 g), and heat gently for about 2 min to remove XS $SO_2$ and to allow the development of the blue color of V(IV). Add a measured XS of std. 0.05 $M$ EDTA and buffer soln. (40 mL). Titrate with std. 0.05 $M$ Hg(II) soln. The det. of V in mixtures with other metals is also reported. | 40 |
| 109. Water hardness | EDTA | Direct | Ag|Ag$^+$ | 1 $M$ $H_3BO_3$ adj. to $p$H 9.2 with concd. NaOH. | Remove Cl$^-$ by passing sample through a 1 in. × 5 in. anion exchange column (e.g., Amberlite IR 400 in nitrate form). Discard the first 50–100 mL and then collect the sample. Take 50–100 mL (depending on hardness), add buffer soln. (5 mL), 0.008 $M$ AgNO$_3$ (0.05 mL), and titr. with std. 0.01 $M$ EDTA to the point when $E_{cell}$ has dropped by 30 mV from its reading taken near the beginning of the titrn. | 1 |

(*continued*)

Table 11.1. (Continued)

| Serial No. Determinand | Chelon | Titrn. type | Electrode | Buffer | Procedures and Remarks | Ref. |
|---|---|---|---|---|---|---|
| 110. Water hardness | EDTA | Direct | $Ca^{2+}$ ISE | 0.03 M Glycine adj. to pH 9.7. | A computer-controlled titrn. To the sample (10 mL) containing a max. of 2.3 mmol Ca + Mg is added 10 mL of a reagent consisting of 0.02 M 3,4-dihydroxybenzoic acid plus 0.03 M glycine (pH 9.7 buffer). This soln. was titrated under computer control with 0.1 M EDTA. The presence of 3,4-dihydroxybenzoic acid increases the ratio of the Ca and Mg conditional stability constants so that two pronounced inflection points are obtained on the titrn. curve. Titrns. were also performed in the presence of 2,4-pentanedione and TRIS buffer. | 41 |
| 111. — | EDTA | Direct | $Cu^{2+}$ ISE ($Ag_2S$–CuS) | $0.2\ M\ H_3BO_3$ + NaOH adj. to pH 9.2 | Add to the water sample (5–50 mL) the buffer soln. (50 mL) and $10^{-3}\ M$ Cu EDTA$^{2-}$ complex (1 mL). Titr. with std. 0.1 or 0.01 M EDTA. Cl$^-$ does not interfere and the inflection is more pronounced than when a $Ca^{2+}$ ISE is used. | 42 |
| 112. Y(III) | EDTA | Indirect with Ca. | Ag\|Ag$^+$ | $H_3BO_3$ (0.1 M) + NaOH (0.1 M) adj. to pH 9.2 | Proceed as in Serial 1. | 1 |
| 113. — | EDTA | Indirect with Zn | Hg\|Hg$^{2+}$ | $NH_3$ pH 9.5–10.0 | Proceed as in Serial 30. | 3 |
| 114. — | CDTA | Indirect with Hg(II) | Hg(Ag)\|Hg$^{2+}$ | 10% Hexamine$^i$ (200 mL) + 0.5 M NaOH (~ 8 mL) pH 10. | Proceed as in Serial 7. | 5 |

| No. | Titrant | Method | Electrode | Medium | Procedure | Ref. |
|---|---|---|---|---|---|---|
| 115. Yb(III) | EDTA | Indirect with Ca | Ag\|Ag$^+$ | $H_3BO_3$ (0.1 $M$) + NaOH (0.1 $M$) adj. to $p$H 9.2 | Proceed as in Serial 1. | 1 |
| 116. Zn | EDTA | Direct | Hg\|Hg$^{2+}$ | 0.5 $M$ Acetate adj. to $p$H 4.6. | Proceed as in Serial 14. | 3 |
| 117. — | EDTA | Direct | Cd$^{2+}$ ISE | Acetate $p$H 5.5 | Adjust the sample soln. to $p$H 5.5 with the acetate buffer, add 0.01 $M$ Cd-EDTA (1 mL) and titr. with std. 0.01 $M$ EDTA added at the rate of 0.1 mL/10 sec. K$^+$, Na$^+$, Mg$^{2+}$, Ba$^{2+}$, Cl$^-$, NO$_3^-$, and SO$_4^{2-}$ did not interfere. | 43 |
| 118. — | EDTA | Direct | Cu$^{2+}$ ISE (Ag$_2$S–CuS) | NH$_3$ | Proceed as in Serial 6. | 4 |
| 119. — | Polyamine Direct TRIEN$^c$ TETREN$^d$ | | Ag\|Ag$^+$ | Saturated borax soln. | To the sample containing 0.002–0.2 mmol Zn(II) add 10$^{-4}$ $M$ Ag–TRIEN complex (1 mL) and satd. borax (50 mL). Dilute the soln. to ~100 mL, adj. the $p$H to 8.0, and titr. with std. 0.001–0.05 $M$ TRIEN. A tenfold XS of Ca and Mg causes no interference and Al(III) in equimolar quantities can be tolerated when the titr. is performed at $p$H 9.1. Fe(III) is masked in 0.01 $M$ sulfosalicylic acid at $p$H 8.8. Similar results are obtained with TETREN titrant. | 17 |
| 120. — | EDTA | Indirect with Ca | Ag\|Ag$^+$ | $H_3BO_3$ (0.1 $M$) + NaOH (0.1 $M$) adj. to $p$H 9.2 | Proceed as in Serial 1. | 1 |
| 121. — | EDTA | Indirect with Cu(II) | Cu$^{2+}$ ISE [Ag$_2$S/CuS (1:1)] | 0.05 $M$ Acetate $p$H 4.7 | Proceed as in Serial 28. | 18 |

(*continued*)

Table 11.1. (*Continued*)

| Serial No. Determinand | Chelon | Titrn. type | Electrode | Buffer | Procedures and Remarks | Ref. |
|---|---|---|---|---|---|---|
| 122. Zn | CDTA | Indirect with Hg(II) | Hg(Ag)|Hg$^{2+}$ | 10% Hexamine$^i$ (200 mL) + 0.5 NaOH (~3 mL) $p$H 8–9. | Proceed as in Serial 7. Zn in pigments has been determined similarly using EDTA (44). | 5 |
| 123. Zr(IV) | EDTA | Direct | Cu$^{2+}$ ISE | HOAc (60 mL) + NaOAc 3H$_2$O (270 g) in 1 L $p$H 5. | Proceed as in Serial 18, but use only 1 drop of Cu–EDTA complex. | 14 |
|  | EDTA | Indirect with Cu(II) | Hg|Hg$^{2+}$ | $p$H 4 | Proceed as in Serial 51. | 3 |

$^a$Chloride must be absent and is removed by passage through 5 in. of Dowex 1X-8 (100–200 mesh) or Amberlite IR400 resin contained in a 1 in. dia. column. The resin should be in the nitrate form.
$^b$n-butyl-1-biguanidide.
$^c$Triethylenetetramine (i.e., 1,4,7,10-tetraazadecane).
$^d$Tetraethylenepentamine (i.e., 1,4,7,10,13-pentaazatridecane).
$^e$NTA. Nitrilotriacetic acid N(CH$_2$COOH)$_3$;
$^f$Ferroin is prepared by dissolving FeSO$_4$ 7H$_2$O (0.028 g) and 1,10 phenanthroline (0.11 g) in H$_2$O (100 mL).
$^g$Addition of Cu(II)–EDTA is not necessary as the Cu$^{2+}$ ISE also responds to Hg(II) (HgS is less soluble than the other metal sulfides in the membrane).
$^h$The electrode is made responsive to Li$^+$ by storing in 0.1 $M$ LiCl for 2 weeks.
$^i$Hexamine is an alternative trivial name for methenamine[*100-97-0*] which is 1,3,5,7-tetraazatricyclo[3.3.1.1(3,7)]decane.
$^j$The titration. medium was a solvent mixture of chlorobenzene (20 mL), water (40 mL), and 2-methyl-2-propanol (130 mL), and yielded homogeneous solutions with up to 1 g of the oil sample.

Readers are reminded that a description of electrodes used as indicator electrodes for complexometric titrations is to be found in Section 5.3, and it should be noted that not many of these had been used routinely at the date of compiling Table 11.1. It was pointed out earlier that the slow response times of liquid membrane electrodes at low concentrations of determinand is an impediment to their use for these routine applications. Thus only two liquid membrane ion selective electrodes are to be found in the survey and of these only the calcium electrode is used frequently.

## 11.2  THE PREPARATION OF TITRANTS AND OTHER REAGENTS

### Polyaminocarboxylic Acids

EDTA solutions are usually prepared by dissolving disodium dihydrogen ethylenedinitrilotetraacetate dihydrate ($M = 372.24$) in an appropriate volume of deionized water. As was previously mentioned, these solutions must be standardized, preferably against a standard solution of the determinand metal ion submitted to the same experimental procedure as is to be used for the sample solutions. Sometimes the disodium salt dihydrate is dissolved in an equimolar amount of sodium hydroxide to minimize the $p$H changes during the titration (14). Although most titrand solutions are strongly buffered with respect to $p$H before the titration, it is advisable to check that the $p$H remains constant during the titration. The glass electrode used for this purpose can be used in conjunction with the same reference half-cell against which the potential of the titration indicator electrode is monitored.

1,2-Cyclohexanediyldinitrilotetraacetic acid (CDTA) is usually supplied as the monohydrate of the free acid ($C_{14}H_{22}N_2O_8 \cdot H_2O$). A 0.05 $M$ solution is prepared by dissolving this solid (18.22 g) in 1 $M$ sodium hydroxide (100 mL) and diluting the solution to 1 L with deionized water. This solution may be standardized against a standard solution of calcium(II) prepared from dried calcium carbonate.

Titrant solutions of DTPA [3,6,9-tris(carboxymethyl)-3,6,9-triazaundecanedioic acid; trivial name diethylenetriaminepentaacetic acid] are prepared similarly from the acid ($C_{14}H_{23}O_{10}N_3$) by dissolving the solid (~ 39.5 g) in water with the addition of sodium hydroxide to adjust the $p$H to 8.5. The resulting solution when diluted to 1 L with water is approximately 0.1 $M$. This solution has been standardized against standard solutions of lead(II) nitrate (26) or iron(III) chloride (36).

EGTA ($C_{14}H_{24}O_{10}N_2$) is supplied as the anhydrous free acid, and a 0.05 $M$ solution is prepared by dissolving the acid (19.02 g) in 1 $M$ sodium hydroxide (100 mL) and diluting the solution to 1 L with deionized water. This solution is standardized against pure calcium carbonate.

## Polyamines

These compounds were first used as complexometric titrants by Reilley and co-workers (45, 46) who showed that calcium(II), magnesium(II), aluminum(III), and lanthanum(III) do not interfere in the determination of transition metals. TRIEN is available commercially as the dihydrated disulfate salt ($C_6H_{18}N_4 \cdot 2H_2SO_4 \cdot 2H_2O$; $M = 378.42$), but TETREN must be purified by precipitation as the disulfate salt as follows (46):

To a well-stirred mixture containing crude TETREN (9.5 g), 2-propanol (150 g), and crushed ice (400 g) is added slowly a cold solution of dilute sulfuric acid (100 mL) previously prepared by adding concentrated sulfuric acid (11 g) to water and diluting to 100 mL. Check that the $pH$ is less than 2, and if it is not, add more sulfuric acid until this condition is met. Filter the cold crystals and wash them immediately with cold 3:1 methanol–water solution (200 mL). Recrystallize the wet crystals immediately by dissolving them in a minimum amount of hot 1:4 methanol–water. Decolorize the near boiling solution with activated charcoal (1–2 g Norit), then boil the mixture and filter it. Allow the filtrate to cool slowly to room temperature and then cool it to 0°C. Filter the crystals and wash them twice with cold 3:1 methanol–water solution (150 mL), twice with methanol (150 mL), and finally twice with anhydrous ether (150 mL). After the compound is dried at 80°C, the yield is about 70% by weight.

A 0.01 M solution of TETREN is prepared by dissolving the disulfate salt (3.85 g) in water, adjusting the $pH$ to 7.0 with concentrated sodium hydroxide and diluting the solution to 1 L with deionized water. Both TRIEN and TETREN titrants may be standardized using a standard copper(II) solution made by dissolving pure copper foil in concentrated nitric acid.

## The Cryptands

The titrants Kryptofix 211 ($C_{14}H_{28}N_2O_4$; $M = 288.39$), Kryptofix 221 ($C_{16}H_{32}N_2O_5$; $M = 332.44$), and Kryptofix 222 ($C_{18}H_{36}N_2O_6$; $M = 376.50$) are marketed in ampoules containing either 1 mL (Kryptofix 211 and 221) or 1 g (Kryptofix 222) by E. Merck, Darmstadt. When the contents of the ampoules are each dissolved and diluted to 200 mL with water, the resulting solutions are such that 1 mL Kryptofix 211 is equivalent to 0.1 mg lithium(I); 1 mL Kryptofix 221 solution is equivalent to 0.3 mg sodium(I), and 1 mL Kryptofix 222 is equivalent to 0.5 mg potassium(I).

## Metal Ion Titrants

Standard solutions of metal ions are required for indirect titrations. In these, a known number of moles of chelon solution is added in an amount expected to be in excess of the number of moles of determinand metal ion in the analyte solution. The number of moles of chelon added in excess is then determined by titration with a standard solution of a metal ion titrant. The number of moles of determinand ion reacted with the chelon is the difference between the total number of moles chelon added and the number of moles determined to be in excess. As is evident from Table 11.1, the metal ion titrants used for these indirect titrations are standard solutions of bismuth(III), calcium(II), iron(III), magnesium(II), mercury(II), or zinc(II).

A 0.05 $M$ solution of bismuth nitrate has been prepared by dissolving the pure metal (10.450 g) in 1 : 1 nitric acid and diluting the solution to 1 L with water (22). Standard solutions of magnesium (16) and zinc (32) are usually also prepared from the pure metals. For example, magnesium metal containing not more than 0.01% of total metal impurities is washed successively in 5 $M$ hydrochloric acid, water, ethanol, and acetone and allowed to dry at room temperature. The required quantity of metal is dissolved in a minimum amount of 2 $M$ sulfuric acid and diluted to the appropriate volume with deionized water. Zinc solutions are prepared similarly by treating granulated zinc with 2 $M$ hydrochloric acid and then washing thoroughly with water, methanol, and, finally, diethyl ether. After drying the granules under vacuum, the required quantity is dissolved in the minimum amount of warm 2 $M$ nitric acid and made up to volume with water. Standard solutions of zinc are sometimes prepared from pure dry zinc oxide. The solid (0.8138 g) when dissolved in a minimum amount of 1 : 1 nitric acid and diluted to 500 mL with water yields a 0.02 $M$ solution.

High purity metallic iron ( > 99.8% pure) is available commercially in sealed ampoules from which it is possible to prepare standard iron(III) solutions directly. Alternatively, an iron(III) solution has been prepared from reagent grade iron(III) chloride which was standardized gravimetrically (26).

Standard solutions of calcium are best prepared by dissolving the required quantity of pure calcium carbonate (dried at 150° for 24 h) in a minimum quantity of 2 $M$ nitric acid and diluting this solution to an appropriate volume with deionized water (32).

Mercury(II) nitrate has been the salt of choice from which to prepare standard solutions of mercury. This solid is dissolved in water aided by the addition of a minimum concentration of concentrated nitric acid. The solution can either be standardized gravimetrically as sulfide, or titrimetrically by indirect titration of excess EDTA with a standard solution of magnesium (29). It has also been standardized against CDTA using methyl thymol blue as indicator, or by titrating it with a standard solution of potassium iodide using a silver amalgam electrode (28).

## Metal Ion–Chelon Complexes

If the indicator electrode responds neither to a determinand ion nor to the metal ion chosen as an indirect titrant, then the indicator ion must be added to the titrand solution either alone or as a metal ion–chelon complex. The role of the mercury(II)–EDTA complex in fulfilling this function when a mercury $p$M electrode is the indicator electrode was discussed in Chapter 5 and the preparation of this complex is found elsewhere in this book. The cadmium(II), copper(II), and silver(I)–chelon complexes are prepared similarly, although it is only necessary to use the silver complex if metal solutions more dilute than 0.002 $M$ are to be titrated using a silver metal electrode (17). The general method of preparation is to mix solutions containing equimolar quantities of the metal ion, chelon, and sodium hydroxide making the final stoichiometric adjustment with the metal ion and chelon solutions using the appropriate electrode to indicate the equivalence point. Finally, it should be noted that the addition of the copper(II)–chelon complex may not be necessary when a copper(II) ion selective electrode based on a silver(I) sulfide/copper(II) sulfide is to be used as the indicator electrode for the titrations of certain metal ions with EDTA, CDTA, or TRIEN (4). In these cases the behavior of the membrane can be interpreted directly in terms of copper(II)–chelon interactions and the silver(I) ion membrane activity can be assumed to remain dependent upon the $Cu^{2+}/S^{2-}/Ag^+$ equilibria.

## REFERENCES

1. J. S. Fritz and B. B. Garralda, *Anal. Chem.* **36**, 737 (1964).
2. W. Kemula, A. Hulanicki, and M. Trojanowicz, *Chem. Anal. (Warsaw)* **14**, 481 (1969).
3. C. N. Reilley, R. W. Schmid, and D. W. Lamson, *Anal. Chem.* **30**, 953 (1958).
4. V. K. Olson, J. D. Carr, R. D. Hargens, and R. K. Force, *Anal. Chem.* **48**, 1228 (1976).
5. H. Khalifa, *Fresenius' Z. Anal. Chem.* **203**, 161 (1964).
6. E. Hopirtean, C. Liteanu, and R. Vlad, *Talanta* **22**, 912 (1975).
7. G. E. Baiulescu, V. V. Cosofret, and F. G. Cocu, *Talanta* **23**, 329 (1976).
8. R. Gallego Andreu and C. Verdejo Morales, *An. Quim.* **73**, 415 (1977); *Anal Abstr.* **33** 4B36; *CA* **88**, 130256r.
9. J. A. King and A. K. Mukherji, *Naturwissenschaften* **53**, 702 (1966).
10. S. L. Tacket, *Anal. Chem.* **41**, 1703 (1969).

# REFERENCES

11. T. P. Hadjiioannou and D. S. Papastathopoulas, *Talanta* **17**, 399 (1970).
12. Z. Pranjic-Anusic, Z. Cimerman, and Z. Stefanac, *Acta Pharm. Jugosl.* **27**, 55 (1977); *CA* **86**, 177409s.
13. J. Ruzicka and E. H. Hansen, *Anal. Chim. Acta* **63**, 115 (1973).
14. E. W. Baumann and R. M. Wallace, *Anal. Chem.* **41**, 2072 (1969).
15. I. E. Lichtenstein, E. Coppola, and D. A. Aikens, *Anal. Chem.* **44**, 1681 (1972).
16. B. Wallen, *Anal. Chem.* **46**, 304 (1974).
17. A. Hulanicki, M. Trojanowicz, and J. Domanska, *Talanta* **20**, 1117 (1973).
18. J. M. Van der Meer, G. Den Boef, and W. E. Van der Linden, *Anal. Chim. Acta* **79**, 27 (1975).
19. M. Taga, M. Mizuguchi, H. Yoshida, and S. Hikime, *Bunseki Kiki* **14**, 230 (1976); *CA* **85**, 201549q.
20. H. Khalifa and B. Ateya, *Microchem. J.* **13**, 247 (1968).
21. J. M. Van der Meer, G. Den Boef, and W. E. Van der Linden, *Anal. Chim. Acta* **76**, 261 (1975).
22. J. Horacek and R. Pribil, *Talanta* **16**, 1495 (1969).
23. M. Taddia, M. T. Lippolis, and L. Pastorelli, *Microchem. J.* **24**, 102 (1979).
24. I. Sekerka, J. Lechner, and B. F. Afghan, *Anal. Lett.* **6**, 977 (1973).
25. W. E. Van der Linden and S. Beijer, *Anal. Chim. Acta* **58**, 472 (1972).
26. R. Pribil and J. Horacek, *Talanta* **16**, 750 (1969).
27. F. Sierra Jimenez, C. Sanchez-Pedreno, and M. Duran Sanz, *An Quim.* **71**, 785 (1975); *Anal. Abstr.* **30**, 4B138; *CA* **85**, 10333v.
28. H. Khalifa, *Anal. Chim. Acta* **30**, 593 (1964).
29. H. Khalifa and M. G. Allam, *Anal. Chim. Acta* **22**, 421 (1960).
30. G. Czerwenka and E. Scheubeck, *Fresenius' Z. Anal. Chem.* **276**, 37 (1975).
31. D. Nonova and M. Gesheva, *God. Sofii Univ., Khim. Fak.* **64**, 213 (1969–70); *CA* **78**, 143424n.
32. J. N. Wilson and C. A. Marczewski, *Analyst (London)* **98**, 42 (1973).
33. R. S. Lambert and R. J. DuBois, *Anal. Chem.* **37**, 427 (1965).
34. D. C. Cormos, I. Haiduc, and P. Stetiu, *Rev. Roum. Chim.* **20**, 259 (1975).
35. H. Khalifa and A. M. Abdallah, *Microchem. J.* **13**, 726 (1968).
36. J. P. Cummings, *Talanta* **17**, 1013 (1970).
37. H. Khalifa, Y. M. Issa, and A. T. Hussein, *Fresineius' Z. Anal. Chem.* **282**, 223 (1976).
38. A. Napoli and M. Mascini, *Anal. Chim. Acta* **89**, 209 (1977).
39. A. Napoli, *Ann. Chim. (Rome)* **68**, 443 (1978).
40. H. Khalifa and A. El-Sirafy, *Fresenius' Z. Anal. Chem.* **227**, 109 (1967).
41. T. F. Christiansen, J. E. Busch, and S. C. Krogh, *Anal. Chem.* **48**, 1051 (1976).
42. M. Mascini, *Anal. Chim. Acta* **56**, 316 (1971).
43. M. Taga, M. Mizuguchi, H. Yoshida, and S. Hikime, *Bunseki Kagaku* **25**, 362 (1976); *CA* **86**, 11469m.
44. H. Khalifa and A. M. Abdallah, *Microchem. J.* **14**, 399 (1969).
45. C. N. Reilley and M. V. Sheldon, *Talanta* **1**, 127 (1958).
46. C. N. Reilley and A. Vavoulis, *Anal. Chem.* **31**, 243 (1959).

CHAPTER

12

# OXIDATION–REDUCTION TITRATIONS

In Chapter 1 it was shown that predictions of the direction of spontaneity for oxidation–reduction reactions are possible in uncomplicated cases by considering the $E°$ values for the component half-reactions. The stoichiometric reaction can be deduced by arranging these data in order of increasing positive $E°$ values:

$$\text{oxidant}_1 + n_1 e \rightleftharpoons \text{reductant}_1 \quad E_1°$$

$$\text{oxidant}_2 + n_2 e \rightleftharpoons \text{reductant}_2 \quad E_2° > E_1°$$

to yield the overall reaction

$$n_2 \text{reductant}_1 + n_1 \text{oxidant}_2 \rightarrow n_2 \text{oxidant}_1 + n_1 \text{reductant}_2$$

The difference ($E_2° - E_1°$) is related to the standard free energy change for the reaction by equation 1.3b. In such uncomplicated cases many parallels can be drawn between acid–base (proton interchange) and oxidation-reduction reactions (electron interchange). For example, just as acid and base strengths are leveled or limited by the solvent, so are the strengths of oxidants and reductants. Thus under conditions of equilibrium in water strong reductants are leveled to that of hydrogen, and oxidants to that of oxygen. For many reagents used in oxidation–reduction titrimetry, however, this type of equilibrium is attained so slowly that this restriction does not apply in practice. This example, therefore, highlights an important difference between the two types of reactions. Acid–base reactions are fast, whereas many oxidation–reduction reactions are slow, and, as was shown in Chapter 1, are often considerably more complicated than would be expected if the $E°$ values for the constituent half-reactions were taken as the sole criteria by which to establish the feasibility of reactions. There are many oxidation–reduction reactions that, although feasible thermodynamically, take place at such a slow rate that they would be useless as analytical reactions unless they were catalyzed. Thus kinetic factors assume an importance in oxidation-reduction reactions far greater than they do in other types of titrimetric reactions. Frequently the mechanisms of oxidation-reduction reactions are complex, and often several reactions can occur simultaneously. Hence, although the majority of oxidants and reductants used as titrants today have been in use for a number of years (1), the main thrust of research during the past two decades has been towards the creation of conditions whereby a single reaction out of a number of competing reactions proceeds selectively, quantitatively, and rapidly. These conditions have been achieved either by altering the medium or by introducing a complexing component into one of the reacting systems that is able to influence favorably the course of the desired reaction. Generally, the net effect of each alternative is to alter the formal oxidation-reduction potentials of the titrants so that their oxidizing or reducing capacities are enhanced. Simultaneously, the potential of the determinand may sometimes be advantageously diminished or increased. Some of these factors are considered in Section 12.1.

Atmospheric oxygen should be regarded as the ubiquitous interferent in oxidation-reduction potentiotitrimetry and it is advisable, therefore, to use either nitrogen or carbon dioxide as a blanket for the titrand solution. To this end the titrand solution is contained in a tall-form beaker which can be sealed by a rubber bung that carries the indicator and reference electrodes, inlet and outlet tubes for the gas, and a further hole to take the tip of the burette. If the reductant is added as the titrant, it is also necessary to prevent air from coming into contact with this solution by the type of arrangement shown in Figure 12.1 used for the preparation and storage of molybdenum(III) solutions (75).

The calomel half-cell is the reference most frequently used in aqueous titrations. In some cases, however, a species present in the titrand solution may react with the chloride emanating from this half-cell causing a marked drift in potentials with time. For example, chromium(VI) in a medium of phosphoric acid reacted slowly with chloride (2). In this particular case a sodium nitrate salt bridge subsequently interposed between the titrand and the reference half-cell was also a source of unstable potentials caused apparently by the reduction of the diffusing nitrate ions by vanadium(IV) which was also present in the titrand solution. The problem was solved by connecting the titration vessel to the calomel reference half-cell by means of two salt bridges: one of the saturated sodium perchlorate and the other of saturated sodium nitrate. Consideration can also be given to the use of a glass electrode as the reference electrode since in many oxidation-reduction titrations the change in hydrogen ion concentration is not great. Ion activity meters, now the most widely used instrument for measurement of $E_{cell}$, are usually operable independently of the actual polarities of the electrodes, and

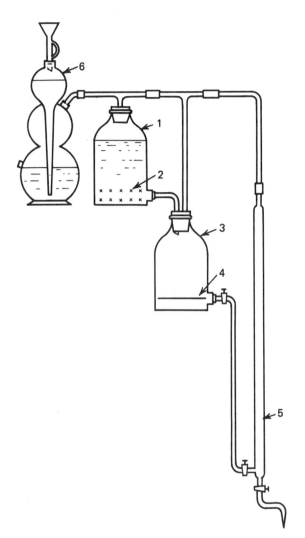

**Figure 12.1.** Apparatus for preparing and storing molybdenum(III) solutions: (1) 2-3 L bottle for reduction of Mo(VI); (2) granulated zinc amalgam; (3) 2-3 L bottle for storing Mo(III) solution prepared in 1; (4) concentrated hydrochloric acid or KCl; (5) 25 or 50 mL burette; (6) Kipp apparatus for preparing carbon dioxide (75).

hence it is possible nowadays to use the glass electrode as a reference without the necessity of modifying the electrode connecting sockets. The indicator electrode of choice for most titrations is the platinum electrode (see Section 5.5). It has been noticed, however, that this electrode cannot be used when vanadium(II) is present in the titrand solution (3). Apparently, the reduction of hydrogen ion by vanadium(II) is catalyzed at the platinum surface leading to unstable potentials. A mercury coated platinum electrode can be used for this application as a result of the high hydrogen overpotential on mercury.

The general shape of the titration curve for an oxidation–reduction reaction was calculated earlier from elementary considerations. A study of the experimentally derived titration curves is essential in the development of any titrimetric method based on these reactions. Frequently, small discontinuities in the curve are indicative of undesirable side reactions that may be eliminated by changing the titration medium. It is also possible to deduce the stoichiometry of the reaction from the curve. Many of the references cited in the tabulated material of this chapter contain useful case studies of this application. Explicit equations pertaining to oxidation–reduction titration curves have been derived in a series of six papers by J. A. Goldman (4).

## 12.1 GENERAL CONSIDERATIONS AND APPLICATIONS

This section deals with some of the methods used to ensure that the determinand is present in the correct oxidation state before the titration. In addition, it outlines the determination of formal potentials that can also be an important preliminary in the development of a titrimetric method. Some guidelines are given in the choice of a medium best suited to a particular determination, and then the catalysts used in oxidation–reduction reactions and their applications are briefly described.

### Preparatory Procedures

The quantitative conversion of the determinand to a particular oxidation state prior to the titration is an essential preliminary for some determinations. The oxidant or reductant chosen for this purpose must react rapidly with the determinand, and after the reaction it must be possible to completely and selectively remove the excess of the reagent from the titrand solution. A well-known example of such a preparatory procedure is often used in the determination of iron(III). The determinand is quantitatively reduced to iron(II) with zinc amalgam and the resulting solution titrated with a standard solution of potassium dichromate. If at all possible, however, it is more economic to dispense with the extra reduction step and to titrate the oxidant directly with a standard solution of a reductant. Thus, for the determination of iron(III), the solution can be titrated directly with a standard solution of hydrazine sulfate in the presence of a small amount (0.2 mL, 2% solution) of osmium tetroxide catalyst (5).

Most of the oxidants used currently to prepare a solution of a determinand in a desired oxidation state prior to its titration with a standard solution of reductant are given in Table 12.1. The solid oxidant sodium bismuthate has not been included in this table because its use has largely been supplanted by the silver-ion catalyzed peroxydisulfate (i.e., persulfate) oxidations. Also excluded are ozone (25), since it does not appear to be widely used these days, and perchloric acid, which is used mainly for the dissolution of solid samples such as ores (26) and in the wet oxidation of organic matter (27). The general usage of solid oxidants in analysis has been reviewed (28).

The use of reductants to lower the valency state of a determinand prior to its titration with a standard solution of oxidant is often more convenient than the reverse procedure previously described. This is because the oxidizing titrants in common use are far more stable in solution than most of the reductant titrants. Prior reduction of the determinand is often achieved by passing the analyte solution through a column (*ca.* 35 cm long, 16–19 mm dia.) containing a metal such as zinc, cadmium, lead, silver, mercury, or bismuth. The selectivity of these metals as reductants is reflected approximately by the magnitude of their $E°$ values in the medium in which they are usually employed. These are

$$Zn^{2+} + 2e \rightleftharpoons Zn \qquad E° = -0.76 \text{ V}$$

$$Cd^{2+} + 2e \rightleftharpoons Cd \qquad E° = -0.40 \text{ V}$$

$$PbCl_2 + 2e \rightleftharpoons Pb + 2Cl^- \qquad E° = -0.26 \text{ V}$$

$$AgCl + e \rightleftharpoons Ag + Cl^- \qquad E° = +0.22 \text{ V}$$

$$Hg_2Cl_2 + 2e \rightleftharpoons 2Hg + 2Cl^- \qquad E° = +0.27 \text{ V}$$

$$BiO^+ + 2H^+ + 3e \rightleftharpoons Bi + H_2O \qquad E° = +0.32 \text{ V}$$

As is to be expected from these data, zinc, the strongest of the reductants listed, is also the least selective. Thus zinc will reduce such oxidants as chromium(III) to chromium(II), europium(III) to europium(II), iron(III)

Table 12.1. Some Oxidants Used to Obtain a Required Oxidation State

| Oxidant | Some Applications | Removal of Excess | Ref |
|---|---|---|---|
| Lead dioxide (solid) | Mn(II) to Mn(III) ipo[c] $P_2O_7^{4-}$ in acid soln. | By filtration | 6 |
| Silver(II) oxide solid[a] | Mn(II) to Mn(VII), Cr(III) to Cr(VI) in acid soln. | Warm for a few minutes | 7 |
| | Pu(IV) to Pu(VI) in 1 $M$ $H_2SO_4$ | | 8 |
| | Pu(IV) to Pu(VI) in 2.5 $M$ $HNO_3$ | | 9 |
| Sodium perxenate (solid) | Mn(II) to Mn(VII) in acid soln. hydroxides of Np(IV) and Pu(IV) to hexavalent states in neutral soln. | Decomposes in acid. soln. | 10 |
| Hydrogen peroxide | Cr(III) to Cr(VI) in NaOH soln. Co(II) to Co(III) in bicarbonate soln. | By boiling the alkaline soln., sometimes ipo[c] catalysts such as $Ni^{2+}$, $I^-$, or Pt. | 11, 13 |
| Periodate (as $KIO_4$ soln.) | Mn(II) to Mn(VII) ipo[c] $H_3PO_4$. Ru(IV) to Ru(VI). | Precipitation with Hg(II). With $H_2O_2$ in alk. soln.; $H_2O_2$ decomposed spontaneously. | 14 15 |
| Permanganae (as $KMnO_4$ soln.) | Ce(III) to Ce(IV) ipo[c] $F^-$ or $P_2O_7^{4-}$. V(IV) to V(V) selectively ipo[c] Cr(III) in acid soln. In alk. soln. Cr(III) → Cr(VI). | Add sodium azide and boil the soln. or add $NaNO_2$ and destroy XS $NO_2^-$ with urea | 16 17 18 |
| Peroxydisulfate[b] ($S_2O_8^{2-}$ as K or $NH_4$ salt) | Ce(III) to Ce(IV). Cr(III) to Cr(VI), W(V) to W(VI), V(IV) to V(V). V in steel. Mn(II) to Mn(VII) ipo[c] $H_3PO_4$. Pu(IV) to Pu(VI) | By boiling | 19, 20 21 22, 24 |

[a] Prepared by adding potassium persulfate (100 mL, 3% w/v) to silver nitrate (10 mL, 10% w/v). Al to stand 1 h. Filter and wash precipitate well with water. Dry *in vacuo* (8).
[b] Used in acid solution in the presence of silver(I) as catalyst.
[c] ipo: in the presence of.

to iron(II), molybdenum(VI) to molybdenum(III), titanium(IV) to titanium(III), uranium(VI) to uranium(III) and (IV), and vanadium(V) to vanadium(II). The lack of selectivity in the reduction of uranium(VI) results in the necessity of an extra oxidation step in the determination of this element by oxidimetric titration. Thus the fraction present in the tervalent state must be oxidized to the tetravalent state by bubbling air through the solution prior to titration with a standard solution of potassium dichromate. When silver is used as the reductant, however, a solution of uranium(VI) in hydrochloric acid is reduced quantitatively to uranium(IV). A similar difference in the selectivities of these two reductants is experienced in the reduction of molybdenum(VI); with zinc this oxidant is reduced to molybdenum(III), whereas with silver it can also be selectively reduced in hydrochloric acid to molybdenum(V) if required. On the other hand, silver will neither reduce chromium(III) nor titanium(IV).

Columns of these solid reductants have become known as *reductors*, the earliest example of which is the Jones zinc reductor (29). The preparation and use of these devices are to be found in standard textbooks (30, 31). Reductants like zinc and cadmium which can reduce hydrogen ion are used as amalgams of low mercury content to control the rate of hydrogen evolution. The amount of mercury present is determined by the strength of the oxidant to be reduced, and whether the oxidant will react also with mercury. For the stronger oxidants that react with mercury, a higher content of mercury should be present in the amalgam to compensate for its loss through reaction. For example, if a Jones reductor is to be used for the reduction of stronger oxidants such as chromium(III) or iron(III) which react with mercury, the zinc amalgam should contain between 1 and 5% mercury. A mercury content in the range 0.1–1% is sufficient to control the rate of hydrogen evolution when weaker oxidants are to be treated. There have been reports that hydrogen peroxide, formed by the reduction of atmospheric oxygen, can cause subsequent interference when using zinc, lead, and silver reductors, particularly in the determination of small quantities of iron(III) (30). It is advisable, therefore, to deoxygenate solutions with either carbon dioxide or hydrogen before passing them through these reductors.

There are, of course, many examples of the applications of zinc, cadmium, lead, silver, mercury, and bismuth reductors in the literature (1). More recent examples have included the determination of europium(III) in mixtures of rare earth metals using a Jones reductor to quantitatively produce europium(II) (32). A departure from the traditional use of the silver reductor for oxidants prepared in hydrochloric acid is its application for the reduction of determinands prepared in solutions containing bromide. The formal potential for the Ag(I)/Ag system is decreased by about 0.13 V in bromide with respect to its value in chloride at the same concentration (33). Thus iron(III) was quantitatively reduced to iron(II) in 0.1–4 $M$ hydrobromic acid; similarly, copper(II) was reduced to copper(I) in $> 1.5$ $M$ hydrobromic acid, vanadium(V) to vanadium(IV), and uranium(VI) to uranium(IV) in $> 0.3$ $M$ hyrobromic acid at room temperature. The selective reduction of uranium(VI) to uranium(IV) in sulfuric acid solution (20%) in the presence of large amounts of niobium was found to be possible using a bismuth metal reductor (34). A bismuth amalgam containing 3% w/w bismuth was used for the reduction of tungsten(VI) to tungsten(V) in hydrochloric acid solution which was followed by titration with cerium(IV) (35). In a Jones reductor, tungsten(VI) was reduced to tungsten(III) in a medium of 9 $M$ hydrochloric acid (36). A lead reductor has been used to simultaneously and quantitatively reduce uranium(VI) to uranium(IV) and iron(III) to iron(II) in a medium of approximately 3 $M$ hydrochloric acid prior to the sequential titration of uranium(IV) and iron(II) with a standard solution of potassium dichromate (37). 3 $M$ hydrochloric acid was also the medium of choice for the reduction of molybdenum(VI) to molybdenum(V) with mercury as a prelude to the titration of the molybdenum(V) with cerium(IV) sulfate (38).

Although soluble reductants are more convenient to use than the solid reductants mentioned, the selective removal of the excess of the soluble reagent introduces a complication that does not arise when using solids. The following $E°$ values of the commonly used soluble reductants are indicative of their relative strengths:

$$H_3PO_3 + 2H^+ + 2e \rightleftharpoons H_3PO_2 + H_2O \qquad E° = -0.50 \text{ V}$$

$$Cr^{3+} + e \rightleftharpoons Cr^{2+} \qquad E° = -0.41 \text{ V}$$

$$H_3PO_4 + 4H^+ + 4e \rightleftharpoons H_3PO_2 + 2H_2O \qquad E° = -0.39 \text{ V}$$

$$N_2 + 5H^+ + 4e \rightleftharpoons N_2H_5^+ \qquad E° = -0.22 \text{ V}$$

$$Sn^{4+} + 2e \rightleftharpoons Sn^{2+} \text{ (1 } M \text{ HCl)} \qquad E°' = 0.14 \text{ V}$$

$$Fe^{3+} + e \rightleftharpoons Fe^{2+} \qquad E° = 0.77 \text{ V}$$

Hypophosphorous acid as $NaH_2PO_2 \cdot H_2O$ is usually employed in acid solution and reduces arsenic(V) and arsenic(III) to elemental arsenic (30, 40), iron(III) to iron(II) (41), tin(IV) to tin(II) (42, 43), and selenium(IV) and tellurium(IV) to the elements (44). A medium of phosphoric acid ($> 4.5$

$M$) was found suitable for the reduction of germanium(IV) to germanium(II), which was accomplished by the addition of solid sodium hypophosphite to the solution followed by a 15 min period of boiling in an atmosphere of carbon dioxide (45).

Chromium(II) chloride is a reductant of similar strength to hypophosphorous acid and will also reduce arsenic(III) to the element (46). It has been used to reduce uranium(VI) to uranium(IV). In this instance the excess reagent was removed by air oxidation using the color change of phenosafraniun [Reg. No. *81-96-6*] from colorless to pink to indicate that the excess chromium(II) had been oxidized (47).

Hydrazine hydrochloride or sulfate is a convenient reductant since its irreversible oxidation produces nitrogen, which is innocuous. It can be used to reduce arsenic(V) to arsenic(III) (48), and as $N_2H_4$ HCl it has been used in hydrocyhloric acid solution for the reduction of molybdenum(VI) to molybdenum(V) (49).

Reduction of molybdenum(VI) to molybdenum(V) (50) and arsenic(V) to arsenic(III) (51) can also be accomplished with tin(II). The reagent is tin(II) chloride dissolved in hydrochloric acid ($\sim 2\ M$) which is added in slight excess to the analyte solution which also contains a high concentration of hydrochloric acid. The excess is destroyed by the addition of mercury(II) chloride

$$Sn^{2+} + 2HgCl_2 \rightarrow Sn^{4+} + Hg_2Cl_2 + 2Cl^-$$

Excessive tin(II) chloride must be avoided since this will reduce mercury(II) chloride to finely divide mercury, which will then react subsequently with the oxidant used as titrant.

### Formal Potentials

The significance of these conditional constants was discussed in Chapter 1. As was shown there, these data are preferable expressed on the same scale as the arbitrary standard chosen for standard reduction potentials; that is, with respect to

$$2H^+ + 2e \rightleftharpoons H_2 \qquad E° = 0.00 \text{ at all temperatures}$$

In the presentation of formal potentials this standard would represent a notional reference half-cell often referred to as a *standard hydrogen electrode* which is an hypothetical rather than a practice entity. Formal potentials are measured typically in a cell such as

$$Hg;\ Hg_2Cl_2|KCl\ (c \geqq 3.5\ M)\|c_{ox} = c_{red}\ (\text{medium stated})|Pt \quad (12.1)$$

in which $c_{ox}$ represents the formal concentration of the oxidized species and $c_{red}$ is the formal concentration of the conjugate reductant. The formal potential $E°'$ is deduced from the measured value of $E_{cell}$ for cell (12.1) by

$$E°' = E_{cell} + E_{ref} \tag{12.2}$$

where $E_{ref}$ is the potential of the reference half-cell. More often than not it is assumed that the potential of the actual reference electrode employed in the cell can be equated to the published value for that reference electrode which was obtained with every care as to the preparation of materials, the assembly of the half-cell, and refinement of technique in the measurement of its value. Sometimes the measured value of $E_{cell}$ is reported as $E°'$ with respect to the actual reference half-cell used which is often a saturated calomel electrode. It has also been usual to assume that the response of the indicator electrode, frequently platinum, is Nernstian. In comparison, therefore, with the equivalent acid–base determination of a mixed $pK_a$ value, an $E°'$ value lacks accuracy. This is a result of a paucity of data upon which the calibration of the electrodes can be based. In a majority of cases, therefore, formal potentials should be regarded as guides to the strength of a particular oxidant–conjugate reductant system under a given set of conditions. Nonetheless, as was pointed out in Chapter 1, they are useful data in drawing conclusions as to the conditions under which an oxidation–reduction reaction may be possible.

The formal potential for the half reaction

$$Fe^{3+} + e \rightleftharpoons Fe^{2+} \; (1 \; M \; H_2SO_4)$$

has been suggested as a suitable calibration standard for cells containing a platinum indicator electrode (52). This oxidation–reduction buffering system, or *redox buffer* as such systems are sometimes called, was prepared as a solution having the following composition:

0.1 $M$ ammonium iron(II) sulfate, $Fe(NH_4)_2(SO_4)_2 \cdot 6H_2O$ (39.31 g $L^{-1}$)
0.1 $M$ ammonium iron(III) sulfate, $FeNH_4(SO_4)_2 \cdot 12H_2O$ (48.22 g $L^{-1}$)
1.00 $M$ sulfuric acid (56.2 mL $L^{-1}$ concentrated $H_2SO_4$)

Although the iron(II) and iron(III) solutions are hydrates, they are commercially available not only in reagent grade quality but also with lot assays confirming them to be within 1% of the nominal composition. When stored in closed suitable glass or plastic storage vessels this solution has been found to be stable indefinitely. The solution is also reported as being quite stable on prolonged exposure to the atmosphere, the potential of a cell containing a platinum electrode and a reference half-cell changing by less than 3 mV over a period of about a week. As befits a buffer solution, minor concentration errors in actual composition do not exert a great influence upon the formal potential of the iron(III)–iron(II) system. For example, a 4% change in the activity ratio iron(III)/iron(II) is required before the potential of the cell changes by 1 mV. This is likely to be a small change in comparison to the change in liquid junction potential when the electrodes are transferred after standardization to a solution of different acidity. Values of $E_{cell}$ for this iron(III)–iron(II) solution when a platinum electrode was used in conjunction with various reference half-cells at 25° are given in the following:

Hg; $HgCl_2$|saturated KCl       430 mV (SCE)

Ag; AgCl|1.00 $M$ KCl       439 mV

Ag; AgCl|4.00 $M$ KCl       475 mV

Ag; AgCl|saturated KCl       476 mV

The calculated formal reduction potential was +675 mV with respect to $2H^+ + 2e \rightleftharpoons H_2$; $E° = 0.00$ V taking the value for the Ag; AgCl|KCl ($4M$) half-cell as +200 mV with respect to the $H^+/H_2$ half-reaction. The pretreatment of the platinum electrode used to obtain these data is not given. However, in the determination of the formal potentials of arylferrocenes in aqueous acetic acid solutions (53), although the rates of attainment of stable $E_{cell}$ values were very dependent upon the pretreatment of the indicator electrodes, the actual equilibrium values of $E_{cell}$ agreed to within 0.5 mV regardless of the pretreatment. Untreated platinum electrodes, stored in water and not subjected to any cleaning, were found to attain steady values very quickly.

Formal potentials are quite frequently determined by measuring the potential of a cell solution containing equal formal concentrations of the oxidant and reductant prepared in the medium of interest. The deoxygenated stock solutions of electroactants from which the cell solution is prepared are standardized by an appropriate titrimetric method immediately before the determination of $E_{cell}$. Usually the concentration of both oxidant and reductant in the cell solution is kept small in comparison with the concentration of the medium. For example, a concentration of 0.01 F with respect to both species suffices when the concentration of the medium, often an acid solution, exceeds 0.25 $M$. This solution is kept under nitrogen or carbon dioxide during the measurement of $E_{cell}$. It is advisable to check the response of the indicator electrode through the constancy of $E°'$ obtained by measuring the e.m.f. values for a series of solutions containing ratios of oxidant to reductant in the range 1 : 4 to 4 : 1. For a cell having the polarity inferred by cell notation (12.1), the formal potential is given by

$$E°' = E_{cell} + E_{ref} + \frac{k}{n}\log\frac{c_{red}}{c_{ox}} \qquad (12.3)$$

where $c_{red}$ and $c_{ox}$ are the formal concentrations of the reductant and oxidant, respectively. A variant of this method is to use a potentiometric titration to progressively and quantitatively convert a standardized solution of reductant to its conjugate oxidant by titration with a standard solution of a strong oxidant such as potassium dichromate or cerium(IV). The value of $E_{cell}$ is measured after the addition of each 0.1 equivalent of titrant within the range 0.2–0.8 equivalent. The equation analogous to (12.3) is

$$E°' = E_{cell} + E_{ref} + \frac{k}{n}\log\frac{(C_t - c_{ox})}{c_{ox}} \qquad (12.4)$$

where $C_t$ is the formal concentration of reductant taken initially, corrected for the volume change caused by the addition of each aliquot of titrant, and $c_{ox}$ in this case is the concentration of titrant in the cell solution also corrected for volume change. The overall spread of the individual values of $E°'$ about the mean value is usually within $\pm 2$ mV. The value of $E_{ref}$ to use in equations (12.3) and (12.4) may be determined using the value of $E°'$ of $+675$ mV for the redox buffer solution, already mentioned in this subsection, in conjunction with the measured value of $E_{cell}$ for this solution provided that the temperature for all the measurements remains at 25° [see equation 12.2]. The reversibility of the system may be checked by back-titrating the 80% oxidized solution with the original standard solution of reductant. Although the internal consistency for a set of results may be satisfactory, there is often a difference between values of formal reduction potentials reported in the literature for the same system. Often the difference exceeds 10 mV, and this possibly reflects the lack of a uniform standardization procedure for the electrodes. The formal reduction potentials for some systems used in potentiotitrimetry in recent years are given in Table 12.2, and a collection of earlier values is to be found in Reference 61.

Table 12.2. Formal Reduction Potentials ($E^{\circ\prime}$ Volt) in Various Media

| Half-Reaction | $E^{\circ\prime}$ Value (Medium) | Ref. |
|---|---|---|
| As(V) + $2e \rightleftharpoons$ As(III) | 0.25 ($p$H 7), 0.60 (2 F HCl), 0.65 (4 F HCl), 0.81 (8 F HCl), 0.71 | 54 |
| $H_3AsO_4 + 2H^+ + 2e \rightleftharpoons HAsO_2 + 2H_2O$ | (20–30% aq. HOAc, KI–KIO$_3$ catalyst). 0.63 (0.5 F HCl in MeOH), 0.30 (3.5 F HOAc in MeOH), 0.48 (2 F HOAc in MeCN), 0.59 (2 F HOAc in EtOH), 0.58 (4 $M$ HOAc in 2-PrOH), 0.29 (1 F HOAc in 4:1 2-PrOH and MeOH). | 55 |
| Dehydroascorbic acid $\rightleftharpoons$ ascorbic acid | 0.19 ($p$H 7). | 56 |
|  | −0.01 ($p$H 8.7), 0.33 ($p$H 1.05). | 57 |
| ($C_6H_6O_6 + 2H^+ + 2e \rightleftharpoons C_6H_8O_6$) | 0.45 (1 F H$_3$PO$_4$), 0.46 (2 F H$_3$PO$_4$), 0.51 (4 F H$_3$PO$_4$), 0.57 (6 F H$_3$PO$_4$), 0.61 (8 F H$_3$PO$_4$), 0.65 (10 F H$_3$PO$_4$), 0.69 (12 F H$_3$PO$_4$). | 58 |
| BrCN + $H^+$ + $2e \rightleftharpoons Br^-$ + HCN[b] | 0.51 ($p$H 3). | 59 |
| Ce(IV) + $e \rightleftharpoons$ Ce(III) | 1.44 (1 F H$_2$SO$_4$), 1.22 (12 F H$_3$PO$_4$). | 58[a] |
|  | 1.24 (6.0 $M$ H$_3$PO$_4$), 1.24 (7.5 $M$ H$_3$PO$_4$), 1.24 (9.0 $M$ H$_3$PO$_4$), 1.24 (10.5 $M$ H$_3$PO$_4$), 1.23 (12.0 $M$ H$_3$PO$_4$), 1.22 (13.5 $M$ H$_3$PO$_4$). | 58a |
| Cr(III) + $e \rightleftharpoons$ Cr(II) in DMF | −0.51 (0.05 F HCl). | 60 |
| $Cr_2O_7^{2-} + 14H^+ + 6e \rightleftharpoons 2Cr^{3+} + 7H_2O$ | 1.13 (0.5 F HClO$_4$), 1.16 (1.0 F HClO$_4$), 1.24 (2.0 F HClO$_4$), 1.37 (5.0 F HClO$_4$), 1.42 (6.0 F HClO$_4$), 1.49 (8.0 F HClO$_4$). | 61a[a] |
|  | 1.11 (2.0 F H$_2$SO$_4$), 1.15 (4.0 F H$_2$SO$_4$), 1.30 (6.0 F H$_2$SO$_4$), 1.35 (8.0 F H$_2$SO$_4$). | 61a[a] |

(*continued*)

Table 12.2. (Continued)

| Half-Reaction | $E^{\circ\prime}$ Value (Medium) | Ref. |
|---|---|---|
| | 0.93 (0.1 F HCl), 0.97 (0.5 F HCl), 1.05 (2.0 F HCl), 1.08 (3.0 F HCl), 1.10 (4.0 F HCl). | 61a[a] |
| | 1.02 (1 F $H_3PO_4$), 1.07 (2 F $H_3PO_4$), 1.11 (3 F $H_3PO_4$), 1.15 (4 F $H_3PO_4$), 1.18 (5 F $H_3PO_4$), 1.22 (6 F $H_3PO_4$), 1.29 (7.5 F $H_3PO_4$), 1.38 (10.5 F $H_3PO_4$), 1.48 (12 F $H_3PO_4$). | 61a[a] |
| $Cu(II) + e \rightleftharpoons Cu(I)$ | 0.70 (0.015 F $H_2SO_4$, 7.24 F $H_3PO_4$). | 62 |
| In acetonitrile | 1.20. | 63[a] |
| in DMF | 0.71 (0.02 F HCl), 0.63 (0.50 F HCl), 0.62 (1.0 F HCl), 0.60 (2.0 F HCl). | 60 |
| In pyridine | 0.43 (0.05 F HCl), 0.41 (0.5 F HCl). | 60 |
| Dibromamine T (R = p-toluenesulfonyl) $(RNBr_2 + 2H^+ + 4e \rightleftharpoons RNH_2 + 2Br^-)$ Dichloramine T (R = p-toluenesulfonyl) | 1.28 (glacial HOAc). | 64 |
| $(RNCl_2 + 2H^+ + 4e \rightleftharpoons RNH_2 + 2Cl^-)$ (Dichloroiodo)benzene | 1.2 (glacial HOAc). | 65 |
| $(C_6H_5ICl_2 + 2e \rightleftharpoons C_6H_5I + 2Cl^-)$ | 1.2 (glacial HOAc). | 66 |
| $Fe(III) + e \rightleftharpoons Fe(II)$ | 0.52 (6.5 F $H_3PO_4$), 0.52 (8.5 F $H_3PO_4$). | 67 |
| In 0.25 F $H_2SO_4 + H_3PO_4$ | 0.68, 0.59 (0.177 F $H_3PO_4$), 0.53 (0.886 F $H_3PO_4$), 0.47 (3.55 F $H_3PO_4$), 0.40 (10.69 F $H_3PO_4$). | 68 |

| | | |
|---|---|---|
| In 0.5 F H$_2$SO$_4$ + H$_3$PO$_4$. | 0.68, 0.63 (0.177 F H$_3$PO$_4$), 0.56 (0.886 F H$_3$PO$_4$), 0.49 (3.55 F H$_3$PO$_4$), 0.44 (7.98 F H$_3$PO$_4$), 0.41 (10.69 F H$_3$PO$_4$). | 68 |
| In H$_2$SO$_4$ + H$_3$PO$_4$. | 0.43 (0.1 F H$_2$SO$_4$, 8.9 F H$_3$PO$_4$), 0.38 (0.065 F H$_2$SO$_4$, 12.3 F H$_3$PO$_4$), 0.36 (0.04 F H$_2$SO$_4$, 13.8 F H$_3$PO$_4$). | 69 |
| In acetonitrile. | 1.57 vs. 0.01 M AgNO$_3$\|Ag. | 70 |
| In DMF. | 0.41 (0.05 F HCl). | 60 |
| In 0.5 F mannitol. | −0.64 (0.5 F KOH), −0.70 (1.0 F KOH), −0.77 (2.0 F KOH), 0.81 (3.0 F KOH), −0.94 (4.0 F KOH). | 71 |
| In 0.5 F triethanolamine. | −0.77 (0.5 F KOH). | 72[a] |
| In pyridine | 0.69 (0.05 F HCl). | 60 |
| Ge(IV) + 2e ⇌ Ge(II) | 0.15 (5 F HCl), 0.15 (5 F HBr), 0.08 (5 F HI), 0.14 (5 F H$_3$PO$_4$). | 45 |
| Hydrazine sulfate N$_2$ + 4H$^+$ + 4e ⇌ N$_2$H$_4$ | 0.65 (1 F HCl), 0.34 (NaHCO$_3$ soln.), −0.55 (1 F NaOH). | 73 |
| I$_2$ + 2e ⇌ 2I$^-$ | 0.42 (8 F HCl). | 59 |
| In methanol. | 0.62 (0.5 F HCl), 0.61 (3.5 F NaOAc). | 55 |
| In acetonitrile | 0.61 (2 F HOAc). | |
| In ethanol | 0.58 (2 F HOAc). | |
| In 2-propanol | 0.57 (4 F HOAc). | |
| 4:1 2-propanol-methanol | 0.55 (1 F HOAc). | |
| ICN + H$^+$ + 2e ⇌ I$^-$ + HCN[b] | 0.40 (pH 3). | 59 |
| 2ICN + 2H$^+$ + 2e ⇌ I$_2$ + 2HCN | 0.56 (8 F HCl). | |
| 3I$_2$ + 2e ⇌ 2I$_3^-$. In acetonitrile | 0.34 vs. 0.01 M AgNO$_3$\|Ag. | 63 |
| I$_3^-$ + 2e ⇌ 3I$^-$. In acetonitrile | −0.25 vs. 0.01 M AgNO$_3$\|Ag. | |

(*continued*)

Table 12.2. (Continued)

| Half-Reaction | $E°$ Value (Medium) | Ref. |
|---|---|---|
| Isoniazid [Reg. No. 54-85-3] (RCOOH + $N_2$ + $4H^+$ + $4e$ ⇌ RCONHNH$_2$ + $H_2O$) | 0.79 (1 F HCL), 0.78 (0.1 F HCL), 0.25 (0.025 F borax), −0.22 (3 M NaOH). | 73 |
| $MnO_4^-$/Mn(III) in 0.28 F NaF. ($MnO_4^-$ + $6H^+$ + $2HF_2^-$ + $4e$ ⇌ $MnF_4^-$ + $4H_2O$) | 1.58 ($p$H 0), 1.49 ($p$H 0.9), 1.46 ($p$H 1.2), 1.38 ($p$H 2.2), 1.34 ($p$H 2.9), 1.14 ($p$H 4.7), 1.08 ($p$H 5.5), 1.03 ($p$H 6.0). | 74 |
| Mn(III)/Mn(II) in 0.28 F NaF ($MnF_4^-$ + $2H^+$ + $e$ ⇌ $Mn^{2+}$ + $2HF_2^-$) | 1.52 ($p$H 0), 1.42 ($p$H 0.9), 1.38 ($p$H 1.2), 1.27 ($p$H 2.2), 1.20 ($p$H 2.9), 0.96 ($p$H 4.7), 0.87 ($p$H 5.5), 0.81 ($p$H 6.0). | 74 |
| Mn(III)/Mn(II) | 1.32 (3.0 M $H_3PO_4$), 1.31 (4.0 M $H_3PO_4$), 1.31 (5.0 M $H_3PO_4$), 1.31 (9.0 M $H_3PO_4$), 1.31 (12.0 M $H_3PO_4$). | 74a |
| Mo(VI) + $e$ ⇌ Mo(V) in 0.2 F HCl | 0.41, 0.42 (1.16 F $H_3PO_4$), 0.45 (4.64 F $H_3PO_4$), 0.46 (5.8 F $H_3PO_4$), 0.50 (6.96 F $H_3PO_4$), 0.57 (9.28 F $H_3PO_4$), 0.61 (10.44 F $H_3PO_4$), 0.65 (11.6 F $H_3PO_4$). | 38 |
| Mo(VI) + $e$ ⇌ Mo(V) | 0.61 (8.5 F HCl). 0.54 (5 F HCl). | 67 |
| Mo(V) + $2e$ ⇌ Mo(III) | 0.19 (5 F HCl). | |
| $[Mo(CN)_8]^{3-}$ + $e$ ⇌ $[Mo(CN)_8]^{4-}$ | 0.97. | 76[a] |
| Mo(V) + $e$ ⇌ Mo(IV). Species prepared from green Mo$_{aq}$(III); solns. contained 0.3 F HCl. | 0.27 (0.17 F $H_3PO_4$), 0.28 (0.34 F $H_3PO_4$), 0.31 (1.0 F $H_3PO_4$), 0.32 (1.35 F $H_3PO_4$), 0.35 (3.9 F $H_3PO_4$). 0.66 (0.495 F HOAc), 0.70 (0.99 F HOAc), 0.70 (1.98 F HOAc). | 77 |

| | | |
|---|---|---|
| Mo(V) + $e$ ⇌ Mo(IV). Species prepared from orange-red MoCl$_{aq}$(III); solns. contained 0.3 F HCl. | 0.11 (0.5 F H$_3$PO$_4$), 0.13 (1.0 F H$_3$PO$_4$), 0.14 (2.0 F H$_3$PO$_4$), 0.17 (3.9 F H$_3$PO$_4$), 0.21 (8 F H$_3$PO$_4$), 0.25 (12 $M$ H$_3$PO$_4$). 0.20 (0.495 F HOAc), 0.20 (0.99 F HOAc), 0.20 (1.98 F HOAc), 0.21 (3.96 F HOAc), 0.23 (11.87 F HOAc). | 77 |
| Mo(IV) + $e$ ⇌ Mo(III). Species prepared from green Mo$_{aq}$(III); solns. contained 0.3 F HCl. | $-0.04$ (0.17 F H$_3$PO$_4$), $-0.03$ (0.34 F H$_3$PO$_4$), 0.00 (1.0 F H$_3$PO$_4$), 0.01 (1.35 F H$_3$PO$_4$), 0.02 (3.9 F H$_3$PO$_4$). 0.03 (0.495 HOAc), 0.06 (0.99 F HOAc), 0.08 (1.98 HOAc). | 77 |
| Mo(IV) + $e$ ⇌ Mo(III). Species prepared from orange-red MoCl$_{aq}$(III); solns. contained 0.3 F HCl. | 0.08 (0.5 F H$_3$PO$_4$), 0.11 (1.0 F H$_3$PO$_4$), 0.12 (2.0 F H$_3$PO$_4$), 0.17 (3.9 F H$_3$PO$_4$), 0.19 (8 F H$_3$PO$_4$), 0.31 (12 F H$_3$PO$_4$). 0.17 (0.495 HOAc), 0.16 (0.99 F HOAc), 0.16 (1.98 F HOAc), 0.19 (3.96 F HOAc), 0.20 (11.87 F HOAc). | 77 |
| Sb(V) + $2e$ ⇌ Sb(III) | 0.75 (3.5 F HCl), 0.82 (6 F HCl). | 61 |
| In methanol | 0.67 (0.5 F HCl), 0.28 (3.5 F HOAc). | 55 |
| In acetonitrile | 0.64 (2 F HOAc). | |
| In ethanol | 0.62 (2 F HOAc). | |
| In 2-propanol | 0.65 (4 F HOAc). | |
| Sn(IV) + $2e$ ⇌ Sn(II) | 0.42 (8.9 F H$_3$PO$_4$). | 69 |
| In methanol | 0.36 (0.5 F HCl). | 55 |
| In acetonitrile | 0.46 (2 F HOAc). | |
| In ethanol | 0.45 (2 F HOAc). | |
| In 2-propanol | 0.34 (4 F HOAc). | |

(*continued*)

**Table 12.2.** (*Continued*)

| Half-Reaction | $E^{\circ\prime}$ Value (Medium) | Ref. |
|---|---|---|
| $Ti(IV) + e \rightleftharpoons Ti(III)$ | $-0.09$ (1 F HCl), 0.24 (6 F HCl), $-0.05$ (1 F $H_3PO_4$), $-0.15$ (5 F $H_3PO_4$), $-0.24$ (0.1 F KSCN), $-0.01$ (0.2 F $H_2SO_4$), 0.12 (2 F $H_2SO_4$), 0.20 (4 F $H_2SO_4$). | 61 |
| In DMF | 0.00 (0.05 F HCl). | 60 |
| In pyridine | $-0.19$ (0.2 F HCl). | |
| $U(VI) + 2e \rightleftharpoons U(IV)$ in 0.25 F $H_2SO_4$ | 0.49, 0.52 (0.177 F $H_3PO_4$), 0.52 (0.866 F $H_3PO_4$), 0.56 (3.55 F $H_3PO_4$), 0.59 (10.69 F $H_3PO_4$). | 68 |
| $UO_2^{2+} + 4H^+ + 2e \rightleftharpoons U^{4+} + 2H_2O$ In 0.50 F $H_2SO_4$. | 0.55, 0.56 (0.177 F $H_3PO_4$), 0.56 (0.866 F $H_3PO_4$), 0.57 (3.55 F $H_3PO_4$), 0.60 (10.69 F $H_3PO_4$). | 68 |
| $V(V) + e \rightleftharpoons V(IV)$ $(VO_2^+ + 2H^+ + e \rightleftharpoons VO^{2+} + 2H_2O)$ | 0.89 (0.25 F $H_3PO_4$), 0.92 (0.50 F $H_3PO_4$), 0.96 (1.0 F $H_3PO_4$), 1.01 (2.5 F $H_3PO_4$), 1.08 (5.0 F $H_3PO_4$), 1.13 (7.5 F $H_3PO_4$), 1.19 (10.0 F $H_3PO_4$), 1.21 (12.0 F $H_3PO_4$). 1.19 (6.5 F $H_3PO_4$). | 78 |
| $V(IV) + e \rightleftharpoons V(III)$ $(VO^{2+} + 2H^+ + e \rightleftharpoons V^{3+} + H_2O)$ | 0.36 (0.5 F $H_2SO_4$), 0.39 (1.0 F $H_3PO_4$), 0.42 (2.0 F $H_3PO_4$), 0.55 (6.0 F $H_3PO_4$), 0.60 (8.0 F $H_3PO_4$), 0.66 (10.0 F $H_3PO_4$), 0.70 (12.0 F $H_3PO_4$). 0.74 (6.5 F $H_3PO_4$). | 67 79 67 |
| In pyridine | 0.20 (0.2 F HCl). | 60 |

[a] Value quoted.
[b] $X^+ + 2e \rightleftharpoons X^-$ (X = Br or I).

## Medium Effects

The data selected for inclusion in Table 12.2 serve to emphasize the effect of medium upon the value of the formal reduction potential for a given oxidation–reduction system. An increase in the value of $E^{\circ\prime}$ for a system as the concentration of a particular medium is increased indicates that the strength of the oxidant species is increased. Conversely, a diminution of the formal reduction potential with increasing medium concentration is indicative that the strength of the reductant species is increased. For example, in a medium of phosphoric acid, potassium dichromate becomes a stronger oxidant as the acid concentration is increased, whereas iron(II) becomes a stronger reductant. Indeed, in some cases the effect of changing the medium is to completely reverse the direction of spontaneity for a reaction between an oxidant and a reductant. Thus in 0.5 $M$ sulfuric acid, iron(III) oxidizes uranium(IV) to uranium(VI)

$$U(IV) + 2Fe(III) \rightarrow 2Fe(II) + U(VI)$$

whereas in concentrated phosphoric acid the reverse reaction occurs. In cases like these either the oxidized or reduced form of the couple is bound into a strong complex by the medium. This complexation effect is also utilized in many other ways: for example, the use of fluoride ion or pyrophosphate ion in titrations with potassium permanganate, and the use of complexants such as glycerol, mannitol, and triethanolamine in alkaline solution when iron(II) is used as the titrant. The result is to greatly enhance the versatility of these well-established titrimetric reagents. A disadvantage is that the stability of a standard solution of a reductant such as iron(II) is very much diminished in the complexing medium as is evident from the difference in the formal reduction potentials for iron(III)/iron(II) in 0.5 F sulfuric acid ($E^{\circ\prime} = 0.68$ V) and in 0.5 F mannitol containing 0.5 F potassium hydroxide ($E^{\circ\prime} = -0.64$ V). In cases like this the titrant is prepared in an acid solution, and sufficient complexant and alkali are added to the titrand solution to compensate for the effect of the acidity of the titrant when it is added to the medium. The titrand solution must be purged of dissolved oxygen before the titrant is added, and it must also be protected by a blanket of inert gas while the titration is in progress.

The types of complexing media used advantageously for oxidation–reduction titrations may be conveniently subdivided according to the $p$H range of the medium as follows:

1. *pH < 0. These are strong acid complexants of which hydrochloric and phosphoric acids are examples.*
2. *pH 3–5. Weakly acid media such as the dilute aqueous solutions of sodium fluoride and hydrofluoric acid.*

3. pH 6–7. *A buffered solution containing the complexant ion; dihydrogen pyrophosphate* $H_2P_4O_7^{2-}$ *is an example of this type.*
4. pH > 9. *These complexing media range from aminoacids which form buffer solutions in the pH range 9–10.5 to polyols such as TEA [triethanolamine, i.e., nitrilotris(2-ethanol)], 1,2,3-propanetriol (glycerol), or mannitol prepared in solutions of strong alkali (NaOH or KOH) having pH values > 12.*

These subdivisions have been used in Table 12.3 in which a survey of some oxidation–reduction titrations performed in the various categories of complexing media is presented. The basic instructions for these titrations are given together with the interferents and the literature references. The use of complexants prepared in organic solvents as the media for the titration of inorganic determinands does not appear to offer any advantage over the aqueous-based methods cited in Table 12.3, and have not, therefore, warranted inclusion in the tables. Only a few organic compounds have been included, since the determination of organic oxidants and reductants in nonaqueous media are considered separately in Section 12.4.

### Catalysts Used in Oxidation–Reduction Titrations

Catalysts are added to accelerate the rate of a reaction, and in some cases they have a further beneficial effect: influencing the course of a reaction so that it follows a single stoichiometric pathway. In some instances where two catalysts can promote the same reaction, one catalyst may be more selective than the other in that it will not catalyze a similar but undesirable side reaction. The principal catalysts used, arranged in alphabetical order of element, are given together with a brief account of the application(s) of each.

**Bismuth(III).** In a medium of nitrilotris(2-ethanol) (0.1–0.15 $M$) and sodium hydroxide (0.15–0.2 $M$) small amounts (0.05–0.2 mg) of bismuth(III) catalyze the reduction of lead(II) to the metal by an excess of a standard solution of iron(II). This excess can be titrated with standard dichromate solution (116).

**Boron Trifluoride.** A dilute solution in glacial acetic acid has been used to catalyze the quantitative esterification of alcoholic −OH groups with acetic acid (117):

$$ROH + CH_3COOH \xrightarrow{BF_3} CH_3COOR + H_2O$$

| Serial | Complexing Medium | Determinand | Titrant | Procedure | Remarks | Ref. |
|---|---|---|---|---|---|---|
| | | | | A. $pH < 0$ | | |
| 1 | 10–11 $M$ HCl | $BrO^-$ | Fe(II) | To the sample was added sufficient HCl to maintain the overall concn. at 10–11 $M$. The soln. was titrd. with std. Fe(II) at 5° with vigorous stirring using a Pt indicator electrode versus SCE. | Air need not be excluded. $BrO^- + 2Fe^{2+} + 10Cl^- + 2H^+ \rightarrow Br^- + 2[FeCl_5]^{2-} + H_2O$ | 80 |
| 2 | 4–6 $M$ HCl in 1:1 $H_2O$–$Me_2CO$ | Cu(II) | Ferrocene[a] in EtOH | To an aliquot of the Cu(II) soln. (0.3–0.5 mg Cu) prepared in concd. HCl was added concd. HCl (15 mL) and acetone (20–30 mL). This soln. was titrd. with 0.01 $M$ ferrocene in EtOH using a Pt indicator electrode versus SCE. | Mo(VI) and Sb(V) interfere. Ni(II), Mn(II), or a fifty-fold XS Fe(III) do not interfere. $(C_5H_5)_2Fe + Cu^{2+} \rightarrow Cu^+ + (C_5H_5)_2Fe^+$ | 81 |
| 3 | $> 8\ M$ HCl | Fe(II) | $K_3Fe(CN)_6$ | Fe(II) (2–100 mg) was determined by direct titrn. Can be used for det. of Fe in its ores, Fe(III) being det. by prior reduction with $SnCl_2$. | V(IV) interferes. Fe(III), Al(III), As(III), Sn(IV), Zn(II), Cd(II), Pb(II), Mn(II), Mg(II) do not interfere. | 82 |
| 4 | $> 8\ M$ HCl | Mo(VI) | $(NH_4)_3MoCl_6$[b] | 8–10 $M$ HCl (70 mL) was added to Mo(VI) soln. and the $O_2$ removed by passage of $CO_2$ through the soln. for 20–25 min. The soln. was titrd. at 50° with Mo(III) using a Pt disk electrode versus SCE. | V(V), U(VI), Cu(II), and Fe(III) interfere. W, Ni, Co, Cd, Al, Bi, and Mg do not interfere. $2MoO_4^{2-} + Mo^{3+} + 16H^+ \rightarrow 3Mo^{5+} + 8H_2O$. | 83 |
| 5 | 8 $M$ HCl | Tl(I) | $Pb(OAc)_4$ in HOAc | Tl(I) (8–102 mg) in 8 $M$ HCl (50 mL) was titrd. with 0.1 N $Pb(OAc)_4$ in glacial HOAc using a Pt electrode versus SCE. | Microdeterminations are also described. $Pb(OAc)_4 + 4H^+ + Tl^+ \rightarrow Tl^{3+} + Pb^{2+} + 4HOAc$. See also Reference 85 for Mo(III) $\rightarrow$ Mo(V). | 84 |

(*continued*)

**Table 12.3.** (*Continued*)

| Serial | Complexing Medium | Determinand | Titrant | Procedure | Remarks | Ref. |
|---|---|---|---|---|---|---|
| 6 | 10–12 $M$ $H_3PO_4$ | Ascorbic acid | $[Fe(CN)_6]^{3-}$ | Ascorbic acid solns. were titrd. directly with hexacyanoferrate(III) using a Pt electrode versus SCE. $C_6H_8O_6 + 2[Fe(CN)_6]^{3-} \rightarrow C_6H_6O_6 + 2H^+ + 2[Fe(CN)_6]^{2-}$. | Ten-twentyfold XS of sucrose, dextrose, citric acid, tartaric acid, malic acid, and oxalic acid did not interfere. Can also be determined using Ce(IV) as titrant. | 86, 58 |
| 7 | 9–13 $M$ $H_3PO_4$ | $BrO_3^-$ | $(NH_4)_2Fe(SO_4)_2$ | Sample of bromate (0.3–0.45 mg) was dissolved in $H_2O$ (10 mL), 85% $H_3PO_4$ (40 mL) added and soln. titrd. with 0.01 or 0.1 N Fe(II). Two inflections observed corresponding to: $2BrO_3^- + 10 Fe^{2+} + 12H^+ \rightarrow Br_2 + 10Fe^{3+} + 6H_2O$ and $Br_2 + 2Fe^{2+} \rightarrow 2Br^- + 2Fe^{3+}$. | $SO_4^{2-}$, $PO_4^{3-}$, or $Br^-$ in great XS do not interfere, and neither does $Cl^-$ up to the concn. ratio 1:1. $NO_3^-$ interferes when the second break is used for evaluation; $I^-$, $ClO_3^-$, $IO_3^-$, and $IO_4^-$ are also reduced. | 87 |
| 8 | 9–13 $M$ $H_3PO_4$ | $N$-Bromo-succinimide | $(NH_4)_2Fe(SO_4)_2$ | $N$-bromosuccinimide (1.5–90 mg) was dissolved in $H_2O$ (10 mL), concd. $H_3PO_4$ (35 mL) was added followed by the addn. of $OsO_4$ soln. (0.2–0.4 mL, 0.01 $M$). The titrn. of this soln. with std. Fe(II) resulted in 2 inflections corresponding to: $2C_4H_4O_2N\text{-}Br + 2Fe^{2+} + 2H^+ \rightarrow 2C_4H_4O_2N\text{-}H + 2Fe^{3+} + Br_2$ and $Br_2 + 2Fe^{2+} \rightarrow 2Br^- + 2Fe^{3+}$. | As for Serial 7. | 87 |

| | | | | | | |
|---|---|---|---|---|---|---|
| 9 | 12–13 $M$ $H_3PO_4$ | Ce(III) | $K_2Cr_2O_7$ | Ce(III) soln. (30–120 mg in 2–10 mL) containing 90% $H_3PO_4$ (40–50 mL) titrd. with std. $K_2Cr_2O_7$, the potentials being monitored with a Pt electrode versus SCE through two salt bridges: one of satd. $NaClO_4$ and the other of satd. $NaNO_3$. $6Ce^{3+} + Cr_2O_7^{2-} + 14H^+ \rightarrow 6Ce^{4+} + 2Cr^{3+} + 7H_2O$ | $Cl^-$ and $NO_3^-$ interfere. $F^-$ interferes by precipitation Fe(II), U(IV), Mo(V), As(III), Sb(III), Mn(II), and lower valences of W also react with Cr(VI). Co(II), Ni, Fe(III), W(VI), U(VI), Cr(III), Ce(IV), Cu(II), Zn, Th(IV), and V(V) do not interfere. Fe(II), V(IV), and Ce(III) can be titrd. sequentially in one aliquot. | 58a |
| 10 | 7 $M$ $H_3PO_4$ | Cu(II) | $(NH_4)_2Fe(SO_4)_2$ | Sufficient syrupy $H_3PO_4$ was taken to ensure a concn. of 7 $M$ in a final vol. of 30 mL. KSCN (0.5 mL, 10%) was added and $CO_2$ passed through the soln. for 10 min after which the soln. of Cu(II) (6–25 mg) was added. The soln. was titrd. with std. Fe(II). | Sb(V) and $NO_2^-$ interfere. Ni, Cd, Ba, Co(II), Mn(II), Zn, Hg(I), Pb(II), Al, As(III), Fe(III), $Cl^-$, $SO_4^{2-}$, $OAc^-$, $B_4O_7^{2-}$, oxalate, tartrate, citrate, and $NO_3^-$ did not interfere. Cu(II) + Fe(II) $\rightarrow$ Cu(I) + Fe(III). | 62 |
| 11 | 8 $M$ $H_3PO_4$ | 1,4-benzo-quinone[c] | Fe(II) | The soln. of determinand in 8 $M$ $H_3PO_4$ was made $O_2$-free by passage of $CO_2$ and then titrd. with std. Fe(II). | Fe(II) can also be titrd. with std. 1,4-benzoquinone soln. $2Fe^{2+} + C_6H_4O_2 + 2H^+ \rightarrow C_6H_4(OH)_2 + Fe^{3+}$ | 88 |
| 12 | 5 $M$ $H_3PO_4$ | Ge(IV) | $KIO_3/KI$[d] | Ge(IV) (20 mL, ~ 0.05 $M$), water (80 mL), $H_3PO_4$ (70 mL, 85%), and $NaH_2PO_2 \cdot H_2O$ (~ 7 g) heated at the boil for 5–10 min under $CO_2$. After cooling quickly to ~ 20° (ice-bath), the soln. of Ge(II) was titrd. with std. 0.05 N $KIO_3/KI$ soln. | Sn(IV) interferes. Det. of Ge in sulfides of Ge and Cd, and in oxides of Zn, Ni, Ge, and Fe are described. $[GePO_4]^- + I_2 + 3H_2O \rightarrow [Ge(OH)_3PO_4]^{2-} + 3H^+ + 2I^-$ | 45 |

(*continued*)

**Table 12.3.** (*Continued*)

| Serial | Complexing Medium | Determinand | Titrant | Procedure | Remarks | Ref. |
|---|---|---|---|---|---|---|
| 13 | 3–6 $M$ $H_3PO_4$ | $IO_4^-$ | $(NH_4)_2Fe(SO_4)_2$ | A sample of $IO_4^-$ (0.09–70 mg) was dissolved in $H_2O$ (30 mL). 85% $H_3PO_4$ (20 mL), 0.01 $M$ $OsO_4$ (0.2 mL) was added and the soln. titrd. with 0.001 or 0.1 $M$ Fe(II) using a Pt electrode versus SCE. | $Br^-$ and $I^-$ interfere. $SO_4^{2-}$, $PO_4^-$, $B_4O_7^{2-}$, and $NO_3^-$ in great XS do not interfere. $F^-$ to fourteenfold and $Cl^-$ to twentyfivefold XS do not interfere. $2IO_4^- + 14Fe^{2+} + 16H^+ \rightarrow I_2 + 14Fe^{3+} + 8H_2O$. | 89 |
| 14 | 3–6 $M$ $H_3PO_4$ | $IO_3^-$ | $(NH_4)_2Fe(SO_4)_2$ | Proceed as in Serial 13. $2IO_3^- + 10Fe^{2+} + 12H^+ \rightarrow I_2 + 10Fe^{3+} + 6H_2O$. | $Br^-$ and $I^-$ interfere. $SO_4^{2-}$, $PO_4^{3-}$, $B_4O_7^{2-}$ in great XS do not interfere. $Cl^-$ to threefold, $F^-$ to thirtyfold and $NO_3^-$ to seventyfold XS do not interfere. Method recommended for standardization of Fe(II) solns. | 90 |
| 15 | 10–11 $M$ $H_3PO_4$ | Fe(II) | $[Fe(CN)_6]^{3-}$ | Sufficient syrupy $H_3PO_4$ to yield a concn. of 10 $M$ in 40 mL was deoxygenated with $CO_2$ for 10 min after which Fe(II) soln. was added. The soln. was titrd. with std. $[Fe(CN)_6]^{3-}$. | $[Fe(CN)_6]^{3-}$ can also be similarly titrd. with std. Fe(II). As(III) and Mn(II) interfere. Ba, Na, Hg(II), K, $SO_4^{2-}$, glucose, and sucrose do not interfere. Method applied to det. of iron in ores. | 91 |
| 16 | 10–11 $M$ $H_3PO_4$ | Fe(II) | $KBrO_3^e$ | Syrupy $H_3PO_4$ (30 mL) was purged of $O_2$ by passage of $CO_2$ for 10 min. A suitable aliquot of Fe(II) was added to the soln. which was then titrd. with std. $KBrO_3$ soln. under $CO_2$ using a Pt indicator electrode versus SCE. | Suitable for application to det. iron in pharmaceutical products since org. subs. such as oxalate, tartrate, citrate, sucrose, and glucose do not interfere. $6Fe^{2+} + BrO_3^- + 6H^+ \rightarrow 6Fe^{3+} + Br^- + 3H_2O$. | 69 |
| 17 | 8–9 $M$ $H_3PO_4$ | Fe(II) | $NaVO_3^f$ | $O_2$ was removed from 10 $M$ $H_3PO_4$ by passage of $CO_2$ for 10 min after which the Fe(II) soln. was added. Soln. was titrd. with std. V(V) or Ce(IV). | Oxalate, tartrate, citrate, methanol, As(III), U(IV), Hg(I), and $Cl^-$ do not interfere. (Some of these interfere in $H_2SO_4$ media.) $VO_3^- + 2Fe^{2+} + 6H^+ \rightarrow 2Fe^{3+} + V^{3+} + 3H_2O$ | 92 |

| | | | | | |
|---|---|---|---|---|---|
| 18 | 12 $M$ H$_3$PO$_4$ | Mn(II) | K$_2$Cr$_2$O$_7$ | Proceed as in Serial 9. Fe(II) and Mn(II) can be titrd. sequentially on the same aliquot. It is essential to exclude O$_2$ during titrn. of Fe(II). | Cl$^-$ and NO$_3^-$ interfere. $6Mn^{2+}$ + Cr$_2$O$_7^{2-}$ + 14H$^+$ → 6Mn$^{3+}$ + 2Cr$^{3+}$ + 7H$_2$O. | 74a |
| 19 | 12–13 $M$ H$_3$PO$_4$ | Mn(III) and Ag(I) in an alloy | Fe(II) | The alloy containing Mn and Ag was dissolved in concd. H$_3$PO$_4$–HNO$_3$ and the soln. heated to remove HNO$_3$. After cooling, concd. H$_3$PO$_4$ was added to the soln. which was then titrd. with std. Fe(II) using Pt–W electrodes. | The two inflections in the titrn. curve are due to Mn(III) → Mn(II) and Ag(I) → Ag. | 93, 94 |
| 20 | 13 $M$ H$_3$PO$_4$ | Mo(VI) | Fe(II)$^g$ | Mo(VI) soln. (2–4 mL, ~ 0.1 $M$) and concd. H$_3$PO$_4$ (30–60 mL) sufficient to give a concn. of 13 $M$ was titrd. with 0.5 $M$ Fe(II) soln. using a Pt indicator electrode versus SCE. | W(VI) accelerates the attainment of stable potentials and does not interfere. Cr(VI) and V(V) interfere. Mo(VI) + Fe(II) → Mo(V) + Fe(III). | 38, 67 |
| 21 | 12 $M$ H$_3$PO$_4$ | Mo(VI) | K$_2$Cr$_2$O$_7$ | Mo(VI) soln. (2–8 mL, ~ 0.1 $M$) in syrupy H$_3$PO$_4$ (40–50 mL) was treated with an XS of Fe(II) until $E_{cell}$ was in the range 150–180 mV. (Cell was: Hg;Hg$_2$Cl$_2$;satd. KCl\|satd. NaClO$_4$\|Mo(V), Fe(II), Fe(II)\|Pt.) On titrn. slowly with std. Cr(VI) the first inflection corresponded to oxidation of XS Fe(II) and the second to Mo(V) → Mo(VI). Volume of titrant between first and second inflection corresponds to volume required for oxidation of Mo(V). | Cl$^-$ and NO$_3^-$ interfere. O$_2$ does not interfere and neither do Mn(II), Mn(VII), Cu(II), Ni, Zn, Co(II), Fe(III), Cr(VI), Ce(III), Ce(IV), Ti(IV), Th(IV), or W(VI). U(VI) behaves like Mo and is titrd. simultaneously. V(V) → V(III) and is then oxidized simultaneously with Mo(V) forming V(IV). An extra inflection to V(V) is observed as more Cr(VI) is added allowing the simultaneous det. of Mo and V in the same aliquot. | 95 |

(*continued*)

Table 12.3. (Continued)

| Serial | Complexing Medium | Determinand | Titrant | Procedure | Remarks | Ref. |
|---|---|---|---|---|---|---|
| 22 | 11–13.5 $M$ $H_3PO_4$ | U(VI) | Fe(II) | Uranyl salt soln. (4–8 mL, ~ 0.025 $M$) in concd. $H_3PO_4$ (65 mL) titrd. under $CO_2$ with std. Fe(II) soln. using a Pt electrode versus SCE. $UO_2^{2+} + 2Fe^{2+} + 4H^+ \rightarrow U^{4+} + 2Fe^{3+} + 2H_2O$. | Fe(III), W(VI) do not interfere and W(VI) is beneficial since it aids the attainment of the stable potentials. V(V), Mo(VI), and Cr(VI) interfere. | 68 |
| 23 | 12–13 $M$ $H_3PO_4$ | U metal | Fe(II) indirect | The metal was dissolved in 85% $H_3PO_4$ by heating. Air was passed through the cooled soln. to ensure U was present as U(III). A weighed quantity of pure $K_2Cr_2O_7$ was added in XS, and this XS titrd. with std. Fe(III) added from a weight burette. | Interferences from Fe, Cr, V, and Mo at comparatively high levels were eliminated. | 96 |
| 24 | 12 $M$ $H_3PO_4$ | U(VI) | $K_2Cr_2O_7$ | Similar to Serial 21 except that U(VI) soln. was ~ 0.05 $M$, and second inflection corresponded to U(IV) → U(VI). | $Cl^-$ and $NO_3^-$ interfere. Mo(VI), V(IV), V(V), Mn(II), and (VII), Ce(III), or (IV) interfere. Mn and Ce interferences can be eliminated by diluting titrand so that concn. $H_3PO_4$ is 8–9 $M$ before titrn. with Cr(VI). | 97, 98 |
| 25 | 8–9 $M$ $H_3PO_4$ | U(VI) | $K_2Cr_2O_7$ | U(VI) was reduced to U(IV) in 8–9 $M$ $H_3PO_4$ by the addition of XS Fe(II) in the presence of $HNO_3$ and $NH_2SO_3H$. The XS Fe(II) was oxidized with $HNO_3$ using Mo(VI) as catalyst. After dilution with water to yield a medium ~ 2.5 $M$ in $H_3PO_4$, the soln. was titrd. with std. $K_2Cr_2O_7$ soln. Davies-Gray method. | A systematic study of interferents given. Method more specific than those used previously. $3U^{4+} + Cr_2O_7^{2-} + 2H^+ \rightarrow 3UO_2^{2+} + 2Cr^{3+} + H_2O$ | 99, 100 |

| | | | | |
|---|---|---|---|---|
| 26 | 12–13 $M$ $H_3PO_4$ | V(IV) | A soln. of V(IV) (3–110 mL, ~0.05 $M$), syrupy $H_3PO_4$ (40–65 mL) was titrd. with 0.1 $M$ Fe(II) using a Pt indicator electrode versus SCE. Note that when V(V) is titrd. 2 distinct endpoints are observed: the first corresponds to titrn. of ~99% V(V) and the second to the stoichiometric equivalence point (±0.5%). Stabilization of potential at first break requires about 10 min. | Pt electrode becomes inert after several titrn. and can be reactivated by heating in a flame. Cr(VI) and V(V) can be titrd. in the same soln. Fe(III), W(VI), Mn(II), Co(II) do not interfere. | 79, 67, 78 |
| 27 | 12 $M$ $H_3PO_4$ | V(IV) | Proceed as in Serial 9. An alternative is to reduce V(IV) with an XS of std. Fe(II). The resulting titrn. gives these inflections: (a) Fe(II) → Fe(III); (b) V(III) → V(IV); (c) V(IV) → V(V). When procedure of Serial 9 is used, $\Delta E/\Delta V$ for inflection V(III) → V(IV) is greater than inflection V(IV) → V(V). | The sequential titrns. of V(IV) and Mn(II) are possible. Applications for det. of V in alloys are given. $Cl^-$ and $NO_3^-$ interfere. Fe(II), U(IV), Mo(V), W(V), Ce(III), As(III), and Mn(II) are also oxidized. Co(II), Ce(IV), Ni, Fe(III), W(VI), Mo(VI), U(IV), and Cr(III) do not interfere. | 101 |

*B. pH 3–5*

| | | | | |
|---|---|---|---|---|
| 28 | 0.35–0.4 $M$ NaF ~ 0.06 $M$ $H_2SO_4$ | Ce(III) Sample | A std. soln. of $KMnO_4$ (1.5–6 mg Mn) was mixed with 1 $M$ $H_2SO_4$ (~4 mL) and 2% NaF (50 mL). The mixture was diluted to 100 mL, heated to ~40° and titrd. with Ce containing sample soln. [21–54 mg Ce(III)]. | A procedure for oxidizing Ce(III) with XS std. $KMnO_4$ followed by titrn. of the XS with Ti(I) is also described. $4Ce^{3+} + MnO_4^- + 2H_2F_2 + 4H^+ \rightarrow MnF_4^- + 4Ce^{4+} + 4H_2O$ | 102 |

(*continued*)

Table 12.3. (Continued)

| Serial | Complexing Medium | Determinand | Titrant | Procedure | Remarks | Ref. |
|---|---|---|---|---|---|---|
| 29 | 0.35–0.4 $M$ NaF ~ 0.06 $M$ $H_2SO_4$ | Hydrazine | $KMnO_4$ | To a soln. of $N_2H_4$ (0.9–3.5 mg) was added 0.5 $M$ $H_2SO_4$ (5 mL), 2% NaF (50 mL) and 0.25 $M$ $CuSO_4$ (5 mL). The mixture was diluted to 100 mL with $H_2O$ and titrd. with std. $KMnO_4$ using a Pt indicator electrode versus SCE. | The second inflection corresponds to the reaction: $N_2H_4 + MnO_4^- + 4F^- + 4H^+ \to N_2 + MnF_4^- + 4H_2O$. Reaction catalyzed by Cu(II). Similar determinations: V(IV) (104) and $HSO_3^-$ (105). | 103 |
| 30 | 0.35–0.4 $M$ NaF ~ 0.06 $M$ $H_2SO_4$ | Mn(II) | $KMnO_4$ | An aliquot (10 mL) containing Mn(II) (5–56 mg) was mixed with 0.5 $M$ $H_2SO_4$ (8.0 mL) and 2% NaF (50–75 mL). The mixture was diluted to 100 mL with $H_2O$ and titrd. with std. $KMnO_4$ using a Pt indicator electrode versus SCE. | $MnO_4^- + 10HF_2^- + 4Mn^{2+} \to 5MnF_4^- + 2H^+ + 4H_2O$. | |

C. pH 6–7

| Serial | Complexing Medium | Determinand | Titrant | Procedure | Remarks | Ref. |
|---|---|---|---|---|---|---|
| 31 | Sodium pyrophosphate | Mn(II) | $KMnO_4$ | The soln. of Mn(II) was added to satd. $Na_4P_2O_7$ soln. (200–400 mL) and the $p$H adj. to 6–7 with either 1:1 $H_2SO_4$ or 5 $M$ NaOH. The Mn(II) was titrd. with std. $KMnO_4$ using a Pt indicator electrode versus SCE. | The influence of diverse ions upon this Lingane–Karplus method, used for determination of Mn in metallurgical products and steels, has been investigated subsequently (107, 108). $4Mn^{2+} + MnO_4^- + 8H^+ 15H_2P_2O_7^{2-} \to 5Mn(H_2P_2O_7)_3^{3-} + 4H_2O$. | 106 |

D. pH > 9

| | | | | | |
|---|---|---|---|---|---|
| 32 | 0.5 M TEA[h] 0.5 M NaOH | Bi(III) | Fe(II) | The sample soln. was added to the TEA and then NaOH was added to yield the required medium. After removing $O_2$ by passage of $N_2$, the soln. was titrd. with Fe(II) at 40–50° using a Pt indicator electrode versus SCE. Bi(III) → Bi. | Fivefold XS of Ni, Cd, Fe(III), Sb(III), Sn(IV), and Pb(II) does not interfere. | 72 |
| 33 | 0.5–2.5 M Glycerol[i] 0.1–2 M KOH | Bi(III), Cr(VI), Cu(II), or V(V) | Fe(II) | The deoxygenated soln. ($N_2$ used) of the respective element was titrd. in the alkaline medium with a std. soln. of Fe(II) under $N_2$ using a Pt indicator electrode versus SCE. [Bi(III) → Bi, Cu(II) → Cu(I) → Cu, V(V) → V(IV), Cr(VI) → Cr(III)]. | For < 5 mg Cu(II) use the second inflection [Cu(II) → Cu(I)], and for larger amounts use the first inflection. The titrn. of V(V) is best performed at 50–60°. Al, Ni(II), Cd, Zn, Fe(III), and Cr(III) in fivefold XS do not interfere and neither do $Cl^-$, $SO_4^{2-}$, or $NO_3^-$. | 109 |
| 34 | 0.04 M $Na_4P_2O_7$ 0.04 M KBr. pH > 9. | $BrO^-$ | Fe(II) | To the soln. of hypobromite (50–70 mL, 1–100 mg) was added $Na_4P_2O_7$ (1–4 g) and KBr (0.5 g). The soln. was titrd. in an inert atmosphere with 0.1 N $FeSO_4$. | $BrO^- + 4P_2O_7^{4-} + 2Fe^{2+} + H_2O \rightarrow 2Fe(P_2O_7)_2^{5-} + Br^- + 2OH^-$. | 110 |
| 35 | 0.05–0.12 M TEA[h], 0.04–0.07 M NaOH | Co(II) | $K_3Fe(CN)_6$ | Co(II) (1.5–60 mg) in the $O_2$-free medium (70–80 mL) was titrd. under $N_2$ with 0.06 M $K_3Fe(CN)_6$ using a Pt indicator electrode versus SCE. Co(II) → Co(III). | Fe(III) does not interfere when present in a fivefold XS. | 111 |
| 36 | 0.2 M Glutamic acid adj. to pH 9.8–10.4 | Co(II) | $K_3Fe(CN)_6$ | Co(II) soln. (1.5–50 mg) in the medium (50 mL) was titrd. with 0.01 or 0.05 M $K_3Fe(CN)_6$. | Method applied to the det. of Co in a steel and an ore. Fivefold XS of Mn(II) does not interfere. | 112 |

(continued)

Table 12.3. (Continued)

| Serial | Complexing Medium | Determinand | Titrant | Procedure | Remarks | Ref. |
|---|---|---|---|---|---|---|
| 37 | 0.3–0.6 $M$ TEA[h], 0.06–0.13 $M$ NaOH | Co(II) + Mn(II) mixtures | $K_3Fe(CN)_6$ | Co(II) (3–9 mg), and Mn(II) (8–25 mg) in the $O_2$-free medium (50–60 mL) was titrd. under $N_2$ with 0.06 $M$ $K_3Fe(CN)_6$ using a Pt indicator electrode versus SCE. First inflection corresponds to Mn(II) → Mn(III) and the second to Co(II) → Co(III). | Fe(III) does not interfere in amounts up to Mn:Fe 1:15 and Co:Fe 1:5. | 111 |
| 38 | 0.25 $M$ TEA[h] 1 $M$ NaOH | Cr(VI) or Cu(II) | Fe(II) | As for serial 32. Cr(VI) → Cr(III), Cu(II) → Cu. | V(V) does not interfere in det. of Cr(VI). Zn(80:1), Ni(5:1), Fe(III)(4:1), Cd(10:1), As(III) (10:1), and Pb(II) do not interfere in titrn. of Cu(II). | 72 |
| 39 | 0.25–0.5 $M$ Mannitol 1 $M$ KOH | Mn(II) | $K_3Fe(CN)_6$ | A soln. of Mn(II) (6.66 mL, 0.09–2.3 × $10^{-2}$ $M$), mannitol (10 mL, 1 $M$) and KOH (3.33 m$\overline{L}$, 6 $M$) was deaerated with $N_2$ for 10 min. and then titrd. with $[Fe(CN)_6]^{3-}$ using a Pt indicator electrode versus SCE. Two inflections corresponding to Mn(II) → Mn(III) and Mn(III) → Mn(IV) were observed. | Method applied to the det. of Mn in slate and in limestone. Cl⁻, $NO^{-3}$, $SO_4^{2-}$, $PO_4^{3-}$, F⁻, Mg, Ca, Sr, Ba, Tl, As(III), Sb(IV), Al, Cd, Pb, Zn, Cu, and Ge(IV) did not interfere. Ni and Cr(III) in concn. in XS of 1:1 to Mn interfere and so does Co(II). The addition of EDTA masks to a certain extent the interference of Co(II). Only 1 inflection, Mn(II) → Mn(IV), occurs when EDTA is added. | 113 |

| | | | | | |
|---|---|---|---|---|---|
| 40 | 0.5 M Mannitol 1–2 M KOH | Mn(II) | Fe(II) | Mn(II) in 0.5 M mannitol and 2–3 M KOH was preoxidized to Mn(III) with air or $O_2$. $PbO_2$ (1–2 g) was added and the oxidation Mn(III) → Mn(IV) was carried out with vigorous stirring for 15–20 min. The soln. was filtered through a G4 sinter, and an aliquot was titrd., after deaeration, with 0.05 M $FeSO_4$ in an inert atmosphere using a Pt indicator electrode versus SCE. Mn(IV) → Mn(III). | An alternative method using $[Fe(CN)_3]^{3-}$ as the initial oxidant is also described. The medium can be used also for the reductions Cr(VI) → Cr(III), Co(III) → Co(II), and Cu(II) → Cu(I). | 71 |
| 41 | 0.5 M TEA[h] 1 M NaOH | Mn(III) or Te(VI) | Fe(II) | As for Serial 32. Mn(III) → Mn(II), $TeO_4^{2-}$ → $TeO_3^{2-}$ | Ni(1:1), Fe(III)(4:1), Cd(20:1), Zn(50:1), Al(200:1), and small amounts of Cr(III)(1:10) do not interfere in titrn. of Mn(III). Fivefold XS of Ni, Cd, Fe(III), Sb(III), Sn(IV), and Pb(II) do not interfere in titrn. of Te(VI). | 72 |
| 42 | 0.5 M TEA[h] 2 M $NH_4OH$ | Hg(II) | $(NH_4)_2 Fe(SO_4)_2$ | The sample soln. was added to the TEA soln. and then $NH_4OH$ was added to yield the required medium. After removing $O_2$ by passage of $N_2$, the soln. was titrd. with Fe(II) using a Pt indicator electrode versus SCE. | 5 mg Hg(II) in 50 mL can be determined with a relative error of 1%. 2Fe(II) + Hg(II) → Hg + 2Fe(III). | (72) |

*(continued)*

Table 12.3. (Continued)

| Serial | Complexing Medium | Determinand | Titrant | Procedure | Remarks | Ref. |
|---|---|---|---|---|---|---|
| 43 | 0.5 M Mannitol, 0.5 M KOH | Pt(IV) | (NH$_4$)$_2$Fe(SO$_4$)$_2$ | To an O$_2$-free soln. (N$_2$ for 5 min) of 1 M mannitol (30 mL) and 5 M KOH (5 mL) was added an aliquot of Pt(IV) soln. (Na$_2$PtCl$_6$). After again passing N$_2$ for 5 min the soln. was titrd. under N$_2$ with 0.1 N Fe(II) using a Pt foil indicator electrode versus SCE. | 0.39 mg Pt(IV) in a volume of 50 mL can be determined with an error $\geq$ 2%. Ir, Pd, and Rh did not interfere. Pt(IV) + 2Fe(II) $\rightarrow$ Pt(II) + 2Fe(III). |  |
| 44 | 0.5 M TEA$^h$ 0.5 M NaOH | Ag(I) | (NH$_4$)$_2$Fe(SO$_4$)$_2$ | The sample soln. and the TEA soln. and the NaOH was added to yield the required medium. After removing O$_2$ by passage of N$_2$, the soln. was titrd. with Fe(II) using a Pt indicator electrode. Ag(I) $\rightarrow$ Ag. | 0.005 mg Ag(I) in 50 mL was determined with a 3% relative error. Cl$^-$ does not interfere and free Ag(I) may be determined in the presence of [Ag(CN)$_2$]$^-$, [Ag(SCN)$_2$]$^-$, AgBr, or AgI. S$_2$O$_3^{2-}$ interferes. | 72 |
| 45 | 0.36 M 1,2-Benzenediol, 1 M NH$_4$Cl, 1 M NH$_3$ in 0.1 M Na$_2$SO$_3$ | U(VI) or V(IV) | Fe(II) | The respective element was titrd. with Fe(II) using a Hg pool electrode versus Ag\|AgCl reference electrode. | U(VI) $\rightarrow$ U(IV), V(IV) $\rightarrow$ V(III). | 115 |

$^a$Ferrocene, (C$_5$H$_5$)$_2$Fe, bis(cyclopentadienyl)iron.
$^b$See Reference 75.
$^c$2,5-Cyclohexadiene-1,4-dione.
$^d$IO$_3^-$ + 5I$^-$ + 6H$^+$ $\rightarrow$ 3I$_2$ + 3H$_2$O.
$^e$Freed of dissolved O$_2$.
$^f$Cerium(IV) sulfate can also be used.
$^g$At least five times more concentrated than Mo(VI) solution.
$^h$TEA is triethanolamine, nitrilotris(2-ethanol).

which permits the indirect determination of ROH by titrating the water formed using the Karl Fischer procedure. Other hydrolytic organic reactions can also be catalyzed by boron trifluoride, the water produced or consumed in the reaction being related to the original concentration of determinand. The catalyst is often prepared as a 10% solution in glacial acetic acid to which is added 0.1–0.2% (v/v) water.

**Chromium(III) Sulfate.** A small amount (10–20 mg) of this solid has been found to catalyze the oxidation of tellurous acid with cerium(IV) (118), and a drop of 1% aqueous solution of chromium(III) sulfate was sufficient to catalyze the oxidation of 1,2-ethanediol and 1,2,3-propanetriol by cerium(IV) sulfate to carbon dioxide and water (119).

**Copper(II) Salts.** Titanium(III) can be oxidized selectively in the presence of iron(II) by adding a trace of copper(II) sulfate (3 mL, $10^{-4}$ $M$) and passing air through the acid solution (10% $H_2SO_4$) (120). Copper(II) (5 mL, $10^{-3}$ $M$ $CuSO_4$) accelerates the reaction between chromium(VI) and iodide (121), and similarly copper(II) acetate (1 mL, 1%) promotes the oxidation of thiosulfate to tetrathionate with vanadium(V) in 2–8 $M$ acetic acid (122). More recently, the copper(II)-catalyzed reaction (2 mL, 0.4 $M$ $CuSO_4$) between standard potassium dichromate solution and thiosulfate in an acetate buffer of $p$H 4.5 has been recommended for the standardization of thiosulfate solutions (123).

In the presence of ammonium arsenate and phosphoric acid, copper(II) chloride can be used to catalyze the reaction between iron(II) and bromate (124). Copper(II) at a concentration of 0.05 $M$ also acts as a catalyst for the oxidation of hydrazine by chromium(VI) in a medium of 5–6 $M$ hydrochloric acid (125). The reaction is

$$2Cr_2O_7^{2-} + 3N_2H_5^+ + 13H^+ \rightarrow 4Cr^{3+} + 3N_2 + 14H_2O \quad (12.5)$$

The oxidation of hydrazine to nitrogen gas by potassium permanganate in a medium containing fluoride is also catalyzed by copper(II) (see Serial 29, Table 12.3). Similarly, the oxidation of isonicotinic acid hydrazide (2′-amino-4-pyridinecarboxamide) by chromium(VI) in hydrochloric acid medium (10–37%) is promoted by the addition of copper(II) (~ 70 mg $CuSO_4 \cdot 5H_2O$ for 1–50 mg of $Cr_2O_7^{2-}$) (73).

The standardization of vanadium(II) solutions by titration into a standard solution of iron(III) sulfate is also catalyzed by copper(II), about 1 mL of 0.1 $M$ copper(II) sulfate being required for every 5 mL 0.1 $M$ iron(III) solution (126).

**Halogen Compounds.** Potassium bromide catalyzes the oxidation of thioethers to the corresponding sulfoxide with lead tetracetate in a medium of 7:3 acetic acid–water (127), and also catalyzes the oxidations of hydrazine to nitrogen, tin(II) to tin(IV), thallium(I) to thallium(III), antimony(III) to antimony(V), and cinnamic acid to dibromocinnamic acid with dichloramine T (128) in a medium of aqueous acetic acid (50–70%). Potassium bromide was also found to be effective in the catalysis of the oxidations of phenol, aniline, and 8-quinolinol with (dichloroiodo)benzene $PhICl_2$ in a medium of acetic acid. However, for the titration of thallium(I) with this reagent, iodine monochloride (5 mL, 0.1 $N$ in 5 $N$ HCl) was added to the sample as catalyst (66).

Iodate or iodine monochloride can be used to catalyze the reduction of potassium permanganate by arsenic(III) $AsO_2^-$ (129). The catalysts are used as 0.0025 $M$ solutions, the iodine monochloride solution being prepared by titrating a potassium iodide solution in 4 $M$ hydrochloric acid with potassium iodate until on shaking with a little carbon tetrachloride no iodine remains in the organic layer. A portion of the aqueous layer is then standardized with sodium thiosulfate solution in the presence of water and more potassium iodide, and another portion is then diluted to the required strength with 4 $M$ hydrochloric acid.

Iodide or iodate can also be used to catalyze the oxidation of arsenic(III) by cerium(IV) even in the presence of hexacyanoferrate(III), a species that renders the catalytic action of osmium(VIII) oxide ineffective (130). This reaction has been investigated more recently, and a solution of catalyst was prepared by mixing 0.1 $M$ potassium iodide (20.8 mL) and 0.1 $M$ potassium iodate (4.16 mL) and diluting the solution to 250 mL (54). The titration was performed in 0.5–1.5 $M$ sulfuric acid and 0.5–1.0 mL of the catalyst solution was added. The catalyst was also effective in a medium of 30% acetic acid. The oxidations of thioureas with cerium(IV), as bis(ethylenediamine)cerium(IV) sulfate, in 7 $M$ sulfuric acid were also catalyzed by iodate. However, when $N$-chloroacetamide was used as the oxidant for the titration of thioureas in a medium of 1–3.5 $N$ sulfuric acid, addition of 0.4–0.5 mL 2% potassium iodide was found satisfactory to catalyze some reactions of the type (131)

$$2RNHC(:NH)SH + CH_3CONHCl \rightarrow$$
$$RHNC(:NH)-S-S-C(:NH)NHR + CH_3CONH_2 + HCl \quad (12.6)$$

A similar reaction occurred when hexacyanoferrate(III) was used as titrant in the presence of potassium iodide (132). Potassium iodide also catalyzes the oxidation of iron(II) in a medium of 2 $N$ hydrochloric acid and 2 $N$ sulfuric acid (133).

Iodine monochloride catalyzes the oxidation of thiocyanic acid (HSCN) in a medium of 1.5–2 $M$ hydrochloric acid by potassium permanganate (134), and has been used as a catalyst in the standardization of cerium(IV) salts against sodium oxalate in a medium of dilute hydrochloric acid. In 0.1–5.0 $N$ sulfuric acid solutions it was found necessary to add iodine monochloride to catalyze the reaction between sulfite and chromium(VI), although no catalyst was necessary when 1–3 $M$ acetic acid or 1–3 $M$ hydrochloric acid was used as the medium (135).

**Manganese(II) Salts.** Manganese(II) sulfate catalyzes the oxidation of oxalic acid (ex. sodium oxalate) by cerium(IV) sulfate in a medium of dilute sulfuric acid [*cf.* ICl (136)], and also catalyzes the reaction between oxalic acid and potassium permanganate at room temperature in dilute sulfuric acid (137). When combined with potassium iodide as a solution containing manganese(II) sulfate (0.005 g) and potassium iodide (0.021 g) in 250 mL, a portion of this solution (5 mL) was found to be effective in catalyzing the titration of chromate in the presence of vanadate with arsenic(III) solution (138). Manganese(II) chloride catalyzes the reduction of bromate in acid solution by chromium(III) [Cr(III) → Cr(VI); (139)], and by hydrogen peroxide in hydrochloric acid solution ($\sim 1\ M$). The latter reaction was reported to yield good results in the direct titration of hydrogen peroxide with standard potassium bromate solution (140). Similarly, manganese(II) sulfate was found to catalyze the reaction between potassium permanganate and hydrogen peroxide in a medium of sulfuric acid (141). Manganese(II) also catalyzes the reduction of molybdenum(VI) with vanadium(II) sulfate (142).

**Molybdenum(VI).** This species, often prepared as a 1% aqueous solution of ammonium molybdate tetrahydrate, has been used to increase the rate of oxidation of hypophosphorous acid by potassium permanganate in dilute sulfuric acid solutions (143), and to catalyze the reaction (144)

$$3Fe^{2+} + NO_3^- + 4H^+ \rightarrow 3Fe^{3+} + NO + 2H_2O \qquad (12.7)$$

This reaction was applied to selectively oxidize an excess of iron(II) remaining after the reduction of uranium(VI) to uranium(IV) by iron(II) in a medium of concentrated phosphoric acid solution containing nitric and sulfamic acids (see serial 25, Table 12.3). Molybdenum(VI) also catalyzes the reduction of chlorate with iron(II) in aqueous hydrochloric acid solution (145), and also the reaction between bromate and iodide before titration with sodium thiosulfate solution (146).

**Osmium(VIII) Oxide.** Small amounts of osmium tetroxide (3 drops 0.01 $M$ $OsO_4$ in 0.1 $N$ $H_2SO_4$, water, or 0.3 $M$ NaOH) have been used predominantly to catalyze titrimetric reactions using reductants such as arsenic(III) (ex. $As_2O_3$), iron(II), hydrazine sulfate, and its derivatives. Its catalytic action is effective for a wide variety of media ranging from 5 $M$ sulfuric acid to 10 $M$ sodium hydroxide, and it is thus convenient to subdivide this summary accordingly into reactions performed in acidic solution and into reactions performed in alkaline solution.

### Acidic Media

Traces of osmium tetroxide catalyze the reaction of arsenious acid (ex. $As_2O_3$) with potassium permanganate in sulfuric acid solution (147), and this reaction has formed the basis for the determination of manganese in ores or alloys (148). A similar method is used for the standardization of cerium(IV) with arsenic(III) (147). Traces of hexacyanoferrate(III), however, render the catalyst ineffective, and in the titration of a mixture of hexacyanoferrate(III) and arsenite with cerium(IV) an iodide or iodate catalyst should be used. The reduction of chlorate ion with arsenite in a medium of 2 $M$ sulfuric acid is also catalyzed by osmium tetroxide, about 1 h at room temperature being required for the reaction (149). However, the reaction between arsenious acid and cerium(IV) sulfate induces the reduction of chlorate by arsenious acid so that an excess of arsenious acid can be titrated immediately with cerium(IV) sulfate at room temperature in the absence of air. It has been reported more recently (150) that chlorate may be determined directly with standard arsenic(III) solution in 2–4 N sulfuric acid in the presence of osmium(VIII). It is suggested that arsenic(III) reduces osmium tetroxide to a compound of indeterminate composition, the reduced form of the catalyst then reducing chlorate to hypochlorous acid, the catalyst being oxidized back to osmium(VIII). The hypochlorous acid, then, reacts with arsenite. The reduction of chlorate by iron(II) is also accelerated by the presence of a small amount of osmium tetroxide (few drops 0.1%), and this permits the determination of chlorate by reaction with excess iron(II) and the titration of this excess with potassium permanganate or potassium bromate (151).

In a medium of 1–5 $N$ sulfuric acid, bromate (152), iodate, and periodate are also reduced by arsenite in the presence of osmium tetroxide (153). Periodate (89) and iodate (90) may also be determined by titration with a standard solution of iron(II) in a medium of 3–6 $M$ phosphoric acid using 0.01 osmium tetroxide (0.2 mL) as catalyst. The reaction with periodate is

$$2IO_4^- + 14Fe^{2+} + 16H^+ \xrightarrow[H_3PO_4]{OsO_4} I_2 + 14Fe^{3+} + 8H_2O \quad (12.8)$$

A medium of 1 $M$ sulfuric acid was used for the titrimetric determinations of vanadium(V) and chromium(VI) with a standard solution of hydrazine sulfate using osmium tetroxide as catalyst (154). The stoichiometry of the vanadium(V) reaction corresponds to

$$4VO_3^- + N_2H_5^+ + 11H^+ \rightarrow 4VO^{2+} + 8H_2O + N_2 \qquad (12.9)$$

An analogous reaction occurs when isonicotinic acid hydrazide is titrated into a known volume of standard vanadium(V) solution in a medium of 0.25–4.0 $M$ sulfuric acid containing phosphoric acid (1 mL, 85%) and a small amount of osmium tetroxide (155). The catalyst has also been used for the reduction of perchlorate to chloride with vanadium(III) in a medium of 7.5 $M$ sulfuric acid. The vanadium(IV) so produced was determined spectrophotometrically and related to the original concentration of perchlorate (156). Alternatively, the chloride formed may be titrated.

At a $p$H in the range 2.5–3.0, osmium tetroxide also markedly catalyzes the reaction

$$N_2H_5^+ + 4Fe^{3+} \rightarrow N_2 + 4Fe^{2+} + 5H^+ \qquad (12.10)$$

which formed the basis for the determination of iron(III) in an alloy (5). Within the $p$H range 1–3 osmium tetroxide was also effective in catalyzing the oxidation of 2-mercaptoacetic acid (thioglycolic acid) with iron(III) (157).

### Osmium(VIII) Oxide in Alkaline Media

Chlorite ($ClO_2^-$) is reduced quantitatively by arsenite in the presence of osmium tetroxide in a bicarbonate medium (158). Chlorite is also reduced in a bicarbonate medium (0.5–1 $M$ $NaHCO_3$) in the presence of potassium iodide and osmium tetroxide with a standard solution of arsenic (III) (ex. $As_2O_3$), the overall reaction being

$$AsO_2^- + 2I^- + ClO_2^- \rightarrow AsO_4^{3-} + I_2 + Cl^- \qquad (12.11)$$

Chlorite does not oxidize iodide in this medium and iodine is formed only after the addition of arsenite titrant (159). The possible reaction sequence is

$$AsO_2^- + OsO_4 + 4OH^- \rightarrow AsO_4^{3-} + OsO_4^{2-} + 2H_2O$$

$$OsO_4^{2-} + ClO_2^- + H_2O \rightarrow OsO_4 + ClO^- + 2OH^-$$

$$2I^- + ClO^- + H_2O \rightarrow I_2 + Cl^- + 2OH^-$$

The same medium (0.2–0.5 $M$ NaHCO$_3$) was used to determine chlorine dioxide and the chlorite ion sequentially (160). Chlorine dioxide in the presence of potassium iodide can be titrated directly with arsenic(III), but on addition of osmium tetroxide both chlorine dioxide and chlorite are titrated.

A saturated solution of disodium hydrogen phosphate having a $p$H in the range 8–10 provided the medium for the titration of iodate with arsenite in the presence of potassium iodide and osmium tetroxide. For this application the catalyst solution must be accurately dispensed (0.50 mL, 0.1%, 161). The reaction between chloramine T and arsenic(III) also occurs in this weakly alkaline medium in the presence of osmium tetroxide, but it is quantitative also in $\leq 6$ $M$ sodium hydroxide (162). The overall reaction is

$$AsO_2^- + RSO_2N^--Cl + 2OH^- \rightarrow AsO_4^{3-} + RSO_2NH_2 + Cl^- \quad (12.12)$$

where R = $p$-tolyl.

Control of $p$H enabled the sequential determinations of chlorine and chlorine dioxide in a single sample (163). A measured amount of 0.1 $N$ sodium arsenite in 0.2–0.5 $M$ alkali was titrated with the sample solution to reduce chlorine to chloride and chlorine dioxide to chlorite. Then sodium bicarbonate was added, and the chlorite formed was titrated with sodium arsenite in the presence of osmium tetroxide catalyst. A similar sequence was adopted in the sequential titration of hyprobromite, bromite, and bromate (152). Hypobromite was titrated with arsenic(III) in 0.2–0.5 $M$ alkali without the catalyst, bromite was then titrated in the presence of catalyst, and, finally, the solution was made acid (0.5–1.0 $N$ H$_2$SO$_4$) and the bromate titrated. Hypochlorites, chlorites, and chlorates may be determined similarly (164). An indirect method for the determination of formaldehyde or benzaldehyde has been developed which involves the use of an excess of a standard solution of hypochlorite. The reaction between the hypochlorite and the aldhyde is carried out in 0.5–1 $M$ sodium hydroxide, and the excess of standard hypochlorite is then titrated with standard arsenic(III) in the presence of osmium tetroxide catalyst (165). An even higher concentration of sodium hydroxide was used for the osmium(VIII)-catalyzed oxidation of chromium(III) with hexacyanoferrate(III) (166). In a medium initially of approximately 5 $M$ sodium hydroxide, which decreases to a final concentration of about 3 $M$ after the titration, this reaction can be represented as

$$Cr^{3+} + 3[Fe(CN)_6]^{3-} + 8OH^- \rightarrow CrO_4^{2-} + 3[Fe(CN)_6]^{4-} + 4H_2O$$

$$(12.13)$$

Osmium tetroxide has also been used in 10 $M$ solutions of strong alkalis (NaOH, KOH) to determine the concentration of reducing impurities in the alkali samples by titration with standard hexacyanoferrate(III) (167). For these applications the catalyst was prepared as a 0.1% solution of osmium tetroxide in 0.3 $M$ sodium hydroxide.

**Oxalic Acid.** This acid catalyzes the oxidation of iron(II), uranium(IV) (168), 1,4-benzenediol (169), and indigo [Reg. No. *482-89-3*] (170) with vanadium(V) titrant. An accurate method for the the determination of titanium(III) in a medium of approximately 1 $M$ hydrochloric acid using vanadium(IV) as titrant also utilized oxalic acid as catalyst (171). When oxalic acid was present at a concentration of 0.06–0.1 $M$, the method gave results accurate to ±0.2%.

**Phosphoric Acid.** Catalyzes the oxidation of vanadium(IV) and iron(II) with permangante (172).

**Potassium Ruthenate ($K_2RuO_4$).** This catalyst has been used as an alkaline solution prepared by the oxidation of metallic ruthenium with potassium peroxydisulfate in potassium hydroxide solution (173). It is described as the universal catalyst for the reduction of periodate with arsenic(III) in acid media ($\geq$ 0.1 $N$ $H_2SO_4$, $HNO_3$, $H_3PO_4$, $HClO_3$, or $CH_3COOH$), in weakly alkaline media ($NaHCO_3$, $Na_2B_4O_7$, or $Na_2HPO_4$) or in strongly alkaline media ($\geq$ 0.1 $N$ NaOH or KOH). It also catalyzes the reaction of periodate with various reductants [formaldehyde, hypophosphate, Cu(II), etc.] and of arsenite with different oxidants [hexacyanoferrate(III), chlorite, Mn(III), Ce(IV), etc.]. Although reaction (12.13) is catalyzed by osmium tetroxide, in the presence of formates osmium tetroxide also promotes the oxidation of formate to carbon dioxide and water by hexacyanoferrate(III). By contrast, potassium ruthenate (0.002 $M$ in 0.5–1 $M$ KOH) catalyzes only reaction (12.13) in the presence of formate thus permitting the direct titration of chromium(III) in formate solutions with standard hexacyanoferrate(III) in a medium of 4–6 $M$ sodium hydroxide (174).

Ruthenium(IV) sulfate has been used in strong acid medium ($\sim$ 50% $H_2SO_4$) to catalyze the reaction between cerium(IV) and hydrazine sulfate. This reaction has been suggested as being suitable for the standardization of cerium(IV) solutions (175), the overall reaction being

$$N_2H_5^+ + 4Ce^{4+} \rightarrow N_2 + 4Ce^{3+} + 5H^+ \tag{12.14}$$

The addition of 3–5 $\mu$mol of ruthenium(IV) sulfate or perchlorate suffices when the hydrazine sulfate titrant concentration is in the range 0.0025–0.025 $M$.

**Silver Salts.** These catalyze oxidations with peroxydisulfate in acid solution (see Table 12.1), including the oxidation of plutonium(IV) to plutonium(VI) (24). Silver sulfate (5 mL, 2%) catalyzes the oxidation of hypophosphite and phosphite by excess cerium(IV) sulfate in 1 $M$ sulfuric acid at the boil (176), and in more concentrated sulfuric acid solutions (17-24 $M$) also catalyzes the oxidations of hypophosphite and phosphite with vandium(V) (177). The oxidation of sodium alginate [Reg. No. *9005-38-3*] by excess cerium(IV) is also catalyzed by silver sulfate (178).

When a catalyst is used it is advisable to ascertain the volume of titrant required to reach the endpoint potential when the titrand medium containing the catalyst and any other reagents is titrated in the absence of determinand. If this volume is significant with respect to the titers obtained in the presence of determinand, it must be deducted to yield the actual titration volume required to react with the determinand.

## 12.2 SOME OXIDANTS USED AS TITRANTS

In this section the oxidants used as titrants are listed together with a brief description, where appropriate, of some of the principal applications of each. They are arranged in alphabetical order of the oxidant species rather than in order of compound name. Oxidants containing chlorine, bromine, or iodine atoms as part of a reducible group are to be found under the heading of *halogen-containing oxidants*.

### *Cerium(IV)*

Cerium(IV) sulfate has been in regular use as a titrant for more than 50 years (179). Its preparation and applications have been fully described (180) and practical directions for its use are included also in some undergraduate texts (see, e.g., Reference 181). The standard reduction potential for the couple Ce(IV)/Ce(III) is 1.44 V and, as is shown in Table 12.2, the strength of the oxidant cerium(IV) diminishes in solutions of phosphoric acid due to complexation, but increases in solutions of perchloric acid (182). Cerium(IV) sulfate must be used in relatively strong acid media because precipitation of the hydrated oxides of cerium(IV) occurs at relatively low $p$H.

Aqueous solutions (0.1 $M$) can be prepared by warming a mixture of cerium(IV) sulfate tetrahydrate (~ 41 g) or ammonium cerium(IV) sulfate dihydrate (64-66 g) and 1:1 sulfuric acid (56 mL) adding water with stirring until the solid dissolves. When the solution is cool it is diluted to 1 L with water. The solution is usually standardized against pure arsenic(III) oxide in the presence of osmium(VIII) oxide catalyst (3 drops, 0.01 $M$) in

1-2 $M$ sulfuric acid. An alternative method (54) used a potassium iodate-iodide catalyst. The reaction is

$$2Ce^{4+} + HAsO_2 + 2H_2O \rightarrow 2Ce^{3+} + H_3AsO_4 + 2H^+ \quad (12.15)$$

Standardized 0.1 $M$ solutions of cerium(IV) are available commercially.

Nonaqueous cerium(IV) solutions (0.05 $M$) in acetonitrile, for example, can be prepared by dissolving ammonium hexanitratocerate(IV) [$(NH_4)_2Ce(NO_3)_6$, $M = 548.23$] in the solvent and standardizing the solution against standard iron(II) solution in the presence of dilute aqueous sulfuric acid (183). Titrations with cerium(IV) in the mixed solvent tributylphosphate:carbon tetrachloride, 1:4 have also been performed using solutions prepared from ammonium hexanitratocerate(IV) (184). Additions of glacial acetic acid to the titrand aided the attainment of stable potentials.

A comparative study of the characteristics of cerium(IV) and chromium(VI) as oxidants for the precise determination of uranium revealed that in aqueous sulfuric acid solutions they were equally suitable (185) as titrants. Among many other recent applications, cerium(IV) sulfate has been used as the oxidant for the microdetermination of plutonium (186) and has been used in very dilute solution ($2.5 \times 10^{-4} - 10^{-3} M$) for the determination of small quantities of iron ($> 5 \times 10^{-5} M$ $Fe^{2+}$) in glass (187).

### *Chromium(VI)*

Potassium dichromate is the most widely used form of this oxidant in acid solution and was first proposed as a titrant in 1850 (188). The salt is available as a standard reference material, the reduction reaction in acid solution being

$$Cr_2O_7^{2-} + 14H^+ + 6e \rightleftharpoons 2Cr^{3+} + 7H_2O$$

for which the standard reduction potential, $E°$ is 1.36 V. The formal potentials are less than this in complexing media (see Table 12.2) but become greater than 1.36 V in fairly concentrated solutions of perchloric acid. The properties and use of potassium dichromate have been described (188) and reviewed more recently (61a). Some applications of this titrant in complexing media are given in Table 12.3, and its use for the standardization of sodium thiosulfate (123) in the presence of copper(II) catalyst was mentioned earlier. The reaction is

$$6S_2O_3^{2-} + Cr_2O_7^{2-} + 14H^+ \rightarrow 3S_4O_6^{2-} + 2Cr^{3+} + 7H_2O \quad (12.16)$$

### Cobalt(III)

This species is a very strong oxidant, the formal potential of the couple cobalt(III)/cobalt(II) in 1 F nitric acid being 1.84 V. It is, therefore, sufficiently powerful to oxidize the very weak reductant cerium(III) in concentrated sulfuric acid solution. However, because cobalt(III) reacts with water liberating oxygen, it is extremely difficult to store solutions of cobalt(III), although these solutions are stable in concentrated sulfuric acid solutions (189). The effect of complexing cobalt(III) as its acetate, prepared in glacial acetic acid by anodic oxidation (190), is to substantially reduce the potential of the cobalt(III)/cobalt(II) system although these solutions are comparatively stable. The use of this solution as a titrant has been reviewed (191). It has been known for some time (192), however, that hexamminecobalt(III) tricarbonatocobaltate(III), $[Co(NH_3)_6][Co(CO_3)_3]$, is an oxidant that combines to a certain degree the advantages of complexed and non-complexed cobalt(III) compounds. The substance, solid or dissolved in a saturated bicarbonate solution, is very stable, but when added to an acid solution, tervalent cobalt in the ionic form is released from the carbonate anion. The complex salt is prepared by mixing the hexamminecobalt(III) cation as its chloride or nitrate, and the tricarbonatatocobalt(III) anion in a saturated solution of sodium bicarbonate. The complete method of preparation is given in Reference 192.

The optimum $pH$ for the stability of hexamminecobalt(III)tricarbonatocobaltate(III) solutions is $pH$ 7.6. Although the hexamminecobalt(III) cation is stable in acid solutions, the tricarbonatocobaltate(III) anion decomposes at $pH < 7$ liberating cobalt(III) ion. At $pH > 8$, however, a brown precipitate $[Co(OH)_3]$ separates from the solution. The titrant solution ($\sim 0.005\ M$) is prepared by adding hexamminecobalt(III)tricarbonatocobaltate(III) (3 g) to water (1 L) that has previously been saturated with sodium bicarbonate. After the solution has been stirred for 2–3 h it is filtered through a sintered glass crucible and the resulting green filtrate standardized against iron(II) solution. The oxidant solution appears to be more stable when stored in the light, and under this condition the slope of the line obtained by plotting concentration (normality) versus time (days) is about $-1.22 \times 10^{-5}$ N day$^{-1}$ over a period from the twentieth day after preparation to the one hundred twentieth day. The oxidant has been used for the titration of cerium(III) in a medium of $\sim 0.5\ M$ sulfuric acid and for the determination of vanadium(IV) in 5 $M$ sulfuric acid (192). Its use has been reinvestigated more recently (193) and applied to the determination of iodide in a medium of 4–5 hydrochloric acid. Chloride (five hundred-fold excess) and tenfold excess of potassium nitrate did not interfere in the determination of 8.3 $\mu$g potassium iodide.

## Copper(II)

In aqueous solution an excess of copper(II) chloride dihydrate (0.5–1 g) has been used to quantitatively oxidize hypophosphite and phosphite in 6 $M$ hydrochloric acid according to the equations

$$H_3PO_2 + 2HCuCl_3 + H_2O \xrightarrow{20-25°} H_3PO_3 + 2HCuCl_2 + 2HCl$$

(12.17)

$$H_3PO_3 + 2HCuCl_3 + H_2O \xrightarrow{boil} H_3PO_4 + 2HCuCl_2 + 2HCl$$

(12.18)

The copper (I) so formed was titrated with a standard solution of cerium(IV), chromium(VI), or vanadium(V). These reactions formed the basis for the determination of hypophosphite and phosphite either separately, or sequentially in mixtures of the two (194).

In nonaqueous solutions copper(II) has been used as the titrant for the direct determinations of thiols in dimethylformamide (60, 195), iodides (63, 196), thiourea alone (63) or in complexes (197), arylamines (198), and potassium ethyl xanthate (196), all in acetonitrile. Detailed instructions for the preparation of anhydrous copper(II) perchlorate solution in acetonitrile are given in Reference 63, and where exclusion of water is not essential the copper(II) perchlorate solution can be prepared directly from the hydrated salt, the white solid which precipitates being removed by filtration [see also (199)]. The copper(II) concentration may be determined in aqueous solution iodometrically (199), by titration with EDTA (197), or in acetonitrile against pure thiourea (63).

### Halogen-Containing Oxidants

**Bromine.** Standard solutions of this oxidant are generally used for bromine substitution titrations under conditions that exclude the application of the commonly used procedure of adding an excess of bromate–bromide reagent to an acidified aqueous solution of the sample followed by the addition of iodide and titration of the liberated iodine with standard thiosulfate solution. For example, amines and phenols of poor aqueous solubility can be titrated in nonaqueous solutions directly with bromine in the presence of a base such as pyridine to accept the protons released in the substitution. A medium of acetic acid–pyridine in the ratio 9:1 has been used for the titration of some phenols directly with 0.15 $M$ bromine in glacial acetic acid (200), and direct titrations of phenols and aromatic amines have also been performed in propylene carbonate (4-methyl-1,3-dioxolan-2-one; 201). The bromine titrant, prepared as a 0.5 $M$ solution in propylene carbonate, was standardized against tetraethylammonium bromide (also dissolved in propylene carbonate) yielding a potential break at a 1 : 1 mole ratio corresponding to the formation of tribromide ion

$$Br^- + Br_2 \rightarrow Br_3^-$$

**Iodine.** Dissolving 12.7 g of iodine in potassium iodide solution (20 g in 40 mL water) and then diluting to 1 L with water yields a 0.1 $N$ solution of this oxidant, the use of which is well established (202). This solution is often standardized against pure arsenic(III) oxide in a solution of bicarbonate, and this medium proved suitable also for the determination of 2-naphthol by titration with 0.1 $N$ iodine solution (203). For the titration of tripotassium oxohydroxotetracyanomolybdate(IV), however, a medium of 0.5 $M$ potassium hydroxide was required (204). The use of iodine as the oxidant for the sequential determination of tin(II), arsenic(III), and antimony(III) in various nonaqueous media has also been described (55).

**Bromine(I) or Iodine(I).** Cations are liberated when bromine cyanide, iodine cyanide, or iodine monochloride, respectively, are dissolved in dilute acid solution or donor nonaqueous solvents. The formal reduction potentials for these cations are

$$I^+ + 2e \rightleftharpoons I^-; \qquad E^{o\prime} = 0.40 \, (pH \, 3)$$

$$Br^+ + 2e \rightleftharpoons Br^-; \qquad E^{o\prime} = 0.51 \, (pH \, 3)$$

Solutions of bromine cyanide and iodine cyanide can be prepared directly from the solids dissolved in water. Their solutions in 0.1 $M$ hydrochloric acid, 0.1 $M$ sulfuric acid, 0.05 $M$ nitric acid, glacial acetic acid, ethanol, and acetonitrile are stable, and their rates of decomposition do not exceed 1% in any of these media over a period of 2 months (59).

It is usually necessary to titrate the common reductants such as thiosulfate, sulfite, sulfide, tin(II) (205), and vanadium(II) with bromine cyanide in dilute hydrochloric acid solutions (0.1–0.2 $M$) containing an excess of potassium iodide (5 mL, 10%), the oxidation taking place through the formation of iodine

$$Br^+ + 2I^- \rightarrow Br^- + I_2 \qquad (12.19)$$

For example, the overall reaction with thiosulfate is

$$2S_2O_3^{2-} + BrCN + H^+ \xrightarrow[{[H^+] > 0.01 M}]{KI} S_4O_6^{2-} + HCN + Br^- \quad (12.20)$$

By contrast, it is not necessary to add potassium iodide when iodine cyanide is the titrant, and iodide can also be dispensed with when alkyl xanthates are oxidized to the corresponding disulfides with bromine cyanide.

Bromine cyanide is a useful oxidant in nonaqueous media such as acetonitrile, glacial acetic acid, 1 : 1 acetic acid–acetic anhydride, methanol, and ethanol (206), and in these media the addition of potassium iodide is unnecessary. Both bromine cyanide and iodine cyanide have also been used in various media for the titrations of phenols that undergo substitution the extent of which depends upon the medium involved (207). Further applications in aqueous solution are given in Reference 208.

When used in aqueous solution, iodine monochloride is more of a reagent than a titrant. Its use is classically associated with oxidations using standard solutions of potassium iodate in a medium of 5–6 $M$ hydrochloric acid (209). In these titrations iodine may be oxidized to the iodine(I) cation in preference to the oxidation of the determinand leading to false endpoints. However, when an excess of iodine monochloride is added to the solution of reductant prior to titration with iodate an equivalent amount of iodine is released

$$2I^+ + 2e \rightarrow I_2 \quad (12.21)$$

This is then oxidized by potassium iodate back to iodine monochloride in the hydrochloric acid solution. The reaction between octacyanomolybdate(IV) and iodine monochloride, for example, is (210)

$$2K_4[Mo(CN)_8] + 2ICl \rightarrow 2K_3[Mo(CN)_8] + I_2 + 2KCl \quad (12.22)$$

The solution is then made 6 $M$ with respect to hydrochloric acid and then titrated with standard potassium iodate

$$2I_2 + KIO_3 + 6HCl \rightarrow KCl + 5ICl + 3H_2O \quad (12.23)$$

The usual method of preparation of iodine monochloride is to dissolve potassium iodide (5.00 g) and potassium iodate (3.22 g) in water (37.5 mL). To this solution is added concentrated hydrochloric acid (37.5 mL) and carbon tetrachloride (5 mL). After shaking the mixture vigorously, the color of the organic layer should be faint pink. If the pink color is pronounced, add a little potassium iodate to convert some of the iodine to iodine monochloride, and if there is no iodine coloration, add a little potassium iodide.

A medium of glacial acetic acid containing anhydrous sodium acetate in the concentration range 0.8–1.6 $M$ has been used for the titration of sulfite, arsenic(III) chloride, antimony(III) chloride, or iron(II) perchlorate with iodine monochloride in glacial acetic acid. The titrant was prepared from the recrystallized solid (211).

The newer oxidants dibromamine T [$NN$-dibromo-4-toluenesulfonamide, 4-$CH_3$-$C_6H_4SO_2N(Cl)_2$] and (diacetoxy)iodobenzene [$PhI(OCOCH_3)_2$] may also be regarded as sources of the bromine(I) and iodine(I) cations, respectively. They too are used in glacial acetic acid, the reduction of these compounds corresponding to the reactions

$$RNBr_2 + 2H^+ + 4e \rightarrow RNH_2 + 2Br^-, \quad (R = CH_3-C_6H_4-SO_2)$$

(12.24)

and

$$PhI(OCOCH_3)_2 + 2e \rightarrow C_6H_5I + 2CH_3COO^- \quad (12.25)$$

Dibromamine T is prepared by the bromination of chloramine T (64), and (diacetoxy)iodobenzene can be prepared by the method described in Reference 212. Titrant solutions are prepared as ~ 0.05 $M$ solutions in anhydrous acetic acid and stored in amber-colored bottles. These solutions can be standardized by adding an aliquot to an aqueous solution of potassium iodide (20 mL, 10%) and, after dilution to 100 mL with water, titrating the liberated iodine with standard thiosulfate solution [compare reactions (12.19) and (12.20)]. Solutions of (diacetoxy)iodobenzene in acetic acid are stable for at least 30 days (213), but dibromamine T is less stable, the normality of a stock solution changing from 0.2388 to 0.2305 $N$ over a period of 15 days. Both can be used for the titration of a number of reductants such as arsenic(III), iodide, iron(II), hexcyanoferrate(III), thallium(I), hydrazine, ascorbic acid, 1,4-benzenediol, phenol, and aniline. Details are given in References 64 and 213.

**Chlorine(I) Cation.** The use of this species as a titrant is traditionally associated with hypochlorite, a strong oxidant in alkaline solution as shown by

$$OCl^- + H_2O + 2e \rightleftharpoons Cl^- + 2OH^- \quad E° = 0.89 \text{ V}$$

A 1 $N$ solution of sodium hypochlorite is prepared by dissolving chlorine (35.5 g) in water (750 mL) containing sodium hydroxide (44 g) and diluting the resulting solution to 1 L. Examples of the use of hypochlorite are to be found in References 214 and 215. Although this solution is a useful oxidant, it is unstable and has been replaced to a large extent by the chlorine analogues of the organic titrants mentioned in the previous paragraph. The most widely used replacement titrant for hypochlorite is Chloramine T (sodium $N$-chloro-4-toluenesulfonamidate, 4-$CH_3$-$C_6H_4SO_2NClNa$). In alkaline solution it reacts analogously to hypochlorite

$$RSO_2N^--Cl + 2H_2O + 2e \rightleftharpoons RSO_2NH_2 + Cl^- + 2OH^-; (R = 4\text{-tolyl})$$

but has the advantage over hypochlorite in that it can be used also in acid solution. When prepared from the pure solid and stored in brown glass bottles, solutions of chloramine T in water have the great additional advantage over hypochlorite in that they are stable over a period of at least two months (216). These solutions may be standardized against arsenic(III) (ex. $As_2O_3$) in the presence of added halide, the type of which determines the conditions of acidity necessary for the titration (217). When chloride at a concentration of 0.5 $M$ or greater is used, the hydrogen ion concentration should be in the range 0.5–5.0 $M$. In the presence of 0.1 $M$ bromide the reaction is quantitative over a hydrogen concentration of $10^{-5}$–5.0 $M$, and in the presence of comparatively small amounts of iodide (0.005 $M$) a $p$H in the range 4–9 suffices.

Chloramine T solutions (10.0–15.0 mL, 0.05 $M$) added to water (30 mL) may also be standardized at 60 ± 3° by titration with a standard solution of hydrazine sulfate (216). The reactions of chloramine T with a number of reductants have been the subject of careful study in a series of publications traceable from Reference 218. Mention was made earlier of the construction of a chloramine T ion selective electrode, and this electrode has been used as indicator electrode for the indirect determinations of isoniazid, sulfide, and thiosulfate (219). An excess of a standard solution of chloramine T was added to the solution containing one of these determinands, and the concentration of determinand was deduced by measuring the concentration of unconsumed reagent with the electrode.

Other titrants similar in characteristics to chloramine T are $N$-chloroacetamide, dichloramine T, and (dichloroiodo)benzene, the reductions of which can be represented respectively as

$$CH_3CONHCl + H^+ + 2e \rightarrow CH_3CONH_2 + Cl^- \tag{12.26}$$

$$RSO_2NCl_2 + 2H^+ + 4e \rightarrow RSO_2NH_2 + 2Cl^-; (R = 4\text{-tolyl}) \tag{12.27}$$

$$PhICl_2 + 2e \rightarrow PhI + 2Cl^- \tag{12.28}$$

$N$-chloroacetamide is prepared as a solution in water to which is added chloroform (1 mL $L^{-1}$ of solution). The titrant is stable for periods in excess of 10 days, and has been used in a medium of 1–3.5 $N$ sulfuric acid and in the presence of potassium iodide (0.4–0.5 mL, 2%) for the titration of thiourea and its derivatives (131). Common reductants such as arsenic(III), iodide, tin(II), and hydrazine derivatives when dissolved in a medium of 2.5–4 $N$ hydrochloric acid can also be titrated with $N$-chloroacetamide (220).

When prepared by the further chlorination of chloramine T (65), dichloramine T ($NN$-dichloro-4-toluenesulfonamide, $CH_3$-$C_6H_4SO_2NCl_2$), is similar in properties to dibromamine T [see reactions (12.27) and (12.24)] and has been used as a titrant solution in anhydrous acetic acid for the titration of arsenic(III), iron(II), and ascorbic acid in both glacial acetic acid and in aqueous solutions (65). When prepared as a solution in acetic acid it is not so stable as solutions of (diacetoxy)iodobenzene, the normality of the dichloramine T solution changing from 0.4603 to 0.4463 $N$ over a period of 15 days. Its applications may be traced through References 64 and 221.

Solutions of (dichloroiodo)benzene ($PhICl_2$) in acetic acid are not so stable as dichloramine T, and their instability (0.1047–0.09625 $N$ in 5 days) was attributed to a rearrangement leading to the formation of 4-chloroiodobenzene. The reagent is prepared by direct chlorination of iodobenzene in dry acetic acid solution (66). It can be used for the same type of titrations as performed with the other organic derivatives of the halogen(I) cations and, in common with them, may be standardized by addition of aqueous potassium iodide (5 mL, 20%) to an aliquot of the titrant in aqueous 1 $N$ sulfuric acid. The liberated iodine is then titrated with standard thiosulfate solution.

**Chlorine(III).** Sodium chlorite ($NaOCl_2$) is the salt most commonly used to prepare aqueous solutions of this titrant (222) which are said to be stable when stored in the dark (133). For the titration of sulfite, satisfactory results were obtained if the $pH$ was kept between $pH$ 4.0 and 4.6, but for the titration of iodide a medium of 2 $N$ sulfuric acid was necessary. The titration of iron(II) is best carried out in a medium containing 2 $N$ hydrochloric acid and 2 $N$ sulfuric acid in the presence of a small amount of potassium iodide as catalyst. Arsenic(III) can be titrated in 5.5 or in 2 $N$ hydrochloric acid using potassium iodate as catalyst (133).

**Halogen(V) Titrants.** These can undergo reductions in acid solution according to the reaction type

$$XO_3^- + 6H^+ + 6e \rightarrow X^- + 3H_2O; \quad (X = Cl, Br, \text{ or } I) \quad (12.29)$$

The $E°$ values for these reactions are 1.45 V when X = Cl, 1.44 V when

X = Br, and when X = I the $E°$ value is 1.09 V. Titrant solutions of these halates are prepared by dissolving their potassium salts in water.

Potassium chlorate is seldom used as a titrant because its reactions with many reductants are kinetically slow despite its high reduction potential. These reactions are, however, strongly catalyzed by hydrogen ion, and thus it has been possible to titrate iron(II) and molybdenum(V) in 7–8 $M$ and 5–6 $M$ sulfuric acid, respectively, after the addition of 85% phosphoric acid (5 mL 50 mL$^{-1}$ titrand solution). Thallium(I) may be determined in a medium of 5–6 $M$ hydrochloric acid (223). By contrast, potassium bromate and potassium iodate are well-established titrimetric reagents [see (224) and (225), respectively], the description and use of which are included in undergraduate textbooks. Potassium bromate in the presence of > 0.15 $M$ potassium bromide and 0.67 $M$ hydrochloric acid has been used typically for the bromometric determination of aniline (226), and has also been used as titrant for the determination of some derivatives of 2-amino and 7-amino-3-phenoxazinones by titration in a medium of 96% ethanol containing a small amount of 5% potassium bromide acidified with hydrobromic acid (227). A typical use of potassium iodate as a titrant is given elsewhere (see reaction 12.23). Under the conditions of that reaction, however, iodine(V) undergoes a four-electron reduction to iodine(I) cation in contradistinction to the six-electron transfer of reaction type (12.29). Potassium iodate has also been used as titrant for the determination of organic sulfides in a medium of 60 : 3 : 5 acetic acid–benzene–1 $M$ hydrochloric acid (228).

### *Gold(III)*

A 0.05 $M$ solution of gold(III) chloride can be prepared by dissolving spectrographically pure gold (4.925 g) in aqua regia. After slow evaporation nearly to dryness the solution is acidified with hydrochloric acid and diluted to 500 mL. In the presence of a fourfold excess of either 1,10-phenanthroline or 2,2′-bipyridine and at a $p$H in the range 1–5, cobalt(II) can be titrated directly with gold(III) at 40–50°, the reaction being

$$Au^{3+} + 3[Co(1,10\text{-phen})_3]^{2+} \rightarrow Au + 3[Co(1,10\text{-phen})_3]^{3+} \quad (12.30)$$

or

$$Au^{3+} + 3[Co(2,2'\text{-bipy})_3]^{2+} \rightarrow Au + 3[Co(2,2'\text{-bipyr})_3]^{3+} \quad (12.31)$$

Either ligand, prepared as a 0.1 $M$ solution in water containing a minimal amount of hydrochloric acid, is added to the cobalt(II) solution so that its concentration will be at least four times higher than that of cobalt(II). The

$p$H is buffered in the range 3–5 (acetate buffer) and the solution warmed to 50° before titration with standard gold(III) solution. The method was applied to the determination of cobalt in high-temperature nickel-base alloys (229). Even small amounts of silver(I), copper(II), palladium(II), mercury(II), and iron(III) interfere.

### *Iron(III)*

Iron(III) chloride has been used as titrant for cobalt(II) in the presence of 1,10-phenanthroline by a method similar to that previously described (230). The titrant was prepared by dissolving the reagent grade solid in water. It has also been used as a titrant in pyridine and dimethylformamide (231), and in methanol (232). As a titrant in acetonitrile (70), iron(III) was prepared from commercially available hydrated iron(III) perchlorate that had been stored over magnesium perchlorate. This was dissolved in dry acetonitrile, and, if precautions were taken to exclude water from the solution, the molarity decreased by about 0.1% per day. When no such precautions were taken, the change in molarity due to hydrolysis of iron(III) was about 1% per day. In either case it is advisable to check the concentration daily by titration with potassium iodide, ferrocene, or 2,3,5,6,-tetrachloro-1,4-benzenediol (70). The latter is oxidized to 2,3,5,6-tetrachloro-2,5-cyclohexadien-1,4-dione.

Solutions of iron(III) chloride or ammonium iron(III) sulfate are often added in excess to solutions of reductants which have formal potentials less than the iron(III)/iron(II) couple in the medium of interest. The iron(II), formed in an amount equivalent to the original concentration of reductant, is then titrated with a strong oxidant such as chromium(VI) or cerium(IV). The determinations of tin(II) (233), europium(II) (32), and phosphate (194) are examples of the application of this method.

Reference to Table 12.3 will confirm that hexacyanoferrate(III) is a versatile oxidant both in strong phosphoric acid medium (see serials 6 and 15), and also in alkaline complexing media (see serials 35, 36, 37, and 39). The titrant solution, prepared by dissolving the potassium salt [i.e., $K_3Fe(CN)_6$] in water, is stable for about six weeks if stored in the dark (234). The purification of potassium hexacyanoferrate(III) by recrystallization is given in Reference 235 and for the accurate determinations of cobalt(II) described therein it was used as a prime standard.

### *Lead(IV)*

Lead(IV) is a strong oxidant, the standard reduction potential for the lead(IV)/lead(II) couples being 1.7 V (236). Lead tetraacetate is the solid most commonly used for the preparation of lead(IV) titrant solutions. These solutions undergo reduction in acid solution according to the equation

$$Pb(CH_3COO)_4 + 4H^+ + 2e \rightleftharpoons Pb^{2+} + 4CH_3COOH \qquad (12.32)$$

The formal potential is, therefore, dependent upon $p$H and at $p$H 4 (acetate buffer) it is 1.27 V (237). Lead tetraacetate is prepared by heating $Pb_3O_4$ with a mixture of glacial acetic acid and acetic anhydride at 60°. The product can be recrystallized from glacial acetic acid, and the titrant solution is usually prepared by dissolution of the pure solid in glacial acetic acid. It can be standardized against standard iron(II) sulfate solution in a medium of 2 $N$ hydrochloric acid (238). This lead(IV) titrant has been used for the determination of osmium after reduction to osmium(III) (239) in a medium of 6 $N$ sulfuric acid for the determination of common reductants such as chromium(II), iron(II), titanium(III), tin(II), antimony(III), arsenic(III) in hydrochloric acid or sulfuric acid media (238), nitrite in a medium of 0.1 or 1 $M$ sodium chloride (240), and bisulfite in acetate buffer at $p$H 4 (237). In 5–7 $M$ hydrochloric acid, molybdenum(III) is quantitatively oxidized to molybdenum(VI) and in 8–10 $M$ hydrochloric acid to molybdenum(V). In 7 $M$ hydrochloric acid the titration curve exhibits two well-defined breaks corresponding to the two products (85). Lead tetraacetate may also be used to titrate organic reductants either in acidic aqueous solutions (241) or in acetic acid solutions (242).

Lead(IV) chloride has been used as a titrant in solutions of anhydrous dimethylformamide (243) and lead tetraformamide $Pb(HCONH)_4$, prepared by heating lead dioxide with formamide under reflux at 120° for 3–4 h, was used to titrate solutions of tin(II), arsenic(III), antimony(III), cerium(III), iron(II), and thiosulfate in formamide (244). Solutions of lead tetraformamide are unstable, but another formamide derivative, trilead octaformamide $[Pb_3(HCONH)_8]$, is said to yield slightly more stable solutions (244).

## *Manganese(III)*

The stability and reactivity of tervalent manganese can be varied widely by means of the formation of complexes of differing stability. The most reactive form of manganese(III) is the hexaquomanganese(III) ion in a noncomplexing medium of perchloric acid, and many reactions involving this ion are rapid and quantitative even in very dilute solutions. A 0.005 $N$ solution of this titrant in 6 $M$ perchloric acid and 0.4 $M$ manganese(II) perchlorate can be prepared by slowly adding with stirring 0.1 $N$ potassium permanganate (10.00 mL) to a mixture of concentrated perchloric acid (100 mL), 2 $M$ managanese(II) perchlorate (40 mL), and water (50 mL; 245). The procedure recommended for its standardization is to use pure potassium iodide (dried at 220° for 6 h) as the primary standard. To an accurately known volume of a standard aqueous solution of potassium iodide is added an equal volume of 2 $M$ perchloric acid, and this solution is then titrated with the manganese(III) solution. The reaction is

$$2Mn^{3+} + 2I^- \rightarrow I_2 + 2Mn^{2+} \tag{12.33}$$

Some uses of this titrant may be traced through Reference 245.

### Manganese(IV)

A 0.05 $M$ solution of this titrant is prepared by dissolving potassium permanganate (7.9 g) in 9 $M$ sulfuric acid with vigorous stirring over a period of 6-8 h. The solution is allowed to stand overnight, and then it is diluted to 1 L with 9 $M$ sulfuric acid (246). It may be standardized against standard iron(II) solution in a medium of 1-5 $M$ sulfuric acid or against sodium oxalate in the same medium at 70°. The standardizations need to be performed daily since the solution deteriorates by 0.1-0.2% per day. Managanese(IV) oxidizes almost all the conventional reductants including cerium(III) (246).

### Manganese(VII)

The use of potassium permanganate as a titrant has been well established for a number of years and much information on its properties and applications is to be found in Reference 247. Some examples of its use in complexing media are given in Table 12.3 and these reflect the modern applications of this important titrant. Examples of oxidations by permanganate that are catalyzed by copper(II), iodate or iodine monochloride, and manganese(II) are given in another section.

### Molybdenum(V)

This oxidant has been used as the complex octacyanomolybdate(V) ion. The formal potential of the octacyanomolybdate(V)/octacyanomolybdate(IV) couple is about 0.97 V and the species $[Mo(CN)_8]^{3-}$ has been used for the indirect titration of cobalt(II) (76) and the direct titration of sulfide (247) in media having $p$H values in the range 9-10. The potassium salt is unstable and hence the titrant solutions have been prepared by the oxidation of the more stable potassium octacyanomolybdate(IV) with lead dioxide in a medium of approximately 0.3 $M$ sulfuric acid (76). Alternatively, they have been prepared from cesium octacyanomolybdate(V) dihydrate by the method given in Reference (248).

### Vanadium(IV)

This species ($VO^{2+}$) has been used as an oxidizing titrant (171) for the determination of titanium(III) in the presence of oxalic acid catalyst, the overall reaction being

$$Ti(III) + V(IV) \rightarrow V(III) + Ti(IV) \quad (12.34)$$

It has also been used as a *reductant* for the determination of copper(II) (249):

$$Cu(II) + V(IV) \rightarrow V(V) + Cu(I) \quad (12.35)$$

For both applications the titrant solutions were prepared from ammonium vanadate by reduction. The reductant was sulfur dioxide for the preparation of the vanadium(IV) oxidant solution, the excess sulfur dioxide being removed by passing carbon dioxide (171). The reduction was performed with hypophosphite in the preparation of the vanadium(IV) solution used subsequently for the determination of copper(II). In this instance pure ammonium metavanadate (11.6980 g) was added to sodium hypophosphite, $NaH_2PO_2 \cdot H_2O$ (6.4 g) dissolved in water (10 mL). Sulphuric acid 1:1 (28–30 mL) was added in small portions, the mixture being gently heated until all the solid dissolved. On dilution to 1 L with water, the solution was 0.100 $N$ in $VOSO_4$, approximately 0.5 $N$ in $H_2SO_4$, and 0.02 $N$ in hypophosphite. The excess reductant did not interfere in the titration of copper(II). Vanadium(IV) solutions have been standardized against permanganate at 60° (171).

### Vanadium(V)

Ammonium metavanadate, sodium metavanadate $NaVO_3$, or sodium orthovanadate $Na_3VO_4$ are the salts commonly used for the preparation of this titrant. When recrystallized and dried, ammonium vanadate is said to be a primary standard (250), but the concentrations of its solutions are usually checked against iron(II) solution that has been standardized previously against potassium dichromate. A 0.1 $N$ solution is prepared by dissolving ammonium vanadate (~ 12 g) in 1:1 sulfuric acid (80 mL) and diluting the solution to 1 L with water.

Vanadium(V) is generally used for reactions in acid solutions and is reduced according to the reaction

$$VO_2^+ + 2H^+ + e \rightarrow VO^{2+} + H_2O \qquad (12.36)$$

The formal potentials for this reaction in phosphoric acid solutions are given in Table 12.2, and these are similar in magnitude to those of chromium(VI) and cerium(IV) in similar media. However, as an oxidant vanadium(V) is less subject to interference from organic compounds than is dichromate or cerium(IV) (251). In acid solution and in the presence of osmium tetroxide catalyst it has been used for the determination of isonicotinic acid hydrazide, and in a medium of 1:1 hydrochloric acid and 1:1 sulfuric acid, phenidone [1-phenyl-3-pyrazolidinon (Reg. No. *92-43-3*)], and 1,4-benzenediol, respectively, could be determined separately in a single sample of developer solution by titrations with vanadium(V) (252).

## 12.3 SOME REDUCTANTS USED AS TITRANTS

The applications of some reductants used for the standardization of solutions of oxidants have been mentioned in Section 12.2. Arsenic(III) features prominently in these applications and, as was mentioned in Chapter 9, arsenic(III) oxide is used as a prime standard for these purposes. Although no iron(II) salt can be regarded as a satisfactory prime standard, iron(II) solutions, standardized against potassium dichromate, are often used for the standardization of oxidant solutions. Reductants such as arsenic(III) and iron(II) are, of course, used also as titrants in their own right, and this section lists these and some other reductants arranged in alphabetical order. However, reductants such as potassium iodide, used for the standardization of manganese(III) solutions, for example, and the prime standards sodium oxalate and thiourea mentioned in Chapter 9 are not included in this section since their solutions are not usually employed as reducing titrants.

### *Antimony(III)*

Antimony potassium tartrate is available as the hemihydrate $C_4H_4KO_7Sb \cdot \frac{1}{2}H_2O$ ($M = 333.93$) after air-drying the recrystallized material at room temperature, and the anhydrous material is obtained by drying it at 100° (253). In either form it is quoted as being suitable for use as a primary standard (253) and reacts similarly to arsenic(III). The titrant solution, when prepared as a solution in water that contains sodium potassium tartrate ($\sim 1$ g $L^{-1}$), is stable, its concentration changing by about 1% over a period of 7 months (254). It can be standardized against potassium bromate in acid medium or iodometrically in a bicarbonate medium. By changing the conditions, the sequential titrations of hypochlorite ($pH > 8$), chlorite ($pH$ 8–12, 3–5 drops 0.1% $OsO_4$), and chlorate (2–4 $N$ $HNO_3$, 1–3 drops 1% $OsO_4$) have been performed with this titrant (254).

### Arsenic(III)

A standard 0.1 $N$ solution of arsenic(III) is prepared by dissolving dry arsenic(III) oxide (4.9460 g) in 1 $N$ sodium hydroxide (60 mL), neutralizing the resulting solution with 1 $N$ hydrochloric acid or 1 $N$ sulfuric acid (59–60 mL) followed by dilution to 1 L with water (255). The solution is stable for at least 2 months when prepared by this method or as a solution in 0.4 $N$ bicarbonate solution but is not stable on prolonged storage when prepared in 0.1 $N$ sodium hydroxide solution. At $pH$ 7 the species in solution is arsenious acid ($HAsO_2$ or $H_3AsO_3$), the $pK$ value for which is 9.3. This can undergo oxidation by the reaction

$$HAsO_2 + 2H_2O \rightarrow HAsO_4^{2-} + 4H^+ + 2e \qquad (12.37)$$

in which the dianion of arsenic acid is formed. Arsenic acid is a triprotic acid having the following $pK_a$ values: $pK_1 = 2.3$, $pK_2 = 4.4$, $pK_3 = 9.2$. The formal reduction potential of the couple As(V)/As(III) is thus very dependent upon $pH$, the strength of the reductant species [As(III)] increasing as the $pH$ is raised.

The use of standard arsenic(III) solutions has now become associated with the catalytic action of osmium(VIII) oxide, and some of the applications of this titrant were outlined when discussing this catalyst. Osmium(VIII) itself may be determined in a medium of 0.1–1 $N$ sodium hydroxide by reduction to osmium(VI) with standard arsenic(III) (256).

### 1,4-Benzenediol

Formerly known as hydroquinone, the use of this reductant is associated with the titration of gold(III) according to the equation (257)

$$2HAuCl_4 + 3C_6H_4(OH)_2 \rightarrow 2Au + 3C_6H_4O_2 + 8HCl$$

When pure 1,4-benzenediol (0.4186 g) is dissolved in water (200 mL) containing hydrochloric acid (10 mL) and the solution is diluted to 500 mL, 1 mL of this titrant solution corresponds to 1 mg gold. The method has been extended more recently to determine gold, platinum, and palladium in mixtures without prior separations. A gold indicator electrode was used in this application (258).

1,4-Benzenediol has also been used as an indirect titrant to react with the added excesses of standard solutions of oxidants used in the determination of some reductants. These have included the determination of excess lead(IV) acetate in the determination of tartaric acid and tartrates in a medium of 80% acetic acid (259), and the determination of the excess standard cerium(IV) above that required for the oxidation of thiourea in a medium of tributyl phosphate–carbon tetrachloride (260). The value of 1,4-benzenediol as a reductant has been reviewed (261).

### *Cerium(III)*

This species is a very weak reductant and is not widely used. The titrant solution is prepared by dissolving cerium(III) sulfate octahydrate in sufficient sulfuric acid to make the resultant stock solution 0.2 $N$ in this acid (262). It has been standardized by the persulfate method (263) and, as mentioned earlier, it could also be standardized against manganese(IV). Cerium(III) does not react with any of the following species when they are present in 30% potassium carbonate solution: arsenate, bromate, iodate, periodate, perchlorate, 2,5-cyclohexadiene-1,4-dione (quinone), hydroxylamine, and hydrazine. Hypohalites react with cerium(III) under these conditions, but the overall reaction is so slow as to be unsuitable for analytical purposes (262). Hexacyanoferrate(III) and permanganate in strong carbonate solutions (3–4 $M$) can be determined by titration with cerium(III), and glucose can be determined indirectly by adding an excess of standard hexacyanoferrate(III) and titrating this with cerium(III) (262).

### *Chromium(II)*

This titrant has been used for the determination of tungsten or tungsten plus vanadium in heteropoly compounds without prior removal of phosphorus (264). The titrand solutions were prepared in a medium of 12 $M$ hydrochloric acid and titrated at room temperature. A wax-impregnated carbon indicator electrode was found to yield more pronounced inflections than a platinum indicator electrode since the latter had a catalytic effect upon the reduction of water by chromium(II). The methods for the preparation and storage of chromium(II) sulfate solutions are given in Reference 265.

### *Copper(I)*

Although solutions of copper(I), prepared by dissolving copper(I) chloride in ~ 1 $N$ hydrochloric acid, have been used as a titrant in aqueous solution

(258), the species has also found application when prepared as a solution in acetonitrile (266). For the latter application copper(I) chloride was dissolved in a 1 : 1 mixture of acetonitrile and glacial acetic acid. These solutions are unstable; in a 0.05 $N$ solution decomposition becomes evident after 2 h, whereas a 0.025 $N$ solution is said to be stable over a period in excess of 12 h. When mixed with an equal volume of 12 $N$ sulfuric acid, aqueous solutions of common oxidants such as chromium(VI), vanadium(V), cerium(IV), and manganese may be titrated with copper(I). Presumably, this titrant solution could also be standardized against potassium dichromate.

## *Hydrazine and Its Derivatives*

Titrant solutions of this reductant are prepared from pure hydrazine sulfate ($N_2H_4H_2SO_4$, M = 130.12) that has been dried at 150° for 1 hr. This solid can be used as a prime standard (2) from which to prepare aqueous solutions of known concentration. In most of its titrimetric reactions hydrazine undergoes irreversible oxidation according to the reaction

$$N_2H_4 \rightarrow N_2 + 4H^+ + 4e \qquad (12.38)$$

As can be deduced from the formal potentials given in Table 12.2, the strength of hydrazine as a reductant increases as the $p$H of the medium is increased, being a strong reductant in 1 $M$ sodium hydroxide.

The catalyst osmium(VIII) oxide is required to promote most of the analytically useful reactions of hydrazine [see reactions (12.9) and (12.10)]. Similarly, the titration of vanadium(V) in a medium of 0.25–0.75 $M$ sulfuric acid with hydrazine also requires the use of this catalyst (267). In a medium of 10–12 $M$ phosphoric acid, however, the catalyst was not required when vanadium(V) and chromium(VI) were titrated sequentially with hydrazine (2). Chromium(VI) can also be titrated with hydrazine in a medium of 5–6 $M$ hydrochloric acid in the presence of copper(II) chloride catalyst as shown by reaction (12.5) (125). In alkaline medium (0.1–1 $N$ NaOH) osmium(VIII) can be titrated directly with standard hydrazine solution (256) or arsenic(III).

Organic derivatives have also been examined as potential titrants in place of hydrazine. These undergo the analogous oxidation reactions to reaction (12.38) and have similar formal reduction potentials (73). Of those tested, 2′-amino-4-pyridinecarboxamide [isoniazid, (Reg. No. *54-85-3*)] produced the most stable solutions both in water and in 0.5–1% sulfuric acid. In water the concentration of the solution remained unchanged over a period of 6 months and the acid solution was stable for a period in excess of 2 months. In alkaline solutions, however, decomposition was detected soon after being prepared in 1% sodium hydroxide solution. Pure isoniazid ($M$ = 137.15, mp

171.4°) is said to be suitable for use as a prime standard (73). It can be purified by recrystallization from ethanol and is dried to constant weight at 105°.

Analogous to the reaction of hydrazine with chromium(VI), the reaction between dichromate and isoniazid is also promoted by the presence of copper(II) sulfate (about 70 mg $CuSO_4 \cdot 5H_2O$ for 1–50 mg $Cr_2O_4^{2-}$) in a medium of 10–37% hydrochloric acid. The conditions under which a number of oxidants can be titrated are given in Reference 73. No work has been sighted, however, where osmium(VIII) oxide has been investigated as a possible catalyst for the reactions of isoniazid.

## *Hydroxylamine and Its Derivatives*

Hydroxylamine shows reducing properties similar to iron(II) in acidic media and acts as a strong reductant in alkaline media as evidenced by the following formal potentials (268)

$N_2O + H_2O + 6H^+ + 4e \rightleftharpoons 2NH_3^+OH$   $E^{\circ\prime} = 0.72$ V $(0.3\ M\ H_2SO_4)$

$N_2O + 5H_2O + 4e \rightleftharpoons 2NH_2OH + 4OH^-$   $E^{\circ\prime} = -0.45$ V $(5.7\ M\ KOH)$ at 25°

$-0.17$ V $(7\ M\ NH_4OH)$ at 70°

$E^{\circ\prime} = 0.02$ V $(0.75\ M\ KOH + 0.045\ M\ Na_4P_2O$ at 50°

$NO_3^- + 5H_2O + 6e \rightleftharpoons NH_2OH + 7OH^-$   $E^{\circ\prime} = 0.13\ V$ (satd. $NaHCO_3$)

The titrant solution is prepared by dissolving analytical reagent grade hydroxylamine hydrochlorine in water. This solution appears to be stable and can be standardized against silver(I) in a concentrated alkaline medium (5 M KOH) containing ammonia (269). To the solution containing silver(I) (20–250 mg) is added ammonia (2 mL, 13 M), sodium thiosulfate (1 mL, 0.1 N) followed by sufficient potassium hydroxide to yield a titrand solution (50 mL) 5.0–5.5 M in potassium hydroxide. When a silver indicator electrode is used in conjunction with a saturated calomel reference half-cell, the equivalence point potential is 0.280 V. The reaction is

$$4Ag(NH_3)_2^+ + 2NH_2OH + 4OH^- \rightarrow 4Ag + N_2O + 5H_2O + 8NH_3$$

(12.39)

Silver may also be determined when it is complexed with thiocyanate in a medium of 5 $M$ potassium hydroxide (269)

$$4Ag(SCN)_4^{3-} + 2NH_2OH + 4OH^- \rightarrow 4Ag + N_2O + 5H_2O + 16SCN^-$$

(12.40)

The equivalence point potential is 0.320 V in this instance. The titrant has been used for the determination of cyclohexanone in a medium of 4 $M$ sodium hydroxide (270), copper (271) in a medium of potassium hydroxide (0.5–2 $M$) containing sodium pyrophosphate (5–10 g in 100 mL), and osmium [Os(VI)–Os(IV)] also in alkaline medium ($\geq$ 0.5 $M$ NaOH) (272). Other applications can probably be traced through Reference 270.

Of the possible derivatives of hydroxylamine, apparently only $N$-phenylbenzohydroxamic acid [PhC(O)N(OH)Ph] has found application for the determination of molybdenum(VI) (273).

### *Iron(II)*

Reference to the formal reduction potentials of the iron(III)/iron(II) couple given in Table 12.2 will give some indication of how the strength of iron(II) as a reductant can be varied by changing the medium in which the titrand solution is prepared. This, of course, is reflected in the versatility of iron(II) as a direct titrant. Thus it has been used when the determinand oxidant solution has been prepared in phosphoric acid (e.g., Table 12.3, serials 7, 8, 10, 11, 13, 14, 19, 20, 22, and 26), and it is also a useful and more powerful reducing titrant when the titrand solution is prepared in a complexing alkaline medium (e.g., Table 12.3, serials 32–34 and 38). Iron(II) also finds many applications for indirect determinations.

Sometimes the excess amount of iron(II) remaining after reduction of a determinand oxidant is determined by titration with a standard solution of oxidant, usually dichromate, as in the determination of lead(II) (116), plutonium (9), and diethyleneglycol dinitrate [(oxybis(2-ethyl nitrate) (Reg. No. *693-21-0*)] (274). It is used also as an indirect titrant to determine the excess of a standard solution of oxidant remaining after reaction with a determinand reductant. Quite frequently the oxidant added in excess is either cerium(IV) as in the determination of bromide and iodide (275) and ruthenium (276), or chromium(VI) as in the determination of uranium (96) and tellurium (277).

Iron(II) solutions are usually prepared from ammonium iron(II) sulfate [$(NH_4)_2Fe(SO_4)_2 \cdot 6H_2O$, Mohr's salt] which is dissolved in an appropriate volume of dilute sulfuric acid (0.5–2 N). This titrant has been stored in an aspirator under xylene (9). It is recommended that the solution be standardized daily against potassium dichromate. Less frequently, standard iron(II) solutions are prepared by dissolving an accurately weighed quantity of pure electrolytic iron in an excess of acid and diluting this solution to an accurately known volume (278).

Iron(II) solutions have also been prepared from potassium hexacyanoferrate(II), $K_4Fe(CN)_6 \cdot 3H_2O$. The addition of a small amount of sodium carbonate (0.2 g $L^{-1}$) is said to stabilize the solution (279), but even without this addition the concentration of a solution of hexacyanoferrate(II) was found to remain invariant over a period of 2 months (280). Some applications can be traced through the second paper of Reference 280.

Ferrocene [bis(cyclopentadienyl)iron(II)], prepared as a 0.01 $M$ solution in ethanol, has been used to determine copper(II) in a medium of 4–6 $M$ hydrochloric acid prepared in 1 : 1 water–acetone (see serial 2, Table 12.3). It has also been used for the determination of molybdenum in steels and alloys (281).

### *Molybdenum(III)*

Solutions of trivalent molybdenum are inherently unstable, although when prepared in solutions of hydrochloric acid the stability increases with increasing acid concentration. It has been reported (75) that solutions of molybdenum(III) in 2 $M$ hydrochloric acid containing potassium chloride (50 g $L^{-1}$) are relatively stable when an atmosphere of carbon dioxide is maintained above them. Nonetheless, it is recommended that the concentration of the solution should be checked every 2–3 days by titration against a standard solution of copper(II) in a medium of 1 : 1 hydrochloric acid. The reaction is

$$2CuCl^{2-} + Mo(III) \rightarrow 2CuCl_2^- + Mo(V) + 2Cl^- \qquad (12.41)$$

Molybdenum(III) solutions are prepared from ammonium molybdate dissolved in 2 $M$ hydrochloric acid by reduction with zinc amalgam either dynamically through a Jones reductor (77), or statically by contact with the reductant (500 g) for a period of about 2 h in an aspirator under carbon dioxide (75). If the latter method is used, the dark green solution must be removed from the zinc and stored under carbon dioxide in another aspirator connected to a burette (see Figure 12.1).

The use of molybdenum(III) for the determination of molybdenum(VI) is outlined in Table 12.3 (serial 4). It has also found application for the determination of some common oxidants and for the sequential titrations of mixtures containing chromium(VI) and iron(III), cerium(IV) and vanadium(V), manganese(VII) and vanadium(V), and manganese(VII) and iron(III). The individual members of some three-component mixtures of these oxidants could also be determined (282). More recently, the properties of molybdenum(III) chloride as a reducing titrant have been independently evaluated (77).

## Sodium Thiosulfate

The properties and uses of this well-established reductant are thoroughly discussed in Reference 283 and are included also in undergraduate textbooks that are devoted to or contain sections on titrimetric analysis. In most of its titrimetric reactions thiosulfate undergoes oxidation according to the reaction

$$2S_2O_3^{2-} \rightarrow S_4O_6^{2-} + 2e$$

It is often standardized against potassium iodate, but as was previously mentioned, it can also be standardized directly against potassium dichromate in the presence of copper(II) catalyst (123). To this end, acetate–acetic acid buffer (30 mL, $p$H 4.45) and copper(II) sulfate solution (2 mL, 0.2 $N$) are added to an aliquot of standard dichromate solution contained in a beaker. The solution is diluted to 50 mL and titrated with sodium thiosulfate using a bright platinum rod electrode versus a saturated calomel reference half-cell as the indicating system.

Sodium thiosulfate is commonly used for the determination of iodine quantitatively liberated by reaction between an oxidant and an excess of potassium iodide. This reaction can also be utilized in nonaqueous media and has been applied, for example, to the determination of heteroaromatic sulfur compounds in a medium of 1 : 3 chloroform–glacial acetic acid (284). The effects of various nonaqueous solvent systems upon reactions of these types have also been investigated using thiosulfate as the titrant (285). A medium of glacial acetic acid was used for the direct titration of lead(IV) with this titrant (286). The same medium was used also to determine the excess of a standard solution of dichloramine T remaining after the oxidation of 3-substituted-2-propen-1-ols (R–CH=CH–CH$_2$OH, R = Me or Ph). After allowing sufficient time for the oxidation of these alcohols to the corresponding aldehydes (45 min when R = Me and 90 min when R = Ph), the excess of dichloramine T was determined by titrating with standard thiosulfate solution the iodine liberated when potassium iodide (10 mL, 20%) was added to the solution (221). A similar procedure, also involving the use of dichloramine T, was adopted for the determination of thiosemicarbazide [Reg. No. *79-19-6*] and its metal complexes in a medium of acetic acid containing 12–15% water (287). Metal cyanides were also determined by a similar reaction sequence carried out in an aqueous acetate buffer at $p$H 4 (288).

*Tin(II)*

Stannous chloride was the first titrant to be used for the determination of iron(III) (289). Titrant solutions are usually prepared by dissolving tin(II) chloride dihydrate under carbon dioxide in an amount of concentrated hydrochloric acid sufficient to ensure that on dilution the solution becomes approximately 1 $N$ with respect to this acid. These solutions are unstable and should, therefore, be standardized against potassium iodate at regular intervals. Complexation with polyols such as 1,2,3-propanetriol (glycerol) and 1,2-ethanediol (ethylene glycol) increases the stability of tin(II) considerably (290).

Tin(II) has been used in the determination of soluble silica by the direct titration of molybdosilicic acid (291), in the determination of selenium and tellurium in refined selenium products (292), and for the determination of tellurium (293).

*Titanium(III)*

Some formal reduction potentials for the titanium(IV)/titanium(III) couple are given in Table 12.2. Titanium(III) undergoes oxidation by the reaction

$$Ti^{3+} + H_2O \rightarrow TiO^{2+} + 2H^+ + e \qquad (12.42)$$

from which it follows that titanium(III) becomes a stronger reductant as the hydrogen ion concentration is decreased. In weakly acid or in alkaline solution titanium(III) can be oxidized by hydrogen ion. Over the years titanium(III) has been used as an indirect titrant for the reduction of organic compounds, particularly nitro compounds. For such determinations standard titanium(III) is added in excess to a solution of determinand and the excess is then titrated with a standard solution of iron(III) (294). The titanium(III), as a solution in hydrochloric acid, can be prepared from solid titanium(III) chloride which is available from some laboratory supply houses.

More recently, titanium(III) prepared as a solution in $NN$-dimethylformamide has been used for the titrations of copper(II), iodine, iron(III), bromine, and antimony in this medium (295). In a study designed to evaluate the performance of indicator electrode systems in $NN$-dimethylformamide by titrating copper(II) with titanium(III), the titrant solution

(0.1–0.2 $M$) was prepared in 25 mL batches by dissolving solid titanium(III) chloride in the solvent under cover of nitrogen contained in a glove box (296). An investigation of the titrations of copper(II), chromium(VI), iron(III), and mercury(II) in $NN$-dimethylformamide with titanium(III) has been made (297). It was found possible to titrate copper(II) and iron(III) sequentially in the medium with titanium(III). This titration was applied to the determination of these metals in an alloy. A detailed description of this titration, presented as a potentiotitrimetric laboratory exercise for students, is available (298). Hydroxylamine and its derivatives have also been determined in $NN$-dimethylformamide by titration with titanium(III) formate complex in 1,2,3-propanetriol-DMF (299).

### *Vanadium(II)*

The standard reduction potential for the vanadium(III)/vanadium(II) couple is quoted as being $-0.255$ V (300) indicating that this species, too, is a strong reductant. The titrant solution can be prepared by dissolving reagent grade vanadium(II) sulfate in distilled water (1 L) containing concentrated sulfuric acid (30 mL; 126). It may also be prepared by the reduction of ammonium metavanadate with sulfur dioxide followed by further reduction with zinc amalgam [(126); see also Reference 3]. In common with other solutions of strong reductants, these solutions must be stored under an inert atmosphere. Vanadium(II) solutions may be standardized against iron(III) in the presence of copper(II) catalyst (2 mL, 0.1 $M$ $CuSO_4$). Under these conditions the reaction is

$$2Fe(III) + V(II) \rightarrow 2Fe(II) + V(IV) \qquad (12.43)$$

Alternatively, titrant solutions may be standardized against iron(III) (2–5 mL, 0.1 $M$) in a medium of syrupy phosphoric acid (5–15 mL). After deoxygenation with carbon dioxide for about 10 min, the solution is titrated with vanadium(II). Under these conditions vanadium(II) is oxidized to vanadium(III). In a medium of 6 $N$ sulfuric acid, however, vanadium(II) can be standardized also against cerium(IV), chromium(VI), or vanadium(IV), the vanadium(II) being oxidized to vanadium(IV) under these conditions (126). Once standardized, binary mixtures of cerium(IV) and iron(III), chromium(VI) and iron(III), and vanadium(V) and iron(III) can be determined sequentially by titration with vanadium(II) [(126); see also Reference 301]. The reductions of molybdenum(VI) (142) and tungsten(VI) (302) with vanidum(II) have also been studied, and the stabilization of vanadium(II) solutions with 1,2,3-propanetriol has been investigated (303).

## Vanadium(IV)

The use of this species as a reductant is mentioned in Section 12.2 under the same heading.

## 12.4 OXIDATION–REDUCTION TITRATIONS OF SOME ORGANIC COMPOUNDS IN VARIOUS MEDIA

Mention has already been made of the determination of ascorbic acid in a medium of 10–12 $M$ phosphoric acid (serial 6, Table 12.3), and the references already cited contain details of the traditional types of determinations of organic compounds using some of the more recently developed titrants. For example, halogen substitution reactions used for the determination of aniline by bromination (226) and the iodometric determination of naphthol (203) have also been applied in aqueous titrand solutions using the newer halogen(I) cation titrants mentioned in Section 12.2 (66, 213). Some substituted anilines, phenol, and 8-quinolinol have been determined with these. Hydrazine derivatives such as isoniazid have also been determined by oxidation with halogen(I) cations (216, 219). This class of compounds includes semicarbazide (1-hydrazinoformamide) derivatives (220). Isoniazid has also been determined by titration with vanadium(V) at room temperature (155). Thiols and thiones (e.g., thiourea, its alkyl and aryl derivatives) have been titrated with halogen(I) cations (131). Complexed iron(III) has also been used as titrant for the oxidation of thiourea (132) and thiols (195) including 2-mercaptoacetic acid (thioglycolic acid). Phenothiazine derivatives (241) have been determined by oxidation with lead tetraacetate, and the determinations of organic nitro, nitroso, and diazonium compounds by reduction with iron(II) in an alkaline medium of triethanolamine have also been investigated (72). Determinations of these types in aqueous solutions are, of course, well established, but it is only comparatively recently that the scope of oxidation–reduction titrations has been extended by using nonaqueous solvents as the medium for the determination of organic compounds. Some of these applications are given in this section which is arranged in alphabetical order of the major component of the medium.

### Acetic Acid Media

These have been used extensively as media for the oxidation of thiols, thiones, and thioethers with a variety of titrants. Thiols have been oxidized with dichloramine T in glacial acetic acid to the corresponding disulfides according to the reaction

$$2RSH \rightarrow R-S-S-R + 2H^+ + 2e \quad (12.44)$$

The determinations of cysteine [Reg. No. *52-90-4*] (304) and 2-mercaptoacetic acid (305) are specific examples of the use of this reaction. A similar reaction was found to occur in glacial acetic acid when lead(IV) acetate was the oxidant (306), although when water was present ( < 70%) both cysteine and 2-mercaptoacetic acid could be oxidized to the corresponding sulfinic acids.

Thioethers in the presence of bromide (1 mol KBr to 100 mol thioether) can be titrated in a medium of 70:30 acetic acid–water with lead(IV) acetate (127). It is postulated that the bromide is oxidized by the initial addition of the titrant and the bromine so formed reacts immediately with the thioether according to the sequence

$$Br_2 + RSR' \rightarrow [RR'S^+ - Br]Br^- \xrightarrow[H_2O]{HOAc} RR'S=O + 2HBr$$

$$(12.45)$$

The hydrobromic acid liberated is continuously oxidized with the lead(IV) titrant, and is immediately used up in reaction (12.45). The oxidation $2Br^- \rightarrow Br_2 + 2e$ is, therefore, established only after all the thioether has undergone reaction. Thus the major inflection in the titration curve was found to be commensurate with the completion of reaction (12.45) for a variety of alkyl and aryl thioethers. The method was applied also to the determination of cephalosporins of pharmaceutical interest. Methods for the determinations of thioethers are also given in References 228 and 242.

A medium of 1:3 chloroform–acetic acid was found to be satisfactory for the determination of 2,2'-dithiobis[benzothiazole] [Reg. No. *120-78-5*] by reduction with excess potassium iodide (10% aqueous) followed by titration of the liberated iodine with standard thiosulfate (284), and a similar indirect method was used for the determination of thiosemicarbazide and its metal complexes (287). In the latter instance, however, thiosemicarbazide was reacted with an excess of standard dichloramine T in an acetic acid medium containing 12–15% water buffered at $p$H 4. The reaction is

$$NH_2NHCSNH_2 + 3RNCl_2 + 6H_2O$$

$$\rightarrow 3RNH_2 + CO_2 + H_2SO_4 + N_2 + NH_4Cl + 5HCl \quad (12.46)$$

After allowing a suitable reaction time (30 min for thiosemicarbazide and 10 min for its metal complexes), the excess dichloramine T was determined by titration with standard thiosulfate after the addition of excess aqueous potassium iodide (10 mL, 20%).

Hydrazine and its 1,1-disubstituted derivatives have been titrated in glacial acetic acid with bromine cyanide in the presence of an excess of anhydrous sodium acetate (3–4 g, 206). Some of these reactions can be represented as

$$2RR'NNH_2 + 2Br^+ \rightarrow RR'NN=NNRR' + 4H^+ + 2Br^- \quad (12.47)$$

when R = Ph and R' = H, and when R = Ph and R' = Me, but when R = H and R' = $Cl_3CCHOH$ as in 2,2,2-trichloro-1-hydrazinoethanol, the reaction was found to be

$$2CCl_3CH(OH)NHNH_2 + 2Br^+ \rightarrow CCl_3CH(OH)NHNHCH(OH)CCl_3$$
$$+ N_2 + 4H^+ + 2Br^- \quad (12.48)$$

The bromine(I) cation emanating from bromine cyanide has also been used for the determination of some phenols in glacial acetic acid and in a 1 : 1 mixture of acetic acid and acetic anhydride (207). The iodine(I) cation was similarly used. The phenols were found to undergo substitution reactions stoichiometrically, the extent of the substitution (i.e., monohalo, dihalo, or trihalo) depending upon the nature of the substituent present in the phenol, the nature of the halogen(I) cation, and the solvent system. The degree of substitution is, of course, deducible from the shape of the titration curve and as little as 0.5 mg of the phenols tested could be determined with $\pm 0.25\%$ error in these media.

A partially aqueous medium (80 : 20 acetic acid–water) was used for the indirect determinations of tartaric acid and tartrates (259) by oxidation with an excess of standard lead(IV) acetate in the presence of aqueous potassium acetate (5 mL, 20%). The reaction requires about 30 min and in the case of tartaric acid, for example, can be represented as

$$COOH(CHOH)_2COOH + 5Pb(OAc)_4 + 2H_2O$$
$$\rightarrow 4CO_2 + 5Pb(OAc)_2 + 10HOAc \quad (12.49)$$

The excess of standard lead(IV) was determined by titration with standard 1,4-benzendiol solution prepared in 1% sulfuric acid solution.

## Acetonitrile Media

Organic sulfur compounds have also been determined in these media by titration with oxidants such as bromine(I) cation, cerium(IV), copper(II), and iron(III). These compounds are often oxidized through a thiol group forming the corresponding disulfide. Thus bromine(I) cation, as a solution of bromine cyanide in acetonitrile, has been used to oxidize the anions of dialkylcarbamodithioc acids $R_1R_2NC(:S)S^-$ (e.g., sodium diethyldithiocarbamate) in a medium of acetonitrile (206). The reaction is

$$2R_1R_2NC(:S)S^- + Br^+ \rightarrow R_1R_2NC(:S)-S-S-C(:S)NR_1R_2 + Br^-$$

(12.50)

The bromine(I) cation, this time emanating from $N$-bromosuccinimide, was used also to determine these dithiocarbamates in pure acetonitrile containing a small amount of iodide (2 mL, 1% KI) (307). Alkyl derivatives of the anion of dithiocarbonic acid $RO-C(:S)S^-$ (e.g., potassium ethylxanthate) were also determined by the same method and undergo similar oxidations

$$2ROC(:S)S^- \rightarrow ROC(:S)-S-S-C(:S)OR + 2e$$

(12.51)

Ammonium hexanitratocerate(IV) has also been used for this oxidation in anhydrous acetonitrile and was applied similarly to the oxidation of monosubstituted thioureas reacting through the thiol tautomer (183)

$$2RNHC(:NH)SH \rightarrow RNHC(:NH)-S-S-C(:NH)NHR + 2H^+ + 2e$$

(12.52)

When titrated with ammonium hexanitratocerate(IV), substituted trithiocarbonates [$RS-C(:S)S^-$] are oxidized analogously to reaction (12.51) (308).

Copper(II) perchlorate has been used as a titrant for the oxidation of thiourea [see reaction (12.52), R = H, (63)] and also for the determination of the thiourea–copper(I) complexes (197). The copper(II) titrant is reduced to copper(I) in these reactions and this titrant has been applied also to the determination of potassium ethyl xanthate by reaction (12.51) (196). Thiols and thiones have been similarly titrated with iron(III) in acetonitrile (70). Some sulfur compounds titrated were benzothiazole-2(3$H$)-thione, 2-naphthalenethiol, phenothiazine [Reg. No. *92-84-2*] and its $N$-phenyl derivative, tetramethylthiourea, thioacetamide, thioacetanilide, $NN$-diphenylthiourea, and thiourea.

Iodide, emanating from tetrabutylammonium iodide, has been titrated with copper(II) in anhydrous acetonitrile and also when the solvent contained 0.1 $M$ water ($\sim$ 0.2%). In both instances two inflections were observed, the first corresponding to the formation of triiodide ($3I^- \rightarrow I_3^- + 2e$) and the second to the oxidation of triiodide to iodine ($2I_3^- \rightarrow 3I_2 + 2e$). However, when 1,4-benzenediol was titrated, an inflection was observed only in the presence of water (0.1 $M$), and this corresponded to the formation of 2,5-cyclohexadiene-1,4-dione (63). Copper(II) was used also for the quantitative oxidation of ascorbic acid to dehydroascorbic acid in a medium of acetic acid and acetonitrile (199). Some arylamines have been titrated with copper(II) in acetonitrile (198), and iron(III) can also be used for these and other titrations (70).

## Alcohols

Methanol has been used as the medium for the oxidation of $N$-monosubstituted thioureas with a methanolic solution of bromine cyanide (206) which proceeds according to reaction (12.52). Methanol (150 mL) acidified with perchloric acid (10 mL, 70%) was the medium for the titration of alkyl and hydroxyalkylferrocenes with a methanolic solution of iron(III) chloride (232).

Ethanol (96%) was the medium of choice for the microdeterminations of substituted 2-amino- and 7-amino-3-phenoxazinones by titration with bromine in the presence of a small amount of hydrobromic acid (0.5 mL, 0.1 $N$). Alternatively, these compounds may be titrated with potassium bromate in the presence of potassium bromide solution (5%) acidified with hydrobromic acid (227). The sodium salts of some dialkylcarbamodithioic acids have been titrated in ethanol with an ethanolic solution of bromine cyanide (206). The oxidations proceed according to equation (12.50).

Azo dyes, prepared as solutions in $NN$-dimethylformamide (10–20 mL, $\sim$ 0.01 $M$), have been titrated with 0.01 $N$ iron(II) solution in an alkaline medium (2 mL, 10% KOH) composed of approximately 1 : 1 glycerol–DMF (20–30 mL) which also contained triethanolamine (2 mL) (309). The titrant was prepared by dissolving ammonium iron(II) sulfate (Mohr's salt) in a minimum amount of water and diluting to the required volume with glycerol (80–85%). A medium containing glycerol has also been used for the determination of dyes containing either azo or nitro groups by reduction with an excess of standard titanium(III) followed by titration of this excess with iron(III).

## Other Media

Copper(II) chloride prepared in $NN$-dimethylformamide (DMF) has been used as an oxidant for the determination of thiols dissolved in DMF (60).

Some 2-mercaptoalkanoic acids and some benzenethiols (i.e., thiophenols) were oxidized to the corresponding disulfides according to the stoichiometry of reaction (12.44), the copper(II) being reduced to copper(I) (see also Reference 195). Similar reactions occurred when these compounds were titrated in pyridine with a solution of iron(III) chloride prepared in pyridine (60).

Propylene carbonate (4-methyl-1,3-dioxolan-2-one) has been suggested as a convenient medium for the determination of phenols and aromatic amines by direct titration with bromine (201). Most phenols and aromatic amines as well as their bromination products are soluble in this solvent, but a base such as pyridine must be present when phenols are being titrated in order to accept the protons released in the substitution. The titration of some aromatic amines is possible, however, without the addition of pyridine. The amount of pyridine present determines the extent of the bromination of phenols and to a lesser extent the bromination of anilines. This, in turn, depends upon the substituent(s) present, and for the optimum conditions necessary for the application of these determinations the original publication should be consulted. The reactions are rapid, and the bromine solution may be conveniently standardized against solutions of tetraethylammonium bromide.

A medium of benzene–tetrahydrofuran has been used for the determination of zinc alkyls by titration with iodine prepared in anhydrous benzene (311). These reactions take place in two steps

$$ZnR_2 + I_2 \rightarrow RI + RZnI$$

$$RZnI + I_2 \rightarrow RI + ZnI_2$$

The use of 1 : 4-tributylphosphate–carbon tetrachloride as a medium for the preparation of cerium(IV) solutions has also been investigated (184, 260).

## REFERENCES

1. I. M. Kolthoff and R. Belcher, *Volumetric Analysis, Volume III—Titration Methods: Oxidation Reduction Reactions*, Interscience, New York, 1957.
2. G. G. Rao and P. V. K. Rao, *Anal. Chim. Acta* **65**, 347 (1973).
3. N. T. Akinchan and S. N. Prasad, *Indian J. Chem.* **10**, 845 (1972).
4. J. A. Goldman, *J. Electroanal. Chem. Interfacial Electrochem.* **11**, 255 (1966); **11**, 416 (1966); **14**, 373 (1967); **16**, 47 (1968); **18**, 41 (1968); **19**, 205 (1968).
5. G. G. Rao, P. V. K. Rao, and G. B. B. Rao, *Bull. Chem. Soc. Jpn.* **45**, 3434 (1972).
6. I. M. Kolthoff and J. I. Watters, *Ind. Eng. Chem., Anal. Ed.* **15**, 8 (1943).

7. J. J. Lingane and D. G. Davis, *Anal. Chim. Acta* **15**, 201 (1956).
8. G. W. C. Milner, A. J. Wood, G. Weldrick, and G. Phillips, *Analyst (London)* **92**, 239 (1967).
9. J. L. Drummond and R. A. Grant, *Talanta* **13**, 477 (1966).
10. R. W. Bane, *Analyst (London)* **90**, 756 (1965): **95**, 722 (1970).
11. E. Schulek and M. Szakacs, *Acta Chim. Acad. Sci. Hung.* **4**, 457 (1954).
12. F. Feigl, K. Klanfer, and L. Weidenfeld, *Fresenius' Z. Anal. Chem.* **80**, 5 (1930).
13. H. A. Laitinen and L. W. Burdett, *Anal. Chem.* **23**, 1268 (1951).
14. H. H. Willard and J. J. Thompson, *Ind. Eng. Chem., Anal. Ed.* **3**, 399 (1931).
15. D. I. Anokhina, *Metody Anal. Kontrolya Proizvod. Khim. Prom-sti.* **2**, 43 (1976); *CA* **85**, 116114z.
16. V. I. Bogovina, Y. I. Usatenko, and V. F. Mal'tsev, *Zh. Anal. Khim.* **23**, 1152 (1968); *J. Anal. Chem. USSR* **23**, 1012 (1968).
17. H. A. Fales and P. S. Roller, *J. Am. Chem. Soc.* **51**, 345 (1929).
18. B. Reinitzer and P. Conrath, *Fresenius' Z. Anal. Chem.* **68**, 81 (1926).
19. N. H. Furman, *J. Am. Chem. Soc.* **50**, 755 (1928).
20. H. H. Willard and P. Young, *J. Am. Chem. Soc.* **50**, 1379 (1928); *Ind. Eng. Chem., Anal. Ed.* **4**, 187 (1932); **5**, 154, 158 (1933).
21. A. Claasen and L. Bastings, *Fresenius' Z. Anal. Chem.* **202**, 241 (1964).
22. H. A. Bright and C. P. Larrabee, *J. Res. Natl. Bur. Stand. (U.S.)* **3**, 573 (1929).
23. R. Lang and F. Kurtz, *Fresenius' Z. Anal. Chem.* **85**, 181 (1931).
24. H. Nakano, K. Onishi, and Y. Ohuchi, *Bunseki Kagaku* **19**, 1293 (1970).
25. H. H. Willard and L. L. Merritt, Jr., *Ind. Eng. Chem., Anal. Ed.* **14**, 486 (1942).
26. W. C. Hoyle and H. Diehl, *Talanta* **18**, 1072 (1971).
27. G. F. Smith, *Anal. Chim. Acta* **8**, 397 (1953); *Talanta* **11**, 633 (1964).
28. W. F. Pickering, *Chemist-Analyst* **53**, 91 (1964).
29. C. Jones, *Trans. Am. Inst. Mining Eng.* **17**, 411 (1889).
30. Reference 1, p. 11.
31. J. Bassett, R. C. Denney, G. H. Jeffery, and J. Mendham, *Vogel's Textbook of Quantitative Inorganic Analysis*, 4th ed., Longman, London, 1978, p. 395.
32. N. G. Serba, L. P. Shul'gin, Y. A. Koz'min, and R. M. Korostyshevskaya, *Zavod. Lab.* **33**, 696 (1967); *CA* **68**, 18350c.
33. F. Pantani, *Anal. Chim. Acta* **31**, 121 (1964).
34. R. J. Jaworowski and W. D. Bratton, *Anal. Chem.* **34**, 111 (1962).
35. A. S. Witwit and R. J. Magee, *Anal. Chim. Acta* **27**, 366 (1962).
36. J. Sevcik and J. Cihalik, *Collect. Czech. Chem. Commun.* **31**, 3140 (1966).
37. W. R. Amos and W. B. Brown, *Anal. Chem.* **35**, 309 (1963).
38. G. G. Rao and S. R. Sagi, *Talanta* **10**, 169 (1963).
39. I. M. Kolthoff and E. Amdur, *Ind. Eng. Chem., Anal. Ed.* **12**, 177 (1940).
40. J. Haslam and N. T. Wilkinson, *Analyst (London)* **78**, 390 (1953).
41. M. N. Sastri and C. Radhakrishnamurti, *Fresenius' Z. Anal. Chem.* **147**, 16 (1935).
42. B. S. Evans, *Analyst (London)* **56**, 171 (1931).

# REFERENCES

43. B. S. Evans and D. G. Higgs, *Analyst (London)* **69**, 201 (1944).
44. B. S. Evans, *Analyst (London)* **68**, 874 (1938).
45. K. L. Cheng, *Anal. Chim. Acta*, **35**, 293 (1966).
46. P. P. Shatko, *Zh. Anal. Khim.* **7**, 242 (1952).
47. W. D. Cooke, F. Hazel, and W. M. McNabb, *Anal. Chim. Acta* **3**, 656 (1949).
48. E. Schulek and P. Von Villecz, *Fresenius' Z. Anal. Chem.* **76**, 81, (1929).
49. R. Khristova and Mac Van Men, *God. Sofii Univ., Khim. Fak.* **67**, 243 (1972–1973); *CA* **87**, 33181j.
50. R. Lang and S. Gottlieb, *Fresenius' Z. Anal. Chem.* **104**, 1 (1936).
51. S. Tribalat, *Anal. Chim. Acta* **1**, 149 (1947).
52. T. S. Light, *Anal. Chem.* **44**, 1038 (1972).
53. J. G. Mason and M. Rosenblum, *J. Am. Chem. Soc.* **82**, 4206 (1960).
54. G. G. Rao, M. Gandikota, and S. G. Viswanath, *Anal. Chim. Acta* **87**, 511 (1976).
55. A. Ramadan, P. K. Agasyan, and S. I. Petrov, *Zh. Anal. Khim.* **29**, 1144 (1974); *J. Anal. Chem. USSR* **29**, 977 (1974).
56. L. Erdey and E. Bodor, *Anal. Chem.* **24**, 418 (1952).
57. G. G. Rao and V. N. Rao, *Fresenius' Z. Anal. Chem.* **147**, 338 (1955).
58. G. G. Rao and S. G. Sastry, *Anal. Chim. Acta* **56**, 325 (1971).
58a. G. G. Rao, P. K. Rao, and S. B. Rao, *Talanta* **11**, 825 (1964).
59. R. Pakash, R. K. Chauhan, and J. Zyka, *Microchem. J.* **17**, 519 (1972).
60. Z. Hladky and J. Vrestal, *Collect. Czech. Chem. Commun.* **34**, 984 (1969).
61. L. Meites in *Handbook of Analytical Chemistry*, L. Meites, Ed., McGraw-Hill, New York, 1963, p. 5–7.
61a. G. G. Rao, *Talanta* **13**, 1473 (1966).
62. K. Krishnamurty and Y. Pullarao, *Anal. Chim. Acta* **73**, 413 (1974).
63. B. Kratochvil, D. A. Zatko, and R. Markuszewski, *Anal. Chem.* **38**, 770 (1966).
64. C. G. R. Nair and P. Indrasenan, *Talanta* **23**, 239 (1976).
65. T. J. Jacob and C. G. R. Nair, *Talanta* **19**, 347 (1972).
66. C. G. R. Nair and P. N. K. Nambisan, *Indian J. Chem.* **12**, 645 (1974).
67. L. I. Veselago, *Zh. Anal. Khim.* **23**, 384 (1968); *J. Anal. Chem. USSR* **23**, 317 (1968).
68. G. G. Rao and S. R. Sagi, *Talanta* **9**, 715 (1962).
69. G. G. Rao and N. K. Murty, *Fresenius' Z. Anal. Chem.* **208**, 97 (1965).
70. B. Kratochvil and R. Long, *Anal. Chem.* **42**, 43 (1970).
71. J. Dolezal and F. J. Langmyhr, *Anal. Chim. Acta* **61**, 73 (1972).
72. J. Zyka and J. Dolezal, *Microchem. J.* **10**, 554 (1966).
73. J. Vulterin, *Talanta* **10**, 891 (1963).
74. I. M. Issa and M. M. Ghoneim, *Talanta* **20**, 517 (1973).
74a. G. G. Rao and P. K. Rao, *Talanta* **10**, 1251 (1963).
75. A. I. Busev and Gyn Li, *Zh. Anal. Khim.* **14**, 668 (1959); *J. Anal. Chem. USSR*, **14**, 741 (1959).
76. B. Kratochvil and H. Diehl, *Talanta* **3**, 346 (1960).
77. S. R. Sagi and P. R. M. Rao, *Talanta* **23**, 427 (1976).
78. G. G. Rao and L. S. A. Dikshitulu, *Talanta* **10**, 1023 (1963).

79. G. G. Rao and L. S. A. Dıkshitulu, *Talanta* **10**, 295 (1963).
80. J. Vulterin and L. Knourkova, *Sb. Vys. Sk. Chem.-Technol. Praze. Anal. Chem.* **1969**, No. 4, 63; *CA* **73**, 116073n.
81. V. I. Ignatov, V. T. Solomatin, and A. A. Nemodruk, *Zh. Anal. Khim.* **33**, 2328 (1978); *J. Anal. Chem. USSR* **33**, 1785 (1978).
82. W. Wisniewski and H. Basinska, *Chem. Anal. (Warsaw)* **12**, 99 (1967); *CA* **67**, 50087n.
83. A. I. Busev and Li Gyn, *Zh. Anal. Khim.* **15**, 191 (1960); *J. Anal. Chem. USSR* **15**, 217 (1960).
84. A. Berka, J. Dolezal, I. Nemec, and J. Zyka, *Anal. Chim. Acta* **25**, 533 (1961).
85. A. Berka, J. Dolezal, I. Nemec, and J. Zyka, *J. Electroanal. Chem. Interfacial Electrochem.* **3**, 278 (1962).
86. G. S. Sastry and G. G. Rao, *Talanta* **19**, 212 (1972).
87. J. Vulterin, *Collect. Czech. Chem. Commun.* **32**, 3349 (1967).
88. N. K. Murty, V. P. Rao, and V. Satyanarayana, *J. Indian. Chem. Soc.* **55**, 686 (1978).
89. J. Vulterin, *Collect. Czech. Chem. Commun.* **31**, 3529 (1966).
90. J. Vulterin, *Collect. Czech. Chem. Commun.* **31**, 2501 (1966).
91. N. K. Murty and V. Satyanarayana, *J. Indian Chem. Soc.* **53**, 712 (1976).
92. N. Krishnamurthy and Y. P. Rao, *Indian J. Chem., Sect. A* **14A**, 213 (1976).
93. S. G. Dyatel and T. N. Rapoport, *Izv. Vyssh. Uchebn, Zaved., Khim. Khim. Tekhnol.* **19**, 1145 (1976); *CA* **86**, 64974d.
94. M. Goto, H. Ota, and D. Ishii, *Nippon Kagaku Zasshi*, **92**, 468 (1971); *CA* **75**, 83862e.
95. U. Muralikrishna and G. G. Rao, *Talanta* **15**, 143 (1968).
96. J. A. Duckitt and G. C. Goode, *Analyst (London)* **87**, 121 (1962).
97. G. G. Rao, P. K. Rao, and M. A. Rahman, *Talanta* **12**, 953 (1965).
98. G. G. Rao and P. K. Rao, *Talanta* **11**, 1031 (1964).
99. W. Davies and W. Gray, *Talanta* **11**, 1203 (1964).
100. L. Pszonicki, *Talanta* **13**, 403 (1966).
101. G. G. Rao and P. K. Rao, *Talanta* **11**, 703 (1964).
102. K. A. Idriss, I. M. Issa, and Y. M. Temerk, *Fresenius' Z. Anal. Chem.* **278**, 364 (1976).
103. I. M. Issa, R. M. Issa, A. M. Hammam, and M. R. Mahmoud, *J. Indian Chem. Soc.* **53**, 698 (1976).
104. I. M. Issa, A. S. Misbah, and M. Hamdy, *U.A.R. J. Chem.* **13**, 127 (1970); *CA* **74**, 119726j.
105. I. M. Issa, K. A. Idriss, and M. M. Ghoneim, *Egypt. J. Chem.* **18**, 633 (1975); *CA*, **89**, 139854n.
106. J. J. Lingane and R. Karplus, *Ind. Eng. Chem., Anal. Ed.* **18**, 191 (1946).
107. W. G. Scribner and R. A. Anduze, *Anal. Chem.* **33**, 770 (1961).
108. W. G. Scribner, *Anal. Chem.* **32**, 966 (1960); **32**, 970 (1960).
109. P. Blanicky, J. Dolezal, and J. Zyka, *Fresenius' Z. Anal. Chem.* **240**, 233 (1968).
110. J. Vulterin, *Collect. Czech. Chem. Commun.* **34**, 1229 (1969).

# REFERENCES

111. M. Herman, Z. Sulcek, and J. Zyka, *Collect. Czech. Chem. Commun.* **31**, 2005 (1966).
112. Y.-H. Chang, J. Dolezal, and J. Zyka, *Zh. Anal. Khim.* **16**, 308 (1960); *J. Anal. Chem. USSR* **16**, 321 (1960).
113. O. Guertler and J. Dolezal, *Fresenius' Z. Anal. Chem.* **233**, 97 (1968).
114. N. Chughtai, J. Dolezal, and J. Zyka, *Microchem. J.* **20**, 363 (1975).
115. J. W. Miller, *Talanta* **4**, 292 (1960).
116. K. N. Raoot and V. G. Vaidya, *Fresenius' Z. Anal. Chem.* **292**, 412 (1978).
117. W. M. D. Bryant, J. Mitchell, Jr., and D. M. Smith, *J. Am. Chem. Soc.* **62**, 1 (1940).
118. H. H. Willard and P. Young, *J. Am. Chem. Soc.* **52**, 553 (1930).
119. N. N. Sharma and R. C. Mehrota, *Anal. Chim. Acta* **13**, 419 (1955).
120. F. S. Grimaldi, R. E. Stevens, and M. K. Carron, *Ind. Eng. Chem., Anal. Ed.* **15**, 387 (1943).
121. B. D. Sully, *J. Chem. Soc.* **1942**, 366.
122. H. S. Gowda, K. B. Rao, and G. G. Rao, *Anal. Chim. Acta* **12**, 506 (1955).
123. V. P. R. Rao and B. V. S. Sharma, *Anal. Chem.* **37**, 1373 (1965).
124. G. F. Smith and H. H. Bliss, *J. Am. Chem. Soc.* **53**, 4291 (1931).
125. S. Syamsunder and T. K. S. Murthy, *Indian J. Chem.* **11**, 669 (1973).
126. K. L. Chawla and J. P. Tandon, *Talanta* **12**, 665 (1965); **13**, 545 (1966).
127. C. Casalini, G. Cesarano, and G. Mascellari, *Anal. Chem.* **49**, 1002 (1977).
128. C. G. R. Nair and V. R. Nair, *Talanta* **20**, 696 (1973).
129. D. E. Metzler, R. J. Myers, and E. H. Swift, *Ind. Eng. Chem., Anal. Ed.* **16**, 625 (1944).
130. R. Lang, *Fresenius' Z. Anal. Chem.* **115**, 103 (1938).
131. B. Singh, S. C. Nistandra, and B. C. Verma, *Fresenius' Z. Anal. Chem.* **257**, 348 (1971).
132. B. Singh, B. C. Verma, and Y. K. Kalia, *Indian J. Chem.* **2**, 124 (1964).
133. J. Minczewski and U. Glabisz, *Talanta* **5**, 179 (1960).
134. G. S. Deshmukh and M. K. Joshi, *Fresenius' Z. Anal. Chem.* **142**, 275 (1954).
135. K. B. Rao and G. G. Rao, *Anal. Chim. Acta* **13**, 313 (1955).
136. V. R. Wheatley, *Analyst (London)* **69**, 207 (1944).
137. G. G. Rao and U. Muralikrishna, *Anal. Chim Acta* **13**, 8 (1955).
138. Z. G. Sabo and L. Csanyi, *Anal. Chem.* **21**, 1144 (1949).
139. I. M. Kolthoff and E. B. Sandell, *Ind. Eng. Chem., Anal. Ed.* **2**, 140 (1930).
140. L. Szebelledy and W. Madis, *Fresenius' Z. Anal. Chem.* **109**, 391 (1937).
141. K. G. Boto and L. F. G. Williams, *Anal. Chim. Acta* **85**, 179 (1976).
142. N. T. Akinchan and S. N. Prasad, *J. Indian Chem. Soc.* **50**, 449 (1973).
143. A. Schwicker, *Fresenius' Z. Anal. Chem.* **110**, 167 (1937).
144. I. M. Kolthoff, E. B. Sandell, and B. Moskowitz, *J. Am. Chem. Soc.* **55**, 1454 (1933).
145. A. J. Boyle, V. V. Hughey, and C. C. Casto, *Ind. Eng. Chem., Anal. Ed.* **16**, 370 (1944).
146. I. M. Kolthoff and H. Yutzy, *Ind. Eng. Chem., Anal. Ed.* **9**, 75 (1937).
147. K. Gleu, *Fresenius' Z. Anal. Chem.* **95**, 305 (1933).

148. H. D. Hillson, *Ind. Eng. Chem., Anal. Ed.* **16**, 560 (1944).
149. K. Gleu, *Fresenius' Z. Anal. Chem.* **95**, 385 (1933).
150. P. K. Norkus, *Zh. Anal. Khim.* **20**, 496 (1965); *J. Anal. Chem. USSR* **20**, 460 (1965).
151. J. H. Van der Meulen, *Chem. Weekblad.* **28**, 348 (1931).
152. P. Norkus and S. Stulgiene, *Zh. Anal. Khim.* **24**, 1565 (1969); *J. Anal. Chem. USSR* **24**, 1263 (1969).
153. P. Norkus and G. Simikeviciute, *Zh. Anal. Khim.* **26**, 1076 (1971); *J. Anal. Chem. USSR* **26**, 964 (1971).
154. P. V. K. Rao and G. G. Rao, *Talanta* **20**, 907 (1973).
155. P. V. K. Rao and G. B. B. Rao, *Analyst (London)* **96**, 712 (1971).
156. D. A. Zatko and B. Kratochvil, *Anal. Chem.* **37**, 1560 (1965).
157. N. K. Murty and K. R. Rao, *Indian J. Chem. Sect. A* **15A**, 569 (1977).
158. E. G. Brown, *Anal. Chim. Acta* **7**, 494 (1952).
159. P. Norkus, *Zh. Anal. Khim.* **19**, 518 (1964); *J. Anal. Chem. USSR* **19**, 478 (1964).
160. P. Norkus, *Zh. Anal. Khim.* **20**, 612 (1965); *J. Anal. Chem. USSR* **20**, 567 (1965).
161. P. Norkus, *Zh. Anal. Khim.* **20**, 88 (1965); *J. Anal. Chem. USSR* **20**, 77 (1965).
162. P. Norkus and S. Stulgiene, *Zh. Anal. Khim.* **22**, 101 (1967); *J. Anal. Chem. USSR* **22**, 83 (1967).
163. J. Jankaukas and P. Norkus, *Zh. Anal. Khim.* **28**, 2257 (1973); *J. Anal. Chem. USSR* **28**, 2009 (1973).
164. P. Norkus, *Liet. TSR Mosk. Akad. Darb. Ser B* **1961**, 151; *CA* **57**, 2850b.
165. P. Norkus, *Zh. Anal. Khim.* **18**, 650 (1963); *J. Anal. Chem. USSR* **18**, 560 (1963).
166. J. Jankaukas and P. Norkus, *Zh. Anal. Khim.* **27**, 1629 (1972); *J. Anal. Chem. USSR* **27**, 1478 (1972).
167. P. Norkus, J. Jankaukas, and G. I. Rozovskii, *Zh. Anal. Khim.* **26**, 1427 (1971); *J. Anal. Chem. USSR* **26**, 1272 (1971).
168. G. G. Rao, G. Aravamudam, and N. C. Venkatamma, *Fresenius' Z. Anal. Chem.* **146**, 161 (1955).
169. G. G. Rao, V. P. Rao, and M. N. Sastri, *Curr. Sci. (India)* **18**, 381 (1949).
170. G. G. Rao, and M. N. Sastri, *Curr. Sci. (India)* **21**, 189 (1952).
171. B. V. S. R. Murty and G. G. Rao, *Talanta* **8**, 547 (1961).
172. L. S. A. Dikshitulu, *Indian J. Chem.* **8**, 371 (1970).
173. P. Norkus, G. Simkevichute, and J. Jankauskas, *Zh. Anal. Khim.* **25**, 1673 (1970); *J. Anal. Chem. USSR* **25**, 1438 (1970).
174. V. G. Rozovskii, P. K. Norkus, R. R. Sharmaitis, and Yu. Yu. Matulis, *Zh. Anal. Khim.* **32**, 757 (1977); *J. Anal. Chem. USSR* **32**, 601 (1977).
175. P. Norkus and J. Jankauskas, *Zh. Anal. Khim.* **30**, 925 (1975); *J. Anal. Chem. USSR* **30**, 778 (1975).
176. K. B. Rao and G. G. Rao, *Fresenius' Z. Anal. Chem.* **147**, 274, 279 (1955).
177. H. S. Gowda, K. B. Rao, and G. G. Rao, *Anal. Chim. Acta* **12**, 506 (1955).
178. A. Gadeke, *Fresenius' Z. Anal. Chem.* **131**, 428 (1950).

## REFERENCES

179. H. H. Willard and P. Young, *J. Am. Chem. Soc.* **50**, 1322, 1334, 1368, 1372, 1379 (1928).
180. Reference 1, p. 121.
181. J. Bassett, R. C. Denney, G. H. Jeffery, and J. Mendham, *Vogel's Textbook of Quantitative Inorganic Analysis*, 4th ed., Longman, London, 1978, p. 363.
182. Reference 1, p. 122.
183. B. C. Verma and S. Kumar, *Talanta* **23**, 241 (1976).
184. H. C. Mruthyunjaya and A. R. V. Murty, *Indian J. Chem.* **5**, 430 (1967).
185. M. V. Ryzhinskii, L. D. Preobrazhenskaya, L. F. Solntseva, and E. A. Gromova, *Zh. Anal. Khim.* **33**, 1738 (1978); *J. Anal. Chem. USSR* **33**, 1331 (1978).
186. W. Helbig, *Fresenius' Z. Anal. Chem.* **182**, 84 (1961).
187. P. T. Close, E. J. Hornyak, T. Baak, and J. F. Tillman, *Microchem. J.* **10**, 334 (1966).
188. Reference 1, p. 169.
189. C. E. Bricker and L. J. Loeffler, *Anal. Chem.* **27**, 1419 (1955).
190. J. Budesinsky, J. Dolezal, B. Sramkova, and J. Zyka, *Microchem. J.* **16**, 121 (1971).
191. B. Sramkova and J. Zyka, *Microchem. J.* **19**, 295 (1974).
192. J. A. Baur and C. E. Bricker, *Anal. Chem.* **37**, 1461 (1965).
193. M. Vasatova and J. Zyka, *Microchem. J.* **22**, 34 (1977).
194. P. Norkus and R. M. Markyavichene, *Zh. Anal. Khim.* **22**, 1527 (1967); *J. Anal. Chem. USSR* **22**, 1282 (1967).
195. K. G. Haeusler, R. Geyer, and S. Rennhak, *Z. Chem.* **13**, 196 (1973).
196. H. C. Mruthyunjaya and A. R. V. Murthy, *Indian J. Chem.* **7**, 403 (1969).
197. H. C. Mruthyunjaya and A. R. V. Murthy, *Anal. Chem.* **41**, 186 (1969).
198. B. Kratochvil and D. A. Zatko, *Anal. Chem.* **40**, 422 (1968).
199. B. C. Verma and S. Kumar, *Talanta* **24**, 694 (1977).
200. C. O. Huber and J. M. Gilbert, *Anal. Chem.* **34**, 247 (1962).
201. R. D. Krause and B. Kratochvil, *Anal. Chem.* **45**, 844 (1973).
202. Reference 1, p. 199.
203. N. A. Shevlyakova, Yu. S. Bryksina, and Yu. V. Lyande, *Zh. Anal. Khim.* **28**, 349 (1973); *J. Anal. Chem. USSR* **28**, 298 (1973).
204. Kabir-ud-Din, A. A. Khan, and M. A. Beg, *J. Indian Chem. Soc.* **52**, 1148 (1975).
205. R. C. Paul, R. K. Chauhan, and R. Parkash, *Indian J. Chem.* **9**, 879 (1971).
206. R. C. Paul, R. K. Chauman, and R. Parkash, *Talanta* **21**, 663 (1974).
207. R. Parkash and J. Zyka, *Microchem. J.* **20**, 193 (1975).
208. R. Parkash, R. K. Chauhan, and J. C. Bhatia, *Acta Cienc. Indica* **4**, 130 (1978); *Anal. Abstr.* **37**, 3B17 (1979).
209. Reference 1, p. 453.
210. P. K. Jain and K. N. Sharma, *Indian J. Chem., Sect. A* **15A**, 369 (1977).
211. G. Piccardi and P. Cellini, *Anal. Chim. Acta.* **29**, 107 (1963).
212. K. H. Pausacker, *J. Chem. Soc.* **1953**, 107.
213. V. S. N. Pillai and C. G. R. Nair, *Talanta* **22**, 57 (1975).
214. J. Kepinski and G. Blaszkiewicz, *Talanta* **13**, 357 (1966).

215. M. Amin and M. Rumeau, *Ann. Chim. (Paris)* **2**, 75 (1977).
216. S. Pinzauti, V. Dal Piaz, and E. La Porta, *J. Pharm. Sci.* **63**, 1446 (1974).
217. E. Bishop and V. J. Jennings, *Talanta* **8**, 22 (1961).
218. E. Bishop and V. J. Jennings, *Talanta* **9**, 679 (1962).
219. M. A. Koupparis and T. P. Hadjiioannou, *Talanta* **25**, 477 (1978).
220. B. Singh, S. C. Nistandara, and B. C. Verma, *Fresenius' Z. Anal. Chem.* **258**, 366 (1972).
221. Rangaswamy and D. S. Mahadevappa, *J. Indian Chem. Soc.* **55**, 620 (1978).
222. Reference 1, p. 644.
223. C. Radhakrishnamurty and G. G. Rao, *Fresenius' Z. Anal. Chem.* **254**, 366 (1971).
224. Reference 1, p. 501.
225. Reference 1, p. 449.
226. N. E. Khomutov and V. I. Eberil, *Zh. Anal. Khim.* **17**, 763 (1962); *J. Anal. Chem. USSR* **17**, 754 (1962).
227. E. Ruzicka, *Mikrochim. Acta* **1968**, 778.
228. K. G. Haeusler, R. Geyer, and S. Rennhak, *Z. Chem.* **16**, 232 (1976).
229. B. V. Rao, S. V. Athavale, T. H. Rao, S. L. N. Acharyulu, and R. V. Tamhankar, *Anal. Chim. Acta* **70**, 169 (1974).
230. F. Vydra and R. Pribil, *Talanta* **5**, 44 (1960).
231. Z. Hladky, *Z. Chem.* **5**, 424 (1965).
232. D. M. Knight and R. C. Schlitt, *Anal. Chem.* **37**, 470 (1965).
233. R. W. Collins and W. H. Nebergall, *Anal. Chem.* **34**, 1511 (1962).
234. H. Kirtchik and F. H. Swearingen, *Analyst (London)* **86**, 188 (1961).
235. J. J. Lingane, *Anal. Chim. Acta* **30**, 319 (1964).
236. A. Berka, V. Dvorak, I. Nemec, and J. Zyka, *J. Electroanal. Chem. Interfacial Electrochem.* **4**, 150 (1962).
237. I. M. Issa, M. S. El-Meligy, and K. A. Idriss, *Indian J. Chem.* **12**, 877 (1974).
238. A. Berka, V. Dvorak, I. Nemec, and J. Zyka, *Anal. Chim. Acta* **23**, 380 (1960).
239. J. Santrucek, I. Nemec, and J. Zyka, *Collect. Czech. Chem. Commun.* **31**, 2679 (1966).
240. A. Morales and J. Zyka, *Collect. Czech. Chem. Commun.* **27**, 1029 (1962).
241. A. Berka, V. Prochazkova, and J. Zyka, *Cesk. Farm.* **13**, 121 (1964); *CA* **61**, 12640b.
242. K. G. Haeusler, R. Geyer, and S. Rennhak, *Z. Chem.* **17**, 146 (1977).
243. Z. Hladky and J. Ruza, *Chem. Zvesti* **23**, 336 (1969).
244. R. C. Paul, K. S. Sooch, O. C. Vaidya, and S. P. Narula, *Anal. Chim. Acta* **46**, 131 (1969).
245. J. Barek, A. Berka, and A. Pokorna, *Microchem. J.* **23**, 104 (1978).
246. S. K. Mandal and B. R. Sant, *Talanta* **23**, 485 (1976).
247. Reference 1, p. 33.
248. S. S. Basson, L. D. C. Bok, and S. R. Grobler, *Fresenius' Z. Anal. Chem.* **268**, 287 (1974).
249. P. Norkus and R. Markeviciene, *Liet. TSR Mokslu Akad. Darb., Ser. B* **1969**, 127; *CA* **71**, 77002a.
250. E. Bishop and A. B. Crawford, *Analyst (London)* **75**, 273 (1950).

251. G. G. Rao, V. B. Rao, and M. N. Sastri, *Curr. Sci.* **18**, 381 (1949).
252. D. Norkiene and G. P. Tomashevich, *Tekh. Kino Telev.* **11**, 20 (1967); *CA* **69**, 16068p.
253. Reference 1, p. 233.
254. P. K. Norkus and Yu. Yu. Yankauskas, *Zh. Anal. Khim.* **28**, 2284 (1973); *J. Anal. Chem. USSR* **28**, 2031 (1973).
255. Reference 1, p. 43.
256. P. K. Norkus and Yu. Yu. Yankauskas, *Zh. Anal. Khim.* **26**, 1837 (1971); *J. Anal. Chem. USSR* **26**, 1641 (1971).
257. Reference 1, p. 662.
258. K. N. Pshenitsyn, S. I. Ginzburg, and I. V. Prokof'eva, *Zh. Anal. Khim.* **17**, 343 (1961); *J. Anal. Chem. USSR* **17**, 346 (1961).
259. A. Berka, *Anal. Chim. Acta* **24**, 171 (1961).
260. H. C. Mruthyunjaya, *Curr. Sci.* **36**, 537 (1967).
261. A. Berka, J. Vulterin, and J. Zyka, *Chemist-Analyst* **51**, 88 (1962).
262. N. H. Furman and A. J. Fenton, Jr., *Anal. Chem.* **32**, 745 (1960).
263. Reference 1, p. 367.
264. D. P. Smith and M. T. Pope, *Anal. Chem.* **40**, 1906 (1968).
265. Reference 1, p. 600.
266. G. J. Misra and J. P. Tandon, *Indian J. Chem.* **5**, 343 (1967).
267. P. V. K. Rao and G. G. Rao, *Indian J. Chem.* **11**, 1309 (1973).
268. J. Vulterin, *Sb. Vys. Sk. Chem.-Technol. Praze, Anal. Chem.* **1970** (6), 77; *CA* **74**, 119626b.
269. J. Vulterin and H. Polacek, *Collect. Czech. Chem. Commun.* **37**, 1999 (1972).
270. J. Vulterin and P. Straka, *Sb. Vys. Sk. Chem.-Technol. Praze, Anal. Chem.* **1976** (H11), 257; *CA* **86**, 133151x.
271. J. Vulterin, *Sb. Vys, Sk. Chem.-Technol. Praze, Anal. Chem.* **1970** (6), 85; *CA* **74**, 119722e.
272. P. Norkus and J. Jankauskas, *Zh. Anal. Khim.* **28**, 127 (1973); *J. Anal. Chem. USSR* **28**, 105 (1973).
273. L. V. Boeva, K. N. Bagdasarov, and V. A. Kimstach, *Zavod. Lab.* **42**, 515 (1976); *CA* **85**, 201578y.
274. A. O. Rucci, E. J. Wood, R. Gonzalez Palacin, and J. A. Blanco, *An. Asoc. Quim. Argent.* **63**, 43 (1975).
275. N. K. Mathur, S. P. Rao, and J. N. Gaur, *J. Sci. Ind. Research (India)* **20B**, 552, (1961).
276. D. K. Sahu, N. Rangachari, and P. Ghosh, *Fert. Technol.* **12**, 344 (1975).
277. V. K. Khakimova and P. K. Agasyan, *Uzbek. Khim. Zh.* **1960** (6), 21; *CA* **56**, 20f.
278. E. J. Zinser and J. A. Page, *Anal. Chem.* **42**, 787 (1970).
279. R. Dewolfs and F. Verbeek, *Fresenius' Z. Anal. Chem.* **246**, 358 (1969).
280. K. Eross, I. Buzas, and L. Erdey, *Fresenius' Z. Anal. Chem.* **255**, 271 (1971); **255**, 275 (1971).
281. P. Ya. Yakovlev, V. T. Solomatin, and L. M. Bakalova, *Zavod. Lab.* **40**, 1046 (1974); *CA* **82**, 51060v.
282. A. M. El-Aggan and S. F. Sidarous, *U.A.R. J. Chem.* **14**, 325, 339, (1971);

CA **77**, 69656a, 69655z.
283. Reference 1, p. 223.
284. N. N. Mikhailova, Z. K. Timokhina, and L. A. Panova, *Zh. Anal. Khim.* **26**, 2248 (1971); *J. Anal. Chem. USSR* **26**, 2015 (1971).
285. A. A. Ramadan, P. K. Agasyan, and S. I. Petrov, *Zh. Anal. Khim.* **28**, 2396 (1973); **29**, 544 (1974); *J. Anal. Chem. USSR* **28**, 2124 (1973); **29**, 464 (1974).
286. K. A. Idriss, I. M. Issa, and M. S. El-Meligy, *Indian J. Chem. Sect. A.* **14A**, 195 (1976).
287. D. S. Mahadevappa and B. T. Gowda, *Curr. Sci.* **44**, 306 (1975).
288. D. S. Mahadevappa and B. T. Gowda, *Talanta* **23**, 601 (1976).
289. R. Fresenius, *Fresenius' Z. Anal. Chem.* **1**, 26, 32 (1862).
290. S. Arribas, R. Moro, M. L. Alvarez, and C. Garcia, *Inform. Quim. Anal. (Madrid)* **21**, 85, 115 (1967); CA **68**, 26501t, 101433c.
291. T. Takahashi and S. Miyake, *Talanta* **4**, 1 (1961).
292. S. Barabas and P. W. Bennett, *Anal. Chem.* **35**, 135 (1963).
293. P. W. Bennett and S. Barabas, *Anal. Chem.* **35**, 139 (1963).
294. Reference 1, p. 617.
295. J. F. Hinton and H. M. Tomlinson, *Anal. Chem.* **33**, 1502 (1961).
296. J. T. Stock and R. D. Braun, *Microchem. J.* **15**, 519 (1970).
297. R. D. Braun and J. T. Stock, *Anal. Chim. Acta* **60**, 167 (1972).
298. R. D. Braun, *J. Chem. Educ.* **53**, 463 (1976).
299. E. Ruzicka, *Chem. Zvesti* **30**, 160 (1976).
300. S. N. Prasad and N. T. Akinchan, *J. Indian. Chem. Soc.* **48**, 21 (1971).
301. K. L. Chawla and J. P. Tandon, *Indian J. Appl. Chem.* **31**, 207 (1968); CA **68**, 119235z.
302. N. T. Akinchan and S. N. Prasad, *Indian J. Chem.* **11**, 37 (1973).
303. S. Arribas Jimeno, M. L. Alvarez Bartholome, and M. A. Aberturas Pastor, *Ion (Madrid)* **34**, 681 (1974); CA **82**, 132560k.
304. N. M. M. Gowda, A. S. A. Murthy, and D. S. Mahadevappa, *Curri. Sci.* **44**, 5 (1975).
305. D. S. Mahadevappa and M. Madaiah, *Curr. Sci.* **42**, 420 (1973).
306. L. Suchamelova and J. Zyka, *J. Electroanal. Chem. Interfacial Electrochem.* **5**, 57 (1963).
307. B. C. Verma and S. Kumar, *J. Indian Chem. Soc.* **52**, 528 (1975).
308. B. C. Verma and S. Kumar, *J. Indian Chem. Soc.* **53**, 1015 (1976).
309. G. N. Nikolaev, R. S. Tsekhanskii, and Yu. A. Fedorov, *Zh. Anal. Khim.* **33**, 999 (1978); *J. Anal. Chem. USSR* **33**, 771 (1978).
310. G. N. Nikolaev, V. G. Stavinchuk, and M. K. Saikina, *Zh. Anal. Khim.* **31**, 2038 (1976); *J. Anal. Chem. USSR* **31**, 1486 (1976).
311. L. Giuffre and E. Losio, *Chim. Ind. (Milan)* **47**, 515 (1965); CA **63**, 4949g.

CHAPTER

13

# PRECIPITATION TITRATIONS

These titrations depend upon the addition of a standardized reagent to a solution containing the determinand that reduces the concentration of the latter to a low level by its precipitation as a salt or a complex. The composition of this insoluble product must be well defined and it must be formed rapidly. Although the energies of activation for these reactions are small when compared with those for some oxidation–reduction reactions, there are other kinetic factors that must be considered in precipitation potentiotitrimetry. These include the physical processes of adsorption and coprecipitation which can each give rise to a precipitate of uncertain composition and thus lead to spurious results unless action is taken to minimize them. This is often achieved by some modification to the medium in which the reaction occurs. Adsorption of the freshly formed precipitate on the surface of the indicator electrode can also cause the resistance of the cell to increase sharply which may result in a marked decrease in the rate of response of the indicator electrode.

In its routine application to equivalence point detection, potentiotitrimetry assumes a greater importance for precipitation reactions than it does in the applications of acid–base, complexometric, or oxidation–reduction reactions to routine analysis. This is because in comparison with each of the latter reactions there are fewer visual indicators available with which to detect the equivalence points for precipitation titrations. Generally, therefore, the potentials are monitored during the titration using a cell with liquid junction in which the indicator electrode could be one of the types described in Chapter 5 (see Section 5.4), or a commercially available ion selective electrode. The application of chloride, fluoride, iodide, lead, nitrate, perchlorate, potassium, silver, and sulfide ion selective electrodes has increased the scope of precipitation titrations considerably.

## 13.1 SOME BASIC CONSIDERATIONS

### Reference Electrodes

The use in precipitation titrations of an unmodified calomel reference half-cell against which to monitor the potential of the indicator electrode is curtailed by the interference that can be caused by the diffusion of potassium and chloride ions across the liquid junction into the titrand solution. These ions cause interference either by reaction with one or more of the species present in the titrand solution or by reaction with a species introduced by the addition of titrant. Thus for the large number of titrations in which silver(I), mercury(I) or (II), cadmium(II), lead(II), halides, perchlorate, or tetraphenylborate(III) feature either as determinands or titrants, electrical contact between the calomel reference half-cell and the titrand solution must be made by means of an intermediate bridge solution composed of innocuous ions. This solution, sometimes immobilized in agar, may be contained in a salt bridge interposed between the reference half-cell and the titrand solution, or it may be contained in the outer compartment of a double junction reference half-cell. Potassium nitrate is the salt most frequently selected for the preparation of these bridge solutions, which range in concentration from 0.05-2.0 $M$. Sometimes these solutions contain other solutes to make the bridge solution more compatible with the titrand solution. For example, potassium nitrate (10%) has been prepared in alkali (1.0 $M$ NaOH) for the titration of sulfide ion in alkaline solution with cadmium(II) (1). For the determination of halides in seawater with silver nitrate titrant, the bridge solution was 0.05 $M$ potassium nitrate in 0.05 $M$ nitric acid, which was compatible with the titrand solution that contained an aliquot of the seawater added to an identical solution (2). Potassium nitrate has also been used as the bridge solution when the titration is to be carried out in partially aqueous media. Thus a saturated solution of this salt in methanol formed the bridge solution when a predominantly methanolic or acetonic medium was used for the microtitration of chloride with silver(I) solution (3). Similarly, a 6% solution of potassium nitrate in 40% 2-propanol in water was the bridge solution for the titration of sulfate with lead(II) in a medium of 80% 2-propanol in water. This expedient minimized the drift attributed to an unstable liquid junction potential when a completely aqueous solution of potassium nitrate was used (4). Sodium nitrate is best used as the bridge solution if the titrant or titrand solutions contain perchlorate (5) and is also favored when lead(II) is used as titrant as in the titration of phosphate (6), sulfide (7), and sulfate (8). Other bridge solutions with which to connect a calomel reference half-cell to the titrand solution have included ammonium nitrate for the titration of perchlorate with tetraphenyl arsenic(V) chloride (9), and 0.1 $M$ ammonium fluoride for the determination of perchlorate with 1,2,4,6-tetraphenylpyridinium acetate (10).

The calomel reference half-cell can, of course, be used without modification for titrations in which there is no possibility that the ions of potassium and chloride will cause interference. These include the titration of fluoride with lanthanum(III) (11) and other titrations that can be monitored with the

## SOME BASIC CONSIDERATIONS

fluoride ion selective electrode in which fluoride features as the titrant (12). The reference half-cell has also been used in titrations such as nitrate with nitron [Reg. No. *2218-94-2*] (13), copper and zinc with sodium pentacyanonitrosylferrate(III) (i.e., sodium nitroprusside; 14), palladium(II) with dithiooxamide [Reg. No. *79-40-3*] (15), and indium with dithiocarbamic acid derivatives (16).

The half-cell $Hg;Hg_2SO_4|satd.K_2SO_4$ or a variant of this mercury–mercurous sulfate electrode has found some application as the reference half-cell in the titration of nitrate with diphenythallium(III) using a nitrate ion selective electrode (17), in the determination of salt in meat products (18), and in the titrations of halides with silver nitrate (19) and quaternary ammonium compounds with tetraphenylborate(III) (20).

The use of the comparatively inert hydrogen ion selective glass electrode as the reference electrode in a cell without liquid junction seems to be confined to those titrations in which the indicator electrode is an oxidation–reduction electrode rather than an ion selective electrode. There appears to be no reason why such a glass electrode cannot be used as a reference against which to monitor the potentials of other ion selective electrodes provided the $p$H remains fairly constant and the titrand solution does not contain a high proportion of a solvent having a low dielectric constant. In the latter circumstance the resistance in the cell may tend to be too large for its potential to be measured with the usual type of ion activity meter. However, in the literature surveyed, the glass electrode has assumed the reference potential only for those titrations in which the potential of a silver indicator electrode has been monitored. These have included the determination of copper and silver with dithiooxamide (21), the titration of silver(I) sulfate (ex. organic sulfur compounds) with potassium iodide (22), the simultaneous determination of hydrogen sulfide and thiols using silver nitrate as titrant (23), and the titration of quaternary compounds with tetraphenylborate(III) (24).

### Titrants and their General Applications

This information is summarized in Table 13.1 in which the titrants are arranged in alphabetical order of the precipitating species. If more than one type of indicator electrode has been used to monitor titrations performed with a given precipitant, the determinands are subdivided according to the type of electrode used. Oxidation–reduction electrodes are given before ion selective electrodes (ISE), and where special electrodes of either kind are used, cross-reference is made to Chapter 5 where these are described.

**Table 13.1. Some Applications of Precipitants Used as Titrants**

| Titrant | Determinand(s) and Reference(s) | Indicator Electrode |
|---|---|---|
| Alkoxymethanedithioates as K salts,[a] ROC(:S)SK | R = Me; Ag(I), Cu(II), Pb(II). R = Pr; the above + Ni(II). R = Bu; the above + Tl(I) and Cd(II). R = Hexyl; the above + Zn(II) and Bi(III). (25) | Ag or Hg |
| Barium as $BaCl_2$ or $Ba(ClO_4)_2$ | $SO_4^{2-}$ (26). $SO_4^{2-}$ (ex. org. sulfur comp.) (27) | Fe-1173 glass Na ISE (glass) |
| 1H-Benzothiazole [95-16-9] Benzothiazole-2(3H)-thione [149-30-4] | Ag(I) in presence of $Cl^-$ or $Br^-$ (28). Cu(II) and Pb(II) in mixtures (29). Pb(II), Ni(II), Cd(II) (30). Ag(I), Hg(II), Bi(III) (31). Theophylline [58-55-9], back titrn. Ag(I) (32) | Ag Hg pool Ag ISE ($Ag_2S$) Not stated.[l] |
| Benzothioamide, $PhC(:S)NH_2$ | Ag(I), Hg(II), Pb(II), Cd(II), Mn(II), Zn(II), Cu(II) (33). | Ag |
| Bromide as KBr | Phenoxyalkanoic acids by precipitation with Hg(I) followed by titrn. of Hg(I) with KBr (34). | Hg(I) ISE[b] |
| Carbamodithioic acid derivatives, >NC(:S)SH | Sodium NN-diethylcarbamodithioate [148-18-5] for titrn. Cu(II) (35), Ag–Cu and Ag–Pb mixtures (36), Ag(I), Cu(II), Cd(II), and Zn(II) (37), In(III) (38). Various derivatives for det. of In(III); sodium piperidine-1-dithiocarboxylate $C_5H_{10}NC(:S)SNa$ gave largest inflection (16). | Ag |

| | | |
|---|---|---|
| Chloride as KCl | Indirect det. of thiourea, XS std. Ag(I) titrd. with KCl (39). | Not stated[1] |
| Cupferron [135-20-6], PhN(NO)ONH$_4$ | Zr(IV) (40). | Au |
| | Th(IV) and Ce(III) (41) | Hg |
| 3-Dimethylaminomethyl-2,6-dimercaptopyran-4(4H)-thione | Tl(III), apparently selective (42). | Not stated[1] |
| Diphenylthallium(III) sulfate | NO$_3^-$ (17) | NO$_3^-$ ISE |
| 2,4-Dithiobiuret [541-53-7], NH(C(:S)NH$_2$)$_2$ | Tl(III) (43) | Pt |
| Dithiooxamide, ethanedithioamide | Ag and Cu(II) in mixtures (21), Pd(II) (15) | Ag, Pt |
| Fluoride as NH$_4$F or NaF | Li (44), Al (12). Th(IV) and total rare earths by titrn. of XS std. F$^-$ with La(III) (45). | F$^-$ ISE |
| Iodide as NaI or KI | Ag$_2$SO$_4$ (ex. org. sulfur comp.) (22) | Ag |
| | Pd(II) (46) | Pt |
| | Hg(II) (47), Pd(II) (48), Ag(I) (49), Hg(II) (ex. org. mercury comp.) (50). Indirect for det. of nitrates and nitramines (51). | I$^-$ ISE |
| Iron(III) as sodium pentacyanonitrosylferrate(III) | Cu(II) and Zn(II) (14). Ag and Cu in alloys (52) | Pt |
| Lanthanum(III) as La(NO$_3$)$_3$ | F$^-$ general (11, 53, 54, 55). F$^-$ in various media (56, 57). F$^-$ in presence of SO$_4^{2-}$ or PO$_4^{3-}$ (58). F$^-$ (ex. org. comps.) (59–61). Titrn. of oxalate comps.) (62). | F$^-$ ISE |

(continued)

Table 13.1. (Continued)

| Titrant | Determinand(s) and Reference(s) | Indicator Electrode |
|---|---|---|
| Lead(II) as $Pb(NO_3)_2$ or $Pb(ClO_4)_2$ | $SO_4^{2-}$ in 70% 2-PrOH (63, 64). $SO_4^{2-}$ in various aq. org. media (65, 5, 4, 66, 67). $SO_4^{2-}$ in seawater (8), $SO_4^{2-}$ (ex. org. sulfur comps.) (7). $PO_4^{3-}$ (6, 68). $PO_4^{3-}$ or $SO_4^{2-}$ in presence of $F^-$ (58). $SO_4^{2-}$, $CrO_4^{2-}$, $WO_4^{2-}$, $P_2O_7^{4-}$, and $C_2O_4^{2-}$ (69). $C_2O_4^{2-}$ in 40% dioxane (70). $S^{2-}$ (71), $S^{2-}$ (ex. org. sulfur comps.) (72). | $Pt^c\|Fe(CN)_6^{4-}, Fe(CN)_6^{3-}$ Pb ISE<br><br><br><br><br><br><br>$Ag/S$ ISE $(Ag_2S)$ |
| Mercury(II) as $Hg(NO_3)_2$ or $Hg(ClO_4)_2$ | Thiols (73). $I^-$ (74). XS std. $I^-$ used in det. of Ag(I) and Cu(II) (75), Pt(IV) (76), Hg(I), Cr(VI)–Cr(III) mixtures (77), Ce(IV) (78) see also (79, 80). Trithiocarbonates (81). Thiols (82). Solutizene [959-79-5] (83). $Cl^-$ and $Br^-$ (ex. volatile org. comps.) (84). Azide and $Cl^-$ sequential (85). Thiols in acetonic media (86). | Hg pool$^d$<br>$Hg(Pt)^e$<br>$Hg(Ag)^e$<br><br><br><br>$Ag^f$<br>$W^g$<br>$Ag_2S$ or AgI ISE<br>$Cl^-$ ISE<br>$Br^-$ ISE $(Ag_2S)$ |
| Molybdenum(VI) as $Na_2MoO_4$ | Pb(II) (also ex. org. lead comps.) (87) | Pb ISE |
| Nitron [2218-94-2] | $NO_3^-$ (13). | $NO_3^-$ ISE. |

614

| | | |
|---|---|---|
| 8-Quinolinethiol (thiooxine) Silver(I) as AgNO₃ or AgClO₄ | Ag(I), apparently selective (88). Cl⁻ in presence of Br⁻ and I⁻ (89), Cl⁻ ex. chloroformate from –OH endgroups (90). I⁻ emanating ex. Cs₂TeI₆ in det. of Cs (91). Halides in seawater (2). Halogens ex. org. comps. (92–94). $PO_4^{3-}$ (95), $VO_4^{3-}$ (96). $S^{2-}$ (97), sequential $S^{2-}$ and thiol (23), thioamides (98). Tetraphenyl borate(III) (99), see also (100). Adenine (101, 102). Mixtures 8-bromoadenine, adenine and Br⁻ (103). | Not stated$^l$ Ag |
| | Cl⁻ (3), Cl⁻ in meats (18). Titrn. I⁻ after treatment of thiols and thiones with I₂ and pyridine (104). Separate titrn. Cl⁻, Br⁻, and I⁻ (105). Sequential Cl⁻, Br⁻, and I⁻ (56), also CN⁻ and Cl⁻ (106), Br⁻ and SCN⁻ (107). Successive $S^{2-}$ and dissolved S (108), also $S^{2-}$, $AsO_2^-$ and $AsO_4^{3-}$ (109). $S_2O_3^{2-}$ (110), SCN⁻ (56). H₂S and thiols (111, 112), thiols (113), thioacetamide (114), dithiooxamide (115). OO-Dialkylphosphorodithioates (116), phenylthiourea and diphenylthiourea (117), mercaptopyrimidines (118). Mixtures Na₂S, cyanamide, and thiourea (119). | Ag ISE (Ag₂S) |

(continued)

Table 13.1. (Continued)

| Titrant | Determinand(s) and Reference(s) | Indicator Electrode |
|---|---|---|
| Silver(I) as AgNO$_3$ or AgClO$_4$ (continued) | $m$-Dinitro compounds by means of XS std. KCN (120). | |
| | $p$-Urazine (121), tetrazoles (122). | |
| | Titrn. azide and Cl$^-$ (59). Cl$^-$ in plant tissue (123) and soil extracts (124). Cl$^-$ in org. materials (125) and ex. RNH$_2$HCl (126). | Cl$^-$ ISE. |
| | Sequential Cl$^-$, Br$^-$, and I$^-$ (127). I$^-$ in det. SO$_3^{2-}$ by reaction with I$_2$ (128), in similar det. of hydrazine and hydroxylamine (129) and also thiones and thiols (104). Det. of HCHO by means of titrn. of XS I$^-$ (130). $m$-Dinitro comps. by means of XS std. KCN (120). | I$^-$ ISE. |
| | Titrn. thiourea in presence of picrate (131). | Picrate ISE[h] |
| Sulfate as Na$_2$SO$_4$ or H$_2$SO$_4$. | Ba(II) (132). | Pt\|Fe(II), Fe(III)[i] |
| | Ba(II) and Sr(II) in aq. dioxane (133). | |
| Sulfide as Na$_2$S. | Ultramicro simultaneous det. of Ag(I) and Cd(II) (134) | S$^{2-}$ ISE (Ag$_2$S) |
| Tetraphenylarsenic(V) chloride [507-28-8]. | ClO$_4^-$ (9). | ClO$_4^-$ ISE |
| Tetraphenylborate(III) as sodium salt [143-66-8]. | K in presence of Na (135). | Graphite |
| | Polyethylene glycols (136), quaternary $N$ surfactants (20). | Ag |
| | Ionic detergents (137). | Hg |
| | Alkali metal cations (138). | Na ISE (glass) |

616

| | | |
|---|---|---|
| Tetraphenylborate(III) (continued) | K (139), N-base cations (140), alkaloid cations (141). | K ISE |
| 1,2,4,6-Tetraphenylpyridinium acetate. | $ClO_4^-$, $ReO_4^-$, $IO_4^-$, $Cr_2O_7^{2-}$, $MnO_4^-$, $PF_6^-$, $PtCl_6^{2-}$, $BF_4^-$ (10) Org. anions, e.g., subs. phenols (142). | $ClO_4^-$ ISE |
| 1,3,4-Thiadiazole-2,6(3H,4H)-dithione. | Cu(II) (143). | Ag |
| Thioacetamide ($S^{2-}$ alk. soln.) | Pb(II), (144), Tl (145). | $S^{2-}$ ISE ($Ag_2S$) |
| Thiocyanate as $NH_4SCN$ | Ag (146). | Ag ISE |
| Thionalide [93-42-5] | Sb(III) (147). | Polarized Pt[j] |
| Thiosulfate | Ag (148). | Ag |
| Thiourea | Ag (149). | Ag |
| | Ag in presence of picrate ion (150). | Picrate ISE[k] |
| Thorium(IV) as $Th(NO_3)_4$ | $F^-$ (range 0.2–1 mM $F^-$) (55). $F^-$ ex. org. comps. (151). | $F^-$ ISE |
| Zephiramine [139-08-2] | Ionic surfactants (137,152). | $I^-$ ISE |
| Zinc(II) as $ZnSO_4$ | Benzotriazole and N-heteroaromatic-2(3H)-thiones (153). | Zn (Hg) |

[a] Potassium alkyl xanthates.
[b] See Chapter 5, p. 288.
[c] See Chapter 5, p. 282.
[d] See Chapter 5, p. 281.
[e] See Chapter 5, p. 281.
[f] See Chapter 5, p. 279.
[g] See Chapter 5, p. 284.
[h] See Chapter 5, p. 287.
[i] In the presence of o-phenanthroline, see Chapter 5, p. 283.
[j] See Chapter 5, p. 291.
[k] See Chapter 5, p. 287.
[l] Not given in the abstract.

## Solubility Products

It was mentioned earlier that the equilibrium constant for a reaction potentially suitable for application as a precipitation titration can be deduced approximately from the value of the solubility product $K_{sp}$. This constant relates specifically to the equilibrium that is set up between a sparingly soluble ionic solid and its ions in a saturated solution of the solid. On shaking an ionic solid $A_xB_y$ in pure water at a specified temperature and pressure until the solution becomes saturated with respect to the ions $A^{z+}$ and $B^{z-}$, and equilibrium is established which can be represented as

$$A_xB_y(\text{solid}) \rightleftharpoons xA^{z+} + yB^{z-} \tag{13.1}$$

The equilibrium constant for this reaction is called the *solubility product* and is defined as

$$K_{sp} = a_{A^{z+}}^x \cdot a_{B^{z-}}^y = m_{A^{z+}}^x \cdot m_{B^{z-}}^y \cdot \gamma_{A^{z+}}^x \cdot \gamma_{B^{z-}}^y \tag{13.2}$$

The solubilities of salts encountered in precipitation titrations are very low, and hence in a saturated solution of the solid in pure water the values of the molal ionic activity coefficients are often assumed to be unity. However, if the saturated solution is prepared in an electrolyte solution containing ions additional to $A^{z+}$ and $B^{z-}$, the magnitude of the activity coefficients in equation (13.2) will, of course, be determined by the total ionic strength of the medium and the composition. Now the value of $K_{sp}$ defined by this equation is a thermodynamic quantity and is thus independent of the ionic strength at the specified temperature. Hence if the ionic strength is increased up to a reasonable limit (1 $m$), the activity coefficients will decrease causing a concomitant increase in the concentrations of the ions $A^{z+}$ and $B^{z-}$. The product $m_{A^{z+}}^x \cdot m_{B^{z-}}^y$ thus becomes larger. This is the situation that pertains to precipitation titrations, and therefore, as in the case of formal potentials $E^{\circ\prime}$, an ionic strength dependent value of the conditional solubility product may be defined as

$$K_{sp}' = m_{A^{z+}}^x \cdot m_{B^{z-}}^y$$

which is more usually expressed in terms of molar concentrations as

$$K_{sp}' = [A^{z+}]^x [B^{z-}]^y \tag{13.3}$$

The solubility $S$ of the precipitate $A_xB_y$ in water or, indeed, any other predominantly aqueous medium is expressed as

$$S = \frac{[A^{z+}]}{x} = \frac{[B^{z-}]}{y}$$

which allows the conditional solubility product to be defined in terms of solubility (moles per liter) at a given temperature and a given ionic strength as

$$K'_{sp} = (xS)^x \cdot (yS)^y \quad (13.4)$$

Conversely, the solubility of the precipitate under the given conditions may be calculated from

$$S = \left( \frac{K'_{sp}}{x^x y^y} \right)^{1/(x+y)} \quad (13.5)$$

If a known excess of anion $B^{z-}$ is added to a saturated solution of the precipitate $A_x B_y$ so that the total concentration of the species $B^{z-}$ in the mother liquor can be assumed to be equal to the concentration of $B^{z-}$ added (i.e., $[B^{z-}]$), then the solubility of the precipitate will be markedly reduced and can be expressed as

$$S = \frac{A^{z+}}{x}$$

By means of equation (13.3) this can be related to the solubility product $K'_{sp}$:

$$K'_{sp} = (xS)^x [B^{z-}]^y$$

from whence the solubility of the precipitate is given by

$$S = \frac{1}{x} \left( \frac{K'_{sp}}{[B^{z-}]^y} \right)^{1/x} \quad (13.6)$$

Similarly, if the cation is present in known excess at a concentration of $[A^{z+}]$, then the solubility of the precipitate $A_x B_y$ becomes

$$S = \frac{1}{y} \left( \frac{K'_{sp}}{[A^{z+}]^x} \right)^{1/y} \quad (13.7)$$

For example, if bismuth(III) sulfide is the precipitate and sulfide is added to the solution adjusted to a suitable $pH$ so that its concentration is known to be $[S^{2-}]$, then the solubility of $Bi_2 S_3$ is given by

$$S = \frac{1}{2} \left( \frac{K'_{sp}}{[S^{2-}]^3} \right)^{1/2} \quad [\text{cf. equation } (13.6)]$$

and, conversely, if bismuth(III) be added in known excess, then

$$S = \frac{1}{3}\left(\frac{K'_{sp}}{[Bi^{3+}]^2}\right)^{1/3} \quad [\text{cf. equation (13.7)}]$$

Equations (13.6) or (13.7) are useful when the concentration of a saturated solution has to be calculated.

Solubility products are sometimes determined by measuring the solubility of the ionic solid in the medium of interest [see equation (13.4)]. A saturated solution is made by shaking an excess of the solid in the solvent at constant temperature for a prolonged period in order to ensure that equilibrium has been attained. The saturated solution is freed from solid, often by centrifugation or filtration at the equilibrating temperature, and then analyzed for the total concentration of one of the species emanating from the solid. If the solubility of the substance is not too low, the concentration of the saturated solution may be determined titrimetrically, but for less soluble salts it is necessary to apply one of the instrumental methods of trace analysis to this purpose. When pure water is the solvent, substitution of the determined molar concentration in equation (13.4) should yield a close approximation to the thermodynamic value.

A more convenient and widely used potentiometric method usually dispenses with the filtration or centrifugation step and relies on the determination of the concentration of one of the species, either $A^{z+}$ or $B^{z-}$, in the presence of a swamping concentration of the other. A calibrated electrode reversible to one of the species, say, $A^{z+}$, is used to measure its concentration at a given temperature in an equilibrated well-stirred suspension of $A_xB_y$ prepared by adding a small amount of $A^{z+}$ (e.g., 0.5 mL of 0.01 $M$) to a large excess of a solution of $B^{z-}$ (e.g., 100 mL of 0.01 $M$). Allowance is made for the diminution of $B^{z-}$ as a result of the precipitation of an amount equivalent to the concentration of $A^{z+}$ added and the accompanying change in volume. It is often assumed that the solubility of $A_xB_y$ will not affect the concentration of $B^{z-}$, but this should be checked by adding $y/x$ times the determined concentration of $A^{z+}$ to the calculated concentration of $B^{z-}$. For most of the precipitates encountered in this branch of titrimetry the latter refinement is unnecessary since it makes negligible difference to the value of the conditional solubility product calculated from

$$K'_{sp} = [A^{z+}]^x[B^{z-}]^y$$

Careful attention must be given to the correct choice of conditions for these determinations to ensure that the equilibrium

$$A_xB_y \rightleftharpoons xA^{z+} + yB^{z-}$$

is the only significant equilibrium process that occurs. This, in turn, requires a knowledge of the acid-base properties of $A^{z+}$ or $B^{z-}$ and an appreciation of the conditions under which these species are likely to form complex ions either with each other or with any other species present in the solution. For example, to minimize the formation of the chlorocomplexes of silver(I) and to ensure that the effect of the junction potential is negligible, the conditional solubility product of silver chloride could be determined at an ionic strength of 0.1 $M$ by using the experimental cell

$$\text{Hg; Hg}_2\text{Cl}_2|3.5\ M\ \text{KCl} \left| \begin{array}{l} \text{KNO}_3(0.09\ M), \\ \text{HNO}_3(0.01\ M) \end{array} \right| \left. \begin{array}{l} \text{KNO}_3(0.08\ M), \\ \text{HNO}_3(0.01\ M), \end{array} \right.$$

$$\text{KCl}(0.01\ M), \text{AgNO}_3(x\ M)|\text{Ag} \qquad (13.8)$$

A suitable cell for the establishment of a reference potential would, in this instance, consist of

$$\text{Hg; Hg}_2\text{Cl}_2|3.5\ M\ \text{KCl} \left| \begin{array}{l} \text{KNO}_3(0.09\ M), \\ \text{HNO}_3(0.01\ M) \end{array} \right| \left. \begin{array}{l} \text{KNO}_3(0.08\ M), \\ \text{HNO}_3(0.01\ M), \end{array} \right.$$

$$\text{AgNO}_3(0.01\ M)|\text{Ag} \qquad (13.9)$$

When the same electrodes are used sequentially in these cells, the unknown concentration of silver(I) in cell (13.8) may be calculated by measuring $\Delta E$, which is defined as the difference in potential between cell (13.9) and cell (13.8). Thus

$$-\log[\text{Ag}^+] = \frac{\Delta E}{\text{slope}} + 2.00$$

where "slope" has been obtained earlier from the usual calibration graph.

A different sort of approach can be applied when the anion $B^{z-}$, for example, is derived from a weak acid. In this instance the $pK_a$ values of the acid at the ionic strength at which the solubility product of $A_xB_y$ is to be determined must be known so that the concentration of the precipitating anion at any given $p$H may be calculated. This type of situation pertained in the determination of the solubility product of silver 1$H$-benzotriazolate, for example (28). The $pK_a$ value of 1$H$-benzotriazole [Reg. No. 95-14-7] was found to be 8.4. A small amount of silver(I) (0.3 mL, 0.01 $M$ AgNO$_3$) was added to a solution of the acid (100 mL, 0.01 $M$) and to this well-stirred suspension was added small portions of 0.5 $M$ sodium hydroxide. The silver(I) concentration in the solution was measured potentiometrically after each addition by a method analogous to that described, and the $p$H was simultaneously determined. The conditional solubility product was calculated for each addition by the usual equation

$$K'_{sp} = [Ag^+][B^-]$$

but in this case the concentration of the anion must be calculated from

$$[B^-] = \frac{C_t}{1 + 10^{(pK-pH)}}$$

where $C_t$ is the stoichiometric concentration of acid taken originally allowing for volume corrections.

Solubility products are also commonly obtained by potentiometric precipitation titrations. In one method the precipitate $A_xB_y$ is formed by titrating a standard solution of one species, for example, $B^{z-}$, with a standard solution of the other, in this case $A^{z+}$ and measuring the concentration of $B^{z-}$ remaining in solution by a calibrated electrode reversible to this species. This method relies upon the potentials recorded *after* the stoichiometric equivalence point to measure the concentration of $B^{z-}$ in the presence of known excesses of the titrant $A^{z+}$, the titration being continued until at least 1.5 equivalents have been added. A worked example of this method as applied to the determination of the conditional solubility product of lanthanum(III) fluoride is to be found in Reference 154. In this instance the fluoride ion selective electrode of known slope was calibrated simply by measuring its potential with respect to a saturated calomel half-cell in the fluoride solution of known concentration immediately before addition of any of the standard lanthanum(III) nitrate titrant. This allowed the constant of the Nernst expression

$$E_{cell} = \text{constant} + \text{slope}\, pF$$

to be evaluated so that $E_{cell}$ values measured subsequently could be interpolated in terms of the fluoride ion concentration. No attempt was made to maintain a condition of constant ionic strength in this particular example.

Another titrimetric method of determining solubility products relied upon the potentials measured before the equivalence point to calculate the concentration of the precipitated ion in the saturated solution during the course of the titration. A linear equation was developed, and the thermodynamic values of the solubility products for silver chloride, silver bromide, and silver iodide were obtained by solving this equation using a graphical method or by computer (155).

The errors inherent in the determination of solubility products are greater than those usually associated with the determinations of equilibrium constants for other types of ionic reactions. This is because solubility equilibria are attained comparatively slowly. Solubilities also depend on the crystal size, and for the best results it is desirable to remove the fine fraction. If, for example, the solubility product is calculated from potentiometric measurements obtained during the early part of a titration, supersaturation may negate the assumption that the activity of the solid is unity and hence cause spurious results. In the region of the equivalence point and afterwards a unidirectional drift in the $E_{cell}$ values is sometimes observed. Although the rate of this drift may settle to what appears to be a negligible rate ($< 0.5$ mV min$^{-1}$), this does not necessarily represent the true state of equilibrium for the aged precipitate, and the value of the solubility product calculated from the apparent equilibrium value of $E_{cell}$ is likely to be larger than the true value. In the determination of the solubility product for lanthanum(III) fluoride an equilibration period of about 10 h was necessary before a stable value of $E_{cell}$ was obtained (55).

Solubility products are useful data from which to predict the general shape of a titration curve under a given set of conditions, and whether the sequential titration of two species forming precipitates with the same titrant is feasible. Values for some of the precipitates formed with some of the titrants listed in Table 13.1 are collected in Table 13.2. Solubility product constants are also included with some of the data in the IUPAC collection *Stability Constants* (Chapter 8, References 2–4) and a list is also to be found in Reference 161 of this chapter.

The concentration of determinand ($A^{z+}$ or $B^{z-}$) remaining in solution at the equivalence point for the reaction

$$xA^{z+} + yB^{z-} \rightarrow A_xB_y \tag{13.10}$$

may be calculated from the appropriate conditional solubility product by the equation

$$[A^{z+}]_{eq} = \left[K_{sp}(x/y)^y\right]^{1/(x+y)} \tag{13.11}$$

or by

$$[B^{z-}]_{eq} = \left[K_{sp}(y/x)^x\right]^{1/(x+y)} \tag{13.12}$$

Table 13.2. Solubility Products for Some Substances Formed in Precipitation Titrations (18–25°).

| Substance | Equilibrium | $K_{sp}$ | Ref. |
|---|---|---|---|
| Barium(II) sulfate | $BaSO_4 \rightleftharpoons Ba^{2+} + SO_4^{2-}$ | $1.1 \times 10^{-10}, I = 0$ | 156 |
| Cadmium(II) | | | |
| benzothiazole-2(3H)-thione | $CdL_2 \rightleftharpoons Cd^{2+} + 2L^-$ | $8.9 \times 10^{-9}, I = ?^a$ | 30 |
| NN-diethylcarbamodithioate | $Cd[Et_2C(:S)S]_2 \rightleftharpoons Cd^{2+} + 2Et_2C(:S)S^-$ | $1.0 \times 10^{-22}, I = 0.1$ | 157 |
| sulfide | $CdS \rightleftharpoons Cd^{2+} + S^{2-}$ | $2 \times 10^{-28}, I = ?^b$ | 161 |
| Calcium fluoride | $CaF_2 \rightleftharpoons Ca^{2+} + 2F^-$ | $3.5 \times 10^{-11}, I = 0$ | 158 |
| | | $3.1 \times 10^{-11}, I = 0$ | 167 |
| | | $3.3 \times 10^{-10}, I = 1.0$ | 167 |
| Copper(II) | | | |
| benzothiazole-2(3H)-thione | $CuL_2 \rightleftharpoons Cu^{2+} + 2L^-$ | $2.5 \times 10^{-21}, I = ?^a$ | 159 |
| NN-diethylcarbamodithioate | $Cu[Et_2C(:S)S]_2 \rightleftharpoons Cu^{2+} + 2Et_2C(:S)S^-$ | $2.5 \times 10^{-30}, I = 0.1$ | 157 |
| pentacyanonitrosylferrate(III) | $Cu[Fe(CN)_5NO] \rightleftharpoons Cu^{2+} + [Fe(CN)_5NO]^{2-}$ | $3.8 \times 10^{-9}, I = ?^a$ | 52 |
| Europium(III) fluoride | $EuF_3 \rightleftharpoons Eu^{3+} + 3F^-$ | $2.2 \times 10^{-17}, I \sim 0.03$ | 154 |
| Indium(III) | | | |
| NN-diethylcarbamodithioate | $In[Et_2C(:S)S]_3 \rightleftharpoons In^{3+} + 3Et_2C(:S)S^-$ | $9.9 \times 10^{-26}, I = 0.1$ | 38 |
| Lanthanum(III) fluoride. | $LaF_3 \rightleftharpoons La^{3+} + 3F^-$ | $2.2 \times 10^{-18}, I = 1.0$ | 55 |
| | | $1.2 \times 10^{-18}, I \sim 0.03$ | 154 |
| Lead(II) | | | |
| benzothiazole-2(3H)-thione | $PbL_2 \rightleftharpoons Pb^{2+} + 2L^-$ | $6.1 \times 10^{-14}, I = ?^a$ | 30 |
| molybdate | $PbMoO_4 \rightleftharpoons Pb^{2+} + MoO_4^{2-}$ (pH 6.0) | $1.2 \times 10^{-13}, I = 0$ | 160 |
| perrhenate | $Pb(ReO_4)_2 \rightleftharpoons Pb^{2+} + 2ReO_4^-$ (pH 8.0) | $6.9 \times 10^{-9}, I = 0$ | 160 |
| phosphate | $Pb_3(PO_4)_2 \rightleftharpoons 3Pb^{2+} + 2PO_4^{3-}$ | $7.9 \times 10^{-43}, I = ?^b$ | 6 |
| sulfate | $PbSO_4 \rightleftharpoons Pb^{2+} + SO_4^{2-}$ | $1.6 \times 10^{-8}, I = ?^b$ | 161 |
| tungstate | $PbWO_4 \rightleftharpoons Pb^{2+} + WO_4^{2-}$ (pH 6.0) | $8.4 \times 10^{-11}, I = 0$ | 160 |

| | | | |
|---|---|---|---|
| Mercury | | | |
| benzothiazole-2(3H)-thione | $HgL_2 \rightleftharpoons Hg^{2+} + 2L^-$ | $4.0 \times 10^{-39}, I = ?^a$ | 159 |
| iodide | $Hg_2I_2 \rightleftharpoons Hg_2^{2+} + 2I^-$ | $4.5 \times 10^{-29}, I = ?^b$ | 161 |
| sulfate | $Hg_2SO_4 \rightleftharpoons Hg_2^{2+} + SO_4^{2-}$ | $8.1 \times 10^{-7}, I = 0$ | 162 |
| Nickel(II) | | | |
| benzothiazole-2(3H)-thione | $NiL_2 \rightleftharpoons Ni^{2+} + 2L^-$ | $4.9 \times 10^{-15}, I = ?^a$ | 30 |
| Nitron nitrate[c] | Nitron nitrate $\rightleftharpoons$ Nitron$^+$ + $NO_3^-$ | $1.8 \times 10^{-6}, I = 0.15$ | 13 |
| Diphenylthallium(III) nitrate | $(Ph)_2TlNO_3 \rightleftharpoons (Ph)_2Tl^+ + NO_3^-$ | $1.0 \times 10^{-7}, I \sim 0$ | 180 |
| Silver(I) | | | |
| adenine [73-24-5][d] | $H^+ + AgAdHNO_3 \rightleftharpoons H_2Ad^+ + NO_3^- + Ag^+$ | $1.4 \times 10^{-6}, I = 0.11$ | 102 |
| arsenate | $Ag_3AsO_4 \rightleftharpoons 3Ag^+ + AsO_4^{3-}$ | $6.5 \times 10^{-23}, I = 0.1$ | 109 |
| arsenite | $Ag_3AsO_3 \rightleftharpoons 3Ag^+ + AsO_3^{3-}$ | $5.4 \times 10^{-32}, I = 0.1$ | 109 |
| azide | $AgN_3 \rightleftharpoons Ag^+ + N_3^-$ | $2.9 \times 10^{-9}, I = ?^b$ | 85 |
| benzimidazole-2(3H)-thione[e] | $AgL \rightleftharpoons Ag^+ + L^-$ (30°) | $5 \times 10^{-12}, I \sim 0.1$ | 163 |
| benzothiazole-2(3H)-thione | $AgL \rightleftharpoons Ag^+ + L^-$ (30°) | $7.9 \times 10^{-13}, I \sim 0.1$ | 163 |
| 1H-benzotriazole[f] | $AgL \rightleftharpoons Ag^+ + L^-$ | $2.2 \times 10^{-14}, I \sim 0.01$ | 28 |
| | | $7.1 \times 10^{-14}, I = 0.1$ | 109 |
| benzoxazole-2(3H)-thione[g] | $AgL \rightleftharpoons Ag^+ + L^-$ (30°) | $5.0 \times 10^{-12}, I \sim 0.1$ | 163 |
| 8-bromoadenine [6974-78-3] | $AgL \rightleftharpoons Ag^+ + L^-$ | $1.4 \times 10^{-12}, I = 0.1$ | 103 |
| bromide | $AgBr \rightleftharpoons Ag^+ + Br^-$ | $1.2 \times 10^{-12}, I = 0$ | 155 |
| | | $8.1 \times 10^{-13}, I = 0.1$ | 109 |
| chloride | $AgCl \rightleftharpoons Ag^+ + Cl^-$ | $2.8 \times 10^{-10}, I = 0.025$ | 164 |
| | | $1.7 \times 10^{-10}, I = 0$ | 164 |
| | | $2.0 \times 10^{-10}, I = 0$ | 155 |
| 5-chloro-1(H)-benzotriazole | $AgL \rightleftharpoons Ag^+ + L^-$ | $5.8 \times 10^{-18}, I = 0.1$ | 109 |
| cyanide | $AgCN \rightleftharpoons Ag^+ + CN^-$ | $3.5 \times 10^{-16}, I = 0.1$ | 109 |
| hexacyanoferrate(III) | $Ag_3[Fe(CN)_6] \rightleftharpoons 3Ag^+ + [Fe(CN)_6]^{3-}$ | $1.3 \times 10^{-28}, I = 0.1$ | 109 |
| iodide | $AgI \rightleftharpoons Ag^+ + I^-$ | $6 \times 10^{-16}, I = 0$ | 155 |
| | | $2.0 \times 10^{-16}, I = 0.1$ | 109 |

(*continued*)

Table 13.2. (Continued)

| Substance | Equilibrium | Procedures and Remarks | Ref. |
|---|---|---|---|
| Silver | | | |
| pentacyanonitrosylferrate (III) | $Ag_2[Fe(CN)_5NO] \rightleftharpoons 2Ag^+ + [Fe(CN)_5NO]^{2-}$ | $1 \times 10^{-16}, I = ?^a$ | 52 |
| phosphate | $Ag_3PO_3 \rightleftharpoons 3Ag^+ + PO_4^{3-}$ | $1.3 \times 10^{-20}, I = ?^b$ | 161 |
| selenite | $Ag_2SeO_3 \rightleftharpoons 2Ag^+ + SeO_3^{2-}$ | $3.5 \times 10^{-16}, I = 0.1$ | 109 |
| thiocyanate | $AgSCN \rightleftharpoons Ag^+ + SCN^-$ | $1.1 \times 10^{-12}, I = 0$ | 165 |
| | | $2.7 \times 10^{-12}, I = 0.1$ | 109 |
| 1,2,4,6-Tetraphenylpyridinium TPP$^+$ | | | |
| fluoroborate | $TPPBF_4 \rightleftharpoons TPP^+ + BF_4^-$ | $< 2.0 \times 10^{-6}, I = 0$ | 178 |
| hexachloroplatinate | $(TPP)_2PtCl_6 \rightleftharpoons 2TPP^+ + PtCl_6^{2-}$ | $< 2.1 \times 10^{-12}, I = 0$ | 178 |
| perchlorate | $TPPClO_4 \rightleftharpoons TPP^+ + ClO_4^-$ | $< 9.0 \times 10^{-8}, I = 0$ | 178 |
| perrhenate | $TPPReO_4 \rightleftharpoons TPP^+ + ReO_4^-$ | $1.4 \times 10^{-10h}$ | 178 |
| tetrachloroaurate | $TPPAuCl_4 \rightleftharpoons TPP^+ + AuCl_4^-$ | $< 4.0 \times 10^{-10}, I = 0$ | 178 |
| Thorium(IV) fluoride | $ThF_4 \rightleftharpoons Th^{4+} + 4F^-$ | $5.0 \times 10^{-26}$ | 166 |
| Zinc(II) | | | |
| NN-diethylcarbamodithioate. | $Zn[Et_2C(:S)S]_2 \rightleftharpoons Zn^{2+} + 2Et_2C(:S)S^-$ | $1.3 \times 10^{-17}, I = 0.1$ | 157 |
| phosphate | $Zn_3(PO_4)_2 \rightleftharpoons 3Zn^{2+} + 2PO_4^{3-}$ | $9.1 \times 10^{-33}, I = ?$ | 161 |

[a] Not stated in abstract.
[b] Value quoted in reference.
[c] Nitrate of the cation of 3,5,6-triphenyl-2,3,5,6-tetraazabicyclo[2.1.1]hex-1-ene, $pK_a$ (proton gained) = 10.34 at 20°, $I = 0.15$ (13).
[d] $pK_a$ values: $pK_1 = 4.2$, $pK_2(-NH) = ?^b$ at 20°.
[e] $pK_a$ (proton lost) = 10.2.
[f] $pK_a = 8.4$.
[g] $pK_a = 6.3$.
[h] In 1.00 $M$ $NH_3$.

As a first approximation, these equations allow an assessment of the minimum error that can be expected when a given concentration of $A^{z+}$ is titrated with $B^{z-}$. For example, when $x = y = 1$ and the concentration of determinand is 0.1 $M$, then an error of 0.01% can be expected if $K_{sp}$ approximates $10^{-10}$ (e.g., AgCl), whereas an error of 1% can be expected if $K_{sp}$ approximates $10^{-6}$ (e.g., nitron nitrate) as a result of the incomplete precipitation of the determinand. Meaningful results can still be obtained in the latter case although the inflection circumjacent to the equivalence points is much reduced. The error will, of course, be increased if the concentration of determinand taken is decreased, and Gran plots will also show increasing errors under this condition. Gran plots are useful when an interfering ion is present for which the indicator electrode has an appreciable selectivity coefficient. If the electrode's primary response is, for example, toward the titrant species, then the presence of the interfering species will cause a distortion of the titration curve before the equivalence point is reached, even though this interferent may not react with the titrant. Under this condition it may not be possible to locate the inflection point, and application of the Gran method to the titration results before the equivalence point would be equally futile. However, when the concentration of the titrant is increased after the equivalence point, the effect of the interferent upon the electrode's response is greatly diminished and a reliable Gran plot becomes possible. For a cell having predetermined values of slope and constant, the potential at the equivalence point will be given approximately by

$$E_{cell} = \text{constant} \pm \text{slope}\, pC_{eq} \quad (13.13)$$

where $C_{eq}$ is the equivalence point concentration for the species $A^{z+}$ or $B^{z-}$, to which the indicator electrode is reversible, calculated by equation (13.11) or (13.12).

The feasibility of the sequential titration of two or more species with the same titrant can also be assessed from the magnitude of the conditional solubility products for each species. Generally, discrete inflections should be observed if the ratio of the solubility products for 2 species is at least $10^3$, but a discernible inflection is possible when the ratio is as small as 10. Smaller ratios than this lead to the simultaneous titration of the two species and only one inflection is observed in the titration curve. When the ratios are favorable, however, the species with the smallest solubility product will be titrated first, followed by the remainder in order of ascending values of their solubility products. The concentration of a particular determinand remaining in solution at a particular equivalence point may be assessed by equation (13.11) or by equation (13.12), and the corresponding value of $E_{cell}$ at this point may be calculated from equation (13.13). In assessing the feasibility of obtaining separate inflections in the titration of a mixture of species, a difference of 100–150 mV between the calculated values of $E_{cell}$ for successive equivalence points should certainly allow accurate determination of the species separately, whereas a minimum difference of 35–40 mV is sufficient to allow a discernible inflection. In such sequential titrations, however, it must be noted that the errors of the individual determinations are accumulative.

The results of these calculations can only form the basis for experimental design since the important kinetic aspects associated with precipitation reactions have been ignored. The time taken for the system to reach equilibrium has already been mentioned, but in the titration of a single species other effects such as supersaturation, the occlusion of ions within the precipitate and the adsorption of ions on its surface also operate. Furthermore, when mixtures are titrated, the additional effects of coprecipitation and induced postprecipitation are operable to an unpredictable extent giving spuriously large values of the equivalence volume. For example, when iodide, bromide, and chloride were titrated sequentially, the large relative errors observed ($> 10\%$) were attributed to coprecipitation. However, in the presence of aluminium nitrate ($5 \times 10^{-5} - 0.1\ M$), which acted as a coagulant, the error was reduced to less than 0.5% (127). Amongst the other coagulants used in the sequential titrations of two or more of these halides, potassium nitrate at a concentration of $0.1\ M$ has also been recommended (168), and this salt has performed a similar function in the sequential titration of silver(I) and copper(II) mixtures with sodium $NN$-diethylcarbamodithioate (sodium diethyldithiocarbamate) (36).

### The Use of Partially Aqueous Media

In the preceding sub-section it was mentioned that the magnitude of the conditional solubility product determined both the error inherent in a particular precipitation titration at a given concentration, and the magnitude of the change of $E_{cell}$ in the vicinity of the equivalent point. A small value of the solubility product is theoretically commensurate with a desirably large inflection circumjacent to the equivalent point, and a concomitantly small error resulting from the finite solubility of the precipitate. On the other hand, if the solubility product of the precipitate is very small, the concentration of determinand in the vicinity of the equivalence point may be reduced below the lower limit of the response range for the indicator electrode which may result in an erratic response of the electrode in this important region of the titration curve. For precipitates having comparatively large solubility products, for example silver chloride ($K_{sp} \sim 10^{-10}$), lead sulfate ($K_{sp}\ 10^{-8}$) and nitron nitrate ($K_{sp}\ 10^{-6}$), any ploy by which the solubility of the substance will be reduced will decrease the lower level of concentration at which a given ion present in the precipitate may be effectively determined.

Two methods are commonly adopted to decrease the solubility of an ionic solid in an aqueous solution. The first consists of lowering the temperature of the solution, and the second relies upon the decrease in solubility observed when the precipitation is carried out in the presence of an organic solvent. The second method is generally the more effective as is evident from the data given in Table 13.3. When using potentiotitrimetry, however, a balance must be struck between the desire to decrease the solubility of the precipitate, and the often adverse effect of an increase in the organic component upon the rate of response of the indicator electrode. The results given in Table 13.4 clearly demonstrate this effect and show also how the inflection is increased as the solubility of the precipitate is decreased by the addition of organic solvent. A less obvious effect is the possible influence that a trace impurity present in the organic solvent can exert upon the performance of the indicator electrode. For example, it has been reported (169) that a minute amount of peroxide present in 1,4-dioxane, dispensed from a bottle that had been opened sometime previously, can poison a lead(II) ion selective electrode. Rather than risk the possibility of this deleterious effect upon the performance of their lead(II) indicator electrode, subsequent workers used 80% 2-propanol–water as the medium for the titration of sulfate in natural waters with lead(II) nitrate solution prepared in the same medium (4).

Table 13.3. Solubility Product of Silver Chloride in Various Media[a]

| Temp (°C) | Medium | $-\log K_{sp}$ | Solubility Product | Chloride Concentration (M) | Dielectric Constant of Pure Solvent |
|---|---|---|---|---|---|
| 25 | Aqueous | 9.75 | $1.75 \times 10^{-10}$ | $1.33 \times 10^{-5}$ | 78.5 |
| 5 | Aqueous | 10.595 | $2.54 \times 10^{-11}$ | $5.04 \times 10^{-6}$ | — |
| 25 | Methanol | 13.05 | $8.91 \times 10^{-14}$ | $2.98 \times 10^{-7}$ | 32.7 |
| 25 | 50% Ethanol | 11.11 | $7.76 \times 10^{-12}$ | $2.79 \times 10^{-6}$ | 24.6 |
| 25 | Acetonitrile | 12.90 | $1.26 \times 10^{-13}$ | $3.55 \times 10^{-7}$ | 37.5 |
| 25 | 20% Dioxane | 10.22 | $6.02 \times 10^{-11}$ | $7.77 \times 10^{-6}$ | 2.2 |
| 25 | 60% Acetone | 11.90 | $1.26 \times 10^{-12}$ | $1.12 \times 10^{-6}$ | 20.7 |
| 25 | 80% Acetone | 13.61 | $2.46 \times 10^{-14}$ | $1.57 \times 10^{-7}$ | — |

[a] Data from Reference 3.

Table 13.4. Influence of the Medium upon the Response of a Lead ISE in the Titration of 1.5 mg of Lead(II) with 0.0025 $M$ Sodium Molybdate.[a]

| Medium | Number of Determinations | $\Delta E/0.05$ ml[b] (mV) | Time Required for Stable Potential[b] (min) |
|---|---|---|---|
| Water | 8 | 30–40 | 2–3 |
| 10% Methanol | 2 | 70–80 | 6 |
| 20% Methanol | 2 | 99–103 | 8 |
| 50% Methanol | 2 | 133–137 | 18–20 |
| 10% Acetone | 2 | 70–85 | 8–9 |
| 20% Acetone | 2 | 106–110 | 6 |
| 50% Acetone | 2 | 161–166 | 15–17 |

[a] Data from Reference 87.
[b] Value at the endpoint.

Because many different types of precipitation titrations have been performed using a variety of mixed aqueous–organic media, it is difficult to associate a particular medium with a particular type of titration. A decision as to the composition of the medium must be based, in the absence of any contrary evidence, on observations of the type given in Table 13.4.

Aqueous acetone in the composition range 40–90% acetone has been used for the titration of chloride (in 40% $Me_2CO$) with silver(I) nitrate (125) and for the titration of halide mixtures in 90% acetone (89, 170). Sulfate has been titrated in 70% acetone–water (27) with barium(II), and 80% acetone–water was the medium of choice for the titration of sulfide and thiols with silver(I) (112). Acetone has provided the medium for the microtitration of chloride with silver(I) (3) and is regarded as the most universally applicable solvent for the titration of thiols with mercury(II) perchlorate (86). In the latter application the addition of pyridine is recommended to inhibit the possible reaction between the carbonyl group and the thiol. Acetone has also been recommended as a medium for the titration of silver(I), copper(II), and cadmium(II) with sodium $NN$-diethylcarbamodithioate (171).

Alcoholic media based mainly on ethanol and 2-propanol have also been used widely for precipitation titrations. Aqueous ethanol solutions having compositions in the range of 60–80% ethanol have been favored for some titrations in which the fluoride ion has been involved. Examples are to be found in the titration of aluminum(III) with sodium fluoride (12), and in the determinations of fluorine in organic compounds by its conversion to fluoride and subsequent titration with thorium(IV) nitrate (151) or lanthanum(III) nitrate (61). A medium of 3 : 1 ethanol–water was found suitable for the titration of oxalate with lanthanum(III) (62), and the same medium was used for the ultramicrotitration of sulfate with lead(II) nitrate (172). More dilute aqueous solutions of ethanol have been used as media for the titrations of some metals in 5–10% ethanol with benzothiazole-2(3$H$)-thione (30), for the precipitation of mercury(I) phenoxyalkanoates in 30% ethanol (34), and 50% ethanolic solutions provided the media for the titration of total halides in seawater (2) and for the titration of dialkyl dithiophosphates (116) with silver(I) nitrate. Fluoride has also been titrated in 2-propanol (60% 2-PrOH-H$_2$O) (56), and approximately 70% 2-propanol in water was the medium for the microtitration of sulfate with lead(II) nitrate (63). As mentioned earlier, 80% 2-propanol provided the medium for the same titration applied to the determination of sulfate in natural waters (4).

By contrast to the objection to the use of the lead(II) ion selective electrode in anhydrous 1,4-dioxane mentioned previously, aqueous solutions containing about 50% dioxane have provided the medium for the titration of sulfate with lead(II) using this electrode (5, 7, 65, 66). Similarly, oxalate was determined by titration with lead(II) in a medium of 40% dioxane (70). A chloride ion selective electrode has been used in 50% and 80% dioxane to monitor the respective titrations of arylamine hydrochlorides (126) and chloride–azide mixtures (85) with silver nitrate.

An anhydrous medium of 3 : 1 2-propanol–benzene was found satisfactory for the determination of trialkoxysilanethiols (RO)$_3$SiSH by titration with silver nitrate (173), and sulfate has also been titrated with lead(II) perchlorate in a ternary medium composed of 1 : 2 : 3 formaldehyde–water–acetone (174).

## 13.2 TITRATION OF INORGANIC ANIONS

The purpose of this section is to summarize some of the applications of precipitation potentiotitrimetry to the determination of halogen anions, anions that contain nitrogen or phosphorus, the anions formed by sulfur, and transition metal anions.

### Halogen Anions

Interest has been centered upon the titrations of the four halides, but perchlorate and periodate have also been determined by precipitation titrations.

## Fluoride

This halide can be determined by titrating the neutral but unbuffered sample solution, contained in a polythene beaker, with a standard solution of lanthanum(III) nitrate utilizing a fluoride ion selective electrode versus a calomel reference half-cell as the indicating system. When attempts were made to buffer the titrand solution at $p\text{H}$ 4.9 with acetic acid–acetate buffer, the change in potential in the vicinity of the equivalence point was diminished with respect to the unbuffered solution (11) as a result of the interference of the carboxylate ion with the precipitation of lanthanum(III) fluoride by forming $\text{LaF}_3-\text{A}_x$ ($\text{A} = \text{RCOO}^-$). This anion also affected the response of the lanthanum(III) fluoride membrane of the electrode (54). Lanthanum(III) nitrate of high purity is available, and solutions of this salt can be standardized against dried sodium fluoride. Interference by sulfate has been masked by its precipitation as barium sulfate, and interference by phosphate has been eliminated by the use of zinc oxide to adsorb the phosphate (175). Small amounts of fluoride were determined in the presence of a tenfold excess of phosphorus (as phosphate) without the necessity of removing the zinc oxide by filtration. A small amount of acetic acid (3 drops, 30%), used in the initial $p\text{H}$ adjustment of the titrand solution ($p\text{H}$ 5.0 ± 0.3), may account for the reported nonstoichiometry of the titration reaction. This impediment was overcome by standardizing the titrant against a sodium fluoride solution having a similar fluoride concentration to that of the sample solution.

## Chloride

It has been known since 1893 that the titration of chloride with silver nitrate can be monitored by a silver electrode (76) and this, of course, has become a very widely used method. In more recent years the silver ion selective electrode ($\text{Ag}_2\text{S}$ membrane) has been applied successfully to this titration, and so has the chloride ion selective electrode. The latter was used to monitor the sequential titrations of chloride and azide in 80% dioxane with aqueous silver nitrate solution (85). However, chloride in the presence of azide was best determined in 80% methanol at $p\text{H}$ 1.2 by titration with mercury(II) perchlorate.

The medium best suited to the determination of chloride at the lower levels of concentration was found to be acetone–acetic anhydride mixed in the ratio 4 : 1, and silver perchlorate solution prepared in acetone was the most suitable titrant. In 50 mL acetone alone (containing < 0.8% v/v $\text{H}_2\text{O}$) two potential breaks were obtained with a silver ion selective electrode for chloride, the first corresponding to dichloroargentate(I) ($\text{AgCl}_2^-$) and the second to silver chloride. If the water content was increased, the break for dichloroargentate(I) became less pronounced and finally disappeared. Generally, however, the preferred medium was 4:1 acetone–acetic anhydride since acetic anhydride removes traces of water in those titrations where water impairs the titration curve. It was found to be more convenient than acetone alone, in which some evaporation occurs because of the rapid stirring of the solution. Furthermore, the use of the mixed solvent gives rise to a single inflection approximately equal in magnitude to the sum of the inflections for dichloroargentate(I) and silver chloride observed in acetone alone. The lowest concentration at which chloride could be usefully determined was $4 \times 10^{-6}$ $M$ and it was thought likely that this limit was imposed by the amount of chloride present in the reagents. A useful survey of the methods used for the microdetermination of chloride is also to be found in Reference 3.

The silver ion selective electrode has been used also to monitor the sequential titration of cyanide and chloride in aqueous solution with standard silver nitrate (106). Three inflections were obtained in the titration curve: the first corresponded to the formation of dicyanoargentate(I) ($Ag(CN)_2^-$), the second to the conversion of this complex ion to the insoluble silver cyanide, and the third to the quantitative precipitation of chloride as silver chloride. In a solution containing a large amount of chloride the second inflection was obscured, and hence the cyanide concentration was calculated from the first inflection and the chloride from the third. The lower detectable limit for cyanide was 0.3 ppm in a $10^5$ excess of chloride relative to cyanide, and hence this method was suggested as being suitable for determining cyanide in river and seawater provided that the influence of other interferents is minimal. A rapid semiautomatic method for the determination of the total halide concentration of seawater by titration with silver nitrate has also been developed (2).

### *Bromide*

This halide can, of course, be determined analogously to chloride by titration with standard silver nitrate using a silver electrode or a silver ion selective electrode as the indicator. It is of interest to note that the bromides and thiocyanates of alkali metals can be determined consecutively by a method based on the solubility of copper(I) bromide and the insolubility of copper(I) thiocyanate in ammoniacal solutions (107). Reference to Table 13.2 will confirm that the solubility products for silver bromide and silver thiocyanate are similar in magnitude. The sample solution was treated with copper(II) acetate followed by ascorbic acid to reduce copper(II) to copper(I). On addition of ammonia solution the precipitate of copper(I) thiocyanate could be quantitatively separated from the soluble copper(I) salt by filtration. The filtrate containing bromide was treated with nitric acid and iron(III) nitrate, and titrated with standard silver nitrate using a silver ion selective electrode as the indicator electrode. The precipitate of copper(I) thiocyanate was then dissolved in nitric acid and iron(III) nitrate, and the thiocyanate titrated similarly to bromide with the same silver nitrate solution.

### Iodide

A mercury-coated platinum foil electrode has been used as the indicator electrode for the titration of iodide with mercury(II) nitrate (74). When the iodide in an acidified sample, 0.01 $N$ with respect to sulfuric acid, was titrated with 0.05 $M$ mercury(II) nitrate, the concentration of mercury(II) in the solution underwent a pronounced increase from $10^{-13}$–$10^{-4}$ $M$ in the vicinity of the endpoint, which corresponded to a change in potential of 463 mV 0.1 mL$^{-1}$ of this titrant. Such a break is sharp enough to permit the accurate determination of minute amounts of iodide or mercury(II). This titration has formed the basis for the indirect determination of some ions (75–80; see Table 13.1).

Reductants such as hydrazine and hydroxylamine have been determined indirectly by reaction with iodine–methanol reagent followed by the titration of the iodide formed with standard silver nitrate solution (129). The iodine–methanol solution (0.1 $M$) was prepared by dissolving pure iodine (~ 12.7 g) in pure methanol (500 mL). An excess of this reagent was added to an aliquot of hydrazine or hydroxylamine (10 mL ~ 0.01 $M$) contained in an amber glass titration cell. After adjusting the volume to 100 mL with pure water, the solution was titrated with standard silver nitrate using an iodide ion selective electrode as the indicator electrode. The whole procedure required only 10 min., the relevant iodide-forming reactions being

$$N_2H_4 + 2I_2 \rightarrow N_2 + 4HI$$

$$NH_2OH + 2I_2 + H_2O \rightarrow HNO_2 + 4HI$$

### Mixtures of Halides

The development of titrimetric methods whereby a mixture of at least two of the three halides chloride, bromide, and iodide could each be determined accurately by a single titration has been given much attention since the first attempt to resolve such a mixture was made in 1893 (176). As was mentioned earlier, coprecipitation leading to the formation of mixed crystals

TITRATION OF INORGANIC ANIONS    635

or solid solutions can give rise to large errors in these determinations. This is particularly noticeable in the sequential titration of bromide and chloride in which an appreciable amount of chloride is precipitated before all the bromide has been titrated. It was suggested, therefore, that separate determinations are advisable to determine the individual halides chloride, bromide, and iodide in mixtures of the three (89). Thus chloride was determined by titration in 90% acetone with silver nitrate after treating the sample solution with hydrogen peroxide and 8-quinolinol to eliminate interferences by bromide and iodide. The bromine and iodine formed by the oxidation of the respective halide with hydrogen peroxide were removed by a halogenation reaction with 8-quinolinol. Iodine was then determined in a separate sample iodometrically. Finally, the total halide was determined by titration with silver nitrate performed this time without the addition of hydrogen peroxide and 8-quinolinol but using the same silver billet electrode. The concentration of bromide was then calculated by difference. The relative errors were < ±0.3%.

More recently, however, an iodide ion selective electrode has been used as the indicator electrode for the sequential titrations of iodide, bromide, and chloride in an aqueous medium containing aluminum nitrate (127). To the sample solution containing approximately 0.01 $M$ iodide, bromide, and chloride (10.0 mL of each) was added 0.01 $M$ aluminum nitrate (10 mL) followed by sufficient water to adjust the total volume to 100 mL. The solution was titrated with a 0.1 $M$ standard silver nitrate solution, the whole procedure requiring about 15 min. When the concentrations of determinands were each 0.001 $M$, relative errors and relative standard deviations of less than 0.4% were reported.

Mention was made previously of attempts to titrate fluoride chloride, bromide, and iodide sequentially with a mixed thorium nitrate–silver nitrate titrant using a platinum indicator electrode.

### *Perchlorate*

This ion has been determined in pyrotechnic mixtures by reduction to chloride with titanium hydride (0.5 g) in acid solution ($\sim 25\%$ $H_2SO_4$) under reflux (15 min). After cooling and filtering the mixture, the filtrate was treated with ammonium peroxydisulfate (2.0 g) and titrated with standard silver nitrate (177). Perchlorate has also been determined directly in aqueous solution having a $p$H in the range 4–7 using tetraphenylarsonium chloride as titrant and a perchlorate ion selective electrode as the indicator electrode (9). The latter type of electrode was used also in the titration of perchlorate with 1,2,4,6-tetraphenylpyridinium acetate (10). This titrant was prepared as a 0.05 $M$ solution in 0.3 $M$ sodium acetate, and the titration is best carried out at constant $pH$ in the range 4–7 although these titrations were found to be feasible in the $pH$ range of 2.2–11.0.

### *Periodate*

This anion has been titrated similarly to perchlorate with 1,2,4,6-tetraphenylpyridinium acetate using the perchlorate ion selective electrode as the indicator electrode (10).

## Anions Containing Nitrogen or Phosphorus

The titration of cyanide with standard silver nitrate solution has already been mentioned and is not described again in this subsection, and neither is the titration of azide. Both these anions are, of course, potentially very hazardous, and heavy metal azides are among the most dangerous chemical compounds known. These azides may be destroyed by acidification and addition of excess sodium nitrite.

### *Cyanate*

A silver–silver sulfate electrode has been used to monitor the titration of cyanate with silver nitrate (179). Methanol (50 vol %) and barium nitrate (~ 50 mg to precipitate $CO_3^{2-}$) were added to a $pH$ 8 solution containing cyanate (~ 0.01 $N$), and the solution was diluted to 50 or 100 mL with water. After cooling to 5–10°, the solution was titrated with 0.1 $N$ silver nitrate. The method was applied to the determination of cyanate in mixtures with carbonate, cyanide, chloride, thiocyanate, and also in a mixture containing cyanide and chloride.

### *Nitrate*

Two cationic titrants, nitron and diphenylthallium(III), have been reported as being suitable for the titrimetric determination of nitrate, and both of these are used in conjunction with a nitrate ion selective electrode. Of the two, diphenylthallium(III) nitrate has the lower solubility product as reference to Table 13.2 will confirm. Diphenylthallium(III) chloride is available commercially, but is much more expensive than nitron. Titrant solutions of nitron are usually prepared in 5% acetic acid and should exceed a concentration of 0.01 $M$. The optimum concentration of nitrate is about 0.01 $M$ and the $pH$ of the solution should be adjusted to $pH$ 2–3 by the addition of sulfuric acid (~ 50% v/v). Interference from nitrite occurs when the concentration of nitrite equals or exceeds that of nitrate and can be eliminated by the addition of 1 $M$ hydroxyammonium sulfate (13 mL) followed 1 min afterward by the addition of 25% phosphoric acid (1 drop). When the evolution of gases has ceased (up to 10 min) the $p$H of the solution is adjusted and the nitrate titrated (13).

Diphenylthallium(III) titrant is prepared as a 0.1 $M$ solution in 0.1 $M$ sulfuric acid by means of the corresponding diphenylthallium(III) oxide which, in turn, is prepared from the commercially available chloride by heating the latter under reflux with alcoholic potassium hydroxide (17). The oxide can also be prepared similarly from the diphenylthallium(III) nitrate precipitates recovered from previous determinations. Dissolution of an appropriate amount of the oxide in 0.1 $M$ sulfuric acid yields the titrant solution. The conditions for titrating nitrate with diphenylthallium(III) sulfate are the same as for the titration with nitron. The nitrate solution (0.01–0.1 $M$) must be adjusted in the $p$H range 2–4 with sulfuric acid. As was mentioned earlier, a mercury–mercury(I) sulfate reference half-cell was used for these titrations. This is because diphenylthallium(III) chloride is insoluble. Interference from chloride, bromide, and iodide must be eliminated by precipitation with a slight excess of silver sulfate (without filtration) prior to the titration, and fluoride did not interfere. Nitrite was removed by prior reduction with hydrazine sulfate at $p$H 4–6. Perchlorate must be absent because of the high selectivity coefficient ($\sim 10^3$) of the nitrate ion selective electrode towards this ion.

## *Thiocyanate*

Mention has already been made of the consecutive determinations of bromide and thiocyanate using a silver ion selective electrode. The same type of electrode was used for the separate titrations of cyanide and thiocyanate with silver nitrate (56).

## *Phosphate*

A microestimation of phosphate in the presence of large amounts of fluoride has been based on the precipitation of the normal orthophosphate as $Pb_3(PO_4)_2$ in buffered aqueous solution at $p$H 8.25–8.75 (68, 6). The titrant was 0.01 $M$ lead(II) perchlorate adjusted to $p$H 5.0, and the potentials were monitored with a lead(II) ion selective electrode used in conjunction with a double junction reference half-cell (6). To a sample containing phosphorus (31 $\mu$g) as orthophosphate ($\sim 10^{-6}$ $M$) was added $p$H 8.9 ammonium acetate buffer solution (4 mL, 0.5 $M$ $NH_3OAc$ + $NH_3$ to $p$H 8.9). The solution was diluted to 50 mL with water and titrated with 0.01 $M$ lead(II) perchlorate. In the presence of > 100 mg fluoride the solutions required $p$H adjustment to 8.9 with ammonia prior to the buffer addition.

Orthophosphate has also been titrated similarly with standard silver nitrate solution at $p$H approximately 9.2 (0.12 $M$ borate, 0.04 $M$ boric acid) using a silver wire indicator electrode (95). In the orthophosphate concentration range 0.03–0.3 $M$ a small but reproducible and linear departure from a stoichiometric reaction was observed as the phosphate concentration was increased. This effect was believed to be due to the adsorption of phosphate ions on the silver phosphate precipitate, and was allowed for by reference to a linear calibration curve or by the use of a multiplication factor. Fluoride when present in a fivefold excess did not interfere, and it was shown that chloride, bromide, or iodide could be determined in a mixture with phosphate by a single titration. Interfering cations (e.g., $Ca^{2+}, Mg^{2+}, Al^{3+}, \ldots$) must be removed prior to the titration by ion exchange separation.

### Pyrophosphate

This ion has been titrated in aqueous solution with standard lead(II) nitrate using a lead(II) ion selective electrode (69).

### Hexafluorophosphate

The perchlorate ion selective electrode is responsive also to hexafluorophosphate ($PF_6^-$) allowing this anion to be titrated similarly to perchlorate with 1,2,4,6-tetraphenylpyridinium acetate (10). The $p$H of the titrand solution was adjusted to 6.0.

## Sulfur Anions

The titration of thiocyanate has already been described in the previous subsection. The titrations of sulfate, sulfite, and thiosulfate are described here.

### Sulfate

Some of the less obvious types of electrodes used as indicators in the determination of this ion by titration with a standard solution of barium(II) or lead(II) were mentioned in Chapter 5. The commercially available lead(II) ion selective electrode, however, is the electrode usually chosen for these titrations. Although not Nernstian in its response to sulfate ion, this electrode has been used to monitor titrations of sulfate with standard barium perchlorate performed at $p$H 5–7 in 75% dioxane if the sulfate concentration was 0.01 $M$ or less, or in 50% dioxane if the concentration was greater than 0.01 $M$ (133). The detection limit was $5 \times 10^{-4}$ mmol, and the standard deviations were $\pm 0.007$ and $\pm 0.04$ mg for 0.5–2 mg and for 5–30 mg sulfate, respectively.

When a lead ion selective electrode is chosen as the indicator it has been more usual to titrate sulfate with either lead(II) nitrate or lead(II) perchlorate, generally in water–dioxane mixtures, although 2-propanol has also been used. The optimum concentration of dioxane was found to be approximately 50% by volume (65). This represents a compromise between the small inflections observed when dioxane was present at $< 25\%$, and the insolubility of water-soluble inorganic salts observed when the solvent concentration was above 70%. The response of the lead ion selective electrode is susceptible to interferences from copper(II), mercury(II), and silver(I), and these should not be present in the titrand solution. Anions that form more insoluble lead salts than lead sulfate should be absent (e.g., phosphate), and chloride, nitrate, and bicarbonate at high concentration (one hundredfold excess) interfere by forming mixed precipitates. These anions also tend to decrease the potential break circumjacent to the equivalence point through complexation with lead(II). Bicarbonate interference can be eliminated by adjusting the titrand solution to $p$H 4. The relative standard deviation was found to be $\pm 0.2$ and $\pm 1.0\%$ at $> 5 \times 10^{-4}$ $M$ and at $10^{-5}$ $M$ sulfate, respectively.

More recent adaptations of this method have relied upon Gran plots based on the readings of $E_{cell}$ obtained after the equivalence point when the lead(II) titrant had been added in excess. This linear plot of the antilogarithm of $E_{cell}$ versus the volume of titrant added to the titrand (prepared in 80% 2-propanol) when extrapolated yielded the equivalence volume. Since data were collected beyond the equivalence point, the electrode response was observed to be more rapid, more stable, and less susceptible to interferences (4). Samples of natural waters containing between 2 and 100 ppm sulfate concentration were analyzed by this method. A similar procedure has also been applied in approximately 50% dioxane (67).

*Sulfide*

The application of the silver metal electrode to the titrations of sulfides and thiols with silver nitrate was mentioned in Chapter 5. It is usual nowadays, however, to employ a commercially available silver sulfide membrane electrode as the indicator electrode for titrations of sulfide with cadmium(II), lead(II), or silver(I) titrants. Thus this sulfide ion selective electrode has been used in the determination of hydrogen sulfide in air at the parts per billion level by titration with a dilute solution of cadmium(II) sulfate ($6 \times 10^{-6}$ $M$) (1). The hydrogen sulfide in the air sample was absorbed in a solution of 1.0 $M$ sodium hydroxide, 0.1 $M$ ascorbic acid (10 mL) contained in a glass midget impinger. The electrode was used also to monitor the titration of sulfide with lead(II) nitrate in a medium 1.0 $M$ with respect to sodium hydroxide presaturated with lead(II) sulfide. The relative standard deviations were found to be approximately 2% for > 90 ng sulfide and < 20% for 6–90 ng sulfide. Halides, sulfate, acetate, cyanide, nitrate, phosphate, and ammonium ion did not interfere (71).

Sulfide and dissolved sulfur in a polysulfide solution have been sequentially determined using the sulfide ion selective electrode as indicator (108). Free sulfur dissolves in an alkaline solution of sulfide to form polysulfides

$$S^{2-} + (x - 1)S \rightarrow S_x^{2-}$$

where one sulfur atom in each polysulfide is denoted as sulfide sulfur ($S^{2-}$) and the remainder of the ion ($S_{x-1}$) as dissolved sulfur. The polysulfide was converted by reaction with an excess of potassium cyanide into thiocyanate and sulfide:

$$S_x^{2-} + (x - 1)CN^- \rightarrow S^{2-} + (x - 1)SCN^-$$

The excess cyanide was masked with formaldehyde, the $p$H being adjusted to 3 with dilute sulfuric acid. The alkalinity of the solution was restored by the addition of ammonia (pale pink to phenolphthalein indicator), and after dilution with water (oxygen-free) to 100 mL, the free sulfide plus the sulfide bound by polysulfide was titrated with standard silver nitrate. The $p$H of the titrand solution was then adjusted to $p$H 3 with dilute sulfuric acid, and the thiocyanate, corresponding to the dissolved sulfur originally present, titrated with the silver nitrate titrant.

The silver sulfide membrane electrode has been used also to monitor the sequential titrations of sulfide, arsenite, and arsenate at $p$H 11 (adjusted with sodium hydroxide) with standard silver nitrate (109). The given sequence of inflections reflects the ascending order of the solubility products, silver sulfide having the smallest value. Sodium sulfide, thiourea, and cyanamide have also been determined in mixtures (119).

## *Sulfite*

A rapid determination of this reductant has been based upon the same type of reaction that was described earlier for the determination of hydrazine and hydroxylamine (128). The solution of sulfite (10 mL, ~ 0.1 $N$) was added to iodine–methanol reagent (10 mL, 0.2 $N$) acidified with sulfuric acid (1 mL, 2 $N$). The solution was diluted to 100 mL with water and the iodide formed titrated with silver nitrate using an iodide ion selective electrode as indicator. Thiosulfate and sulfide interfere with the titration, but large amounts of chloride, bromide sulfate, and carbonate do not. The method was applied to the determination of sulfur dioxide in chimney gases.

### *Thiosulfate*

A procedure especially suitable for the determination of thiosulfate in the presence of sulfite has been based upon the titration of thiosulfate (100 mL, 0.0005–0.005 $M$) at $p$H 1.5–10.5 with 0.1 $M$ silver nitrate at room temperature ($> 20°$) using a silver ion selective electrode as indicator (110). When large amounts of sulfite were present, the titrand solution was adjusted to $p$H 1.5–2.0 with dilute sulfuric acid at approximately 60° before the titration. Small amounts of thiosulfate (0.025%) in sodium sulfite can be determined with a relative error of +0.6% and a relative standard deviation of 0.3% when titrated slowly over 50 min.

### Transition Metal Anions

The determinations of chromate, dichromate, hexacyanoferrate(II), hexachloroplatinate, permanganate, perrhenate, and tungstate are not usually associated with precipitation titrations. The availability of the lead and perchlorate ion selective electrodes now makes these titrations feasible.

### *Chromate, Hexacyanoferrate(II), and Tungstate*

These have each been determined by precipitation titration with lead(II) nitrate using a lead(II) ion selective electrode as indicator (69).

### *Dichromate, Hexachloroplatinate, Permanganate, and Perrhenate*

These have each been determined by titration with 1,2,4,6-tetraphenylpyridinium acetate using the perchlorate ion selective electrode as indicator (10). The titrations are carried out under the same conditions as those described for the titration of perchlorate with this titrant.

### *Orthovanadate*

Silver nitrate was used as titrant for the determination of sodium orthovanadate which was prepared by dissolving calculated amounts of vanadium pentoxide in boiling sodium hydroxide of the required strength. The potentials were monitored with a cell consisting of a silver coated platinum gauze electrode connected to a saturated calomel half-cell by means of a potassium nitrate salt bridge. The temperature of this cell was maintained at $40 \pm 1°$ during the titration. The $p$H of the titrand solution was initially 11.5, and a marked change in $E_{cell}$ was observed in the vicinity of the equivalence point which corresponded to the precipitation of $Ag_3VO_4$ in the $p$H range 8–9 (96).

## 13.3 TITRATION OF INORGANIC CATIONS

Precipitation potentiotitrimetry has been applied widely to the determination of metal cations in recent years. Once again, the development of ion selective electrodes has increased the versatility of this method greatly, and the development of organic precipitants has increased the specificity of some determinations considerably. Some of the latter are chelating agents, and hence the degree to which an insoluble chelate is formed with a given metal ion depends upon the $p$H of the solution, just as it does in the case of the soluble EDTA complexes. Careful control of $p$H, therefore, allows the selective precipitation of some ions in the presence of others that could form complexes with the precipitant. The cations in this section are arranged in the order: monovalent cations (alkali metals), alkaline earths, group 3a metals, transition metals, group 5a metals, lanthanides, and actinides.

### Monovalent Cations (Alkali Metals and $NH_4^+$)

The determination of potassium, rubidium, and cesium is possible by titration with tetraphenylborate(III). These ions, together with the ammonium ion, have been titrated with calcium tetraphenylborate(III) in a medium of 0.1 $M$ calcium acetate using a cation-sensitive glass electrode as indicator (138). The calcium tetraphenylborate(III) titrant replaced the more usual sodium tetraphenylborate(III) in this application because the indicator electrode responds also to sodium ion. This conversion was achieved by passing the sodium salt through cation exchange resin previously converted to the calcium form. The solubilities of these four tetraphenylborates are so similar that the titration of a mixture of them yields the total concentration of the four.

### *Potassium*

This ion has also been determined by titration at 25–30° with sodium tetraphenylborate(III) using a potassium ion selective electrode in a solution buffered at $p$H 4-5 with glycine and hydrochloric acid (139). Sodium(I) and lithium(I) up to concentrations equivalent to that of potassium(I) ($\sim 2 \times 10^{-3}$ $M$) did not interfere, but magnesium(II) and calcium(II) gave a negative error whereas ammonium ion gave a positive error.

As mentioned previously, the silver metal electrode responds to changes in phenylborate(III) concentration, and this has formed the basis for the indirect determination of potassium (99). A known volume of a sodium tetraphenylborate(III) solution of known concentration (10 mL, 0.1 $N$) was added to a known volume (10 mL) of the test solution containing potassium(I) (4-20 mg) buffered at $p$H 5 with acetate buffer. A few drops of 2% aluminum(III) nitrate were added and the suspension diluted to 25 mL in a volumetric flask. The precipitate was collected on a filter paper and 10 mL of the filtrate titrated with 0.05 $N$ silver nitrate. Another method for the determination of potassium using a silver metal electrode is given in Reference 100.

## Cesium

Mention has already been made of the titration of this ion with calcium tetraphenylborate(III). Cesium(I) has also been determined in the presence of lithium, sodium, potassium, iron, and magnesium by its precipitation as cesium hexaiodotellurate(IV) ($Cs_2TeI_6$) and the subsequent titration of the iodide present in this precipitate with silver nitrate (91). The analyte solution containing 20 mg cesium was evaporated to dryness and the residue dissolved in glacial acetic acid (4 mL). The precipitate was formed by adding a solution (3 mL) of tellurium dioxide (1.88 g) in aqueous hydriodic acid (100 mL, 20-30% HI), and then it was digested on a steam bath for 30 min. After collecting it on a filter paper and washing it with small portions of glacial acetic acid (25 mL total), it was then dissolved in sodium hydroxide (10 mL, 10%). The filter was washed with water (50 mL) and the total volume of the solution containing the dissolved precipitate adjusted to 100 mL. This solution was then acidified with concentrated nitric acid (2 mL) and the iodide titrated with standard silver nitrate ($\sim 0.1$ $N$) using a silver metal electrode as indicator. The method allows the determination of 5-40 mg cesium provided that no rubidium is present.

## Lithium

This ion has been determined by titration in a medium of 95% ethanol with ammonium fluoride by using a fluoride ion selective electrode as the indicator electrode (44). The method, developed primarily for the analysis of the concentrated solutions of lithium(I) used in the Tramex process for the separation of lanthanides from actinides, can be applied to determine > 5 mg lithium(I), although more satisfactory results can be expected with > 15 mg lithium(I). The range of utility is determined by the solubility of lithium fluoride in 95% ethanol, and hence the amount of water in the medium must be minimized to avoid excessive solubility of lithium fluoride. Interference by hydrogen ion was eliminated by adjusting the $p$H of the medium with ammonia so that it was slightly alkaline, and interferences due to tin(II) and lanthanum(III) were masked with ammonium sulfide.

### Alkaline Earths

Apart from barium(II), these cations are apparently not determined by precipitation potentiotitrimetry. Barium ($\sim 0.1\ M$), however, was titrated with standard sulfate using a platinum electrode in the presence of iron(II), iron(III)-$o$-phenanthrolines, or the thallium(III), thallium(I) oxidation-reduction systems (132). The sharp potential change at the equivalence point was caused by the complexing of iron(III) or thallium(III) with sulfate.

### Group 3a Metals

The cations of this subgroup determined by precipitation titration are aluminum, indium, and thallium.

#### *Aluminum*

An ethanolic medium has been used for the titration of aluminum(III) at $p$H 5 (acetate buffer) with sodium fluoride by a method similar to that described for lithium (12). A sample solution (> 1 mL) containing approximately 10 $\mu$mol aluminum(III) was acidified with 1 $M$ nitric acid, if necessary, and mixed with 1 mL acetate buffer solution (60 mL HOAc, 270 g NaOAc $\cdot$ 3H$_2$O in 1 L water) and 95% ethanol (15–20 mL). The stirred solution was titrated with 0.1 $M$ sodium fluoride, the potentials being monitored with a fluoride ion selective electrode used in conjunction with a saturated calomel reference half-cell. Interfering metal ions were removed before acidification by precipitation in 1 $M$ sodium hydroxide. The relative standard deviation was 1–2%.

#### *Indium*

A silver metal electrode has been found suitable as the indicator electrode for the titration of indium(III) with sodium carbamodithioate derivatives [R$_1$R$_2$NC(:S)S$^-$Na$^+$]. Thus indium(III) (2–12 mg) was determined by titration with 0.03 $M$ sodium $NN$-diethylcarbamodithioate (R$_1$ = R$_2$ =

ethyl) at $p$H 4.4 (acetate buffer) (38). In a search for other suitable carbamodithioates, sodium $N$-cyclohexylcarbamodithioate ($R_1$ = H, $R_2$ = $C_6H_{11}$), ammonium $N$-phenylcarbamodithioate ($R_1$ = H, $R_2$ = $C_6H_5$), and sodium piperidine-1-dithiocarboxylate [$C_5H_{10}$NC(:S)S$^-$Na$^+$] all gave satisfactory results (average relative error $\pm 0.9\%$) in the determination of indium(III) over a concentration range of approximately 3.5–69 × $10^{-4}$ $M$. Of the three titrants, sodium piperidine-1-dithiocarboxylate gave the largest inflection when titrated into the indium solution having a concentration within this range and maintained at $p$H 6.5 with 0.5 $M$ sodium tartrate solution (16). Aluminum, molybdenum, tungsten, alkali metal, and alkaline earth metals do not interfere, but tin(IV), bismuth, and antimony(III) must be absent.

### Thallium

A titrant apparently quite specific to the determination of thallium has been reported (42). This titrant, 3-diethylaminomethyl-2,6-dimercaptopyran-4(4$H$)-thione, has been used for the potentiotitrimetric determination of thallium(III) in strong acid media (e.g., 0.1–2 $N$ HCl) or in acetate buffer solution ($p$H 3.7–4.3). The relative error reported in the *amperometric titration* was < 2%, and the sensitivity was 10 $\mu$g 50 mL$^{-1}$. Zinc(II), tin(IV), copper(II), lead(II), aluminum(III), barium(II), nickel(II), iron(III), cobalt(III), tellurium(IV), manganese(II), tin(II), and chromium(III) in 200–2000-fold excess do not interfere.

Aqueous solutions of thallium(I) in the $p$H range 2.3–8.0 can be titrated with potassium hexyloxymethanedithioate [potassium hexylxanthate, $C_6H_{13}$OC(:S)SK] or the corresponding nonyloxy or decyloxy-derivatives (25). These titrant solutions were prepared in ethanol, and the inflection points of the titrations were detected with either a silver or a mercury electrode. Thus thallium(I) could be determined in the presence of silver(I), copper(II), manganese(II), nickel(II), and chromium(III), but zinc(II) and cadmium interfered.

### Transition Metals

Cations of this classification that have been determined by precipitation potentiotitrimetry have included cadmium, chromium, copper, lead, manganese, mercury, nickel, palladium, platinum, silver, zinc, and zirconium.

### Cadmium(II)

This cation has been determined by precipitation as the sulfide either directly with sodium sulfide titrant (134) or with benzothioamide

[PhC(:S)NH$_2$] when the solution was warmed in the presence of hydrazine (33). A silver ion selective electrode was the indicator electrode for the ultramicroanalytical simultaneous determination of cadmium(II) and silver(I) with the sodium sulfide titrant, and the same type of electrode was preferred over a silver metal wire electrode when the titrant was benzothioamide. Cadmium(II) can also be titrated with potassium hexyloxymethanedithioate and its higher alkoxy homologues in the presence of silver(I), copper(II), lead(II), manganese(II), zinc(II), and chromium(III) under similar conditions to those given for thallium(I) (25). Nickel(II) and thallium(I) interfere in this case.

### Chromium

Mention was made earlier that the titration of iodide with mercury(II) has formed the basis of the indirect determination of some metal ions. Thus chromium(VI) in chromium(VI)–chromium(III) mixtures was determined by this method, the total chromium(III) in another aliquot being determined indirectly by EDTA titration (77).

### Copper(II)

A variety of precipitants are available for the titrimetric determination of copper(II), among which organic sulfur compounds feature prominently. One category of these can be thought of as being derivatives of carbamodithioic acid since they contain the arrangements >N—C(:S)—S—. This category includes the simple structure Et$_2$NC(:S)S$^-$ as in the $NN$-diethylcarbamodithioate anion (structure I, Figure 13.1), and also the more complex heteroaromatic type structures given as structures II and III in Figure 13.1. The second category is really a variant of structure I in that it, too, is related to the same acid (i.e., methanedithioic acid HC(:S)SH), but in this case the compounds are alkoxy derivatives (structure IV) rather than the dialkylamino derivative of structure I. These alkoxymethanedithioates, already mentioned for the determination of thallium(I) and cadmium(II), are examples of this category. The third category also bears a relation to the other two in that these can be regarded as amides of thioic acids, an example of which is dithiooxamide (structure V). These thioamides sometimes form 1:1 chelates which are stable at room temperature, but in other cases unstable complexes or salts are formed, and these decompose to yield the metal sulfide. Sometimes a catalyst is used to hasten this decomposition as in the titration of copper(II) with benzothioamide in the presence of hydrazine catalyst (33).

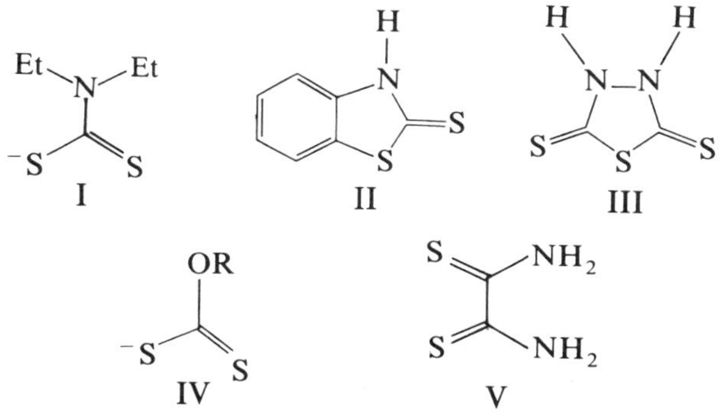

Figure 13.1. Some organic precipitants used for the titration of copper(II): I, $NN$-diethylcarbamodithioate anion; II, benzothiazole-2-($3H$)thione; III, 1,3,4-thiadiazole-2,5($3H,4H$)-dithione; IV, an alkoxymethanedithioate anion; V, dithiooxamide (ethanedithioamide).

The carbamodithioic acid derivative sodium $NN$-diethylcarbamodithioate has been used for the titration of copper(II) in the presence of iron(III) (35). The iron(III) was masked with ammonium sulfosalicylate buffered at $p$H 9 with ammonia–ammonium buffer. A medium containing water (50 mL), potassium nitrate (2 g), ethanol (40 mL), and diethyl ether (100 mL) was used for the sequential titration of silver(I), copper(II), lead(II), and zinc(II) with this titrant (37). An aqueous medium buffered in the range $p$H 3.5–5.8 was used for the titration of copper(II) with benzothiazole-2($3H$)-thione (29). When prepared as a solution in 0.1 or 0.05 $M$ sodium hydroxide, the titrant solution was reported to be stable for several months and is thus possibly more stable than sodium $NN$-diethylcarbamodithioate. An aqueous medium also buffered at about $p$H 4 and containing sodium fluoride (0.5 g) was used for the titration of copper(II) with 1,3,4-thiadiazole-2,6($3H,4H$)-dithione, but silver, lead, zinc, and gold were also precipitated under these conditions. Iron, nickel, and cobalt did not interfere (143).

Copper(II), nickel(II), and zinc(II) have been titrated sequentially with potassium hexyloxymethanedithioate (25). Potassium methoxymethanedithioate can be used as titrant for copper(II) in the presence of five- to sixfold amounts of thallium(I), zinc(II), nickel(II), lead(II), manganese(II), cadmium(II), and chromium(III). As noted earlier, these titrants are prepared in ethanol, and the solutions are stable for 10–12 days (25).

Dithiooxamide forms a stable 1:1 chelate with copper(II) which is thought to be polymeric (21). This titrant has been used to determine copper(II) and silver(I) in mixtures ranging from 1–99% silver(I) with a relative standard deviation ranging from 0.16–2.00% for silver(I) and from 0.20–1.34% for copper(II). The titration medium contained potassium hydrogen phthalate (1 g) and sodium fluoride (0.5 g) in about 100 mL water.

Mercury(II) and silver(I) react simultaneously with dithiooxamide, and gold and platinum are reduced to the metallic state by this titrant and must, therefore, be absent. Under the conditions employed, iron(III), cobalt(II), nickel(II), and zinc(II) in excess did not interfere with the sequential titrations of silver(I) and copper(II). An aqueous solution of the titrant ($\sim 0.001$ $M$) appeared stable for at least one month.

Other titrants used for the determination of copper(II) by precipitation titrimetry, not covered by the foregoing three categories, include pentacyanonitrosylferrate(III) (sodium nitroprusside) and tetraphenylborate(III). Sodium pentacyanonitrosylferrate(III) has been used for the sequential titrations of silver(I) and copper(II) at concentrations as small as $2.5 \times 10^{-4}$ $M$ silver(I) and $5 \times 10^{-4}$ $M$ copper(II) with respective precisions of 0.35 and 0.41% (52). This method was applied to the analysis of a 72:28 Ag/Cu alloy. The titrant has also been used for the sequential titrations of zinc(II) and copper(II) within the $p$H range 3.0–5.0 using a platinum indicator electrode (14). A sharper potential jump for zinc(II) is observed, however, if this ion is titrated at $p$H 12. Satisfactory results with an error of not more than 1.0% were obtained for copper(II) contents of not less than $2.7 \times 10^{-6}$ $M$ and zinc(II) contents of not less than $2.2 \times 10^{-6}$ $M$. A tenfold excess of iron(III), a sevenfold excess of iron(II), a sixfold excess of aluminum(III), a sixtyfold excess of barium(II), a fortyfold excess of calcium(II), a 1.8-fold excess of manganese(II), a 1.6-fold excess of lead(II), or a 1.5-fold excess of bismuth caused no interference to the determination of copper(II). However, cadmium(II), nickel(II), and cobalt interfere with the determination. Copper(II) has also been determined, after reduction to copper(I) with ascorbic acid, by precipitation with sodium tetraphenylborate(III) in calcium acetate and magnesium acetate–acetic acid buffers (181).

### *Lead(II)*

This cation has been determined using the same type of sulfur-containing organic titrants as were mentioned for copper(II). Thus in the first category sodium $NN$-diethylcarbamodithioate and benzothiazole-2(3$H$)-thione have been applied to these determinations, and in the second category alkoxymethanedithioates have also found application. The thioamide category is represented by the use of thioacetamide and benzothioamide.

Lead(II), in company with silver(I), copper(II), and zinc(II), has been determined sequentially by titration with sodium $NN$-diethylcarbamodithioate in a medium of water (50 mL), potassium nitrate (2 g), ethanol (40 mL), and diethyl ether (100 mL) (37).

A similar medium was used also for the sequential determination of silver(I) and lead(II) with this titrant utilizing a silver indicator electrode

(36). The titration of lead(II) with benzothiazole-2(3$H$)-thione was accomplished in a medium containing 5–10% ethanol buffered in the $p$H range 3–6 (30). A mercury metal electrode was used as the indicator, and the error was less than 1% for the determination of $9 \times 10^{-5}$ $M$ lead(II).

Potassium methoxymethanedithioate has been used for the titration of lead(II) in the presence of five- to sixfold amounts of thallium(I), zinc(II), nickel(II), manganese(II), cadmium(II), chromium(III), copper(II), and silver(I), and separate determinations of the following combinations of metals with lead(II) in their mixtures were found to be possible with alkoxymethanedithioates because of the differences in the solubilities: copper(II) and lead(II), zinc(II) and lead(II), and combinations of copper(II), lead(II), nickel(II), and zinc(II) (25).

Lead(II) has been determined in iron yttrium garnets containing 0.1–10% lead(II) by titration with 0.01 or 0.05 $M$ thioacetamide in a medium of 1 $M$ hydrazine in the presence of citrate at $p$H 10 using a silver sulfide membrane electrode as the indicator (144). Lead(II) sulfide has been similarly precipitated by titration with benzothioamide in alkaline solution (33).

Other titrants used for the precipitation of lead(II) have included chromate (182) and molybdate (87). A 0.0025 $M$ solution of sodium molybdate was employed as titrant for the microdetermination of lead(II) contained in a predominantly aqueous unbuffered medium adjusted to $p$H 5–5.5 (87). Standardization of the titrant against 1.5 mg lead nitrate overcomes the observed slight departure of the reaction from the expected stoichiometry, and accurate results were obtained for 1–2 mg lead(II). Under these conditions the absolute error did not exceed 3 $\mu$g lead(II), and the relative standard deviation was 0.05%. As is shown in Table 13.4, the potential break of this titration, monitored with a lead ion selective electrode, is increased in water–organic solvent mixtures with respect to that observed in aqueous media, but the electrode response becomes very slow. Chloride in less than thousandfold excess and bromide in hundredfold excess do not interfere, but the presence of iodide or sulfate causes interference.

### Manganese(II)

This cation can be precipitated as manganese(II) sulfide with benzamide in the presence of hydrazine (33).

### Mercury(II)

This cation has been titrated with benzothiazole-2(3$H$)-thione at $p$H 9–10 using a silver sulfide membrane electrode as indicator (31). The titrant solution was prepared in a 9:1 mixture of ethanol–methanol.

Mercury(II) sulfide has been precipitated by titration in alkaline medium with thioacetamide (183). To a solution containing mercury (5–15 mg) and 0.1 $M$ disodium EDTA (30 mL) was added 2 $M$ sodium hydroxide (20 mL) and 1.2% gelatin (20 mL). The resulting solution was titrated with 0.005 $M$ thioacetamide prepared in $p$H 5 buffer [0.1 $M$ potassium hydrogen phthalate and 0.05 $M$ trisodium phosphate mixed in the ratio 50 : 24 (v/v)]. Silver and platinum interfere because they are precipitated in the presence of EDTA, but none of the common anions such as nitrate, phosphate, sulfate, acetate, or halides interfere. Cyanide interferes with the functioning of the silver metal billet electrode but does not affect the precipitation of mercury(II) sulfide. This solid was also precipitated by titration with benzothioamide which was monitored with a silver wire electrode (33). Dithiooxamide can also be used to precipitate mercury(II) sulfide, this time from a solution buffered to $p$H approximately 4.5 with potassium hydrogen phthalate (1 g 100 mL$^{-1}$ solution). This titration was monitored with a silver metal indicator electrode, and mercury(II) in the range 0.1–40 ppm was determined with 0.26–1.5% relative standard deviation with approximately 0.3% relative error (184). Copper(II) and silver(I) interfere with this determination.

Mercury(II) ($\geq 10^{-6}$ $M$) in a medium of 0.1 $M$ nitric acid has been titrated with standard sodium iodide solution using an iodide ion selective electrode as the indicator electrode (47). The relative standard deviation was $\pm 0.4\%$ for the determination of 1 mg mercury(II) in the presence of hydrogen peroxide, iron(III), iron(II), DTPA, sodium(I), or aluminum(III). Mercury(II)-complexing ligands can be masked with metals.

## *Nickel(II)*

Benzothiazole-2(3$H$)-thione has been used as titrant for the determination of nickel(II) at $p$H 3–6 in a medium containing 5–10% ethanol (30). A mercury metal electrode was the indicator electrode for this determination which was applied also to the analysis of nickel(II) in electrolytes.

Potassium alkoxymethanedithioates have also been used as titrants for this determination (25). Nickel(II) has been determined in the presence of copper(II), silver(I), lead(II), manganese(II), and chromium(III) with hexyloxymethanedithioate and its higher homologues, but cadmium interfered. The different solubilities of these metal alkoxymethanedithioates allowed the determination of the following combinations of metals in their mixtures: copper and nickel, nickel and zinc, copper, nickel, and zinc by titration with potassium hexyloxymethanedithioate. A silver metal or mercury metal indicator electrode was used to monitor the potentials against a saturated calomel reference half-cell.

### Palladium(II)

When this ion in acetate buffer ($pH$ 4.5) was titrated into a standard solution of dithiooxamide, the jump in potential of the platinum indicator electrode corresponded to a palladium/dithiooxamide mole ratio of 1:2 (15). The standard deviation was ≤ 0.16 for determining 2.66–19.95 mg palladium, but the results of the titration of palladium(II) with standard dithiooxamide solution were not reproducible.

Palladium(II) in the presence of excess platinum and copper has been successfully titrated with iodide (46). The determination of palladium in catalysts was performed by titration of the dissolved residue after ashing (500°) with a standard solution of potassium iodide using an iodide ion selective electrode (48). Palladium (1–20 mg) was determined with a relative standard deviation of 0.4%.

### Platinum(IV)

This cation can be determined indirectly by its precipitation as platinum(IV) iodide with a known volume of standard potassium iodide added in excess, followed by the titration of this excess with standard mercury(II) nitrate using a silver amalgam indicator electrode (76). Masking of copper(II), cobalt(II), mercury(II), nickel(II), or iron(III) was achieved by the addition of EDTA which allowed the analysis of binary mixtures of each of these cations with platinum(IV). Thus platinum was determined by the method outlined, whereas the other metal was determined by an appropriate complexometric or other method. The relative errors in the determination of platinum(IV) were ≤ ±1.2 and ≤ 3.9% for the determination of 1.6–29 mg and 24–808 µg platinum, respectively.

### Silver(I)

There are a variety of titrants available for the titrimetric determination of silver(I) by precipitation. In addition to the types of organic sulfur compounds employed for the precipitation of copper(II), which in this case also include 8-quinolinethiol and thiourea, there are the halide titrants, principally potassium bromide and potassium iodide, and other titrants such as sodium thiosulfate, ammonium thiocyanate, tetraphenylborate(III) (138), and 1,2,3-benzotriazole (28).

Mention has already been made of the sequential titrations of silver(I), copper(II), lead(II), and zinc(II) with sodium $NN$-diethylcarbamodithioate and the determination of silver(I) and copper(II) in binary mixtures by titration with either dithiooxamide or sodium pentacyanonitrosylferrate(III). Similarly, the determination of silver(I) and lead(II) in mixtures by titration with sodium $NN$-diethylcarbamodithioate was referred to earlier, and binary mixtures of silver(I) with either thallium(I) and cadmium(II) can be sequentially titrated with hexyloxymethanedithioate (25). Silver(I) and cadmium(II) have also been titrated sequentially on the ultramicro scale with sodium sulfide (134). Silver has also been determined as silver(I) sulfide by titration with benzothioamide (33) and with sodium thiosulfate (148).

Thiourea has been used as titrant for the determination of silver(I) in photographic fixing baths. The sample solution (25 mL) containing silver(I) ($< 100$ mg) in the presence of sodium thiosulfate (up to 0.5 $M$) was heated to 90°, mixed with 1 $M$ sodium hydroxide (5 mL), and titrated with 0.05 or 0.1 $M$ thiourea using a silver wire indicator electrode. The titration of silver(I) with thiourea in the presence of picrate ion using a picrate ion selective electrode as indicator was mentioned in Chapter 5. It has been reported that 8-quinolinethiol (thiooxine) is an especially good titrant for determining silver(I) in multicomponent mixtures (88), and silver(I) has also been titrated at $p$H 9-10 with benzothiazole-2(3$H$)-thione solution, prepared in 9:1 ethanol-methanol mixture using a silver sulfide membrane electrode as indicator (31). A similar electrode was used for the titration of silver(I) with ammonium thiocyanate (146). This method was applied to the determination of silver in silver alloys and solders.

Although it was recommended sometime ago that silver(I) be determined by direct titration with potassium iodide (185), improved results have been obtained by the indirect titration of silver(I) using an excess of iodide (75). The amount of standard potassium iodide solution added in excess of that required to precipitate silver(I) was determined by titration with a standard solution of mercury(II) nitrate. Binary mixtures of silver(I) and copper(II) have been analyzed utilizing this method by masking copper(II) with EDTA during the determination of silver(I). Copper(II) was then determined iodometrically in another aliquot of the analyte solution. A similar approach was used in the analysis of ternary mixtures of silver(I) with lead(II) and zinc(II), cadmium(II), nickel(II), cobalt(II), or copper(II). This indirect method is reported as having advantages over the direct determination in that both the high relative error ($+3.2\%$) and the time taken for a single titration (1-1.5 h) reported for the latter method are much reduced in the indirect method. Thus the errors observed for the indirect determination of silver(I) alone ranged from $-1.3- +0.43\%$, and even in the determination of micro amounts of silver(I) ($\geq$ 6.5 µg) the time required for a single determination did not exceed 15 min.

### Zinc(II)

As was previously mentioned, zinc(II) in company with silver(I), copper(II), and lead(II) can be titrated sequentially with sodium $NN$-diethylcarbamodithioate, and in mixtures with copper(II) and nickel(II) by titration with potassium hexyloxymethanedithioate. Again, mention has already been made of the sequential titrations of copper(II) and zinc(II) within the $p$H range 3.0–5.0 with sodium pentacyanonitrosylferrate(III). Zinc(II), similar to copper(II), can also be precipitated as the sulfide by reaction with benzothioamide in alkaline solution in the presence of hydrazine (33).

### Zirconium(IV)

This cation has been successfully titrated with cupferron, $PhN(NO)O^-NH_4^+$ [Reg. No. *135-20-6*] in a medium of 2 $M$ sulfuric acid by using a gold indicator electrode (40). Fourfold tin(II) and iron(III), tenfold nickel(II) and chromium(III), and all concentrations of thorium(VI), lanthanum(III), manganese(II), cobalt(II), indium(III), and uranium(VI) did not interfere. The average error in the determination of 2–30 mg zirconium(IV) in solutions was $\leq$ 0.95%. The method was used to analyze a synthetic mixture corresponding in its composition to Zircalloy.

### Group 5a Metals

Apparently, only antimony(III) of this subgroup has been determined by precipitation potentiotitrimetry. The titrant used was thionalide, 2-mercapto-$N$-(2-naphthyl)acetamide [Reg. No. *93-42-5*], and the titration was carried out in a medium of 1 $N$ sulfuric acid at 25° using the pair of polarized platinum electrodes mentioned earlier (147). Lead(II), iron(III), copper(II), silver(I), bismuth(III), tin(IV), aluminum(III), arsenic(III), sodium(I), potassium(I), and chloride do not interfere, but zinc(II) and cadmium(II) do.

### Lanthanides and Actinides

Lanthanum(III) can be titrated in an unbuffered solution with a standard solution of fluoride using a fluoride ion selective electrode. Similarly, the total concentration of thorium(IV) and lanthanides in analyte solutions prepared from samples of monazite sand have been determined indirectly by the addition of excess standard fluoride solution (45). The excess fluoride was then determined by back-titration with standard lanthanum(III). In another aliquot of analyte solution, the lanthanides were separated from thorium(IV) by precipitation of the lanthanide as oxalate. These oxalates were then dissolved, 30–40 vol % ethanol was added, and the lanthanides were determined directly by titration with standard fluoride. The thorium(IV) concentration was calculated by difference. The relative standard deviations were 0.8 and 1.0% for the determination of the lanthanides and thorium, respectively. Cerium(III) and thorium(IV) have been determined at optimum $p$H values of 4–6 for cerium(III) and 3.0–3.5 for thorium(IV) using cupferron titrant and the mercury electrode mentioned previously (41). Thorium(IV) and cerium(III) may also be determined in their mixtures with a relative error of < 2% by this method by first titrating their total concentrations, and then in another aliquot titrating thorium(IV) alone at $p$H 2 in the presence of sulfosalicylic acid. The inflection points of these titrations corresponded to mole ratios of cupferron : thorium(IV) and cupferron : cerium(III) of 4 : 1 and 3 : 1, respectively.

Cerium(IV) has been determined in a medium of 0.2–2 $N$ sulfuric acid by the addition of excess standard potassium iodide, and the titration of this excess with standard mercury(II) using a silver amalgam indicator electrode (78). This method formed the basis for the determination of cerium(IV) in binary and ternary mixtures of other ions. In another aliquot cerium(IV) was reduced to cerium(III), and the other cation or the total cations were determined by back-titration of excess EDTA with mercury(II). The cation mixtures were resolved by separations or masking followed by titration with EDTA. Cerium(IV) in the range 42 $\mu$g–35 mg 20 mL$^{-1}$ aliquot could be determined by this method. A method based upon the titration of excess iodide with mercury(II) has also been developed for thorium(IV) (80).

## 13.4 APPLICATIONS TO ORGANIC ANALYSIS

Precipitation titrations play an important role both in the fields of organic elemental analysis and in the determination of some types of organic compounds through the reaction of particular functional groups with a precipitant. The application of ion selective electrodes to organic analysis in general is the subject of a book (186) in which the application of these electrodes to precipitation potentiotitrimetry has been included. The treatment in this section is not confined to the use of ion selective electrodes but is restricted to a summary of the general applications of precipitation titrimetry to organic elemental analysis and to the determination of organic

APPLICATIONS TO ORGANIC ANALYSIS 655

compounds. When the latter summary is combined with the information given in Chapter 10, a large number of potentiotitrimetric methods applied to the determination of organic compounds becomes available.

### Elemental Analysis

The principal elements determined in organic compounds by precipitation titrimetry are the halogens, sulfur, selenium, tellurium, arsenic, lead, and mercury. Once converted into the appropriate inorganic ion, usually by preliminary oxidative or reductive decomposition of the original compound, the actual titration is similar to the analogous method mentioned in Sections 13.2 or 13.3.

#### *Fluorine*

In one method (59) the sample containing 2–4 mg was combusted in a Vycor oxygen flask (> 96% $SiO_2$) by the standard Schoniger technique (187), the gaseous products being absorbed in 0.1 $N$ sodium hydroxide (10 mL). After transferring the contents to a plastic beaker and adjusting the $p$H of the solution (50–75 mL) to $p$H 4.4 with dilute hydrochloric acid, the solution was freed from carbon dioxide by boiling and the fluoride titrated with standard lanthanum(III) nitrate in the usual way. If the sample contained sulfur, the sulfate formed was masked by a quantitative precipitation with barium nitrate and the titration of fluoride performed in the $p$H range 5–7 without separating the barium sulfate precipitate. The other halogens and nitrogen do not interfere. However, phosphate emanating from the oxidation of organic phosphorus must be separated by ion exchange, or alternatively, the fluoride freed from phosphate by distillation as fluorosilicic acid. In another method (60) fluorine and sulfur were determined in one sample by combustion of the sample (5–10 mg) in the presence of glucose (4–6 mg) in an oxygen flask containing 1 $N$ sodium hydroxide (5 mL), water (15 mL), and hydrogen peroxide (1 mL, 30%). The sulfate was titrated at $p$H 5.5–6.6 with standard barium perchlorate visually using methylsulfonazo III indicator, and the fluoride by titration with standard lanthanum(III) nitrate using a fluoride ion selective electrode as indicator. Other methods for the determination of fluorine are given in References 61 and 151.

#### *Other Halogens*

The oxygen flask combustion method, referred to in the preceding, is often applied also to the conversion of chloro, bromo, or iodo substituents to the corresponding halides, the solution of one of the latter being titrated with one of the titrants available for that ion. This method has been applied to the determination of chlorine in organomercury compounds, the combustion products in this instance being absorbed in an aqueous medium containing EDTA to mask mercury(II) (188). It has also been applied to the determination of residues of the herbicide Lasso, 2-chloro-2′,6′-diethyl-$N$-methoxymethylacetanilide [Reg. No. *15972-60-8*] (189).

This Schoniger method when applied to highly halogenated volatile organic compounds, however, often yields low results because of incomplete combustion (190, 191). Chlorine or bromine have been determined in such compounds in a specially designed combustion tube apparatus (84). The samples were introduced into the combustion system by injection and burned in an oxygen stream at 1000° over platinum and quartz. The combustion gases were absorbed directly in a water-cooled titration vessel containing 80% v/v acetic acid, some hydrogen peroxide, nitric acid, and mercury(II) chloride or mercury(II) bromide. This solution was titrated with mercury(II) using a silver sulfide membrane electrode or an iodide ion selective electrode as indicator.

The convenient oxygen combustion flask method fails also in the analysis of some halogen-containing organometallic derivatives. In such instances the sample can be decomposed by fusion with sodium peroxide in a sealed container often referred to as a bomb. Thus an electrically fired calorimetric bomb was used for the fusion of gelatin encapsulated samples (20–30 mg) of chlorinated organorhenium, -ruthenium, and -osmium compounds in an atmosphere of oxygen (94). The solutions of the rhenium and ruthenium halogen compounds were acidified with $HNO_3$ (1 : 1) to $pH$ 3–4, and those of osmium-containing compounds with sulfuric acid (1 : 1) to $pH$ 1–2 and transferred quantitatively to 100 mL volumetric flasks. Aliquots containing 1–2 mg of the halide were titrated with 0.01 $N$ silver nitrate using a silver metal indicator electrode. Halogenated platinum and palladium compounds were determined similarly except that the alkaline residue was acidified with concentrated nitric acid (4 mL) and heated until it dissolved. After cooling, the $pH$ of the solution was adjusted to $pH$ 1.8–2.5. A two- to fivefold excess of EDTA was added to aliquots of these solutions which were then boiled and cooled before titration with silver nitrate. The standard deviation was 0.17%, and the method was used to determine chlorine and bromine in compounds of the type $(R_3NH)_2[MX_6]$, $(RNH_3)_2[MNOX_5]$, and $MX_2(PR_3)_2$ where M = Re, Ru, Os, Pt, or Pd, and X = Cl or Br. Similar methods were used also for the determination of halogen in halogenated organoiron and organotin compounds (93). Conversion of halogens to halide ions has also been accomplished by fusion with potassium metal in a bomb at 800–850° (92).

APPLICATIONS TO ORGANIC ANALYSIS  657

An alternative method to the combustion and fusion methods for the determination of chlorine, bromine, or iodine in some organic compounds has involved the dehalogenation of the compound with sodium borohydride in the presence of a palladium catalyst (192). Water-soluble or alkali-soluble compounds (1.8 meq) were dissolved in water (10 mL) or 0.2 $N$ sodium hydroxide. Sodium borohydride solution (5 mL of 3% w/v $NaBH_4$ in 2 $N$ NaOH) was added, followed by the addition of palladium(II) nitrate solution (1 mL of a solution of 1.0 g $Pd(NO_3)_2 \cdot 2H_2O$ in 100 mL $H_2O$) to the vigorously stirred solution. Dehalogenation of iodo compounds occurred within 15 min at room temperature, but dehalogenation of the chloro and bromocompounds was accomplished by boiling the solution. After decomposition of the sodium borohydride with acetone (10 mL) and the addition of 2 $N$ sulfuric acid (20 mL), the halide ions were titrated with standard silver nitrate using a silver–silver chloride electrode as indicator. Water insoluble compounds were dissolved in dimethylformamide (10 mL), treated with palladium on charcoal catalyst (100 mg of 5% Pd on charcoal), and then the sodium borohydride solution (5 mL) was added before proceeding as outlined. The same palladium–charcoal catalyst replaced the palladium(II) solution for the dehalogenation of the more stable halogenated compounds, which in this case was accomplished by boiling the reaction mixture under reflux until evolution of hydrogen ceased. It was not necessary to add acetone if this procedure was adopted. The error for the method was $\pm 0.5\%$, and the volatility of the sample and the poisoning of the catalyst are quoted as possible interferences.

In special cases when chlorine is bound to boron, as in compounds such as bis(chlorovinyl)chloroborane [(ClCH=CH)$_2$ BCl], for example, it was found possible to discriminate between the chlorine bound to carbon and chlorine bound to boron by esterification of the latter bond in nitric acid–methanol medium (193). The hydrochloric acid so formed was titrated with standard silver nitrate solution.

*Sulfur*

The Schoniger oxygen flask combustion method has been used also for the oxidation of sulfur-containing organic compounds (7). In this instance the gaseous products from the combustion of 2–4 mg of the compound were absorbed in distilled water (10 mL) containing sodium nitrite (10 mg). This absorption and oxidation of the sulfur-containing species to sulfate with sodium nitrite required 30–45 min after which the solution was transferred to a 150 mL beaker. The volume of the absorbing solution and the washings was reduced to 20 mL by boiling, and the solution cooled to room temperature. The $p$H was adjusted to between 4 and 6.5 with dilute sodium hydroxide, 1,4-dioxane (30 mL) was added, and the solution titrated with 0.01 $M$ lead(II) perchlorate using a lead ion selective electrode as indicator. Fluoride can be masked by the addition of a 10% excess of boric acid to the absorbing solution, but when both fluoride and chloride were present this addition of boric acid was unnecessary. Instead, the combined absorbing solution and washings were acidified with 70% perchloric acid (3-4 drops) before boiling. The absolute errors were ±0.3% sulfur. Phosphorus interferes and must be separated prior to the sulfur determination. The oxidation of organic sulfur to sulfate has also been accomplished by an "empty tube" combustion method in which solid silver sulfate was formed (22). This solid was dissolved in water and titrated with potassium iodide.

A rapid determination of sulfur in organic compounds has been devised in which the sample was heated at approximately 1050° in a hydrogen stream over quartz wool to give hydrogen sulfide which was absorbed in 1 $M$ potassium hydroxide–1 $M$ hydroxylamine solution (194). The sulfide was then titrated immediately with 0.0002 $M$ lead(II) nitrate using a silver sulfide membrane electrode in an automatic titration apparatus. The total time for the analysis was approximately 5-6 min, and the relative standard deviation was < 3% for the determination of > 5 ppm sulfur. Nitrogen-containing compounds interfered only at high concentrations.

### *Selenium, Tellurium, and Arsenic*

The conversion of these elements to selenite, tellurite, or arsenate, respectively, and the titrations of these ions at $p$H 8-8.5 with standard silver nitrate solution using a platinum titrant stream reference electrode together with a silver indicator electrode dipping in the titrand solution forms the basis for these analyses (195). The organic compounds (4-8 mg) containing selenium, tellurium, or arsenic were decomposed by oxygen flask combustion and absorption in aqueous sodium hydroxide, or by wet digestion (Kjeldahl flask) with sulfuric acid and nitric acid. The selenite, tellurite, or arsenate solution thus formed was adjusted to $p$H 8-8.5 (borate buffer) and made approximately 33% in acetone. This titrand solution was then titrated with 0.1 $M$ silver nitrate added from an Agla Micrometer Syringe modified by the sealing of a platinum wire into the barrel of the syringe near the exit. The tip of the syringe was dipped into the titrand solution in the usual way. Full experimental details are given in the paper, and these include a good description of the oxygen flask combustion technique. The interference of chlorine or bromine was eliminated by the addition of dilute nitrous acid to the absorption solution, and 0.3% sodium borohydride in 2 $M$ sodium hydroxide was used to reduce iodine. The halides thus formed could be determined by titration in a slightly acidic medium. The interference of

APPLICATIONS TO ORGANIC ANALYSIS 659

metal ions was masked with iminodiacetic acid. The errors were $\leq 0.3\%$, and the systematic error of approximately 1% observed in the titration of arsenate was compensated for by standardization of the titrant with a standard solution of arsenate.

## Mercury

Precipitation of mercury(II) as the sulfide in alkaline solution by titration with 0.005 $M$ thioacetamide solution using a silver indicator electrode is one method that is used for this determination (183). This titration can be applied directly to those compounds which are soluble in alkaline EDTA by treating the solution containing 5–15 mg of mercury with 0.1 $M$ EDTA (30 mL disodium salt), 2 $M$ sodium hydroxide (20 mL), and 1.2% gelatin solution (20 mL), and then titrating the resulting solution with thiocetamide. For insoluble organomercury compounds the same procedure was applied to a solution prepared by the combustion of the sample (5–15 mg Hg) in a Schoniger oxygen flask in the presence of concentrated nitric acid (10 mL). After heating the contents of the flask under reflux for 15 min, the solution was cooled, transferred to a beaker, and neutralized with sodium hydroxide. The relative error of the method was found to be 1%. A similar combustion of 3–5 mg organic substance in the presence of concentrated nitric acid (4 mL) followed by reflux for 6 min sufficed for the preparation of the mercury(II) solution, which was titrated with 0.005 $M$ potassium iodide after adjustment of the titrand to $p$H 0.8–1.0 with 30% potassium hydroxide solution (50). An iodide ion selective electrode was used as the indicator electrode and the results were within the usual $\pm 0.3\%$ required for elemental analysis.

## Lead

The precipitation of lead(II) by titration with molybdate, mentioned earlier, also formed the basis for the microdetermination of lead in organic lead compounds (87). The lead was converted to lead(II) by a wet mineralization process since the oxygen flask combustion method has been reported as being unsuitable for this application owing to the formation of Pt–Pb alloy. The sample (3–4 mg), containing 1.4–1.7 mg of lead, was digested in a micro Kjeldahl flask with 65% nitric acid followed by additions of 30% hydrogen peroxide and, finally, 70% perchloric acid. After the organic matter had been destroyed the digest solution ($\sim 1$ mL) was allowed to cool, diluted with water (10 mL), and boiled vigorously for 3 min to expel any oxidizing gases present (e.g., chlorine dioxide) which cause interference with the response of the lead ion selective electrode. It is important also to ensure that all hydrogen peroxide be removed since this poisons the membrane of the electrode. The cooled solution was then transferred to a 100 mL beaker and its $p$H adjusted to 5 with sodium hydroxide before titration with 0.0025 $M$ sodium molybdate. A potential break of 8–10 mV/0.05 mL was observed circumjacent to the equivalence point.

### The Determination of Organic Anions

Organic compounds that can exist as anions in solution may sometimes be determined through the reaction of the anionic species with metal ions or other cations to form sparingly soluble salts or complexes. Silver(I) or mercury(II) are the titrants used predominantly for these precipitations and are discussed here under the subheadings argentometric and mercurimetric methods, respectively. Other cations have included lanthanum(III), lead(II), and zinc(II), and the application of 1,2,4,6-tetraphenylpyridinium ion for the precipitation of organic anions provides an example of the use of an organic precipitant.

#### *Argentometric Methods*

Some nitrogen-containing heteroaromatic compounds that possess a hydrogen atom attached to a ring nitrogen atom as part of a weakly acidic imino group ($p\text{K}_a$ values generally in the range 8–10) have been determined by precipitation as silver salts. These have included derivatives of 9$H$-purine, 1$H$-triazole, and 1$H$-tetrazole in which the acidity of the >N–H group is not impaired by substitution. Some of these types of compounds, often too weakly acidic to be titrated with alkali at low concentrations, have been titrated successfully with standard silver nitrate using an electrode reversible to silver ion as the indicator electrode. Thus adenine (6-purinylamine, $p$K 9.9) has been titrated at concentrations of $2.5 \times 10^{-3}$ $M$ with 0.01 $M$ silver(I) at $p$H 9.8 ± 0.2 (101). The buffer solution was prepared by mixing 1 $M$ sodium acetate (100 mL), 1 $M$ ammonium acetate (100 mL), and 0.1 $M$ sodium hydroxide (800 mL), and the 0.01 $M$ silver(I) titrant [i.e., $Ag(NH_3)_2^+$] was prepared by tenfold dilution of a 0.1 $M$ solution of silver nitrate with this buffer. When the indicator electrode and the calomel reference half-cell were dipped into the buffered solution of adenine, the value of $E_{cell}$ required 30 min to equilibrate. During the titration the potentials were recorded 1 min after each addition, and the equivalence point corresponded to 1:1 stoichiometry. 8-Bromoadenine has been similarly titrated (103). Similar titrations with silver nitrate of 1$H$-benzotriazole ($p\text{K}_a$ 8.4) and its 4 and 5-methyl derivatives in a medium of 1.5% aqueous sodium bicarbonate solution ($p$H 8–8.5) have also been reported (196; see also 163). Determination of the three compounds in extracts of antirust paper, antirust fluid, and automotive antifreeze are given as an example. For the analogous titration of theophylline [1,3-dimethylpurine-2,6(1$H$, 3$H$)-dione, $pK_a$ 8.7], the best results were obtained at $p$H 11 (32). In this instance, however, more precise results were obtained by precipitation of the silver salt with excess standard silver nitrate followed by the back-titration of this excess with benzothiazole-2(3$H$)-thione. More recently, a number of derivatives of tetrazole, including bitetrazole, have been determined by argentometric titration (122).

Although many thiols have been determined successfully by titration with silver nitrate, mention must be made at this stage of the advantages given by the application of the mercurimetric method for these determinations. Nonetheless, thiols are often titrated with silver nitrate in alkali media containing ammonia to prevent the precipitation of silver oxide. Thus water-soluble alkanethiols have been titrated in 1 $M$ sodium hydroxide containing 0.05 $M$ ammonia, but samples not soluble in this medium were titrated with ethanolic silver nitrate in a medium of ethanolic sodium acetate (23). Mixtures containing hydrogen sulfide and an alkanethiol were titrated sequentially, satisfactory results being obtained for binary mixtures of methanethiol, ethanethiol, propanethiol, 2-propanethiol, 2-butanethiol, 2-methyl-2-propanethiol, or 2-methylpropanethiol with hydrogen sulfide. However, marked coprecipitation was obtained when butanethiol and pentanethiol were titrated similarly. With the exception of the latter two compounds, two inflections were observed in the titration curve: the first inflection corresponding to hydrogen sulfide and the second to the alkanethiol. A silver electrode, prepared as described previously, was used as indicator electrode. A similar method has been described for the separate determination of thiol, disulfide, and inorganic sulfide in a single sample (197). The silver ion selective electrode has also been applied to equivalence point detection of similar titrations. For example, thiols and hydrogen sulfide have been determined argentometrically at $p$H 1–2 in a medium containing 1% gelatin (111). Solutions of volatile samples were added to known volumes of standard silver nitrate by means of a micrometer burette fitted with a tight Teflon plunger. A $p$H in the range 2.5–9 sufficed for the titration of an aqueous solution of 2-mercaptoethanol, L-cysteine, or glutathione with silver nitrate using a silver ion selective electrode (113). In the vicinity of the equivalence point the response of this electrode was slow, and up to 2 min was necessary for the attainment of a stable potential after each addition of titrant. A somewhat different approach has been applied to the determination of thiols including cysteine (104). A peroxide-free dioxane solution containing 70 µg to 5 mg of the sample was reacted with iodine (1 mL, 1 $N$, $CCl_4$ soln.) in carbon tetrachloride (15 mL) in the presence of pyridine (1 mL). One equivalent of iodide was formed for each thiol group reacted, and, after extraction into water (15 mL, followed by 3 × 10 mL portions), the iodide was titrated with 0.005 $N$ silver nitrate using a silver ion selective electrode. The mean absolute error ranged from 0.2–0.6% in determining these small quantities of thiols.

The latter method has been applied also to the determination of thiones including compounds containing thiocarbonyl groups. In these cases four equivalents of iodide was formed for each thione group reacted. Thioamides are more commonly titrated directly with silver nitrate, often in alkaline solution. For example, the important analytical reagent thioacetamide may be titrated in the concentration range 0.001–0.1 $M$ in 0.1 or 1 $M$ sodium hydroxide, in 0.01 $M$ ammonia, or in distilled water using a sulfide ion selective electrode versus a saturated calomel reference half-cell to monitor the potentials (114). In such media silver sulfide precipitated, but in a medium of 0.5 $M$ nitric acid a precipitate of silver thioacetamide [AgS–C(:NH)CH$_3$] was formed by the reaction of the thiol tautomer [HS–C(:NH)CH$_3$]. Similar well-defined titration curves were obtained for the titrations of $N$-phenylthiourea [Ph NHC(:S)NH$_2$] and $NN'$-diphenylthiourea [PhNHC(:S)NHPh] in 1 $M$ sodium hydroxide. Both compounds reacted with two equivalents silver(I) to give silver sulfide and $N$-phenylurea or $NN'$-diphenylurea, respectively (117). Thiourea has also been determined in slightly acid media indirectly by titration of excess standard silver nitrate with standard potassium chloride solution (39). A method has also been developed that allows the sequential determinations of thiourea, sodium sulfide, and cyanamide (119). Therapeutically active thioamides have been analyzed in pharmaceutical formulations by titration with standard silver nitrate in 90% 2-propanol, the thioamide being dissolved in a medium of 90% ethanolic 0.1 $M$ sodium acetate (98). A silver electrode was used in these titrations. A rapid and accurate determination of dithiooxamide [H$_2$N–C(:S)—C(:S)—NH$_2$] is possible by titration with silver nitrate in a titrand medium of 1 $M$ sodium hydroxide (115), the reaction products being silver sulfide, oxalic acid, and nitric acid. Methods have also been devised for the determinations of $OO$-dialkylphosphorodithioates in a medium of 1 : 1 water–ethanol (116) or in acetic acid–acetone–water medium (198).

A miscellany of other types of compounds has been determined argentometrically. Thus chlorinated hydrocarbons have been determined quickly and accurately by splitting off the chlorine atoms with potassium hydroxide in 1,2-ethanediol at 170° and then titrating the chloride with silver nitrate (199). A titration of chloride has also formed the basis for the determination of terminal hydroxy groups in polymers. The hydroxy group was reacted with phosgene to form chloroformate which was then hydrolyzed to yield chloride in amounts equivalent to the original hydroxy content (90). Interfering compounds included amines, some carboxylic acids, unsaturated

APPLICATIONS TO ORGANIC ANALYSIS 663

compounds, and hydroxy groups on aromatic nucleii. $m$-Dinitro compounds have also been determined indirectly by reacting them with 0.2 $M$ potassium cyanide followed by titration of the excess cyanide with silver nitrate. One mole of potassium cyanide was quantitatively consumed for each mole of $m$-dinitro compound, but in those cases in which the compound contained three nitro groups in the meta position to each other two moles potassium cyanide per mole of compound were required (120). Grignaud reagents (MeMgCl and $Et_2Mg$) have been titrated with silver perchlorate in tetrahydrofuran (200), and formaldehyde has been determined using its reaction with iodine in alkali medium to quantitatively produce iodide which was then titrated (130).

*Mercurimetric Methods*

As previously mentioned, these methods can be applied advantageously to the determination of thiols because some argentometric methods give rise to coprecipitation phenomena when the silver sulfide or silver thiolate is formed (73). Some doubts also exist on the specificity and stoichiometry of the reaction between thiols and silver(I), and these are based mainly upon the variable stoichiometry between cysteine and silver(I) (113). Furthermore, mercury thiolates have been found to be considerably more stable than silver thiolates (201), and mercury(II) titrants, unlike silver nitrate, are not light sensitive (86). The reaction of thiols with mercury(II) proceeds according to the equation

$$Hg^{2+} + 2RSH \rightarrow (RS)_2Hg + 2H^+$$

in neutral acetone or aqueous solution (73). Acetone is the most universally applicable solvent for thiols although several are soluble only in water. It is recommended (73) that the acetonic solution be adjusted to $pH$ 5-7 and buffered with pyridine, since at low $pH$ the carbonly group may react with the thiol group to form a dithioacetal,

$$>C=O + 2RSH \rightarrow >C(SR)_2 + H_2O$$

Addition of pyridine also improves the character of the mercury(II) thiolate and enhances the stability of the e.m.f. readings. The latter can be monitored by a J-shaped mercury pool electrode or with a commercially available bromide ion selective electrode used in conjunction with suitable reference half-cells (73, 86). Mercury(II) perchlorate was used as the titrant (86), which was prepared as a 0.01 $N$ solution by dissolving the nonahydrate (2.81 g) in water (950 mL), adjusting the $pH$ of this solution to 1.2-1.3 with perchloric acid, and then diluting it to 1 L with water. This solution was standardized against an aliquot of potassium bromide containing 8–10 mg Thiol samples in the 0.01–0.1 mM range were titrated with the 0.01 M mercury(II) perchlorate by weighing the sample directly into the titration vessel and diluting it to about 40 mL with acetone, ethanol, or dioxane. When acetone was used, pyridine (2 mL) was added and the solution titrated under an atmosphere of nitrogen with mercury(II). A variety of different types of thiols was determined successfully by this method including some heteroaromatic thiols. Many of the latter can undergo thiol–thione tautomerism, and in water it is often the thione tautomer that predominates (202). It seems likely that a similar situation exists in acetone since it was impossible to analyze some of these compounds by this method. An example of the latter was "8-mercaptopurine" which probably exists overwhelmingly in the thione tautomer configuration named for convenience as purine-8(1$H$)-thione. Cysteine and albumin have also been determined by titration with mercury(II) (82) using a silver electrode.

The titration of nickel(II) bis($OO$-diethylphosphorodithioate) in a medium of 0.01 $M$ perchloric acid with silver(I) and mercury(II) perchlorate solutions was investigated as a means of determining $OO$-diethylphosphorodithioate. Better results were obtained by the mercurimetric method (203). Mercury(II) has also been the titrant for the determination of trithiocarbonate in the presence of other sulfidic anions (81). This titration, used to determine the trithriocarbonate concentration in viscose, was performed in an air-free solution containing the viscose sample (1 g) in a medium of 0.1 $N$ sodium hydroxide–0.9 $N$ sodium acetate (100 mL) using an amalgamated silver electrode. Two inflections were observed, the first corresponding to the titration of sulfide, and the second to that of trithiocarbonate.

Mention has been made of the indirect determination of phenoxyalkanoic acids in pesticides (34) using a mercury(I) responsive membrane electrode. An indirect method has also been devised for the determination of organic nitrates or nitramines in which the sample (2–5 mg) was treated with a mixture of concentrated sulfuric acid (2–3 mL) and mercury (3 drops) in an atmosphere of nitrogen. The mercury(I), formed by the reaction

$$2R-O-NO_2 + 3H_2SO_4 + 6Hg \rightarrow 3Hg_2SO_4 + 2NO + 2R-OH + H_2O$$

was titrated with standard potassium iodide solution using an iodide ion selective electrode as indicator (51). Several compounds used as high explosives, industrial intermediates, and vasodilators were analyzed successfully.

APPLICATIONS TO ORGANIC ANALYSIS    665

## Other Metal Ion Titrants

Oxalate has been determined both by titration with lead(II) perchlorate (70) and with lanthanum(III) nitrate (62) using a lead(II) or a fluoride ion selective electrode, respectively. The medium chosen for the titration with lead(II) was 40% dioxane ($p$H 3.5–10.5), whereas an unbuffered medium of 3:1 ethanol–water at $p$H 5.5 was satisfactory for the lanthanum(III) titration. When lead(II) was used for the titration of oxalate (1–25 mg) an average recovery of 100.06% (0.25 absolute standard deviation) was obtained, and acetate, formate, propanoate, and phthalate could be tolerated in 18-, 24-, 19- and 3-fold excesses, respectively, relative to oxalate. Species that complex or precipitate lead(II) interfered. In the lanthanum(III) titrations 0.025–2.50 mM oxalate were determined with standard deviations of 0.0003–0.01 mM. Bismuth(III), calcium, sodium, and potassium interfered. The interference of the latter two were attributed to the formation of double salts of the type $La_2(C_2O_4)_3(Na$ or $K)_2C_2O_4$.

Benzimidazole-2(3$H$)-thione ($pK_a$10.2), 1$H$-benzotriazole ($pK_a$8.4), and benzothiazole-2(3$H$)-thione ($pK_a$6.9) have all been titrated with zinc(II) sulfate using a zinc amalgam electrode (153).

### 1.2.4.6-Tetaphenylpyridinium Acetate

Mention has already been made of the use of this titrant for the determination of large inorganic anions. The titrations of organic anions are similar to these in that the potentials observed during the titrations are monitored with a perchlorate ion selective electrode used in conjunction with a double junction reference half-cell (142). Nitrate and fluoroborate selective electrodes also could be used in some cases. The classes of compounds that can be determined are nitroform and some nitroform yielding compounds, phenylborates, 2,4,6-trinitro compounds, some dinitrophenols and nitrophenols, and some halogenated phenols. As a result of titrating a significantly representative number of compounds within these categories, the requirements for the successful titrations of organic compounds with 1,2,4,6-tetraphenylpyridinium acetate were summarized as the following:

1. The compound must be reasonably soluble in water or in dilute alkali.
2. It must form a sufficiently insoluble precipitate with the titrant.
3. The perchlorate ion selective electrode must respond to the compound or, alternatively, the fluoroborate or nitrate electrodes must respond to it.

Direct potentiometric determinations were studied for picric acid, which can be determined by titration, and 2,6-dinitro-4-methylphenol, which does not give an analytically useful titration curve. Both gave near-Nernstian responses as recorded with a perchlorate ion selective electrode in the concentration range $10^{-4}$–$10^{-2}$ for picric acid and $10^{-3}$–$10^{-2}$ $M$ for 2,6-dinitro-4-methylphenol.

## The Determination of Organic Cations

Tetraphenylborate(III), the principal species used for the precipitation of alkali metals and ammonium cations, has also found wide application as titrant for the determination of organic cations. The general applications of sodium tetraphenylborate(III) $NaB(Ph)_4$ to analytical chemistry has been reported and reviewed (204), and the applications of ion selective electrodes specifically to titrations with this compound have been reviewed more recently (205).

A variety of nitrogen bases has been titrated with sodium tetraphenylborate(III), the potentials being monitored with a potassium ion selective electrode (valinomycin type) used in conjunction with a saturated calomel reference half-cell (140). The titrant solution was prepared by dissolving sodium tetraphenylborate(III) (25 g) in water (500 mL), and then shaking the resulting solution with chromatographic grade alumina (5 g). The suspension was left overnight, then filtered, and the volume of the filtrate adjusted to 1 L. The solution was standardized against thallium(I) nitrate. Solutions of the conjugate acid species $BH^+$ (5 mL 0.1 $M$, or 50 mL 0.01 M) were diluted to about 70 mL in the titration beaker and titrated with the 2.5% sodium tetraphenylborate(III) solution. The larger the cation the lower the solubility of its tetraphenylborate(III), and the greater was the potential jump observed circumjacent to the equivalence point. The magnitude of the latter depended also upon the selectivity of the indicator electrode to the species $BH^+$, and the symmetry of the overall charge distribution in the cation was also significant. Thus the inflections obtained for methylene blue, methylene green, and crystal violet were very pronounced being 70–80, 20–30, and 30–40 mV mL$^{-1}$ titrant, respectively, whereas for 2-methylpropylamine the inflection was only 1.4 mV mL$^{-1}$.

A selection of alkaloids including aconitine nitrate [Reg. No. *6509-18-8*], brucine hydrochloride [Reg. No. *5786-96-9*], and papaverine hydrochloride [Reg. No. *61-25-6*] have also been titrated similarly using a valinomycin electrode (141), and surfactant quaternary ammonium salts used alone or in pharmaceutical preparations have also been determined (20). The preparation of the silver wire electrode for the latter application was described earlier. The titrations of sodium tetraphenylborate(III) with cetylpyridinium chloride [Reg. No. *123-03-5*] and Zephiramine (*NN*-dimethyl *N*-tetradecylbenzylammonium chloride) [Reg. No. *139-08-2*] in a medium of 5 mM sodium chloride were monitored satisfactorily using a J-shaped mercury cup indicator electrode (137).

Oxonium cations, formed by the complexation of polyethyleneglycols with barium(II) acetate buffer (*p*H 4.6), can be precipitated by titration with sodium tetraphenylborate(III) (136). A cell consisting of a silver indicator electrode and a silver–silver chloride reference half-cell ($NaNO_3$ bridge) was used to monitor the potentials during these titrations which formed the basis for the determination of polyethyleneglycols. Other related compounds such as the Tweens, nonylphenol adducts, and polypropyleneglycols were also titrated. Organic bases, polyvinylpyrrolidone, ethylene oxide, lithium, potassium, and ammonium ions are quoted as being possible interferents. Anions that form insoluble silver salts should be absent since they have a deleterious effect upon the performance of the silver electrode.

## REFERENCES

1. D. L. Ehman, *Anal. Chem.* **48**, 918 (1976).
2. D. Jagner and K. Aren, *Anal. Chim. Acta* **52**, 491 (1970).
3. W. Selig, *Microchem. J.* **21**, 291 (1976).
4. E. P. Scheide and R. A. Durst, *Anal. Lett.* **10**, 55 (1977).
5. W. Eysenbach, B. Suttkus, and G. Heller, *Fresenius' Z. Anal. Chem.* **277**, 183 (1975).
6. W. Selig, *Mikrochim Acta* **1970**, 564.
7. W. Selig, *Mikrochim Acta* **1970**, 168.
8. M. Mascini, *Analyst (London)* **98**, 325 (1973).
9. R. J. Baczuk and R. J. Dubois, *Anal. Chem.* **40**, 685 (1968).
10. W. Selig, *Mikrochim. Acta* **1978(2)**, 75.
11. J. J. Lingane, *Anal. Chem.* **39**, 881 (1967).
12. E. W. Baumann, *Anal. Chem.* **42**, 110 (1970).
13. A. Hulanicki and M. Maj, *Talanta* **22**, 767 (1975).
14. O. N. Rusina, P. N. Kovalenko, and Z. I. Ivanova, *Zh. Anal. Khim.* **21**, 257 (1966); *J. Anal. Chem. USSR* **21**, 227 (1966).
15. B. Stankovic, A. Stefanovic, and M. Dugandzic, *Mikrochim. Acta* **1977(1)**, 395.
16. Z. I. Ivanova, T. V. Sazhneva, V. K. Chebotarev, and O. E. Shelepin, *Zh. Anal. Khim.* **29**, 466 (1974); *J. Anal. Chem. USSR* **29**, 398 (1974).
17. J. S. DiGregorio and M. D. Morris, *Anal. Chem.* **42**, 94 (1970).
18. M. Kapel and J. C. Fry, *Analyst (London)* **99**, 608 (1974).
19. D. A. Katz and A. K. Mukherji, *Microchem. J.* **13**, 604 (1968).

20. S. Pinzauti and E. La Porta, *Analyst* (*London*) **102**, 938 (1977).
21. L. H. Kalbus and G. E. Kalbus, *Anal. Chim. Acta* **53**, 225 (1971).
22. J. P. Dixon, *Talanta* **4**, 221 (1960).
23. M. W. Tamele, L. B. Ryland, and R. N. McCoy, *Anal. Chem.* **32**, 1007 (1960).
24. W. J. Kirsten, A. Berggren, and K. Nilsson, *Anal. Chem.* **30**, 237 (1958); **31**, 376 (1959).
25. P. N. Kovalenko, Z. I. Ivanova, V. K. Chebotarev, and V. D. Dionis'ev, *Zh. Anal. Khim.* **24**, 1810 (1969); *J. Anal. Chem. USSR* **24**, 1468 (1969).
26. R. Jasinski and I. Trachtenberg, *Anal. Chem.* **44**, 2373 (1972).
27. N. Akimoto and K. Hozumi, *Anal. Chem.* **46**, 766 (1974).
28. J. Havir, *Collect. Czech. Chem. Commun.* **32**, 130 (1967).
29. O. N. Rusina, P. N. Kovalenko, and Z. I. Ivanova, *Zh. Anal. Khim.* **20**, 44 (1965); *J. Anal. Chem. USSR* **20**, 38 (1965).
30. E. I. Galkina, K. N. Bagdasarov, and V. I. Man'ko, *Fiz. Khim. Metody. Anal. Kontrolya Proizvod., Mezhvuz. Sb.* **2**, 21 (1976); *CA* **87**, 161124e.
31. H. Hopkala and L. Przyborowski, *Chem. Anal.* (*Warsaw*) **20**, 785 (1975); *CA* **84**, 69037u.
32. L. Przyborowski, *Acta Pol. Pharm.* **29**, 247 (1972); *CA* **77**, 172397r.
33. L. M. Andreasov, E. I. Vail, V. A. Kremer, Z. N. Chernyaeva, and S. A. Reznichenko, *Izv. Vyssh. Ucheb. Zaved., Khim. Tekhnol.* **10**, 743 (1967); *CA* **67**, 113456e.
34. N. Ciocan and G. E. Baiulescu, *Anal. Chem.* **50**, 1407 (1978).
35. A. Hulanicki, *Chem. Anal.* (*Warsaw*) **5**, 881 (1960); *CA* **55**, 14169a.
36. I. M. Bhatt and K. P. Soni, *Indian J. Appl. Chem.* **34**, 8 (1971); *CA* **79**, 142544k.
37. A. M. Trivedi, K. P. Soni, and I. M. Bhatt, *Indian J. Chem.* **4**, 328 (1966).
38. R. Staroscik and H. Siaglo, *Chem. Anal.* (*Warsaw*) **10**, 265 (1965); *CA* **64**, 1341g.
39. L. A. Sales and J. A. De Morgan, *Arch. Bioquim., Quim. Farm.* **18**, 3 (1973); *CA* **83**, 125876k.
40. Z. I. Ivanova and V. Ya. Rivina, *Zh. Anal. Khim.* **27**, 189 (1972); *J. Anal. Chem. USSR* **27**, 156 (1972).
41. P. N. Kovalenko, Z. I. Ivanova, and I. F. Poyarkova, *Tr. Kom. Anal. Khim., Akad. Nauk SSSR* **17**, 381 (1969); *CA* **72**; 117410m.
42. L. F. D'yachenko and Yu. I. Usatenko, *Vopr. Khim. Khim. Tekhnol.* **34**, 25 (1974); *CA* **83**, 71006f.
43. A. S. Sukhoruchkina, V. A. Postnikov, and Yu. I. Usatenko, *Zavod. Lab.* **39**, 917 (1973); *CA* **80**, 22291z.
44. E. W. Baumann, *Anal. Chem.* **40**, 1731 (1968).
45. F.-C. Chang. H.-T. Tsai, and S.-C. Wu, *J. Chin. Chem. Soc.* (*Taipei*) **22**, 309 (1975); *CA* **85**, 13389p.
46. S. Barabas and J. Vinaric, *Anal. Chem.* **36**, 2365 (1964).
47. R. F. Overman, *Anal. Chem.* **43**, 616 (1971).
48. B. Schreiber, *Fresenius' Z. Anal. Chem.* **278**, 343 (1976).
49. P. Hartmann, U. Gruenke, and H. Berge, *Z. Chem.* **17**, 452 (1977).
50. A. Campiglio, *Mikrochim. Acta* **1977(2)**, 71.

51. S. S. M. Hassan, *Talanta* **23**, 738 (1976).
52. P. N. Kovalenko, Z. I. Ivanova, and L. V. Sekretova, *Zavod. Lab.* **35**, 409 (1969); *CA* **71**, 45496y.
53. T. Anfalt and D. Jagner, *Anal. Chim. Acta* **47**, 483 (1969).
54. T. Anfalt and D. Jagner, *Anal. Chim. Acta* **50**, 23 (1970).
55. T. Eriksson and G. Johansson, *Anal. Chim. Acta* **52**, 465 (1970).
56. W. E. Bazzelle, *Anal. Chim. Acta* **54**, 29 (1971).
57. E. Heckel and P. F. Marsh, *Anal. Chem.* **44**, 2347 (1972).
58. W. Selig, *Mikrochim. Acta* **1974**, 515.
59. W. Selig, *Fresenius' Z. Anal. Chem.* **249**, 30 (1970).
60. M. Gachon, A. Gehenot, and G. Maire, *Bull. Soc. Chim. Fr.* **1975**, 2442.
61. L. Helesic, *Collect. Czech. Chem. Commun.* **37**, 1514 (1972).
62. A. Cedergren and G. Sundin, *Anal. Chim. Acta* **94**, 467 (1977).
63. G. C. Cortellessa and C. A. Napoli, *Analyst (London)* **93**, 546 (1968).
64. T. A. Kokina, M. N. Petrikova, and I. P. Alimarin, *Zh. Anal. Khim.* **26**, 2237 (1971); *J. Anal. Chem. USSR* **26**, 2003 (1971).
65. J. W. Ross, Jr., *Anal. Chem.* **41**, 967 (1969).
66. S. Iga, K. Nakamura, K. Tanaka, and M. Iwaida, *Eisei Shikenjo Hokoku* **92**, 54 (1974); *CA* **83**, 7287w.
67. R. Delmas, *Mikrochim. Acta* **1978(1)**, 219.
68. W. Selig, J. W. Frazer, and A. M. Kray, *Mikrochim. Acta* **1975(2)**, 581.
69. E. H. Hansen and J. Ruzicka, *Anal. Chim. Acta* **72**, 365 (1974).
70. W. Selig, *Microchem. J.* **15**, 452 (1970).
71. J. Slanina, E. Buysman, J. Agterdenbos, and B. Griepink, *Mikrochim. Acta* **1971**, 657.
72. J. Slanina, P. Vermeer, J. Agterdenbos, and B. Griepink, *Mikrochim. Acta* **1973**, 607.
73. J. S. Fritz and T. A. Palmer, *Anal. Chem.* **33**, 98 (1961).
74. H. Khalifa, *Microchem. J.* **18**, 529 (1973).
75. H. Khalifa and B. Ateya, *Microchem. J.* **12**, 440 (1967).
76. H. Khalifa and B. N. Barsoum, *Microchem. J.* **15**, 224 (1970).
77. H. Khalifa and Y. M. Issa, *Microchem. J.* **18**, 436 (1973).
78. H. Khalifa and B. N. Barsoum, *Microchem. J.* **18**, 428 (1973).
79. H. Khalifa, M. M. Khater, and M. A. Zayed, *Egypt. J. Chem.* **17**, 161 (1974); *CA* **86**, 150022y.
80. H. Khalifa, B. N. Barsoum, Y. M. Issa, and M. S. Okda, *Egypt. J. Chem.* **18**, 679 (1975); *CA* **89**, 139855p.
81. H. Dautzenberg, B. Philipp, and J. Schumann, *Faserforsch. Textiltech.* **23**, 372 (1972).
82. T. Y. Toribara and L. Koval, *Talanta* **17**, 1003 (1970).
83. V. G. Belikov, K. N. Bagdasarov, V. A. Kimstach, and N. A. Sorokoumova, *Farmatsiya (Moscow)*; *CA* **85**, 10484e.
84. W. Potman and E. A. M. F. Dahmen, *Microchim. Acta* **1972**, 303.
85. W. Selig, *Mikrochim. Acta* **1971**, 46.
86. W. Selig, *Mikrochim. Acta* **1973**, 453.
87. A. Campiglio, *Mikrochim Acta* **1979(1)**, 267.

88. V. I. Suprunovich, V. V. Velichko, and Yu. I. Shevchenko, *Zavod. Lab.* **44**, 525 (1978); *CA* **89**, 156841y.
89. T. S. Prokopov, *Anal. Chem.* **42**, 1096 (1970).
90. D. G. Bush, L. J. Kunzelsauer, and H. M. Stewart, *Anal. Chem.* **35**, 1250 (1963).
91. G. M. Serebrennikova, L. A. Sazikova, V. E. Phjushcheva, and B. D. Stepin, *Zh. Anal. Khim.* **23**, 446 (1968); *J. Anal. Chem. USSR* **23**, 371 (1968).
92. V. A. Klimova and M. D. Vitalina, *Zh. Anal. Khim.* **15**, 339 (1960); *J. Anal. Chem. USSR* **15**, 387 (1960).
93. M. P. Strukova and I. I. Kashiricheva, *Zh. Anal. Khim.* **24**, 1244 (1969); *J. Anal. Chem. USSR* **24**, 999 (1969).
94. M. P. Strukova, I. I. Kashiricheva, R. G. Abdulina, and L. K. Kalashnikova, *Zh. Anal. Khim.* **25**, 1198 (1970); *J. Anal. Chem. USSR* **25**, 1036 (1970).
95. D. M. McColl and T. A. O'Donnell, *Anal. Chem.* **36**, 848 (1964).
96. R. S. Saxena and O. P. Sharma, *Talanta* **11**, 863 (1964).
97. M. W. Tamele, V. C. Irvine, and L. B. Ryland, *Anal. Chem.* **32**, 1002 (1960).
98. I. E. Davidson, *J. Pharm. Pharmacol.* **27**, Suppl., 32P (1975).
99. E. Siska and E. Pungor, *Fresenius' Z. Anal. Chem.* **257**, 12 (1971).
100. T. S. Prokopov, *Anal. Chem.* **43**, 793 (1971).
101. J. Roemer and E. Mittag, *Mikrochim. Acta* **1974**, 879.
102. J. Roemer, *Mikrochim. Acta* **1976(1)**, 321.
103. J. Roemer, *Mikrochim. Acta* **1977(2)**, 479.
104. S. S. Hassan, *Mikrochim. Acta* **1977(1)**, 405.
105. W. Krijgsman, J. F. Mansveld, and B. F. A. Griepink, *Fresenius' Z. Anal. Chem.* **249**, 368 (1970).
106. F. J. Conrad, *Talanta* **18**, 952 (1971).
107. J. E. Burroughs and A. I. Attia, *Anal. Chem.* **40**, 2052 (1968).
108. S. Ikeda, M. Satake, T. Hisano, and T. Terazawa, *Talanta* **19**, 1650 (1972).
109. E. E. Chao and K. L. Cheng, *Anal. Chem.* **48**, 267 (1976).
110. S. Ikeda and H. Satake, *Bunseki Kagaku* **25**, 544 (1976); *CA* **86**, 50194n.
111. F. Peter and R. Rossett, *Anal. Chim. Acta* **70**, 149 (1974).
112. V. I. Nazarova, *Koks Khim.* **1978**, 51; *CA* **89**, 122490t.
113. L. C. Gruen and B. S. Harrap, *Anal. Biochem.* **42**, 377 (1971); **42**, 398 (1971).
114. M. K. Papay, K. Toth, V. Izvekov, and E. Pungor, *Anal. Chim. Acta* **64**, 409 (1973).
115. N. M. Sheina, V. P. Izvekov, M. K. Papay, K. Toth, and E. Pungor, *Anal. Chim. Acta* **92**, 261 (1977).
116. I. Haiduc and V. Munteau, *Rev. Chim. (Bucharest)* **28**, 375 (1977); *CA* **87**, 110974z.
117. M. K. Papay, V. P. Izvekov, K. Toth, and E. Pungor, *Anal. Chim. Acta* **69**, 173 (1974).
118. M. T. Neshkova, V. P. Izvekov, M. K. Papay, K. Toth, and E. Pungor, *Anal. Chim. Acta* **75**, 439 (1975).
119. T. V. Kramareva and V. Shul'man, *Zh. Anal. Khim.* **23**, 750 (1968); *J. Anal. Chem. USSR* **23**, 648 (1968).
120. S. S. M. Hassan, *Anal. Chem.* **49**, 45 (1977).

121. V. P. Izvekov, M. Kucsera-Papay, K. Toth, and E. Pungor, *Analyst* (*London*) **97**, 634 (1972).
122. W. Selig, *Mikrochim. Acta* **1979(1)**, 53.
123. R. L. La Croix, D. R. Keeney, and L. M. Walsh, *Commun. Soil Sci. Plant Anal.* **1**, 1 (1970).
124. B. W. Hipp and G. W. Langdale, *Commun. Soil Sci. Plant Anal.* **2**, 237 (1971).
125. G. W. Heunisch, *Anal. Chim. Acta* **101**, 221 (1978).
126. S. S. M. Hassan, *Fresenius' Z. Anal. Chem.* **270**, 125 (1974).
127. J. Motonaka, S. Ikeda, and N. Tanaka, *Anal. Chim. Acta* **105**, 417 (1979).
128. S. Ikeda, J. Hirata, and H. Satake, *Nippon Kagaku Kaishi* **1973**, 1473; *CA* **79** 100155d.
129. S. Ikeda and J. Motonaka, *Anal. Chim. Acta* **90**, 257 (1977).
130. S. Ikeda, *Anal. Lett* **7**, 343 (1974).
131. E. P. Diamandis and T. P. Hadjiioannou, *Mikrochim. Acta* **1977(2)**, 255.
132. E. M. Zingel, V. G. Korsakov, and I. A. Kedrinskii, *Fiz. Khim.* **1974**, 44; *CA* **84**, 83695p.
133. C. Harzdorf, *Fresenius' Z. Anal. Chem.* **262**, 167 (1972).
134. P. Hartmann, U. Gruenke, and H. Berge, *Z. Chem.* **19**, 71 (1979).
135. R. Geyer and P. Stein, *Acta Chim.* (*Budapest*) **67**, 1 (1971); *CA* **74**, 106841t.
136. R. J. Levins and R. M. Ikeda, *Anal. Chem.* **37**, 671 (1965).
137. T. Kiba and T. Kambara, *J. Electroanal. Chem. Interfacial Electrochem.* **44**, 129 (1973).
138. G. A. Rechnitz, S. A. Katz, and S. B. Zamochnick, *Anal. Chem.* **35**, 1322 (1963).
139. M. Kawamura, F. Hiratsuka, and T. Kashima, *Kyoritsu Yakka Daigaku Kenkyu Nempo* **1973**, 21; *CA* **80**, 152493c.
140. K. Vytras, *Collect. Czech. Chem. Commun.* **42**, 3168 (1977).
141. K. Vytras, *Cesk. Pharm.* **26**, 9 (1977); *CA* **87**, 90766c.
142. W. Selig, *Mikrochim Acta* **1978(2)**, 359.
143. Y. G. Ha, *Taehan Hahak Hoechi* **20**, 277 (1976); *CA* **85**, 201694h.
144. A. P. Mirnaya, A. M. Bulgakov, and L. A. Egorova, *Zavod. Lab.* **36**, 1045 (1970); *CA* **74**, 38036g.
145. A. P. Mirnaya and A. K. Timchenko, *Metody Anal. Galogenidov Shchelochin. Shchelochnozemel. Metal. Vys. Chist.* **1971**, No. 2, 199; *CA* **77**, 121774d.
146. M. Geissler, *Anal. Chim. Acta* **90**, 249 (1977).
147. B. Kh. Gorbatkova and O. N. Rusina, *Zh. Anal. Khim.* **31**, 2147 (1976); *J. Anal. Chem. USSR* **31**, 1569 (1976).
148. R. E. Humphrey, R. M. Maniscalco, and W. Hinze, *Microchem. J.* **16**, 410 (1971).
149. R. Soloniewicz, *Chem. Anal.* (*Warsaw*) **22**, 177 (1977); *CA* **88**, 83062n.
150. E. P. Diamandis and T. P. Hadjiioannou, *Mikrochim. Acta* **1977(2)**, 255.
151. T. S. Light and R. F. Mannion, *Anal. Chem.* **41**, 107 (1969).
152. N. Ishibashi, K. Kina, and K. Tamura, *Anal. Lett* **8**, 867 (1975).
153. S. N. Prajapati, I. M. Bhatt, and K. P. Soni, *J. Electrochem. Soc. India* **25**, 133 (1976); *CA* **87**, 126681n.

154. J. J. Lingane, *Anal. Chem.* **40**, 935 (1968).
155. D. C. Moore and G. Davies, *Anal. Chem.* **47**, 2477 (1975).
156. C. C. Templeton, *J. Chem. Eng. Data* **5**, 514 (1960).
157. A. Hulanicki, *Acta Chim. Acad. Sci. Hung.* **27**, 411 (1961).
158. R. M. H. Verbeeck, R. A. Khan, and H. P. Thun, *Bull. Soc. Chim. Belg.* **86**, 503 (1977).
159. O. N. Rusina, *Mater. Nauch.-Tekh. Konf. Sev.-Kavkaz.-Gornomet. Inst.* **1968**, 69; *CA* **78**, 66491c.
160. E. E. Chao and K. L. Cheng, *Talanta* **24**, 247 (1977).
161. L. Meites (Ed.), *Handbook of Analytical Chemistry*, McGraw-Hill, New York, 1963, p. 1–13.
162. L. Sharma and B Prasad, *J. Indian Chem. Soc.* **47**, 193 (1970).
163. S. N. Prajpati, I. M. Bhatt, and K. P. Soni, *J. Inst. Chem.*, Calcutta **48** Pt. 3, 123 (1976); *CA* **85**, 167493e.
164. D. G. Peters and J. J. Lingane, *Anal. Chim. Acta* **26**, 75 (1962).
165. I. Leden and R. Nilsson, *Svensk. Kem. Tidskr.* **66**, 126 (1954).
166. N. S. Nikalaev and Yu. A. Lukyanichev. *Zh. Neorg. Khim.* **8**, 1786 (1963).
167. J. B. Macaskill and R. G. Bates, *J. Phys. Chem.* **81**, 496 (1977).
168. W. Selig, *Microchem. J.* **20**, 388 (1975).
169. W. Selig and A. Solomon, *Mickrochim. Acta* **1974**, 663.
170. T. S. Prokopov, *Mikrochim. Acta* **1968(2)**, 401.
171. K. G. Sultanova and V. M. Stepanov. *Fiz.-Khim. Metody. Anal. Kontrolya Proizvod.* **1975**, 80; *CA* **87**, 110863n.
172. T. A. Kokina, M. N. Petrikova, and I. P. Alimarin, *Zh. Anal. Khim.* **26**, 2237 (1971); *J. Anal. Chem. USSR* **26**, 2003 (1971).
173. W. Wojnowski and K. Kwiatkowska-Sienkiewicz, *Z. Anorg. Allg. Chem.* **396**, 333 (1973).
174. E. J. Lokka, *Pap. Puu* **60**, 441 (1978).
175. W. Selig, *Mikrochim. Acta* **1970**, 229.
176. R. Behrend, *Z. Phys. Chem.* **11**, 466 (1893).
177. B. J. Alley and H. W. H. Dykes, *Anal. Chem.* **36**, 1124 (1964).
178. T. C. Chadwick, *Anal. Chem.* **48**, 1201 (1976).
179. M. Higashiura and S. Mii, *Kagaku To Kogyo (Osaka)* **43**, 506 (1969); *CA* **75**, 157901r.
180. J. S. DiGregorio and M. D. Morris, *Anal. Lett.* **1**, 811 (1968).
181. S. Lal and G. D. Christian, *Indian J. Chem.* **10**, 123 (1972).
182. F. Frenzel and E. Hartwagner, *Fresenius' Z. Anal. Chem.* **289**, 366 (1978).
183. B. Coulter and D. G. Bush, *Anal. Chim. Acta* **51**, 431 (1970).
184. G. E. Kalbus, R. D. Wesley, and L. H. Kalbus, *Analyst (London)* **96**, 488 (1971).
185. I. M. Kolthoff and J. J. Lingane, *J. Am. Chem. Soc.* **58**, 2457 (1936).
186. G. E. Baiulescu and V. V. Cosofret, *Applications of Ion Selective Membrane Electrodes to Organic Analysis*, Ellis Horwood, Chichester, U.K., 1977.
187. W. Schoniger, *Mikrochim. Acta* **1955**, 123; **1956**, 869.
188. L. Srp and V. Chromy, *Chem. Zvesti* **31**, 83 (1977); *CA* **88**, 114883n.

189. Y. P. Handa, I. S. Bains, D. S. Chahal, and S. L. Chopra, *J. Indian Chem. Soc.* **50**, 333 (1973).
190. H. Soep and P. Demoen, *Microchem. J.* **4**, 77 (1960).
191. E. Debal and R. Levy, *Mikrochim. Acta* **1964**, 285.
192. R. A. Egli, *Fresenius' Z. Anal. Chem.* **247**, 39 (1969).
193. H. G. Nadeau, D. M. Oaks, Jr., and R. D. Buxton, *Anal. Chem.* **33**, 341 (1961).
194. J. Slanina, P. Vermeer, J. Agterdenbos, and B. Griepink, *Mikrochim. Acta* **1973**, 607.
195. M. R. Masson, *Mikrochim. Acta* **1976(1)**, 399.
196. Y. Morita, S. K. Komatsu, and N. A. Yukitoshi, *Bunseki Kagaku* **22**, 1036 (1973); *CA* **80**, 55589f.
197. N. A. Kolchina and G. M. Kondrat'eva, *Zh. Anal. Khim.* **24**, 1884 (1969); *J. Anal. Chem. USSR* **24**, 1530 (1969).
198. J. L. Sabot and D. Bauer, *Analusis* **6**, 441 (1978); *CA* **90**, 114582e.
199. A. Balint-Novotny, G. Gyorgy, and A. Vegh, *Acta. Pharm. Hung.* **44**, 262 (1974); *CA* **82**, 80149m.
200. K. Kham, C. Chevrot, J. C. Folest, M. Troupel, and J. Perichon, *Bull. Soc. Chim. Fr.* **1977**, 243.
201. I. M. Kolthoff and J. Eisenstadter, *Anal. Chem. Acta* **24**, 83 (1961).
202. A. Albert and G. B. Barlin, *J. Chem. Soc.* **1959**, 2384; **1962**, 3129.
203. V. M. Shul'man, S. V. Larionov, and L. A. Podol'skaya, *Zh. Anal. Khim.* **22**, 1165 (1967); *J. Anal. Chem. USSR* **22**, 983 (1967).
204. H. Flaschka and A. J. Barnard in *Tetraphenylboron as an Analytical Reagent. Advances in Analytical Chemistry and Instrumentation*, Vol. 1, C. N. Reilley, Ed., Interscience, New York, 1960, pp. 1–117.
205. K. Vytras, *Int. Lab.* **9**, 35 (1979).

# APPENDIX

**Table A.1. Values of the Nernst Factor $k$ in Volts from 0 to 100 °C**[a,b]

| $T\,°C$ | $k = (RT \ln 10)/F$ (V) | $T\,°C$ | $k = (RT \ln 10)/F$ (V) |
|---|---|---|---|
| 0  | 0.054197 | 55  | 0.065110 |
| 5  | 0.055189 | 60  | 0.066102 |
| 10 | 0.056181 | 65  | 0.067094 |
| 15 | 0.057173 | 70  | 0.068086 |
| 20 | 0.058165 | 75  | 0.069078 |
| 25 | 0.059157 | 80  | 0.070070 |
| 30 | 0.060149 | 85  | 0.071062 |
| 35 | 0.061142 | 90  | 0.072054 |
| 38 | 0.061737 | 95  | 0.073046 |
| 40 | 0.062134 | 100 | 0.074038 |
| 45 | 0.063126 | —   | —        |
| 50 | 0.064118 | —   | —        |

[a] $\ln 10 = 2.30259$.
[b] $R = 8.31433$ J K$^{-1}$ mol$^{-1}$; $F = 96{,}487.0$ C mol$^{-1}$; $T$ (in kelvins) $= T(°C) + 273.150$: International Union of Pure and Applied Chemistry. *Pure Appl. Chem.* **9**, 453 (1964).

## Table A.2. Values of A and B in the Debye–Huckel Equation from 0 to 100 °C for Water[a,b]

| $T\,°C$ | Unit Volume of Solvent | | Unit Weight of Solvent | |
|---|---|---|---|---|
| | A | B | A | B |
| 0 | 0.4918 | 0.3248 | 0.4918 | 0.3248 |
| 5 | 0.4952 | 0.3256 | 0.4952 | 0.3256 |
| 10 | 0.4989 | 0.3264 | 0.4988 | 0.3264 |
| 15 | 0.5028 | 0.3273 | 0.5026 | 0.3272 |
| 20 | 0.5070 | 0.3282 | 0.5066 | 0.3279 |
| 25 | 0.5115 | 0.3291 | 0.5108 | 0.3286 |
| 30 | 0.5161 | 0.3301 | 0.5150 | 0.3294 |
| 35 | 0.5211 | 0.3312 | 0.5196 | 0.3302 |
| 38 | 0.5242 | 0.3318 | 0.5224 | 0.3306 |
| 40 | 0.5262 | 0.3323 | 0.5242 | 0.3310 |
| 45 | 0.5317 | 0.3334 | 0.5291 | 0.3318 |
| 50 | 0.5373 | 0.3346 | 0.5341 | 0.3326 |
| 55 | 0.5432 | 0.3358 | 0.5393 | 0.3334 |
| 60 | 0.5494 | 0.3371 | 0.5448 | 0.3343 |
| 65 | 0.5558 | 0.3384 | 0.5504 | 0.3351 |
| 70 | 0.5625 | 0.3397 | 0.5562 | 0.3359 |
| 75 | 0.5695 | 0.3411 | 0.5623 | 0.3368 |
| 80 | 0.5767 | 0.3426 | 0.5685 | 0.3377 |
| 85 | 0.5842 | 0.3440 | 0.5750 | 0.3386 |
| 90 | 0.5920 | 0.3456 | 0.5817 | 0.3396 |
| 95 | 0.6001 | 0.3471 | 0.5886 | 0.3404 |
| 100 | 0.6086 | 0.3488 | 0.5958 | 0.3415 |

[a] For values of the ion-size parameter $a_i$ in angstrom units.
[b] Data from R. A. Robinson and R. H. Stokes, *Electrolyte Solutions*, 2nd ed. revised, Butterworths, London, 1970, p. 468. The values for unit weight of solvent were obtained by multiplying the corresponding values for unit volume by the square root of the density of water at the appropriate temperature.

Table A.3. Values of the Ion Size Parameter $a_i$ and Ion Activity Coefficients in Water at 25°[a]

| Ion | Ion-Size Parameter $a_i$ cm × $10^8$ | $\gamma_i$ at Ionic Strengths | | | |
|---|---|---|---|---|---|
| | | 0.005 | 0.01 | 0.05 | 0.1 |
| $H^+$ | 9 | 0.933 | 0.914 | 0.860 | 0.830 |
| $(C_3H_7)_4N^+$ | 8 | 0.931 | 0.912 | 0.850 | 0.820 |
| $(C_3H_7)_3NH^+$, $\{OC_6H_2(NO_3)_3\}^-$ | 7 | 0.930 | 0.909 | 0.845 | 0.810 |
| $Li^+$, $C_6H_5COO^-$, $(C_2H_5)_4N^+$ | 6 | 0.929 | 0.907 | 0.835 | 0.800 |
| $CHCl_2COO^-$, $(C_2H_5)_3NH^+$ | 5 | 0.928 | 0.904 | 0.830 | 0.790 |
| $Na^+$, $IO_3^-$, $HSO_3^-$, $(CH_3)_3NH^+$, $C_2H_5NH_3^+$ | 4–4.5 | 0.927 | 0.901 | 0.815 | 0.770 |
| $K^+$, $Cl^-$, $Br^-$, $I^-$, $CN^-$, $NO_2^-$, $NO_3^-$ | 3 | 0.925 | 0.899 | 0.805 | 0.755 |
| $Rb^+$, $Cs^+$, $NH_4^+$, $Tl^+$, $Ag^+$ | 2.5 | 0.924 | 0.898 | 0.800 | 0.750 |
| $Mg^{2+}$, $Be^{2+}$ | 8 | 0.755 | 0.690 | 0.520 | 0.450 |
| $Ca^{2+}$, $Cu^{2+}$, $Zn^{2+}$, $Mn^{2+}$, $Ni^{2+}$, $Co^{2+}$ | 6 | 0.749 | 0.675 | 0.485 | 0.405 |
| $Sr^{2+}$, $Ba^{2+}$, $Cd^{2+}$, $H_2C(COO)_2^{2-}$ | 5 | 0.744 | 0.670 | 0.465 | 0.380 |
| $Hg_2^{2+}$, $SO_4^{2-}$, $CrO_4^{2-}$ | 4 | 0.740 | 0.660 | 0.445 | 0.355 |
| $Al^{3+}$, $Fe^{3+}$, $Cr^{3+}$, $La^{3+}$ | 9 | 0.540 | 0.445 | 0.245 | 0.180 |
| $\{Co(en)_3\}^{3+}$ | 6 | 0.520 | 0.415 | 0.195 | 0.130 |
| $Citrate^{3-}$ | 5 | 0.510 | 0.405 | 0.180 | 0.115 |
| $PO_4^{3-}$, $Fe(CN)_6^{3-}$, $\{CO(NH_3)_6\}^{3+}$ | 4 | 0.505 | 0.395 | 0.160 | 0.095 |
| $Th^{4+}$, $Zr^{4+}$, $Ce^{4+}$ | 11 | 0.350 | 0.255 | 0.100 | 0.065 |
| $Fe(CN)_6^{4-}$ | 5 | 0.310 | 0.200 | 0.048 | 0.021 |

[a] Data from J. Kielland, *J. Am. Chem. Soc.* **59**, 1675 (1937).

Table A.4. Thermodynamic Values of the Ionic Product of Water $K_w$ from 0 to 60 °C[a]

| $T$ °C | $K_w \times 10^{14}$ | $pK_w$ |
|---|---|---|
| 0  | 0.1117 | 14.952 |
| 5  | 0.1820 | 14.740 |
| 10 | 0.2877 | 14.541 |
| 15 | 0.4446 | 14.352 |
| 20 | 0.6714 | 14.173 |
| 25 | 0.9908 | 14.004 |
| 30 | 1.4390 | 13.842 |
| 35 | 2.0420 | 13.690 |
| 40 | 2.8510 | 13.545 |
| 45 | 3.9170 | 13.407 |
| 50 | 5.2970 | 13.276 |
| 55 | 7.0790 | 13.150 |
| 60 | 9.3110 | 13.031 |

[a] Data from A. K. Covington, M I. A. Ferra, and R. A. Robinson, *J. Chem. Soc. Faraday Trans. 1* **73**, 1721 (1977).

# INDEX

Acetamide, $pK_B(CH_3COOH)$, $pK_B(H_2O)$, 382
Acetamidine, $pK_a$ $(H_2O)$, 308
Acetaminophen determination in mixtures, 485
Acetanilide, $pK_B(CH_3COOH)$, $pK_B(H_2O)$, 382
Acetic acid:
   acid base equilibria in, 378–382
   autoprotolysis constant, 366
   determination in mixtures, 484
   medium:
      for acid base titrations, 482
      for oxidation-reduction titrations, 594
   physical properties, table, 369
   $pK_a(DMF)$, $pK_a(H_2O)$, 394
   $pK_a(DMSO)$, $pK_a(H_2O)$, 398
   $pK_a(H_2O)$, 308
      determination, table, 315
   $pK_a$(2-methyl-2-propanol), $pK_a(H_2O)$, 405
   $pK_a$(N-methyl-2-pyrrolidinone), 393
   preparation of anhydrous, 378
   reference half cells for, 116
Acetic anhydride, 378
   medium for acid-base titration, 482
   reference half cells for, 116
Acetone:
   acid-base equilibria in, 407
   autoprotolysis constant, 372
   dielectric constant, 407
   medium for precipitation titrations, 630
   reference half cells for, 114, 116
Acetonitrile, 571
   acid-base equilibria in, 383–390
   autoprotolysis constant, 366, 367
      determination, 383
   Debye-Huckel constants, 372
   effect on calomel, 115
   medium:
      for acid-base titrations, 483
      for oxidation-reduction titrations, 597
   physical properties, table, 370
   purification, 386
   reference half cells for, 116

   relative acidity constant for, 375
Acetyl acetone, *see* 2,4-Pentanedione
Acetylcholine:
   determination, 176
   in membrane preparation, 153
Acetylcholinesterase, 176
Acetylsalicylic acid determination in mixtures, 486. *See also* Aspirin
   determination in mixtures
Acid:
   concept of, 305–308, 365
   dissociation constant, 306. *See also* Ionization constant
   strength, 306
   leveling of, 367
Acid-base:
   equilibria:
      in acetic acid, 378–382
      in acetonitrile, 383–390
      in dimethylformamide, 390–393
      in dimethylsulfoxide, 397–401
      in ketones, 406–409
      in 2-methyl-2-propanol, 401–406
      in N-methyl-2-pyrrolidinone, 393–397
      in mixed solvents, 410–426
   non-aqueous titrimetry:
      in acetic acid, 482
      in acetic anhydride, 482
      in acetonitrile, 483
      in alcohols, 483
      in amides, 485
      applications and solvents, table, 481–488
      in dimethylsulfoxide, 485
      in ketones, 485
      in nitromethane, 487
      in phenols, 487
      in pyridine, 487
      solvent contamination in, 481
      in tetraalkoxysilanes, 488
      in tetrahydrofuran, 488
      in tetramethylguanidine, 488

679

Acid-base: non-aqueous titrimetry (*Continued*)
titrants:
acid, 488–490
base, 490–493
titrimetry:
in aqueous sodium acetate, 468
in aqueous sodium sulfite, 468
cell design, 476
effect of carbon dioxide, 476
electrodes for, 255–264
enhancement of acidity in, 477–479
enhancement of basicity in, 479–481
indirect titration of divalent metals, 479
linearization of curves:
equations, 465
methods, 468
refinements, 469
multiparametric curve fitting in, 470, 475
primary standards, 449
titrants:
aqueous, 476
non-aqueous, 488–493
two-phase titrations, 478
Acidity:
constant relative, 374
function $p(a_D\gamma_{Cl})$, 202
$p(a_H\gamma_{Cl})$, 195
*see also* $p(a_H\gamma_{Cl})$
functions, 309
relative scale, 372–375
effect of water on, 375
histogram, 374
scales, 309
Acids:
determination, 263
enhancements of acidity, 477–479
$pK_a(H_2O)$ values, 307
polyfunctional, 337–357
self association of, 371
titrations in presence of polyelectrolytes, 478
Aconitine nitrate determination, 666
Acridines determination, 483
Acrylamide in enzyme immobilization, 169
Actinides, *see* Lanthanides and actinides determination
Activation energy, 70
of titrimetric reactions, 447
Activity coefficient:
mean ionic, 15

calculation of, 18, 204
defined, 16
determination, 19–26
for hydrochloric acid in salt solution, 335
for sodium fluoride solution, 193
for zinc sulfate solution, 22
for hydrochloric acid solutions, 94
molal and molar, 14
single ion, 79
calculation, 18
for calcium(II), 204
for chloride, 196, 202, 203
in 2-methyl-2-propanol, 403
determination for hydrogen ion, 435
equations for, table, 19
in non-aqueous solvents, 371
Activity, mean ionic, 15
Activity of electrolyte, 15, 94
Activity standards:
pD, 202
pH, 194–201. *See also* pH
pX, 203–208
Acylamidines determination, 487
Adenine:
determination, 660
$pK_a(H_2O)$ values, 626
5'-Adenosine monophosphate determination, 174
5'-Adenylic acid deaminase, 174
Adsorption:
effects, 609, 638
ions on vessel walls, 214
Albumin determination, 664
Alcohol oxidase, 177
Alcohols:
for acid-base titrations, 483
determination, 168, 177, 550
for oxidation-reduction titrations, 598
Aldehydes determination, 484
Aliquat 336S, 143, 144, 273
Alkali blue electroactant, 147
Alkali metals determination, 642
Alkaloid bases determination, 259
Alkaloids determination, 666
Alkanethiols determination, 661
Alkoxy groups determination, 487
Alkoxymethanedithioates:
as precipitants, 646, 649
as titrants, 612

see also Potassium(I),
  alkoxymethanedithioates as titrants
Alkoxy(phenylamino)silanes determination,
  483
Alkyl borates determination, 486
Alkyldithiocarbonates determination, 597
Alkylferrocenes determination, 598
Alkylnaphthalene detergents, 282
Alkylpyridinium iodide, 487
Alkyl sulfates determination, 484
Alkyl thiols, 279
Alkyl xanthates, oxidation of, 575
Aluminium, determination of chloride in, 290
Aluminium(III) determination:
  complexometric, 266, 274, 504
  precipitation, 644
Aluminium(III) nitrate coagulant, 286, 628, 635, 643
Amberlite XAD-2, 478
Amides as titration media, 485
Amines:
  determination, 263, 573
  primary, secondary, tertiary mixtures, 483
  tertiary in mixtures, 484
  hydrochlorides determination, 174
  oxides determination, 482, 483, 484
  sympathomimetic, titrations in various solvents, 424
  titrants, 493
Aminium picrates preparation, 389
Aminoacetic acid, $pK_a(H_2O)$, 308. See also Glycine, $pK_a(H_2O)$
L-Aminoacid oxidase, 177
Aminoacids determination, 482
L-Aminoacids determination, 168, 177
3-Aminobenzoic acid, $pK_a(DMSO)$, $pK_a(H_2O)$, 398
4-Aminobenzoic acid:
  determination in mixtures, 486
  $pk_a(DMSO)$, $pK_a(H_2O)$, 398
4-Aminobenzoic acid-formaldehyde-4-bromophenyl copolymer determination, 487
4-Amino-5-bromopyrimidine-2(1H)-one, $pK_a(DMF)$, $pK_a(H_2O)$, 394
Aminocarboxylic acids, microscopic ionization constants, 347
2-Aminoethanol, physical properties, table, 370
Amino groups determination, 485
  of terminal, 485

2-Amino-2-hydroxymethyl-1,3-propanediol (TRIS):
  as buffer, 169
  $pK_a(H_2O)$ determination, table, 325
  primary standard:
    in acetic acid, 451
    impediments, 450
    in 2-methoxyethanol, 451
1-Amino-4-methylpiperazine determination in mixtures, 485
4-Amino-5-methylpyrimidin-2(1H)-one, $pK_a(DMF)$, $pK_a(H_2O)$, 394
Aminophenols determination, 482
2-Amino-3-phenoxazinone derivatives determination, 579, 598
7-Amino-3-phenoxazinone derivatives determination, 579, 598
2'-Amino-4-pyridinecarboxamide, 587. See also Isoniazid
5-Aminopyrimidine-2,4(1H,3H)-dione, $pK_a(DMF)$, $pK_a(H_2O)$, 394
4-Aminopyrimidin-2(1H)-one, $pK_a(DMF)$, $pK_a(H_2O)$, 394
Aminopyrine hydrochloride determination, 483
Ammonia:
  $pK_a(CH_3CN)$, 389
  $pK_a(DMF)$, 395
4-Ammoniomethylpyridine $pK_a(DMF)$, 395
Ammonium:
  arenesulfonates determination, 488
  cerium(IV) sulfate, 570
  fluoride titrant, 285, 643
  hexanitratocerate(IV) titrant, 571
  in acetonitrile, 597
  ion determination with calcium tetrophenylborate(III), 287, 642
  iron(II) sulfate titrant (Mohr's salt), 552
  standardization, 589
  see also Iron(II), titrant
  iron(III) sulfate as oxidant, 254, 580
  molybdate catalyst, 565
  perchlorate as standard, 286
  peroxydisulfate oxidant, 452, 635. See also Peroxydisulfate oxidant
  N-phenylcarbamodithioate titrant, 645
  sulfosalicylate masking agent, 647
  thiocyanate titrant, 651, 652
  vanadate:
    primary standard, 583
    titrant, 583

# 682　　INDEX

Amphiprotic solvents, 365. *See also* Solvents, amphiprotic
Analgesics determination, 485
Anilides determination, 488
Anilines:
　bromination of, 599
　determination:
　　acid-base, 485
　　bromometric, 594
　　of mixtures, 488
　　oxidation-reduction, 576, 579
　oxidation of, 564
　$pK_a(CH_3CN)$, 389
　$pK_a(DMF)$, $pK_a(H_2O)$, 395
　$pK_a(DMSO)$, $pK_a(H_2O)$, 399
　$pK_a(H_2O)$, 256, 308
　titration of, 256
Antibiotic A23187 as electroactant, 150
Antidiuretic hormone determination, 173, 176
Anti-freeze, 661
Antimony determination in dimethylformamide medium, 592
Antimony(III):
　chloride determination in non-aqueous medium, 576
　determination:
　　oxidation-reduction, 574, 581
　　precipitation, 291, 653
　iodide determination in mixtures, 487
　oxidation of, 564
　titrant preparation and properties, 584
Antimony(V) formal reduction potential, 547
Antimony potassium tartrate:
　primary standard, 584
　titrant, 584
Antioxidants, 216
Anti-rust:
　fluid, 661
　paper, 661
Argentometric methods, 660. *See also* Silver(I), titrant
L-Arginine, 172
　determination, 171, 175
Arginine deaminase, 171, 172
Aromatic amines determination, 599
Arsenate determination, 658
Arsenic:
　determination, 486
　　in organic compounds, 658
　　sequential, 640
　formation of, 538

Arsenic(III), 568
　chloride determination in non-aqueous medium, 576
　determination:
　　oxidation-reduction, 574, 576, 578, 581
　　precipitation, sequential, 640
　formation of, 539
　iodide determination, 486
　　in mixtures, 487
　oxidation, 564
　oxide, 574, 577, 585
　　as primary standard, 453, 584
　　in standardization cerium(IV), 570
　standard solution, 567
　titrant, 568
　　preparation and properties, 585
Arsenic acid:
　determination, 486
　$pK_a(H_2O)$, 585
Arsenic(V) formal reduction potentials, 543
Arsenious acid, $pK_a(H_2O)$, 585
Arylamines determination, 573, 598
Ascorbic acid determination:
　oxidation-reduction, 293, 552, 576
　in acetonitrile, 598
　potential, 543. *See also* Dehydroascorbic acid
L-Asparaginase, 174
L-Asparagine determination, 174
L-Aspartase, 171
L-Aspartate determination, 171, 175
Aspirin determination in mixtures, 485, 486. *See also* Acetylsalicylic acid determination in mixtures
Asymmetry potential, 127
Automatic titration system, 470
Autoprotolysis, 365
　constant, 256, 366, 372–375, 475
　determination, 383
　　in aqueous dimethylsulfoxide, 416
　effect of water on, 410
　for N-methyl-2-pyrrolidinone, 393
　constants, table as $pK_{SH}$, 369
　values for aqueous dimethylsulfoxide, table, 417
　values for aqueous ethanol, table, 411
　values for aqueous methanol, table, 411
Avagadro's number value, 455
Azide:
　determination, 632
　hazardous nature and disposal, 636

INDEX 683

Azo-dyes, titration in dimethylformamide, 598
Azoles substituted determination, 483
*Azotobacter vinelandii,* 174

Background solutions, 215–222, 226, 235
　dilutions with, 218
　quality of water for, 218
　*see also* LIPB; TISAB
Bacterium:
　*cadaveris,* 175
　in bioselective sensor, 171
　*sarcina flava,* 174
Barbital, *see* 5,5-Diethylpyrimidine-
　　2,4,6(1$H$,3$H$,5$H$)-trione, $pK_a$(2-
　　methyl-2-propanol), $pK_a(H_2O)$
Barbiturates determination, 485
Barium(II):
　chloride titrant, 284
　determination:
　　complexometric, 266, 270, 504
　　precipitation, 283, 644
　　nitrate masking agent, 655
　　perchlorate titrant, 655
　　　in aqueous dioxane, 639
　　sulfate solubility product, 624
　titrant applications, 612
Base concept of, 305–308
Bases:
　enhancement of basicity of, 479
　ionization in acetic acid, 379
　organic determination, 263
　　with hexacyanoferrate(II), 291
　polyfunctional, 337–357
Bates-Guggenheim convention, 19, 196, 202,
　　312, 315, 327, 339, 372, 414
Bemegride determination, 488
Benzaldehyde determination, 568
Benzamide titrant, 649
Benzanilides determination, 487
1,4-Benzenedicarboxylic acid determination in
　　mixtures, 485
1,2-Benzenedicarboxylic acid (phthalic acid),
　　$pK_a$(2-methyl-2-propanol),
　　$pK_a(H_2O)$, 405
1,2-Benzenediol, titration of, 478
1,3-Benzenediol, titration of, 478
1,4-Benzenediol (hydroquinone):
　determination, 294, 576, 584
　　in acetonitrile medium, 598
　oxidation of, 569

titrant, 596
　preparation and properties, 585
　titration of, 478
1,3-Benzenedisulfonylchloride-L-lysine
　　copolymer, 352
Benzenehexacarboxylic acid (mellitic acid):
　buffers for quinhydrone reference half cell,
　　64
　$pK_a$ values, 64, 345
Benzenethiols determination, 599
Benzene-tri and -tetracarboxylic acids,
　　titrations of, 422
Benzethonium chloride, *see* Hyamine 1622
Benzimidazole and derivatives determination,
　　484
Benzimidazole-2(3$H$)-thione:
　determination, 284, 665
　$pK_a(H_2O)$, 626
Benzodiazepine derivatives determination,
　　482
Benzoic acids:
　determination, 259, 485, 488
　　in mixtures, 488
　half-neutralization potentials, 401
　standard, 419
　$pK_a(CH_3CN)$, 389
　$pK_a(DMF)$, $pK_a(H_2O)$, 394
　$pK_a(DMSO)$, $pK_a(H_2O)$, 398
　$pK_a(H_2O)$, 308
　　determination, table, 324
　$pK_a$(2-methyl-2-propanol), $pK_a(H_2O)$, 405
　$pK_a(N$-methyl-2-pyrrolidinone), 393
　primary standard, 490
　purification, 419
　2-,3-, and 4-substituted:
　　titrations in dimethylsulfoxide, 421
　　titrations in pyridine, 421
　3- and 4-substituted, titrations in various
　　solvents, 421
Benzophenone determination, 484
1,4-Benzoquinone determination, 553
Benzothiazole-2(3$H$)-thione:
　determination, 284, 665
　in acetonitrile, 597
　titrant, 281, 647, 648, 649, 650, 652, 661
　applications, 612
1H-Benzothiazole titrant applications, 612
Benzothioamide:
　precipitant, 653
　titrant, 645, 646, 648, 649, 652
　applications, 612

1$H$-Benzotriazole:
  determination, 284, 660, 665
  $pK_a(H_2O)$, 622, 626
  titrant, 651
Benzylamine, $pK_B(CH_3COOH)$, $pK_B(H_2O)$, 382
6-Benzylaminopurine determination, 484
$N$-Benzyl $NN$-dimethyl 1-hexadecylammonium:
  chloride as titrant, 289
  nitrate as electroactant, 145
$N$-Benzyl $NN$-dimethyl N-octadecylammonium:
  chloride as electroactant, 144
  thiocyanate as electroactant, 147
$N$-Benzyl $NN$-dimethyl N-tetradecylammonium:
  perchlorate as electroactant, 146
  thiocyanate as electroactant, 147
Benzylpenicillin determination, 176
4-Benzylpyridine, $pK_a(DMF)$, $pK_a(H_2O)$, 395
Beryllium(II) bromide determination, 486
Bias potential, 60, 83, 89, 230
  between hydrogen electrodes, 85
  between silver-silver chloride electrodes, 95, 96
Bimetallic electrode systems, 65
  carbon-tungsten, 284
  platinum-platinized platinum, 65
  platinum-silver, 282
  platinum-tungsten, 65
    exchange current densities for, 67
  silver of different sizes, 65
Biological fluids:
  calibration standards for, table, 220
  cell calibration for, 219
Bipotentiometric titration, 70. *See also* Potentiotitrimetry, differential, electrolytic
2,2'-Bipyridine, 579
Bis(aryl and alkylamine)tetrabromotellurium adducts, 485
Bis(1-butyl-2,6-dimethylpyridin-4(1$H$)-one) perchlorate, primary standard, 451
Bis(chlorovinyl)chloroborane, 657
Bis(cyclopentadienyl)iron(II), 590. *See also* Ferrocene titrant
Bis(ethylenediamine)cerium(IV) titrant, 564. *See also* Cerium(IV), titrant
$OO$-Bis(2-ethylhexyl) ethylphosphonate, in membrane preparation, 153
Bis(2-ethylhexyl)hexanedioate, solvent in membrane preparation, 159

Bis(2-ethylhexyl)phosphate:
  electroactant, 153
  in membrane preparation, 159
Bis(1,2-dihydro-1,5-dimethyl-2-phenylpyrazol-3(3$H$)-one) perchlorate, primary standard, 451
Bis(dimethylglyoxine)-1,10-phenanthroline cobaltate(III):
  dodecylbenzenesulfonate, 289
dodecylsulfate:
  electroactant, 149
  preparation, 289
Bismuth:
  metal reductor, 538
  primary standard, 452
  reductant, 535
Bismuth(III):
  catalyst, 550
  determination:
    complexometric, 266, 505
    sequential, 505
    oxidation-reduction, 559
  iodide determination in mixtures, 487
  nitrate titrant preparation, 527
  phosphate membrane electroactant, 133
  titrant, 527
Bis(2-nitrophenyl)ether [2-(2'-Nitrophenoxy)nitrobenzene] in membrane preparation, 150
Bis(phenols) determination, 485
Bisthiosemicarbazones determination, 479
Bitetrazole determination, 661
Bomb combustion method, 656
Borax, *see* Sodium (I), tetraborate
Boric acid:
  determination, 486
  in mixtures, 486
  titration, 477
Boron:
  determination in:
    amineboranes, 477
    boron silicides, 477
    organoboron compounds, 477
    titanium boride, 477
  trifluoride catalyst, 550
Bovine serum albumin, in preparation of enzyme sensors, 169
Bridge solution, 104, 378
  equitransferent, 106
  for precipitation titrations, 610
  requirements in non-aqueous media, 114
Brilliant green perrhenate as electroactant, 146

# INDEX

Bromate determination, 552, 566, 568
Bromide:
  determination:
    general methods, 633
    oxidation-reduction, 589
    precipitation, 283, 286, 289
    sequential, 633
    titrant, 288
    applications, 612
Bromine:
  cyanide:
    formal reduction potentials, 543
  titrant:
    in acetic acid, 596
    in ethanol, 598
    in methanol, 598
    preparation and properties, 575
  determination:
    in dimethylformamide medium, 592
    in organic compounds, 656, 657
  titrant, 573
    preparation in non-aqueous media, 574
    in propylene carbonate, 599
    standardization, 573
Bromine(I) cation titrants, 574
Bromite determination, 568
8-Bromoadenine determination, 660
3-Bromobenzoic acid:
  $pK_a(CH_3CN)$, 389
  $pK_a(DMSO)$, $pK_a(H_2O)$, 398
  $pK_a(2\text{-methyl-2-propanol})$, $pK_a(H_2O)$, 405
4-Bromophenol:
  $pK_a(DMSO)$, $pK_a(H_2O)$, 398
  $pK_a(2\text{-methyl-2-propanol})$, $pK_a(H_2O)$, 405
$N$-Bromosuccinimide:
  determination, 552
  titrant, in acetonitrile medium, 597
Brucine hydrochloride determination, 666
Buffers:
  EDTA titrations, 267
  $p(a_H\gamma_{Cl})$, table, 318
  pH, table, 197
Buformin determination, 506
2-Butanamine, $pK_B(CH_3COOH)$, $pK_B(H_2O)$, 382
Butanedioic acid:
  $pK_a(DMF)$, $pK_a(H_2O)$, 395
  $pK_a(H_2O)$, 344
  $pK_a(2\text{-methyl-2-propanol})$, 405
2,3-Butanediyldioxybis[$N$-ethyl $N$-($\alpha$-methylbenzyl)acetamide] as electroactant, 154

2-Butanethiol determination, 661
Butanol physical properties, table, 369
2-Butanone:
  autoprotolysis constant, 367, 372
  physical properties, table, 370
($Z$)-Butenedioic acid (maleic acid), $pK_a(2\text{-methyl-2-propanol})$, $pK_a(H_2O)$, 405
tert-Butyl alcohol, see 2-Methyl-2-propanol
Butylamine:
  $pK_B(CH_3COOH)$, $pK_B(H_2O)$, 382
  $pK_a(DMSO)$, $pK_a(H_2O)$, 399
4-Butyl-$\alpha,\alpha,\alpha$-trifluoroacetophenone in membrane preparation, 144

Cadmium(II):
  benzothiazole-2($3H$)-thione solubility product, 624
  chelon complexes preparation, 528
  determination:
    acid-base, via EDTA, 479
    complexometry, 266, 271, 272, 273, 274, 508
    precipitation, 282, 645
    sequential, 652
    ultramicro, 646
  $NN$-diethylcarbamodithioate solubility product, 624
  sulfate titrant, 640
  sulfide:
    as membrane material, 130
    solubility product, 624
Cadmium reductant, 535
Caffeine:
  determination, 480
  $pK_a(H_2O)$, 480
Calcium(II):
  bis(decyl)phosphate as electroactant, 150, 151
  bis(2-ethylhexyl)phosphate as electroactant, 151
  bis(4-octylphenyl)phosphate as electroactant, 150
  bis[4-(1,1,3,3-tetramethylbutyl)phenyl]phosphate as electroactant, 150, 159
  carbonate, 527
    primary standard, 452
  chloride, basicity enhancement with, 480
  fluoride solubility product, 624
  determination:
    acid-base, via EDTA, 479

Calcium(II): determination (*Continued*)
  complexometry, 266, 268, 274, 276, 506, 508
  sequential, 507
  diethyl *r*-13, *t*-17, *c*-18, *c*-22-tetramethyl-14,21-dioxo-16,19,13,22-diazatetratriacontanedioate as electroactant, 158
  hydroxide pH reference, 197
  iodide determination, 486
  nitrate titrant, 270, 504
  tetraphenylborate(III) titrant preparation, 642
  titrant, 527
    preparation, 527
Calibration:
  of cells without diffusion, 20, 192–194
  standards, 191. See also $p(a_H\gamma_{Cl})$; pH
Calomel electrode, see Reference half cell
Caprolactam determination, 482
Carbamates determination, 483
  alkyl derivatives, 488
  phenyl derivatives, 488
  pyridyl-substituted, 486
Carbamic acid, 172
  determination, alkyl esters, 483
Carbamodithioic acid derivatives:
  as precipitants, 646
  as titrants, 612
Carbamoyl phosphate, 172
Carbon acids, $pK_a(DMSO)$, $pK_a(H_2O)$, 398
Carbonate buffer solutions:
  pD reference, 203
  pH reference, 197
Carbonyl compounds determination, 484
Carbonyl group determination with fluorene, 293
Carboxy groups determination in polymers, 486, 487
Carboxylates determination, 259
Carboxylic acid chlorides determination, 488
Carboxylic acids:
  aromatic determination, 484
  determination, 484, 485
  in mixtures, 486
  naphthyl esters determination, 487
  $pK_a(DMF)$, $pK_a(H_2O)$, 394
  $pK_a(DMSO)$, $pK_a(H_2O)$, 398
  $pK_a(2$-methyl-2-propanol), 405
  titrations:
    in N-methyl-2-pyrrolidinone, 422

  in polyelectrolyte solutions, 478
  in two phases, 478
9-Carboxymethylfluorene, $pK_a(DMSO)$, $pK_a(H_2O)$, 398
Carisoprodal determination, 483
Carius procedure, micro, 477
Catalysts, 447
  use, 550, 563–570
Cationic acids:
  $pK_a(DMF)$, $pK_a(H_2O)$, 395
  $pK_a(DMSO)$, $pK_a(H_2O)$, 399
CDTA, 501
  barium complex stability constant, 502
  calcium complex stability constant, 502
  titrant, 269, 504, 508, 509, 511, 513, 514, 515, 518, 519, 520, 522, 524
    preparation, 525
Cell:
  calibration, 10, 11, 28, 83, 192
  in biological fluids, 219
  curve, 12
    effects of interferents on, 235, 238
    technique for obtaining, 218
  factor $E^°_{cell}$, 20
  in seawater, 220
  solutions for, 29
  technique for measuring $E^°_{cell}$, 218
  design for use in non-aqueous solvents, 377
  with diffusion (with liquid junction):
    compared, 56
    precision of measurement of $E_{cell}$, 255
    response time, 244
    stability of, 245
  without diffusion:
    calibration, 20, 192–194
    design features, 21–26
    determination of mean activity coefficient with, 22
    diffusion potential in, 24
  notation, 6, 47–50
    abbreviated form, 49
  gas sensor, 160
  glass electrode-calomel reference half cell, 50
  ion selective electrodes, 119
  IUPAC convention, 47
  potentials during Fe(II)-Ce(IV) reaction, table, 45
  quasi-thermodynamic, 26–28, 79
  thermodynamic, 28
  *see also* Chemical cell; Concentration, cells

Cellulose acetate membrane matrix, 139
Cephalosporins determination, 176, 482, 595
Ceramic sinter membrane matrix, 139
Ceresin wax, 294
Cerium(III):
    determination:
        complexometry, 509
        oxidation-reduction, 553, 557, 572
        sequential, 593
        precipitation, 281, 654
    oxidation of, 572
    sulfate titrant, 586
        preparation and properties, 586
Cerium(IV):
    determination, 587
        indirect in mixtures, 654
        sequentially, 590
    formal reduction potentials, 543
    formation of, 536
    sulfate, 260
        preparation and use, 570
        titrant, 538, 570
    titrant, 573
        comparison with chromium(VI), 571
        in non-aqueous media, 571
        in tributylphosphate medium, 599
        preparation, 570
        standardization, 565, 566, 569, 570
            in non-aqueous media, 571
    see also Ammonium, cerium(IV) sulfate; Ammonium, hexanitratocerate(IV) titrant
Cesium(I):
    determination:
        general methods, 643
        with tetraphenylborate(III), 287
    dimsyl titrant, 400
    hexaiodotellurate(IV) precipitate, 643
    tetraphenylborate(III) preparation, 287
Cetylpyridinium chloride determination, 666
Cetyltrimethylammonium hydroxide titrant, 263
Chalcogenides, electroactive, 134
Chelate definition, 431
Chelaton(II), see EDTA
Chelaton(III), see EDTA
Chelon, 501
    determination, 511
    metal ion complexes preparation, 528
Chemical cell:
    defined, 57
    with diffusion, 61
        application to titrimetry, 62–65
        role of reference electrode in, 82
    without diffusion, role of reference electrode in, 82
Chloramine T (sodium N-chloro-4-toluenesulfonamidate), 568
    titrant, 295
    properties, 577
    standardization, 577
Chlorate:
    determination, 566, 568, 585
    reduction of, 565, 566
Chloride:
    determination, 283, 284, 286, 288, 289, 635
    conditions, 632
    by differential potentiometry, 232
    micro, 290, 633
    sequential, 633
    effect on pM electrode response, 273
    titrant applications, 613
Chlorinated hydrocarbons determination, 662
Chlorine(I) cation titrants, 576
Chlorine determination, 568
    in organic compounds, 656, 657
    in organo-boron compounds, 657
    in organo-osmium compounds, 656
    in organo-rhenium compounds, 656
    in organo-ruthenium compounds, 656
Chlorine dioxide determination, 568
Chlorine(III) titrant, 578
Chlorite:
    determination, 568, 585
    reduction of, 567
N-Chloroacetamide titrant, 577
Chloroacetic acid:
    $pK_a$(2-methyl 2-propanol), $pK_a(H_2O)$, 405
    poly(vinyl alcohol) derivatives titration, 349
4-Chloroaniline, $pK_a$(DMSO), $pK_a(H_2O)$, 399
Chlorobenzene in membrane preparation, 145
2-Chlorobenzoic acid, $pK_a$(DMSO), $pK_a(H_2O)$, 398
4-Chlorobenzoic acid, $pK_a$(DMSO), $pK_a(H_2O)$, 398
Chlorocyclohexane in membrane preparation, 153
4-Chloro-2,6-dinitrophenol:
    buffer:
        in dimethylformamide, 391
        in dimethylsulfoxide, 397

4-Chloro-2,6-dinitrophenol (*Continued*)
   $pK_a(DMF)$, $pK_a(H_2O)$, 394
   $pK_a(DMSO)$, $pK_a(H_2O)$, 398
Chloroform in membrane preparation, 151
4-Chloromercuribenzenesulfonate titrant, 280
4-Chloro-2-nitroaniline, $pK_a(DMSO)$,
   $pK_a(H_2O)$, 399
4-Chloro-3-nitrobenzoic acid:
   $pK_a(DMSO)$, $pK_2(H_2O)$, 398
   $pK_a$(2-methyl-2-propanol), $pK_a(H_2O)$, 405
3-Chlorophenol, $pK_a(DMF)$, $pK_a(H_2O)$, 394
4-Chlorophenol:
   $pK_a(DMF)$, $pK_a(H_2O)$, 394
   $pK_a$(*N*-methyl-2-pyrrolidinone), 393
Chromate determination, 641
Chromazurol S indicator, 274
Chromium(II):
   chloride reductant, 539
   determination, 581
   formation, 535
   sulfate titrant, 586
   titrant properties, 586
Chromium(III):
   determination:
      complexometric, 266, 509
      oxidation-reduction, 569
   formal reduction potentials, 543
   oxidation of, 568
   sulfate catalyst, 563
Chromium(VI):
   determination, 559, 560, 567, 587
      in Cr(VI)-(III) mixtures, 646
      sequential, 590, 593
   formal reduction potentials, 543
   formation of, 536
   titrant:
      preparation, 571. See also Potassium, dichromate
Cinnamic acid, oxidation of, 564
Citric acid:
   buffer solution for quinhydrone reference half-cell, 64
   $pK_a(H_2O)$, 64
*Citrobacter freundii*, 176
Citrulline, 172
Cobalt(II) determination:
   acid-base, via EDTA, 479
   complexometric, 509
   oxidation-reduction, 293, 559, 579, 582
      in mixtures, 560

   precipitation sequential, 652
Cobalt(III):
   formation, 536
   titrant:
      preparation and use, 572
      see also Hexamminecobalt(III)-tricarbonatocobaltate(III) titrant
Codeine determination, 486
   in mixtures, 483
Combustion tube apparatus, 656
Complexometric titrations, table, 504
   electrodes for, 264–278
   equations for linearization of curves, 466
   primary standards, 451
   titrants, 525–528
Complexone(II), *see* EDTA
Complexone(III), *see* EDTA
Computer programs:
   ACBA, 470
   CFT3, 470
   COMICS, 441
   Curve-fitting on-line, 470
   DALSFEX, 440
   EQUIL, 441
   HALTAFALL, 441, 468, 469
   LETAGROP VRID, 440
   MINIQUAD 75, 440
   PSEUDO PLOT, 440
   SGOGS, 440
Concentration:
   cells:
      defined, 57
      with diffusion, role of reference electrode in, 83
      without diffusion, 57, 58, 61
         application to titrimetry, 65–66
   dependent $pK_a$ values, 331. See also $pK_a$ values, conditional
   hysteresis, 248
   ionization constant, 332. See also $pK_a$ values, conditional
   standards, 208
   units, 454–456
      normality, 455
Conjugate acid-base, 305
Conjugate acids, $pK_a(H_2O)$, 307
Constant ionic strength, 28–29, 434
   in stability constant determinations, 435
Controlled current, potentiometric titrations, *see* Potentiotitrimetry, electrolytic

INDEX 689

with two indicator electrodes, *see* Potentiotitrimetry, differential, electrolytic
Copper:
determination in alloy, 593
primary standard, 452
Copper(I):
chloride titrant in acetonitrile, 586
titrant properties, 586
Copper(II):
benzothiazole-2(3$H$)-thione solubility product, 624
chelon complexes preparation, 528
chloride:
catalyst, 587
titrant in dimethylformamide, 598
determination:
acid-base, via EDTA, 479
complexometric, 271, 272, 273, 274, 510, 511
sequential, 510
oxidation-reduction, 551, 553, 559, 560, 583, 589, 590
in dimethylformamide medium, 292, 592
precipitation, 280, 281, 282, 646, 647
sequential, 647, 648, 650, 652
*NN*-diethylcarbamodithioate solubility product, 624
formal reduction potential, 544
pentacyanonitrosylferrate(III) solubility product, 624
perchlorate titrant:
in acetonitrile medium, 597
preparation and standardization, 573
-*N*-phenylbenzohydroxamic acid as electroactant, 151
salts as catalysts, 563
sulfate catalyst, 588, 593
titrant, 509, 513
in non-aqueous media, 573
standardization and use, 573
Coprecipitation, 609
Creatinase, 174
Creatinine determination, 174
Crown ethers, 155, 372
Cryptands, 372, 503
as titrants, 526
Cryptates, 503
Crystal violet electroactant, 145, 295

CTAB, *see NNN*-Trimethyl-1-hexadecylammonium, bromide titrant (CTAB)
Cupferron titrant, 281, 654
applications, 613
Cyanamide determination:
in mixtures, 640
sequential, 662
Cyanate determination, 636
Cyanide:
determination, 633, 637
sequential, 633
in determinations of metals, 598
Cyanoacetic acid, p$K_a$(2-methyl-2-propanol), p$K_a$($H_2O$), 405
4-Cyanomethylpyridine, p$K_a$(DMF), 395
Cyanuric acid determination, 485
Cyclohexanecarboxylic acid:
p$K_a$($H_2O$), 308
p$K_a$(2-methyl-2-propanol), p$K_a$($H_2O$), 405
Cyclohexanone:
determination, 589
in membrane preparation, 141
Cyclohexa[b]pyrrole determination, 483
Cyclohexylamine titrant, 493
Cyclopenta[b]pyrroles, determination of 1,5-alkylated, 483
Cysteine:
desulfhydrase, 177
determination, 595, 664
L-Cysteine determination, 177, 661
Cytosine, *see* 4-Aminopyrimidin-2(1$H$)-one, p$K_a$(DMF), p$K_a$($H_2O$)

Davies equation, 19
Davies-Gray method, 556
DCTA, *see* CDTA
Debye-Huckel equation, 17
in calculation of p$K_a$($H_2O$) values, 312, 327
constants (A and B), 17
calculation of, 371
values in aqueous dimethylsulfoxide, table, 417
values in aqueous ethanol, table, 411
values in aqueous methanol, table, 411
values in deuterium oxide, 202
values in water, table, 676
forms of, 17, 18
ion size parameter, 17
allowance for, 21

Debye-Huckel equation, ion size parameter (*Continued*)
  effect on calculation of $pK_a(H_2O)$ values, 316
  estimates of values, table, 677
  limiting in acetonitrile, 385
  for use in 2-methyl-2-propanol, 403
Decanol in membrane preparation, 144
Dehydroascorbic acid, formal reduction potential, 543. *See also* Ascorbic acid determination
Density:
  values:
    for amphiprotic solvents, table, 369
    for aqueous dimethylsulfoxide, table, 417
    for aqueous-ethanol, table, 411
    for aqueous-methanol, table, 411
  and volumetric measurements, 376
Derivative titration, defined, 55
Detergents, determination of inorganic sulfate in, 282
Determinand definition, 3
Deuterium oxide:
  glass electrode in, 202
  ionization constants of weak acids in, 202
(Diacetoxy)iodobenzene:
  oxidant, 576
  preparation, 576
  titrant:
    stability in acetic acid, 576
    standardization, 576
Dialkylcarbamodithioates determination in acetonitrile medium, 597
Dialkyl hydrogen phosphates as electroactants, 159
*OO*-Dialkylphosphorodithioates determination, 662
Diamines determination, 483
  in mixtures with monoamines, 483
Diazabicyclooctane determination, 480
Diazonium compounds determination, 594
Diborane determination, 477
Dibromamine T:
  formal reduction potential, 544
  oxidant, 576
  preparation, 576
  titrant:
    stability in acetic acid, 576
    standardization, 576
2,3-Dibromopropanoic acid, $pK_a$(2-methyl-2-propanol), $pK_a(H_2O)$, 405

Dibutylamine, $pK_a$(DMSO), $pK_a(H_2O)$, 399
*NN*-Dibutyl-1-butanamine, $pK_B(CH_3COOH)$, $pK_B(H_2O)$, 382
Dibutyldecanedioate in membrane preparation, 152, 159
Dibutylferrocene, 172, 177
2,6-Di-*tert*-butyl-4-nitrophenol, $pK_a$(DMSO), $pK_a(H_2O)$, 398
Dibutylphthalate in membrane preparation, 144, 150
Dicarboxylic acids determination, 484, 485, 488
  in mixtures, 484, 486
Dichloramine T:
  formal reduction potential, 544
  titrant, 564, 577
  in acetic acid medium, 594
  properties, 578
Dichloroacetic acid, $pK_a$(2-methyl-2-propanol), $pK_a(H_2O)$, 405
2,5-Dichloroaniline, $pK_B(CH_3COOH)$, $pK_B(H_2O)$, 382
Dichloroargentate(I) formation, 632
*o*-Dichlorobenzene in membrane preparation, 144
3,4-Dichlorobenzoic acid:
  $pK_a$(DMSO), $pK_a(H_2O)$, 398
  $pK_a$(2-methyl-2-propanol), $pK_a(H_2O)$, 405
3,5-Dichlorobenzoic acid, $pK_a$(DMSO), $pK_a(H_2O)$, 398
1,2-Dichloroethane in membrane preparation, 154
2,6-Dichloro-4-(4-hydroxyphenyl)imino-2,5-cyclohexadien-1-one titrant, 293
(Dichloroiodo)benzene:
  formal reduction potential, 544
  titrant, 564, 577
  properties, 578
Dichlorophenolindophenol, *see* 2,6-Dichloro-4-(4-hydroxyphenyl)imino-2,5-cyclohexadien-1-one titrant
3,5-Dichlorophenol, $pK_a$(2-methyl-2-propanol), $pK_a(H_2O)$, 405
3,5-Dichloro-2,4,6-trinitrophenol, $pK_a$(2-methyl-2-propanol), $pK_a(H_2O)$, 406
Dichromate determination, 641. *See also* Chromium(VI)
Dicyclohexyl-18-crown-6 electroactant, 151
*OO*-Didecyl dithiophosphate electroactant, 153

Dielectric constant, 363, 368–372
  effect on:
    electrode response, 368
    membrane selectivity, 158
  values:
    amphiprotic solvents, table, 369
    in aqueous dimethylsulfoxide, table, 417
    in aqueous ethanol, table, 411
    in aqueous methanol, table, 411
Diethylamine:
  determination, 486
  $pK_B(CH_3COOH)$, $pK_B(H_2O)$, 382
  $pK_a(DMSO)$, $pK_a(H_2O)$, 399
  $pK_a(H_2O)$, 308
3-Diethylaminomethyl-2,6-dimercaptopyran-4(4$H$)-thione titrant, 645
$NN$-Diethylaniline, $pK_B(CH_3COOH)$, $pK_B(H_2O)$, 382
$NN$-Diethylcarbamodithioate precipitant, 646
Diethyleneglycol dinitrate determination, 589
$NN$-Diethylethanamine, 382. *See also* Triethylamine
$OO$-Diethylphosphorodithioates determination, 664
5,5-Diethylpyrimidine-2,4,6(1$H$,3$H$,5$H$)-trione, $pK_a$(2-methyl-2-propanol), $pK_a(H_2O)$, 406
Diethyl $r$-13, $t$-17, $c$-18, $c$-22-tetramethyl-14,21-dioxo-16,19,13,22-dioxadeazatetratriacontanedioate electroactant, 150. *See also* Calcium(II)
Differential titration, defined, 54
Diffusion potential, 24, 326. *See also* Liquid junction, potential
Dihydrazides determination, 480
Dihydro-1,4-benzodiazepines titrations in nonaqueous solvents, 425
5,6-Dihydro-5-methylpyrimidine-2,4(1$H$,3$H$)-dione, $pK_a(DMF)$, $pK_a(H_2O)$, 394
5,6-Dihydropyrimidine-2,4(1$H$,3$H$)-dione, $pK_a(DMF)$, $pK_a(H_2O)$, 394
2,3-Dihydro-2-thioxopyrimidin-4(1$H$)-one, $pK_a(DMF)$, $pK_a(H_2O)$, 394
3,4-Dihydro-4-thioxopyrimidin-2(1$H$)-one, $pK_a(DMF)$, $pK_a(H_2O)$, 394
2,6-Dihydroxybenzoic acid:
  $pK_a(DMF)$, $pK_a(H_2O)$, 394
  $pK_a(DMSO)$, $pK_a(H_2O)$, 398
  $pK_a(H_2O)$, 316
Dihydroxydiethylpiperazine determination, 480

β-Diketones determination, 484
Dimethylamine, $pK_a(CH_3CN)$, 389
3-Dimethylaminomethyl-2,6-dimercaptopyran-4(4$H$)-thione titrant, 613
$NN$-Dimethylaniline, $pK_B(CH_3COOH)$, $pK_B(H_2O)$, 382
3,4-Dimethylbenzoic acid, 404
$pa_H$ values in 2-methyl-2-propanol, 403
$pK_a(DMSO)$, $pK_a(H_2O)$, 398
$pK_a$(2-methyl-2-propanol), $pK_a(H_2O)$, 405
Dimethyldibenzo-30-crown-10, 155
  as electroactant, 151, 158
Dimethyldicyclohexyl-18-crown-6 electroactant, 151
5,5-Dimethyl-3,7-dioxanonanedioic acid, 158
2,2-Dimethyl-1,3-diyldioxybis($N$-heptyl-$N$-methylacetamide) as electroactant, 151
Dimethylformamide:
  acid-base equilibria in, 390–393
  autoprotolysis constant, 367, 393
  effect on calomel, 115
  physical properties, table, 370
  purification, 390
  reference half-cells for, 116
  relative acidity constant, 375
1,1-Dimethylhydrazine, $pK_B(CH_3COOH)$, $pK_B(H_2O)$, 382
Dimethylpropanedioic acid (dimethylmalonic acid), $pK_a(DMF)$, $pK_a(H_2O)$, 395
3,5-Dimethylpyrazole determination in mixtures, 487
Dimethylsulfate determination, 487
Dimethysulfoxide:
  acid-base equilibria in, 397–401
  effect on calomel, 115
  physical properties, table, 370
  purification, 401
  reference half-cells for, 116
  titration medium, 485
Dimsyls as titrants, 400
2,4-Dinitroaniline, $pK_a(DMSO)$, $pK_a(H_2O)$, 399
2,4-Dinitrobenzoic acid, $pK_a(DMSO)$, $pK_a(H_2O)$, 398
3,5-Dinitrobenzoic acid:
  $pK_a(CH_3CN)$, 389
  $pK_a(DMF)$, $pK_a(H_2O)$, 394
  $pK_a(DMSO)$, $pK_a(H_2O)$, 398
  $pK_a$(2-methyl-2-propanol), $pK_a(H_2O)$, 405
$m$-Dinitrocompounds determination, 663

2,4-Dinitrodiphenylamine, $pK_a$(DMSO),
    $pK_a(H_2O)$, 399
2,6-Dinitro-4-methylphenol determination, 666
2,4-Dinitrophenol:
    $pK_a(CH_3CN)$, 389
    $pK_a(DMF)$, $pK_a(H_2O)$, 394
    $pK_a$(N-methyl-2-pyrrolidinone), 393
2,6-Dinitrophenol:
    buffer:
        in dimethylformamide, 391
        in dimethylsulfoxide, 397
    $pK_a(DMF)$, $pK_a(H_2O)$, 394
    $pK_a(DMSO)$, $pK_a(H_2O)$, 398
    purification, 391
3,5-Dinitrophenol:
    $pK_a(DMF)$, $pK_a(H_2O)$, 394
    $pK_a(DMSO)$, $pK_a(H_2O)$, 398
    $pK_a$(2-methyl-2-propanol), 406
Dinitrophenols determination, 259, 665
Dinonylphthalate:
    in membrane preparation, 152
    plasticizer component, 158
Dioctyl adipate in membrane preparation, 152
$OO$-Dioctylphenylphosphonate in membrane preparation, 150, 159
Dioctylphthalate in membrane preparation, 147
Diols, relative acidity constants, 375
Diorganodithiophosphoric acids determination, 485
1,4-Dioxane medium for precipitation titrations, 631
3,6-Dioxaoctanedioic acid, 158
Dipentylphthalate in membrane preparation, 151
Diphenylamine:
    determination in mixtures, 485
    $pK_B(CH_3COOH)$, $pK_B(H_2O)$, 382
    $pK_a(H_2O)$, 308
Diphenyl esters determination, 484
Diphenyl ether as plasticizer component, 158
1,3-Diphenylguanidine:
    half-neutralization potential standard, 419
    $pK_a(CH_3CN)$, 389
    $pK_B(CH_3COOH)$, $pK_B(H_2O)$, 382
    $pK_a(DMF)$, $pK_a(H_2O)$, 395
    $pK_a$(N-methyl-2-pyrrolidinone), 393
    primary standard, 489
    purification, 419
    titrant, 493
Diphenylthallium(III):
    chloride titrant, 636

nitrate solubility product, 625
sulfate titrant application, 613
titrant preparation, 637
Diphenylthiocarbazone, see Dithizon (diphenylthiocarbazone)
NN-Diphenylthiourea determination, 597, 662
Dipicrylamine:
    -vitamin B1 electroactant, 154
    -vitamin B6 electroactant, 154
Dipiperidylbenzaminals determination, 487
Dipropylamine, $pK_a$ (DMSO), $pK_a(H_2O)$, 399
Diprotic acids:
    ionization in dimethylformamide, 392
    $pK_a(DMF)$, $pK_a(H_2O)$, 395
    $pK_a$(2-methyl-2-propanol), $pK_a(H_2O)$, 405
Discriminative titration, 54
Disodium hydrogen phosphate, 320
Dissociation constant:
    hydrogen chloride in dimethylformamide, 376
    ion pair, 381
    in acetic acid, table, 382
    overall in acetic acid, 378
    tetraalkylammonium salts in 2-methyl-2-propanol, 402
    tetrabutylammonium picrate in 4-methyl-2-pentanone, 408
    tetraethylammonium 3,5-dinitrobenzoate in 4-methyl-2-pentanone, 408
    tetraethylammonium salts in 2-methyl-2-propanol, table, 405
Disulfide determination, 661
2,2'-Dithiobis[benzothiazole] determination, 595
2,4-Dithiobiuret titrant, 613
Dithiocarbamates determination, 483
Dithiocarbazates determination, 483
Dithiooxamide:
    determination, 662
    titrant, 280, 283, 647, 650, 651
    applications, 613
    see also Ethanedithioamide titrant, applications
Dithiophosphates determination, 485
Dithizon (diphenylthiocarbazone), 288
    -mercury(I) complex as electroactant, 151
    -metal complexes as cation exchangers, 143
    -palladium complex as electroactant, 151
Dodecalactam determination, 483
Dolomite, determination of calcium in, 507

Double junction reference electrode, 104. *See also* Reference half-cells, double junction
DTPA (3,6,9-tris(carboxymethyl)-3,6,9-triazaundecanedioic acid):
  determination, 511
  -silver(I) complex stability constant, 270
  titrant, 270, 512, 518
  preparation of solutions, 525
Dyes, azo- and nitro-, determination, 254

$E°_{cell}$, significance of, 29
Edetic acid, 501. *See also* EDTA
EDTA, 266, 279, 451
  cadmium(II) complex, 437
  copper(II) pM buffer, 221
  determination of, 511
  iron(III) complex, 272
  lead(II) complex, 437
  masking of, 217
  masking agent, 651, 652, 656
  mercury(II) complex, 276
    as indicator, 254
    preparation, 266
    stability constant, 265
  metals determined by, table, 267
  metals determined indirectly, 269
  $pK_a(H_2O)$, 337
  silver(I) complex stability constant, 269
  titrant, 271, 272, 273, 274, 276, 501, 504–524
  preparation of solutions, 525
EGTA:
  calcium(II) complex stability constant, 502
  magnesium(II) complex stability constant, 502
  mercury(II) complex, 268
  titrant, 266, 268, 271, 507, 516
  preparation of solutions, 525
Electrodes:
  air gap, 165. *See also* Sensors, gas
  auxiliary, 69
  bimetallic, *see* Bimetallic electrode systems
  characteristics, 83
  glass hydrogen ion responsive, 107, 118, 124–128, 160, 176, 244, 309
    asymmetry potential, 127
    calibration of, 198
      in acetic acid, 379
      in acetone, 407
      in acetonitrile, 384
    in aqueous alcohol solutions, 416
    in dimethylformamide, 391
    in dimethylsulfoxide, 397
    in 4-methyl-2-pentanone, 407
    in 2-methyl-2-propanol, 402
    in $N$-methyl-2-pyrrolidinone, 395
  conditioning, 317
    in acetic acid, 379
    in acetonitrile, 384
    and storage, 242
  in determination of stability constants, 434
  effect of light on, 249
  low temperature type, 126
  measurements with:
    in acetic acid, 380
    in acetonitrile, 384
    in aqueous dimethylsulfoxide, 415
    in deuterium oxide, 202
    in dimethylformamide adaptation for use, 390
    in dimethylsulfoxide adaptation for use, 397, 401
    in 2-methyl-2-propanol response range, 402
    in $N$-methyl-2-pyrrolidinone cell, 395
    in non-aqueous solvents, 113, 366
      cell design, 377
      range of potentials, 372
    in measurements of $p(a_H\gamma_{Cl})$, 310, 317–321, 337
      calibration for, 317
  mechanism of response, 126–128
  mercury-filled, 126
  rate of response of,
    in acetic acid, 379
    in dimethylformamide, 391
    in dimethylsulfoxide, 402
    in 2-methyl-2-propanol, 402
    in 4-methyl-2-pentanone, 409
    in non-aqueous solvents, 377
  as reference electrode, 102, 518, 532, 611
  reference half-cell for:
    external, 125
    internal, 125
  response:
    effect of enzyme layer on, 170
    transitory peaks in, 244
    stability of response, 245
    temperature hysteresis, 248

Electrodes: glass hydrogen ion responsive
(*Continued*)
*see also* Membranes, electroactive
glass sodium ion responsive, 173, 193, 219
effect of pH on response, 233
in enzyme sensor, 176
reproducibility, 245
indicator, defined, 13
ion selective, 80
anion responsive:
acetate, 147
alkylbenzenesulfonate, 149
benzenesulfonate or naphthalene-1-sulfonate, 149
benzoate or salicylate, 148
bicarbonate, 144
bile salt anion, 149
bromide, 137, 144, 233
butanoate, 148
carbonate, 144, 234
chloride, 135, 144, 233, 234
chlorometallate, 144
cyanide, 131, 135, 166
2,4-dichlorophenoxyacetate, 147
dodecylsulfate, 149
effect of pH on response, 233, 234
formate, 147
fluoroborate, 234
fluoride, 118, 129, 137, 138, 166, 193, 219, 230, 233, 241, 244, 246, 253
for microdetermination, 230
as reference electrode, 219
halide, 130, 132, 135, 137, 138, 246
hydrogen phthalate or salicylate, 148
iodide, 131, 138, 145, 168, 173, 233
in enzyme sensor, 177
leucinate, 148
maleate or phthalate, 148
nitrate, 143, 145, 234, 241, 243, 248
oxalate, 148
perchlorate, 143, 146, 234, 241
perrhenate, 146
phenobarbital, 148
phenylalanate, 148
phosphate, 135, 138
propanoate, 148
sulfate, 136, 138, 146
sulfide, 135, 137, 166, 246
sulfite, 137
surfactant anionic, 149
tartrate, 137
tetrafluoroborate, 147
thiocyanate, 147, 233
trifluoroacetate, 147
calibration in biological fluids, 219
cation responsive:
ammonium, 150, 169, 170, 234, 241
in enzyme sensor, 175
antipyrine, 153
atropine, 153
barium, 150
brucine, 153
cadmium(II), 130, 135, 138, 234
calcium(II), 150, 219, 234, 241
cesium(I), 138
choline, 153
cinchonine, 153
codeine, 153
copper(II), 128, 130, 134, 135, 138, 151, 221, 234, 246
divalent cation, 151, 159, 234, 241
effect of pH on response, 233
ephedrine, 153
hydrogen(I), *see* Electrodes, glass hydrogen ion responsive
iron(III) (glass), 128
lead(II), 130, 134, 136, 138, 234
lithium(I), 151
magnesium(II), 151
mercury(I), 151
mercury(II), 131, 134, 135, 234
narcotine, 153
onium ions, large, 153
papaverine, 153
potassium(I), 138, 151, 158, 219, 241
propranolol, 154
pyramidon, 154
quinine, 154
silver(I), 131, 152, 166, 234, 246
sodium(I), 153. *See also* Electrodes, glass hydrogen ion responsive
strontium(II), 153
strychnine, 154
thallium(I), 153
uranium(VI), 153
vitamin B1, 154
vitamin B6, 154
water hardness, 151
cell notation, 119
classification of membrane materials, *see also* Membranes, electroactive, 118, 122–124.

coated graphite (Selectrode®), 120, 137
  conditioning and storage, 243
coated wire, 120, 136, 141, 144, 150
  storage of, 141
commercially available, 118
conditioning and storage, 242
definition of, 13
drifts in potential, 141
effects of:
  light, 249
  pH, 233–235
  polarizing current, 291
  stirring, 259
in enzyme sensors, 168
enzyme type, 123. See also Sensors, enzyme
gas sensing, 123. See also Sensors, gas
glass membrane, 123. See also Electrodes, glass hydrogen ion responsive
inorganic salt membrane, 128–139. See also Membranes, electroactive
internal reference half-cell in, 83, 120
ISFET type, 123, 179
liquid membrane, 123, 139, 144, 150
lower limit of response, 235–236
  effect of temperature on, 248
Nernst equation for, 119
organic membrane, 139–159. See also Membranes, electroactive
precision of measurement with, 229
preparation and assessment, 134
properties of, 232–233
pX(S) values in calibration, 203–208. See also pX(S) values
response and detection limits, 235–236
response time, 243–245, 272
  effects of solvents on, table, 630
selectivity, 158, 191
selectivity coefficient, 236–241
  defined, 237
  determination of, 237–241
  values, table, 241
solid membrane:
  heterogeneous, 122
  homogeneous, 122
solid state, 120
stability and reproducibility, 245–246
stabilization of response at low concentrations, 221
temperature:
  coefficients, 247
  control, 246
  effects on, 244, 246
  thermal equilibrium in, 247
  use in non-aqueous media, 113
  see also Membranes, electroactive
membrane, 79. See also Electrodes, glass hydrogen ion responsive; Electrodes, ion selective
oxidation-reduction:
  amalgam type, 7, 9, 81
  cadmium amalgam, 437
  categories of, 80–81
  first kind, 81
  hydrogen, 5, 7, 84–91
    bias potentials between, 85
    cell assembly for, 87
    depth effects on, 88
    in dimethylsulfoxide, 397
    gas supply for, 87
    Hildebrand type, figure, 7, 86
    Nernst equation for, 87
    palladium type, 89
    in $p(a_H\gamma_{Br})$ measurements, 321
    in $p(a_H\gamma_{Cl})$ measurements, 310, 316
    poisoning, 89
    test for, 89
    preparation, 85–87
    pressure corrections for, 88, 90–91
    standard potential of, 84
    storage, 242
  kind, definition, 80
  lead amalgam, 24, 437
  lead amalgam - lead(II) sulfate, 22, 84
    effect of solubility lead(II) sulfate, 23–25
    lower limits of concentration, 25
  mercury-mercury(II) oxide, 103, 107
    standard potential, 103
  mercury-mercury(I) picrate, 103
  platinum:
    as base for enzyme sensor, 177
    calibration, 541
    net, 172
    see also Electrodes, potentiotitrimetric
  quinhydrone, 107
  second kind, 81
  silver-silver(I) bromide in $p(a_H\gamma_{Br})$ measurements, 321

Electrodes: oxidation-reduction (*Continued*)
  silver-silver(I) chloride, 83, 84, 92–98, 194, 245
    calibration, 93–94
    conditioning and storage, 242
    in dimethylsulfoxide, 415
    effect of impurities, 93
    flow-through type, 98
    limit of detection with, 236
    lower limits of concentration, 92
    in $p(a_H\gamma_{Cl})$ measurements, 310, 316–321
    preparation:
      bielectrolytic method, 96
      electrolytic method, 96
      thermal-electrolytic method, 95
    reactivity, 139
    in reference half-cell, *see* Reference half-cells
    response time, 96
    solubility effects, 92
    standard potential, 84
    upper limit of concentration, 92
    use in sea water, 93, 98
    third kind, 81
    zinc amalgam, 7, 9, 22, 24
  potentiotitrimetric
    acid-base titrations, 255–264
      aluminium, 257
      antimony, 256
        -platinum pair, 488
        polarized, 263
      bismuth, 257
      carbon:
        epoxide membrane, 260
        fiber, 260
        graphite, 259, 260
        silicone rubber membrane, 260
      gallium, 257
      germanium, 257
      glass pH, 255–256, 257, 261, 488
        in non-aqueous media, 256
      gold, 258
      hydrogen, 488
        $\alpha$-palladium type, 261
      indium, 257
      manganese dioxide, 261
      metal and metalloid, 256–262
      molybdenum, 258
      platinum, 258
        -platinum pair, 488
        polarized, 262
        preparation, 259
        quinhydrone polarized, 263
      silicon, 257
      tungsten, 258, 259, 488, 491
        bronze, 262
    adsorption polarized, 66
    complexometric titrations, 264–278
      anion-responsive coated wire, 273
      bismuth(V) oxide anodically polarized, 277
      cadmium(II) ISE, 507, 508, 509, 515, 517, 521, 523
      calcium(II) ISE, 506, 522
      chlorometallate, 273
      Chromazurol S, 273, 274
      copper(II), ISE, 505–509, 511–515, 517–520, 522, 523, 528
      Eriochrome Black T ISE, 273
      8-hydroxyquinoline-5-sulfonate ISE, 275
      ion selective in, 272–275
      lead(II) ISE, 518
      (lead)$PbHg(SCN)_4$type, 273
      manganese(IV) oxide anodically polarized, 277
      mercury(II) ISE, 273, 505, 509, 512
      mercury pM, 254, 264–269, 276, 504, 505, 506, 508, 509, 510, 513, 514, 515, 516, 517, 519, 520, 522, 523, 524
      coated gold, 268, 275
      coated platinum, 268
      coated silver, 268–269, 505, 507, 509, 511, 513, 514, 515, 518–522, 524
      metals determined with, table, 267
      metal oxide anodically polarized, 277
      $M_x(N(II)L_4)$ type, 273
      platinum, 271–272, 504, 511, 517, 520
      hexachloroplatinate, 503
      potassium(I) ISE, 514
      silver, 269–271, 504–512, 514–523
      coated platinum, 270
      metals determined with, 269
      silver(I) ISE, 273
      silver(I) sulfide:
        -cadmium(II) sulfide, 272
        -copper(II) sulfide, 272
        -lead(II) sulfide, 272

-metal sulfide, 272
sodium(I) ISE, 515, 516
thallium(III) oxide anodically polarized, 277
tungsten bronze, 275
oxidation-reduction titrations, 292–296
carbon:
  glassy polarized, 296
  graphite impregnated, 292, 294, 586
  graphite silicone membrane, 294
  rod (ex-dry cell), 294
  -silver pair, 294
  -tungsten pair, 294
Chloramine T ISE, 294, 577
gold-gold pair polarized, 295
mercury:
  coated platinum, 534
  polarized drop, 296
perbromate ISE, 295
platinum, 254, 255, 534
  -antimony pair, 293
  calibration, 541
  -gold pair, 292
  -graphite pair, 292
  -molybdenum pair, 292
  -platinum pair polarized, 295
  polarized, anodically and cathodically, 72, 295
  preparation and conditioning, 292
  -stainless steel pair, 292
  -talantum pair, 292
  -titanium pair, 292
  -tungsten pair, 292
  use in various media, 292
tungsten polarized, 296
precipitation titrations, 278–292
  bromide ISE, 285, 614, 663
  carbon (graphite), 616
  -calomel pair polarized, 290
  -carbon pair polarized, 290
  -silver pair polarized, 290
  -tungsten pair, 284
  cesium ISE, 287
  chloride ISE, 614, 616
  fluoride ISE, 253, 285, 613, 643, 644, 653, 655, 665
  fluoroborate ISE, 287, 665
  gold, 613, 653
  iodide ISE, 286, 613, 616, 617, 635, 641, 650, 651, 656, 659
  iron(III) (glass), 284, 612

lead amalgam, 284
lead(II) ISE, 614, 637, 638, 641, 649, 658, 659, 665
  poisoning of, 629
  polarized, 291
  response in various solvents, table, 630
mercury, 281–282
  coated platinum, 614, 634
  coated silver, 281, 614, 651, 654
  -mercury(II) alkoxymethanedithioate, 280
  -mercury(II) benzothiazole-2(3$H$)-thiolate, 281
  -mercury(II) cupferronate, 281
  -mercury(II) sulfide, 279, 281
  -mercury(II) thiolate, 281
mercury(I) ISE, 254, 288, 612, 664
mercury(II) ISE, 288
nitrate ISE, 287, 613, 614, 636, 665
perchlorate ISE, 286, 616, 635, 638, 641, 665
phosphate ISE, 286
picrate ISE, 287, 616, 617, 652
platinum, 281, 282–283, 613, 614, 616, 644, 648, 651
  needle polarized, 291, 617
  -platinum pair polarized, 291, 653
  -silver pair, polarized, 282, 290
potassium(I) ISE, 617, 642, 666
silver, 278–281, 612–617, 632, 643, 645, 648, 650, 656, 658, 661, 664, 666
  coated platinum, 642
  needle polarized, 290
  polarized, 291
  -silver(I) alkoxymethanedithioates, 280
  -silver chloride, pair polarized, 290, 657
  -silver(I) dithiocarbamates, 280
  -silver(I) ethanedithioamidate, 280
  -silver(I) halide, 253
  -silver(I) iodide, 281
  -silver pair polarized, 289
  -silver(I) sulfide, 279–280, 282
  -silver(I) tetraphenylborate(III), 280
  -silver(I) thiolate, 279
silver(I) ISE, 612, 614, 615, 617, 641, 646, 656, 661
  polarized, 291

Electrodes: potentiotitrimetric, precipitation titrations (*Continued*)
  sodium ISE (glass), 284, 612, 616
  sulfide ISE, 285, 616, 617, 639, 649, 658, 662
  surfactant ISE, 289
  tetraphenylborate(III) ISE, 287
  tungsten, 284, 614
  zinc amalgam, 284, 617, 665
  requirements for, 253
  response range, 191
  slope, 79
  temperature coefficient, 247
  transference between solutions, 323
  *see also* Reference electrode; Reference half-cell
Electrolytes soluble in organic solvents, 116
Electromotive efficiency, 199, 319
Elemental analysis, 655–660
EMF, measurement of, 6, 218
Empty tube combustion method, 658
Endpoint, 456. *See also* Equivalence volume
Enzyme:
  action of, 166
  immobilization, 167
  immunoassays, 173
  sensors, *see* Sensors, enzyme
Ephedrine hydrochloride determination in mixtures, 486
Equivalence point, 456
  detection, 46, 47
  by differential electrolytic potentiotitrimetry, 71
  by differential potentiotitrimetry, 65–66, 69–71, 73–76
  by electrolytic potentiotitrimetry, 71–73
  with ion selective electrodes of different selectivities, 65
  Pinkhof-Treadwell method, 64
  retarded electrode method, 66–69
  titrant stream monitor method, 64
  relation to inflection point, 459
Equivalence volume, 456
  assessment, 457–470
  arithmetical:
    Fortuin's method, 463
    Hahn's method, 463
    Kolthoff's method, 462
  curve fitting methods, 469
  effect of interfering ions, 464
  graphical methods, 459–462

differentiation, 486
Tubb's, 460
linearization methods, 465–469
Gran, 465. *See also* Gran's method
other, 468
methods available, 457
Equivalent, definition, 454, 455
Eriochrome Black T indicator, 273
$pK_a(H_2O)$, 274
*see also* Electrodes, potentiotitrimetric, complexometric titrations
1,2-Ethanediamine:
  autoprotolysis constant, 367
  $NN$-disubstituted determination, 486
  physical properties, table, 370
  reference half-cells for, 116
1,2-Ethanediol (ethyleneglycol), 477
  physical properties, table, 369
Ethanedithioamide titrant, applications, 613
Ethanethiol:
  determination, 661
  $pK_a(H_2O)$, 308
Ethanol:
  medium:
    oxidation-reduction titrations, 598
    precipitation titrations, 630
  physical properties, table, 369
  reference half-cells for, 116
Ethanolamine:
  determination, 483
  *see also* 2-Aminoethanol, physical properties, table
4-Ethoxybenzoic acid, $pK_a(DMSO)$, $pK_a(H_2O)$, 398
N-(4-Ethoxyphenyl)acetamide, $pK_B(CH_3COOH)$, $pK_B(H_2O)$, 382
Ethylacetoacetate determination, 484
Ethylamine:
  $pK_a(CH_3CN)$, 389
  $pK_a(DMSO)$, $pK_a(H_2O)$, 399
  $pK_a(H_2O)$, 308
$N$-Ethylaniline, $pK_B(CH_3COOH)$, $pK_B(H_2O)$, 382
Ethylenedinitrotetraacetic acid, *see* EDTA
Ethylenedioxybis($NN$-dibenzylacetamide) electroactant, 150
Ethyleneglycol, *see* 1,2-Ethanediol (ethyleneglycol)
$N$-Ethylethanamine, *see* Diethylamine
4-Ethylpyridine, $pK_a(DMF)$, $pK_a(H_2O)$, 395

Europium(II):
  determination, 580
  formation of, 535
Europium(III):
  determination, 537
  fluoride solubility product, 624
$E°$ values, see Reduction potentials, standard
Exchange current density, 67–69, 71, 83
  estimates, table, 68
Exchangers, methacrylic type, titration of, 351

Ferrocenes determination, 598
Ferrocene titrant, 551, 580
  uses, 590
Ferroin preparation, 524
Fischer titration, 76–77
Flavinduline $O$ electroactant, 144
Fluoradene, $pK_a(DMSO)$, 398
Fluorene:
  $pK_a(DMSO)$, $pK_a(H_2O)$, 398
  titrant, 293
Fluorene-9-carboxylic acid, $pK_a(DMSO)$, $pK_a(H_2O)$, 399
Fluoride:
  determination:
    by differential potentiometry, 59, 230–232
    in seawater, 227
    titrimetric, 283, 468
      conditions, 632
  masking agent, 270, 549
  masked with boric acid, 658
  precipitant, 654
  titrant, 653
    applications, 613
Fluorine determination in organic compounds, 655
Fluorosulfonic acid titrant, 490
Fluorosilicic acid, 655
Formaldehyde:
  determination, 568, 663
  masking agent, 640
  $pK_a(H_2O)$, 308
Formality, defined, 38
Formal potential, 265, 374, 539–548
  definition, 38
  determination, 539–542
  general treatment, 38–42
  medium effects on, table, 543
  use of, 41–42

values (reduction):
  chromium(VI), effect of acidity, table, 40
  cobalt(III), 572
  halogen(I) cations, 574
  hydroxylamine, 588
  iron(III):
    calibration standards, 541
    effect of acidity, table, 40
    lead(IV), 580
    octacyanomolybdate(V), 582
    table of, 543
Formamide, physical properties, table, 370
Formate, oxidation of, 569
Formation constant:
  electrolytes, 368
  stepwise, 432
Formic acid:
  autoprotolysis constant, 366
  physical properties, table, 369
  $pK_a(H_2O)$, 308, 316
Fortuin's method nomogram, 464
Fowler-Bright method, 453
Fructose, 477
Fulvic acids, potentiometric titrations, 350
2-Furoic acid, $pK_a(H_2O)$, 308

Gallium(III) determination, 512, 513
  in mixtures, 485
Garnets, determination of lead in, 649
Gas sensor:
  ammonia in enzyme sensor, 170
  carbon dioxide in enzyme sensor, 170
  see also Sensors, gas
Gelatin, 279, 282
Gentian violet electroactant, 143, 147
Germanium(II) formation, 539
Germanium(IV) determination, 553
Glass:
  determination of:
    calcium in, 507
    iron in, 571
    lead in, 518
  electrode, see Electrodes, glass hydrogen ion responsive
  hydrophobized matrix, 139
D-Gluconate determination, 175
Gluconate kinase, 175
Glucose, 477
  determination, 167, 586
Glucose oxidase, 167
Glutaminase, 174

L-Glutamine determination, 174
Glutaraldehyde, see Pentanedial in enzyme sensor preparation
Glutathione determination, 661
Glycerol complexant, 549, 550
Glycine, $pK_a(H_2O)$, 308
Gold(III):
  chloride titrant preparation, 579
  determination, 585
  titrant, 579
Gran's method:
  graph paper, 226, 467
  multistandard addition, 225
  titration curve linearization, 465–468, 475, 478, 627, 639
Grignaud reagents determination, 663
Group II fluorides as electroactants, 132
Group IIIB metals determination, 486
Guaiacyl-2-hydroxybenzoate nicotinate hydrochloride determination, 483

Hafnium(IV) determination, 513
Hahn's method, 463
Half-cells defined, 6
Half-neutralization potential, 418–426
  applications, table, 421
  benzoic acids, 401
  definition, 373
  determination, 419
  in $pK_a$ determination, 400
  relative ($\Delta HNP$), 419
    measurement significance, table, 421
    relation to $pK_a(H_2O)$, 420
    for phenols, 420
Halides determination:
  nanogram amounts, 76, 290
  in seawater, 633
  sequential determinations of, 278, 283, 286, 634
Halogen:
  anions determination, 631–642
  compounds as catalysts, 564
Halogen(I) cations as titrants, 574
Halogens determination in halogenated:
  organic compounds, 655
  organoiron compounds, 656
  organotin compounds, 656
  phenols, 665
Halogen(V) titrants, 578
HEDTA, 451
  cadmium(II) complex, 437

  as primary standard, 452
  lead(II) complex, 437
Hepatitis B surface antigen, 177
  determination, 173
Heteroaromatic acids:
  $pK_a(DMF)$, $pK_a(H_2O)$, 394
  $pK_a$(2-methyl-2-propanol), $pK_a(H_2O)$, 406
Heteroaromatic nitrogen compounds determination, 660
Heteroaromatic sulfur compounds determination, 591
Heteroaromatic thioamide precipitants, 646
Heteroaromatic thiols determination, 664
Heterocyclic amines determination, 482
Heterocyclic bases (nitrogen), 482
  N-oxides determination, 483
Heterovalent reactions, 459
Hexachloroplatinate determination, 286, 641
Hexacyanoferrate(II):
  determination, 641
  -hexacyanoferrate(III) in enzyme sensor, 172
  titrant, 282, 291
  properties, 590
  see also Potassium(I), hexacyanoferrate(II) titrant
Hexacyanoferrate(III):
  determination, 576, 586
  titrant, 293, 552, 554, 559, 560, 569. See also Potassium(I), hexacyanoferrate(III) titrant
N-Hexadecyl NNN-trimethylammonium nitrate electroactant, 145
Hexafluorophosphate determination, 286, 638
Hexamminecobalt(III)tricarbonatocobaltate(III) titrant, 572
  preparation and properties, 572
2-Hexanone physical properties, table, 370
Hexaphenyldisiloxanes determination, 487
Hexitols determination, 478
4-Hexylnitrobenzene in membrane preparation, 151
Homoconjugation, 364, 371
  allowance for, 387
  complexes, 364
  formation constant, 388, 389, 394, 407
    in dimethylformamide, 392
  test for, 364, 386, 392, 400
Horseradish peroxidase, 173, 177
Humic acids, potentiometric titrations of, 350
Hyamine 1622, 143

INDEX                                                                701

electroactant, 149
Hydration number, 204, 205
Hydrazine:
  catalyst, 646
  derivatives, 578
  determination in acetic acid, 596
  determination, 482, 558, 576, 634
  hydrochloride reductant, 539
  oxidation of, 563, 564
  physical properties, table, 370
  $pK_B(CH_3COOH)$, $pK_B(H_2O)$, 382
  sulfate:
    formal reduction potentials, 545
    prime standard, 587
    reductant, 539
    titrant, 535, 567, 587
    preparation and properties, 587
Hydrobromic acid, $pK_a(CH_3CN)$, 383
Hydrochloric acid:
  complexant, 549
  determination in mixtures, 487
  $0.01m$ as $p(a_H\gamma_{Cl})$ standard, 320
  $pK_a(CH_3CN)$, 383
  $pK_a(DMF)$, 394
  titrant, 476
  in organic solvents, 489
Hydrofluoric acid:
  complexant, 549
  determination in mixtures, 484, 487
Hydrogen electrode, see Electrodes, oxidation-
    reduction, hydrogen
Hydrogen ion concentration, 332
  calculation of, 313
  see also $pc_H$
Hydrogen peroxide:
  -iodide reaction, 178
  oxidant, 536
  reduction by, 565
Hydrogen sulfide determination, 279, 285, 661
  in air, 639
Hydroquinone, 585. See also 1,4-Benzenediol
    (hydroquinone)
Hydroxamic acids, 485
Hydroxyalkylferrocenes determination, 598
2-Hydroxybenzoic acid:
  $pK_a(CH_3CN)$, 389
  $pK_a(DMF)$, $pK_a(H_2O)$, 394
  $pK_a(DMSO)$, $pK_a(H_2O)$, 398
  $pK_a(N$-methyl-2-pyrrolidinone), 393
3-Hydroxybenzoic acid, $pK_a(DMSO)$,
    $pK_a(H_2O)$, 398

4-Hydroxybenzoic acid:
  $pK_a(CH_3CN)$, 389
  $pK_a(DMSO)$, $pK_a(H_2O)$, 398
  $pK_a(H_2O)$, 337
2-Hydroxyethyl methacrylate-methacrylic acid
    copolymer potentiometric titration,
    351
Hydroxy groups, terminal, determination in
    polymers, 662
Hydroxylamine:
  and derivatives determination, 593
  determination, 634
  formal reduction potentials, 588
  hydrochloride titrant standardization, 588
  titrant preparation and properties, 588
Hydroxyl ion concentration, calculation in
    $pK_a(H_2O)$ determinations, 314
4-Hydroxymethylpyridine, $pK_a(DMF)$,
    $pK_a(H_2O)$, 395
8-Hydroxyquinoline, see 8-Quinolinol
8-Hydroxyquinoline-5-sulfonic acid, 275
  titrant, 275
Hyoscyamine sulfate determination, 479
Hypobromite determination, 551, 559, 568
Hypochlorite:
  determination, 568, 585
  oxidant, 576
  titrant preparation, 576
Hypophosphite:
  determination, 573
  oxidation of, 570
Hypophosphorous acid:
  oxidation of, 565
  reductant, 538
Hysteresis in electrodes, 248

Imidiazoles determination, 483
Iminodiacetic acid masking agent, 659
Immunoassays enzyme, see Enzyme
Immuno-gamma globulin, potentiometric
    titration of, 356
Indene, $pK_a(DMSO)$, 399
Indigo, oxidation of, 569
Indirect titration, 254
Indium(III):
  determination, 514, 644
  $NN$-diethylcarbamodithioate, solubility
    product, 624
  in mixtures, 485
Indole, $pK_a(H_2O)$, 308
Indomethacin determination, 485

Inert electrodes, 80. *See also* Zeroth-kind electrodes
Inflection point, 459, 460. *See also* Equivalence volume
Infra-red effect of irradiation in gold electrodes, 258
Iodate:
 catalyst, 564
 determination, 554, 566, 568
 titrant, 553. *See also* Potassium(I), iodate titrant
Iodide:
 catalyst, 564. *See also* Potassium(I), iodide
 determination:
  by oxidation-reduction, 572, 573, 576, 578, 589
  by precipitation, 283, 286, 289, 634
 titrant applications, 613. *See also* Potassium(I), iodide
Iodine:
 determination:
  in dimethylformamide medium, 592
  in organic compounds, 657
  with thiosulfate, 468
 formal reduction potentials, 545
 -methanol reagent, 641
 preparation, 634
 -thiosulfate reactions, 591
 titrant:
  preparation and use, 574
  standardization, 574
Iodine(I) cation as titrant, 574
Iodine cyanide:
 formal reduction potentials, 545
 titrant preparation and properties, 574
Iodine monochloride, 574
 catalyst, 564, 565
 preparation, 575
 reagent, 575
 titrant, 576
Ion exchange resin, anionic, 273
Ionic associate compounds, 451
Ionic product, *see* Autoprotolysis; $pK_w$
Ionic strength:
 calculation of, 19
  in $pK_a(H_2O)$ determinations, 313–316, 328, 339, 341, 344
 constant concept of, 28–29
 defined, 17
 relation to concentration, 16
 in titrimetry, 458

Ionization constant:
 calculation from emf data, 311–316
 defined, 306
 determination:
  in aqueous alcohol, 415
  in aqueous dimethylsulfoxide, 415–418
  of bases in acetic acid, 379. *See also* $pk_B$ values
 limits for Gran plots, 467
 microscopic, 346
 for poly-electrolytes, 348
 relation to titration error, 459
 weak acids in deuterium oxide, 202
 *see also* $pK_a$ values
Ion pairs, 378
 formation, 371
Ion selective electrodes, *see* Electrodes, ion selective
Ion size parameter, *see* Debye-Huckel equation
Iron:
 determination:
  in alloy, 567, 593
  in glass, 571
 primary standard, 452
Iron(II):
 determination, 295, 551, 554, 555, 576, 579, 581
 formation of, 537, 538
 -iron(III) *o*-phenanthroline couple as indicator, 283
 oxidation of, 565, 569
 perchlorate determination in non-aqueous medium, 576
 titrant, 551, 553, 554, 555, 556, 557, 559, 560, 561
  in dimethylformamide medium, 598
  properties and preparation, 589
  stability of, 549
  *see also* Ammonium, iron(II) sulfate titrant (Mohr's salt)
Iron(III):
 chloride titrant, 511, 516, 536
 in methanol, 598
 preparation, 580
 in pyridine, 599
 determination:
  complexometric, 255, 271, 274, 511, 512
  in presence aluminium(III), 512
  oxidation-reduction, 534
  in dimethylformamide medium, 592
  sequentially, 590, 591, 593

formal reduction potentials, 544
perchlorate titrant in acetonitrile, 580
titrant, 527, 580, 592, 613
   in acetonitrile medium, 597
   preparation, 527
Irreversible oxidation-reduction couples, 71, 73
Irving-Rossotti equation, 437
ISE, see Electrodes, ion selective
ISFET, see Electrodes, ion selective, ISFET type
Isoniazid:
   determination, 577, 584
   formal reduction potentials, 546
   oxidation of, 563
   prime standard, 587
   titrant, 567
      preparation and properties, 587
Isonicotinic acid hydrazide, see Isoniazid
Isoquinoline, $pK_a(H_2O)$, 308
Isovalent reactions, 459

Jones reductor, 271, 511
   effect of amalgamation, 537
Junction potential, see Liquid junction, potential

Ketones:
   determination, 484
   -hydroxyphenyl determination, 485
   as titration media, 485
Kjeldahl digestion, 658, 659
Kolthoff's method, 462
Kryptofix 211, 503
   titrant, 515
      preparation of solutions, 526
Kryptofix 221, 503
   titrant, 516
      preparation of solutions, 526
Kryptofix 222, 503
   titrant, 514
      preparation of solutions, 526

Lactate:
   dehydrogenase, 172, 177
   determination, 172, 177
*Lactobacillus arabinosus,* 176
β-Lactoglobulin, potentiometric titration of, 356
Lanthanide precipitation, 654
Lanthanides and actinides determination, 653

Lanthanum(III):
   determination, 484, 486, 514, 653
   fluoride:
      -carboxylate complex, 632
      solubility product, 624
      determination, 634
   nitrate titrant, 655, 665
   standardization, 632
   titrant, 285, 286, 654
      applications, 613
Laser, effect of irradiation on gold electrode, 258
Lasso determination of residues, 656
LAST DP-4 stabilizer, 482
Laurates determination, 482
Lead:
   determination in organic compounds, 659
   reductant, 535
   reductor, 538
Lead(II):
   benzothiazole-2(3H)-thione solubility product, 624
   chloride primary standard, 452
   determination:
      acid-base, via EDTA, 479
      complexometric, 272, 273, 274, 517, 518
      oxidation-reduction, 589
      precipitation, 282, 648
      sequential, 647, 648, 652
   halide electroactants, 132
   hydrogen phosphate preparation, 286
   molybdate solubility product, 624
   nitrate titrant, 282, 638, 641, 658
   perchlorate titrant, 637, 658, 665
   perrhenate solubility product, 624
   phosphate:
      electroactants, 133
      solubility product, 624
   sulfate:
      electroactant, 132
      solubility product, 624
   titrant, 291
      applications, 614
   tungstate solubility product, 624
Lead(IV):
   acetate, 260
   titrant, 294, 551, 594
      preparation and standardization, 580
   chloride titrant in dimethylformamide, 581
   determination, 591
   formamide titrant preparation, 581

Lead(IV) (*Continued*)
   oxide oxidant, 536
   titrant preparation and properties, 580
Lead dioxide, *see* Lead(IV), oxide oxidant
Lead tetraacetate, *see* Lead(IV), acetate
Leveling effects:
   acid-base, 367
   oxidants and reductants, 531
Lewis acid titrants, 490
Lidocaine determination, 483
Ligand:
   concentration calculation, 438
   definition, 431
   interference preventive buffers, 215. *See also* LIPB
   number average, 437
   stepwise protonation formation constants, 438
Light effect on electrodes, 248
Lignosulfonic acid, determination of phenol and sulfo-groups, 263
Lingane-Karplus method, 558
LIPB, 215, 217. *See also* Background solutions
   composition for copper(II), 218
Liquid junction:
   formation of, 108–113
      adjustable screw type, 113
      capillary U-tube, 112
      ceramic plug, 109
      crack, 111
      ground sleeve, 109
      J-shaped, 112
      wick, 111
   potential, 9, 79, 98, 104, 105–108, 211, 213, 224, 309, 381, 414, 435
      magnitude, 107
      medium effects, 107
      minimization, 108
      in non-aqueous media, 118
      relative error due to, 105
Lithium(I):
   acetate:
      $pK_B(CH_3COOH)$, 379, 382
      purification, 381
   chloride in basicity enhancement, 480
   determination, 285, 515, 643
   diisopropylamide titrant, 491
   perchlorate ion-pair dissociation constant exponent, 382
Lyate ion, 366

Lyonium ion, 366
L-Lysine:
   decarboxylase, 175
   determination, 175

Macrotetrolides, 155
Magnesium(II):
   determination, 268, 274, 486, 515
      automated, 270
   nitrate titrant, 271
   titrant, 527
   preparation of solutions, 527
Malachite green electroactant, 147
Maleic acid, *see* (Z)-Butenedioic acid (maleic acid)
Malonic acid, *see* Propanedioic acid
Manganese determination in alloys, 555, 566 or ores, 566
Manganese(II):
   chloride catalyst, 565
   determination:
      acid-base, via EDTA, 479
      complexometric, 515, 516
      oxidation-reduction, 555, 558, 560
      precipitation, 649
   salts as catalysts, 565
   sulfate catalyst, 453, 565
Manganese(III):
   determination, 561
   formal reduction potentials, 546
   formation, 536
   titrant preparation and standardization, 581
Manganese(IV) titrant preparation and properties, 582
Manganese(VII):
   determination, 586
      sequential, 591
   formal reduction potentials, 546
   formation of, 536
   titrant:
      applications and properties, 582
      *see also* Potassium, permanganate
Mannitol, 477
   complexant, 549, 550
   determination, 478
Masking agent, 215, 270
Melamine determination, 485
Mellitic acid, *see* Benzenehexacarboxylic acid (mellitic acid)
Membrane dialysis, 174

INDEX  705

Membranes:
electroactive:
classification, 118, 122–124
common properties of, 119
glass, 123, 124–128
effect of composition on response, 124
gel layers, 126
mechanism of response, 126–128
stability, 245
heterogeneous, 122
homogeneous, 122
inorganic salt, 123, 128–139, 234
electrical connections to, 121, 128
heterogeneous, 133–134
polyethylene matrix, 133
properties of matrix, 133
properties of salt, 133
silicone matrix, 133
homogeneous, 129–133
$Ag_2Hg\ I_4$, 133
binders for, 129
compression of materials, 132
effect of pH, 233
fluoride, 129
preparation of solids, 131
primary response, 130
sealing of, 132
silver sulfide type, 130, 131
suitable salts, 132
reactivity, 139
stability, 245
deterioration in, 246
storage, 243
liquid, 123
organic, 123, 139–159
anion exchangers, 143, 144
cation responsive, table, 150
coated wire, 141
conditioning, 243
construction, 140, 141, 142
effect:
on calcium(II) response, 159
of pH, 234
influence of plasticizer, 158
ion exchangers for, 143
matrices for, 139
neutral carriers, 155
effect of solvent on selectivity, 158
polyethylene in, 141
PVC, 140
containing dibutylferrocene, 173
solid, 144, 150
relative merits, 141
role of solvent, 158
silicone rubber, 141
solvents for electroactants, 139
stability, 246
in enzyme sensors, 169
gas permeable, 159, 163
homogeneous, 164
microporous, 164
see also Sensors, gas
Memory effects, 243, 246, 248
Meprobamate determination, 483
2-Mercaptoacetic acid, 567
determination, 594, 595
2-Mercaptoalkanoic acids determination, 599
2-Mercaptobenzothiazole, see Benzothiazole-2(3H)-thione
2-Mercaptoethanol determination, 294, 661
2-Mercapto-N-(2-naphthyl)acetamide titrant, 653
8-Mercaptopurine determination, 664
Mercurimetric methods, 660. See also Mercury(II), titrant
Mercury:
determination in organic compounds, 659
reductant, 535
Mercury(I):
iodide solubility product, 625
precipitant for phenoxyalkanoic acids, 288
sulfate solubility product, 625
Mercury(II):
benzothiazole-2(3H)-thione solubility product, 625
chalcogenide matrices, 134
-chelon complex, 266. See also EDTA, mercury(II) complex
chloride titrant, 280, 281
determination:
acid-base, via EDTA, 479
complexometric, 513
oxidation-reduction, 561
precipitation, 279, 280, 282, 634, 649
-EDTA complex, 528
halides as membrane electroactants, 132
nitrate titrant, 269, 284, 289, 291, 527, 634, 651, 652
oxide as primary standard, 452
perchlorate titrant, 281, 285
preparation and standardization, 663

Mercury(II) (Continued)
  sulfide:
    membrane binder, 129
    precipitation, 650
    thiolates stability, 663
    titrant, 268, 505, 508, 513, 519, 527, 656, 660
      advantages over silver(I), 663
      applications, 614
      preparation, 527
Metal:
  determinations in oil, 516
  ion titrants preparation, 527–528
  and metalloid electrodes, 256–262
  perchlorates preparation, 435
Metallic junctions, thermal-electric effects of, 48
Methanesulfonic acid, $pK_a(H_2O)$, 308
Methanethiol determination, 661
Methanol:
  medium for oxidation-reduction titrations, 598
  physical properties, table, 369
  $pK_a(H_2O)$, 308
  reference half-cell for, 117
L-Methionine determination, 174
Methioninelyase, 174
2-Methoxyaniline, $pK_B(CH_3COOH)$, $pK_B(H_2O)$, 382
4-Methoxyaniline, $pK_B(CH_3COOH)$, $pK_B(H_2O)$, 382
Methoxycarbonylacetic acid, $pK_a$(2-methyl-2-propanol), $pK_a(H_2O)$, 405
2-Methoxycarbonylbenzoic acid, $pK_a$(2-methyl-2-propanol, 405
3-Methoxycarbonylpropanoic acid, $pK_a$(2-methyl-2-propanol), $pK_a(H_2O)$, 405
(Z)-3-Methoxycarbonylpropenoic acid, $pK_a$(2-methyl-2-propanol), $pK_a(H_2O)$, 405
4-Methoxyphenol, $pK_a(DMSO)$, $pK_a(H_2O)$, 398
N-Methylacetamide, reference half-cells for, 117
Methylamine:
  determination, 486
  $pK_a(CH_3CN)$, 389
3-Methylbenzoic acid, $pK_a(DMSO)$, $pK_a(H_2O)$, 398

4-Methylbenzoic acid:
  determination in mixtures, 485
  $pK_a(DMSO)$, $pK_a(H_2O)$, 398
4-Methyl-1,3-dioxolan-2-one (propylene carbonate):
  effect on calomel, 115
  reference half-cells for, 117
NN'-Methylenebisacrylamide in enzyme immobilization, 169
Methylene blue:
  determination, 666
  perchlorate electroactant, 146
Methylene green determination, 666
4,5-Methylenephenanthrene, $pK_a(DMSO)$, $pK_a(H_2O)$, 399
9-Methylfluorene, $pK_a(DMSO)$, $pK_a(H_2O)$, 399
Methyl hydrogen maleate, see (Z)-3-Methoxycarbonylpropenoic acid $pK_a$(2-methyl-2-propanol), $pK_a(H_2O)$
Methyl hydrogen malonate, see Methoxycarbonylacetic acid $pK_a$(2-methyl-2-propanol), $pK_a(H_2O)$
Methyl hydrogen phthalate, see 2-Methoxycarbonylbenzoic acid, $pK_a$(2-methyl-2-propanol)
Methyl hydrogen succinate, see 3-Methoxycarbonylpropanoic acid $pK_a$(2-methyl-2-propanol), $pK_a(H_2O)$
Methyl isobutyl ketone, see 4-Methyl-2-pentanone $pK_a$(2-methyl-2-propanol), $pK_a(H_2O)$
4-Methyl-2-pentanone:
  dielectric constant, 407
  purification, 409
4-Methylphenol, $pK_a(DMSO)$, $pK_a(H_2O)$, 398
Methylpiperazine determination in mixtures, 485
2-Methyl-2-propanamine, $pK_B(CH_3COOH)$, $pK_B(H_2O)$, 382
2-Methylpropanethiol, 661
2-Methyl-2-propanethiol determination, 661
2-Methyl-2-propanol:
  acid-base equilibria in, 401–406
  autoprotolysis constant, 366
    determination, 372
  Debye-Huckel constants for, 372
  discriminative effect in, 373

discriminative solvent for phenols, 420
physical properties, table, 369
purification, 404
reference half cell for, 114, 115
2-Methylpropylamine, 666
4-Methylpyridine, $pK_a$(DMF), $pK_a(H_2O)$, 395
5-Methylpyrimidine-2,4($1H,3H$)-dione, $pK_a$(DMF), $pK_a(H_2O)$, 394
$N$-Methyl-2-pyrrolidinone:
  acid-base equilibria in, 393, 395–397
  autoprotolysis constant, 393
  general properties, 393
  physical properties, table, 370
Methylsulfonazo III indicator, 655
Micrometer syringe in potentiometric titrations, 326
Millipore filter as matrix material, 139
Mineral acids:
  determination, 484, 493
  in mixtures, 486
  titrants, 488
  see also Strong acids determination
Mixed constants, 331. See also $pK_a$ values, conditional
Mixed ligand complex, 431
Mixed solution method for selectivity coefficients, 238
Mohr's salt, see Ammonium, iron(II) sulfate titrant (Mohr's salt)
Molality, mean ionic, 16
Molecular sieve drying agent, 378
Mole definition, 454
Molybdenum determination in alloys and steels, 590
Molybdenum(III):
  chloride titrant, 591
  formation of, 537
  oxidation of, 581
  titrant, 551
  preparation and properties, 590
Molybdenum(V):
  determination, 579
  formal reduction potentials, 546
  formation of, 537, 539
  titrant preparation and properties, 582
Molybdenum(VI):
  catalyst, 168, 565
  determination, 516, 551, 555, 589
  formal reduction potentials, 546
  reduction of, 565, 593

titrant, 614
Molybdenum(IV) formal reduction potentials, 547
Molybdosilicic acid titration, 592
Monactin electroactant, 150
Monazite sand, 654
Monensin electroactant, 152
Monoamine oxidase, 174
Monosodium glutamate determination, 482
Monovalent cation determinations, 642
Multidentate ligand, 431
Multiparametric curve fitting, 469–470

2-Naphthalenethiol determination, 597
Naphthol determination, 594
2-Naphthol determination, 574
Neodymium(III) determination, 517
Neptunium(VI), formation of, 536
Nernst equation, 8, 87, 192
  adaptations for:
    differential potentiometry, 231
    Gran addition method, 225
    liquid junction potential, 10
    mercury pM electrode, 265
    multistandard addition, 227
    non-aqueous solvents, 378
    selectivity coefficient, 237
  derived, 8
  general form, 79
  numerical factor for (k), 79, 675
  use of E° values in formulating, 35
Neutral carriers in membrane preparation, 155–159
Neutral salts, enhancement of basicity by, 479
Nickel:
  -base alloys determination of cobalt in, 580
  as primary standard, 452
Nickel(II):
  benzothiazole-2($3H$)-thione solubility product, 625
  determination:
    acid-base, via EDTA, 479
    complexometric, 517
    precipitation, 650
    sequential, 647, 650, 652
Nicotinic acid determination, 176
Nikolsky equation, 237
Nitramines determination, 664
Nitrate determination, 174
  general methods, 636

Nitric acid:
    determination in mixtures, 484, 486, 487
    $pK_a(CH_3CN)$, 383
Nitrilotriacetic acid:
    -copper(II) pM buffer, 221
    determination, 511
    masking of, 217
Nitrite:
    determination, 175
    reductase, 175
β-Nitroalcohols determination in mixtures, 485
Nitroalkanes determination, 482
2-Nitroaniline:
    buffer for use in acetonitrile, 384
    $pK_B(CH_3COOH)$, $pK_B(H_2O)$, 382
    $pK_a(H_2O)$, 307
    purification, 387
3-Nitroaniline:
    determination, 480
    $pK_a(H_2O)$, 480
4-Nitroaniline, $pK_a(DMSO)$, $pK_a(H_2O)$, 399
Nitrobenzene:
    determination, 487
    in membrane preparation, 144, 150
2-Nitrobenzoic acid, $pK_a(DMSO)$, $pK_a(H_2O)$, 398
3-Nitrobenzoic acid, $pK_a(DMSO)$, $pK_a(H_2O)$, 398
4-Nitrobenzoic acid:
    $pK_a(CH_3CN)$, 389
    $pK_a(DMF)$, $pK_a(H_2O)$, 394
    $pK_a(DMSO)$, $pK_a(H_2O)$, 398
    $pK_a(2\text{-methyl-2-propanol})$, $pK_a(H_2O)$, 405
    $pK_a(N\text{-methyl-2-pyrrolidinone})$, 393
Nitro compounds:
    determination, 594
    reduction, 592
4-Nitrocymene, 294
    in membrane preparation, 145
4-Nitrocymol in membrane preparation, 153
4-Nitrodiphenylamine, $pK_a(DMSO)$, $pK_a(H_2O)$, 399
Nitroethane, $pK_a(DMSO)$, $pK_a(H_2O)$, 399
Nitroform:
    determination, 665
    yielding compounds, determination, 665
    see also Trinitromethane determination
Nitrogen:
    acids, $pK_a(DMSO)$, $pK_a(H_2O)$, 399
    bases titrations in:
        acetophenone and nitrobenzene, 424

        nitromethane, 424
    compounds in:
        asphaltenes, 424
        petroleum, 424
    determination in amineboranes, 477
Nitroglycerin determination, 517
Nitroguanidine determination, 486
Nitromethane:
    $pK_a(DMSO)$, $pK_a(H_2O)$, 399
    reference half-cells for, 117
    titration medium, 487
Nitron:
    nitrate solubility product, 625
    $pK_a(H_2O)$, 626
    titrant:
        applications, 614
        preparation of solutions, 636
2-Nitrophenol:
    $pK_a(CH_3CN)$, 389
    $pK_a(DMSO)$, $pK_a(H_2O)$, 398
3-Nitrophenol:
    $pK_a(DMF)$, $pK_a(H_2O)$, 394
    $pK_a(2\text{-methyl-2-propanol})$, $pK_a(H_2O)$, 406
    $pK_a(N\text{-methyl-2-pyrrolidinone})$, 393
4-Nitrophenol:
    $pK_a(CH_3CN)$, 389
    $pK_a(DMF)$, $pK_a(H_2O)$, 394
    $pK_a(DMSO)$, $pK_a(H_2O)$, 398
    $pK_a(H_2O)$, 308
    $pK_a(2\text{-methyl-2-propanol})$, $pK_a(H_2O)$, 406
    $pK_a(N\text{-methyl-2-pyrrolidinone})$, 393
2-Nitrophenol octyl ether, see 2-Octyloxynitrobenzene (o-nitrophenyl octyl ether)
Nitrophenols determination, 665
4-Nitrophenylhydrazones determination, 487
5-Nitropyrimidine-2,4(1H,3H)-dione, $pK_a(DMF)$, $pK_a(H_2O)$, 394
Nitroso compounds determination, 594
4-Nitro-3-trifluoromethyl phenol, $pK_a(2\text{-methyl-2-propanol})$, $pK_a(H_2O)$, 406
Nitrous acid determination in mixtures, 486
3-Nitro-o-xylene as electroactant, 151
Nonactin electroactant, 150
Nonactinic acid, 155
Nonylphenol adducts determination, 667
Nonylphenoxypoly(ethyleneoxy)ethanol, 478
    barium complex-tetraphenylborate(III) electroactant, 150
    strontium complex electroactant, 153

Normality, 454
Normal solution, definition, 455
Nuclepore® filters as matrix material, 139

Octacyanomolybdate(IV) determination, 575
Octanol in membrane preparation, 145
2-Octyloxynitrobenzene (o-nitrophenyl octyl ether), in membrane preparation, 145, 150, 159
Oils, determination of metals in, 516
Organic compounds:
  anionic determination, 660–666
  basic determination, 487
    as picrates, 488
  cationic determination, 666–667
  elemental analysis, 655–660
  hydroxylic determination, 487
  ion exchangers, types and uses, 143
  nitrated determination, 664
  sulfidic determination, 597
Organo compounds:
  -germanium mixtures determination, 486
  -metallic, determination of halogens in, 656
  -palladium, determination of halogens in, 656
  -phosphinic acids determination, 484
  -platinum, determination of halogens in, 656
Ornithine, 172
  trans-carbamylase, 172
Orotic acid determination in mixtures, 486
Orthovanadate determination, 641
Osmium(VIII):
  determination, 585, 587
  oxide catalyst, 535, 564, 566, 570, 585, 587
    in acidic media, 566–567
    in alkali media, 567–569
Osmium(III) determination, 581
Osmium(VI) determination, 589
Osmium tetroxide, see Osmium(VIII), oxide catalyst
Ovalbumin, potentiometric titration of, 357
Overlapping $pK_a$ values, 342–345
Oxalate determination, 285, 665
Oxalic acid:
  catalyst, 569
  oxidation of, 565
Oxazole, $pK_a(H_2O)$, 308
Oxidant:
  halogen containing as titrants, 573–579

for obtaining required oxidation state, table, 536
  primary standard, 452
  titrants, 570–584
Oxidation-reduction:
  buffering system, 540
  characteristics of reactions, 531–534
  titrations:
    apparatus, 532
    catalysts, 550, 563–570
      titre corrections for, 570
    complexing media for, 551–562
    curves, 534
      linearization of, 466, 468
    electrodes, 292–296
    medium effects, 549–562
    organic compounds, 594–599
    preparatory procedures, 534–539
      oxidants used, table, 536
      reductants used, 535, 537–539
    primary standards for, 452–453
    reference half-cells, 532
    titrants, 570–594
    stability, 549
Oxonium cations determination, 667
Oxygen:
  dissolved, 266, 270, 532
  flask combustion, 655–659

$pa_H$:
  calculation:
    in acetone, 407
    in acetonitrile, 384
    in dimethylformamide, 391
    in dimethylsulfoxide, 397
    in 2-methyl-2-propanol, 402
  measurement:
    in acetic acid, 380
    in acetone, 407
    in acetonitrile, 384
    in 4-methyl-2-pentanone, 407
  values:
    in aqueous ethanol, table, 414
    in aqueous methanol, table, 412
    in 4-methyl-2-pentanone, table, 408
    in 2-methyl-2-propanol, table, 403
  see also $pc_H$; pH
$p(a_H\gamma_{Br})$, 321
$p(a_H\gamma_{Cl})$, 309
  definition, 195
  derivation of pH from, 195

p($a_H\gamma_{Cl}$) (Continued)
  in determination of p$K_a$($H_2O$), 311–316
  for diprotic acids, 343
  measurement:
    in aqueous alcohol, 410
    in aqueous dimethylsulfoxide, 415
    in formamide, 376
    in water, 316–321
      with glass and silver-silver chloride electrodes, 245, 317
        calibration, 320
      with hydrogen and silver-silver chloride electrodes, 316, 334, 339
      solutions for calibration, table, 318
    meter, 320
    solutions for calibration, 320
  values:
    for aqueous ethanol, table, 413
    for aqueous methanol, table, 412
    for 0.01m hydrochloric acid, table, 335
    for 0.01m potassium hydroxide, table, 335
Palladium catalyst, 657
Palladium(II) determination, 283, 651
Pallmann effect, 107
o-PAN (1-(2-Pyridylazo)-2-naphthol)-mercury(II) complex preparation, 288
Papain, potentiometric titration of, 357
Papverine hydrochloride determination, 482, 666
p($a_H\gamma_{Cl}$), 323
p$c_H$:
  in aqueous dimethylsulfoxide, 416
  buffers in dimethylformamide, table, 392
  measurement of, 209–214, 332
  in multiparametric curve fitting, 470
  relation to p($a_H\gamma_{Cl}$), 210
  relation to pH, 211
  in stability constant determinations, 434
  see also Hydrogen ion concentration
p$c_X$, measurement of, 214–222
pD:
  relation to pH, 202
  scale of, 202
pD(S) values for buffers, table, 203
Pectic acid, potentiometric titration of, 351
Penicillinase in enzyme sensors, 170
Penicillin β-lactamase, 176
Penicillins determination, 176
Pentachlorophenol, 479

Pentacyanonitrosylferrate(III), 282
  determination, 291
Pentadecylbenzenesulfonate in membrane preparation, 149
Pentaerythritol tetranitrate determination, 487
Pentanedial in enzyme sensor preparation, 169
2,4-Pentanedione:
  masking agent, 270
  p$K_a$(DMSO), p$K_a$($H_2O$), 399
  p$K_a$($H_2O$), 308
Pepsinogen, potentiometric titration of, 357
Perchloric acid:
  buffer component:
    in acetic acid, 381
    in 4-methyl-2-pentanone, 407
  determination, 286, 635
  in mixture, 487
  p$K_B$($CH_3COOH$), 381
  polymerization in acetonitrile, 383
  reduction of, 567, 635
  titrant, 259, 263, 322, 420, 451
    in 4-methyl-2-pentanone, 407
    in organic solvents, 489
    standardization, 489
    precautions, 489
  preparation:
    in acetic acid (0.1M), 381
    in acetonitrile, 387
Perfluoroalkanedicarboxylic acids determination, 487
Periodate:
  determination, 286, 554, 566, 636
  oxidant, 536
  reduction of, 569
Permanganate:
  determination, 286, 641
  oxidant, 536
  see also Manganese (VII); Potassium, permanganate
Peroxydisulfate oxidant, 477, 536, 570. See also Ammonium peroxydisulfate oxidant
Perrhenate determination, 286, 641
Persulfate, see Peroxydisulfate oxidant
Pesticides determination, 254, 288
pH, 309
  derivation of activity standards for, 194–201
  measurements:
    in aqueous alcohols, 410–415
    for multiparametric curve fitting, 470
  operational, definition, 197

scale:
  in countries other than US, 201
  NBS, 194
  accuracy, 200
  reference solutions, table, 197
  in stability constant determinations, 434
  titration, 326, 434
Phenacetin, $pK_B(CH_3COOH)$, $pK_B(H_2O)$, 382
1,10-Phenanthroline, 579, 580
  -iron(II) as electroactant, 143, 146
  -nickel(II) as electroactant, 143, 145
Phenazinduline $O$ electroactant, 144
Phenidone determination, 584
Phenobarbital determination in mixtures, 486
Phenolic compounds determination, 485, 486, 487
Phenolic groups determination, 263
Phenols:
  bromination, 599
  determination, 172, 177, 261, 286, 484, 485, 488, 573, 575
    in acetic acid media, 596
    alkyl and aryl, 485
    in mixtures, 484, 487
    in propylene carbonate medium, 599
  enhancement in acidity of, 478
  medium for basicity enhancement, 480
  oxidation of, 564
  $pK_a(CH_3CN)$, 389
  $pK_a(DMF)$, $pK_a(H_2O)$, 394
  $pK_a(DMSO)$, $pK_a(H_2O)$, 398
  $pK_a$(2-methyl-2-propanol), 405
  2-substituted $pK_a(H_2O)$-$\Delta HNP$ relation, 420
  titration, 478
    as media, 487
    in non-aqueous solvents, 422, 423
Phenosafrin, 539
Phenothiazines:
  determination, 594, 597
  titrations in various solvents, 424
Phenoxyacetic acid, tetrazolium salts as electroactants, 147
Phenoxyalkanoic acids determination, 254, 288, 664
L-Phenylalanine:
  ammonia lyase, 175
  determination, 175
$N$-Phenylbenzohydroxamic acid titrant, 589
$N$-Phenylbenzylamine, $pK_B(CH_3COOH)$, $pK_B(H_2O)$, 382
Phenylborates determination, 286, 665

1-Phenyl-1,3-butanedione, $pK_a(DMSO)$, $pK_a(H_2O)$, 399
Phenylbutazone determination, 485
Phenylenediamines determination, 480
1-12-$o$-Phenylenedihydropleiadene, $pK_a(DMSO)$, 399
1,2-Phenylenedioxybis($N$-benzyl $N$-phenylacetamide) electroactant, 152
9-Phenylfluorene, $pK_a(DMSO)$, $pK_a(H_2O)$, 399
2-Phenylphenol determination, 259
$N$-Phenylphenothiazine determination, 597
Phenyls, alkyl-determination, 485
$N$-Phenylureas:
  determination, 662
  -substituted, titrations in butylamine, 423
Phosphate:
  buffer:
    as pD reference, 203
    as pH reference, 197
  determination, 286, 291, 484, 580, 637
Phosphine oxides determination, 482
Phosphines-substituted titrations in nitromethane, 424
Phosphite:
  determination, 573
  oxidation of, 570
6-Phospho-D-gluconate dehydrogenase, 175
Phosphoric acid:
  alkyl and aryl esters as exchangers, 143
  catalyst, 569
  complexant, 549
  diorganodithio- determination, 483
Phosphorus determination in polymer, 484
pH(S):
  defined, 198
  effect of temperature on, 200
  values of various temperatures, table, 201
Phthalic acid, see 1,2-Benzenedicarboxylic acid (phthalic acid)
Picric acid, see 2,4,6-Trinitrophenol (picric acid)
Pinkhof-Treadwell method, 64, 457
Piperidine:
  determination in mixtures, 487
  $pK_a(CH_3CN)$, 389
  $pK_a(DMSO)$, $pK_a(H_2O)$, 399
  $pK_a(H_2O)$, 308
  titrant, 493
$pK_a$, defined, 306
$pK_{SH}$, see Autoprotolysis, constant

pK$_w$, 332
  conditional, pK$_w^c$, 332
  determination, 334–337
  values:
    in potassium nitrate, 337
    in potassium and sodium chlorides, table, 336
    in seawater, 337
  determination, 333
  values, table, 678
pK$_a$ values:
  in acetic acid for perchloric acid, 381
  in acetonitrile, table, 389
  determination, 388
  analytical applications, 308
  calculation:
    in 4-methyl-2-pentanone, 408
    from p(a$_H$γ$_{Cl}$) data, 311–316, 322–326
    from pH data, 326–332
  conditional, 329, 331–332, 416
  in dimethylformamide, table, 394
  in dimethylsulfoxide, table, 398
  in 2-methyl-2-propanol, table, 405
  in $N$-methyl-2-pyrrolidinone, 393
  determination, 396
  in mixed solvent, variation with dielectric constant, 410
  in water:
    for acids, range of, table, 307
    for acids and bases, table, 308
    for benzenediols, 478
    calculation, effect of ion size parameter on, 316
    determination:
      acetic acid, table, 315
      2-amino-2-hydroxymethyl-1,3-propanediol, table, 325
      ampholytes, 342
      benzoic acid, tables, 324, 330
      diprotic acids:
        overlapping ionizations, 342–345
        separate ionizations, 338–341
      diprotic conjugate acids, 340
      general methods, 309
      polyprotic acids, 345
    for ethoxylated nonylphenol-phenol adducts, 478
    for hexitols, 478
    prediction of, 331
    sources of, 307

pK$_B$ values:
  in acetic acid, table, 382
  determination, 379
  in water, table, 382
Plasticizer in membrane preparation, 158
Platinum(IV):
  determination, 562, 651
  oxidation of, 570
Platinum black preparation, 85–86
Platinum electrodes, see Electrodes, oxidation-reduction, platinum
Plutonium determination, 589
  micro, 571
Plutonium(VI) formation, 536
pM buffers in calibration of electrodes, 221
pm$_H$ for seawater, 221
Polarized electrode, definition, 66
Poly acids potentiometric titration, 347
Polyacrylamide gel, 172
  in enzyme sensors, 169
Poly(acrylic acid) potentiometric titration, 349
Poly(adenylic acid) potentiometric titration, 352
Poly(L-alanine glutamic acid) potentiometric titration, 352
Polyamines as titrants, 526
Polyaminocarboxylic acids, 501
  as titrants, 525
Poly(4-amino-L-phenylalanine) potentiometric titration, 352
Polyampholytes potentiometric titration, 347
Poly(aspartic acid) potentiometric titration, 352
Poly bases potentiometric titration, 347
Poly(bisphenol-A-carbonate) in membrane preparation, 152
Poly(caprolactam), 486
Poly(S-carboxyethyl-L-cysteine) potentiometric titration, 353
Poly carboxylic acids aromatic determination, 484
Poly(S-carboxymethyl-L-cysteine) potentiometric titration, 353
Poly(dimethylaminoethylmethacrylate), use in titrations, 478
Poly(dimethylsiloxane) in membrane preparation, 152
Polyelectrolytes ionization, 345–357
Polyethylene:
  as matrix material, 133
  in membrane preparation, 141

# INDEX

Polyethyleneglycol-barium(II) acetate complexes, 667
Polyethyleneglycols determination, 667
Polyethylenepolyamines determination, 486
Poly(ethylenimine) potentiometric titration, 357
Poly(L-glutamic acid) potentiometric titration, 353
  of derivatives, 354
  of sodium salt, 354
Poly($\alpha$-DL1-glutamic acid) potentiometric titration, 354
Poly(guluronic acid) potentiometric titration, 350
Poly(L-histidine) potentiometric titration, 354
Poly(L-lysine) potentiometric titration, 354
Poly(maleic acid) potentiometric titration, 350
Poly(mannuronic acid) potentiometric titration, 351
Poly(methacrylic acid) potentiometric titration, 351
Polynuclear aromatics determination, 487
Polynuclear complex, 431
  test for formation, 436
Polyols as complexants, 477, 592
Polypropyleneglycols determination, 667
Polysiloxane, 133
Poly(styrylphosphonic acid), 484
Polysulfides formation, 640
Poly(L-tyrosine) potentiometric titration, 356
Poly(vinylamine) potentiometric titration, 357
Poly(2-vinylpyridine), atactic potentiometric titration, 357
Porcine kidney tissue, 174
Potassium, fusion with, 656
Potassium(I):
  acetate p$K_B$(CH$_3$COOH), 382
  alkoxymethanedithioates as titrants, 612, 650
  bromate titrant, 554, 579
    in ethanolic medium, 598
  bromide:
    catalyst, 564
    titrant, 254, 651
  *tert*-butoxide, *see* Potassium(I), 2-methyl-2-propanolate ionization
  butoxymethanedithioate titrant, 612
  chlorate titrant, 579
  chloride primary standard, 453
  determination, 514, 642
  dichromate, 260
    as primary standard, 452
    titrant, 534, 553, 555, 556, 557, 571
    properties and use, 571
    *see also* Chromium(VI)
  dihydrogen citrate:
    p($a_H\gamma_{Cl}$) reference, 320
    pH reference, 197
  dihydrogen citrate-d$_2$, pD reference, 203
  dihydrogen phosphate, 320
  ethyl xanthate determination, 573
  hexacyanoferrate(II) titrant, 590
  hexacyanoferrate(III) titrant, 551
    preparation, 580
    *see also* Hexacyanoferrate(III)
  hexyloxymethanedithioate titrant, 645, 646, 647, 650
    applications, 612
  hydrogen iodate primary standard, 452
  hydrogen phthalate:
    growth of large crystals, 449
    pH reference, 197
    primary standard, 449, 489
  hydrogen tartrate pH reference, 197
  hydroxide titrant, 259, 261
    carbonate-free, 476
    in non-aqueous solvents, 491
  iodate titrant, 575, 579. *See also* Iodate
  iodide, 178
    catalyst, 578
    in membrane preparation, 145
    precipitant, 652
    prime standard, 582
    titrant, 580, 651, 658, 659, 664
    *see also* Iodide
  methoxymethanedithioate titrant, 647, 649
    uses, 612
  2-methyl-2-propanolate ionization, 372
  nitrate:
    background solutions, 436
    bridge solutions, 610
    coagulant, 628
  perbromate titrant, 295
  permanganate, 260
    reduction of, 564
    titrant, 549, 558
    standardization, 453
    *see also* Manganese(VII)
  peroxydisulfate, 569
    in enzyme immobilization, 169

Potassium(I): peroxydisulfate (*Continued*)
see also Peroxydisulfate oxidant, 569
propoxymethanedithioate titrant, 612
ruthenate catalyst, 569
tetrakis(4-chlorophenyl)borate electroactant, 151, 158
tetraphenylborate electroactant, 151
tetroxalate pH reference, 197
Potentiometric titration:
  curve fitting, 458
  defined, 55
  with one polarized electrode, 70. See also Potentiotitrimetry, electrolytic
  $p(a_H\gamma_{Cl})$, 322–326
  pH, 326–330, 434–437
  of polyelectrolytes, 347–357
  and potentiotrimetry contrasted, 55
  with two polarized electrodes, 70
  see also Potentiotitrimetry, differential
Potentiometry:
  defined, 3
  differential, 59–60
    for chloride determination, 232
    defined, 57
    for fluoride determination, 229–232
  direct, 3, 55
  see also Electrodes, ion selective; Electrodes, oxidation-reduction; Reference half-cell
Potentiotitrimetry:
  advantages, 47, 255
  definition, 3, 55
  differential, 57, 60–61, 284, 457
    electrolytic, 69, 73–75, 276, 457
      with alternating current, 76
      apparatus and conditions, 75–76
      in complexometry, 277
      ultramicro, 282
  electrolytic, 69, 71–73, 457
    apparatus and conditions, 75–76
    in complexometry, 275–278
    in microdeterminations of halides, 76
    in oxidation-reduction titrations, 295–296
    in precipitation titrations, 289–292
  requirements of indicator electrode, 253
  see also Electrodes, potentiotitrimetric
Practical $pK_a$ values, 331. See also $pK_a$ values, conditional
Praseodymium(III) determination, 519

Precipitation titrations:
  coagulants for, 628
  electrode response in, 629
  electrodes for, 278–292
  equations for linearization of curve, 466
  errors, 627
  kinetic aspects, 628
  in partially aqueous media, 628–631
  primary standards, 453
  reference half-cells for, 609
  solvents used, 630–631
  titrants, table, 612
Primary standards, 447, 448–454
  for acid-base titrimetry, 449–451
  for complexometry, 451–452
  for oxidation-reduction titrimetry, 452–453
  for precipitation titrimetry, 453–454
  recrystallization, 448
  requirements, 448
L-Proline-poly(styrene) sorbent, potentiometric titration, 355
Propanedinitrile, $pK_a$(DMSO), $pK_a$(H$_2$O), 399
Propanedioic acid:
  $pK_a$(DMF), $pK_a$(H$_2$O), 395
  $pK_a$(2-methyl-2-propanol), $pK_a$(H$_2$O), 405
1,2-Propanediol, physical properties, table, 369
Propanethiol determination, 661
2-Propanethiol determination, 661
1,1,1-Propanetriyltris-4(*N*-heptyl *N*-methyl 3-oxabutanamide) electroactant, 152
1,1,1-Propanetriyltris-4-(3-oxabutanoic acid), 158
Propanol, physical properties, table, 369
2-Propanol, 256
  Debye-Huckel constants, 372
  discriminative solvent for phenols, 420
  medium for precipitation titrations, 630
  physical properties, table, 369
  reference half-cells for, 114
2-Propen-1-ols, 3-substituted determination, 591
Propylene carbonate, 599. See also 4-Methyl-1,3-dioxolan-2-one (propylene carbonate)
  titration medium, 574
Propylene glycol, see 1,2-Propanediol, physical properties, table
Protein antigen, 175
Proteins, potentiometric titrations:
  of globular, 356

of human apo A-II, 356
of tobacco mosaic virus, 357
*Proteus morganii*, 177
Protogenic solvent, 367. *See also* Solvents, amphiprotic, protogenic
Protonation constants, stepwise, 438. *See also* Formation constant
9$H$-Purine:
  determination of derivatives, 660
  $pK_a(H_2O)$, 308
Purine-8(1$H$)-thione determination, 664
PVC:
  glue, 141
  in membrane preparation, 140
pX buffers (X = Ca,Cd,Cu,Pb), 221
pX(S) values:
  divalent cations, 207
  NBS (X = K,Cl,Na,Cl,F), table, 206
  in standardization of pX electrodes, 207
  values (X = Li,Rb,Cs,$NH_4^+$), table, 207
pX values, blood serum standards, table, 208
Pyridine:
  bases determination, 487
  determination, 486
  $pK_a(CH_3CN)$, 389
  $pK_B(CH_3COOH)$, 379, 382
  $pK_a(DMF)$, $pK_a(H_2O)$, 395
  $pK_a(DMSO)$, $pK_a(H_2O)$, 399
  $pK_a(H_2O)$, 308
  reference half-cells for, 114, 117
  as titration medium, 487
Pyridinecarboxylic acids determination in mixtures, 486
2-Pyridone, $pK_a(H_2O)$, 308
1-(2-Pyridylazo)-2-naphthol, *see o-*Pan (1-(2-Pyridylazo)-2-naphthol)-mercury(II) complex preparation
Pyrimidine, $pK_a(H_2O)$, 308
Pyrimidine-2,4(1$H$,3$H$)-dione, $pK_a(DMF)$, $pK_a(H_2O)$, 394
Pyrimidine-2(1$H$)-one, $pK_a(DMF)$, $pK_a(H_2O)$, 394
4-Pyrone, $pK_a(H_2O)$, 308
Pyrophosphate:
  complexant, 549, 550
  determination, 638
Pyrrole, $pK_a(H_2O)$, 308
Pyrrolidine, $pK_a(CH_3CN)$, 389

Quasi-thermodynamic cells, 26–28, 191
Quaternary nitrogen cations, titration, 479
Quinhydrone:
  electrode, 62. *See also* Reference half-cell, quinhydrone
  purification, 102
Quinine hydrochloride determination in mixtures, 486
Quinoline:
  bases determination, 487
  determination, 483
  $pK_a(H_2O)$, 308
8-Quinolinethiol titrant, 651, 652
  application, 615
8-Quinolinol, 479
  oxidation of, 564

Radioimmunoassays, 173
Rare earths determination, 486
  as acetates, 482
  fluorides as electroactants, 132
  *see also* Lanthanides and actinides determination
Rat liver microsomes, 177
Recrystallization, removal of solvent in, 448
Redox electrodes, 80. *See also* Zeroth-kind electrodes
Reductants:
  effects of amalgamation on, 537
  primary standard, 453
  selectivities of, 537
  used to obtain required oxidation-state, 535, 537–539
  used as titrants, 584–594
Reduction potentials:
  formal, *see* Formal reduction potential
  standard, 30, 31–32
    predictive use of, 34
    significance, 32–38
    use in formulating Nernst equation, 35
    values, table, 33
Reductors, 537. *See also* Jones reductor
Reference electrode:
  essential characteristics, 83
  platinum titrant stream, 658
  role of, 82
Reference half-cell:
  calomel (mercury-mercury(I) chloride), 98–100, 125, 215, 278, 281, 284, 290, 291, 292, 644, 651, 662, 666
  double junction, 434
  manufacture of, 99

Reference half-cell: calomel (mercury-
mercury(I) chloride (*Continued*)
   modification for oxidation-reduction
     titrations, 532
   modifications for precipitation titrations,
     609
   potentials, table, 100
   temperature coefficients, 247
   temperature hysteresis, 99, 248
   toxicity of materials, 97
   use in dimethylformamide, 391
  defined, 13
  double junction, 102, 104, 107, 266, 284,
    291, 637, 665
  hydrogen, 247
  internal reference, 120
  mercury-mercury(I) acetate, 261
  mercury-mercury(I) carboxylate, 103
  mercury-mercury(II) oxide, 103
  mercury-mercury(I) sulfate, 101–102, 268,
    280
   potentials, 101
   in precipitation titrations, 611
  quinhydrone, 63, 102
  requirements for, 14
  silver-silver(I), 116
   in acetone, 407
   in dimethylsulfoxide, 397
   in 2-methyl-2-propanol, 402
   in N-methyl-2-pyrrolidinone, 395
   in non-aqueous media, 377
  silver-silver(I) chloride, 94, 97, 667
   in aqueous dimethylsulfoxide, 415
   in dimethylsulfoxide, 400
   double junction, 295
   effect of light, 249
   industrial type, 98, 113
   in N-methyl-2-pyrrolidinone, 395
   potentials of 3.5M KCl type, 98
   temperature coefficient, 232
  stability, 245
  storage, 242
  temperature coefficients, 247
  thallium-thallium(I) chloride (Thalamid®
    type), 100
  thermal equilibrium in, 247
  triple junction, 114
  use in non-aqueous solvents, 113–118
  use in precipitation titrimetry, 609–611
  Wilhelm type, 112, 434
Refractive index, table, 369

Reversible cell example, 4
Reversible oxidation-reduction couples, 71, 73
Response time:
  ion selective electrodes, 243–245
  silver-silver(I) chloride electrodes, 96
Rhodium determination, 519
Riboflavine in enzyme immobilization, 169
Ribonuclease A-3'-cytosine complex,
    potentiometric titration, 357
Ruhemann's purple:
  electroactant, 152
  -mercury(II) complex, 289
Ruthenium(VI), formation of, 536
Ruthenium determination, 589
Ruthenium(IV) sulfate catalyst, 569

Safranine *O* in membrane preparation, 144
Salicylaldoxime, 479
Salicylamide determination, 485, 486
Salicylate masking agent, 270
Salicylic acid, *see* 2-Hydroxybenzoic acid
Salt bridge, 7, 230, 278, 290, 402
Salts of weak acids determination, 482, 487
Samarium(III) determination, 519
Scandium(III) determination, 484, 519
Schoniger combustion method, 655. *See also*
    Oxygen, flask combustion
Seawater:
  calibration of cells, 220
  composition of artificial, 220
  determination:
   of alkalinity, total, 468
   of cyanide in, 633
   of halides in, 633
  pH buffers in, 214
  p$K_w^c$ of, 337
  scales of hydrogen ion concentration, 221
Second kind electrodes, *see* Electrodes,
    oxidation-reduction, second kind
Selectivity coefficient, 236–241. *See also*
    Electrodes, ion selective
Selectrodes®, *see* Electrodes, ion selective,
    coated graphite (Selectrode®)
Selenides as membrane electroactants, 132
Selenite determination, 658
Selenium:
  determination, 592
   in organic compounds, 658
  formation of, 538
Semicarbazide derivatives determination,
    594

INDEX 717

Sensors:
 bioselective, 171-178
  antidiuretic hormone, 173
  in L-arginine determination, 171
  in L-aspartate determination, 171
  hepatitis B surface antigen, 173
 enzyme, 123, 166-178
  amperometric glucose, 167
  based on:
   glass electrode, 176
   iodide ISE, 177
   platinum electrode, 177
   zeroth electrodes, 172
  immobilization of enzyme in, 169
  kinetically based, 178
  penicillin, 171
  urea, 169
 gas, 123, 159-166
  air gap, 165
  in enzyme sensors, 170
  ammonia, 160, 163
   in bioselective sensors, 171
   in enzyme and biological sensors, 174
   in enzyme immunoassay, 178
   temperature coefficients, 247
  calibration, 162
  carbon dioxide, 163
   in enzyme sensor, 175
   and oxygen, 163
   temperature coefficient, 247
  cell electrodes, table, 163
  cell notation, 160
  cell solutions, table, 163
  diethylamine, 163
  in enzyme sensors, 168
  homogenous membrane, 164
  hydrocyanic acid, 163
  hydrogen sulfide, 163
   in enzyme sensor, 177
  microporous membrane, 164
  nitrogen oxides, 163
  osmotic drift in, 165
  permeable membranes for, 163
  response time characteristics, 164
  storage of, 242
  sulfur dioxide, 163
  see also Membranes, gas permeable
Sequestrene, see EDTA
Sevron L in membrane preparation, 144
Sevron red GL in membrane preparation, 144
Silanes, trialkoxy(phenylamino) titrations, 425

Silica determination, 592
Silicone rubber as membrane material, 133, 141
Silver:
 determination in alloys and solders, 652
 as primary standard, 453
 reductant, 535
Silver(I):
 adenine solubility product, 625
 arsenate solubility product, 625
 arsenite solubility product, 625
 azide:
  as membrane electroactant, 133
  solubility product, 625
 benzimidazole-2(3$H$)-thione solubility product, 625
 benzothiazole-2(3$H$)-thione solubility product, 625
 benzotriazole, solubility product, 625
  determination of conditional, 621
 benzoxazole-2(3$H$)-thione solubility product, 625
 bromide solubility product, 625
 8-bromoadenine solubility product, 625
 chalcogenides as matrices, 134
 -chelon complexes preparation, 528
 chloride, solubility product, 625
  determination of conditional, 621
  in various media, table, 629
 5-chloro-1(1$H$)benzotriazole solubility product, 625
 chromate electroactant, 133
 coulometrically generated prime standard, 453
 cyanide:
  electroactant, 133
  solubility product, 625
 determination, 280, 281, 282, 555, 562, 651
  in fixing baths, 652
  sequential, 647, 648, 652
  ultramicro, 646
 halides as electroactants, 132
 hexacyanoferrate(III) solubility product, 625
 iodide:
  electroactant, 131
  solubility product, 625
 nitrate:
  in membrane preparation, 152
  precipitant, 661
  -thorium(IV) nitrate mixed titrant, 283

Silver(I) nitrate (*Continued*)
  titrant, 280, 285, 287, 289, 290, 291, 656, 658, 660
  pentacyanonitrosylferrate solubility product, 626
  perchlorate titrant, 632
    in tetrahydrofuran, 663
  phosphate:
    electroactant, 133
    solubility product, 626
  salts as catalysts, 570
  selenide membrane binder, 129
  selenite solubility product, 626
  sulfate catalyst, 570
  sulfide in membrane preparations, 129, 130, 134
  thiocyanate:
    electroactant, 133
    solubility product, 626
  thiolates stability, 663
  titrant, 660
  applications, table, 615
Silver-copper alloy determination of silver and copper, 648
Silver(II) oxide:
  oxidant, 536
  preparation, 536
Single point titrations, 223
Soaps, 289
Sodium(I):
  acetate:
    medium for acid-base titrations, 468
    in membrane preparation, 147
    $pK_B(CH_3COOH)$, 379, 382
    purification, 381
  alginate, oxidation of, 570
  benzoate in membrane preparation, 148
  bicarbonate buffer component, 321
  borohydride for dehalogenations, 657
  bromide in membrane preparation, 144
  carbonate:
    buffer component, 321
    fusion with, 477
    primary standard, 450
  chloride:
    in membrane preparation, 144
    $p(a_H\gamma_{Cl})$ titrations in 0.01M, 322
    primary standard, 453
  chlorite titrant, 578
  *N*-cyclohexylcarbamodithioate titrant, 645
  deoxytaurocholate in membrane preparation, 149
  determination, 516
  dialkylcarbamodithioates determination in ethanolic medium, 598
  *NN*-diethyldithiocarbamate:
    determination in acetonitrile medium, 597
    titrant, 644, 647, 648
    applications, 612
  dimsyl titrant, 397, 491
  fluoride:
    as complexant, 549
    mean activity coefficient determination, 193
    as standard, 632
    titrant, 644
  hydroxide titrant, 310, 322, 327
    carbonate-free, 476
  hypophosphite, 539
  iodide:
    in membrane preparation, 145
    titrant, 650
  leucinate in membrane preparation, 148
  metavanadate titrant, 555, 583
  methoxide titrant, 491
  molybdate titrant, 660
    standardization, 649
  nitrate bridge solution, 610
  orthovanadate titrant, 583
  oxalate:
    in membrane preparation, 148
    primary standard, 453, 584
  pentacyanonitrosylferrate(III) titrant, 648
    applications, 613
  perchlorate,
    background electrolyte, 435
    ion-pair dissociation constant exponent, 382
    in membrane preparation, 146
    purification, 381
  peroxide, fusion with, 656
  perxenate oxidant, 536
  phenylalanate in membrane preparation, 148
  piperidine-1-dithiocarboxylate titrant, 645
    applications, 612
  salicylate in membrane preparation, 148
  sulfide:
    determination, 640, 662
    titrant, 645
  sulfite medium for acid-base titrations, 468

taurocholate in membrane preparation, 149
tetraborate (borax):
   growth of large crystals, 449
   pH reference, 197
tetraphenylborate(III), 280, 666
   in membrane preparation, 151
   titrant, 642, 648
      applications, 616, 666
      preparation and standardization, 666
      see also Tetraphenylborate(III)
   thiosulfate titrant, 576, 595, 651
   properties and standardization, 591
   standardization, 571
Soil extracts, determination of sulfate, 284
Solasodine determination, 486
Solubility, calculation from solubility product, 619
Solubility product:
   conditional:
      definition, 618
      determination, 620, 622
      predictions based on, 627
   definition and determination, 618-631
   effect of value on electrode response, 23
   errors in determination, 623
   limits for Gran plots, 467
   relation to titration error, 460, 627
   silver(I) chloride in various media, table, 629
   values, table, 624
Solutizene determination, 284
Solvents:
   amphiprotic, 365
      ampholytic, 367
   apparatus for titrations in, 375
   coefficient of cubic expansion, 376
   density, 376
   ideal properties, 368
   ionic product, 366. See also
      Autoprotolysis, constant
   ion pair formation in, 371
   physical properties, table, 369
   protogenic, 367
   protophilic, 367
   range of potentials available in, 372
   relative acidity constant, 374
   relative scale of acidity, 373, 374
   toxicity, 375
   vapor pressure, 375
aprotic, 368

classification, 365-368
purification, 481
Sorbitol, 477
   determination, 478
Speakman equation, 342, 345
Stability constant:
   calculations of:
      basic, 437-439
      computerized, 439-441
   compilations of values, 433
   definition, 431
   determination by pH titration, 433, 434-437
   DPTA-silver(I), 270
   EDTA-mercury(II), 265
   EDTA-silver(I), 269
   HEDTA-cadmium(II), 437
   HEDTA-lead(II), 437
   relation to titration error, 460
   stepwise formation, 432
Stability and response of ion selective electrodes, 245-246
Standard:
   addition:
      differential potentiometry, 229
      method, 53, 59, 108, 222-232
      subtraction methods, 222-232
   additions multiple, 225-229
   reduction potential, see Reduction potential standard
   solution, 447
   subtraction method, 54, 108, 222
Stannous chloride, 592. See also Tin(II), chloride (stannous chloride)
Stirring, effect on ISE response, 244
Stokes-Robinson equation, 204
*Streptococcus faecium*, 171, 175
*Streptomyces cinnamonensis*, 152
*Streptomyces fulvissmus*, 155
Strong acids determination, 482. See also Mineral acids
Strontium(II) determination, 519
Strychnine nitrate determination, 479
Succinic acid, $pK_a(H_2O)$, 308. See also Butanedioic acid
Sugars as media for basicity enhancement, 480
Sulfa drugs determination, 485
Sulfamic acid:
   growth of large crystals of, 449
   primary standard, 450
Sulfanilamides determination, 483

Sulfate:
    determination, 128, 282, 284, 291, 655
        general method, 638
        indirect, 482
        masking agent, 270
        titrant, 283, 616
Sulfides:
    determination, 574, 577, 582, 661
        general method, 639
        sequential, 640
    as membrane electroactants, 132
    titrant application, 616
Sulfinic acids formation, 595
Sulfite determination, 574, 576
    general method, 640
Sulfo groups determination, 263
Sulfolane, see Tetrahydrothiophene-$SS$-dioxide (sulfolane)
Sulfonic acids:
    aryl purification, 489
    $pK_a$(2-methyl-2-propanol), 405
    titrants, 489
Sulfosalicylate masking agent, 270
Sulfur:
    anions determination, 638–641
    determination, 640
        in organic compounds, 655, 657, 658
    dioxide determination in chimney gases, 641
Sulfuric acid:
    determination, 483, 484, 486
        in mixtures, 487
    $pK_a$($CH_3CN$), 383
    $pK_a$(DMF), 397
Surfactants determination:
    of anionic, 289
    of quaternary nitrogen salts, 280, 666

Tartaric acid determination, 586, 596
Teflon as matrix material, 139
Tellurides as electroactants, 132
Tellurite determination, 658
Tellurium:
    determination, 561, 589, 592
        in organic compounds, 658
    dioxide, 543
    formation of, 538
Tellurous acid, oxidation of, 563
Temperature:
    coefficients of electrodes, 247
    effects on ISEs, 246. See also Electrodes, ion selective
    hysteresis, 248
Tetraalkoxysilanes as titration media, 488
Tetraalkylammonium titrants, 420, 451, 491–493
    effect of structure on, 492
    preparation, 493
Tetrabutylammonium:
    hydroxide titrant, 257, 259, 263, 397, 400, 426, 491
    availability, 493
    preparation in dimethylsulfoxide, 401
    stability, 492
    iodide determination, 597
    perchlorate as buffer component in 4-methyl-2-pentanone, 407
    picrate:
        dissociation in 4-methyl-2-pentanone, 408
        formation constants, 368
        preparation, 409
2,3,5,6-Tetrachloroaniline, $pK_a$(DMSO), $pK_a$($H_2O$), 399
2,3,5,6-Tetrachloro-1,4-benzenediol titrant, 580
1,1,2,2-Tetrachloroethane in membrane preparation, 146, 151
Tetracycline hydrochlorides determination, 483, 485
Tetracyclines determination, 482, 485
Tetradecylphosphonium:
    tetrachlorothallate electroactant, 144
    trichloromercurate electroactant, 144
Tetraethylammonium:
    bromide, 574, 599
    carboxylates preparation, 388
    3,4-dimethylbenzoate preparation, 404
    3,5-dinitrobenzoate dissociation in 4-methyl-2-pentanone, 408
    hydroxide:
        as standard, 396
        titrant, 395, 416, 491
            effect of methanol on, 426
            stability, 492
    perchlorate:
        as bridge solution, 384
        preparation, 387
    picrate:
        buffer component in acetone, 407
        purification, 387
Tetraethyl 2,4-dinitrophenolate preparation, 391

Tetraethylenepentamine, 503. See also
    TETREN titrant (1,4,7,10,13-
    pentaazatridecane)
Tetrahepylammonium hydroxide titrant
    stability, 492
Tetrahexylammonium hydroxide titrant
    stability, 492
Tetrahydrofuran:
    in membrane preparation, 141, 154
    $pK_a(H_2O)$, 308
    reference half-cells for, 117
    as titration medium, 488
Tetrahydrothiophene-$S,S$-dioxide (sulfolane),
    reference half-cells for, 117
Tetrakis(decyl)ammonium:
    acetate electroactant, 147
    butanoate electroactant, 148
    formate electroactant, 147
    nitrate electroactant, 145
    propanoate electroactant, 148
Tetramethylammonium hydroxide titrant
    stability, 492
Tetramethylguanidine:
    basic ionization constant in acetonitrile, 383
    $pK_a(CH_3CN)$, 389
    $pK_a(DMF)$, $pK_a(H_2O)$, 395
    $pK_a(DMSO)$, $pK_a(H_2O)$, 399
    as titration medium, 488
Tetramethylguanidinium ion, ionization
    constant in acetonitrile, 383
Tetramethylthiourea determination, 597
Tetranactin electroactant, 150
Tetraoctylammonium nitrate electroactant,
    145
Tetraoctylphosphonium:
    perchlorate electroactant, 146
    perrhenate electroactant, 146
$NNNN$-Tetrapentylammonium picrate
    preparation, 287
Tetraphenylarsenic(V):
    chloride titrant, 616, 635
    thiocyanate electroactant, 147
Tetraphenylborate(III):
    anion exchangers, 143
    determination, 290
    titrant, 280, 287, 648, 651
        applications, 616
    -vitamin B6 electroactant, 154
    see also Sodium(I), tetraphenylborate(III)
Tetraphenylphosphonium:
    perchlorate electroactant, 146
    perrhenate electroactant, 146
1,2,4,6-Tetraphenylpyridinium:
    acetate titrant, 286, 638, 641
    applications, 617
        for organic determinations, 665
    preparation of solutions, 635
    fluoroborate solubility product, 626
    hexachloroplatinate solubility product, 626
    perchlorate solubility product, 626
    perrhenate solubility product, 626
    tetrachloroaurate solubility product, 626
$NNN'N'$-Tetraphenyl-3,6,9-
    trioxaundecanediamide
    electroactant, 150
Tetrapropylammonium bromide determination,
    479
1$H$-Tetrazole derivatives determination, 660
TETREN titrant (1,4,7,10,13-
    pentaazatridecane), 266, 271, 503,
    508, 510, 523
    purification, 526
Thalamid® electrode, 100–101. See also
    Reference half-cell, thallium-
    thallium(I) chloride (Thalamid®
    type)
Thallium(I):
    determination, 551, 576, 579
    halides as electroactants, 132
    nitrate, 666
    oxidation of, 564
Thallium(III):
    determination, 485, 520, 645
    -thallium(I) as indicator, 283
THAM, see 2-Amino-hydroxymethyl-1,3-
    propanediol (TRIS)
Theobromine determination, 482
Theopaverine determination, 482
Theophylline determination, 486, 661
Thermodynamic:
    $pK_a$ value, 329. See also $pK_a$ value
    $pK_w$, 333. See also $pk_w$
    reversibility condition for, 4–6, 70
1,3,4-Thiadiazole-2,6(3$H$,4$H$)-dithione titrant,
    647
    application, 617
Thiazole, $pK_a(H_2O)$, 308
Thioacetamide:
    determination, 285, 597, 662
    titrant, 279, 282, 648, 649
    applications, 617
    preparation, 650

Thioamides:
  determination, 482, 662
  as precipitants, 646
Thiocarbamates determination, 483
  pyridyl-substituted, 487
Thiocarbonyl groups determination, 662
Thiocyanate:
  determination, 289, 637
  sequential, 633
  titrant applications, 617
Thiocyanic acid, oxidation of, 565
Thioethers:
  determination, 482, 594
  oxidation of, 564
Thioglycolic acid (see 2-Mercaptoacetic acid)
Thiols determination:
  acid-base, 484, 485
  oxidation-reduction, 573, 594, 597, 598
  precipitation, 281, 285, 661, 663
  in mixtures, 285
Thionalide (2-mercapto-$N$-(2-naphthyl)acetamide) titrant, 291, 653
  application, 617
Thiones determination, 594, 597, 662
Thiooxine titrant, 615. See also 8-Quinolinethiol titrant
Thiophenol, $pK_a(H_2O)$, 308
Thiophenols determination, 599
Thiosemicarbazide determination, 591
  of complexes, 595
Thiosulfate:
  determination, 72, 574, 577, 641
  oxidation of, 563
  titrant, 617. See also Sodium, thiosulfate titrant
2-Thiouracil, see 2,3-Dihydro-2-thioxopyrimidin-4(1$H$)-one, $pK_a(DMF)$, $pK_a(H_2O)$
4-Thiouracil, see 3,4-Dihydro-4-thioxopyrimidin-2(1$H$)-one, $pK_a(DMF)$, $pK_a(H_2O)$
Thioureas:
  -copper(I) complexes determination, 597
  determination, 594, 597, 598
  acid-base, 483
  oxidation-reduction, 573, 578, 586, 597
  precipitation, 287, 662
  in mixtures, 640, 662
  oxidation of, 564
  $pK_B(CH_3COOH)$, $pK_B(H_2O)$, 382

primary standard, 453, 454, 584
  titrant, 651, 652
  applications, 617
Third kind electrodes, see Electrodes, oxidation-reduction, third kind
Thorium(IV):
  determination, 281, 520, 653, 654
  in monazite sand, 654
  fluoride solubility product, 626
  nitrate-silver(I) nitrate titrant, 283
  titrant application, 617
Thymine, see 5-Methylpyrimidine-2,4(1$H$,3$H$)-dione, $pK_a(DMF)$, $pK_a(H_2O)$
Thyroxine determination, 177
Tin(II):
  chloride (stannous chloride):
    preparation of solution, 539
    titrant preparation and standardization, 593
  determination, 574, 578, 580, 581
  formation of, 538
  oxidation of, 564
  reductant, 539
  titrant preparation and properties, 592
Tin(IV) formal reduction potentials, 547
TISAB, 215–217. See also Background solutions
  antioxidants in, 216
  compositions of, 216
Titanium(III):
  chloride titrant preparation and properties, 592
  determination, 569, 581, 583
  formate complex as titrant, 593
  formation of, 537
  hydride reductant, 635
  oxidation of, 563
  sulfate reductant in glycerol, 254
  titrant, 292, 592
    properties and standardization, 592
Titanium(IV) formal reduction potentials, 548
Titrants:
  acid-base:
    aqueous, 476
    non-aqueous, 488–493
  alkali storage, 476
  complexometric, 525–526
  oxidation-reduction, 570–594
  precipitation, table, 611

INDEX 723

Titration curve, 457
  asymmetrical, 459
  calculation of simple, 42–47
  differential with polarized electrodes, 73–74
  graphical representation, 46, 459–465
  linearization of, 465–469
  multiparametric fitting of, 469
  pH of ligand and complex, 433
  with polarized electrodes, 72
  shape, 42
  symmetrical, 459
Titration error, 457, 459
  effect of $pK_a$, 475
Titrations:
  general classifications, 53–55
  indirect, 254
Titrimetry:
  catalyst, 447
  classifications, 53
  criteria for suitable reactions, 445–448
  defined, 53
  equilibrium constants for suitable reactions, 446
  estimating equivalence volume, 457–470
  high precision, 448
  measurement by volume, 447
  measurement by weight, 448
  primary standard in, 447
  requirements for reaction, 447
  standard solution for, 447
  stoichiometry of reactions, 456
  units of concentration, 454–456
Titriplex III, 501. *See also* EDTA
Titron X-100 in gas sensors, 163
Toad bladder, 173, 176
4-Toluenesulfonic acid:
  standard in $N$-methyl-2-pyrrolidinone, 395
  titrant, 396
3-Toluidine, $pK_B(CH_3COOH)$, $pK_B(H_2O)$, 382
4-Toluidines determination, 483
Total ionic strength adjustment buffers, 215.
  *See also* TISAB
Toxicity of solvents, 375
Tramex process, 643
Transition metals:
  anions determination, 641
  determination, 645
$1H$-Triazole derivatives determination, 660
Tribenzylamine, $pK_B(CH_3COOH)$, $pK_B(H_2O)$, 382
Tributylamine, $pK_a(DMSO)$, $pK_a(H_2O)$, 399

$NNN$-Tributyl $N$-hexadecylammonium bromide electroactant, 149
Tributylphosphate, 571
2,2,2-Trichloro-1-hydrazinoethanol determination, 596
Tridecylammonium bicarbonate electroactant, 144
$NNN$-Tridodecyl $N$-hexadecylammonium nitrate electroactant, 145
TRIEN (1,4,7,10-tetrazadecane):
  determination, 486
  masking agent, 217
  titrant, 266, 271, 503, 508, 510, 523
  preparation of solution, 526
Triethanolamine:
  complexant, 549, 550
  masking agent, 270
Triethylamine:
  $pK_a(CH_3CN)$, 389
  $pK_a(DMF)$, $pK_a(H_2O)$, 395
  $pK_a(DMSO)$, $pK_a(H_2O)$, 399
  $pK_a(N$-methyl-2-pyrrolidinone), 393
  $pK_B(CH_3COOH)$, $pK_B(H_2O)$, 382
Triethylenetetramine, 503. *See also* TRIEN (1,4,7,10-tetrazadecane)titrant
Triflic acid, *see* Trifluoromethane sulfonic acid (triflic acid)
Trifluoroacetic acid, $pK_a(H_2O)$, 307
Trifluoromethane sulfonic acid (triflic acid), 404
  $pa_H$ in 2-methyl-2-propanol, 403
  $pK_a(2$-methyl-2-propanol), 405
  titrant, 490
3-Trifluoromethylphenol:
  $pK_a(DMF)$, $pK_a(H_2O)$, 394
  $pK_a(2$-methyl-2-propanol), $pK_a(H_2O)$, 406
Trifluoroperoxyacetic acid, 477
4,4,4-Trifluoro-1-(2-thienyl)butane-1,3-dione in membrane preparation, 153
Trilead octaformamide, 581
Trilon B, 501. *See also* EDTA, titrant
Trimethylamine:
  $pK_a(CH_3CN)$, 389
  $pK_a(DMSO)$, $pK_a(H_2O)$, 399
$NNN$-Trimethyl-1-hexadecylammonium:
  bromide titrant (CTAB), 289
  hydroxide titrant, 263
2,2,4-Trimethylpentylphthalate in membrane preparation, 151
2,4,6-Trinitrobenzenesulfonic acid:
  $pK_a(2$-methyl-2-propanol), 405

2,4,6-Trinitrobenzenesulfonic acid (*Continued*)
   titrant, 489
   purification, 489
2,4,6-Trinitrobenzoic acid, $pK_a(H_2O)$, 307
2,4,6-Trinitro compounds determination, 286, 665
2,4,7-Trinitrofluoren-9-one complexes determination, 487
Trinitromethane determination, 286. *See also* Nitroform
2,4,6-Trinitrophenol (picric acid):
   buffer component:
      in acetone, 407
      in acetonitrile, 384
   determination, 666
   $pK_a(DMF)$, $pK_a(H_2O)$, 394
   $pK_a(DMSO)$, $pK_a(H_2O)$, 398
   $pK_a$(2-methyl-2-propanol), $pK_a(H_2O)$, 406
   purification, 386
3,6,9-Trioxaundecanedioic acid, 158
Triphenylamine dyes determination, 487
Triphenylmethane, $pK_a(DMSO)$, 399
Triphenyltin compounds determination, 486
TRIS, 197. *See also* 2-Amino-2-hydroxymethyl-1,3-propanediol (TRIS)
Tris(bathophenanthroline)iron(II) trifluoroacetate electroactant, 147
Tris(bathophenanthroline)nickel(II) nitrate electroactant, 294
Tris(2-ethylhexyl)phosphate in membrane preparation, 150, 158
Trithiocarbonate determination, 281, 664
Triton X-100, 294
TTHA, 502
Tubb's method, 460
Tubocin determination, 483
Tungstate determination, 284, 641
Tungsten(VI):
   formation of, 536
   reduction of, 593
Tungsten determination, 586
Tyrosinase, 172, 177
Tweens' determination, 667

Unidentate ligand, 431
Uracil, *see* Pyrimidine-2,4(1*H*,3*H*)-dione, $pK_a(DMF)$, $pK_a(H_2O)$
Uranium(IV):
   formation, 537, 539
   oxidation of, 569

Uranium(VI):
   determination, 556
   formal reduction potentials, 548
   formation, 539
Uranium determination, 589
Uranium(III) formation, 537
Ureas:
   determination, 174, 175, 176
   phenyl determination, 488
   $pK_a(H_2O)$, 308, 480
   $pK_B(CH_3COOH)$, 379, 382
   $pK_B(H_2O)$, 382
   purification, 381
   pyridyl-substituted determination, 487
   titration, 480
Urease, 169, 174, 175, 176
Uric acid:
   determination, 175
   $pK_a(H_2O)$, 308
Uricase, 175

Val, *see* Equivalent definition
Valinomycin electroactant, 152, 155, 158
Vanadium(II):
   determination, 574
   formation, 537
   sulfate titrant, standardization, 593
   titrant:
      preparation and standardization, 593
      standardization, 563
Vanadium(IV):
   determination, 520, 557, 572
   formal reduction potentials, 548
   oxidation of, 569
   titrant, 569
   preparation and properties, 583
Vanadium(V):
   determination, 521, 559, 567, 587
      sequential, 591, 593
   formal reduction potentials, 548
   formation of, 536
   titrant, 569, 573
   preparation and properties, 583
Vanadium(III) standard reduction potential, 593
Veronal, *see* 5,5-Diethylpyrimidine-2,4,6(1*H*,3*H*,5*H*)-trione, $pK_a$(2-methyl-2-propanol), $pK_a(H_2O)$
Versene, 501. *See also* EDTA, titrant
Victawet-12 wetting agent in gas sensors, 163
Viscose, 281, 664

Viscosity:
  of titration medium, 373
  values, table, 369
Vitamin B1 in membrane preparation, 154
Volumetric analysis, see Titrimetry
Volumetric apparatus, solvent properties and calibration, 376

Water:
  determination of, 76–77
    cyanide, 633
    sulfate, 284, 639
  dielectric constant, 368
  -dimethylsulfoxide properties, table, 417
  -ethanol properties, table, 411
  hardness determination, 521, 522
  ionic product, 307, 314, 329, 332, 438
    determination, 332–337
    values, table, 678
  -methanol properties, table, 411
  relative acidity constant, 375
  standard state for, 306

Ytterbium(III) determination, 523
Yttrium(III) determination, 484, 522

Zeroth-kind electrodes, 35, 80
  in enzyme sensor, 172
Zephiramine, 143, 274, 275

determination, 667
  as electroactant, 149
  titrant applications, 617
Zinc:
  alkyls determination, 599
  amalgam reductant, 534
  as primary standard, 452
  reductant, 535
Zinc(II):
  determination:
    acid base, via EDTA, 479
    complexometric, 271, 273, 274, 523
    precipitation, 282, 653
      sequential, 647, 648, 650, 652
  $NN$-diethylcarbamodithioate solubility product, 626
  oxide, 527
  as primary standard, 452
  phosphate solubility product, 626
  sulfate:
    determination of mean activity coefficients, 22–26
    titrant, 665
  titrant, 266, 284, 504, 514, 520, 527
    application in precipitations, 617
    preparation of solutions, 527
Zircalloy, determination of zirconium in, 653
Zirconium(IV) determination, 524, 653
Zwitterionic equilibria microscopic constants, 347

Vol. 60. **Quality Control in Analytical Chemistry.** By G. Kateman and F. W. Pijpers
Vol. 61. **Direct Characterization of Fineparticles.** By Brian H. Kaye
Vol. 62. **Flow Injection Analysis.** By J. Ruzicka and E. H. Hansen
Vol. 63. **Applied Electron Spectroscopy for Chemical Analysis.** Edited by Hassan Windawi and Floyd Ho
Vol. 64. **Analytical Aspects of Environmental Chemistry.** Edited by David F. S. Natusch and Philip K. Hopke
Vol. 65. **The Interpretation of Analytical Chemical Data by the Use of Cluster Analysis.** By D. Luc Massart and Leonard Kaufman
Vol. 66. **Solid Phase Biochemistry: Analytical and Synthetic Aspects.** Edited by William H. Scouten
Vol. 67. **An Introduction to Photoelectron Spectroscopy.** By Pradip K. Ghosh
Vol. 68. **Room Temperature Phosphorimetry for Chemical Analysis.** By Tuan Vo-Dinh
Vol. 69. **Potentiometry and Potentiometric Titrations.** By E. P. Serjeant